冯士筰文集

本书编委会　编

冯士筰院士

从教五十五周年

科学出版社

北京

内 容 简 介

冯士筰院士是我国著名的物理海洋学家和环境海洋学家，本书沿其从事海洋科教事业55年以来的学术发展脉络，选取截至2017年能体现其各个时期学术贡献的35篇文章结集而成。其中既包括他关于潮汐和海洋环流基本认识的早期文章，也包括关于风暴潮动力——数值预报和产品研发工作的部分成果，以及有关拉格朗日余环流和物质长期输运过程的论文。

具有大学数学、力学基础，并具备初步海洋动力学知识的大学生、研究生及青年海洋科技工作者是本书的主要读者群，从事环境科学、近海工程等教学研究的专业人士亦可从中得到启发。

图书在版编目(CIP)数据

冯士筰文集：冯士筰院士从教五十五周年/本书编委会编. —北京：科学出版社, 2017.9

ISBN 978-7-03-054721-7

Ⅰ. ①冯…　Ⅱ. ①本…　Ⅲ. ①海洋学–文集　Ⅳ. ①P7–53

中国版本图书馆CIP数据核字(2017)第238709号

责任编辑：孟美岑　胡晓春／责任校对：张小霞

责任印制：肖　兴／封面设计：北京美光设计制版有限公司

科学出版社 出版

北京东黄城根北街16号

邮政编码：100717

http://www.sciencep.com

中国科学院印刷厂 印刷

科学出版社发行　各地新华书店经销

*

2017年9月第　一　版　开本：787×1092　1/16

2017年9月第一次印刷　印张：37 1/2

字数：890 000

定价：298.00 元

(如有印装质量问题，我社负责调换)

冯士筰简介

冯士筰（1937～　），天津市人，中国海洋大学教授，中国科学院院士。1962年毕业于清华大学工程力学数学系，分配至山东海洋学院（今中国海洋大学）任教。曾任物理海洋研究所所长、海洋环境学院院长、副校长，浙江海洋学院（今浙江海洋大学）校长，全国政协委员、青岛市政协副主席，民主建国会中央常委、青岛市主委。历任国家教委科学技术委员会委员，国务院学位委员会"海洋科学"评议组组长，教育部高等学校"海洋科学与工程"教学指导委员会主任，中国力学学会常务理事，中国海洋湖沼学会常务理事，中国风暴潮及海啸分会理事长，中国海洋学会理事，"国际地圈生物圈计划（IGBP）"中国委员会常委等。

中国风暴潮研究的开拓者之一，与其合作者创建了超浅海风暴潮动力学理论及其动力－数值预报模型；"浅海风暴潮动力机制和预报方法的研究"获国家自然科学三等奖；出版了我国唯一一部关于风暴潮的理论专著——《风暴潮导论》，获全国优秀科技图书一等奖；主持完成了中国第一代和第二代风暴潮数值预报产品的研制，其中国家"七五"攻关专题获"国家七五科技攻关重大成果奖"。在浅海环流和长期输运的研究方面，依据流体物质面守恒原理，扬弃了传统的欧拉时均，发展了拉氏时均场的新理论。首先发展出弱非线性系统动力学模型，导出了风生－潮致－斜压浅海环流湍封闭基本方程组。"拉格朗日余流和长期输运过程的研究——一种三维空间弱非线性理论"获国家自然科学三等奖。其后将浅海环流的弱非线性理论发展为一般非线性理论，从而建立了近海拉格朗日时均环流及其长期输运理论体系。该理论为近海污染物物理自净、悬移质输运、海洋环境预测和近海生态系统动力学等诸多方面，提供了海洋环境流体动力学基础。

冯士筰教授出生于教育世家，始终强调教育是国家和民族的基石，人才培养是高等学校之本质功能。主持完成的教学改革项目"面向21世纪海洋科学专业的教学改革与实践"及"海洋科学类人才培养模式的改革与实践"分别获国家级优秀教学成果二等奖。主编的《海洋科学导论》，已成为全国所有涉海专业的入门经典，并以繁体字在台湾出版，获全国普通高等学校优秀教材一等奖。尤其作为我国最早的物理海洋学博士生导师之一，我国环境海洋学学位点主要创建人，培养物理海洋学和环境海洋学博士、硕士研究生数十名，桃李满天下。

编者的话

今年是冯士筰院士从教五十五周年及八十华诞，几年前他的学生们就觉得要做点什么，以表达对先生的热烈祝贺与深深敬意。经过反复酝酿，大家决定编写两本书——《冯士筰文集》与《学海兰舟》，以分别体现冯先生的学术成就及在人才培养方面的突出贡献。为此成立了编委会，对这两本书的选题定位进行策划，将冯先生发表的学术论文进行梳理精选，广泛联络冯先生培养的研究生，组织他们提供素材、撰写文章，最终形成了这两本书。《冯士筰文集》精选并收录了冯先生发表于不同刊物、时间跨度为40年的多篇学术论文，代表了他在风暴潮、拉格朗日余流和长期物质输运及海洋生态动力学方面的主要工作，展示了冯先生的学术思想发展脉络。我们请冯先生的师长文圣常院士为此书作序，此序使我们对冯先生的学术成就有了更加深刻的认识。《学海兰舟》则以照片和文字相结合的形式，再现了冯先生求学成长、学术交流、人才培养、海洋科教等方面的场景，也包括了学生心得、亲情友情，以及他参加社会活动的方方面面。我们希望这两本书在10月与大家见面，届时能为参加冯士筰先生从教五十五周年研讨会的同仁们奉上这份珍贵的礼物。正如于志刚校长在《学海兰舟》序中所说："我无时不在领略着先生谦逊低调的君子风范、不断求索的科学精神，更为先生的人文情怀、教育情怀所深深感染，先生是我们追随的楷模"。

在这两本书的编辑和出版过程中，得到了各位师兄、师弟、师姐、师妹们的积极响应与大力支持，也得到了科学出版社的鼎力相助，浅海室的学生们为文集的录入校对做了大量工作，在此表示衷心感谢！感谢陈鹭为第二本书取了一个富有诗意的名字——《学海兰舟》！吴德星、孟伟、王辉、闫菊、鹿有余、魏皓、郭新宇、柴扉等为这两本书的筹划付出了辛勤的劳动，江文胜负责《冯士筰文集》的论文筛选和编辑，高会旺负责《学海兰舟》的组稿和编辑，张平负责照片收集、筛选，以及与出版社和各位同门的联络沟通等工作。在此过程中，我们充分体会了“眼高手低”和“文字乏力”的真正含义，原以为很快就可以完成的书稿，在准备过程中却出现了这样或那样的问题，如图的分辨率不够、素材没有足够的代表性、文字先后安排缺乏逻辑性，等等。在我们的共同努力下，希望已经较好地解决了这些问题，若仍有不当，万望见谅！收录到这两本书的文章多数已经得到原版权单位的同意，同时，我们对个别地方做了删改，或是改正原文中的笔误，或是使表达更为简洁，在此特别说明。个别文章没有得到及时反馈，如果有异议可与作者联系。

由于水平所限，书中难免有不少遗漏和缺憾，如我们没有能够联系上冯先生指导过的每一位研究生，无法将研究生们的简历格式统一起来，也尚未收录所有研究生的“感悟”文章，更未能全面反映学生们想表达、而未能以文字书写的感情。冯先生桃李满天下，《学海兰舟》中只将他指导的研究生收录在内，他们只是其学生中的代表，冯先生从教五十五周年，培养过的本科生、研究生何止于此，他们的成长曾得益于冯先生的谆谆教诲和润物无声的启迪，想必他们也有很多话要说，恕我们考虑不周，未能更广泛地征求和吸纳他们的想法！在这个特殊的时刻，也让我们代表工作于全国各地、世界各地的学友们表达你们对冯先生的敬意与祝贺！

本书编委会

2017 年 5 月 20 日

序

1962 年夏天，士筰和他的三位同班同学，从清华大学工程力学数学系流体力学专业毕业，一起分配到山东海洋学院海洋水文气象系工作。当时系里的海洋专业有两个教研室，我是动力海洋学教研室主任，他们四位就自然而然地进入到了我的教研室。他们是时任山东海洋学院教务长、海洋水文气象系主任的赫崇本先生专门从国内名校要来的毕业生，赫先生希望这些数理基础好的年轻人能够成为我国物理海洋教学与科研的后备军。

赫先生的这种布局，顺应了物理海洋学的发展规律，即用数学力学的基础理论来解决物理海洋学问题。士筰在清华大学的学习经历，为其后从事物理海洋学研究打下了深厚的数理基础，开启了科学思维的心扉。自 20 世纪 70 年代中期以来，他在物理海洋学和环境海洋学领域，特别是在风暴潮与浅海动力学研究方面取得了令人瞩目的成就，并于 1997 年当选为中国科学院院士。

士筰及其合作者在风暴潮动力学研究中创建了超浅海风暴潮理论，并将风暴潮动力学数值预报方法系统化，相关成果“浅海风暴潮动力机制和预报方法的研究”获“国家自然科学奖”三等奖，其专著《风暴潮导论》获全国优秀科技图书一等奖。他主持完成了我国第一代风暴潮数值预报产品的研制，主持的国家“七五”、“八五”和“九五”有关风暴潮数值预报重点科技攻关专题，获得了鉴定专家的高度评价，认为其部分成果已达国际领先水平。其中“七五”攻关专题与其他专题共同获得“国家七五科技攻关重大成果奖”。这些工作也促进了我国相关学科人才培养和风暴潮预报的进步。

士筰在浅海动力学方面的研究成果尤为突出。在这一近海环境海洋学研究焦点中，他提出的拉格朗日余环流和输运方程独具特色，物理意义明确，受到国际同行重视，且已被美国的研究人员成功应用于切萨皮克湾盐度的数值模拟和富营养化问题的预测。其研究成果“拉格朗日余流和长期输运过程的研究——一种三维空间弱非线理论”获“国家自然科学奖”三等奖。他建立的斜压浅海环流方程组成功地应用于渤海、黄海、东海环流和泥沙输运的数值模拟和机制研究。近年来，他指出拉格朗日余流的核心是物质面守恒，以此为基础，他把拉格朗日余流理论从对流弱非线性推广到一般非线性，并且从理论、数值、实验室模拟和现场观测诸方面均取得了进展。

士笮在浅海动力学上的成果成为环境海洋学的重要基础，自 20 世纪 90 年代以来，他推动了海洋科学内物理、化学、生物方向的交叉研究，创建了环境海洋学博士点，其理论在海洋生态动力学研究中得到了应用，促进了学科的发展。

今年是士笮从事海洋科教事业 55 周年，他的学术成果优选出版，作为阶段性总结，我在此表示祝贺，并祝愿他在科研、教学和人才培养等方面取得新进展。

约半个世纪以来，我和士笮相处，对他献身一个新的学术领域，成果丰硕，甚为钦佩，谨以为序。

文圣常

2017.2.26.

前 言

冯士筰院士是我国著名的物理海洋学家和环境海洋学家，他在学术上的贡献主要体现在两大方面，一方面是风暴潮的动力学研究，他与合作者提出了超浅海风暴潮理论和数值预报模型，主持完成了国家“七五”、“八五”和“九五”重点科技攻关风暴潮专题，为我国风暴潮数值预报的发展作出了突出贡献；另一方面是在浅海动力学方面的贡献，他提出了一种拉格朗日余流和长期物质输运的理论模型，建立了以拉氏时均速度的最低阶近似——物质输运速度来体现浅海环流速度基本场的新理论框架，导出了浅海潮生–风生–热盐环流基本方程组，建立了一种新型的长期输运方程。这些成果获得了国家自然科学奖和“国家七五科技攻关重大成果奖”，促进了风暴潮理论及预报和浅海环境动力学的发展。

截止到 2016 年底，冯老师共发表学术文章 96 篇，撰写、主编学术著作 5 部，从这些已经发表的文章中，我们沿着其学术发展脉络选取了能体现其学术贡献的 35 篇组成《冯士筰文集》。文集中的文章大致可以分为三个部分，第一部分是冯老师早期的文章，主要是关于潮汐和海洋环流的基本认识，此部分成果的形成开始于 20 世纪 60 年代中期，是他大学毕业刚刚进入物理海洋领域时，在学习过程中的一些体会和发现，这些成果显示了力学思维在发轫于地学的物理海洋学中的重要性。第二部分是关于风暴潮动力学的相关研究，展示的是冯老师及其合作者有关风暴潮动力——数值预报和产品研发工作的部分成果。冯老师在 1982 年出版的《风暴潮导论》是世界上第一部系统论述风暴潮的专著，主要包括风暴潮的基本概念、动力学机制和风暴潮数值预报和经验预报等内容，考虑到这部分成果已经以专著形式出版，本书不再收入。另外，冯老师有关风暴潮研究的大部分成果是在“七五”、“八五”和“九五”国家重点科技攻关“风暴潮数值预报和产品研制”专题中完成的，按照当时的规定是不允许发表论文的，只有相关的技术报告，本论文选集中也不予录入。第三部分是有关拉格朗日余环流和物质长期输运过程的论文，这是冯老师将物理海洋与海洋科学其他分支交叉出环境海洋学这一新学科的基础，这一部分选入的文章最多，也是目前冯老师仍然在研究的问题。在这一部分中，冯老师等编著的《渤海生态动力学导论》是环境海洋学研究的重要成果，但该书也已独立出版，本文集不再收录。

文集中论文的编排是按时间排序的，这三部分文章前后顺序有所交错，为了方便读者阅读，下面将简要地将这三部分文章进行介绍。

第一部分　潮和海洋环流

在这一部分，我们选入的首篇文章是冯老师有关非线性潮汐的论文（Feng, 1977），论文阐明在东中国海广阔大陆架上的海水运动是半日或全日潮周期运动占优的，潮汐对于与其非线性耦合在一起的风暴潮、时均余环流都或多或少地产生一定的影响，故对于非线性潮动力学及其数值模型的探讨是必要的。

第二篇文章建立了一个三维空间变湍黏性系数的浅海动力——数值模型（Feng, 1984），它既可应用于超浅海风暴潮和非线性潮流的模拟和预报，也可用于探讨风生–斜压–欧拉潮余流的机制和模拟（Feng *et al*., 1984）。但应特别指出，在控制欧拉潮余流的表面运动学边界条件中出现了虚假的源汇项，这表示欧拉时均潮余流是值得商榷的。

本部分最后三篇文章是关于大洋环流传统的解析理论（冯士筰，1979，1984）及 f–和 β–坐标系（冯士筰，1982）的探讨。尽管大洋环流方面，冯老师没有后继文章发表，形不成体系，但从历史的观点着眼，主要形成于20 世纪 60 年代中期的这些思想，发表出来作为历史的遗痕，还是蛮有趣味的，特别由此流露出了冯老师在学习和科研道路上求索的神貌和风格，及其从开端到形成之过程。

第二部分　风暴潮

我国是一个风暴潮灾害严重的国家，故风暴潮预报、特别其动力–数值预报的研究是十分重要的。尽管20世纪60年代已开始有人接触到了这一现象，但直到70年代中期，我国才真正开始了风暴潮动力学的研究（Chin and Feng, 1975），该文提出了超浅海风暴潮的概念，并建立了其动力学理论模型，为数值模拟和预报提供了动力学基础。

作为超浅海风暴潮动力–数值模型的首次数值实验，孙文心、冯士筰和秦曾灏（1979）设计了几种理论模型风应力场作为驱动力初步成功地模拟了渤海风潮。这不仅对该理论是一个有力的检验，而且对进一步探讨渤海风潮的机制和预报也是一个有益的参考。

伴随着风暴潮或在风暴潮过境后，陆架上可能出现一类特殊波动——边缘波。边缘波的研究，对于风暴潮预报，无疑是有意义的。在陆架上，尤其在陆架近岸浅水域中，底摩擦导致的阻尼效应，对边缘波产生明显的耗散作用——这种阻尼模型的探讨，与之前的不考虑底摩擦的非耗散模型相较，更贴近实际情况（冯士筰，1979，1981）。

第三部分　拉格朗日余环流及物质长期输运过程

我们将着重导读这最后一部分主要的文章，它们是有关以潮周期运动为主的近海/河口环流和长期输运的研究。其研究的特色为：扬弃了传统的欧拉时均模型，而系统地发展了拉格朗日时均模型（简称拉氏时均模型）——这个问题是浅海环境动力学研究的焦点，而直到目前为止，就我们的认识，它对于海洋界仍旧是一个挑战。

（一）弱非线性系统

这一系列研究从浅海弱非线性动力系统开始，首先表明了传统的物质输运速度（MTV）（欧拉时均速度与斯托克斯漂移速度之和）是拉氏余流（LRV）的一阶近似，特别导出了其二阶近似——拉格朗日漂移速度（LDV），依赖于标识流体微团被释放的时间，它表征了拉氏余流独特的性质，并提出了余流椭圆的概念（Cheng *et al.*, 1986；Feng *et al.*, 1986a）。进而建立了一个新型的潮际输运方程，它不仅是浅海环流场中温、盐的长期对流扩散方程，而且也是在海洋生态系统中耦合余流和生化变量的纽带方程（Feng *et al.*, 1986b），与相应的欧拉时均方程比较，其特点在于其对流输运速度是物质输运速度并消除了多余的"潮弥散"项。鉴于拉格朗日时均问题应成立于一个三维空间的动力学系统中（Feng, 1987），故上述模型和结论自然应由上面的深度平均的二维空间问题推广到一个更具物理意义的三维空间问题（Feng, 1986）。

冯老师在进一步的文章（Feng, 1987）中，开始进入了动力学探讨：建立了物质输运速度满足的动力学方程组，导出并命名了潮致体积力（TBF），这是潮非线性产生的对余环流的强迫力；并建议扬弃欧拉时均速度，应以物质输运速度来近似体现浅海环流速度场。又进一步把上述正压单频潮致拉氏余流理论模型推广到斜压情形（Feng *et al.*, 1990），进而推广到更加完备和有更多实际应用前景的多频潮致–风生–斜压理论模型（Feng, 1990）。

在冯士筰和孙文心于 1992 年主编的《物理海洋数值计算》第九章"浅海环流物理及数值模拟"中，综述了上面提出的理论模型，并建立了相应的全流方程组和给出了拉氏余涡的概念和分析（Feng, 1991）。

上述理论模型中，湍粘性系数是作为已知的常数或空间坐标函数这种传统假设来处理的，即它未考虑湍粘性的非线性效应，而在浅海这一耗散系统中这也是可以产生余流的，所以这一假定是有相当局限性的，有很大的改善余地。到目前为止物理海洋学的动力学方程中，对于湍流的处理和其他学科一样，仍然处于半经验的湍封闭模型水平，在这一认识水平上，Feng 和 Lu（1993）在对流为主的弱非线性假设下，建立了一个考虑湍非线性的 k-ε 湍封闭拉氏余流控制方程组，如果仅考虑其最简化的形式，这一湍封闭可以蜕化为湍动能齐次方程（魏皓等，2000）。与上述湍粘性系数为常数的模型相比，湍封闭模型果然在潮致体积力中增添了一个余流的产生项。

进一步，为了解决存在强流的陆架海环流问题，比如东中国海中就存在着黑潮这样的强流，一个潮流和准定常流共同占优的弱非线性系统中风生–潮生–斜压拉氏余环流热力–动力学的湍封闭模型也被建立了（Feng and Wu, 1995；Feng, 1998）。

以上文章反映的是弱非线性近海/河口系统中拉氏余流和潮际输运理论模型探讨之过程和贡献。

下面我们选进了冯老师参与发表的关于渤海环流、潮际输运和渤海生态模拟和分析之数值实验的代表性文章和论著，作为上述理论模型在实际海域中的应用：如 Wang 等（1993），Feng 等（1994），Feng 和 Xi（1997），Gao 等（1998），Sündermann 和 Feng（2004），Wei 等（2004）以及 Hainbucher 等（2004）。

在浅海中由于余流的量级基本处于观测误差的范围，因此有必要采取其他手段进行研究，因此在一个狭长模型河口/海湾中开展了拉氏余环流的解析研究，区别于传统的相关解法，直接求解上述具有明显物理意义的控制物质输运速度的封闭方程组，解出了解析解，从而给出并分析了潮致余环流之新流型，特别分解出了非线性二次底摩擦产生余环流的分量（Jiang and Feng, 2011, 2014）。

开始展开实验室研究，首次利用 PIV 技术基于拉氏余环流的定义，直接测得潮致拉氏余环流场，并可作为上述解析解的实验验证（Wang *et al.*, 2013）。

（二）一般非线性系统

显然，由上述弱非线性问题向一般非线性问题的探讨，不仅是实际上的需求，也是合乎逻辑之发展（Feng *et al.*, 2008）。

在此文章中：① 给出了拉氏余流速度的一个严格的定义；② 证明了该余流速度满足流体物质面守恒；③ 导出了作为对流输运速度的拉氏余流速度满足的潮际输运方程：该方程控制的“浓度”，不再是弱非线性系统中控制的欧拉时均浓度，而是首次被揭示和命名的“拉氏潮际浓度（Lagrangian Inter-tidal Concentration）”；④ 拉氏余流速度场和拉氏潮际浓度场皆为对应于每个标识流体微团释放时的初始时间之无穷个场。

作为实际海域的应用，分别对渤海和胶州湾的环流作了初步的数值试验，在渤海除部分水域表现为一般非线性性质外，基本为一弱非线性系统（Ju *et al.*, 2009；Mao *et al.*, 2016）；而胶州湾基本表现为一般非线性系统（Liu *et al.*, 2012）。

由于文集中的文章时间跨度为40年，发表的刊物也不相同，因此文章的风格、体例、格式各异，比如早期的中文作者的拼音方式都与现在不同。为了保持原貌，各篇文章基本保持原样，对部分错误进行了修订。文中的公式样式进行了统一，即矢量采用黑斜体而不是采取加箭头的方式。对文章中的图件进行了重新描绘，并按新的格式要求进行了修改。多数收录入文集的文章已经得到原版权单位的书面同意，个别文章因为年代久远无法联系，如果有异议可与作者联系。

本书编委会

2017年5月20日

目 录

A Preliminary Study on the Mechanism of Shallow Water Storm Surges*

Chin Tseng-Hao (秦曾灏) and Feng Shih-Zao (冯士筰)

I. Introduction

The so-called "whole current method" (or "total transport method") has been prevalent in the theoretical study of storm surges for a long time. The storm surges as a three-dimensional problem can be transformed into a two-dimensional one by means of this method, hence the mathematical analysis is simplified. The solution of the storm-induced current thus obtained does not, however, yield its vertical distribution. Besides this, the method embodies another shortcoming, i.e., some subjective assumptions[1–5] have to be introduced by many authors in order to keep the governing equations in a closed form. Some of them have devoted themselves to the study of the bottom stress[6]. Hence it complicates the problem.

It is evident that some real dynamical mechanism about storm surges could not be revealed, at least, by the "whole current method", as discussed in Section V of this paper. To investigate directly the three-dimensional storm surges instead of using the "whole current method" seems more reasonable. Moreover, it is necessary for both theoretical and practical purposes. A preliminary attempt has been made early by Welander[7] to approach this problem. A numerical method has been suggested lately by Heaps[8] for computing the three-dimensional time-dependent wind-induced current structure, in addition to sea level elevations. A tentative scheme has been carried out in a closed rectangular basin with uniform wind. In this paper, an effort is made to investigate analytically the three-dimensional shallow water storm surges by the fundamental hydrodynamical equations. The theory of shallow water surge is subdivided into two parts, namely, the ordinary shallow water theory and the ultra-shallow water theory. The dynamical criteria of the classification are given. An exact solution of the storm-induced current and the corresponding equation of the storm-induced sea level elevation for the ultra-shallow sea are obtained. As an example, the authors give and then analyze these exact solutions for sea areas of uniform depth on the continental shelf with constant coefficient of eddy viscosity.

It is quite an important subject to reveal the mechanism of storm surges as a theoretical basis for numerical analysis and prediction of storm surges.

---

* Chin T-H, Feng S-Z. 1975. A preliminary study on the mechanism of shallow water storm surges. Science in China, Ser. A, XVII(02): 242–261

Rossby[9] was notably the first who had proposed the concept about geostrophic adaptation between the pressure field and the velocity distribution in the ocean. A series of investigations on the problem of the adaptation have been successively made with reference to various scales of atmospheric motion since 1938. The preliminary analysis shows that the adaptation and the evolution also existed for the ultra-shallow water storm surges. The distinction between the adaptation and evolution, as well as their physical meanings, is discussed for the ultra-shallow sea, say, the Pohai Sea, in order to offer a key to further studies on the mechanism of storm surges.

II. The Governing Equations and the Determination of Characteristic Parameters

The basic equations governing the shallow water storm surges may, with good approximation, be expressed as

$$\frac{\partial \boldsymbol{q}}{\partial t}+\boldsymbol{q}\cdot\nabla\boldsymbol{q}+w\frac{\partial \boldsymbol{q}}{\partial z}=f\boldsymbol{q}\times\boldsymbol{e}_3-g\nabla\zeta-\frac{1}{\rho}\nabla p_{\mathrm{a}}+\frac{\partial}{\partial z}\left(\nu\frac{\partial \boldsymbol{q}}{\partial z}\right)+\nu_{\mathrm{L}}\Delta\boldsymbol{q} \tag{1}$$

$$\nabla\cdot\boldsymbol{q}+\frac{\partial w}{\partial z}=0 \tag{2}$$

In these equations t denotes the time, z the vertical coordinate, $\boldsymbol{e}$ the unit vector along z-axis pointing upward, ∇ the horizontal Del operator, $\nabla\cdot$ the horizontal divergence operator, Δ the two-dimensional Laplace operator, $\boldsymbol{q}$, w the horizontal velocity vector and vertical component of current velocity, respectively, ζ the storm-induced elevation of the sea surface, p_{a} the surface atmospheric pressure, f the Coriolis parameter, ρ the density of sea water, g the acceleration of the earth's gravity, ν, ν_{L} the kinematic coefficients of vertical and lateral eddy viscosity, respectively.

Here, f, ρ and g are assumed to be constants. The boundary conditions have to be specified at the surface and the bottom. The surface conditions require that

$$\rho\nu\frac{\partial \boldsymbol{q}}{\partial z}=\boldsymbol{\tau} \tag{3}$$

$$w=\frac{\partial \zeta}{\partial t}+\boldsymbol{q}\cdot\nabla\zeta \tag{4}$$

at $z=\zeta$; the bottom conditions may be taken as

$$\boldsymbol{q}=0,\quad w=0 \tag{5}$$

at $z=-h$; where h denotes the depth, $\boldsymbol{\tau}$ the wind stress vector at the sea surface.

The wind stress is usually the main forcing function when the severe storm surges have been developed over shallow waters[1]. Therefore, it seems reasonable to assume $g\nabla\zeta=\frac{\partial}{\partial z}\left(\nu\frac{\partial \boldsymbol{q}}{\partial z}\right)$ under the action of the wind stress $\boldsymbol{\tau}$ on the surface, as given by Equation (3) .

By introducing the following dimensionless variables: $\bar{z}=z/h^*$, $(\bar{\nabla},\bar{\nabla}\cdot)=L(\nabla,\nabla\cdot)$, $\bar{\Delta}=L^2\Delta$, $\bar{t}=t/T_0$, $\bar{h}=h/h^*$, $\bar{\boldsymbol{q}}=\boldsymbol{q}/q_0$, $\bar{w}=w/(q_0h^*/L)$, $\bar{\zeta}=\zeta/Z$, $q_0=\dfrac{gh^{*2}}{\nu_0 L}Z$, $Z=\dfrac{L}{\rho g h^*}\tau_0$, $T_0=\dfrac{L^2\nu_0}{gh^{*3}}$, $\bar{\nu}=\nu/\nu_0$, $\bar{\boldsymbol{\tau}}=\boldsymbol{\tau}/\tau_0$, $\bar{p}_{\mathrm{a}}=p_{\mathrm{a}}/\delta p_{\mathrm{a}0}$, in which L is the characteristic horizontal scale, h^* the characteristic value of the depth, ν_0, τ_0 and $\delta p_{\mathrm{a}0}$ are the corresponding characteristic values, Eqs. (1), (2) and (3)–(5) may then be written in the dimensionless form as follows:

$$\Xi\frac{\partial\bar{\boldsymbol{q}}}{\partial\bar{t}}+\varepsilon\left(\bar{\boldsymbol{q}}\cdot\bar{\nabla}\bar{\boldsymbol{q}}+\bar{w}\frac{\partial\bar{\boldsymbol{q}}}{\partial\bar{z}}\right)=D(\bar{\boldsymbol{q}}\times\boldsymbol{e}_3)-\bar{\nabla}\bar{\zeta}-P\bar{\nabla}\,\bar{p}_{\mathrm{a}}+\frac{\partial}{\partial\bar{z}}\left(\bar{\nu}\frac{\partial\bar{\boldsymbol{q}}}{\partial\bar{z}}\right)+\bar{\nu}_L\bar{\Delta}\bar{\boldsymbol{q}} \qquad (1)'$$

$$\bar{\nabla}\cdot\bar{\boldsymbol{q}}+\frac{\partial\bar{w}}{\partial\bar{z}}=0 \qquad (2)'$$

with the boundary conditions:

$$\frac{\partial\bar{\boldsymbol{q}}}{\partial\bar{z}}=\bar{\boldsymbol{\tau}} \qquad (3)'$$

$$\bar{w}=\frac{\partial\bar{\zeta}}{\partial\bar{t}}+\kappa\bar{\boldsymbol{q}}\cdot\bar{\nabla}\bar{\zeta} \qquad (4)'$$

at $\bar{z}=\kappa\bar{\zeta}$; and

$$\bar{\boldsymbol{q}}=0,\quad \bar{w}=0 \qquad (5)'$$

at $\bar{z}=-\bar{h}$, where

$$\Xi=\omega_0^2/\omega_\nu^2\quad(\omega_0=\sqrt{gh^*}/L,\ \omega_\nu=\nu_0/h^{*2})$$

$$\varepsilon=\kappa\Xi,\quad D=2\pi^2h^{*2}/d^2\quad(d=\pi\sqrt{2\nu_0/f})$$

$$P=(\delta p_{\mathrm{a}0}/\tau_0)(h^*/L),\quad \bar{\nu}_{\mathrm{L}}=(\nu_{\mathrm{L}}/\nu_0)(h^*/L)^2,\quad \kappa=Z/h^*$$

Although the physical meaning of foregoing dimensionless variables is evident, it seems necessary to elucidate with emphasis two points as follows.

Firstly, the storm surge appears oscillatory. For example, the whole body of water oscillates usually in a closed sea, a semi-closed sea, a bay or over the continental shelf-seiches. The inertial oscillations are gradually damped out by the eddy viscosity of water.

The dimensionless parameter Ξ is a measure of the inertial effect in storm surges, i.e., Ξ gives the criteria measuring the relative magnitude of the effect of inertial oscillation in storm surges. Actually, the above reasoning is obvious, because ω_0 and ω_ν are respectively in proportion to the characteristic frequencies of the water and the coefficient of eddy viscosity. The magnitude of Ξ is proportional to h^{*5}, and it may be expected that the effect of inertial oscillation is negligible in the zero-order model of shallow water surge as the depth diminishes to a certain threshold value. A vivid physical picture would be drawn with further discussion of the characteristic period T_0 of the surge and the criteria Ξ together.

At certain instant, if the oscillations are excited in a storm surge such that the order of magnitude of $\varXi$, i.e., $O(\varXi)$ is unity at least, then the characteristic period $T_0 = \dfrac{1}{\omega_\nu \varXi}$ shows that the greater $\varXi$ is, i.e., the greater the frequency of the excited oscillations is, the less T_0 will result. Moreover, the greater the wind stress τ_0 and the smaller h^* are, the smaller T_0 would be; this implies that the excited oscillations are damped out more quickly by the eddy viscosity. In a shallow sea with small depth such as the Pohai Sea, it may generally be expected that the excited initial oscillations in a storm surge under the action of strong wind, no matter how severe their intensities are, will be dissipated by the eddy vis cosity in a short time. In fact, if assuming $O(\varXi)=1$ at least for the Pohai Sea, the duration of decay for the above-mentioned oscillations under typically strong winds will be less than several hours. Obviously, this is comparatively smaller than several days—the duration of surge process for the Pohai Sea.

Secondly, the dangerous surges develop in very shallow sea and cause the catastrophe there, therefore the nonlinear effects become a specially important problem. Unfortunately, studies on the nonlinear problem have not yet been sufficient and efficient because of mathematical difficulties. There are two nonlinear dimensionless parameters ε and κ involved in the foregoing system of Eqs. (1)′–(2)′ and (3)′–(5)′, which are physically different in essence. The parameter ε represents nonlinear effect of convective acceleration comprised partially in the inertial terms of the equation of motion, while the parameter κ defined by the ratio of the characteristic amplitude of the surge to the characteristic depth implies the nonlinear mutual interaction between the storm-induced currents and the variations of sea surface elevations. In general, $O(\varepsilon)$ and $O(\kappa)$ are unequal, so these nonlinear effects should be treated separately. Especially, in the case of very shallow waters such as the Pohai Sea, κ has to be introduced under certain order of approximation in which ε might be well neglected. This is due to the fact that $\varXi$ is subjected to the severe damping effect of eddy viscosity within the whole sea so that ε would not attain the same order of magnitude as κ. The equality $O(\varepsilon)=O(\kappa)$ holds only when $O(\varXi)$ equals to unity. Consequently, ε and κ are distinct in dynamically nonlinear character and a proposal for subdivision of the theory of shallow water surge into two parts is suggested: the ordinary shallow water surge theory which satisfies the dynamic condition $O(\varXi)=1$ or its equivalence $O(\varepsilon)=O(\kappa)$, and the ultra-shallow water theory which satisfies the dynamic condition $O(\varXi)<1$ or its equivalence $O(\varepsilon)<O(\kappa)$.

In the case of ultra-shallow waters as the Pohai Sea, the foregoing dimensionless variables under the surface wind stress τ_0 ranging from 10.5–22.0 (C.G.S.) (corresponding to the surface wind speed W_a from 18 m/s to 26 m/s) amount to① $O(\varXi)=10^{-1}$; $O(\kappa)=10^{-1}$; $O(\varepsilon)=$

① 原文注: The wind stress formula $\tau_0 = 2.6\times10^{-3}\rho_a W_a^2$ (C.G.S.) is used in estimation, where ρ_a is air density near the sea surface.

Value of ν_0 in estimation comes from $\nu_0 = r_1/3$, where r_1 is cited from W. Hansen (1952).

10^{-2}; $O(D)=1$; $O(P)=10^{-2}$; $O(\nu_L)=10^{-2}$ (corresponding to $\nu_L/\nu_0=10^7$).

The previous dynamic condition for the ultra-shallow sea is obviously satisfied, hence studies on the storm surges occurring in the Pohai Sea belong to the category of the ultra-shallow water theory. The exclusion of inertial oscillations in the zero-order dynamic model is an important characteristic of the ultra-shallow water surges. The zero-order model for the ultra-shallow sea, say, the Pohai Sea, is a linear quasi-equilibrium one in which the horizontal pressure-gradient force expressed as the surface slope is approximately balanced by the horizontal Coriolis force and the horizontal frictional force due to the vertical turbulent transfer (the inertial force can be neglected in the first approximation). As to the first-order model, the equations of motion contain also the time derivative terms and the κ-nonlinear effect in addition to those terms contained in the zero-order model, but the ε-nonlinear effect can be dropped in this model. This is an important characteristic for the ultra-shallow water surges. Thus the previous subdivision of the theory of shallow water surge not only becomes necessary theoretically, but also provides a convenient approach to the problem of ultra-shallow water surges.

It is noteworthy that the nonlinear effect may also be caused by the nonlinear coefficient of eddy viscosity.

Finally, we give the order of magnitude of terms involved in Eqs. (1) & (2) which have been omitted from Reynolds' equations (of incompressible fluid) in complete form. By inserting the typical values of previous variables for the Pohai Sea into the various terms, it is found that only Coriolis terms have the magnitude of 10^{-5}, the rest of them are all less than 10^{-11}. Hence the approximation of the system of Eqs. (1) & (2) to Reynolds' equations is valid with high accuracy. This is apparent because the characteristic depth is much smaller than the characteristic horizontal scale in a shallow sea.

III. The Zero-order Dynamic Model for the Ultra-shallow Water Storm Surges

1. General case

If the study on the ultra-shallow water surges is restricted to the sea area such as the Pohai Sea, the zero-order dynamic model may then be expressed mathematically as follows:

$$\frac{\partial}{\partial z}\left(\nu\frac{\partial u_0}{\partial z}\right)+fv_0=g\frac{\partial \zeta_0}{\partial x} \tag{6}$$

$$\frac{\partial}{\partial z}\left(\nu\frac{\partial v_0}{\partial z}\right)-fu_0=g\frac{\partial \zeta_0}{\partial y} \tag{7}$$

$$\frac{\partial u_0}{\partial x}+\frac{\partial v_0}{\partial y}+\frac{\partial w_0}{\partial z}=0 \tag{8}$$

The boundary conditions require that
at surface $z = 0$:

$$\rho\nu\frac{\partial u_0}{\partial z}=\tau_x,\quad \rho\nu\frac{\partial v_0}{\partial z}=\tau_y \tag{9}$$

$$w_0=\frac{\partial \zeta_0}{\partial t} \tag{10}$$

at bottom $z=-h$:

$$u_0=v_0=w_0=0 \tag{11}$$

along the coasts C_1:

$$Q_{n0}\equiv\cos\alpha'\int_{-h}^{0}u_0\mathrm{d}z+\cos\beta'\int_{-h}^{0}v_0\mathrm{d}z=0 \tag{12}$$

along the open sea boundary C_2:

$$\text{either } Q_{n0}=M \text{ or } \zeta_0=N \tag{13}$$

The initial condition is

$$\zeta_0\big|_{t=0}=\zeta^*(x,y) \tag{14}$$

where x, y, z consist of right-handed Cartesian coordinate system, u_0, v_0, w_0 the corresponding current velocity components, τ_x, τ_y the two horizontal components of the eddy stress, $\cos\alpha'$, $\cos\beta'$ the two direction-cosines of coastal normal, respectively, and M, N, as well as $\zeta^*(x,y)$, are known functions.

The kinematic coefficient of eddy viscosity is assumed to be $\nu=\nu(x,y)$.

By introducing a complex velocity $q_0=u_0+\mathrm{i}v_0$ ($\mathrm{i}=\sqrt{-1}$), an analytical solution of Eqs. (6) & (7) satisfying the boundary conditions (9) & (11) is easily given by[7]

$$u_0=q_{0\mathrm{R}} \tag{15}$$

$$v_0=q_{0\mathrm{I}} \tag{16}$$

where $q_0=\dfrac{\tau}{k\rho\nu}\dfrac{\mathrm{sh}[k(z+h)]}{\mathrm{ch}(kh)}+\dfrac{\Gamma_0 g}{k^2\nu}\left[\dfrac{\mathrm{ch}(kz)}{\mathrm{ch}(kh)}-1\right]$, $\Gamma_0=\dfrac{\partial\zeta_0}{\partial x}+\mathrm{i}\dfrac{\partial\zeta_0}{\partial y}$, $\tau=\tau_x+\mathrm{i}\tau_y$, $k=(1+\mathrm{i})\pi/d$.

The subscripts R and I denote respectively the real and the imaginary parts of a variable.

Integrating (8) with respect to z from bottom to surface and using (10) & (11), we get

$$\frac{\partial \zeta_0}{\partial t}+\frac{\partial U_0}{\partial x}+\frac{\partial V_0}{\partial y}=0 \tag{17}$$

where

$$U_0=\int_{-h}^{0}u_0\mathrm{d}z,\quad V_0=\int_{-h}^{0}v_0\mathrm{d}z$$

Integrating (15) & (16) vertically and substituting the results into (17) lead to

$$\frac{\partial \zeta_0}{\partial t} = \frac{g}{\nu} A_R \Delta \zeta_0 + \left[\frac{\partial}{\partial x}\left(\frac{g}{\nu} A_R \right) + \frac{\partial}{\partial y}\left(\frac{g}{\nu} A_I \right) \right] \frac{\partial \zeta_0}{\partial x} + \left[\frac{\partial}{\partial y}\left(\frac{g}{\nu} A_R \right) - \frac{\partial}{\partial x}\left(\frac{g}{\nu} A_I \right) \right] \frac{\partial \zeta_0}{\partial y} + F(x,y,t) \tag{18}$$

which is the required storm-induced elevation in the zero-order model subjected to the conditions (12)–(14), where

$$F(x,y,t) = \frac{1}{\rho\nu}\left[B_R \nabla \cdot \boldsymbol{\tau} - B_I \nabla \times \boldsymbol{\tau} \right] + \left[\frac{\partial}{\partial x}\left(\frac{B_R}{\rho\nu} \right) + \frac{\partial}{\partial y}\left(\frac{B_I}{\rho\nu} \right) \right] \tau_x + \left[\frac{\partial}{\partial y}\left(\frac{B_R}{\rho\nu} \right) - \frac{\partial}{\partial x}\left(\frac{B_I}{\rho\nu} \right) \right] \tau_y$$

$$Q_{n0} = -\left[\frac{g}{\nu} A_R \frac{\partial \zeta_0}{\partial n} + \frac{g}{\nu} A_I \frac{\partial \zeta_0}{\partial s} + \frac{B_R}{\rho\nu} \tau_n + \frac{B_I}{\rho\nu} \tau_s \right]$$

$$A = \frac{1}{k^2}\left[h - \frac{1}{k}\mathrm{th}(kh) \right], \quad B = \frac{1}{k^2}\left[\mathrm{sech}(kh) - 1 \right]$$

s, n refer to the tangent and the normal in the right-handed natural coordinates.

An analytical solution of Eq. (18) is not generally obtainable when the initial elevation (14), the lateral boundary conditions (12) & (13) and $\nu(x, y)$ as well as wind stress are given. However, the approximate solution is always possible by the numerical integration technique, hence the future states of both the elevation and the current can be determined.

It should be noted that the assumption $\nu = \nu(x, y)$, i.e., ν is independent of the vertical structure of the storm-induced currents, is a weakness of the present analysis.

2. Determination of storm surges on the sea areas of continental shelf

It is significant to investigate the storm surges on the sea areas of continental shelf in order that the mechanism of three-dimensional storm surge and its prediction may be studied.

To obtain an exact solution of Eq. (18) analytically, we consider a comparatively simple case in which both ν and h are assumed to be constant.

Taking $0 \leqslant x < \infty$, $-\infty < y < \infty$ to be the region of the sea, and $x = 0$ to be the coastline (Fig. 1), the storm-induced elevation in the zero-order dynamic model is expressed as follows:

$$\frac{\partial \zeta_0}{\partial t} - \beta \Delta \zeta_0 = -\gamma \nabla \cdot \boldsymbol{\tau} - \delta \nabla \times \boldsymbol{\tau} \tag{19}$$

with boundary conditions

$$\frac{\partial \zeta_0}{\partial x} = \frac{\alpha}{\beta} \frac{\partial \zeta_0}{\partial y} + \frac{\gamma}{\beta} \tau_x + \frac{\delta}{\beta} \tau_y \tag{20}$$

at $x = 0$, and all field variables must remain bounded as both x and $|y|$ tend to infinity. (21)

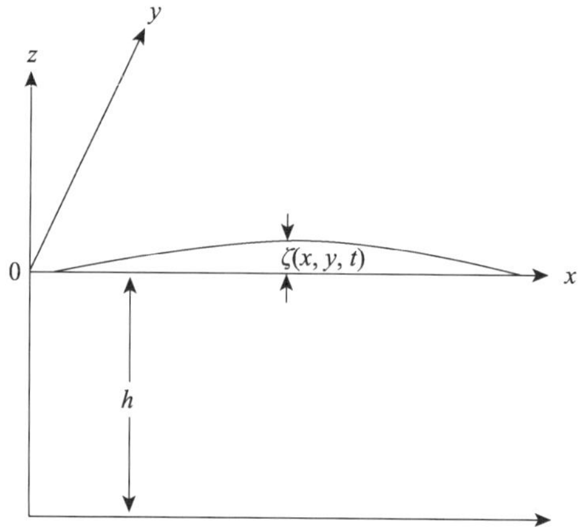

Fig. 1

The initial condition (14) remains unaltered, where

$$\alpha = \frac{g}{f}\left\{h + \frac{r}{2a}\left[\operatorname{sh}(2ah) + \sin(2ah)\right]\right\}$$

$$\beta = \frac{gr}{2fa}\left[\operatorname{sh}(2ah) - \sin(2ah)\right], \quad \gamma = \frac{2r}{f\rho}\operatorname{sh}(ah)\sin(ah)$$

$$\delta = \frac{1}{f\rho}\left[1 - 2r\operatorname{ch}(ah)\cos(ah)\right]$$

$$r = \left[\operatorname{ch}(2ah) + \cos(2ah)\right]^{-1}, \quad a = \pi / d$$

The equation can be solved by the Fourier integral transform with the kernel $K(m,x) = \sqrt{\frac{2}{\pi}}\frac{\beta m\cos(mx) - \mathrm{i}\alpha n\sin(mx)}{\sqrt{\beta^2 m^2 - \alpha^2 n^2}}$. After using the related theorems about the Fourier integral transform, we get an exact solution of Eq. (19) satisfying the conditions (20), (21) and (14)

$$\begin{aligned}\zeta_0(x,y,t) = &\int_0^\infty\int_{-\infty}^\infty \zeta^*(x',y')G_0(x,y;x',y';t)\mathrm{d}x'\mathrm{d}y' \\ &-\int_0^t\int_0^\infty\int_{-\infty}^\infty\left[\gamma\nabla\cdot\boldsymbol{\tau} + \delta\nabla\times\boldsymbol{\tau}\right]G_1(x,y;x',y';t-t')\mathrm{d}t'\mathrm{d}x'\mathrm{d}y' \\ &-\int_0^t\int_{-\infty}^\infty\left[\frac{\gamma}{\beta}\tau_x(0,y',t') + \frac{\delta}{\beta}\tau_y(0,y',t')\right]G_2(x,y;y';t-t')\mathrm{d}t'\mathrm{d}y'\end{aligned} \tag{22}$$

where Green's functions are

$$\begin{aligned}G_0(x,y;x',y';t) = \frac{1}{4\pi}\Bigg\{&\frac{1}{\beta t}\mathrm{e}^{-\frac{(x-x')^2+(y-y')^2}{4\beta t}} + \frac{\beta^2-\alpha^2}{\beta(\alpha^2+\beta^2)t}\mathrm{e}^{-\frac{(x+x')^2+(y-y')^2}{4\beta t}} \\ &-\frac{\sqrt{\pi}\alpha\beta\left[\alpha(x+x') - \beta(y-y')\right]}{\left[\beta(\alpha^2+\beta^2)t\right]^{3/2}}\mathrm{e}^{\frac{\left[\alpha(x+x')+\beta(y-y')\right]^2}{4\beta(\alpha^2+\beta^2)t}}\times\operatorname{erf}\left(\frac{\beta(x+x') - \alpha(y-y')}{2\sqrt{\beta(\alpha^2+\beta^2)t}}\right)\Bigg\}\end{aligned}$$

$$G_1(x,y;x',y';t-t') = G_0(x,y;x',y';t-t')$$

$$G_2(x,y;y';t-t') = \frac{\alpha^2[\alpha x+\beta(y-y')]}{2\pi\beta^2(\alpha^2+\beta^2)(t-t')^2}\mathrm{e}^{-\frac{[\alpha x+\beta(y-y')]^2}{4\beta(\alpha^2+\beta^2)(t-t')}}$$

$$\times F\left(-\frac{3}{2},\frac{3}{2};\frac{[\alpha x+\beta(y-y')]^2}{4\beta(\alpha^2+\beta^2)(t-t')}\right)+\frac{1}{\pi[\beta(t-t')]^{3/2}}\mathrm{e}^{-\frac{\alpha x+\beta(y-y')}{4\beta(\alpha^2+\beta^2)(t-t')}}$$

$$\times \operatorname{erfc}\left(\frac{\alpha(y-y')-\beta x}{2\sqrt{\beta(\alpha^2+\beta^2)(t-t')}}\right)\sum_{k=0}^{\infty}\frac{(-3/2)_k}{(3/2)_k}$$

$$\times\sum_{j=0}^{2k+1}\binom{2k+1}{j}\frac{(-1)^{2k-j+1}\beta^{2(j+1)}}{(\sqrt{\alpha^2+\beta^2})^{2j+1}}\left(\frac{\alpha x+\beta(y-y')}{2\alpha\sqrt{\beta(t-t')}}\right)^{2k+1}$$

$$+\frac{1}{[\pi\beta(t-t')]^{3/2}}\mathrm{e}^{-\frac{x^2+(y-y')^2}{4\beta(t-t')}}\cdot\sum_{k=0}^{\infty}\frac{(-3/2)_k}{(3/2)_k}$$

$$\times\sum_{j=0}^{2k+1}\binom{2k+1}{j}\frac{(-1)^{2k-j+1}\beta^{j+2}}{(\sqrt{\alpha^2+\beta^2})^{j+1}}\cdot\sum_{i=1}^{j}\binom{j}{i}\left(\frac{\alpha x+\beta(y-y')}{2\alpha\sqrt{\beta(t-t')}}\right)^{2k-i+1}$$

$$\times\left\{\left[\frac{\alpha(y-y')-\beta x}{2\sqrt{\beta(\alpha^2+\beta^2)(t-t')}}\right]^{i-1}+\sum_{p=1}^{\infty}\left(\frac{i-1}{2}\right)\right.$$

$$\left.\times\left(\frac{i-3}{2}\right)\cdots\left(\frac{i-2p-1}{2}\right)\left(\frac{\alpha(y-y')-\beta x}{2\sqrt{\beta(\alpha^2+\beta^2)(t-t')}}\right)^{i-2p-1}\right\}$$

and $\operatorname{erf}(x)=\frac{2}{\sqrt{\pi}}\int_0^x \mathrm{e}^{-\xi^2}\mathrm{d}\xi$ is error function, its supplementary function is denoted by $\operatorname{erfc}(x)$, $F(\mu,\nu,x)$ the hypergeometrical function in degenerate case, $(\lambda)_k=\lambda(\lambda+1)(\lambda+2)\cdots(\lambda+k-1)$, $(\lambda)_0=1$.

Eq. (22) shows that the storm-induced elevation change in the zero-order model at a certain point (x,y) and at a certain instant t depends on the contributions of three factors:

(1) The initial elevation at same instant and at points on the semi-infinite sea areas;

(2) The past time history of the horizontal divergence and the vertical vorticity of the wind stress at points on the semi-infinite sea areas;

(3) The past time history of the wind stress at points along the coasts.

The effects of an arbitrary point on a prescribed point are determined by Green's functions G_0, G_1 and G_2. These Green's functions are not symmetrical with respect to horizontal coordinates, i.e., the influence of a point source with unit intensity on the elevation at a prescribed point depends not only upon the distance between them but also on the polar angle. They decrease rapidly, however, both with increasing time t (or $t-t'$) and with increasing horizontal distance between them. This implies that the influences of above-mentioned three physical factors at the point of interest on elevation ζ_0 at a prescribed point will decrease rapidly with increasing time and distance from that point.

These features of Green's functions enable us to approximate the infinite integrals on the right side of (22) by integrals over finite length or finite region, and the theoretical values for the weighted functions can be calculated more conveniently.

It follows also from (22) that there may be a noticeable time delay of the influence at arbitrary point from the wind stress acting on the sea surface and coasts relative to elevation at a prescribed point. This has been verified by the results of a preliminary analysis of the sea surface elevation data for certain port on the west coast of the Pohai Sea[10].

It can be comprehended from the dynamical point of view that the storm surge is generated by the wind field governed by certain storm weather situation in which the scale of the storm is finite, and that the motion of the water is damped somewhat by eddy viscosity and bottom friction. Hence, the time delay has a definite maximal limit, which, for instance, is beyond twenty hours for the Pohai Sea. There is no time delay of initial elevation at the point of interest relative to the elevation at a prescribed point if the inertial oscillations have not been taken into account. It may be expected that the influence of the initial field would be gradually lost.

By integrating Eq. (22) by parts, the resulting formula shows that the wind stress is the direct and unique cause for developing the storm surge. Furthermore, the integrals in the resulting formula can be approximately calculated by conventional method. This enables us to predict the storm-induced elevation numerically.

IV. The First-order Dynamic Model for the Ultra-shallow Water Storm Surges

1. General case

The first-order dynamic model for the ultra-shallow water surges, say, the Pohai Sea surges, can be represented by Eqs. (6)–(14), provided that the subscripts "0", are replaced by "1", for u, v, w and ζ and $z=0$ is replaced by $z=\zeta_1$ in (9) and (12), and a term of $\boldsymbol{v}_1 \cdot \nabla\zeta_1$ is added to the right side of (10), and two additional terms $\frac{\partial u_1}{\partial t}$ and $\frac{\partial v_1}{\partial t}$ are imposed respectively on the right side of (6) and (7) .

The system of equations under consideration will lead to a nonlinear integro-differential equation for ζ_1. Fortunately, it is advantageous to adopt the following approximation by utilizing the dynamical characteristics for the ultra-shallow sea to make the problem considerably simplified and readily generalized.

Since $O\left(\dfrac{\partial u_1}{\partial t}\right)=O(\Xi)$, $O(\boldsymbol{v}_1\cdot\nabla\zeta_1)\leqslant O(\kappa)$ and $O(\Xi)=O(\kappa)=1$, the solutions of the first-order model may be written as $u_1=u_0+u_1^{(1)}$, $v_1=v_0+v_1^{(1)}$, $w_1=w_0+w_1^{(1)}$, $\zeta_1=\zeta_0+\zeta_1^{(1)}$ and $Q_{n1}=Q_{n0}+Q_{n1}^{(1)}$, where quantities with subscripts "0" denote the solutions of the zero-order dynamic model, namely, the fundamental solutions.

Substituting the foregoing expressions into the system of equations and its related conditions in the first-order surge model and neglecting the terms comprising $O(\Xi)=O(\kappa)$ of higher order, we get the system of equations for the perturbation quantities $u_1^{(1)}$, $v_1^{(1)}$, $w_1^{(1)}$, $\zeta_1^{(1)}$ and $Q_{n1}^{(1)}$ as follows:

$$\frac{\partial}{\partial z}\left(\nu\frac{\partial u_1^{(1)}}{\partial z}\right)+fv_1^{(1)}=g\frac{\partial\zeta_1^{(1)}}{\partial x}+\Pi_1(x,y,z,t) \tag{23}$$

$$\frac{\partial}{\partial z}\left(\nu\frac{\partial v_1^{(1)}}{\partial z}\right)-fu_1^{(1)}=g\frac{\partial\zeta_1^{(1)}}{\partial y}+\Pi_2(x,y,z,t) \tag{24}$$

$$\frac{\partial u_1^{(1)}}{\partial x}+\frac{\partial v_1^{(1)}}{\partial y}+\frac{\partial w_1^{(1)}}{\partial z}=0 \tag{25}$$

with the corresponding boundary and initial conditions:

$$\frac{\partial u_1^{(1)}}{\partial z}=\frac{\partial v_1^{(1)}}{\partial z}=0 \tag{26}$$

and

$$w_1^{(1)}=\frac{\partial\zeta_1^{(1)}}{\partial t}+\Theta \tag{27}$$

at surface $z=0$;

$$u_1^{(1)}=v_1^{(1)}=w_1^{(1)}=0 \tag{28}$$

at bottom $z=-h$;

$$Q_{n1}^{(1)}\equiv\cos\alpha'\int_{-h}^{0}u_1^{(1)}\mathrm{d}z+\cos\beta'\int_{-h}^{0}v_1^{(1)}\mathrm{d}z=0 \tag{29}$$

along the coasts C_1, and

$$\text{either } Q_{n1}^{(1)}=0 \text{ or } \zeta_1^{(1)}=0 \tag{30}$$

along the open sea boundary C_2, and

$$\zeta_1^{(1)}\Big|_{t=0}=0 \tag{31}$$

where

$$\Pi_1=\frac{\partial u_0}{\partial t},\quad \Pi_2=\frac{\partial v_0}{\partial t}\quad\text{and}\quad \Theta=(\boldsymbol{v}_0\cdot\nabla\zeta_0)_{z=0}$$

It follows that the initial-boundary value problem for the first-order surge model is transformed into a system of Eqs. (23)–(31) ; hence a nonlinear integro-differential equation for $\zeta_1^{(1)}$ under the prescribed boundary and the initial conditions has been reduced to a simple differential equation for $\zeta_1^{(1)}$. Moreover, the first-order dynamic model can be generalized such that the known functions Π_1, Π_2 and Θ may consist of more terms such as the nonlinear convective acceleration, etc. Then we may write Π_1, Π_2 and Θ in their generalized forms:

$$\Pi_1 = \Pi_1(x,y,z,t), \quad \Pi_2 = \Pi_2(x,y,z,t), \quad \Theta = \Theta(x,y,t)$$

We obtain then without difficulty

$$u_1^{(1)} = q_{1\mathrm{R}}^{(1)} \tag{32}$$

$$v_1^{(1)} = q_{1\mathrm{I}}^{(1)} \tag{33}$$

where

$$q_1^{(1)} = \frac{\Gamma_1^{(1)} g}{k^2 \nu}\left[\frac{\mathrm{ch}(kz)}{\mathrm{ch}(kh)} - 1\right] + \frac{1}{k}\mathrm{sh}(kz)\int_0^z \frac{\Pi}{\nu}\mathrm{ch}(kz')\mathrm{d}z'$$
$$-\frac{1}{k}\mathrm{ch}(kz)\left[\int_{-h}^z \frac{\Pi}{\nu}\mathrm{sh}(kz')\mathrm{d}z' + \mathrm{th}(kh)\int_{-h}^0 \frac{\Pi}{\nu}\mathrm{ch}(kz')\mathrm{d}z'\right]$$

$$\Gamma_1^{(1)} = \frac{\partial \zeta_1^{(1)}}{\partial x} + \mathrm{i}\frac{\partial \zeta_1^{(1)}}{\partial y}$$

The governing equation for $\zeta_1^{(1)}$ may be written in the form

$$\begin{aligned}\frac{\partial \zeta_1^{(1)}}{\partial t} = \frac{g}{\nu}A_\mathrm{R}\Delta\zeta_1^{(1)} + \left[\frac{\partial}{\partial x}\left(\frac{g}{\nu}A_\mathrm{R}\right) + \frac{\partial}{\partial y}\left(\frac{g}{\nu}A_\mathrm{I}\right)\right]\frac{\partial \zeta_1^{(1)}}{\partial x} \\ + \left[\frac{\partial}{\partial y}\left(\frac{g}{\nu}A_\mathrm{R}\right) - \frac{\partial}{\partial x}\left(\frac{g}{\nu}A_\mathrm{I}\right)\right]\frac{\partial \zeta_1^{(1)}}{\partial y} + E(x,y,t)\end{aligned} \tag{34}$$

with boundary and initial conditions (29)–(31), where

$$E(x,y,t) = \frac{\partial}{\partial x}\int_{-h}^0 \left[D_\mathrm{R}(z')\Pi_1(x,y,z',t) - D_\mathrm{I}(z')\Pi_2(x,y,z',t)\right]\mathrm{d}z'$$
$$+\frac{\partial}{\partial y}\int_{-h}^0 \left[D_\mathrm{I}(z')\Pi_1(x,y,z',t) + D_\mathrm{R}(z')\Pi_2(x,y,z',t)\right]\mathrm{d}z' - \Theta(x,y,t)$$

$$D(z) = \frac{1}{k^2\nu}\left[1 - \frac{\mathrm{ch}(kz)}{\mathrm{ch}(kh)}\right]$$

$$Q_{n1}^{(1)} = -\left[\frac{g}{\nu}A_\mathrm{R}\frac{\partial \zeta_1^{(1)}}{\partial n} + \frac{g}{\nu}A_\mathrm{I}\frac{\partial \zeta_1^{(1)}}{\partial s} + \int_{-h}^0 D_\mathrm{R}(z')\Pi_n \mathrm{d}z' + \int_{-h}^0 D_\mathrm{I}(z')\Pi_s \mathrm{d}z'\right]$$

$$\Pi_n = \Pi_1\cos\alpha' + \Pi_2\cos\beta', \quad \Pi_s = -\Pi_2\cos\alpha' + \Pi_1\cos\beta'$$

Equation (34) for $\zeta_1^{(1)}$ may, of course, be solved by finite difference method for an arbitrary sea area. The storm-induced current $u_1(x,y,t)$ and $v_1(x,y,t)$ can thus be obtained.

2. Determination of storm surges on the sea areas of continental shelf

We again consider the storm surges on the sea areas over a continental shelf as given in Part 2 of Section III. The storm-induced current velocities are immediately found to be

$$u_1 = q_{0\mathrm{R}} + q_{1\mathrm{R}}^{(1)} \tag{35}$$

$$v_1 = q_{0\mathrm{I}} + q_{1\mathrm{I}}^{(1)} \tag{36}$$

where

$$\begin{aligned} q_1^{(1)} = & \frac{1}{8\rho k^3 \nu^2 \mathrm{ch}(kh)} \{4kh\mathrm{ch}(kz) - \mathrm{sh}(2kh)\mathrm{e}^{-kz} - 2\mathrm{e}^{kh}\mathrm{ch}(kh)\mathrm{sh}(kz) \\ & + 4k\mathrm{ch}(kh)z\mathrm{ch}[k(z+h)] - 2\mathrm{ch}(kh)\mathrm{sh}[k(z+h)]\}\frac{\partial \tau}{\partial t} \\ & - \frac{g}{2k^4\nu^2\mathrm{ch}^2(kh)}\{[kh\mathrm{sh}(kh) + 2\mathrm{ch}(kh)]\mathrm{ch}(kz) - k\mathrm{ch}(kh)z\mathrm{sh}(kz) \\ & - 2\mathrm{ch}^2(kh)\}\frac{\partial \Gamma_0}{\partial t} + \frac{g\Gamma_1}{k^2\nu\mathrm{ch}(kh)}[\mathrm{ch}(kz) - \mathrm{ch}(kh)] \end{aligned}$$

It is readily seen that the bottom stress is a complicated function of time and spatial coordinates in analogy to that of the zero-order model.

The dependent variable $\zeta_1^{(1)}$ under consideration is now governed by the differential equation,

$$\frac{\partial \zeta_1^{(1)}}{\partial t} + C_\mathrm{R}\Delta\zeta_1^{(1)} = -A_\mathrm{R}'\frac{\partial}{\partial t}(\nabla\cdot\boldsymbol{\tau}) + A_\mathrm{I}'\frac{\partial}{\partial t}(\nabla\times\boldsymbol{\tau}) - B_\mathrm{R}'\frac{\partial}{\partial t}\Delta\zeta_0 - \Theta \tag{37}$$

which is subjected to the conditions:

$$\frac{\partial \zeta_1^{(1)}}{\partial x} = \frac{C_\mathrm{I}}{C_\mathrm{R}}\frac{\partial \zeta_1^{(1)}}{\partial y} + \frac{1}{C_\mathrm{R}}\left(A_\mathrm{I}'\frac{\partial \tau_y}{\partial t} - A_\mathrm{R}'\frac{\partial \tau_x}{\partial t}\right) + \frac{1}{C_\mathrm{R}}\left(B_\mathrm{I}'\frac{\partial^2\zeta_0}{\partial t\partial y} - B_\mathrm{R}'\frac{\partial^2\zeta_0}{\partial t\partial x}\right) \tag{38}$$

at $x = 0$, and (21) & (31), where

$$A' = \frac{1}{4\rho k^4\nu^2\mathrm{ch}(kh)}\{2kh\mathrm{th}(kh) + (1-\mathrm{e}^{kh})\mathrm{sh}(kh) + (3-\mathrm{e}^{kh})[1-\mathrm{ch}(kh)]\}$$

$$B' = \frac{g}{2k^5\nu^2}\{3[kh - \mathrm{th}(kh)] - kh\mathrm{th}^2(kh)\}$$

$$C = \frac{g}{k^3\nu}[\mathrm{th}(kh) - kh]$$

$$\begin{aligned} \Theta(x,y,t) = (\boldsymbol{v}_0\cdot\nabla\zeta_0)_{z=0} = & \frac{r}{2\rho a\nu}\{[\sin(2ah) + \mathrm{sh}(2ah)]\cdot(\boldsymbol{\tau}\cdot\nabla\zeta_0) + [\sin(2ah) - \mathrm{sh}(2ah)] \\ & \cdot\left(\tau_x\frac{\partial\zeta_0}{\partial y} - \tau_y\frac{\partial\zeta_0}{\partial x}\right) - \frac{2gr}{f}\sin(ah)\mathrm{sh}(ah)\left[\left(\frac{\partial\zeta_0}{\partial x}\right)^2 + \left(\frac{\partial\zeta_0}{\partial y}\right)^2\right] + 2\frac{\partial\zeta_0}{\partial x}\cdot\frac{\partial\zeta_0}{\partial y} \end{aligned}$$

As before, the solution of Eq. (37) satisfying (38), (21) and (31) may be obtained straightforward by the Fourier method.

$$\zeta_1^{(1)}(x,y,t)=\int_0^t\int_0^\infty\int_{-\infty}^{+\infty}\left[A_{\mathrm{I}}'\frac{\partial}{\partial t'}(\nabla\times\boldsymbol{\tau})-A_{\mathrm{R}}'\frac{\partial}{\partial t'}(\nabla\times\boldsymbol{\tau})\right.$$
$$\left.-B_{\mathrm{R}}'\frac{\partial}{\partial t'}\nabla\zeta_0-\Theta\right]G_3(x,y;x',y';t-t')\mathrm{d}t'\mathrm{d}x'\mathrm{d}y'-\int_0^t\int_{-\infty}^\infty\left[\frac{A_{\mathrm{I}}'}{C_{\mathrm{R}}}\frac{\partial\tau_y}{\partial t'}\right. \tag{39}$$
$$\left.-\frac{A_{\mathrm{R}}'}{C_{\mathrm{R}}}\frac{\partial\tau_x}{\partial t'}+\frac{B_{\mathrm{I}}'}{C_{\mathrm{R}}}\frac{\partial^2\zeta_0}{\partial t'\partial y'}-\frac{B_{\mathrm{R}}'}{C_{\mathrm{R}}}\frac{\partial^2\zeta_0}{\partial t'\partial x'}\right]G_4(x,y;y';t-t')\mathrm{d}t'\mathrm{d}y'$$

The Green functions G_3 and G_4 are equal to G_1 and G_2 with $-C_{\mathrm{I}}$ and $+C_{\mathrm{R}}$ in replacement of coefficients α and β, respectively. Thus, the characteristics of G_1 and G_2 hold also for them.

Eq. (39) shows that the elevation $\zeta_1^{(1)}$ at a certain point and at a certain instant depends upon the contributions of the following physical factors: (1) the past time history of the rate of change of horizontal divergence and vertical vorticity of the wind stress field, (2) the past time history of the rate of change of Laplacian of the elevation and of the advective effects of the elevations in the zero-order surge model at points on the semi-infinite sea areas, (3) the past time history of the rate of change of the wind stress components and the elevation gradient in the zero-order surge model at points along the coasts.

After integrating (39) by parts and combining with (22), it is found that the elevation $\zeta_1^{(1)}$ depends also on the past time history of the initial elevations and the wind stresses at points over the semi-infinite sea areas and at points along the coast as well as the nonlinear interaction between the fields.

By comparing (22) with (39), it will be seen that the initial elevations which affect $\zeta_1^{(1)}$ are also delayed in time relative to $\zeta_1^{(1)}$ owing to the inclusion of the terms $\frac{\partial u_0}{\partial t}$ and $\frac{\partial v_0}{\partial t}$ involved in the first-order surge model. This demonstrates that the inertial oscillation is the essential reason for the time delay of the initial elevation relative to the predicting elevation.

As mentioned above, the inertial oscillations will be damped out quickly by friction, especially for the ultra-shallow water surges. Accordingly, the effects of the time history of elevations on the predicting elevations will disappear rapidly, and the time delay will no longer hold on.

The zero-order dynamic model is the fundamental and principal one for the ultra-shallow water surges, while higher order approximation plays only an additional and amendatory role.

V. A Simple Analysis of the Mechanism of the Ultra-shallow Storm Surges

Rossby[9] pointed out that in the large-scale motion of ocean currents, there would have been the so-called geostrophic adaptation between the pressure and the velocity field. Is there a process of adaptation between the fields in the case of ultra-shallow sea? If so, what is the physical nature of the adaptation? These questions will be answered in this section.

We limit our discussion to the simple case of the linear model, i.e., two-dimensional storm surges occurring in the shallow channel of both uniform depth and of finite length, in order to give a comparison with Harris' results by the whole current method.

Let L be the length of the shallow channel, $0 \leqslant x \leqslant L$ be the extent of water. Assume $\tau = \tau_x$, let both ν and h be constants and neglect the rotation effect of the earth. Then the equations for the storm surge become

$$\frac{\partial u}{\partial t} = -g\frac{\partial \zeta}{\partial x} + \nu\frac{\partial^2 u}{\partial z^2} \tag{40}$$

$$\frac{\partial \zeta}{\partial t} + \frac{\partial U}{\partial x} = 0 \tag{41}$$

The boundary conditions are

$$\rho\nu\frac{\partial u}{\partial z} = \tau \tag{42}$$

at surface $z = 0$;

$$u = 0 \tag{43}$$

at bottom $z = -h$;

and

$$U\big|_{x=0} = U\big|_{x=L} = 0 \tag{44}$$

The initial conditions are

$$U\big|_{t=0} = 0, \quad \zeta\big|_{t=0} = 0 \tag{45}$$

The storm-induced current velocity is readily found to be

$$\begin{aligned} u(x,z,t) = &\sum_{n=0}^{\infty}\cos\left(n+\frac{1}{2}\right)\frac{\pi z}{h}\int_0^t \frac{2g(-1)^{n+1}}{\left(n+\frac{1}{2}\right)\pi} e^{-\nu\left(n+\frac{1}{2}\right)^2\frac{\pi^2}{h^2}(t-t')}\frac{\partial \zeta(x',t')}{\partial x}\mathrm{d}t' \\ &+\sum_{n=0}^{\infty}\cos\left(n+\frac{1}{2}\right)\frac{\pi z}{h}\int_0^t \frac{2}{\rho h}\tau(x,t')e^{-\nu\left(n+\frac{1}{2}\right)^2\frac{\pi^2}{h^2}(t-t')}\mathrm{d}t' \end{aligned} \tag{46}$$

It is easily seen from Eq. (46) that the bottom friction is not so simple as that presumed on the basis of experience.

Introduce an image function $\varphi(x,t)$ defined by

$$U = \frac{\partial \varphi}{\partial t}, \quad \zeta = -\frac{\partial \varphi}{\partial x} \tag{47}$$

such that φ satisfies Eq. (41) .

Integrating (46) with respect to z from bottom to surface, we get, with the relations (47), an integro-differential equation for $\varphi(x,t)$:

$$\frac{\partial\varphi}{\partial t}=\int_0^t\frac{\partial^2\varphi(x,t')}{\partial x^2}G(t-t')\mathrm{d}t'+F(x,t) \tag{48}$$

where

$$G(t-t')=\frac{2gh}{\pi^2}\sum_{n=0}^{\infty}\frac{1}{\left(n+\frac{1}{2}\right)^2}\mathrm{e}^{-\nu\left(n+\frac{1}{2}\right)^2\frac{\pi^2}{h^2}(t-t')}$$

$$F(x,t)=\frac{2}{\rho\pi}\int_0^t\sum_{n=0}^{\infty}\frac{(-1)^n}{n+\frac{1}{2}}\mathrm{e}^{-\nu\left(n+\frac{1}{2}\right)^2\frac{\pi^2}{h^2}(t-t')}\tau(x,t')\mathrm{d}t'$$

and the boundary and initial conditions (44), (45) become

$$\varphi(0,t)=\varphi(L,t)=0 \tag{49}$$

and

$$\varphi(x,0)=0 \tag{50}$$

After some manipulations, the problem (48)–(50) may be reduced to a Volterra integral equation of the second kind as follows:

$$\overline{\varphi}_p(t)=\lambda_p\int_0^t\mathscr{K}(t-t')\overline{\varphi}_p(t')\mathrm{d}t'+\mathscr{F}(t) \tag{51}$$

where

$$\varphi(x,t)=\sum_{p=1}^{\infty}\overline{\varphi}_p(t)\sin\left(\frac{p\pi}{L}x\right) \tag{52}$$

$$\mathscr{K}(t-t')=\frac{6}{\pi^4}\sum_{n=0}^{\infty}\frac{1}{\left(n+\frac{1}{2}\right)^4}\mathrm{e}^{-\nu\left(n+\frac{1}{2}\right)^2\frac{\pi^2}{h^2}(t-t')}-1$$

$$\mathscr{F}(t)=\frac{2}{L}\int_0^L\sin\left(\frac{p\pi}{L}x'\right)\mathrm{d}x'\int_0^t\mathscr{H}(t-t')\tau(x,t')\mathrm{d}t'$$

$$\mathscr{H}(t-t')=\frac{2h^2}{\rho\nu\pi^3}\sum_{n=0}^{\infty}\frac{(-1)^{n+1}}{\left(n+\frac{1}{2}\right)^3}\mathrm{e}^{-\nu\left(n+\frac{1}{2}\right)^2\frac{\pi^2}{h^2}(t-t')}+\frac{h^2}{2\rho\nu}$$

$$\lambda_p=\frac{gh^3}{3\nu}\left(\frac{p\pi}{L}\right)^2$$

The solution of Eq. (51) may be written as

$$\overline{\varphi}_p(t)=\int_0^t W_p(t-t')Y_p(t')\mathrm{d}t' \tag{53}$$

where

$$Y_p(t')=\frac{2}{L}\int_0^L\tau(x',t')\sin\left(\frac{p\pi}{L}x'\right)\mathrm{d}x'$$

$$W_p(t-t')=\mathscr{H}(t-t')+\lambda_p\int_{t'}^{t}\mathscr{H}(s-t')\Gamma(t-s;\lambda_p)\mathrm{d}s$$

$$\Gamma(t-s;\lambda_p)=\sum_{m=1}^{\infty}\lambda_p^{m-1}\mathscr{K}_m(t-s)$$

$$\mathscr{K}_m(t-s)=\int_s^t\mathscr{K}(t-\vartheta)\mathscr{K}_{m-1}(\vartheta-s)\,\mathrm{d}\vartheta$$

$$\mathscr{K}_1(t-s)=\mathscr{K}(t-s)$$

Since the kernel $\mathscr{K}(t-t')$ is bounded and $\mathscr{F}(t)$ is a continuous function for t, a sequence of the approximate solution should be absolutely and uniformly convergent for arbitrary values of λ_p.

Combining (47), (52) with (53), we get the required solution for the surge elevation

$$\zeta(x,t)=-\frac{\pi}{L}\sum_{p=1}^{\infty}p\int_0^t W_p(t-t')Y_p(t')\cos\left(\frac{p\pi}{L}x\right)\mathrm{d}t' \tag{54}$$

With the aid of the solution (54), it is possible to reveal an important mechanism of the storm surges which cannot be attained by the solution of the system of total transport equations.

1. Distinguishability of adaptation and evolution

It has been demonstrated previously that the term of the time derivative is one order of magnitude smaller than the main terms in the equation of motion for the ultra-shallow water surges. Hence the motion of the water particles in the ultra-shallow sea is always in quasi-balanced state of forces characterized by the zero-order dynamic model. If, however, this quasi-balanced state is disturbed in some way, there must undergo a vigorous change. Accordingly, a mechanism of restoration has to be brought about to maintain a quasi-balanced state of forces constantly. The motion of water particles proceeds slowly under the condition of quasi-balanced forces until the quasi-steady state is ultimately reached.

Therefore, it may be concluded that the ultra-shallow water surges develop through many repetitions of the process of set-up and breakdown of quasi-balanced state of forces.

The foregoing two different stages may also be distinguished by time. Analysis of the solution (54) will give respectively their magnitudes below.

If the motion proceeds such that the term comprising the factor of $\mathrm{e}^{-\nu\left(n+\frac{1}{2}\right)^2\frac{\pi^2}{h^2}(t-t')}$ in Eq. (54) becomes negligibly small, it follows that

$$\mathscr{K}_m(t-s)=-\frac{(s-t)^{m-1}}{(m-1)!},\quad \Gamma(t-s;\lambda_p)=-\mathrm{e}^{-\lambda_p(t-s)}$$

$$W_p(t-t')=\frac{h}{2\rho\nu}\mathrm{e}^{-\lambda_p(t-t')}$$

Hence, Eq. (54) is reduced just to the solution for the storm-induced elevation under quasi-balanced state,

$$\zeta(x,t)=\int_0^t\int_0^L \tau(x',t)G(x,x';t-t')\mathrm{d}t'\mathrm{d}x' \tag{54)'$$

where

$$G(x,x';t-t')=-\frac{\pi h^2}{\rho\nu L}\sum_{p=1}^{\infty} p\mathrm{e}^{-\left(\frac{p\pi}{L}\right)^2\frac{gh^3}{3\nu}(t-t')}\cos\left(\frac{p\pi}{L}x\right)\sin\left(\frac{p\pi}{L}x'\right)$$

It means that the elevation approaches firstly its quasi-balanced value at speed $\geqslant \mathrm{e}^{-\frac{\nu\pi^2 t}{4h^2}}$, and subsequently approaches its quasi-steady value at speed $\geqslant \exp\left[-\left(\frac{\pi^2}{3}\frac{gh}{L^2}\Big/\frac{\nu}{h^2}\right)t\right]$. Strictly speaking, both the former adaptation process and the latter evolution process could only be completed as t approaches infinity, but we can define two characteristic values T_1 and T_2 by $\mathrm{e}^{-\frac{\pi^2\nu}{4h^2}T_1}\ll 1$ and $\exp\left[-\left(\frac{\pi^2}{3}\frac{gh}{L^2}\Big/\frac{\nu}{h^2}\right)T_2\right]\ll 1$, respectively, such that the adaptation process ends beyond T_1 and the evolution process ends beyond T_2. The ratio of T_1 to T_2 is equal to $\frac{4}{3}\varXi$, provided that the magnitude of $\mathrm{e}^{-\pi}$ is very small compared with unity. As an example, it is known that $O(\varXi)=10^{-1}$ for the Pohai Sea, thus $T_1\approx 10^4$ sec and $T_2\approx 10^5$ sec.

Similarly, it may be expected that, as the interval of time becomes sufficiently large, the solution for the storm-induced current (46) also approaches its quasi-balanced value at a first asymptotical speed and then approaches its quasi-steady value at a second asymptotical speed.

Evidently the storm surges develop over the shallow water by the surface wind stress. If the quasi-balanced state of forces, i.e., the approximate balance between the horizontal pressure-gradient force expressed as the surface slope and the horizontal frictional force due to the vertical turbulent transfer, is disturbed by storms, the excited inertial oscillations would be damped out in a few hours by the eddy viscosity and bottom friction, while the mutual adjustment between the pressure and the velocity field proceeds with characteristic time of 10^4 sec or less so as to restore the quasi-balanced state of forces. Subsequently, the fields evolute slowly under the condition of quasi-balanced forces with characteristic time of 10^5 sec. Such a phenomenon is in accordance with the observed fact. Hence, the two stages of motion are distinguished by time.

Although the foregoing discussion is based on a simple linear dynamic model, the conclusions thus obtained may be regarded as an illustration to the analysis of the governing equations of motion in Section II.

The foregoing analysis of the surge problem shows that, for $T_1 \ll T_2$ or $O(\Xi) \ll 1$, we have the solution (54)′ for the quasi-balanced state. This conclusion is of prime importance and justifies initial assumptions analytically.

It should be noted that Harris' solution by the whole current method for the same dynamic model does not offer us the above physical mechanism because his solution approaches its quasi-steady value for one time only at the speed of $e^{-\frac{r_1}{2}t}$.

For a non-ultra-shallow sea, the time derivative is not necessarily less than the main terms involved in the equation of motion. Hence, no distinction between the adaptation and the evolution by time process is feasible. It implies once more the theoretical and practical significance for subdividing the theory of shallow water surge into the ordinary shallow water theory and the ultra-shallow water theory.

2. The physical nature of adaptation and evolution

For the problem of the ultra-shallow water surges, the adaptation and the evolution of the pressure and the velocity field may be distinguished explicitly by their characteristic time as mentioned above. It is expected of course that these two stages would be different in their physical nature.

It is readily seen from solutions (52) and (53) or (54) that the adaptation process is oscillatory. In fact, it may be found by expanding the solutions that they consist of a series of damped "simple waves", their damping factors being proportional to $e^{-1/\Xi}$. The more the quasi-balanced state is disturbed, i.e., the greater the order of magnitude of Ξ is, the more intense these 'waves' are. The 'waves' will, however, be vanished in a comparatively short interval of time because of the strong damping by eddy viscosity. Moreover, it follows, as mentioned above, that the more the quasi-balanced state is disturbed, the shorter the duration of the adaptation is. When $O(\Xi) < 1$, the quasi-balanced state of forces is ultimately estab lished, i.e., the adaptation stage is transformed into the evolution stage and the solution (54) degenerates into (54)'.

The solution (54)' for the quasi-balanced state shows no oscillatory feature. It follows that the chief distinction by the physical nature between the adaptation and the evolution is the oscillatory feature.

It is also seen from (46) that the horizontal velocity divergence is related to the distribution of the divergence of the wind stress and the storm-induced elevation during the adaptation. Hence, the divergence (or convergence) of the wind stress should be the source (or sink) of the horizontal velocity divergence (or convergence).

If the quasi-balanced state of forces is in some way broken down severely under the influence of the surface wind stress, the intensive horizontal velocity divergence will be excited. The condition of continuity will lead to strong upwelling and downwelling and hence the set-up and falling of the sea surface. Variations of surface slope will adjust the pressure field to a new state of balance of forces rapidly. After the adaptation stage is accomplished, the motion of sea water then transits into slow evolution.

VI. Conclusions

The conclusions from the foregoing analysis may be summarized as follows:

(1) On the basis of the dynamical characteristics of shallow water surges, three basic dimensionless parameters, Ξ , κ and ε , are derived, which measure the relative importance of inertial oscillation and nonlinearity, respectively. The dynamical conditions for subdividing the theory of shallow water storm surge into two parts, namely, the ordinary shallow water theory and the ultra-shallow water theory, are obtained.

(2) Studies on the Pohai Sea surges belong to the category of the ultra-shallow water theory. The zero- and first-order three-dimensional dynamic models for the Pohai Sea surges are derived. It is shown that the zero-order linear dynamic model is of prime importance for the ultra-shallow water surges such as the Pohai Sea surges.

(3) For the ultra-shallow sea, say, the Pohai Sea, the structure of the storm-induced currents in analytical form and the equations capable for numerical prediction of storm surge amplitude both in the zero- and first-order dynamic models are given. Moreover, by considering the sea areas over continental shelf, an exact solution for the elevation, which may probably be used for numerical analysis and prediction of the ultra-shallow water surges, is derived. The present method does not require method of finite difference in numerical computations.

(4) The bottom stress is certainly a complicated function of time and spatial coordinates. The assumptions of the bottom stress in the studies of storm surges by the whole current method seem questionable.

(5) There is a quasi-balanced state between the pressure and the velocity field for the ultra-shallow water surges just as the large-scale quasi-geostrophic horizontal motion in the ocean. For example, in the Pohai Sea, the turbulent frictional force, the Coriolis force and the pressure-gradient force due to the surface slope should be in quasi-balanced state.

(6) For the ultra-shallow water surges, if the quasi-balanced state of forces is disturbed by the meteorological forcing functions, the quasi-balanced state is restored rapidly owing to the damping effects of eddy viscosity and bottom friction. Subsequently, the motion proceeds slowly under the condition of quasi-balanced forces until it transits to a quasi-steady state. These two processes are different in their physical features, and hence the former is called adaptation while the latter is called evolution. An analysis of linear surge model for two-dimensional shallow channel of uniform depth shows that it is oscillatory in the adaptation stage but non-oscillatory in evolution stage. The ultra-shallow water storm surges develop through many re-petitions of the process of set-up and breakdown of the quasi-

balanced state of forces.

(7) The processes of adaptation and evolution could be distinguished by their characteristic times. For the Bohai Sea, the characteristic time of adaptation is 10^4 sec (~several hours), and the characteristic time of evolution is 10^5 sec (~several days) .

The authors are indebted to Prof. C. B. Liu (刘智白) for her kind assistance they have received in preparing this paper. We would like especially to express our deep gratitude to Prof. C. P. Ho (赫崇本), who reviewed very carefully the original manuscript and gave us encouragement during the course of this study.

References

[1] Welander, P.: *Advances in Geophysics*, 8, pp. 315–379, 1961.
[2] Groen, P. & Groves, G. W.: *The Sea*, 1, "Surges", 1962.
[3] Harris, D. L.: *Mon. Wea. Rev.*, 90, 331–340, 1962.
[4] 宫崎正衛: 海洋科学基础讲座 3, 物理海洋 III, 第三篇, 第三章, 高潮, 1970.
[5] Isozaki, I.: *Papers in Meteor. and Geophys.*, 21, 421–448, 1970.
[6] Jelesniaski, C. P.: *Mon. Wea. Rev.*, 98, 462–478, 1970.
[7] Welander, P.: *Tellus*, 9, 43–52, 1957.
[8] Heaps, N. S.: *Mémoires Société Royale des Sciences de Liege, 6ᵉ série*, tome I, 143–180, 1971.
[9] Rossby, C. G.: *J. Mar. Res.*, 1, 239–263, 1937–1938.
[10] 山东海洋学院风暴潮研究小组: 《渤海风暴潮的机制和预报》(I), (III), 1974.

A Three-dimensional Nonlinear Model of Tides*

Feng Shih-Zao (冯士筰)

Considerable progress has been made on tidal theory and tidal calculations during the recent twenty five years or so[1-4]. Walter Hansen has developed a so-called tidal "boundary-value problem" which is a notable one and its application to North Sea and other seas has been fairly successful. Later L. Sgibneva and A. Felsenbaum extended the theory to a three-dimensional one, but it is also linear[5]. Apparently such theories should not be applied to shallow water area, where non-linear shallow water constituents may become pronounced. Since our country faces vast shallow seas, the study aiming at establishing some suitable relations for our shallow water seas becomes essential. The main idea of this analysis is different from that of "hydrodynamic-numerical method" that has been widely adopted; we would rather follow the line of Hansen's "boundary-value problem".

Considering the fact that tidal waves are gravitational long-waves by nature, we take up the problem of tides in nondimensional forms as follows:

$$\frac{\partial u}{\partial x}+\frac{\partial v}{\partial y}+\frac{\partial w}{\partial z}=0$$

$$\frac{\partial u}{\partial t}+\chi\left(u\frac{\partial u}{\partial x}+v\frac{\partial u}{\partial y}+w\frac{\partial u}{\partial z}\right)-\Omega v=-\frac{\partial \zeta}{\partial x}+\frac{\partial}{\partial z}\left(\nu\frac{\partial u}{\partial z}\right)+\omega_x$$

$$\frac{\partial v}{\partial t}+\chi\left(u\frac{\partial v}{\partial x}+v\frac{\partial v}{\partial y}+w\frac{\partial v}{\partial z}\right)+\Omega u=-\frac{\partial \zeta}{\partial y}+\frac{\partial}{\partial z}\left(\nu\frac{\partial v}{\partial z}\right)+\omega_y \tag{1}$$

The boundary conditions are
at the sea surface $z=\chi\zeta$:

$$\begin{cases} w=\dfrac{\partial \zeta}{\partial t}+\chi\left(u\dfrac{\partial \zeta}{\partial x}+v\dfrac{\partial \zeta}{\partial y}\right) \\ \dfrac{\partial u}{\partial z}=\dfrac{\partial v}{\partial z}=0 \end{cases}$$

at the sea bottom $z=-h$:

$$u=v=w=0$$

along the shore boundary C_1:

$$\cos\alpha_x\int_{-h}^{\chi\zeta}u\mathrm{d}z+\cos\alpha_y\int_{-h}^{\chi\zeta}v\mathrm{d}z=0$$

along the open boundary C_2:

$$\zeta=S$$

---

* Feng S-Z. 1977. A Three-dimensional nonlinear model of tides. Science in China, Ser. A, XX(04): 436–446

where

$$(x,y)=\frac{(\tilde{x},\tilde{y})}{L},\quad z=\frac{\tilde{z}}{h_0},\quad t=\frac{\tilde{t}}{L/\sqrt{gh_0}},\quad h=\frac{\tilde{h}}{h_0}$$

$$(u,v)=\frac{(\tilde{u},\tilde{v})}{\left(R\sqrt{g/h_0}\right)},\quad w=\frac{\tilde{w}}{\left(R\sqrt{gh_0}/L\right)},\quad \zeta=\frac{\tilde{\zeta}}{R}$$

$$\Omega=\frac{\tilde{\Omega}L}{\sqrt{gh_0}},\quad \nu=\frac{\tilde{\nu}}{h_0^2}\cdot\frac{L}{\sqrt{gh_0}},\quad (\omega_x,\omega_y)=\frac{(\tilde{\omega}_x,\tilde{\omega}_y)}{(gR/L)},\quad S=\frac{\tilde{S}}{R}$$

$\chi=R/h_0$—a non-linear parameter.

In the above $\tilde{x},\tilde{y},\tilde{z}$ constitute a Cartesian co-ordinate system at the right-hand side, the plane $\tilde{x},\tilde{y}$ coincides with the undisturbed sea surface, and $\tilde{z}$ is positive upward; $\tilde{t}$ denotes time; $\tilde{u},\tilde{v}$ denote the components of the tidal current in $\tilde{x},\tilde{y}$ directions respectively, and $\tilde{w}$ represents the vertical component; $\tilde{\zeta}$ is the elevation of the tide measured from the undisturbed sea surface; $\tilde{\Omega}$ is the Coriolis parameter; $\tilde{\nu}$ is the coefficient of vertical eddy viscosity; g is the gravitational acceleration; $\tilde{\omega}_x,\tilde{\omega}_y$ denote two components of tide-generating forces; $\tilde{S}$ denotes the tidal elevation along the open boundary; h_0, L and R are the vertical, horizontal characteristic lengths and the characteristic amplitude of the tidal elevation respectively; $\tilde{h}$ is the depth; $\cos\alpha_x,\cos\alpha_y$ denote the direction—cosines of the boundary normal $\boldsymbol{n}$, and assume

$$\alpha_x=\alpha_x(x,y),\quad \alpha_y=\alpha_y(x,y)$$

Evidently the problem (1) described above is a nonlinear one. Nevertheless, by means of the perturbation method, problem (1) can be reduced to a series of linear sets. Taking χ as a perturbation quantity, we have the solutions:

$$(\zeta,u,v,w)=\sum_{j=0,1,\cdots}\chi^j(\zeta_j,u_j,v_j,w_j)\tag{2}$$

Similarly, the boundary condition can be expressed as

$$S=\sum_{j=0,1,\cdots}\chi^jS_j,\quad S_j=\begin{cases}S & (j=0)\\ 0 & (j=1,2,\cdots)\end{cases}$$

where ζ_j, u_j, v_j, w_j satisfy the following problem:

$$\begin{gathered}\frac{\partial u_j}{\partial x}+\frac{\partial v_j}{\partial y}+\frac{\partial w_j}{\partial z}=0\\ \frac{\partial u_j}{\partial t}-\Omega v_j=-\frac{\partial\zeta_j}{\partial x}+\frac{\partial}{\partial z}\left(\nu\frac{\partial u_j}{\partial z}\right)+\begin{cases}\omega_x & (j=0)\\ -{}_xE_{j-1} & (j=1,2,\cdots)\end{cases}\\ \frac{\partial v_j}{\partial t}+\Omega u_j=-\frac{\partial\zeta_j}{\partial y}+\frac{\partial}{\partial z}\left(\nu\frac{\partial v_j}{\partial z}\right)+\begin{cases}\omega_y & (j=0)\\ -{}_yE_{j-1} & (j=1,2,\cdots)\end{cases}\end{gathered}\tag{3}$$

at the sea surface $z=0$:

$$\begin{cases} w_j = \dfrac{\partial \zeta_j}{\partial t} + \begin{cases} 0 & (j=0) \\ F_{j-1} & (j=1,2,\cdots) \end{cases} \\ \dfrac{\partial u_j}{\partial z} = \begin{cases} 0 & (j=0), \\ {}_x\Gamma_{j-1} & (j=1,2,\cdots), \end{cases} \quad \dfrac{\partial v_j}{\partial z} = \begin{cases} 0 & (j=0) \\ {}_y\Gamma_{j-1} & (j=1,2,\cdots) \end{cases} \end{cases}$$

at the sea bottom $z=-h$:

$$u_j = v_j = w_j = 0$$

along the shore boundary C_1:

$$\cos\alpha_x\left[\int_{-h}^{0} u_j \mathrm{d}z + \begin{cases} 0 & (j=0) \\ {}_xH_{j-1} & (j=1,2,\cdots) \end{cases}\right] + \cos\alpha_y\left[\int_{-h}^{0} v_j \mathrm{d}z + \begin{cases} 0 & (j=0) \\ {}_yH_{j-1} & (j=1,2,\cdots) \end{cases}\right] = 0$$

along the open boundary C_2:

$$\zeta_j = S_j$$

where

$${}_xE_{j-1} = \sum_{m=0}^{j-1}\left\{\frac{\partial}{\partial x}(u_m u_{j-1-m}) + \frac{\partial}{\partial y}(v_m u_{j-1-m}) + \frac{\partial}{\partial z}(w_m u_{j-1-m})\right\}$$

$${}_yE_{j-1} = \sum_{m=0}^{j-1}\left\{\frac{\partial}{\partial x}(u_m v_{j-1-m}) + \frac{\partial}{\partial y}(v_m v_{j-1-m}) + \frac{\partial}{\partial z}(w_m v_{j-1-m})\right\}$$

$$\begin{aligned} F_{j-1} = & \sum_{m=0}^{j-1}\left\{u_m \frac{\partial \zeta_{j-1-m}}{\partial x} + v_m \frac{\partial \zeta_{j-1-m}}{\partial y}\right\} \\ & + \sum_{n=2}^{j}\left\{\frac{1}{(n-1)!}\left(\prod_{l_1=1}^{n} \sum_{m_{l_1}=0}^{j-n-\sum_{k=l_1+1}^{n} m_k}\right)\left(\prod_{l=2}^{n}\zeta_{m_l}\frac{\partial^{(n-1)} u_{m_1}}{\partial z^{(n-1)}}\cdot\frac{\partial \zeta_{j-n-\sum_{k=1}^{n} m_k}}{\partial x}\right)\right\} \\ & + \sum_{n=2}^{j}\left\{\frac{1}{(n-1)!}\left(\prod_{l_1=1}^{n} \sum_{m_{l_1}=0}^{j-n-\sum_{k=l_1+1}^{n} m_k}\right)\left(\prod_{l=2}^{n}\zeta_{m_l}\frac{\partial^{(n-1)} v_{m_1}}{\partial z^{(n-1)}}\cdot\frac{\partial \zeta_{j-n-\sum_{k=1}^{n} m_k}}{\partial y}\right)\right\} \\ & + \sum_{n=1}^{j}\left\{\frac{1}{n!}\left(\prod_{l_1=1}^{n} \sum_{m_{l_1}=0}^{j-n-\sum_{k=l_1+1}^{n} m_k}\right)\times\left(\prod_{l=1}^{n}\zeta_{m_l}\frac{\partial^{(n-1)}}{\partial z^{(n-1)}}\left[\frac{\partial u_{j-n-\sum_{k=1}^{n} m_k}}{\partial x} + \frac{\partial v_{j-n-\sum_{k=1}^{n} m_k}}{\partial y}\right]\right)\right\} \end{aligned}$$

$$\left.\begin{matrix} {}_x\Gamma_{j-1} \\ {}_y\Gamma_{j-1} \end{matrix}\right\} = -\sum_{n=1}^{j}\left\{\frac{1}{n!}\left(\prod_{l_1=1}^{n} \sum_{m_{l_1}=0}^{j-n-\sum_{k=l_1+1}^{n} m_k}\right)\left(\prod_{l=1}^{n}\zeta_{m_l}\left\{\begin{matrix} \dfrac{\partial^{(n+1)} u_{j-n-\sum_{k=1}^{n} m_k}}{\partial z^{(n+1)}} \\ \dfrac{\partial^{(n+1)} v_{j-n-\sum_{k=1}^{n} m_k}}{\partial z^{(n+1)}} \end{matrix}\right.\right)\right\}$$

$$\left.\begin{array}{l} {}_xH_{j-1} \\ {}_yH_{j-1} \end{array}\right\} = \sum_{n=1}^{j}\left\{\frac{1}{n!}\left(\prod_{l_1=1}^{n}\sum_{m_{l_1}=0}^{j-n-\sum\limits_{k=l_1+1}^{n} m_k}\right)\left(\prod_{l=1}^{n}\zeta_{m_l}\left\{\begin{array}{l}\dfrac{\partial^{(n-1)} u_{j-n-\sum\limits_{k=1}^{n} m_k}}{\partial z^{(n-1)}} \\ \dfrac{\partial^{(n-1)} v_{j-n-\sum\limits_{k=1}^{n} m_k}}{\partial z^{(n-1)}}\end{array}\right\}\right)\right\}$$

The symbols:

$$\left(\prod_{l_1=1}^{n}\sum_{m_{l_1}=0}^{j-n-\sum\limits_{k=l_1+1}^{n} m_k}\right) = \sum_{m_n=0}^{j-n}\sum_{m_{n-1}=0}^{j-n-m_k}\sum_{m_{n-2}=0}^{j-n-\sum\limits_{k=n-1}^{n} m_k}\cdots\sum_{m_2=0}^{j-n-\sum\limits_{k=3}^{n} m_k}\sum_{m_1=0}^{j-n-\sum\limits_{k=2}^{n} m_k}$$

$$\sum_a^b = \begin{cases} 0 & (b<a), \\ \sum\limits_a^b & (b \geqslant a), \end{cases} \qquad \prod_a^b = \begin{cases} 0 & (b<a) \\ \prod\limits_a^b & (b \geqslant a) \end{cases}$$

The problem (3) is just to describe the j-th order constituents, in other words, the j-th order constituents are the very solutions which satisfy problem (3). $j=0$ corresponds to the astronomical tides and the linear response of the sea region under consideration to the shallow water constituents coming from the outer adjacent sea; $j=1,2,\cdots$ correspond to the shallow water constituents produced by the nonlinear coupling among lower orders of constituents in the region under consideration. In fact, the problem (3) has explained the mechanism of the formation of tide constituents, particularly, those of shallow water constituents of different orders[6].

Because of the linearity of the dimensionless problem (3) the solution of any tide constituent with the circular frequency $\tilde{\sigma}$ can be solved respectively (in the following, for brevity, the right subscript j of the different quantities will be omitted), thus

$$X = X'\cos\sigma t + X''\sin\sigma t = \mathrm{Re}\left[\bar{X}e^{-i\sigma t}\right],\ \bar{X} = X' + iX'' \tag{4}$$

where X denotes u, v, w, ζ, ω_x, ω_y and S:

$$i = \sqrt{-1},\ \sigma = \frac{\tilde{\sigma}L}{\sqrt{gh_0}}$$

Re[Y] denotes the real part of the complex variable, Y. Furthermore, noting that

$$\begin{aligned} {}_xE_{j-1} = & \sum_{m=0}^{j-1}\sum_{p,q}\mathrm{Re}\left[\left\{\frac{1}{2}\frac{\partial}{\partial x}(\bar{u}_{m,p}\bar{u}_{j-1-m,q}) + \frac{1}{2}\frac{\partial}{\partial y}(\bar{v}_{m,p}\bar{u}_{j-1-m,q})\right.\right. \\ & \left.\left. + \frac{1}{2}\frac{\partial}{\partial z}(\bar{w}_{m,p}\bar{u}_{j-1-m,q})\right\}\exp\{-i(\sigma_{m,p}+\sigma_{j-1-m,q})t\}\right] \\ & + \sum_{m=0}^{j-1}\sum_{p,q}\mathrm{Re}\left[\left\{\frac{1}{2}\frac{\partial}{\partial x}(\bar{u}_{m,p}\bar{u}^*_{j-1-m,q}) + \frac{1}{2}\frac{\partial}{\partial y}(\bar{v}_{m,p}\bar{u}^*_{j-1-m,q})\right.\right. \end{aligned}$$

$$+\frac{1}{2}\frac{\partial}{\partial z}(\overline{w}_{m,p}\overline{u}^{*}_{j-1-m,q})\bigg\}\exp\{-\mathrm{i}(\sigma_{m,p}-\sigma_{j-1-m,q})t\}\bigg]$$

$$_{y}E_{j-1}=\sum_{m=0}^{j-1}\sum_{p,q}\mathrm{Re}\bigg[\bigg\{\frac{1}{2}\frac{\partial}{\partial x}(\overline{u}_{m,p}\overline{v}_{j-1-m,q})+\frac{1}{2}\frac{\partial}{\partial y}(\overline{v}_{m,p}\overline{v}_{j-1-m,q})$$

$$+\frac{1}{2}\frac{\partial}{\partial z}(\overline{w}_{m,p}\overline{v}_{j-1-m,q})\bigg\}\exp\{-\mathrm{i}(\sigma_{m,p}+\sigma_{j-1-m,q})t\}\bigg]$$

$$+\sum_{m=0}^{j-1}\sum_{p,q}\mathrm{Re}\bigg[\bigg\{\frac{1}{2}\frac{\partial}{\partial x}(\overline{u}_{m,p}\overline{v}^{*}_{j-1-m,q})+\frac{1}{2}\frac{\partial}{\partial y}(\overline{v}_{m,p}\overline{v}^{*}_{j-1-m,q})$$

$$+\frac{1}{2}\frac{\partial}{\partial z}(\overline{w}_{m,p}\overline{v}^{*}_{j-1-m,q})\bigg\}\exp\{-\mathrm{i}(\sigma_{m,p}-\sigma_{j-1-m,q})t\}\bigg]$$

$$F_{j-1}=\sum_{m=0}^{j-1}\sum_{p,q}\mathrm{Re}\left[\left\{\frac{1}{2}\overline{u}_{m,p}\frac{\partial\overline{\zeta}_{j-1-m,q}}{\partial x}+\frac{1}{2}\overline{v}_{m,p}\frac{\partial\overline{\zeta}_{j-1-m,q}}{\partial y}\right\}\times\exp\{-\mathrm{i}(\sigma_{m,p}+\sigma_{j-1-m,q})t\}\right]$$

$$+\sum_{m=0}^{j-1}\sum_{p,q}\mathrm{Re}\left[\left\{\frac{1}{2}\overline{u}_{m,p}\frac{\partial\overline{\zeta}^{*}_{j-1-m,q}}{\partial x}+\frac{1}{2}\overline{v}_{m,p}\frac{\partial\overline{\zeta}^{*}_{j-1-m,q}}{\partial y}\right\}\times\exp\{-\mathrm{i}(\sigma_{m,p}-\sigma_{j-1-m,q})t\}\right]$$

$$+\sum_{n=2}^{j}\left(\prod_{l_1=1}^{n}\sum_{m_{l_1}=0}^{j-n-\sum\limits_{k=l_1+1}^{n}m_k}\right)\left(\prod_{l=1}^{n}\sum_{p_l}\right)\sum_{q}\frac{1}{(n-1)!}\frac{1}{2^n}$$

$$\times\left\{\mathrm{Re}\left[\prod_{l=1}^{n}\overline{u}'_{m_l,p_l}\frac{\partial}{\partial x}\overline{\zeta}_{j-n-\sum\limits_{k=1}^{n}m_k,q}\cdot\exp\left\{-\mathrm{i}\left(\sum_{l=1}^{n}\sigma_{m_l,p_l}+\sigma_{j-n-\sum\limits_{k=1}^{n}m_k,q}\right)t\right\}\right]\right.$$

$$+\mathrm{Re}\left[\prod_{l=1}^{n}\overline{u}'_{m_l,p_l}\frac{\partial}{\partial x}\overline{\zeta}^{*}_{j-n-\sum\limits_{k=1}^{n}m_k,q}\cdot\exp\left\{-\mathrm{i}\left(\sum_{l=1}^{n}\sigma_{m_l,p_l}-\sigma_{j-n-\sum\limits_{k=1}^{n}m_k,q}\right)t\right\}\right]$$

$$+\sum_{\lambda=1}^{n-1}\left(\prod_{\theta=1}^{\lambda}\sum_{r_\theta=1}^{n-(\lambda-\theta+1)-\sum\limits_{k=1}^{\theta-1}r_k}\right)\mathrm{Re}\left[\prod_{l=1}^{n}{}^{(\lambda)}\overline{u}'_{m_l,p_l}\prod_{\mu=1}^{\lambda}\overline{u}^{*'}_{m_{\sum\limits_{k=1}^{\mu}r_k},p_{\sum\limits_{k=1}^{\mu}r_k}}\frac{\partial}{\partial x}\times\overline{\zeta}_{j-n-\sum\limits_{k=1}^{n}m_k,q}\right.$$

$$\left.\cdot\exp\left\{-\mathrm{i}\left(\sum_{l=1}^{n}{}^{(\lambda)}\sigma_{m_l,p_l}-\sum_{\mu=1}^{\lambda}\sigma_{m_{\sum\limits_{k=1}^{\mu}r_k},p_{\sum\limits_{k=1}^{\mu}r_k}}+\sigma_{j-n-\sum\limits_{k=1}^{n}m_k,q}\right)t\right\}\right]+\sum_{\lambda=1}^{n-1}\left(\prod_{\theta=1}^{\lambda}\sum_{r_\theta=1}^{n-(\lambda-\theta+1)-\sum\limits_{k=1}^{\theta-1}r_k}\right)$$

$$\times \mathrm{Re}\Bigg[\prod_{l=1}^{n}{}^{(\lambda)}\overline{u}'_{m_l,p_l}\prod_{\mu=1}^{\lambda}\overline{u}^{*'}_{m_{\sum_{k=1}^{\mu}r_k},\,p_{\sum_{k=1}^{\mu}r_k}}\frac{\partial}{\partial x}\overline{\zeta}^{*}_{j-n-\sum_{k=1}^{n}m_k,q}$$

$$\times\exp\Bigg\{-\mathrm{i}\Bigg(\sum_{l=1}^{n}{}^{(\lambda)}\sigma_{m_l,p_l}-\sum_{\mu=1}^{\lambda}\sigma_{m_{\sum_{k=1}^{\mu}r_k},\,p_{\sum_{k=1}^{\mu}r_k}}-\sigma_{j-n-\sum_{k=1}^{n}m_k,q}\Bigg)t\Bigg\}\Bigg]\Bigg]\Bigg\}$$

$$+\sum_{n=2}^{j}\Bigg(\prod_{l_1=1}^{n}\sum_{m_{l_1}=0}^{j-n-\sum_{k=l_1+1}^{n}m_k}\Bigg)\Bigg(\prod_{l_1=1}^{n}\sum_{p_l}\Bigg)\sum_{q}\frac{1}{(n-1)!}\frac{1}{2^n}$$

$$\times\Bigg\{\mathrm{Re}\Bigg[\prod_{l=1}^{n}\overline{v}'_{m_l,p_l}\frac{\partial}{\partial y}\overline{\zeta}_{j-n-\sum_{k=1}^{n}m_k,q}\cdot\exp\Bigg\{-\mathrm{i}\Bigg(\sum_{l=1}^{n}\sigma_{m_l,p_l}+\sigma_{j-n-\sum_{k=1}^{n}m_k,q}\Bigg)t\Bigg\}\Bigg]$$

$$+\mathrm{Re}\Bigg[\prod_{l=1}^{n}\overline{v}'_{m_l,p_l}\frac{\partial}{\partial y}\overline{\zeta}^{*}_{j-n-\sum_{k=1}^{n}m_k,q}\cdot\exp\Bigg\{-\mathrm{i}\Bigg(\sum_{l=1}^{n}\sigma_{m_l,p_l}-\sigma_{j-n-\sum_{k=1}^{n}m_k,q}\Bigg)t\Bigg\}\Bigg]$$

$$+\sum_{\lambda=1}^{n-1}\Bigg(\prod_{\theta=1}^{\lambda}\sum_{r_\theta=1}^{n-(\lambda-\theta+1)-\sum_{k=1}^{\theta-1}r_k}\Bigg)\mathrm{Re}\Bigg[\prod_{l=1}^{n}{}^{(\lambda)}\overline{v}'_{m_l,p_l}\prod_{\mu=1}^{\lambda}\overline{v}^{*'}_{m_{\sum_{k=1}^{\mu}r_k},\,p_{\sum_{k=1}^{\mu}r_k}}\frac{\partial}{\partial y}\times\overline{\zeta}_{i-n-\sum_{k=1}^{n}m_k,q}$$

$$\cdot\exp\Bigg\{-\mathrm{i}\Bigg(\sum_{l=1}^{n}{}^{(\lambda)}\sigma_{m_l,p_l}-\sum_{\mu=1}^{\lambda}\sigma_{m_{\sum_{k=1}^{\mu}r_k},\,p_{\sum_{k=1}^{\mu}r_k}}+\sigma_{j-n-\sum_{k=1}^{n}m_k,q}\Bigg)t\Bigg\}\Bigg]$$

$$+\sum_{\lambda=1}^{n-1}\Bigg(\prod_{\theta=1}^{\lambda}\sum_{r_\theta=1}^{n-(\lambda-\theta+1)-\sum_{k=1}^{\theta-1}r_k}\Bigg)\times\mathrm{Re}\Bigg[\prod_{l=1}^{n}{}^{(\lambda)}\overline{v}'_{m_l,p_l}\prod_{\mu=1}^{\lambda}\overline{v}^{*'}_{m_{\sum_{k=1}^{\mu}r_k},\,p_{\sum_{k=1}^{\mu}r_k}}\frac{\partial}{\partial y}\overline{\zeta}^{*}_{j-n-\sum_{k=1}^{n}m_k,q}$$

$$\times\exp\Bigg\{-\mathrm{i}\Bigg(\sum_{l=1}^{n}{}^{(\lambda)}\sigma_{m_l,p_l}-\sum_{\mu=1}^{\lambda}\sigma_{m_{\sum_{k=1}^{\mu}r_k},\,p_{\sum_{k=1}^{\mu}r_k}}-\sigma_{j-n-\sum_{k=1}^{n}m_k,q}\Bigg)t\Bigg\}\Bigg]\Bigg]\Bigg\}$$

$$+\sum_{n=1}^{j}(-1)\Bigg(\prod_{l_1=1}^{n}\sum_{m_{l_1}=0}^{j-n-\sum_{k=l_1+1}^{n}m_k}\Bigg)\Bigg(\prod_{l=1}^{n}\sum_{p_l}\Bigg)\sum_{q}\frac{1}{n!}\cdot\frac{1}{2^n}$$

$$\times\left\{\mathrm{Re}\left[\prod_{l=1}^{n}\bar{\zeta}_{m_l,p_l}\frac{\partial^{(n)}}{\partial z^{(n)}}\bar{w}_{j-n-\sum_{k=1}^{n}m_k,q}\cdot\exp\left\{-\mathrm{i}\left(\sum_{l=1}^{n}\sigma_{m_l,p_l}+\sigma_{j-n-\sum_{k=1}^{n}m_k,q}\right)t\right\}\right]\right.$$

$$+\mathrm{Re}\left[\prod_{l=1}^{n}\bar{\zeta}_{m_l,p_l}\frac{\partial^{(n)}}{\partial z^{(n)}}\bar{w}^*_{j-n-\sum_{k=1}^{n}m_k,q}\cdot\exp\left\{-\mathrm{i}\left(\sum_{l=1}^{n}\sigma_{m_l,p_l}-\sigma_{j-n-\sum_{k=1}^{n}m_k,q}\right)t\right\}\right]$$

$$+\sum_{\lambda=1}^{n-1}\left(\prod_{\theta=1}^{\lambda}\sum_{r_\theta=1}^{n-(\lambda-\theta+1)-\sum_{k=1}^{\theta-1}r_k}\right)\mathrm{Re}\left[\prod_{l=1}^{n}{}^{(\lambda)}\bar{\zeta}_{m_l,p_l}\prod_{\mu=1}^{\lambda}\bar{\zeta}^*_{m_{\sum_{k=1}^{\mu}r_k},p_{\sum_{k=1}^{\mu}r_k}}\times\frac{\partial^{(n)}}{\partial z^{(n)}}\bar{w}_{j-n-\sum_{k=1}^{n}m_k,q}\right.$$

$$\left.\cdot\exp\left\{-\mathrm{i}\left(\sum_{l=1}^{n}{}^{(\lambda)}\sigma_{m_l,p_l}-\sum_{\mu=1}^{\lambda}\sigma_{m_{\sum_{k=1}^{\mu}r_k},p_{\sum_{k=1}^{\mu}r_k}}+\sigma_{j-n-\sum_{k=1}^{n}m_k,q}\right)t\right\}\right]+\sum_{\lambda=1}^{n-1}\left(\prod_{\theta=1}^{\lambda}\sum_{r_\theta=1}^{n-(\lambda-\theta+1)-\sum_{k=1}^{\theta-1}r_k}\right)$$

$$\times\mathrm{Re}\left[\prod_{l=1}^{n}{}^{(\lambda)}\bar{\zeta}_{m_l,p_l}\prod_{\mu=1}^{\lambda}\bar{\zeta}^*_{m_{\sum_{k=1}^{\mu}r_k},p_{\sum_{k=1}^{\mu}r_k}}\frac{\partial^{(n)}}{\partial z^{(n)}}\bar{w}^*_{j-n-\sum_{k=1}^{n}m_k,q}\right.$$

$$\left.\left.\times\exp\left\{-\mathrm{i}\left(\sum_{l=1}^{n}{}^{(\lambda)}\sigma_{m_l,p_l}-\sum_{\mu=1}^{\lambda}\sigma_{m_{\sum_{k=1}^{\mu}r_k},p_{\sum_{k=1}^{\mu}r_k}}-\sigma_{j-n-\sum_{k=1}^{n}m_k,q}\right)t\right\}\right]\right\}$$

$$\begin{aligned}&{}_x\Gamma_{j-1}=-{}_x\mathscr{D}_{j-1}(n+1),\quad {}_y\Gamma_{j-1}=-{}_y\mathscr{D}_{j-1}(n+1)\\&{}_xH_{j-1}={}_x\mathscr{D}_{j-1}(n-1),\quad {}_yH_{j-1}={}_y\mathscr{D}_{j-1}(n-1)\end{aligned}\tag{5}$$

$$\left.\begin{matrix}{}_x\mathscr{D}_{j-1}(d)\\{}_y\mathscr{D}_{j-1}(d)\end{matrix}\right\}=\sum_{n=1}^{j}\left(\prod_{l_1=1}^{n}\sum_{m_{l_1}=0}^{j-n-\sum_{k=l_1+1}^{n}m_k}\right)\left(\prod_{l=1}^{n}\sum_{p_l}\right)\sum_{q}\frac{1}{n!}\cdot\frac{1}{2^n}\times\left\{\mathrm{Re}\left[\prod_{l=1}^{n}\bar{\zeta}_{m_l,p_l}\frac{\partial^{(d)}}{\partial z^{(d)}}\left\{\begin{matrix}\bar{u}_{j-n-\sum_{k=1}^{n}m_k,q}\\\bar{v}_{j-n-\sum_{k=1}^{n}m_k,q}\end{matrix}\right\}\right.\right.$$

$$\left.\times\exp\left\{-\mathrm{i}\left(\sum_{l=1}^{n}\sigma_{m_l,p_l}+\sigma_{j-n-\sum_{k=1}^{n}m_k,q}\right)t\right\}\right]+\mathrm{Re}\left[\prod_{l=1}^{n}\bar{\zeta}_{m_l,p_l}\times\frac{\partial^{(d)}}{\partial z^{(d)}}\left\{\begin{matrix}\bar{u}^*_{j-n-\sum_{k=1}^{n}m_k,q}\\\bar{v}^*_{j-n-\sum_{k=1}^{n}m_k,q}\end{matrix}\right\}\right.$$

$$\left.\cdot\exp\left\{-\mathrm{i}\left(\sum_{l=1}^{n}\sigma_{m_l,p_l}-\sigma_{j-n-\sum_{k=1}^{n}m_k,q}\right)t\right\}\right]+\sum_{\lambda=1}^{n-1}\left(\prod_{\theta=1}^{\lambda}\sum_{r_\theta=1}^{n-(\lambda-\theta+1)-\sum_{k=1}^{\theta-1}r_k}\right)$$

$$\times \mathrm{Re}\left[\prod_{l=1}^{n}{}^{(\lambda)}\bar{\zeta}_{m_l,p_l}\prod_{\mu=1}^{\lambda}\bar{\zeta}^{*}_{m_{\sum_{k=1}^{\mu}r_k},p_{\sum_{k=1}^{\mu}r_k}}\times\frac{\partial^{(d)}}{\partial z^{(d)}}\begin{Bmatrix}\bar{u}_{j-n-\sum_{k=1}^{n}m_k,q}\\ \bar{v}_{j-n-\sum_{k=1}^{n}m_k,q}\end{Bmatrix}\right.$$

$$\left.\cdot\exp\left\{-\mathrm{i}\left(\sum_{l=1}^{n}{}^{(\lambda)}\sigma_{m_l,p_l}-\sum_{\mu=1}^{\lambda}\sigma_{m_{\sum_{k=1}^{\mu}r_k},p_{\sum_{k=1}^{\mu}r_k}}+\sigma_{j-n-\sum_{k=1}^{n}m_k,q}\right)t\right\}\right]$$

$$+\sum_{\lambda=1}^{n-1}\left(\prod_{\theta=1}^{\lambda}\sum_{r_\theta=1}^{n-(\lambda-\theta+1)-\sum_{k=1}^{\theta-1}r_k}\right)\times\mathrm{Re}\left[\prod_{l=1}^{n}{}^{(\lambda)}\bar{\zeta}_{m_l,p_l}\prod_{\mu=1}^{\lambda}\bar{\zeta}^{*}_{m_{\sum_{k=1}^{\mu}r_k},p_{\sum_{k=1}^{\mu}r_k}}\frac{\partial^{(d)}}{\partial z^{(d)}}\begin{Bmatrix}\bar{u}^{*}_{j-n-\sum_{k=1}^{n}m_k,q}\\ \bar{v}^{*}_{j-n-\sum_{k=1}^{n}m_k,q}\end{Bmatrix}\right.$$

$$\left.\times\exp\left\{-\mathrm{i}\left(\sum_{l=1}^{n}{}^{(\lambda)}\sigma_{m_l,p_l}-\sum_{\mu=1}^{\lambda}\sigma_{m_{\sum_{k=1}^{\mu}r_k},p_{\sum_{k=1}^{\mu}r_k}}-\sigma_{j-n-\sum_{k=1}^{n}m_k,q}\right)t\right\}\right]\Bigg\}$$

where

$$\bar{u}_{m_l,p_l}=\begin{cases}\dfrac{\partial^{(n-1)}\bar{u}_{m_l,p_l}}{\partial z^{(n-1)}} & (l=1)\\ \bar{\zeta}_{m_l,p_l} & (l=2,3,\cdots n)\end{cases},\quad \bar{v}'_{m_l,p_l}=\begin{cases}\dfrac{\partial^{(n-1)}\bar{v}_{m_l,p_l}}{\partial z^{(n-1)}} & (l=1)\\ \bar{\zeta}_{m_l,p_l} & (l=2,3,\cdots,n)\end{cases}$$

$\sum^{(\lambda)}$ and $\prod^{(\lambda)}$ denote the processes of summation and multiplication, excluding the quantities $l=r_1, l=r_1+r_2,\cdots, l=r_1+r_2+\cdots+r_\lambda$. The asterisk denotes the complex conjugate. The expressions under $\Sigma\Sigma\cdots\Sigma$ in (5) are abbreviated as ψ_1, ψ_2, f, γ_1, γ_2, η_1 and $\eta_2(\rightarrow {}_xE_{j-1}, {}_yE_{j-1}, F_{j-1}, {}_x\Gamma_{j-1}, {}_y\Gamma_{j-1}, {}_xH_{j-1}$ and ${}_yH_{j-1})$, any of which satisfies expression (4).

For simplicity let us suppose

$$\nu=\nu(x,y) \tag{6}$$

Now we may combine Eqs. (4), (5), and (6) so that the time variable is eliminated. The dimensionless problems (3) will be then reduced to that of (7), which, as involved only the space coordinates, is satisfied by a certain constituent with the non-dimensional frequency σ:

$$\frac{\partial\bar{u}}{\partial x}+\frac{\partial\bar{v}}{\partial y}+\frac{\partial\bar{w}}{\partial z}=0$$

$$\nu\frac{\partial^2\bar{u}}{\partial z^2}+\mathrm{i}\sigma\bar{u}+\Omega\bar{v}=\frac{\partial\bar{\zeta}}{\partial x}+\bar{\psi}_1,\quad \nu\frac{\partial^2\bar{v}}{\partial z^2}+\mathrm{i}\sigma\bar{v}-\Omega\bar{u}=\frac{\partial\bar{\zeta}}{\partial y}+\bar{\psi}_2 \tag{7}$$

at the sea surface $z=0$:

$$\bar{w}=-\mathrm{i}\sigma\bar{\zeta}+\bar{f},\quad \frac{\partial\bar{u}}{\partial z}=\bar{\gamma}_1,\quad \frac{\partial\bar{v}}{\partial z}=\bar{\gamma}_2$$

at the sea bottom $z=-h$:

$$\overline{u}=\overline{v}=\overline{w}=0$$

along the shore boundary C_1:

$$\cos\alpha_x\left(\int_{-h}^{0}\overline{u}\mathrm{d}z+\overline{\eta}_1\right)+\cos\alpha_y\left(\int_{-h}^{0}\overline{v}\mathrm{d}z+\overline{\eta}_2\right)=0$$

along the shore boundary C_2:

$$\overline{\zeta}=\overline{S}$$

For the zero order model, in the problem (7):

$$\overline{f}=\overline{\gamma}_1=\overline{\gamma}_2=\overline{\eta}_1=\overline{\eta}_2=0,\quad \overline{\psi}_1=\overline{\omega}_x,\quad \overline{\psi}_2=\overline{\omega}_y,\ \text{or}\ \ \overline{\psi}_1=\overline{\psi}_2=0$$

To solve the dimensionless problem (7) we shall not encounter any substantial theoretical difficulty. Finally, we have:

$$\overline{u}=\frac{1}{2}(q+\hat{q}),\quad \overline{v}=\frac{1}{2\mathrm{i}}(q-\hat{q}) \tag{8}$$

where

$$q=q(\kappa,G,\psi,\gamma)=\frac{G}{\kappa^2}\frac{1}{\nu}\left\{\frac{\mathrm{ch}(\kappa z)}{\mathrm{ch}(\kappa h)}-1\right\}+\frac{1}{\kappa}\mathrm{sh}(\kappa z)\int_0^z\frac{\psi}{\nu}\mathrm{ch}(\kappa\zeta)\mathrm{d}\zeta$$
$$-\frac{1}{\kappa}\mathrm{ch}(\kappa z)\left\{\int_{-h}^{z}\frac{\psi}{\nu}\mathrm{sh}(\kappa\zeta)\mathrm{d}\zeta+\mathrm{th}(\kappa h)\int_{-h}^{0}\frac{\psi}{\nu}\mathrm{ch}(\kappa\zeta)\mathrm{d}\zeta\right\}+\frac{\gamma}{\kappa}\frac{\mathrm{sh}\left[\kappa(z+h)\right]}{\mathrm{ch}(\kappa h)}$$

$$\hat{q}=q(\hat{\kappa},\hat{G},\hat{\psi},\hat{\gamma})$$

$$\begin{cases}\kappa=(1+\mathrm{i})\sqrt{\dfrac{\Omega-\sigma}{2\nu}}\\ \hat{\kappa}=(1-\mathrm{i})\sqrt{\dfrac{\Omega+\sigma}{2\nu}}\end{cases},\quad \begin{cases}G=\dfrac{\partial\overline{\zeta}}{\partial x}+\mathrm{i}\dfrac{\partial\overline{\zeta}}{\partial y}\\ \hat{G}=\dfrac{\partial\overline{\zeta}}{\partial x}-\mathrm{i}\dfrac{\partial\overline{\zeta}}{\partial y}\end{cases},\quad \begin{cases}\psi=\overline{\psi}_1+\mathrm{i}\overline{\psi}_2\\ \hat{\psi}=\overline{\psi}_1-\mathrm{i}\overline{\psi}_2\end{cases},\quad \begin{cases}\gamma=\overline{\gamma}_1+\mathrm{i}\overline{\gamma}_2\\ \hat{\gamma}=\overline{\gamma}_1-\mathrm{i}\overline{\gamma}_2\end{cases}$$

Eliminating $\overline{u}$, $\overline{v}$, and $\overline{w}$, we get the following boundary value problem of $\overline{\zeta}$.

$$\frac{1}{\nu}\frac{A+\hat{A}}{2}\Delta\overline{\zeta}+\left[\frac{\partial}{\partial x}\left(\frac{1}{\nu}\frac{A+\hat{A}}{2}\right)+\frac{\partial}{\partial y}\left(\frac{1}{\nu}\frac{A-\hat{A}}{2\mathrm{i}}\right)\right]\frac{\partial\overline{\zeta}}{\partial x}$$
$$+\left[\frac{\partial}{\partial y}\left(\frac{1}{\nu}\frac{A+\hat{A}}{2}\right)-\frac{\partial}{\partial x}\left(\frac{1}{\nu}\frac{A-\hat{A}}{2\mathrm{i}}\right)\right]\frac{\partial\overline{\zeta}}{\partial y}+\mathrm{i}\sigma\overline{\zeta}=\overline{f}-F \tag{9}$$

along the shore boundary C_1:

$$Q_n=\overline{\eta}_n$$

along the open boundary C_2:

$$\overline{\zeta}=\overline{S}$$

where

$$F=\frac{\partial}{\partial x}\int_{-h}^{0}\left(\frac{B+\hat{B}}{2}\overline{\psi}_1-\frac{B-\hat{B}}{2\mathrm{i}}\overline{\psi}_2\right)\mathrm{d}\zeta+\frac{\partial}{\partial y}\int_{-h}^{0}\left(\frac{B-\hat{B}}{2\mathrm{i}}\overline{\psi}_1+\frac{B+\hat{B}}{2}\overline{\psi}_2\right)\mathrm{d}\zeta$$
$$+\frac{\partial}{\partial x}\left(\frac{C+\hat{C}}{2}\overline{\gamma}_1-\frac{C-\hat{C}}{2\mathrm{i}}\overline{\gamma}_2\right)+\frac{\partial}{\partial y}\left(\frac{C-\hat{C}}{2\mathrm{i}}\overline{\gamma}_1+\frac{C+\hat{C}}{2}\overline{\gamma}_2\right)$$

$$Q_n = \frac{1}{\nu}\frac{A+\hat{A}}{2}\frac{\partial\bar{\zeta}}{\partial n} + \frac{1}{\nu}\frac{A-\hat{A}}{2\mathrm{i}}\frac{\partial\bar{\zeta}}{\partial\tau} + \int_{-h}^{0}\left(\frac{B+\hat{B}}{2}\bar{\psi}_n + \frac{B-\hat{B}}{2\mathrm{i}}\bar{\psi}_\tau\right)\mathrm{d}\zeta + \frac{C+\hat{C}}{2}\bar{\gamma}_n + \frac{C-\hat{C}}{2\mathrm{i}}\bar{\gamma}_\tau$$

$$A = A(\kappa) = \frac{1}{\kappa^2}\left\{h - \frac{1}{\kappa}\mathrm{th}(\kappa h)\right\},\quad \hat{A} = A(\hat{\kappa})$$

$$B = B(\kappa, \zeta) = \frac{1}{\nu\kappa^2}\left\{1 - \frac{\mathrm{ch}(\kappa\zeta)}{\mathrm{ch}(\kappa h)}\right\},\quad \hat{B} = B(\hat{\kappa}; \zeta)$$

$$C = C(\kappa) = \frac{1}{\kappa^2}\{\mathrm{sech}(\kappa h) - 1\},\quad \hat{C} = C(\hat{\kappa})$$

$$\frac{\partial\bar{\zeta}}{\partial n} = \frac{\partial\bar{\zeta}}{\partial x}\cos\alpha_x + \frac{\partial\bar{\zeta}}{\partial y}\cos\alpha_y,\quad \frac{\partial\bar{\zeta}}{\partial\tau} = -\frac{\partial\bar{\zeta}}{\partial y}\cos\alpha_x + \frac{\partial\bar{\zeta}}{\partial x}\cos\alpha_y$$

$$\bar{\psi}_n = \bar{\psi}_1\cos\alpha_x + \bar{\psi}_2\cos\alpha_y,\quad \bar{\psi}_\tau = -\bar{\psi}_2\cos\alpha_x + \bar{\psi}_1\cos\alpha_y$$

$$\bar{\gamma}_n = \bar{\gamma}_1\cos\alpha_x + \bar{\gamma}_2\cos\alpha_y,\quad \bar{\gamma}_\tau = -\bar{\gamma}_2\cos\alpha_x + \bar{\gamma}_1\cos\alpha_y$$

$$\bar{\eta}_n = \bar{\eta}_1\cos\alpha_x + \bar{\eta}_2\cos\alpha_y$$

in the above τ denotes the tangent at the boundary which forms a right-hand Cartesian coordinate system with the normal n.

Finally, we have

$$\bar{w} = -\mathrm{i}\sigma\bar{\zeta} + \bar{f} - \frac{\partial}{\partial x}\left[\frac{P(z)+\hat{P}(z)}{2}\right] - \frac{\partial}{\partial y}\left[\frac{P(z)-\hat{P}(z)}{2\mathrm{i}}\right] \tag{10}$$

where

$$P(z) = P(z,\kappa,G,\psi,\gamma) = \frac{G}{\kappa^2}\frac{1}{\nu}\left\{\frac{1}{\kappa}\frac{\mathrm{sh}(\kappa z)}{\mathrm{ch}(\kappa h)} - z\right\} - \frac{1}{\kappa^2}\int_0^z\frac{\psi(\zeta)}{\nu}\mathrm{d}\zeta$$
$$+\frac{1}{\kappa^2}\mathrm{ch}[\kappa(z+h)]\int_0^z\frac{\mathrm{ch}(\kappa\zeta)}{\mathrm{ch}(\kappa h)}\frac{\psi(\zeta)}{\nu}\mathrm{d}\zeta - \frac{1}{\kappa^2}\frac{\mathrm{sh}(\kappa z)}{\mathrm{ch}(\kappa h)}$$
$$\times\int_{-h}^{z}\mathrm{sh}[\kappa(\zeta+h)]\frac{\psi(\zeta)}{\nu}\mathrm{d}\zeta + \frac{\bar{\gamma}}{\kappa^2}\left\{\frac{\mathrm{ch}[\kappa(z+h)]}{\mathrm{ch}(\kappa h)} - 1\right\}$$

$$\hat{P}(z) = P(z,\hat{\kappa},\hat{G},\hat{\psi},\hat{\gamma})$$

The dimensionless problem (7) is now reduced to the boundary value problem (9) of the elliptic differential equation for tidal elevation $\bar{\zeta}$, and the analytical expression (8) for the vertical distribution of tidal currents $\bar{u}$, $\bar{v}$, as well as the formula (10) for $\bar{w}$ are derived. Thus, a basic model for further analysis and numerical calculation is obtained.

In case of $\sigma = \Omega$, we have a special solution. Transforming the complex solutions into real ones and the dimensionless solutions into dimensional forms does not involve any special difficulty in calculation.

The physical hypothesis about $\nu = \nu(x, y)$ is not related with the tidal current structure, so it constitutes the main weakpoint of the proposed model.

It is evident that the linear three-dimensional model given by L. Sgibneva and A. Felsenbaum (1965)[5] is a zero-order approximation of our model.

Thanks are due to Professor C. P. Ho (赫崇本) for his examination of this article.

References

[1] Hansen, W.: The Sea, Vol. I (1962), Ch. 23 "Tides"

[2] Hansen, W.: Recent Development in Tidal Theory, itt. des Ins. für Meer. der Uni. Hamburg (1968), No. X, pp. 1–7

[3] Hendershott, M. & Munk, W.: Tides, Annual Review of Fluid Mechanics, 2 (1970), pp. 205–224

[4] Сгибнева, Л.: Госу, океан. инс. труды (1972), Вып. 112, pp. 5–43

[5] Сгибнева, Л. и Фельзенбаум, А.: ДАН СССР, 164 (1965), No. 2, pp. 315–318

[6] 山东海洋学院: 《海洋科学文集》(海洋水文专辑) (1972), 22–55 页

大洋风生–热盐环流模型*

冯士筰

绪言

二十五年前，W. Munk 发表了他的风生大洋环流的经典论文，阐释了大洋表层定常环流的主要特征。这一理论成就极大地鼓舞了研究者们去探索洋流理论中另一个更加困难的领域——热盐环流的机制；因为正如 H. Stommel 等人所指出的，Munk 风生洋流的理论值与观测值的分歧是由于没有考虑热盐作用的后果；另外，如跃层的形成、深层洋流的机制等重要课题也是联系于热盐环流的。虽然，二十余年来，热盐环流以及温跃层的研究出现了不少可观的成果；但是，迄今为止，还达不到像风生洋流那样的成就。这主要是因为，一方面，洋面上受热和冷却的知识不如洋面上风应力分布的知识多；另一方面，热盐环流的课题，会带来非线性的困难。由此可见，对热盐环流及其机制，同时提出几种不同的解释，互补长短，看来还是必要的。本文提出了一种热盐环流机制和风生–热盐环流的模型，尝试着解释了洋流中的若干现象。

一、基本方程组

洋流的复杂性就在于是斜压流体，且处于行星尺度空间中作湍流运动，洋流是对于洋面上作用着的风应力和热盐因子的一种非线性反应。对于它的一种特殊情形，所谓正压海洋，毫无疑义，其方程组即通常的常密度不可压缩湍流的 Reynolds 方程组。而对于描述风生–热盐环流的斜压模型，则必须在给出其动力学方程组的同时，还要补充以适当的热力学规律。对于斜压海洋定常流的一种最古典的处理方法，是在给出动力学方程组的同时，再假设某种密度模型作为密度的已知分布。无疑，这种办法，主观随意性太大，往往是直接违反热力学定律的；这是因为，洋流微团，作为一个“动力–热力系统”，其流场和密度场，显然是相互制约、相互调整的。一种比较严格的途径，是补充以热力学方程。这方面较早的尝试，是补充一个等容过程的所谓“不可压缩性”方程（П. А. Киткин, 1953）[1]。遗憾的是，这个“不可压缩性方程”不能充分地反映洋流中密度变化的物理过程，因而，没有获得广泛的应用。描述洋流的经典的完整方程组是考虑了洋水微弱压缩性的湍流方程组（N. P. Foffonof, 1962）[2]；该型方

* 冯士筰. 1979. 大洋风生–热盐环流模型. 山东海洋学院学报, 9(2): 1–14

程组是 П. С. Пинейкин（1955）以稍许不同的形式首次引出的[3]，其主要特征为补充引入了一个密度扩散方程。应该指出，具有广泛应用价值的是该方程组引进了 Boussinesq 假设后的近似形式——其特点为：密度变化除了当然反映在密度扩散方程中以外，只仅仅保留于准静力方程中[4~12]。特别由下列这样一个有趣的事实——在 β 平面上，由 Boussinesq 近似后的方程组所导出的大洋斜压层深度的解，蜕化到 f-平面上的时候，与由完整的上述之 Пинейкин 方程组所得到的解是一致的[4]；由此可见，洋流中这点微弱的压缩性的考虑未必是必要的。这就启发我们，能否合理地建立这样一组方程：它在重视湍流扩散效应的同时，却忽略了洋流的微弱压缩性；而已被证实具有广泛应用价值的 Boussinesq 近似形式的方程组仅作为该方程组的一种局部情形被包含，特别，这一方程组应比 Boussinesq 形式的方程组具有更广泛的应用范围。这就是本节要回答的问题。

事实上，如果我们假设洋流为非均匀的不可压缩湍流，则可以下列方程组来描述之：

$$\rho\frac{\mathrm{d}\boldsymbol{q}}{\mathrm{d}t}=\rho\cdot 2\boldsymbol{q}\times\boldsymbol{\Omega}-\nabla p-\rho g\nabla\chi+\nabla\cdot(\overline{-\rho\boldsymbol{q}'\boldsymbol{q}'}) \tag{1}$$

$$\nabla\cdot\boldsymbol{q}=0 \tag{2}$$

$$\frac{\mathrm{d}\Theta}{\mathrm{d}t}=\nabla\cdot(\overline{-\boldsymbol{q}'\Theta'}) \tag{3}$$

$$\rho=\rho_0\left[1-\alpha\left(\Theta-\Theta_0\right)\right] \tag{4}$$

其中，$\frac{\mathrm{d}}{\mathrm{d}t}=\frac{\partial}{\partial t}+\boldsymbol{q}\cdot\nabla$，$\nabla$ 为梯度算子，$\nabla\cdot$ 为散度算子；$\boldsymbol{q}$ 为流速，ρ 和 p 分别为海水密度和压强，$\boldsymbol{\Omega}$ 为地转角速度，g 为重力加速度，$g\chi$ 为重力势；Θ 为海水温度；α, ρ_0, Θ_0 为常数；$\boldsymbol{q}'$ 与 Θ' 分别表示流速和温度的湍流脉动，其上的一横，表示 Reynolds 平均。应指出，若把 Θ 理解为表观温度，则方程（3）和（4）也包含了盐度效应[2]；t 表示时间。

我们要着重强调的是，若把（4）方程微分，再把（3）式代入之，则有密度扩散方程；把密度扩散方程与（2）方程合并则有

$$\frac{\partial\rho}{\partial t}+\nabla\cdot(\rho\boldsymbol{q})=\nabla\cdot(\overline{-\boldsymbol{q}'\rho'}) \tag{5}$$

方程（5）正是 Reynolds 平均后的质量守恒方程[13]；其右端附加的脉动项 $(\overline{-\boldsymbol{q}'\rho'})$ 可解释为“表观质量流”；正如（3）方程表达了平均温度扩散一样，（5）方程表达了平均质量的湍扩散。

至于 Reynolds 方程中其他的脉动项都被略去了，这是因为，它们或者为三阶相关，或者，虽是二阶相关，但与 Mach 数同阶[14]。

最终，当我们引入了广义 Newton 摩擦定律和广义的 Fourier 定律和扩散定律以后，

就获得了最终描述洋流的完整方程组（1）~（4）、（6）、（7）或（1）、（2）、（5）、（6）、（8）。其中，方程（6）、（7）、（8）分别为：

$$\overline{-\rho \boldsymbol{q}'\boldsymbol{q}'} = \nabla(\mu \boldsymbol{q}) \tag{6}$$

$$\overline{-\boldsymbol{q}'\Theta'} = \nabla(\lambda \Theta) \tag{7}$$

$$\overline{-\boldsymbol{q}'\rho'} = \nabla(k \rho) \tag{8}$$

式中，μ、λ 和 k 分别表示湍摩擦系数、湍导热系数和湍扩散系数。

顺便指出，方程（5）的应用也有着流体力学上的兴趣；因为，通常，经 Reynolds 平均后的质量守恒方程保持原始的形式，而为此，必须引进一个假设[15]。

二、Ekman 漂流的密度机制

首先应用上述洋流方程组来阐释一个经典问题——Ekman 漂流的密度机制。

Ekman（1905）求解其漂流的时候，曾假定了密度 ρ 为常数。这其实是一个最简单的密度模型。迫使 Ekman 预先对密度作出这一假定，是由于他无视洋流的热力学因素、无视流场和密度场的相互制约和调整，没有找出一个适当的热力学方程所致。其实，只要我们合理地假定了海洋表面的热盐状况，Ekman 漂流中的密度分布将作为一个理论结果，由上述洋流的完整方程组而推出。事实上，在众所周知的 Ekman 漂流问题中所作的假设条件下（当然，不包括 ρ 为常数这一假设），附加上海面、无限深处以及初始状态等定解条件（10）~（12），求解简化了的扩散方程（9）即可获得 Ekman 漂流中的密度解（13）：

$$\frac{\partial \rho}{\partial t} = k \frac{\partial^2 \rho}{\partial z^2} \tag{9}$$

$$\rho = \rho_0(z) \ (t = 0) \tag{10}$$

$$\rho = \rho^* = 常数 \ \ (t \geqslant 0 及 z = 0) \tag{11}$$

$$\rho 有界 \ (t \geqslant 0 及 z \to \infty) \tag{12}$$

解

$$\begin{aligned} \rho(z,t) = &\frac{1}{2\sqrt{\pi}} \int_0^{\infty} \frac{1}{\sqrt{kt}} \left\{ \exp\left[-\frac{(z-\zeta)^2}{4kt} \right] - \exp\left[-\frac{(z+\zeta)^2}{4kt} \right] \right\} \rho_0(\zeta) \mathrm{d}\zeta \\ &+ \rho^* \left\{ 1 - \frac{1}{\sqrt{\pi}} \left[\frac{z}{\sqrt{kt}} \exp\left(-\frac{z^2}{4kt} \right) + \frac{1}{4} \int_0^{\frac{z^2}{kt}} \zeta^{\frac{1}{2}} \mathrm{e}^{-\frac{\zeta}{4}} \mathrm{d}\zeta \right] \right\} \end{aligned} \tag{13}$$

其中，扩散系数 k 设为常数，z 为重力方向的空间坐标，原点设在静止海面上。

显然，当 $t \to \infty$ 时，解（13）蜕化成定常 Ekman 漂流的密度分布

$$\rho(z,\infty)=\rho^{*}=\text{常数} \tag{14}$$

结论十分清楚了：当 Ekman 漂流发展到定常状况时，密度为常值，证明了 Ekman 的密度模型是正确的；这也阐明了，海洋表层——即 Ekman 层之密度所以均匀，确由湍扩散之结果。

但，非定常的发展阶段，密度不为常值而以（13）式描述之。故，在经典问题中，把密度设为常数求解 Ekman 非定常漂流是违反了热力学规律的，除非再引入 Boussinesq 近似。

强调指出，由此可见，不论热盐要素初始分布如何，当风作用于洋面一段足够长的时间以后，在海洋表层，如密度等要素必被风海流调整为均匀分布。这是早被观测所肯定的事实了。

三、风旋度-热盐梯度方程

由于洋流流场为一薄层流场，其流场的铅垂尺度远小于其水平尺度，于是描述洋流的上述完整方程组可化简如下：

$$\rho\frac{\mathrm{d}u}{\mathrm{d}t}-\rho fv=-\frac{\partial p}{\partial x}+\mu_{\mathrm{H}}\left(\frac{\partial^2 u}{\partial x^2}+\frac{\partial^2 u}{\partial y^2}\right)+\mu_{\mathrm{V}}\frac{\partial^2 u}{\partial z^2} \tag{15}$$

$$\rho\frac{\mathrm{d}v}{\mathrm{d}t}+\rho fu=-\frac{\partial p}{\partial y}+\mu_{\mathrm{H}}\left(\frac{\partial^2 v}{\partial x^2}+\frac{\partial^2 v}{\partial y^2}\right)+\mu_{\mathrm{V}}\frac{\partial^2 v}{\partial z^2} \tag{16}$$

$$\rho g=-\frac{\partial p}{\partial z} \tag{17}$$

$$\frac{\partial u}{\partial x}+\frac{\partial v}{\partial y}+\frac{\partial w}{\partial z}=0 \tag{18}$$

$$\frac{\partial \rho}{\partial t}+\frac{\partial(\rho u)}{\partial x}+\frac{\partial(\rho v)}{\partial y}+\frac{\partial(\rho w)}{\partial z}=\frac{\partial}{\partial z}\left(k_{\mathrm{V}}\frac{\partial \rho}{\partial z}\right)+k_{\mathrm{H}}\left(\frac{\partial^2 \rho}{\partial x^2}+\frac{\partial^2 \rho}{\partial y^2}\right) \tag{19}$$

或等价的，

$$\frac{\partial \Theta}{\partial t}+\frac{\partial(\Theta u)}{\partial x}+\frac{\partial(\Theta v)}{\partial y}+\frac{\partial(\Theta w)}{\partial z}=\frac{\partial}{\partial z}\left(\lambda_{\mathrm{V}}\frac{\partial \Theta}{\partial z}\right)+\lambda_{\mathrm{H}}\left(\frac{\partial^2 \Theta}{\partial x^2}+\frac{\partial^2 \Theta}{\partial y^2}\right) \tag{20}$$

其中，u、v、w 为流速 $\boldsymbol{q}$ 在 x、y、z 三坐标上的分量；μ_{V}、μ_{H} 表示铅垂、水平湍摩擦系数；k_{V}、k_{H}——铅垂、水平湍扩散系数；λ_{V}、λ_{H}——铅垂、水平湍导热系数；$f=2\Omega\sin\varphi$ 表示 Coriolis 参数，其中，Ω 表示 $|\boldsymbol{\Omega}|$，φ 地理纬度；x、y、z 为空间右手正交坐标系，z 铅垂座标，向上为正，原点置于静止海面上，而 x 和 y 可理解为准纬圈和准经圈——这是三维边界层座标使用于洋流问题上的一种特定形式。应指出，在这

种坐标系中，“f–平面”是其零阶近似，“β–平面”为其一阶近似①。

我们进一步局限于定常洋流的粘性理论，即局部微商项和相对加速度是略而不计的，则洋流方程组（15）~（20）化简如下

$$\mu_{\mathrm{H}}\left(\frac{\partial^2 u}{\partial x^2}+\frac{\partial^2 u}{\partial y^2}\right)+\frac{\partial}{\partial z}\left(\mu_{\mathrm{V}}\frac{\partial u}{\partial z}\right)+f\rho v=\frac{\partial p}{\partial x} \tag{21}$$

$$\mu_{\mathrm{H}}\left(\frac{\partial^2 v}{\partial x^2}+\frac{\partial^2 v}{\partial y^2}\right)+\frac{\partial}{\partial z}\left(\mu_{\mathrm{V}}\frac{\partial v}{\partial z}\right)-f\rho u=\frac{\partial p}{\partial y} \tag{22}$$

$$-\rho g=\frac{\partial p}{\partial z} \tag{23}$$

$$\frac{\partial u}{\partial x}+\frac{\partial v}{\partial y}+\frac{\partial w}{\partial z}=0 \tag{24}$$

$$\frac{\partial(\rho u)}{\partial x}+\frac{\partial(\rho v)}{\partial y}+\frac{\partial(\rho w)}{\partial z}=\frac{\partial}{\partial z}\left(k_{\mathrm{V}}\frac{\partial \rho}{\partial y}\right)+k_{\mathrm{H}}\left(\frac{\partial^2 \rho}{\partial x^2}+\frac{\partial^2 \rho}{\partial y^2}\right) \tag{25}$$

或

$$\frac{\partial(\Theta u)}{\partial x}+\frac{\partial(\Theta v)}{\partial y}+\frac{\partial(\Theta w)}{\partial z}=\frac{\partial}{\partial z}\left(\lambda_{\mathrm{V}}\frac{\partial \Theta}{\partial z}\right)+\lambda_{\mathrm{H}}\left(\frac{\partial^2 \Theta}{\partial x^2}+\frac{\partial^2 \Theta}{\partial y^2}\right) \tag{26}$$

显然，上述方程组为非线性的。这反映了热盐环流或风生–热盐洋流问题的一个重要特色、亦是难点。这曾迫使某些作者，为了获得解析解，或者采用“扰动法”，或者仅限于描述那种具有“相似性解”的局部情形[7~9]；否则，放弃解析解转而采用数值方法[10~12]。不过，若当我们仅着眼于描绘洋流体积运输的分布[18, 19]，则可采用全流技术，在较普遍的情形下，将这一非线性问题转化为线性问题，从而给出其解析解而没有根本性的困难。

B. Б. Штокман（1946）[20]和 H. U. Sverdrup（1947）[21]分别独立地导出了斜压海洋中的风生全流方程。他们引入全流概念的目的是为了自然消除密度非均匀的铅垂结构，从而简化了斜压海洋中的风生洋流问题。为此，只需由斜压层底部至海面对（21）~（24）诸方程施以积分即可。十年后，H. Stommel（1957）[19]把这一全流理论推广去研究洋流中的热盐机制，并导出了热盐环流为“内模式”的著名结论[19,22]，然而，明显的是，Штокман-Sverdrup 的全流理论仅适于风生洋流；这是由于该理论仅考虑了运动方程和连续方程，而未引入扩散方程（25）或导热方程（26），故不适于风生–热盐环流的研究。为了正确地推广全流理论到风生–热盐环流的情形，必须对（25）方程或（26）方程也施以上述积分。从而有全流方程组为

① 原文注：在这一种正交坐标系中由基本方程组（1）~（4）导出洋流方程组的简化形式（15）~（20）的过程，虽不复杂，却颇占篇幅，故从略之；其精神可参考文献[16]

$$\mu_{\mathrm{H}}\left(\frac{\partial^2 S_x}{\partial x^2}+\frac{\partial^2 S_x}{\partial y^2}\right)+\tau_x+fM_y=\frac{\partial P}{\partial x} \tag{27}$$

$$\mu_{\mathrm{H}}\left(\frac{\partial^2 S_y}{\partial x^2}+\frac{\partial^2 S_y}{\partial y^2}\right)+\tau_y-fM_x=\frac{\partial P}{\partial y} \tag{28}$$

$$-gQ=p\big|_{z=\zeta}-p\big|_{z=-\mathrm{H}} \tag{29}$$

$$\frac{\partial S_x}{\partial x}+\frac{\partial S_y}{\partial y}=0 \tag{30}$$

$$\frac{\partial M_x}{\partial x}+\frac{\partial M_y}{\partial y}=\Gamma_0 \tag{31}$$

或

$$\frac{\partial \Theta_1}{\partial x}+\frac{\partial \Theta_2}{\partial y}=\Lambda_0 \tag{32}$$

其中

$$(S_x,S_y)=\int_{-H}^{\zeta}(u,v)\,\mathrm{d}z$$

$$(M_x,M_y)=\int_{-H}^{\zeta}(\rho u,\rho v)\,\mathrm{d}z$$

$$P=\int_{-H}^{\zeta}p\mathrm{d}z,\quad Q=\int_{-H}^{\zeta}\rho\mathrm{d}z$$

$$(\Theta_1,\Theta_2)=\int_{-H}^{\zeta}\Theta(u,v)\,\mathrm{d}z$$

τ_x，τ_y 为洋面风应力 $\boldsymbol{\tau}$ 在 x、y 上的分量；

$$\Gamma_0=\left(k_{\mathrm{V}}\frac{\partial\rho}{\partial z}\right)\bigg|_{z=\zeta},\quad \Lambda_0=\left(\lambda_{\mathrm{V}}\frac{\partial\Theta}{\partial z}\right)\bigg|_{z=\zeta}$$

在导出的过程中，假设了斜压层底部（$-H$）为常数，且在 $z=-H$ 上，有 u, v, ρ, Θ 对 z 的导数和 p 对 x、y 的导数和二阶导数以及 w 皆为零；在水面上，$z=\zeta$ 处，有 w 为零而 p 为常数（或 p 远小于斜压底的 p），且因为 $|\zeta|\ll H$ 而以零代之[18~21, 23]。

消去 P 并引入流函数 ψ 满足（30），有

$$(S_x,S_y)=\left(-\frac{\partial\psi}{\partial y},\frac{\partial\psi}{\partial x}\right) \tag{33}$$

并设 Coroilis 参数为常数，则导出了在 f–平面上的风旋度–热盐梯度方程

$$\mu_{\mathrm{H}}\nabla^4\psi=-\,\mathrm{Curl}\boldsymbol{\tau}+f\Gamma_0 \tag{34}$$

式中，

$$\nabla^4 = \frac{\partial^4}{\partial x^4} + 2\frac{\partial^4}{\partial x^2 \partial y^2} + \frac{\partial^4}{\partial y^4}$$

$$\mathrm{Curl}\,\boldsymbol{\tau} = \frac{\partial \tau_y}{\partial x} - \frac{\partial \tau_x}{\partial y}$$

$$\varGamma_0 = \left(k_{\mathrm{V}} \frac{\partial \rho}{\partial z} \right)\bigg|_{z=\zeta} = -\alpha\rho_0 \left(\frac{k_{\mathrm{V}}}{\lambda_{\mathrm{V}}} \right)\bigg|_{z=\zeta} \varLambda_0$$

由（34）式显然看出，经典的Штокман-Sverdrup 风旋度方程

$$\mu_{\mathrm{H}} \nabla^4 \psi = \mathrm{Curl}\,\boldsymbol{\tau} \tag{35}$$

乃是风旋度–热盐梯度方程在不考虑热盐效应（ $\varGamma_0 = 0$ ）时之特殊情形。

下面讨论一下风旋度–热盐梯度方程（34）的物理意义。

（1）全流函数 ψ 取决于洋面的风应力旋度和热盐梯度 $\varGamma_0$（或 $\varLambda_0$）， $\varLambda_0$ 可视为通过海面的“表观热流”。海面条件 $\varGamma_0$ 或 $\varLambda_0$ 的提法是等价的，并分别一致于 Линейкин（1957）[17]和 Fofonoff（1962）[2]。当 $\mathrm{Curl}\,\boldsymbol{\tau}$ 为零时，就获得了纯热盐的水平环流；当 $\varGamma_0 = 0$ 时，就获得了众所周知的风生环流。由于（34）的线性，以及边界条件的线性，故二者可以叠加而获得风生–热盐环流的解。

（2）由于热盐环流乃取决于海洋外部的热盐因子 $\varGamma_0$ 或 $\varLambda_0$；因而，本模型所引入的热盐效应将为所谓“外模式”。这一结论与 H. Stommel（1957）[19]引入的“内模式”热盐效应相反。我们认为，这是由于 Stommel 没有引入适当的热力学方程所致。

（3）在（34）方程中热盐梯度前有 f Coriolis 参数这一乘子。这说明了，在不转动的封闭大洋中，将不会产生纯热盐的水平环流［当然，垂直环流仍可能存在（M. E. Stern, 1975）[24]］。故，在赤道附近热，盐效应对产生水平环流将没有显著贡献；赤道流主要是风生的。的确，赤道流的（质量）全流的风生理论计算值与实测值符合较好（Sverdrup, 1947; Reid, 1948）[21, 25]。

（4）对于我们考查的定常环流的情况，海面表观热流 $\varLambda_0$ 在整个的大洋面上的积分值应为零；否则，收支不抵。这进一步得知，在整个洋面上 $\varLambda_0$ 有正值区和负值区。但当 $\varLambda_0$ 为负值时，出现了密度的不稳定结构，所以方程只能适用于对大洋加热为主的区域，或者净降水量足以补偿在致冷区域造成的密度可能为不稳定结构的差额，致使其仍显现为稳定结构。幸而，实际大洋中，不稳定结构的区域相对的小，仅局限于高纬处[2]。对于海洋中热盐要素的稳定结构，$\varLambda_0$ 为正，$\varGamma_0$ 为负；由（34）可见，纯热盐环流，在封闭大洋中，为气旋式的。

（5）作为一例，设有一个矩形海洋 x（$0,a$），y（$0,b$）；在海而上不存在风应力，只有净受热 $\varLambda_0 = F(y) > 0$（或 $\varGamma_0 = F_1(y) < 0$）

引入粘性侧边界条件：

$$\psi=\frac{\partial\psi}{\partial x}=0(x=0,a)$$
$$\psi=\frac{\partial\psi}{\partial y}=0(x=0,b) \tag{36}$$

在（36）边界条件下求解（34）方程，即获得纯热盐环流的流函数ψ的分布。其实为了获得定性结果，只需利用一个在边缘被镶嵌的平板（(0,a)、(0,b)）上加载($f\Gamma_0$)后得出的小挠度弯曲（相当于ζ）和其等值线的分布（相当于ψ），有$\zeta<0$，且ψ为气旋式；南部密集，北部稀疏[26]。

显然，当f=0时，$\frac{\partial\psi}{\partial x}=\frac{\partial\psi}{\partial y}=0$，即$S_x=S_y=0$——此时，仅有经截面（$x$=常数）中的垂直环流，而无水平环流了；若$f\neq0$，则有水平环流。这个物理图画是明显的：由于海而的热流Λ_0仅对y是不均匀的，故产生了仅在y方向的压强梯度，从而，在y向产生了流动；若f = 0，则这一流动将最终形成在经面中的定常垂直环流，而，由于水体守恒的要求，必有$S_y=\int_{-H}^{\zeta}v\mathrm{d}z=0$；反之，若$f\neq0$，则由于 Coriolis 力的作用，上述流动必然偏转，从而形成了水平环流。

四、大洋风生-热盐环流的两层模型

（I）问题的提法

我们仍局限于考查定常的风生-热盐洋流的粘性理论，于是描述这一现象的洋流方程组，即（21）~（25）（或 26）诸方程。

我们把大洋抽象成一个两层的模型，中间以一个“薄层”——“准不连续面”把它分开。设该薄层即压缩了的密度跃层或表观温度的跃层；作出这一假设的依据是，大洋是密度层化的，且不论其密度的分布之具体形式如何，跃层的存在乃一主要的特征（高维度地区跃层的消失，可作为过渡到极限之情形）。设跃层的平均位置为$-h$，并为了简单，假设h为常数。我们又可以进一步认为，$-h$还具有斜压层下界的含义，即h具有斜压层深度的含义。这样，斜压层底部的性质几乎都可以赋予$-h$处了（参考上一节中$-H$处的诸性质）；但有两点修正：其一，密度（或表观温度）的铅垂梯度，即$\left(\frac{\partial\rho}{\partial z}\right)_{z=-h}$或$\left(\frac{\partial\Theta}{\partial z}\right)_{z=-h}$不为零，而由于$-h$亦为跃层的所在位置，故上述梯度应为足够的大；其二，依据 H. Stommel（1957）[19]，该处为“无运动层”——即无水平速度的辐散面，于是，按照连续方程，有铅垂速度的极大值 w_i，和相应的 $\rho_i w_i$ 存在，由此意味着：上层大洋乃斜压层以上洋流的流场，且将其诸变量以下标“1”标之，下层大洋乃指深层洋流之流场了，并以下标“2”表征其诸变量。由上一节的导出方式，可以立刻导出上层大洋中洋流的有关全流的封闭方程组如下：

$$\frac{\partial S_{x1}}{\partial x}+\frac{\partial S_{y1}}{\partial y}=w_i \tag{37}$$

$$\beta S_{y1}+\mu_{\mathrm{H}}\left(\frac{\partial^3 S_{x1}}{\partial x^2\partial y}+\frac{\partial^3 S_{x1}}{\partial y^3}-\frac{\partial^3 S_{y1}}{\partial y^2\partial x}-\frac{\partial^3 S_{y1}}{\partial x^3}\right)-\mathrm{Curl}\boldsymbol{\tau}+f(\varGamma_0-\varGamma_1+\rho_i w_i)=0 \tag{38}$$

其中

$$\varGamma_i=\left(k_{\mathrm{V}}\frac{\partial\rho}{\partial z}\right)\Bigg|_{z=-h}=-\left(\alpha\rho_0\frac{k_{\mathrm{V}}}{\lambda_{\mathrm{V}}}\right)\Bigg|_{z=-h}\varLambda_i,\quad \varLambda_i=\left(\lambda_{\mathrm{V}}\frac{\partial\varTheta}{\partial z}\right)_{z=-h}$$

应指出，在导出上封闭方程组时，考虑了 Coriolis 参数 f 随纬度的变化 $\beta=\mathrm{d}f/\mathrm{d}y$，且令 β 为常数——即该问题立于“β-平面”上；又引入了 $(\mathrm{d}f/\mathrm{d}y)M_{y1}\approx\beta S_{y1}$（密度采用单位值）。假设了 w_i、$\rho_i w_i$ 和 $\varGamma_i$（或 $\varLambda_i$）也为已知强迫函数，则该方程组封闭。

为了在下层大洋中导出一组同样简单的封闭方程组，还需要补充以洋底的下列诸假设：大洋深度 D 设为常数，且 $z=-D$ 处，有 $\partial p/\partial x=\partial p/\partial y=0$，略掉洋底湍摩擦；以及引用下列底边界条件——$w$、$\partial p/\partial z$ 和 $\partial\varTheta/\partial z$ 在洋底为零。这样，就导出一个形如（37）和（38）方程组的下层大洋中洋流方程组

$$\frac{\partial S_{x2}}{\partial x}+\frac{\partial S_{y2}}{\partial y}=-w_i \tag{39}$$

$$\beta S_{y2}+\mu_{\mathrm{H}}\left(\frac{\partial^3 S_{x2}}{\partial x^2\partial y}+\frac{\partial^3 S_{x2}}{\partial y^3}-\frac{\partial^3 S_{y2}}{\partial y^2\partial x}-\frac{\partial^3 S_{y2}}{\partial x^3}\right)+f(\varGamma_i-\rho_i w_i)=0 \tag{40}$$

由上述两个方程组（37）~（38）和（39）~（40），可以得出的有益结论，除去了上节中由（34）方程的讨论所得出的那些以外，尚可立即得到下述的一个重要结论：下层大洋中——即深层大洋中水平环流的形成，一部分是由于借助 $\rho_i w_i$（和 w_i）沟通上层大洋的垂直对流所致，另一部分是由于通过大洋中跃层这一准不连续面所形成的强大的湍流表观热流 $\varLambda_i$（或 $\varGamma_i$）的湍流扩散所致；并且，显然，地球的旋转是形成深层水平环流的必要条件。深层环流是纯热盐的。当然，跃层的形成又是由于洋流与热盐要素相互调整的结果，而上层洋流——即表层洋流既然本质上是主要取决于海面风应力，故风应力对深层洋流也有贡献，但却是间接的。

对于大洋的中央部分，若不考虑侧向湍摩擦，即令 $\mu_{\mathrm{H}}=0$，则上述二方程组化简

如下，

上层大洋：

$$\frac{\partial S_{x1}}{\partial x}+\frac{\partial S_{y1}}{\partial y}=w_i \tag{41}$$

$$\beta S_{y1}=\mathrm{Curl}\,\boldsymbol{\tau}-f(\varGamma_0-\varGamma_i+\rho_i w_i) \tag{42}$$

下层大洋：

$$\frac{\partial S_{x2}}{\partial x}+\frac{\partial S_{y2}}{\partial y}=-w_i \tag{43}$$

$$\beta S_{y2}=-f(\varGamma_i-\rho_i w_i) \tag{44}$$

显然，当不考虑湍流热扩散的时候—— $\varGamma_0=\varGamma_i=0$ ，即蜕化为 H. Stommel 于 1957 年给出的热盐效应的模型[19]。因而，Stommel 的热盐环流模型是纯对流式的。其实，若从不考虑湍流扩散的 Киткин（1953）方程组出发[1]，依上述同样的处理手段，可直接获得方程组（41）~（44）所描述的 $\varGamma_0=\varGamma_i=0$ 的 Stommel 模型。

基于（37）~（40）两个方程组的线性性质，并且由下面我们即将看出，边界条件也是线性的；故类似于 Stommel 模型那样的对流式的热盐效应可以线性地滤掉；因本文将侧重探讨湍扩散的热盐模型。于是描述它的方程组为：（37）~（40）方程组中当不考虑 w_i 和 $\rho_i w_i$ 时的局部情况；从而，由上一节的分析，可以立刻导出上层大洋中和下层大洋中洋流的风旋度–热盐梯度方程式如下，

上层大洋：

$$\mu_{\mathrm{H}}\nabla^4\psi_1-\beta\frac{\partial\psi_1}{\partial x}=-\,\mathrm{Curl}\,\boldsymbol{\tau}+f(\varGamma_0-\varGamma_i) \tag{45}$$

下层大洋：

$$\mu_{\mathrm{H}}\nabla^4\psi_2-\beta\frac{\partial\psi_2}{\partial x}=f\varGamma_i \tag{46}$$

（II）矩形大洋中的近似解

我们把兴趣集中在以赤道为中心的 ±60° 纬度以内的洋区。可假设矩形大洋东西有界（$x=0,a$），而南北无界，即实际上被两条经线所包围。于是，有粘性侧边界条件如下：

岸界（$x=0$ 和 $x=a$）

$$\psi_n=\frac{\partial\psi_n}{\partial x}=0\qquad n=(1,2) \tag{47}$$

假设风应力分布只有沿东西向分量且仅依 y 而变化，即 $\boldsymbol{\tau}=\boldsymbol{i}\tau_x(y)$ ，于是

$$-\operatorname{Curl}\boldsymbol{\tau}=\frac{\mathrm{d}\tau_x(y)}{\mathrm{d}y} \tag{48}$$

由于大洋表层强烈的湍混合形成了密度比较均匀的铅垂分布(见第二节);而($-h$)处乃一强大的密度跃层,虽然跃层限制了铅直湍的发展,但足够大的密度梯度可以认为足以使$\varGamma_i$为有限值,且使$\varGamma_i \gg \varGamma_0$。又以大洋的幅员来考查,由于盐度的保守性,故设密度分布主要取决于温度分布,即密度跃层的分布主要取决于温跃层的分布,于是自然可以合理地假设$\varGamma_i$仅随纬度而变化,即$\varGamma_i$(或$\varLambda_i$)随y的增加而单调上升(下降)(注意,对于密度的稳定结构,$\varGamma_i(y)\leqslant 0$,而$\varLambda(y)\geqslant 0$)

$$\varGamma_i=\varGamma_i(y) \text{ 或 } \varLambda_i=\varLambda_i(y) \tag{49}$$

在上述边界条件(47)和海面风应力表达式(48)以及跃层分布(49)式下解方程(45)和(46),得出上、下层洋流之流函数分别为

$$\psi_1=\frac{a}{\beta}X\left\{\frac{\mathrm{d}\tau_x(y)}{\mathrm{d}y}-f\varGamma_i(y)\right\} \tag{50}$$

$$\psi_2=\frac{a}{\beta}Xf\varGamma_i(y) \tag{51}$$

其中表达式

$$X=-\left(\frac{2}{\sqrt{3}}-\frac{\sqrt{3}}{\gamma a}\right)\mathrm{e}^{-\frac{\xi}{2}}\cos\left(\frac{\sqrt{3}}{2}\xi+\frac{\sqrt{3}}{2\gamma a}-\frac{\pi}{6}\right)+1-\frac{1}{\gamma a}(\xi-\mathrm{e}^{-\gamma a+\xi}+1) \tag{52}$$

$$\xi=\gamma x,\quad \gamma=\sqrt[3]{\frac{\beta}{\mu H}},\quad \frac{\mathrm{d}X}{\mathrm{d}\xi}=\left(\frac{2}{\sqrt{3}}-\frac{\sqrt{3}}{\gamma a}\right)\mathrm{e}^{-\frac{\xi}{2}}\sin\left(\frac{\sqrt{3}}{2}\xi+\frac{\sqrt{3}}{2\gamma a}\right)-\frac{1}{\gamma a}(\xi-\mathrm{e}^{-\gamma a+\xi}) \tag{53}$$

依 W. Munk(1950)[18],给出(50)、(51)诸流函数没有任何困难。不过应指出的是,该近似解的成立应满足一定条件;从下面我们选择的强迫函数的形式以及有关数据来看,满足上述条件是显然的。另外,把原为在南北方向为有界大洋的解移到这里的南北方向为无界大洋的情形,只需把原为南北方向的界长理解为强迫函数的基本谐波波长即可;特别,在 Munk 模型中在南北向边界上边界条件应被满足的疑难,此处却不复存在了[18]。

取风应力分布如 Munk(1950)[18],而$\varGamma(y)$的分布如下:

$$\varGamma_i(y)=-\varGamma_m\cos(\kappa\varphi) \tag{54}$$

其中κ为一常数,为了相应于 Munk 风生洋流之情形,可取$\kappa=1.5$;这意味着“跃层”($\varGamma_i(y)$)消失于纬度60°处,$\varGamma_m$为幅度,$\varGamma_m=\left(k_{\mathrm{V}}\left|\frac{\mathrm{d}\rho}{\mathrm{d}z}\right|_{\max}\right)_{\kappa=-n}$,依 Пинейкин(1957)[4],

$\left|\frac{d\rho}{dz}\right|_{max} = 5\times 10^{-6}(g/cm^4)$，$k_V$ 取 15（g/(cm·sec)）是为了使 Γ_m 与西风幅度在不同比例下有相接近的值[10]，于是有 $\Gamma_m = 0.75\times 10^{-4}$（c.g.s.）。从而，

$$f\Gamma_i = -2\Omega\sin\varphi\Gamma_m\cos(\chi\varphi) \tag{55}$$

代入以上各数据及 Ω 取 0.73×10^{-4}（1/sec），有

$$f\Gamma_i = -1.1\times 10^{-8}\sin\varphi\cdot\cos(1.5\varphi) \tag{56}$$

大洋几何尺度依 Munk[18]。

依上面的公式和数据进行了 ψ 的计算，讨论如下。

（III）结论和讨论

由解的结构可以看出，我们利用了经典的 Munk 大洋风生环流的近似解（Munk, 1950）[18]。当不考虑热盐因素时，即令 $\Gamma_i = 0$（或 $\Lambda_i = 0$）时，解蜕化为描述大洋风生环流的经典的 Munk 的解（Munk, 1950）[18]。

由（50）和（51）所描绘的大洋环流之图画讨论如下。

（1）上层大洋中的洋流为风生–热盐环流，且环流概貌大体地一致于 Munk 纯风生环流；只是亚热带流涡在南北方向有所扩大，亚寒带流涡在南北方向有所压缩；南赤道流略北移、北赤道流略南移，赤道逆流也略有偏移。这说明了，尽管“表层环流一半是风生的、一半为热盐过程形成的”，又何以“这一环流的总趋势，如流涡的边界等等，竟和大气环流如此吻合”（Munk, 1950）[18]的机理。

（2）由式（52）和（53）可见，正如 Munk 所指出的[18]，近似解可分为三区，而上层大洋的中部有解

$$S_{y1} = \frac{\partial\psi_1}{\partial x} = \beta^{-1}(\mathrm{Curl}\,\boldsymbol{\tau} + f\Gamma_i) \tag{57}$$

由（57）式可见，当不考虑热盐因子时（即 $\Gamma_i = 0$），（57）蜕化为经典的 Sverdrup 解。

在下层大洋的中部有解

$$S_{y2} = \frac{\partial\psi_2}{\partial x} = -\beta^{-1}f\Gamma_i \tag{58}$$

由此可见，在深层洋中区，在同一半球上，南北半部之间，有广阔而缓慢的、向两极方向的水量迁移。

（3）由于式（52）、（53）独立于强迫函数（不论是风应力旋度抑或热盐效应抑或二者皆存），因而，被 Munk[18]所指出的它们的特征，在本模型中完全被保留了。特别是，如西向强化，在西部主流以东有一逆流存在［其量值为主流的 17%，与 Must（1936）

和 Iselin（1936）依据实测资料分析的结果 19%相近]，西部 X 曲线之形式与据水文资料得出的湾流及黑潮的流量函数曲线一致等正确结论完全被保留下来了。

（4）但，强调指出，本模型引入了热盐效应后，亚热带流涡的两边界流加强了，把 Munk 风生洋流的理论计算值与实测值相差一半的流量给补上了（见下表）。

依西边界流流量公式

$$\psi_{1_{\mathrm{WC}}} = 1.17 a\beta^{-1}\left\{\frac{\mathrm{d}\tau_x}{\mathrm{d}y} - f\Gamma_i\right\} \tag{59}$$

计算 35°N 处之湾流和黑潮之流量。

	a/km	$10^{10} f\Gamma$ (cgs)	$10^{-12}\psi_1$（热盐）/(cm³/sec)	$10^{10}\frac{\mathrm{d}\tau_x}{\mathrm{d}y}$ (cgs)	经向风因子[17]	$10^{-12}\psi_1$（风生）/(cm³/sec)	$10^{-12}\psi_1$（合成）/(cm³/sec)	实测 $10^{-12}\psi_1$ /(cm³/sec)	实测 ψ_1 与理论 ψ_1 之差/(cm³/sec)	相对误差/%
黑潮	10000	38	24	50	1.25	39	63	65	2	3
湾流	6500	38	15	70	1.30	36	51	74	23	31

其中，$\sin\varphi = \sin 35° = 0.57$，$\cos 52.5° = 0.61$。

在此，指出三点。

首先，Munk 风生环流理论值仅为实测值之半，确因没计入热盐效应之结果。

其次，上述论断尽管早被 H. Stommel 所指出[19, 22]；但，Stommel 的热盐模型不同于本模型。Stommel 模型至多只适于大西洋（湾流）而不适用于太平洋（黑潮）。这是因为太平洋无下沉的深水源[2]。我们提出的这个热盐模型显然可以用于北太平洋（黑潮），且由表中计算可见，理论值和实测值吻合较好，相对误差仅为 3%。

第三，本模型的理论值与实测值比较，对湾流来说，有差额较著。这是由于本模型没考虑大西洋高纬处下沉水所致；因在其中滤掉了对流式的热盐效应。设想，若再补充以 Stommel 正压大洋热盐模型[22]，会有改善。若进一步假设，Stommel 模型的西边界流与我们的模型西边界热盐流量同量阶，且认为，譬如，在 35°N 处也取值为 15×10^{12} cm³/sec，则理论值与实测值的相对误差可缩减到 11%了。并由此推论，南大西洋上层大洋之西边界流中两种热盐洋流有对消趋势；因而，在南半球，西边界流相对的弱，且主要为风生流——前者，与 Stommel 结论一致，后者，与其结论相反[22]。这可能是由于 Stommel 的热盐模型不包含湍扩散效应所致。

（5）由下层大洋中的解（51）可见，深层洋流为热盐性质的。且，在半球上仅为一个简单的气旋式的大流涡。因而，与上层流系相比较，深层环流的图形要简单得多——这一结论，与数值模拟结果一致[10~12]。

（6）深层环流也是西向强化的。但，由于是气旋式的环流，故伏于黑潮或湾流之

下必有一股强大的热盐逆流存在。依公式

$$\psi_{2_{\mathrm{WC}}} = 1.17 a\beta^{-1} f \Gamma_i \tag{60}$$

计算 35°N 处的这一股热盐逆流之流量，伏于黑潮之下的为 24×10^{12} cm^3/sec，伏于湾流之下的为 30×10^{12} cm^3/sec（已叠加上了 Stommel 热盐逆流之值）。应指出，H. Stommel（1957）[19]以他提出的热盐模型首先预言了湾流下有一股强大的热盐逆流存在，并已被观测所证实。综合本模型和 Stommel 的模型，可进一步推论出，在南大西洋西边界流下，若有逆流，大概也颇弱，且只能由北半球通过赤道到达 20°S 以内；这是因为，上述两股热盐逆流，特别在南大西洋亚热带流涡西边界上有对消趋势，而在 20°S 以内，本模型之热盐逆流较弱（在西边界向北流）而 Stommel 之热盐逆流却能保持其强度（在西边界向南流）之故吧[19]。

（7）假设（$-h$）位于海面下 1000 m 处，距海底为 5000 m。如上所述，热盐洋流的水量迁移与风生洋流之水量迁移，在上层大洋中为同量阶；因而，深层洋流水量迁移约为上层之半。故，深层大洋洋流流速约为上层的 10%。这是一个正确量阶。譬如，西边界流为 100 cm/sec，而热盐逆流为 10 cm/sec 之量阶[19, 27]。

（8）由解（51）、（52）、（53）还可以预言一个现象，即在深层的热盐逆流之东侧有一股反逆流存在，其量值为热盐逆流的 17%，在北半球，其流向北方。这一现象，若被观测所证实，将也是对本模型的一个有力支持。

最后，指出这个大洋环流的两层模型的弱点和问题。

首先，由于它浅性地滤掉了对流热盐模型，不能考虑大西洋高纬度处的下沉水，它对大西洋环流的解释肯定是不完备的和不充分的；有待进一步给出一个“对流–湍扩散模型”的完整解。再有，底摩擦和底形的忽略，对深层环流带来的影响可能是不能无视的。最终，跃层的常深度分布和热盐梯度的简单形式之假设也带有过大的随意性质。

感谢赫崇本教授、景振华先生、奚盘根和张淑珍二位同志提出的宝贵意见。由于这是作者对洋流问题的最初探讨，更由于水平所限，错误在所难免，概由作者自己负责；并希望有识者指出，作者将深表感谢！

参考文献

[1] П. А. Киткин, 1953, ДАН СССР, том. 92, No2, pp293–296.
[2] N. P. Fofonoff, 1962, The Sea, vol. l, pp323–395.
[3] П. С. Инейкин, 1955, ДАН СССР, том. 101, No3, pp461–464.
[4] П. С. Λллинвйкин, 1957, ДАН СССР, том. 117, No6, pp971–914.
[5] A. R. Robinson, 1965, Research frontiers in fluid dynamics, pp504–531.
[6] 景振华, 1966, 海流原理, 科学出版社.

[7] A. R. Robinson & H. Stommel, 1959, Tellus, 11, No3, pp295–308.
[8] P. Welander, 1959, Tellus, 11, pp309–318.
[9] A. R. Robinson & P. Welander, 1963, J. Mar. Res. 21, No1, pp25–38.
[10] K. Bryan & M. D. Cox, 1967, Tellus, 19, pp54–80.
[11] K. Bryan & M. D. Cox, 1968, J. Atmos. Sci. 25, pp945–967.
[12] K. Bryan & M. D. Cox, 1968, J. Atmos. Sci. 25, pp968–978.
[13] J. O. Hinze, 1959, Turbulence.
[14] C. C. Lin, 1959, Turbulent Flow and Heat Transfer.
[15] Г. Н. Абрамович, 1960, Теория Турбулентных Струй.
[16] Н. Е. Кочен, И. А. Кибель, Н. В. Poze, 1963, Теоретическая Гидромеханика, частЬ вторая, госуд. изд. физ —матем, литература москва.
[17] П. С. Пинейкин, 1957, Основные Вопросы Динамической Теории бароклинного Слоя Моря.
[18] W. H. Munk, 1950, J. Meteorol. 7, 79.
[19] H. Stommel, 1957, The Gulf Stream.
[20] В. Б. Штокман, 1946, ДАН СССР, 54, No5.
[21] H. U. Sverdrup, 1947, Proc. Nat. Acad. Sci. USA, 33, 318.
[22] H. Stommel, 1957, Deep-Sea Res. 4, No3, 1.
[23] А. И. фелЬЗенбаум, I960, Теоретические Основы и Методы Расчета Yстановившихся Морских Течений.
[24] M. E. Stern, 1975, Ocean Circulation Physics.
[25] R. O. Reid, 1948, J. Mar. Res. 7, No2.
[26] В. Б. Штокман, 1946, ДАН СССР, 54, No8.
[27] A. M. King, 1964, Oceanography for Geographers.

f–平面上的宽陆架诱导阻尼波*

冯士筰

一、引言

由于陆架区资源的开发和利用以及污染、地震海啸和风暴潮预报等等的实际需要，近廿年来，随着浅海动力学中的陆架动力现象越来越受到人们的关注，陆架诱导波的研究也得到了相应的进展。特别是，我国东临宽广的浅水陆架区域，它蕴藏着丰富的石油和水产等资源，又不断遭到风暴潮的袭击，显然，宽陆架诱导阻尼波的探讨是具有特殊重要意义的。

早在 19 世纪的中期，Stokes 就导出了一类特殊波动的解，由于该波的显著部分仅局限于离边界一个波长的距离以内，故 Lamb 称之为“边缘波”（H. Lamb, 1932）[1]。但直至廿年前，由于 Munk，Snodgrass 和 Carrier 的工作，边缘波才开始被海洋界所重视（W. Munk, F. Snodgrass and G. carrier, 1956）[2]。Ursell（1952）[3]推广了 Stokes 的解；而 Eckart（1951）首先研究了浅水边缘波。Reid[4]和 Kajiura[5]（1958）探讨了 Coriolis 力对边缘波的影响，首次在边缘波的问题中引入了这一地球物理学的特征。Greespan（1956）[6]给出了强迫边缘波的解。其后的理论工作，譬如有 Mysak（1968）[7]和 Ball（1967）[8]；更近代的理论工作，譬如，Clarke（1974）[9]和 Grimshaw（1974）[10]。边缘波确实被观测到了，如 Munk 等（1956），Huntley 和 Bowen（1973），Kajura（1972），Fuller（1975），Lemon（1975）；边缘波在实验室中也实现了（Galvin, 1965；P. LeBlond and L. Mysak, 1977）[11]。

陆架波的发现是 20 世纪 60 年代以后的事情（B. Hamon, 1962, 1966）[12, 13]。Robinson（1964）[14]给予 B. Hamon 的发现以理论解释，并命名其为“陆架波”。其后的一系列理论进展有，譬如，Mysak（1967）[15]，V. Buchwald 和 J. Adams（1968）[16]，Adams 和 Buchwald（1969）[17]，A. Gill 和 E. Schumann（1974）[18]，J. Allen（1976）[19]。各地的观测进一步肯定了陆架波的存在（如，Mooers and Smith, 1968[20]; Mysak and Hamon, 1969[21]; Cutchin and Smith, 1973[22]; Isozaki, 1968[23]）。陆架波的模型实验在实验室中也实现了，如 Caldwell 等（1972）[24]。

上述的诸理论几乎都是局限于不考虑陆架底摩擦影响的无阻尼的陆架诱导波范畴［有个别的例外，但讨论也是很局限的，如 Mysak（1967）[25]，Gill 和 Schumann（1974）；

---

* 冯士筰. 1979. *f*–平面上的宽陆架诱导阻尼波. 海洋学报, 1(2): 177–192

对于所谓“宽陆架”波的探讨，实际上没有引起人们的注目，个别的例外，也仅是由于模型实验的要求而已（Caldwell *et al.*，1972 和 V. Buchwald（1973）[26]］。鉴于上述理由，本文期望能对宽陆架诱导阻尼波的某些问题进行一些探讨和分析。

二、基本方程组及其解

在 f–平面上，取正交曲线坐标系($\bar{X}\ \bar{Y}\ \bar{Z}$)，$\bar{X}$ 为理想化了的岸界 AB 之法线，朝向大海为正，$\bar{Y}$ 为 AB 之弧长，($\bar{X}\,O\,\bar{Y}$)含在未扰动的静止海面内，$\bar{Z}$ 垂直向上，并与($\bar{X}\ \bar{Y}$)组成右手坐标系（图 1），在这个正交曲线坐标系中，描述陆架诱导阻尼波的无因次全流线化方程组如下：

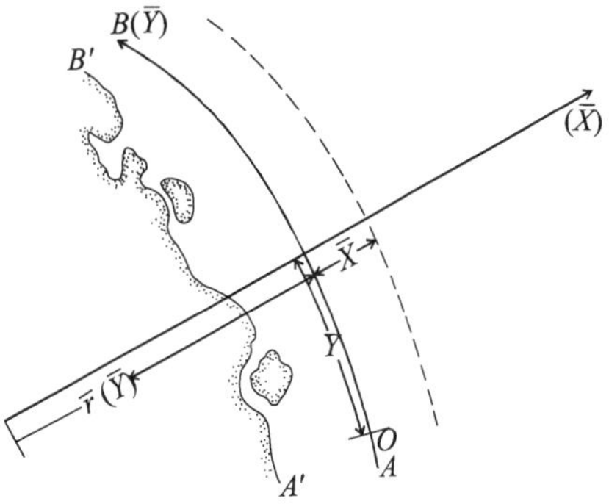

图 1

$$\gamma^2\frac{\partial\eta}{\partial t}+\frac{1}{H}\left\{\frac{\partial(HU)}{\partial x}+\frac{\partial V}{\partial y}\right\}=-\gamma^2P\frac{\partial\xi}{\partial t}\tag{2-1}$$

$$\varepsilon^2\frac{\partial U}{\partial t}-\varepsilon FV+\varepsilon^2BU=-h\frac{\partial\eta}{\partial x}+\varepsilon T\tau_x\tag{2-2}$$

$$\frac{\partial V}{\partial t}+\varepsilon FU+BV=-\frac{h}{H}\frac{\partial\eta}{\partial y}+T\tau_y\tag{2-3}$$

式中

$$H=1+(\varepsilon/k_0r_0)\frac{X}{r}\tag{2-4}$$

$$\gamma^2=\frac{(\omega_0/k_0)^2}{gd}\tag{2-5}$$

$$\varepsilon=lk_0\tag{2-6}$$

$$F=f/\omega_0\tag{2-7}$$

$$B=\beta/\omega_0\tag{2-8}$$

$$P=\xi_0/\eta_0\tag{2-9}$$

$$T=\frac{\tau_0}{\rho gk_0d\eta_0}\tag{2-10}$$

其中

$$X=\bar{X}/l,\quad Y=\bar{Y}/k_0^{-1},\quad r=\bar{r}/r_0,\quad t=\bar{t}/\omega_0^{-1},\quad h=\bar{h}/d$$

$$U=\bar{U}/U_0,\quad V=\bar{V}/V_0,\quad \eta=\bar{\eta}/\eta_0,\quad \tau_x=\bar{\tau}_x/\tau_0,\quad \tau_y=\bar{\tau}_y/\tau_0$$

$$\xi=\bar{\xi}/\xi_0,\quad V_0=gd\eta_0k_0/\omega_0,\quad U_0=V_0\varepsilon$$

$\bar{t}$ 为时间坐标；$\bar{r}(\bar{Y})$ 为理想岸界 AB 的曲率半径，一般为 $\bar{y}$ 的函数，r_0 为其特征量。以一条光滑曲线的理想岸界 AB 代替真实海岸 $A'B'$ 是方便的。这是因为若顾及到陆架底部和岸线的细微结构，目前看来，无法给出解析解。进而，我们又把 $A'B'$ 外的复杂的大陆架，代之以一个 AB 外的理想大陆架，这个大陆架的结构可以描摹真实大陆架的主要特征，即水深 $\bar{h}$ 仅为法线 $\bar{X}$ 的增函数——$\bar{h}=\bar{h}(\bar{x})$。对于 AB 和 $A'B'$之间的狭带视而不见，与我们略掉非线性效应和侧向湍流摩擦力的动力学模型也是相一致的。d 和 l 分别为陆架最大深度和其宽度。k_0、ω_0 为波数和圆频率的特征量。$\bar{U}$ 和 $\bar{V}$ 分别为相应于 $\bar{X}$ 和 $\bar{Y}$ 坐标轴的全流分量；$\bar{\eta}=\bar{\zeta}-\bar{\xi}$，$\bar{\zeta}$ 为波面坐标，$\bar{\xi}$ 为表达了海面大气压强的作用的强迫函数。$\bar{\tau}_x$ 和 $\bar{\tau}_y$ 分别为海面风应力相应于 $\bar{X}$ 和 $\bar{Y}$ 坐标的分量。τ_0 和 ξ_0 分别为风应力和大气压强的特征量。η_0 为 η 的特征量。g 为重力加速度。ρ 为海水密度，设为常数——这就意味着排除了斜压波和内波，f 为 Coriolis 参数①，设为常数——仅局限于 f –平面上，这就意味着排除了 Rossby 行星波。陆架底摩擦的表达式是引入了与全流正比的线性假设，比例系数为 ρ 和所谓“阻尼系数” β 之积——且 β 设为常数，这就排除了非线性底摩擦效应——使其符合于整个问题的线化模型。陆架诱导波的边界条件为

$$(\bar{U},\bar{V},\bar{\zeta})\to 0\,(\bar{X}\to\infty) \tag{2-11}$$

条件（2-11）要求大气扰动力仅局限于岸边附近一个有限区域内，且意味着排除了 Poincare 波。至于岸界边界条件为

$$\bar{U}=0 \quad (\bar{X}=0) \tag{2-12}$$

若进一步采用 $h(0)=0$，岸边不存在垂直壁；这就排除了 Kelvin 边界波（尽管，一般来说，它是可能存在的[27]）。下面，在强迫力 $P=T=0$ 和 $H=1$ 的假设下，给出陆架诱导阻尼自由波的解。令其解为

$$(U,V,\eta)=(U(x),V(x),\eta(x))\exp\{\mathrm{i}(ky+\omega t)\} \tag{2-13}$$

其中：$k=\bar{k}/k_0$，$\omega=\bar{\omega}/\omega_0$，$\bar{k}$、$\bar{\omega}$ 分别为波数和圆频率，且令 $\bar{k}>0$，$\bar{\omega}=\mathrm{Re}(\bar{\omega})+\mathrm{i}I_m(\bar{\omega})$；从而周期 $\bar{T}=\dfrac{2\pi}{|\mathrm{Re}(\bar{\omega})|}$ 和波速 $C=-\dfrac{\mathrm{Re}(\bar{\omega})}{\bar{k}}$。

把（2-13）代入（2-1）~（2-3），消去变量 y 和 t，有仅含变量 X 的全流表达式以及波面所满足的微分方程如下：

$$U=\frac{\mathrm{i}h\left[(\omega-\mathrm{i}B)\eta'+\varepsilon Fk\eta\right]}{\varepsilon^2(\omega-\mathrm{i}B)^2-(\varepsilon F)^2} \tag{2-14}$$

① 原文注：本文令 $f>0$（北半球）并不失去其一般性；对于 $f<0$（南半球），效应相反即可。

$$V=-\frac{h\left[\varepsilon F\eta'+\varepsilon^2(\omega-\mathrm{i}B)k\eta\right]}{\varepsilon^2(\omega-\mathrm{i}B)^2-(\varepsilon F)^2} \tag{2-15}$$

和

$$h\eta''+h'\eta'+(\kappa-k^2\varepsilon^2h)\eta=0 \tag{2-16}$$

其中$\kappa=\frac{1}{\omega-\mathrm{i}B}\left\{h'\varepsilon Fk+\gamma^2\omega\left[\varepsilon^2(\omega-\mathrm{i}B)^2-(\varepsilon F)^2\right]\right\}$，变量右上角的一撇表示对$X$求导。

假设陆架的坡度是均匀的（图2），其宽度为l，与外洋相接处深为d，且设外洋为无限深；有陆架深度分布为

$$h=X\ (0\leqslant X\leqslant 1) \tag{2-17}$$

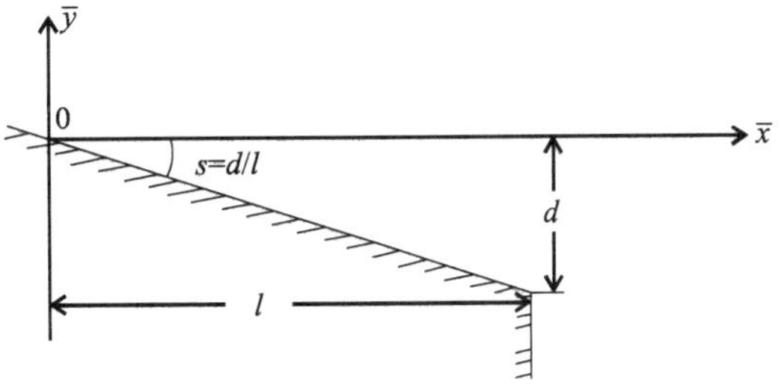

图 2

由（2-17）、（2-14）二式可见：岸界条件（2-12）自然满足，只要$\eta(0)$有界。由于X=1处，与无限深大洋相接，故相应于（2-11）边界条件者为

$$\eta(1)=0 \tag{2-18}$$

于是，方程（2-16）的解为

$$\eta(x)=A\mathrm{e}^{-\varepsilon kx}L_\nu(2\varepsilon kx) \tag{2-19}$$

式中A为任意常数；Laguerre 函数$\mathrm{L}_\nu(z)$为

$$\mathrm{L}_\nu(z)=1-\nu z+(-\nu)_2z^2/(2!)^2+\cdots$$

而$(-\nu)_n=-\nu(-\nu+1)(-\nu+2)\cdots(-\nu+n-1)$，$(-\nu)_0=1$，它满足微分方程

$$\left[z-\frac{\mathrm{d}^2}{\mathrm{d}z^2}+(1-z)\frac{\mathrm{d}}{\mathrm{d}z}+\nu\right]\mathrm{L}_\nu(z)=0$$

把解（2-19）代入方程（2-16），导出了频率–波数方程如下：

$$\omega^3-2\mathrm{i}B\omega^2-\left[F^2+B^2+(2\nu+1)\frac{k}{\varepsilon\gamma^2}\right]\omega+\frac{k}{\varepsilon\gamma^2}\left[F+\mathrm{i}B(2\nu+1)\right]=0 \tag{2-20}$$

且利用边界条件（2-18）给出了ν与k的关系

$$\mathrm{L}_\nu(2\varepsilon k)=0 \tag{2-21}$$

方程（2-20）之解为

$$\omega_j=\sigma_j+\mathrm{i}\frac{2}{3}B\quad(j=1,2,3) \tag{2-22}$$

其中

$$\sigma_1=\sigma_++\sigma_-,\quad \sigma_2=\theta\sigma_++\bar{\theta}\sigma_-,\quad \sigma_3=\bar{\theta}\sigma_++\theta\sigma_-$$

而 $\sigma_{\pm}=\left[-\frac{q}{2}\pm\sqrt{\left(\frac{q}{2}\right)^2-\left(\frac{p}{2}\right)^3}\right]^{1/3}$，$\theta=-\frac{1}{2}+\mathrm{i}\frac{\sqrt{3}}{2}$，$\bar{\theta}=-\frac{1}{2}-\mathrm{i}\frac{\sqrt{3}}{2}$，

$$p=-\frac{1}{3}B^2+F^2+(2\nu+1)\frac{k}{\varepsilon\gamma^2}$$

$$q=\frac{k}{\varepsilon\gamma^2}F+\mathrm{i}B\left[\frac{1}{3}(2\nu+1)\frac{k}{\varepsilon\gamma^2}-\frac{2}{3}F^2-\frac{2}{27}B^2\right]$$

显然，当 B=0 时，解（2-22）蜕化为 Mysak（1968）之解[7]。

三、非弥散陆架波

（一）一般分析

在讨论非弥散陆架波的解以前，先给以一般分析。

1. 陆架波存在的必要条件是

$$\varepsilon F=1 \tag{3.1-1}$$

由此有 $\omega_0/k_0=fl$ ——这给出了陆架波波速是正比于 fl 的表达式；说明了不存在 Coriolis 力将导致陆架波的消失。

2. 把（3.1-1）式代入（2-5）式，有

$$\gamma^2=\frac{f^2l^2}{gd} \tag{3.1-2}$$

由此可见，γ^2 正比于陆架宽度的平方；从而，γ^2 可以作为区分所谓“宽陆架波”和普通的“(狭)陆架波”的动力学准则

$$\left.\begin{array}{ll}O(\gamma^2)<1 & \text{为“(狭)陆架波”}\\ O(\gamma^2)=1 & \text{为“宽陆架波”}\end{array}\right\} \tag{3.1-3}$$

引入实际海洋中的数据量阶 $O(g、f、d)$=（10^3、10^{-4}、10^4）①，计算如下表（表 1）。

表 1

l/km	50	500
$O(\gamma^2)$	10^{-2}	1

由此不难看出，“宽陆架波”不仅仅为实验室中的产物[24]，在实际陆架上（如宽为 400~500 km 以上的陆架）也完全可能出现（如，中国东临的广阔陆架区）；从而，区分“宽陆架波”和普通“(狭)陆架波”的无因次准则（3.1-3）就具有了重要的实际意义了。

① 原文注：若不加特别指明，本文中的单位一律为 c.g.s 制。

3. 非弥散陆架波是陆架波的低频段的极限情形，其满足如下条件：

$$O(F^{-1})<1 \tag{3.1-4}$$

引入典型的频率 $\omega=10^{-5}$ [14, 15, 18]，有 $O(F^{-1})=10^{-1}$；可见，分析非弥散陆架波具有一定的实际意义。

由（3.1-4）式可以导出一个非弥散陆架波的重要推论［注意（3.1-1）式］：

$$O(\varepsilon)<1 \tag{3.1-5}$$

顺便指出，如 Robinson（1964）、Gill 和 Schumann（1974）等，把（3.1-4）和（3.1-5）作为两个独立的假设是没有必要的。

由（2-6）式进一步可知，非弥散的宽陆架波将为“超长波”。

4. 由（2-8）式可见，陆架湍流底摩擦的重要与否取决 $O(B)$的大小；若陆架底摩擦是必要的，应有下式成立：

$$O(B)=1 \tag{3.1-6}$$

事实上，阻尼系数 β 的典型数量阶是 $10^{-5}\sim10^{-4}$，从而有 $O(B)=1$。由此可见，非弥散陆架波的湍阻尼效应是显著的，必须计及。

5. 由于问题（2-1）~（2-3）是线性的，故气压效应和风应力效应可以分别探讨。先假定不存在海面风应力而仅给出气压陆架波。此时，应有下列量阶等式成立：

$$O(\gamma^2 P)=1 \tag{3.1-7}$$

由（3.1-7）看出，对于普通的（狭）陆架波，因为 $O(\gamma^2)<1$，故 $O(P^{-1})<1$——气压效应与“静压效应”相比微不足道（与 Robinson 结论是一致的）[14]；由表 1 知，对于宽为 50 km 的陆架，$O(P^{-1})=10^{-2}$。但是，对于宽陆架波，由（3.1-3）式可见，$O(\gamma^2)=1$，因而，$O(P^{-1})=1$——气压效应与静压效应同量阶；与普通（狭）陆架波相比，这是宽陆架波的重要特征之一。

6. 假定不存在气压效应，给出风生陆架波。由（2-3）方程和（2-10）式可知

$$O(\eta_0)=O\left(\frac{\tau_0 l}{\rho g d\varepsilon}\right) \tag{3.1-8}$$

由（3.1-8）式可见，风生陆架波的幅度正比于陆架宽度。这是自然的，为了保持非弥散陆架波的存在假设，宽陆架将意味着相对大的幅员之风应力效应的出现。

若取 $O(\tau_0)=1$, $l=50$ km，有 $O(\eta_0)=1\sim10$，这个风生陆架波的估值与 Gill 和 Schumann（1974）是一致的，但对于宽陆架上的风生陆架波一旦发生，其幅度将可高达数十厘米之多（如，l =500 km，有 $\eta_0\approx50$ cm）。

一个有兴趣的结论是：风生陆架流的大小与陆架宽度无关；事实上，依据沿岸存在一列波的假定以及（3.1-8）式和（3.1-1）式，有

$$O(V_0 / d) = O\left(\frac{\tau_0}{\rho f \varepsilon d}\right) \tag{3.1-9}$$

利用上面引入的诸数据估值，有 $O(V_0 / d) = 10$，该量阶与 Gill 和 Schumann（1974）是一致的。

最后指出，气压效应 η_{0a} 与风应力效应 η_{0w} 相较正比于陆架宽度，即

$$O(\eta_{0a} / \eta_{0w}) = O\left(\frac{\rho \varepsilon f^2 \xi_0}{\tau_0} l\right) \tag{3.1-10}$$

对于普通（狭）陆架，$O(\eta_{0a} / \eta_{0w}) = 10^{-2}$[17, 18]①；对于宽陆架，依据我们的估值，显然有 $O(\eta_{0a} / \eta_{0w}) = 10^{-1}$——与气压效应相比，风应力效应为主。

7. 综合以上分析，特别是由（3.1-5）式［注意（3.1-1）式］看出，若略去 $O(\varepsilon)$ 以上诸项，则导出了在我们所选用的这一特定正交线坐标系中的非弥散陆架波方程组如下：

$$\gamma^2 \frac{\partial \eta}{\partial t} + \frac{\partial U}{\partial x} + \frac{\partial V}{\partial y} = -\gamma^2 p \frac{\partial \xi}{\partial t} \tag{3.1-11}$$

$$-V = -h \frac{\partial \eta}{\partial x} \tag{3.1-12}$$

$$\frac{\partial V}{\partial t} + U + BV = -h \frac{\partial \eta}{\partial y} + T \cdot \tau_y \tag{3.1-13}$$

在这里：我们假定了 $O(k_0^{-1} / r_0)_{\max} = 1$。

（3.1-11）~（3.1-13）方程组在形式上与直角坐标下（y 轴沿平直的岸）的方程形式没有区别——这一点指出了，与假定海岸为平直的陆架模型相比较，考虑了真实海岸的大致曲线轮廓的陆架模型，不会给非弥散陆架波的问题带来任何一点附加困难。因而，任何有关平直海岸非弥散陆架波的成果，都可由此而直接应用到曲线海岸的模型中去。

由（3.1-12）方程看出，沿岸的法线方向上，非弥散陆架波呈地转反应，无论风应力的向岸分量存在与否；显然，只有当我们期望提高一个量阶的精度时——即保留 $O(\varepsilon)$ 诸项而略去 $O(\varepsilon^2)$ 以上诸项的时候，风应力的向岸分量的出现才破坏了这一地转关系［参考（2-2）方程］，从而，Allen（1976）的论证[19]才是有意义的。不过为了提高这一个量阶的精度，若又希望保持方程组（3.1-11）~（3.1-13）（再附加上一个 τ_x 项）这一简单的形式，必须假定 $O(k_0^{-1} / r_0)_{\max} = O(\varepsilon)$；否则，必须考虑海岸曲率的影响。无疑，对于宽陆架比之对于普通（狭）陆架，要做到上述之点是较困难的，因为 $O(k_0^{-1} / r_0) = O(l / r_0 \varepsilon)$（对于确定的平均曲线率半径 r_0 而言）。譬如，$O(\varepsilon) = 10^{-1}$，$r_0 = 2 \times 10^3$ km[15]，

① 原文注：相应于 ξ_0=1 cm (ΔP_a=1 mb)

$l=50\ \text{km}$，有 $O(l/r_0\varepsilon)=10^{-1}(O(\varepsilon))$［可见，Mysak（1967）的理论模型大概可以很大地化简］；若取 $l=500\ \text{km}$（宽大陆架情形），有 $O(l/r_0\varepsilon)=1$。

（二）自由波解的讨论

在非弥散陆架阻尼波的情况，解（2-19）蜕化为零阶 Bessel 函数。即非弥散陆架阻尼波解为

$$\eta(x)=\text{A}J_0(2\sqrt{\mu x}) \tag{3.2-1}$$

式中 $\mu=\dfrac{k}{\text{Re}[\omega]}=-\gamma^2$；

波速 C_j 为

$$C_j=-\frac{4}{\kappa_j^{\ 2}}\left(1+\frac{4\gamma^2}{\kappa_j^{\ 2}}\right)^{-1} \tag{3.2-2}$$

以及

$$\text{Im}\left[\omega_j\right]=B\left\{1-\frac{4\gamma^2}{\kappa_j^{\ 2}}\left(1+\frac{4\gamma^2}{\kappa_j^{\ 2}}\right)^{-1}\right\} \tag{3.2-3}$$

其中 κ_j 为方程 $J_0(2\sqrt{\mu})=0$ 的诸根，依顺序 $\kappa_0<\kappa_1<\cdots<\kappa_j<\cdots$ 排列；j 为波型的“阶数”（$j=0,1,\cdots$）。

由（3.2-1）~（3.2-3）可得出有关非弥散陆架阻尼波的下列诸结论。

1. 令解（3.2-1）~（3.2-3）中的 $\gamma^2=B=j=0$ 即蜕化为无阻尼的非弥散普通（狭）陆架波的零阶解（Robinson, 1974）[14]；故 Robinson 波为本模型解的特殊情形。

2. 由（3.2-3）式可见，由于陆架底摩擦的影响，对于 Robinson 波，湍阻尼衰亡因子特别简单，即为 e^{-Bt}，若认为 $\text{e}^{-Bt_{\infty\text{R}}}=\text{e}^{-\pi}\ll 1$，则 Robinson 波由于陆架底的湍阻尼而衰亡的时间 $t_{\infty\text{R}}\approx\pi B^{-1}$。譬如，取 $\beta=10^{-5}$，则这一衰亡时间约为三天半，与周期同量阶。

3. 由（3.2-1）~（3.2-3）解看出，陆架宽度对陆架波的效应是取决于因子 $4\gamma^2/\kappa_j^{\ 2}$ 的大小，我们将证明，对于高阶陆架波（j=1,2,⋯），与零阶陆架波相较，该因子迅速递减，并利用实际数据进一步计算表明，陆架宽度效应对高阶陆架波的影响是微不足道的；从而，下面我们将只分析零阶宽陆架波。

为了与零阶陆架波比较，我们只取到了一阶陆架波即足以说明问题了。计算如下表（表 2）。

表 2　$\kappa_0 = 2.40$

j/κ_j	1/5.52	2/8.66	3/11.8
$(\kappa_0/\kappa_j)^2$	0.19	0.08	0.04

表 2 中的 κ_j 取自 Friedrich（1960）[28]。由表 2 不难看出，随着波型的阶数 j 的增长，陆架宽度效应迅速递减。

若引入 $f = 7.3\times10^{-5}$（相当于北纬 30°），$g = 980$，$d = 2\times10^4$，有计算表如下（表 3）。

表 3

l/km	50	100	200	300	400	500
$(4\gamma^2/\kappa_j^2)$/%	0.09	0.36	1.43	3.22	5.71	8.93

由表 3 可见，陆架宽度确对高阶陆架波影响不大（$j\geqslant1$）。

4. 尽管高阶陆架波（$j\geqslant1$）的宽度效应对阻尼衰亡因子影响不大，但零阶陆架波的宽度效应却对阻尼衰亡因子有明显影响（见表 4）。

表 4

l/km	50	100	200	300	400	500
$\left(\dfrac{t_\infty - t_{\infty R}}{t_{\infty R}}\right)$/%	0	2	8	18	30	47

t_∞ 和 $t_{\infty R}$ 分别为宽陆架波和普通（狭）陆架波的由于湍阻尼而衰亡的时间。

由上表可见，随着陆架宽度的增加，陆架底摩擦所引起的波幅衰亡的时间也加长了，亦即衰亡的速率减慢了；这可能是由于宽陆架对应的非弥散陆架波具有较长的波长之结果。从而若以普通（狭）陆架波的阻尼衰亡因子去估值宽陆架（如 l=500 km）波阻尼衰亡的时间的话，则仅为真实值的约 2/3。

5. 由表达式（3.2-2）可见，一个独特的结论是：陆架底的湍阻尼对波速没有影响。另外，确如 3 中所推断的，γ^2 对于高阶波（$j\geqslant1$）影响不大；而对于零阶陆架波的波速之影响列于下表（表 5）。

表 5

l/km	50	100	200	300	400	500
$\left\lvert\dfrac{C_{0R} - C_0}{C_0}\right\rvert$/%	0	2	8	18	30	47

C_{0R} 和 C_0 分别为普通（狭）陆架波和宽陆架波之波速。

由表 5 可见，若用普通（狭）陆架波的公式去计算宽陆架波的波速（如取 l=500 km），其相对误差可高达 47%，由此可见，用（狭）陆架波波速公式去计算宽陆架波波速时将夸张了真实的数值。更为重要的是，对于宽陆架波，由于宽度因子γ^2效应，此时，陆架波已部分地含有了重力波的性质——这反映了陆架波由于宽度的增加所引起的量变到质变。

四、边缘波

（一）一般分析

1. 边缘波存在的必要条件是

$$\gamma^2 = 1 \tag{4.1-1}$$

把（2-5）式代入上式，有$(\omega_0 / k_0)^2 = gd$——这说明了边缘波为重力波的性质。

2. 把（4.1-1）代入（2-8）式，有

$$B = \frac{\beta k_0^{-1}}{\sqrt{gd}} \tag{4.1-2}$$

对于千公里波长的边缘波，有$O(B)=1$——这说明了对于陆架上典型边缘波的动力学问题[2]，陆架底摩擦的阻尼效应是不容忽视的。不过这里应当说明的是，我们在这里引入的β为 10^{-4}；既然β几乎正比于最大流速，而边缘波的流速，从量阶上看，比陆架波流速约大一个量阶，故取β之上限是自然的，显然，对于小于百公里长之波长的边缘波，有$O(B) \leqslant 10^{-1}$。

3. 对于$O(\varepsilon)<1$，以及$O(F) \leqslant 1$（相应于高频振情形）[27]，此时，有$O(\varepsilon F) \leqslant O(\varepsilon)<1$。若略掉所有$O(\varepsilon)$以上的项，则（2-2）变为$\partial\eta / \partial x = 0$——这违背了陆架诱导波的存在条件（2-11）式；因而，在这一情况下不可能出现边缘波（当然，更一般说，不可能存在陆架诱导波）。这个论断，与一些局部情形上的具体计算的结果是一致的（例如，参考文献[7]和[11]）。

4. $O(\varepsilon)=1$的情形：此时，Coriolis 力效应的显著与否取决于参数F之大小，而

$$F = \frac{f k_0^{-1}}{\sqrt{gd}} \tag{4.1-3}$$

显然，对于千公里波长的边缘波，有$O(F)=1$，即 Coriolis 力的影响不能忽略；而对于百公里波长之边缘波，$O(F)=10^{-1}$——这个结论，与 Reid（1958）[4]、Kajiura（1958）[5]等的特殊模型下的计算结果是一致的。

若把$O(\varepsilon)=1$代入（4.1-2）式，有$B = \dfrac{\beta}{\sqrt{gd}} O(l)$——由此可见，宽陆架边缘波的阻尼效应要比普通（狭）陆架边缘波的阻尼效应更为显著，故对于宽陆架边缘波，陆架

底摩擦效应更为重要。这点从物理上看是自然的，因为若假定陆架外缘的深度 d 不变，则较宽陆架相当于一个较小的海底平均坡度 $s(=d/l)$，从而，相应于一个具有较宽广的浅水域的陆架，当然，底摩擦效应就是更为显著的了（一个具体的数值结果请参考§4.2）。

5. 由（2-4）式可见，对于 $O(\varepsilon)=1$ 的情形，H=1 的条件是 $O(k_0^{-1}/r_0)<1$。譬如，边缘波为千公里波长，r_0 取 2×10^3 km [15]，则 $O(k_0^{-1}/r_0)=10^{-1}$——可以作为一个小量。可见，对于如波长小于千公里这种典型的陆架上边缘波的动力学问题，以一个直角坐标系来近似沿岸的曲线坐标系是完全允许的。

（二）自由波解的讨论

对于一个形如图 2 上所描绘的均匀坡度的陆架模型，自由边缘波的幅度正比于 e^{-kx}［令（2-19）式中 $O(\varepsilon)=1$，$\gamma^2=1$］，可见，一个宽为半波长以上的陆架上的边缘波在陆架外缘以内已仅存在边界处边缘波幅度的 4%了。譬如，在一个 500 km 宽的宽陆架上，千公里以下波长的边缘波的显著部分完全被限制在陆架以上——陆架与大洋相接处的边界条件不再起显著作用；因而，作为一个初步近似，可把这种相对宽的陆架看作一个均匀坡度（s）的半无界模型[4, 5, 11]。此时，$\varepsilon=1$，陆架的宽度 l 消失了，应以 k_0^{-1} 代替这个特征量；另外，陆架外缘深度 d 也消失了，应代之以陆架坡度 s 和 k_0^{-1} 的乘积；且 $U_0=V_0$；$\omega_0=k_0\sqrt{gsk_0^{-1}}$；且强调指出，描述陆架深度分布的表达式（2-17）中 x 定义域为 $[0,\infty)$。于是，解（2-19）、（2-20）中有 $\nu=n\ (n=0,1,2,\cdots)$，其形式蜕化如下：

$$\eta(x)=A\mathrm{e}^{-kx}\mathrm{L}_n(2kx) \tag{4.2-1}$$

以及频率–波数方程之解如下：

$$\omega_j=\sigma_j+\mathrm{i}\frac{2}{3}B \quad (j=1,2,3) \tag{4.2-2}$$

其中的符号同（2-22）解中的符号表达式，只是此处要以 n 代替彼处的 ν 而已。显然，当 B=0 时，解（4.2-2）蜕化 Reid（1958）的解[4]。

上面仅就最基本的，最具有实际意义的零阶边缘波 $(n=0)$ 进行讨论[2]。此时，解（4.2-2）蜕化为

$$\omega_j=-\frac{F}{2}+(-1)^j\frac{1}{2}a\cos\psi+\mathrm{i}\left(\frac{B}{2}+(-1)^{j-1}\frac{1}{2}a\sin\psi\right) \quad (j=1,2) \tag{4.2-3}$$

$$\omega_3=F+\mathrm{i}B$$

式中 $a=\left\{(F^2-B^2+4k)^2+4F^2B^2\right\}^{1/4}$，$\psi=\dfrac{1}{2}\mathrm{tg}^{-1}\dfrac{2FB}{F^2-B^2+4k}$

零阶波实际上只有两种可能，即 $\omega_j\ (j=1,2)$的情形；至于对应于 ω_3 的情况是不足

道的。

对应于 $\omega_j\ (j=1,2)$，其周期 T_j 和波速 C_j 分别表达如下：

$$T_j = 2\pi\left|-\frac{F}{2}+(-1)^j\frac{a}{2}\cos\psi\right|^{-1} \tag{4.2-4}$$

$$C_j = \frac{F}{2k}-(-1)^j\frac{a}{2k}\cos\psi \quad (j=1,2) \tag{4.2-5}$$

由（4.2-3）~（4.2-5）诸式可得出边缘阻尼波的如下各结论。

1. 当 $F=B=0$ 时，相应于底为小坡度的 Stokes 边缘波[1]。

2. 当 $B=0$ 而 $F\neq 0$ 时，相应于 f–平面上无阻尼边缘波的零阶波型（Reid, 1958）[4]

$$\omega_j = -\frac{F}{2}+(-1)^j\frac{1}{2}\sqrt{F^2+4k} \tag{4.2-6}$$

$$C_j = \frac{F}{2k}+(-1)^{j+1}\sqrt{(F/2k)^2+k^{-1}} \quad (j=1,2) \tag{4.2-7}$$

3. 当 $F=0$ 时而 $B\neq 0$ 时，给出了 Stokes 边缘波的一个阻尼效应的修正；此时，解（4.2-3）~（4.2-5）蜕化成如下形式：

$$\omega_j = (-1)^j\sqrt{k-(B/2)^2}+\mathrm{i}B/2 \tag{4.2-8}$$

$$T_j = 2\pi\left[k-(B/2)^2\right]^{-1/2} \tag{4.2-9}$$

$$C_j = (-1)^{j+1}\sqrt{k^{-1}-(B/2k)^2} \quad (j=1,2) \tag{4.2-10}$$

由（4.2-8）式可见，此时的阻尼修正非常简单，但，与无阻尼的 Stokes 波相比却是质的改变。我们排除 $k\leqslant(B/2)^2$ 的情形，因为那时将出现“临界阻尼”和“过阻尼”的现象——边缘波不可能存在，或一出现即消失；这样，就仅局限于 $k>(B/2)^2$ 的情形。显然，湍阻尼效应的显著与否取决于比值 $(B/2)^2/k$ 与 1 相比的大小。若取 $g=10^3$，$k=10^{-8}$，$\beta=10^{-4}$，计算如下表（表 6）。

表 6

S	4×10^{-2}	4×10^{3}	4×10^{-4}
$(B/2)^2/k$	0.6%	6.3%	62.5%

由表 6 看出，特别对于相对宽的陆架上之边缘波，湍流阻尼效应确是不可忽视的，$O((B/2)^2/k)=1$。进一步，由（4.2-9）和（4.2-10）两式不难看出，湍阻尼效应，与 Stokes 无阻尼波相比，使周期加大了，波速减慢了；并且，ω_j 的虚部揭示了一个简单的湍阻尼衰亡因子 $\mathrm{e}^{-(B/2)t}$。若取 $\beta=10^{-4}$，则对于略去 4%误差的近似程度来讲，边缘波衰亡时间为 $2\pi/\beta$，约为十数小时。

4. 当 $F\neq 0, B\neq 0$ 时就给出了一个 f–平面上无阻尼边缘波的湍阻尼修正，这是最具有意义的一种情形。依（4.2-3）~（4.2-5）诸式，引入 $g=10^3$，$k=10^{-8}$，$\beta=10^{-4}$，$f=10^{-4}$，计算如表 7。

表 7

S	j	4×10^{-2}	4×10^{-3}	4×10^{-4}	备注
Δ	1	2‰	3%	7%	沿正 y 向传播的边缘波
	2	3‰	4%	39%	沿负 y 向传播的边缘波

表中 $\Delta=\dfrac{T_j-T_{0j}}{T_{0j}}=\dfrac{|C_{0j}|-|C_j|}{|C_j|}=\dfrac{|\omega_{0j}|-|\mathrm{Re}[\omega_j]|}{|\mathrm{Re}[\omega_j]|}$

其中，右下角带下标“0”者为 f–平面下无阻尼边缘波相应的量。

表 8

S	j	4×10^{-2}	4×10^{-3}	4×10^{-4}
δ	2	0.9	0.8	0.3
	1	1.1	1.2	1.7

表中 $\delta=1+(-1)^{j-1}\alpha\sin\psi/B$

由表 7 和表 8 我们可以得出下列结论。

首先，像 Stokes 波的湍阻尼效果一样，此时也是，陆架越宽（即坡度越小）湍阻尼效应越大。其次，这或许是最重要的一个结论，就是，虽然如上所述，只有在相对宽的陆架上（坡度相对小）湍阻尼效应才是显著的，但是，此时湍阻尼对反向传播的边缘波的影响却显著不对称：如果我们在北半球上背陆面海来观察边缘波，则湍阻尼对于向右传播的边缘波周期有显著影响（周期增大，波速减小），波的阻尼衰亡却相对不显著；而湍阻尼对于向左传播的边缘波周期却影响不显著，几如不考虑湍阻尼时 f–平面上的 Reid 边缘波的结果——这也顺便指出了，Reid 的边缘波模型，尽管为无阻尼的，何以能解释 Munk 等的观测结果[4]。

最后，指出下列一点可能是有意义的，我国东海陆架宽而广，但由于台风过境所诱发的边缘波为北向的（沿正 y 向传播）①，故看来，实际上，湍阻尼效应对周期的修正未必重要——这就给研究提供了方便，只需利用无阻尼边缘波的模型来计算周期，大概即可给出一个良好的近似结果吧。

感谢文圣常教授及其领导的山东海洋学院动力海洋学教研室的同志们审阅了本文；感谢赵茂祥同志和高新生同志帮助进行了计算。

五、结语

给出了 f–平面上陆架诱导阻尼波的无因次方程组及其自由波的解；并就非弥散陆架波和边缘波两种情形分别进行了一般分析，特别指出了非弥散宽陆架阻尼波的特征和宽陆架上边缘波的湍阻尼效应。

① 原文注：多蒙中国科学院海洋研究所刘凤树同志首先向作者指出了这一点。

本文物理假设中的最大弱点在于陆架底摩擦的假设带有过大的经验性质。

最后指出，把本文中提出的这一线性模型推广到相应的非线性模型或许是更有意义的①。

参考文献

[1] Lamb, H., 1932. Hydrodynamics, 6th. Ed. Cambridge Univ. Press, 738.
[2] Munk, W., F. Snograss and G. Carrier, 1956. Science, 123: 127–132.
[3] Ursell, F., 1952. Proc. Roy. Soc. (A), 214: 79–97.
[4] Reid, R., 1958. J. Mar. Res., 16 (2): 109–144.
[5] Kajiura, K., 1958. J. Mar. Res., 16 (2): 145–157.
[6] Greespan, H., 1956. J. Fluid Mech., 1: 574–592.
[7] Mysak, L., 1968. J. Mar. Res., 26: 24–33.
[8] Ball, F., 1967. Deep-Sea Res., 14: 79–88.
[9] Clarke, D., 1974. Dtsch. Hydrogr. Z., 27: 1–8.
[10] Grimshaw, R.,1974. J. Fluid Mech., 62: 775–791.
[11] LeBlond, P. and L. Mysak, 1977. The Sea, 6: 459–495.
[12] Hamon, B., 1962. J. Geophys. Res., 67: 5147–5155.
[13] Hamon, B., 1966. J. Geophys. Res., 71: 2883–2893.
[14] Robinson, A., 1964. J. Geophys. Res., 69: 367–368.
[15] Mysak, L., 1967. J. Mar. Res., 25: 205–227.
[16] Buchwald, V. and J. Adams, 1968. Proc. Roy. Soc. (A), 305: 235–250.
[17] Adams, J. and V. Buchwald, 1969. J. Fluid Mech., 35: 815–826.
[18] Gill, A. and E. Schumann, 1974. J. Phys. Oceanogr., 4: 83–90.
[19] Allen, J., 1976. J. Phys. Oceanogr., 6: 426–431.
[20] Mooers, C. and R. Smith, 1968. J. Geophys. Res., 73, 549–557.
[21] Mysak, L. and B. Hamon, 1969. J. Geophys. Res., 74: 1397–1405.
[22] Cutchin, D. and R. Smith, 1973. J. Phys. Oceanogr., 3: 73–82.
[23] Isozaki, I., 1968. J. Oceanogr. Soc. of Japan, 24 (4): 32–44.
[24] Caldwell, D., D. Cutchin and M. Longuet-Higgins, 1972. J. Mar. Res., 30: 39–55.
[25] Mysak, L., 1967. J. Geophys. Res., 72: 3043–3047.
[26] Buchwald, V., 1973. J. Mar. Res., 31: 105–115.
[27] Huthnance, J., 1975. J. Fluid Mech., 69: 689–704.
[28] Friedrich L., 1960. Tables of higher function, New York Toronto London.
[29] Smith, R., 1972. J. Fluid Mech., 52: 379–391.

① 原文注：无阻尼陆架波的非线性模型已被给出，例如，参考 R. Smith (1972)[29]。

超浅海风暴潮的数值模拟（一）
——零阶模型对渤海风潮的初步应用*

孙文心　冯士筰　秦曾灏

一、引言

由我国北方的寒潮或冷空气所引起的渤海风潮，按文献[1]的分析，该属于“超浅海风暴潮”类型。以渤海风潮为例，按超浅海风暴潮理论进行数值计算，不仅对该理论是一有力的检验，且对探讨渤海风潮的机制和预报也是一个有益的参考。

将冷锋所造成的大风风场重现在计算网格上，在现场海面风资料极度贫乏的情况下，显然是一较难的课题。本文仅就发生渤海风潮时，渤海上空风场的几个主要力学特征，设计了几个理论模式风应力场，对渤海风潮进行了数值计算，以期获得全过程概貌及主要特征。计算结果达到了这一预期目的。

二、微分模型及其差分格式

（一）超浅海风暴潮零阶模型潮位方程为[1]

$$\frac{\partial\zeta}{\partial t}=a^{*}(x,y)\nabla^{2}\zeta+b(x,y)\frac{\partial\zeta}{\partial x}+c(x,y)\frac{\partial\zeta}{\partial y}+F(x,y,t) \qquad (1)$$

岸界条件：

$$Q_{n}=-\left[\frac{g}{\nu}A_{\mathrm{R}}\frac{\partial\zeta}{\partial n}+\frac{g}{\nu}A_{\mathrm{I}}\frac{\partial\zeta}{\partial s}+\frac{B_{\mathrm{R}}}{\rho\nu}\tau_{n}+\frac{B_{\mathrm{I}}}{\rho\nu}\tau_{s}\right]=0 \qquad (2)$$

水界条件：

$$\zeta(x_{\mathrm{w}},y_{\mathrm{w}},t)=0 \qquad (3)$$

初始条件：

$$\zeta(x,y,0)=0 \qquad (4)$$

上列诸式中：

$$a^{*}(x,y)=\frac{g}{\nu}A_{\mathrm{R}}$$

---

* 孙文心，冯士筰，秦曾灏. 1979. 超浅海风暴潮的数值模拟（一）——零阶模型对渤海风潮的初步应用. 海洋学报, 1(2): 193–211

$$b(x,y)=\frac{\partial}{\partial x}\left(\frac{g}{\nu}A_{\mathrm{R}}\right)+\frac{\partial}{\partial y}\left(\frac{g}{\nu}A_{\mathrm{I}}\right)$$

$$c(x,y)=\frac{\partial}{\partial y}\left(\frac{g}{\nu}A_{\mathrm{R}}\right)-\frac{\partial}{\partial x}\left(\frac{g}{\nu}A_{\mathrm{I}}\right)$$

$$F(x,y,t)=\frac{\partial}{\partial x}\left[\frac{1}{\rho\nu}\left(B_{\mathrm{R}}\tau_x-B_{\mathrm{I}}\tau_y\right)\right]+\frac{\partial}{\partial y}\left[\frac{1}{\rho\nu}\left(B_{\mathrm{I}}\tau_x+B_{\mathrm{R}}\tau_y\right)\right]$$

$$A_{\mathrm{R}}=\frac{1}{4a^3}\cdot\frac{\mathrm{sh}(2ah)-\sin(2ah)}{\mathrm{ch}(2ah)+\cos(2ah)}\geqslant 0$$

$$A_{\mathrm{I}}=\frac{1}{4a^3}\left[\frac{\mathrm{sh}(2ah)+\sin(2ah)}{\mathrm{ch}(2ah)+\cos(2ah)}-2ah\right]\leqslant 0$$

$$B_{\mathrm{R}}=-\frac{1}{a^2}\cdot\frac{\mathrm{sh}(2ah)\sin(2ah)}{\mathrm{ch}(2ah)+\cos(2ah)}$$

$$B_{\mathrm{I}}=-\frac{1}{a^2}\left[\frac{\mathrm{ch}(ah)\cos(ah)}{\mathrm{ch}(2ah)+\cos(2ah)}-\frac{1}{2}\right]$$

$$a=\sqrt{f/2\nu}$$

其中两个不等号是因为恒有 $h\geqslant 0$，而 $(x_{\mathrm{w}},y_{\mathrm{w}})$ 为水界处坐标。

风暴潮潮流的零阶解析表达式为：

$$u=C_{\mathrm{R}}\frac{\partial\zeta}{\partial x}-C_{\mathrm{I}}\frac{\partial\zeta}{\partial y}+G_{\mathrm{R}}\tau_x-G_{\mathrm{I}}\tau_y$$

$$v=C_{\mathrm{I}}\frac{\partial\zeta}{\partial x}+C_{\mathrm{R}}\frac{\partial\zeta}{\partial y}+G_{\mathrm{I}}\tau_x+G_{\mathrm{R}}\tau_y \tag{5}$$

其中

$$C_{\mathrm{R}}=\frac{g}{a^2\nu}\cdot\frac{\mathrm{ch}(ah)\cos(ah)\mathrm{sh}(az)\sin(az)-\mathrm{sh}(ah)\sin(ah)\mathrm{ch}(az)\cos(az)}{\mathrm{ch}(2ah)+\cos(2ah)}$$

$$C_{\mathrm{I}}=\frac{g}{a^2\nu}\left[\frac{1}{2}-\frac{\mathrm{ch}(ah)\cos(ah)\mathrm{ch}(az)\cos(az)+\mathrm{sh}(ah)\sin(ah)\mathrm{sh}(az)\sin(az)}{\mathrm{ch}(2ah)+\cos(2ah)}\right]$$

$$G_{\mathrm{R}}=\frac{1}{\rho a\nu}\Big\{\big[\mathrm{ch}(ah)\cos(ah)+\mathrm{sh}(ah)\sin(ah)\big]\mathrm{ch}\big[a(h+z)\big]\sin\big[a(h+z)\big]+\big[\mathrm{ch}(ah)\cos(ah)$$
$$-\mathrm{sh}(ah)\sin(ah)\big]\mathrm{sh}\big[a(h+z)\big]\cos\big[a(h+z)\big]\Big\}/\big[\mathrm{ch}(2ah)+\cos(2ah)\big]$$

$$G_{\mathrm{I}}=\frac{1}{\rho a\nu}\Big\{\big[\mathrm{ch}(ah)\cos(ah)-\mathrm{sh}(ah)\sin(ah)\big]\mathrm{ch}\big[a(h+z)\big]\sin\big[a(h+z)\big]+\big[\mathrm{ch}(ah)\cos(ah)$$
$$+\mathrm{sh}(ah)\sin(ah)\big]\mathrm{sh}\big[a(h+z)\big]\cos\big[a(h+z)\big]\Big\}/\big[\mathrm{ch}(2ah)+\cos(2ah)\big]$$

以上诸式中之ζ, u, v即为文献[1]相应的ζ_0, u_0, v_0，而其余未注明之符号均与文献[1]中之相应符号意义相同。

（二）方程（1）中的空间微商以中心差商取代，时间微商以前差商取代，遂成显式差分方程：

$$\zeta_{n,l}^{(m+1)} = (1-4\lambda)\zeta_{n,l}^{(m)} + \left(\lambda + \frac{\Delta t}{2\Delta s}b\right)\zeta_{n+1,l}^{(m)} + \left(\lambda - \frac{\Delta t}{2\Delta s}b\right)\zeta_{n-1,l}^{(m)} + \left(\lambda + \frac{\Delta t}{2\Delta s}c\right)\zeta_{n,l+1}^{(m)} + \left(\lambda - \frac{\Delta t}{2\Delta s}c\right)\zeta_{n,l-1}^{(m)} + \Delta t F_{n,l}^{(m)} \quad (6)$$

其中Δt为时间步长，Δs为空间网格间距，

$$\zeta_{n,l}^{(m)} = \zeta(n\Delta s, l\Delta s, m\Delta t), \quad \lambda = \frac{a^*\Delta t}{(\Delta s)^2}$$

为简便，式中未标出b, c, λ诸量之空间标号。

岸界条件（2）中的法向导数以前差商取代，切向导数以后差商取代，其相应的差分形式为

$$\zeta_{i,0}^{(\mathrm{m}+1)} = p_{i,0}\zeta_{i,1}^{(\mathrm{m}+1)} + q_{i,0}\zeta_{i-1,0}^{(\mathrm{m}+1)} + \Delta s\varphi_{i,0}^{(\mathrm{m}+1)} \quad (7)$$

其中第一个角标之i为岸界切线上网格点之切向数标，自水界沿逆时针方向数至第一个岸界点为$i=1$，此后沿逆时针方向i数值增加；第二个角标为沿岸界内法线方向之数标，其为0即表岸界点，其为1即表与岸界点相邻之内网格点。其中：

$$p = \frac{A_\mathrm{R}}{A_\mathrm{R} - A_\mathrm{I}}, \quad q = \frac{-A_\mathrm{I}}{A_\mathrm{R} - A_\mathrm{I}}, \quad \varphi = \frac{B_\mathrm{R}\tau_n + B_\mathrm{I}\tau_s}{\rho g(A_\mathrm{R} - A_\mathrm{I})}$$

水界条件（3）和初始条件（4）可分别离散化为

$$\zeta_{N,L}^{(m)} = 0 \quad (8)$$

$$\zeta_{n,l}^{(0)} = \zeta_{i,0}^{(0)} = 0 \quad (9)$$

(N, L)为水界网格点数标。

方程（6）~（9）即构成了本文的差分数值模型。网格设计见图1。为适应风场之特点，网格节线取为西南至东北和东南至西北方向。网格图中之四类朝外拐角点的网格边界已做了45°的削平。

（三）稳定性问题与平滑处理。方程（6）这种显式格式，有可能导致计算出现不稳定。但可以证明，只要保证方程（6）是正型的（即右端各ζ项之系数不小于0），则差分问题（6）~（9）是稳定的。这里我们只给出一种最简单的证明，即差分格式指数判定法[2]，依该方法，若网格内点以及并非第一类边界条件的边界点的函数值均能以该时刻以前若干点上的函数值的线性组合表示（可以是非齐次的），而其各线性项之系数模之和的极大值J（差分格式指数）若有$J \leqslant 1 + c\Delta t$（$c$为不依赖于$\Delta t$和$\Delta s$的任意数），

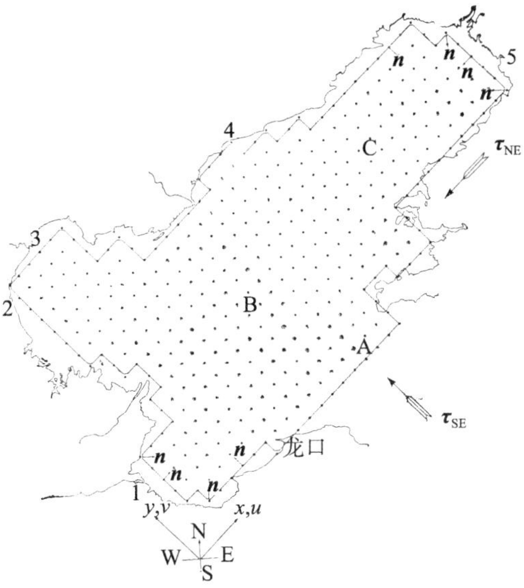

图 1　计算网格

1. 羊角沟；2. 歧口；3. 塘沽；4. 秦皇岛；5. 营口

则差分问题是依初始条件稳定的。我们试证明如下。

若方程（6）是正型的，则（6）式即有 $J=1$。问题只在方程（7）。为简便，去掉（7）中的角标 0，并记 $\Delta s\varphi_{i,0}^{(m+1)}=\Phi_i$。递推方程（7）并利用（6），即可将（7）式也化为 m 时刻的线性组合形式，为此将（6）式简记为

$$\zeta_{n,l}^{(m+1)}=R_{n,l}^{(m)}+\Delta tF_{n,l}^{(m)}$$

于是（7）式化为（注意 $\zeta_0^{(m)}$ 已为水界水位 0）

$$\begin{aligned}\zeta_i^{(m+1)}&=p_i\zeta_{i,1}^{(m+1)}+q_i\left[p_{i-1}\zeta_{i-1,1}^{(m+1)}+q_{i-1}\zeta_{i-2}^{(m+1)}+\Phi_{i-1}^{(m+1)}\right]+\Phi_i\\&=\cdots\cdots\cdots\cdots\cdots\cdots\\&=p_i\zeta_{i,1}^{(m+1)}+\Phi_i+\sum_{k=1}^{i-1}\left[\left(p_k\zeta_{k,1}^{(m+1)}+\Phi_k\right)\prod_{j=k+1}^{i}q_j\right]\end{aligned}$$

令 $\prod_{j=i+1}^{i}q_j=1$，则最终有

$$\zeta_i^{(m+1)}=\sum_{k=1}^{i}\left[\left(p_k\zeta_{k,1}^{(m+1)}+\Phi_k\right)\prod_{j=k+1}^{i}q_j\right]=\sum_{k=1}^{i}\left(p_kR_{k,1}^{(m)}\prod_{j=k+1}^{i}q_j\right)+\psi_i \qquad (7)'$$

其中，$\psi_i=\sum_{k=1}^{i}\left[\left(p_k\Delta tF_{k,1}^{(m)}+\Phi_k\right)\prod_{j=k+1}^{i}q_j\right]$ 为该方程之自由项。

依 A_R 和 A_I 的性质可知 $0\leqslant p_k\leqslant1$； $0\leqslant q_k\leqslant1$， $p_k+q_k=1$，并注意到方程（6）

之 $J=1$。则（7）′之

$$J=\sum_{k=1}^{i}\left(p_k\prod_{j=k+1}^{i}q_i\right)<\sum_{k=1}^{i}\left(p_k\prod_{j=k+1}^{i}q_j\right)+\prod_{j=k+1}^{i}q_j$$
$$=p_i+q_i(p_{i-1}+q_{i-1}(p_{i-2}+q_{i-2}(\cdots\cdots(p_2+q_2(p_1+\underbrace{q_1))\cdots\cdots)}_{i-1\text{层}}$$
$$=1$$

从而在全部网格点上均有 $J\leqslant 1$，则差分问题是依初始条件稳定的。

由（6）式的正型即可得到稳定充分条件为

$$\Delta s\leqslant\min\left(\frac{2a^*}{|b|},\frac{2a^*}{|c|}\right);\quad \Delta t\leqslant\frac{(\Delta s)^2}{4a^*}\tag{10}$$

当然我们也可以直接用稳定性定义，稍加繁琐地证明在同样条件（10）下，该差分问题还依自由项而稳定。

条件（10）的右端是一空间变量，故需取全部网格点上之最小值来设计网格，这就会使 Δs 和 Δt 逼得过小，而使计算量过大。本文最终取 Δs =14 km；Δt =450 sec。在 419 个网格点上，尚有 4 个网格点不满足条件（10），即仍然存在误差短波的传播与发展，而导致计算不稳定。故采用了如下的空间平滑算子，并于每一时刻进行平滑运算[6]。

$$\bar{\zeta}_{n,l}^{\mu}=\bar{\zeta}_{n,l}^{\mu-1}+\frac{s_\mu}{2}(1-s_\mu)\left(\bar{\zeta}_{n+1,l}^{\mu-1}+\bar{\zeta}_{n-1,l}^{\mu-1}+\bar{\zeta}_{n,l+1}^{\mu-1}+\bar{\zeta}_{n,l-1}^{\mu-1}-4\bar{\zeta}_{n,l}^{\mu-1}\right)$$
$$+\frac{s_\mu^2}{4}\left(\bar{\zeta}_{n+1,l+1}^{\mu-1}+\bar{\zeta}_{n+1,l-1}^{\mu-1}+\bar{\zeta}_{n-1,l+1}^{\mu-1}+\bar{\zeta}_{n-1,l-1}^{\mu-1}-4\bar{\zeta}_{n,l}^{\mu-1}\right)\tag{11}$$

其中 μ=1,2，μ 为平滑次数；$\bar{\zeta}^\mu$ 为第 μ 次平滑后的水位；s_μ 为第 μ 次的平滑因子，且

$$s_1=0.5;\quad s_2=-0.5$$

在使用了平滑算子（11）之后，计算获得了稳定。但需说明的是，若 Δs 和 Δt 过大，从而不满足条件（10）的网格点过多时，即使采用平滑算子（11）也不能获得稳定性。

三、风应力场模型及湍粘系数

我国北方的寒潮或冷空气所形成的风场，无外乎以其强、长、广、向四个主要力学特征的综合作用引起渤海海面的大幅度升降和海水的大量转移而形成了渤海风潮。所谓“强”，即风力大，其局部最大风速可达 20 m/sec，甚至更高；所谓“长”，即大风持续的时间长，可达两天左右；所谓“广”，即该种强风的风场水平尺度是大的，达上千公里；所谓“向”，即风向适于岸边增水，主要是东北风，而东南风转东北风尤为典型。

渤海的水平尺度为上百公里，故本文首先就其“广”的特征，将渤海网格上的风

场设计为空间均匀的，只随时间而变。这不妨可以看做一种零阶近似。

至于风场随时间的变化，只简单地考虑其兴衰，而给以半周期内的正弦函数形式。

第一模型（标准模型）风应力场①：

$$\begin{cases} \tau_x = -\tau_{\mathrm{NE}} = -9.25\sin\left[\dfrac{2\pi}{48}(t-21)\right](\mathrm{c.g.s}),\ (21 \leqslant t \leqslant 45)h \\ \tau_y = \tau_{\mathrm{SE}} = 3.2\sin\left(\dfrac{2\pi}{96}t\right)(\mathrm{c.g.s}),\ (0 \leqslant t \leqslant 48)h \end{cases}$$

该模型即摘取了东南转东北的典型风向，除上述时空特征外，其风速相当于 $W_{\mathrm{SEmax}} = 10\ \mathrm{m/sec}$；$W_{\mathrm{NEmax}} = 17\ \mathrm{m/sec}$。东南风持续两天，东北风滞后 21 时之后起风，持续一天。

第二模型风应力场：单纯东北风。

$$\begin{cases} \tau_x = -9.25\sin\left[\dfrac{2\pi}{48}t\right](\mathrm{c.g.s}),\ (0 \leqslant t \leqslant 24)h \\ \tau_y = 0 \end{cases}$$

第三模型风应力场：单纯东南风。

$$\begin{cases} \tau_x = 0 \\ \tau_y = 3.2\sin\left(\dfrac{2\pi}{96}t\right)(\mathrm{c.g.s}),\ (0 \leqslant t \leqslant 48)h \end{cases}$$

第四模型风应力场：常东北风。

$$\begin{cases} \tau_x = -9.25(\mathrm{c.g.s}) \\ \tau_y = 0 \end{cases}$$

本文中之湍粘系数 ν 选取最简模式，即 $\nu = \mathrm{const}$ 在计算各方案时均取 $\nu = 300(\mathrm{c.g.s})$，同时，也以 $\nu = 300 \sim 1000(\mathrm{c.g.s})$ 做了几个计算试验，以考查 ν 值的影响。

四、计算结果及分析

本文将计算结果整理为七类图：图 2 为标准模型风应力场作用下，自东南风起后 15 小时至东北风消失后 6 小时的期间内等水位线在不同时刻的分布图；图 3 为同一风场下自东南风起后 20 小时至东北风消失后 9 小时的期间内之表面流速图；图 4 为渤海沿岸六港口在标准模型风场作用下水位全过程时变曲线；图 5、6、7 为在第二~第四模型风应力场作用下六港口之水位时变曲线；图 8 为标准模型风应力场作用下 *A*、*B*、*C* 三网格点之分层流速矢量端迹图。

① 原文注：本文选取之坐标为：x，指向东北；y，指向西北。

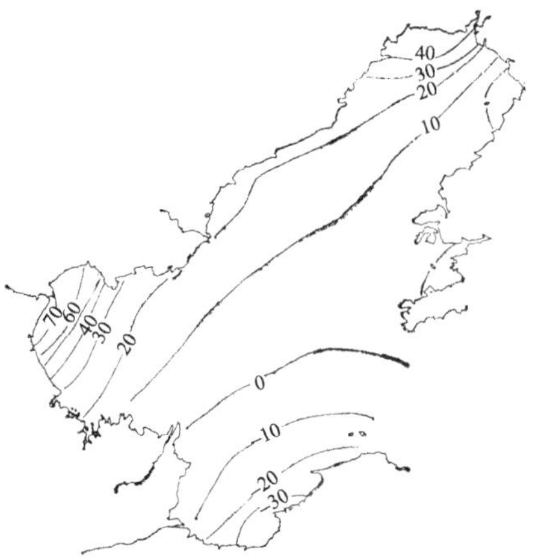

图 2（a） 第 15 小时水位分布图

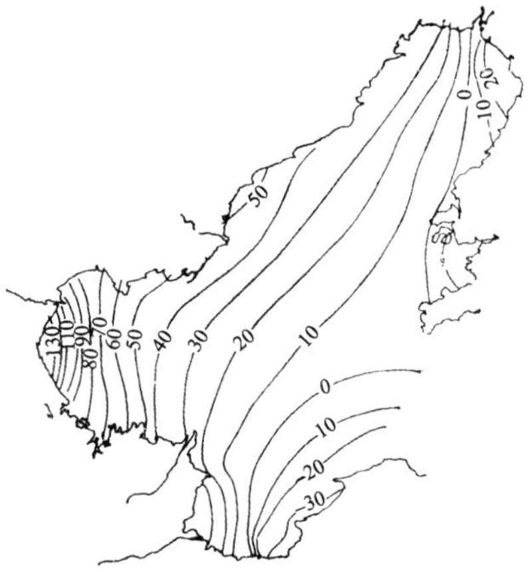

图 2（b） 第 24 小时水位分布图

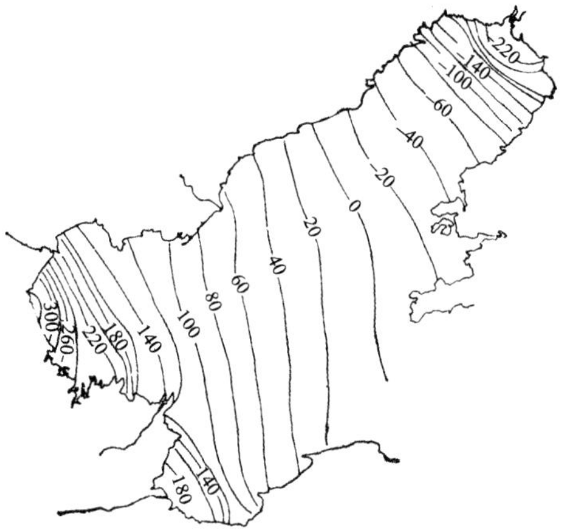

图 2（c） 第 33 小时水位分布图

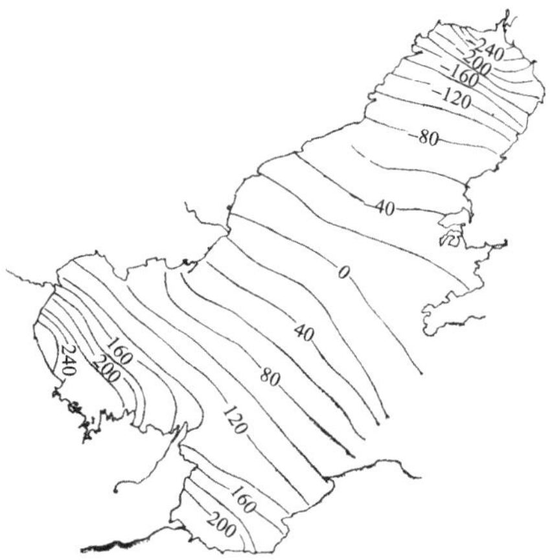

图 2（d） 第 39 小时水位分布图

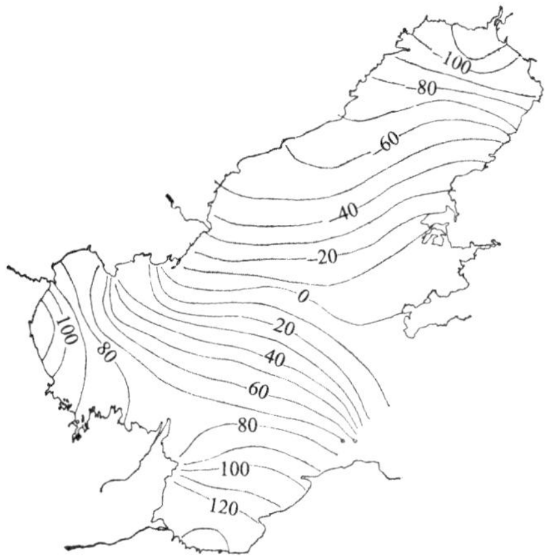

图 2（e） 第 45 小时表面水位分布图

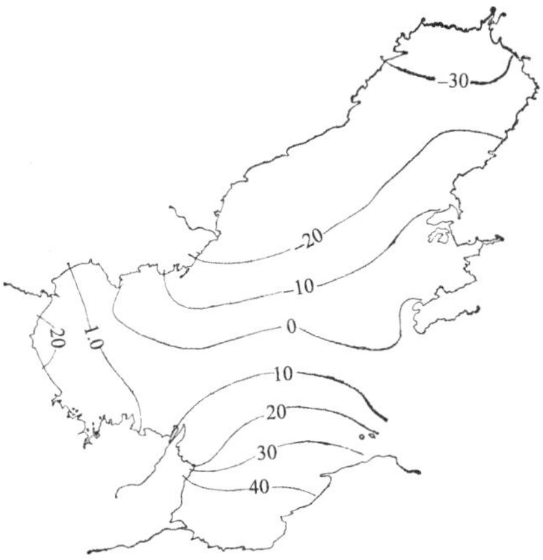

图 2（f） 第 51 小时水位分布图

图 3（a） 第 20 小时表面流速分布图

图 3（b） 第 24 小时表面流速分布图

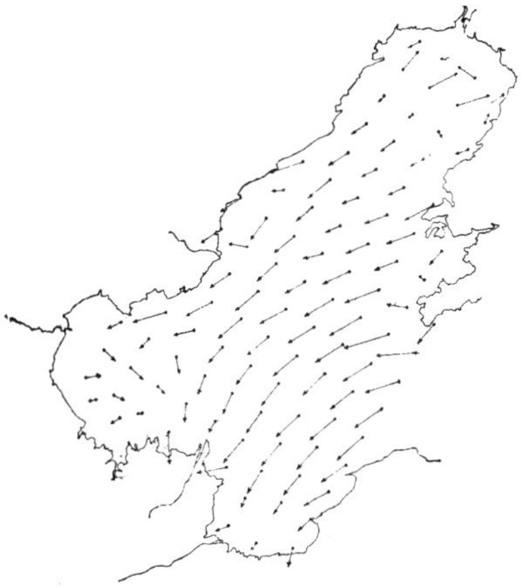

图 3（c） 第 33 小时表面流速分布图

图 3（d） 第 39 小时表面流速分布图

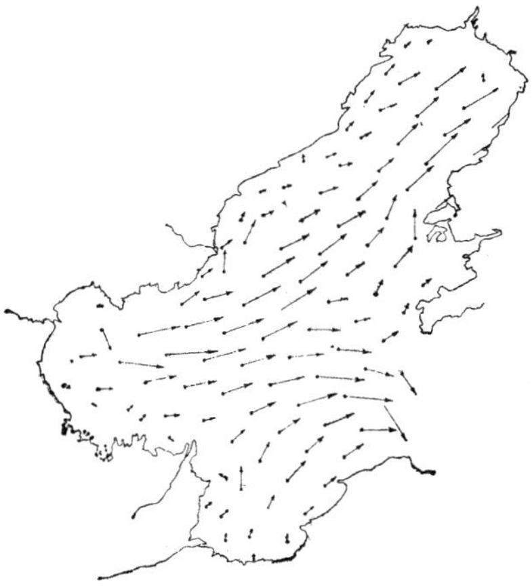

图 3（e） 第 45 小时表面流速分布图

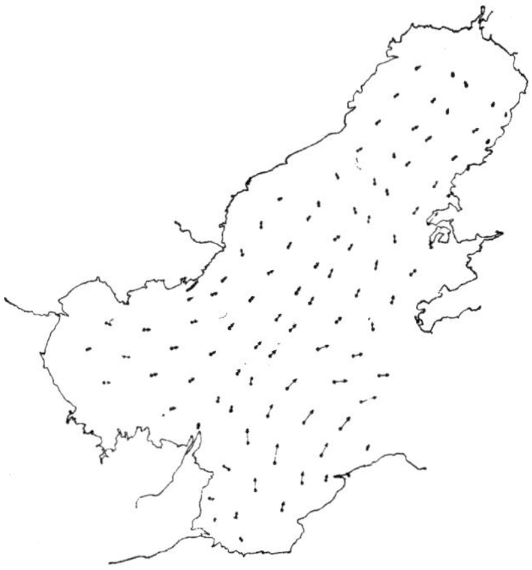

图 3（f） 第 54 小时表面流速分布图

（一）图 2、图 3 清晰地呈现了渤海风潮发生、发展和消衰的全过程：开始，东南风将黄海水经海峡吹入渤海，分别向渤海湾和辽东湾运移。同时，背风的东南沿岸海水也被迁往对岸，造成整个渤海西北岸增水、东南岸减水的势态，而辽东湾和莱州湾又各自独立地形成一类似系统，唯渤海湾处处增水。强东北风起后，一面因柯氏力的作用继续将黄海水迁入，一面又使渤海内部海水自东北向西南转移。因而，辽东湾内的海水迅速倾出，逐渐变为全湾减水，而渤海湾和莱州湾则迅速灌入，变为全湾增水，渤海湾尤为强烈。正因为增水强烈，水面坡度逐增，以至渤海湾内局部地区的倾斜流较快地发展至强于风海流，而出现了表层流的辐聚和水平环流，并在柯氏效应下使进入渤海湾的海水一部分又转入了莱州湾。同样，辽东湾由于猛烈减水，倾斜流也在局部地区首先占据优势，出现了表层流的辐散。风停之后，倾斜流逐渐使水量返回，渤海西南部海水逐渐向东北回迁，一部分经海峡返回黄海，一部分回灌入辽东湾，水面逐渐复原。但由于渤海湾倾出的水一部分继续进入莱州湾，因而莱州湾的复原较渤海湾缓慢，致使风停后莱州湾的水位转而高于渤海湾，直至整个渤海全部复原。

（二）由上述全过程可知：危险增水港必分布于渤海湾内（如塘沽、歧口）和莱州湾内（如羊角沟），而危险的减水港必出现于辽东湾内（如营口），而处于渤海长轴中部的港口（如秦皇岛）增减水均不会强烈。这与渤海实际风潮是一致的。

图 4 曲线表明：除歧口没有实测资料核对外，其它五港在全过程时间长度、水位极值、水风滞后间隔等重要参数，以及塘沽大增水同时对应营口大减水等重要特征也与实际风潮基本一致。

由图 4~图 7 可知，无论是单纯东北风，抑或是东南转东北风，使整个渤海增水最为强烈的就数歧口港。这是由于前期东南风已使迎风的渤海湾沿岸铺上了一层高达一米左右的增水，强东北风起后，在风应力和柯氏力的联合作用下，将海水向风的右前方输送，从而在风的右前方迎风迎流的角落造成最严重的水量堆积，而歧口正处于这一全渤海最为不利的位置，从而具有全渤海的最大增水。

塘沽增水机制与歧口略同，唯东北风为顺岸风，只在柯氏力作用下形成水量堆积，故其水位远低于歧口。

前期东南风对营口海面的升降作用甚微，唯当强东北风起后，因处于背风岸，水位骤减，其减水极值和极值的出现时间都与塘沽增水相映。

羊角沟水位的风效应恰与营口相似，但又相反：相似者，东南风皆无作用，唯东北风有效应；相反者，东北风效应是一增水一减水。

龙口的增水别具特点，因其所处位置，无论东南风还是东北风均不可能在此直接地造成异常增水。正如图 4~图 7 所示，两种方向的风的直接作用反可使该港略有减水。但龙口确属增水港。由图 4 可知，龙口的增水极值是在东北风已接近完全消失时才达到的，此时渤海湾，甚至莱州湾的水位均已回降。这说明龙口港的增水不是强东南和强东北风的直接效应，而是由于柯氏力的作用，在全渤海风潮的消衰阶段，海水逆时针方向流动，最后堆积此处，遂造成了较高的水位。

龙口港水位计算值比实际值相对偏低，一个原因是该处的一个迎流小湾——龙口湾在网格图上被节线削掉了。

（三）图 2、3 并参考图 4 可知：渤海风潮也表现为一逆时针方向的旋转系统。

（1）如图 3 所示，自黄海经海峡进入渤海的水沿逆时针方向几乎遍游了整个渤海，最后又经海峡返回黄海，完成了海水逆时针方向的运移。

（2）海湾水位最大值的逆时针方向的转移：如图 2 所示，在东北风停风前后，水位最大值由渤海湾转移至莱州湾。

（3）如图 2 所示，风潮水位等值线与天文潮波峰线（借助同潮时线）是类似的，也是做逆时针方向的旋转。由风停之后这一旋转仍在继续和由单纯东北风模型也可造成这一现象（本文未给出此种等值线图）说明这一旋转并非以风的转向为前提。

（4）图 4、5 所示各港口增水极值出现的时间呈秦皇岛—塘沽、歧口—羊角沟—龙口的顺序，说明港口水位极值出现的时间也在做逆时针序向的转移。其实这与上述水位等值线的逆时针方向旋转实为两个孪生现象。这从方程（1）~（4）得到直观的解释：将我们的问题（1）~（4）对时间做一次微商，由于问题是线性的，且诸系数均与时间无关，故水位时变率 $\dfrac{\partial\zeta}{\partial t}$ 所满足的方程将与水位 ζ 本身所满足的方程（1）~（4）是一样的，只不过强迫函数中的 τ 代换为其时变率 $\dfrac{\partial\tau}{\partial t}$。由上述 $\zeta=\text{const}$，并非以 τ 之转向而旋转，则可知 $\dfrac{\partial\zeta}{\partial t}=\text{const}$，也可在任何风向下做逆时针方向旋转，这当然包含了 $\dfrac{\partial\zeta}{\partial t}=0$ 线的旋转。

以上诸现象虽早已由 Groen 和 Groves 等人[4]以模型海区阐明过，且已为一些预报机构的实际经验所证实。但需注意的是，本文的结果是按不具线性惯性项的准平衡（但非准定常）模型计算出来的。

（四）注意到本文采用的是均匀风场，各主要增水港水位极值出现时间相对风应力极值时间的滞后量如下表。

由于准定常模型港口水位极值出现时间与风应力极值时间是一致的，故表中的滞后量显然也就是本文所用零阶模型相对准定常模型结果的滞后量。

Groen 和 Groves 曾认为“滞后”是具线性惯性项模型相对准定常模型的一种“摄动”，但本文的这一结果说明惯性效应绝非是这一“滞后”的必要条件。这一点还可以用一个长渠形海湾的解析解说明：考虑一等深、封闭长渠海湾，选用与本文数值计算

渤海沿岸五个增水港水位极值滞后量表

水位极值滞后量/小时 \ 港口 / 风应力模型	秦皇岛	塘沽	歧口	羊角沟	龙口
第一模型	−6	3	3	4.5	9.5
第二模型	−6	4.5	3	4.5	8.5
第三模型	5	5.5	6	26.5	27

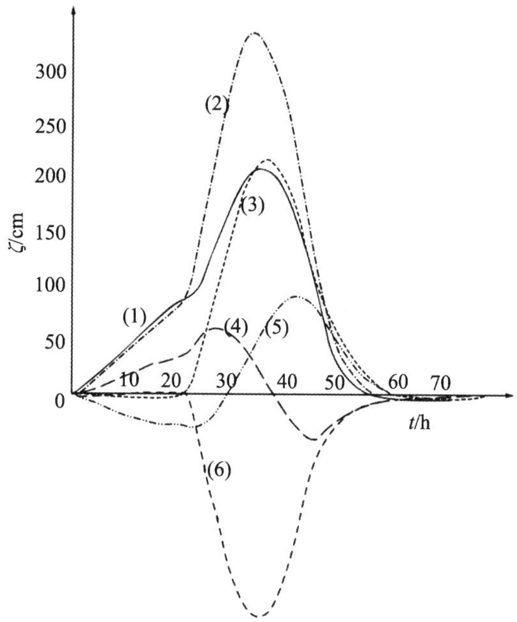

图 4　六港口标准模型风场下水位过程曲线
（1）塘沽；（2）歧口；（3）羊角沟；（4）秦皇岛；（5）龙口；（6）营口

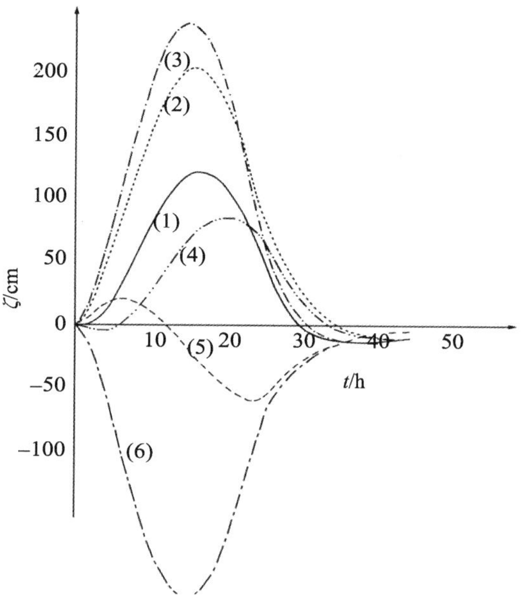

图 5　六港口在第Ⅱ模型风场下水位过程曲线
（1）塘沽；（2）羊角沟；（3）歧口；（4）龙口；（5）秦皇岛；（6）营口

同一超浅海风暴潮零阶模型、常湍粘系数、时间正弦函数的风应力，初始没有任何扰动，所不同的只是不考虑柯氏力，其精确解为

$$\zeta(x,t)=\frac{2\tau_0 h^2}{\rho\nu L}\sum_{k=0}^{\infty}\frac{1}{f_k^2+\omega^2}\left\{\omega e^{-f_k t}+f_k\sin(\omega t)-\omega\cos(\omega t)\right\}\cos\frac{(2k+1)\pi x}{L}$$

其中：L——海湾长度；$\tau_x=-\tau_0\sin(\omega t)$——风应力；

$$f_k=\frac{(2k+1)^2\pi^2 a^2}{L^2}\ ;\ \left(a^2=\frac{gh^3}{3\nu}\right)$$

其余诸符号与本文数值计算同。

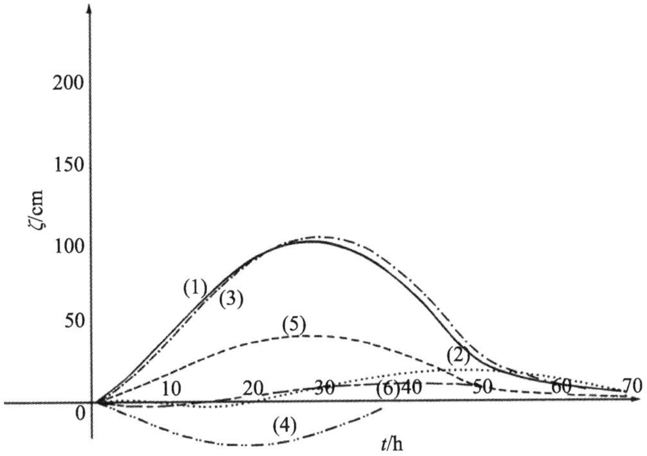

图 6　六港口在第Ⅲ模型风场下水位过程曲线

（1）塘沽；（2）羊角沟；（3）歧口；（4）龙口；（5）秦皇岛；（6）营口

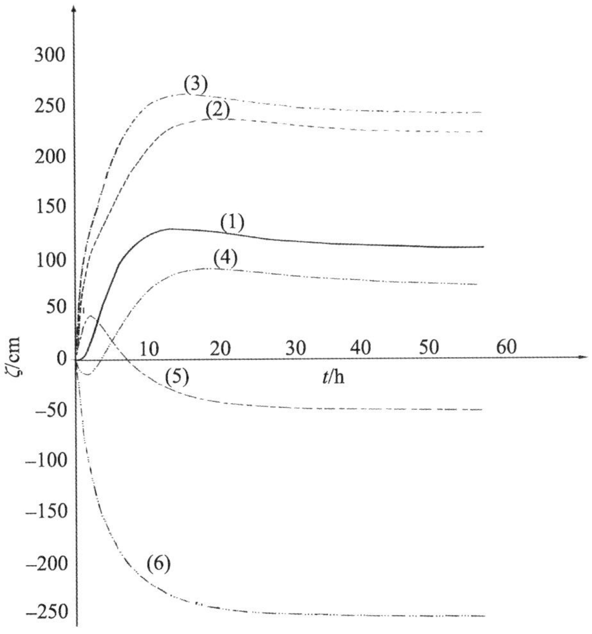

图 7　六港口在第Ⅳ模型风场下水位过程曲线

（1）塘沽；（2）羊角沟；（3）歧口；（4）龙口；（5）秦皇岛；（6）营口

当满足条件

a）$e^{-f_k t} \ll 1$（即运动达准定常阶段）

b）$\omega^2 \ll f_k^2$（即风周期足够长）

可得解：$\zeta(x,t) = A\sin\left[\omega(t-t_l)\right]$

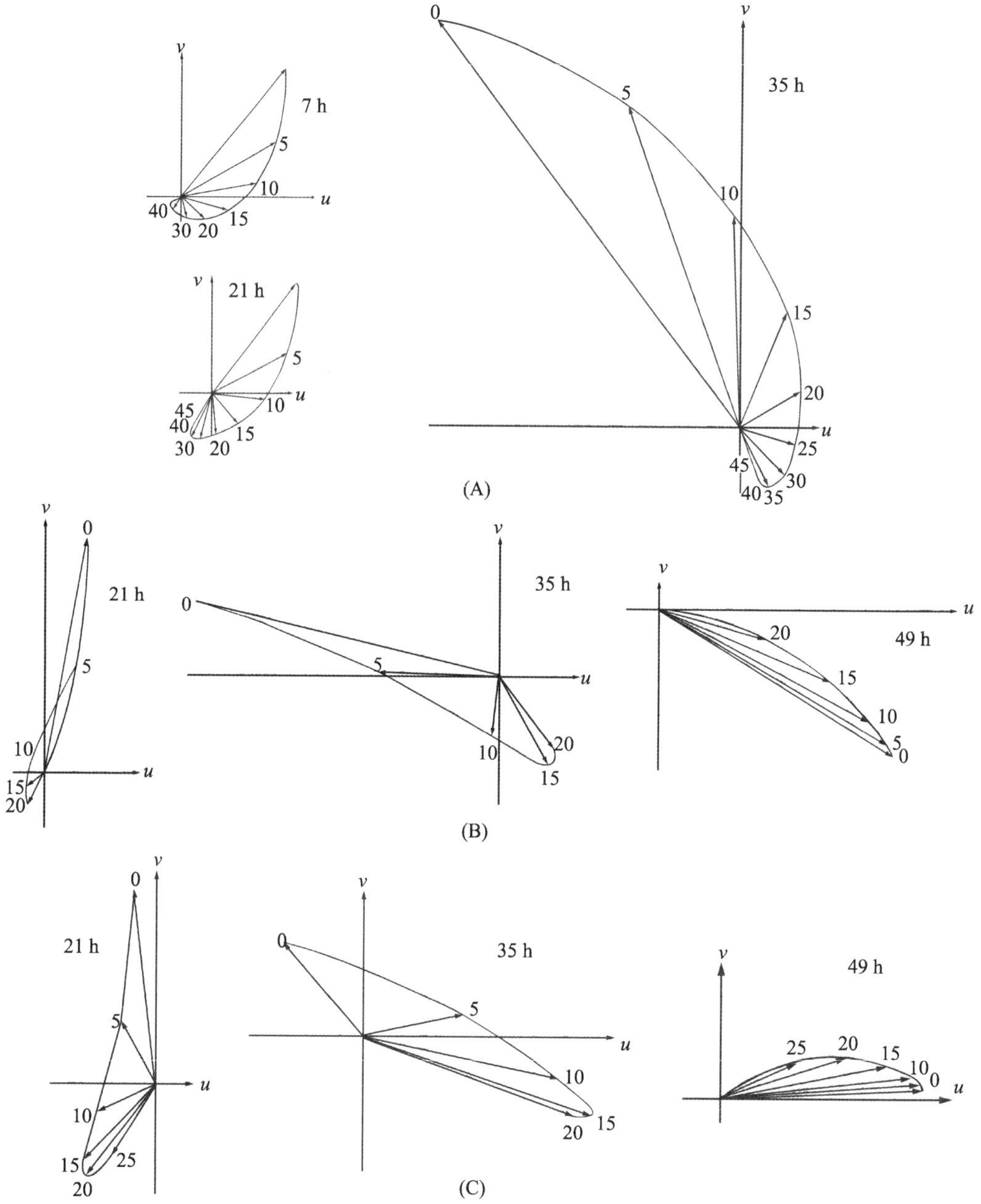

图 8　A、B、C 三网格点不同时刻的风暴潮流垂直结构图

其中：

$$A = \frac{3\tau_0}{4\rho gh}\left(1 - \frac{2x}{L}\right)L$$

$$t_l = \frac{\nu L^2}{4gh^3} \cdot \frac{1 - (6x/L)^2 + 4(x/L)^3}{1 - 2x/L}$$

这一“滞后”的出现不仅说明惯性效应不是其必要条件，而且地转效应也不是必要的；地转效应不仅对“滞后”的产生不是必要的，而且对滞后量随地点而不同这一现象的产生也不是必要的。可以认为这一滞后的出现，是湍摩擦在由准平衡向准定常演变阶段的一种“调节”效应。当然，单纯的湍摩擦也并非是充分的，这正是这种可以反映演变阶段的非准定常模型的机能。

（五）Groen 和 Groves 还发现了具线性惯性项的模型相对准定常模型的所谓“二次效应”，即“过顶”（Overshooting）和“落低”（Undershooting）。而本文依据不具线性惯性项的超浅海风暴潮零阶模型计算的结果，却也出现了这一现象。图 4 和图 5 所示之各增水港水位过程曲线，在最后阶段先降至平均海面之下，然后才恢复为 0，这显然就是“Undershooting”，而图 7 之各增水港水位曲线并非单调地趋向于定常值，而是首先超过定常值，然后才趋向它，这也就是“Overshooting”。在本文的计算结果中，这一“过顶”可达 20 cm，而东北风效应主要是减水的秦皇岛，竟然先行增水近半米，而后才减水半米。显然这种“过顶”现象是不容忽视的。本文的这一结果又说明了惯性效应也不是“过顶”或“落低”出现的必要条件。参考（四）中的解析结果，可以认为这可能是柯氏力在演变阶段的一种调节效应。

（六）由于我们的问题是线性的，风应力场是空间均匀的，故差分问题（6）~（9）可归结为

$$\zeta_{n,l}^{(m)} = \sum_{k=0}^{m}\left[b_{1_{n,l}}^{(k)} \tau_x^{(m-k)} + b_{2_{n,l}}^{(k)} \tau_y^{(m-k)} \right]$$

这就得到了各网格站上的水位预报方程，其系数 $b_1^{(k)}$ 和 $b_2^{(k)}$ 可以用本文的数值方法确定，也可以用统计方法确定，后一种方法，即是依据风应力场来确定回归系数 $b_1^{(k)}$ 和 $b_2^{(k)}$。再注意图 7 所示各港水位趋于常值，且这一结果是在常东北风应力场下得到的，即 $\tau_x = \text{const}$； $\tau_y = 0$。于是

$$\lim_{m\to\infty} \zeta_{n,l}^{(m)} = \left(\lim_{m\to\infty} \sum_{k=0}^{m} b_{1_{n,l}}^{(k)} \right) \tau_x = \text{const}$$

从而可知序列 $\left\{ b_{1_{n,l}}^{(k)} \right\}$ 是收敛的，因为：

$$\lim_{m\to\infty} \sum_{k=0}^{m} b_{1_{n,l}}^{(k)} = \text{const}$$

而有

$$\lim_{m\to\infty} \left[\lim_{k\to m} b_{1_{n,l}}^{(k)} \right] = 0$$

这一结果揭示了下列两点：（i）$k \to m$ 各项是反映了初始阶段的贡献，这一结果说明离初始阶段较晚的时刻 $(m \to \infty)$ 的水位受初始状态的影响是可以忽略的。这一问题已由

Welander[5]精辟地论证过。（ii）$k=0$项是预报时刻风应力的贡献，而随k之增大表示了预报时刻前离预报时刻愈来愈远的风应力之贡献。这一结果说明了离预报时刻愈近的因子的贡献愈大，愈远的愈小。故若以实测风做预报时，欲提高时效而人为地砍去离预报时刻相近的因子的贡献，必然使预报精度大为降低。

（七）湍粘系数ν在正确量阶下，其数值的变化对水位的影响是微弱的。本文试验将$\nu=300$(c.g.s)变为$\nu=600$(c.g.s)时，塘沽的水位降低了不足 20 cm。但试验表明ν值的变化对流速的影响是较大的，几乎是反比关系，且ν值还决定了流速的变化阶段。

（八）从风暴潮流的角度看，超浅海风暴潮零阶模型实为风海流、倾斜流与底流的合成流模型。这一点也可以由图 8 得到证实：处于海峡的A点，由于水深很大，在整个有风的过程中，主要以风海流占优势；风停之后（49 小时），B、C两处的流速矢端图明显地展示了倾斜流与底流的合成，而从 21 小时和 35 小时B、C两处的流速矢端曲线可看到三种流同时存在的结果。

由于无实测资料，计算结果无法与实际风暴潮流比较，但其量阶则与 Heaps[6]和Davies[7,8]的计算结果一致。

（九）本文的水界取在海峡，且其边界条件取为$\zeta=0$，无疑是很粗糙的。为考察外部潮的影响，经计算，若在海峡处出现半米的外部潮，则渤海内几个主要增水港之增水只达 20 cm 左右。

本文的计算是初步的。为更深入探求超浅海风暴潮机制，并使最终获得一简单而尽可能精确的预报方案，尚需做进一步的数值计算。为了更有力地证实渤海风潮的超浅海风暴潮的特征，以超浅海风暴潮一阶模型以及采用实际风场对其进行数值计算是必要的。

参考文献

[1] 秦曾灏, 冯士筰, 1975. 浅海风暴潮动力机制的初步研究, 中国科学, (1): 64–78.

[2] Рябеньский, В. С. и А. ф. фильппов, 1961. Об Устойчибссти Разносетных Уравнений, 差分方程的稳定性, 吴文达、王宗皓译, 科学出版社.

[3] Shapiro, R., 1970. Smoothing, filtering and boundary effects, Rev. of Geophys and Space Phys., May, 8(2): 359.

[4] Groen, P. and G. W Groves, 1962. Surges, The Sea, Ch. 17.

[5] Welender, P., 1961. Numerical prediction of storm surges, Advances in Geophysics, 8: 316–377.

[6] Heaps, N. S., 1971. On the numerical solution of the three dimensional hydrodynamical equations for tides and storm surges, Mem. Soc. R. Sci. Liege, 6, Ser, 2: 143–180.

[7] Davies, A. M., 1977. The numerical solution of the three-dimensional hydrodynamic equations, using a B-spline representation of the vertical current profile, Bottom Turbulence, 1–26.

[8] Davies, A. M., 1977. Three-dimensional model with depth-varying eddy viscosity, Bottom Turbulence, 27–28.

常底坡有限宽陆架诱导阻尼波的一种模型*

冯士筰

在文献[1]中，作者曾给出了 f–平面上陆架诱导阻尼波的无因次方程组及相应的诸无因次参数，就一个底坡均匀的有限宽陆架上自由波模型给出了它的解，并仅对非弥散陆架波和半无限宽陆架上的零阶边缘波这两种特殊情形下的弥散关系作了分析。作为推广，本文就上述陆架几何模型上的边缘波和陆架波的一般情形，对其弥散关系式进行了讨论。

依据文献[1]，并在形式上稍加改变后，其弥散关系式为

$$\sigma_j = \bar{\sigma}_j + \mathrm{i}\frac{2}{3}\mu\ ;\quad (j=1,2,3)$$

式中

$$\bar{\sigma}_1 = \lambda\sigma_+ + \bar{\lambda}\sigma_-\ ,\quad \bar{\sigma}_2 = \sigma_+ + \sigma_-\ ,\quad \bar{\sigma}_3 = \bar{\lambda}\sigma_+ + \lambda\sigma_-$$

$$\sigma_\pm = \left[-\frac{q}{2} \pm \sqrt{\left(\frac{q}{2}\right)^2 - \left(\frac{p}{3}\right)^3}\right]^{1/3}\ ,\quad \lambda = -\frac{1}{2} + \mathrm{i}\frac{\sqrt{3}}{2}$$

$$\bar{\lambda} = -\frac{1}{2} - \mathrm{i}\frac{\sqrt{3}}{2}\ ,\quad p = 1 + (2\nu_n + 1)\frac{\kappa}{\delta} - \frac{1}{3}\mu^2$$

$$q = \frac{\kappa}{\delta} + \mathrm{i}\mu\left[\frac{1}{3}(2\nu_n + 1)\frac{\kappa}{\delta} - \frac{2}{3} - \frac{2}{27}\mu^2\right]\quad (n=0,1,2,\cdots)$$

$$\sigma_j = \omega_j / f\,,\ \kappa = kl,\ \mu = \beta / f\,,\ \delta = f^2 l^2 / (gd)$$

其中，k 和 ω_j 分别为波数（$k>0$）和复圆频率（称其实部 $\mathrm{Re}[\omega_j]$为圆频率），d 和 l 分别为陆架外缘水深和陆架宽度，f 和 g 分别为 Coriolis 参数和重力加速度，β 为湍阻尼系数，$\nu_n = \nu_n(\kappa)$满足方程 $\mathrm{L}_{\nu_n}(2\kappa) = 0$（L 为 Laguerre 函数），$\mathrm{i} = \sqrt{-1}$。

引入下列数据：$g = 981$ 厘米/秒^2，$f = 0.73\times10^{-4}$ 秒$^{-1}$（相应于纬度 $\varphi = 30°\mathrm{N}$），$\beta = 10^{-4}$ 秒$^{-1}$，$d = 200$ 米，以及 $\delta = 0.027$（相应于 $l = 100$ 千米或陆架底坡 $\alpha = 2\times10^{-3}$）——记作 $\mathscr{O}(\delta) < 1$ 和 $\delta = 0.679$（相应于 $l = 500$ 千米或陆架底坡 $\alpha = 4\times10^{-4}$）——记作 $\mathscr{O}(\delta) = 1$；依据弥散关系式，并利用 $\nu_n = \nu_n(\kappa)$（$n = 0,1,2$）函数表[2]，作了计算，绘制成诸曲线图（见图 1~图 13），并讨论如下。

---

* 冯士筰. 1981. 常底坡有限宽陆架诱导阻尼波的一种模型. 海洋与湖沼, 12(1): 1–8

边缘波$\left(\left|\mathrm{Re}[\sigma_j]\right|>1,\ j=1,2\right)$

1. 对比图 1 和图 2，或图 4a 和图 5a，或图 7a 和图 8a 可知：当$\mathscr{O}(\delta)<1$时，有$\mathscr{O}(\mathrm{Re}[\sigma])>1$，而当$\mathscr{O}(\delta)=1$时，有$\mathscr{O}(\mathrm{Re}[\sigma])=1$。这是由边缘波乃惯性重力波的性质所决定的，故有一般性意义；事实上，对于边缘波，有$\mathscr{O}(\delta\,\mathrm{Re}[\sigma]^2/\kappa^2)=1$[1]，故有$\mathscr{O}(\mathrm{Re}[\sigma])=\mathscr{O}(\kappa/\sqrt{\delta})$。特别，当$f=0,\beta=0$时，相对圆频率$\sigma$精确地反比于$\delta$的平方根，因为此时$\sigma_j=(-1)^j\left[(2\nu_n+1)\kappa/\delta\right]^{1/2}$，（$j=1,2$）。

2. 由图 6 可见，随着相对波数κ的增大，互为反向传播的两列边缘波之间由于 Coriolis 力的影响所产生的非对称性是先增而后减，它揭示了必然存在一个对 Coriolis 力效应最敏感的波数区间，在该区间中上述的非对称性表现得最严重；顺便指出，它也统一了 Mysak 模型[2]和 Reid 模型[3]在这一点上的矛盾。

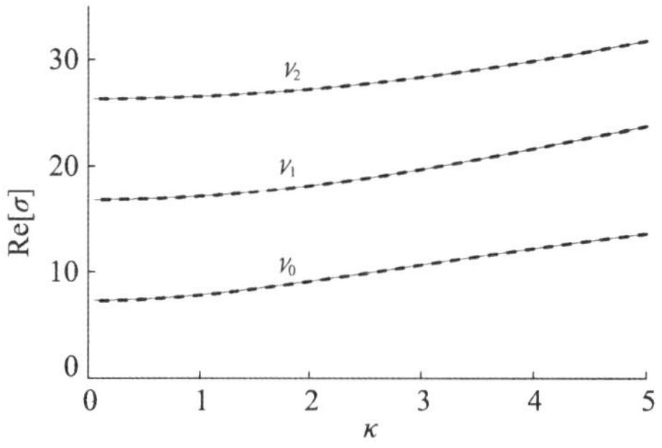

图 1　Re[σ]-κ曲线

$f=0$；$\delta=0.027$；—— $\beta>0$；------ $\beta=0$

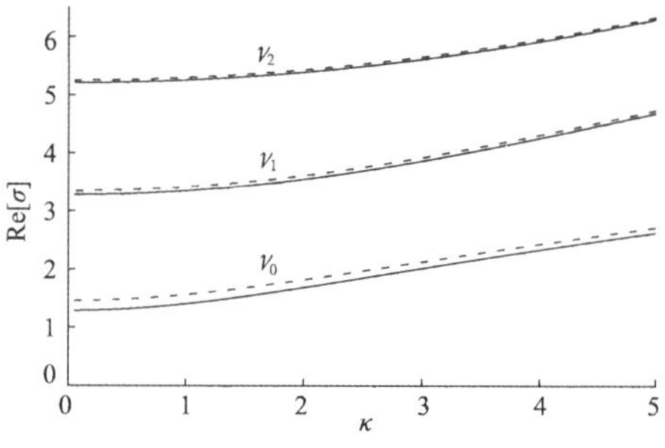

图 2　Re[σ]-κ曲线

$f=0$；$\delta=0.679$；—— $\beta>0$；------ $\beta=0$

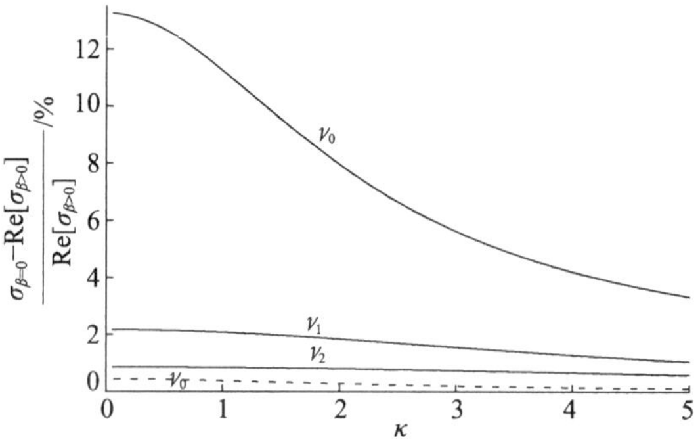

图 3　阻尼波和无阻尼波的相对差异

$f=0$；$\delta=0.679$；- - - -　$\delta=0.027$

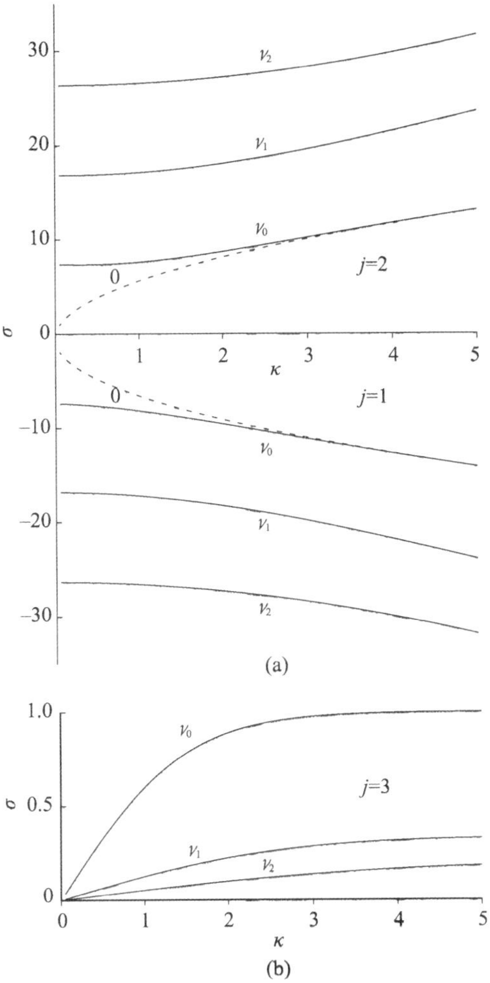

图 4 σ-κ曲线

$f = 0.73\times10^{-4}\ \mathrm{sec}^{-1}$；$\delta = 0.679$；$\beta = 0$；

---- 半无限宽陆架模型

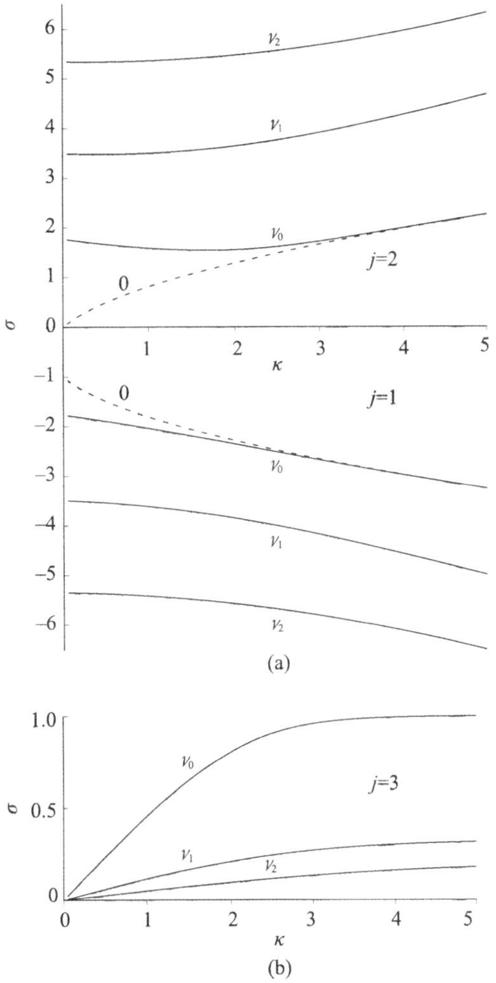

图 5 σ-κ曲线

$f = 0.73\times10^{-4}\ \mathrm{sec}^{-1}$；$\delta = 0.027$；$\beta = 0$ ；

---- 半无限宽陆架模型

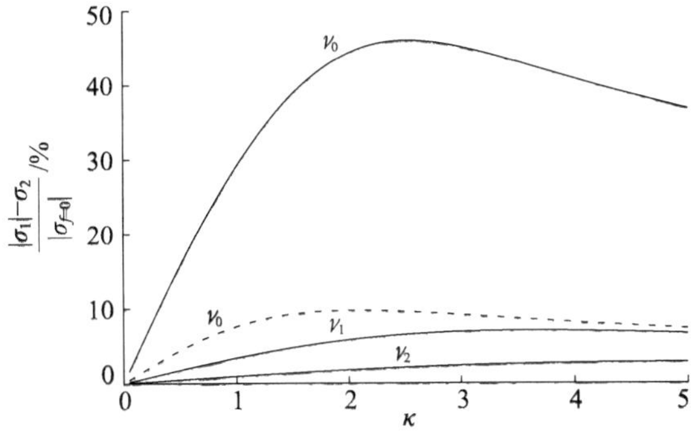

图 6 f–平面上边缘波圆频率的不对称性

$f = 0$; $\beta = 0$; —— $\delta = 0.679$; ---- $\delta = 0.027$

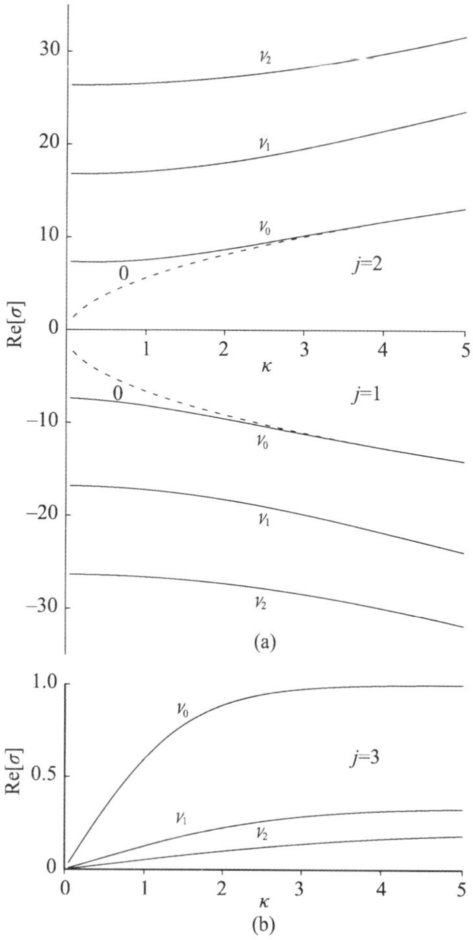

图 7　Re[σ]-κ曲线

$f=0.73\times10^{-4}\ \text{sec}^{-1}$；$\delta=0.027$；$\beta>0$；

---- 半无限宽陆架模型

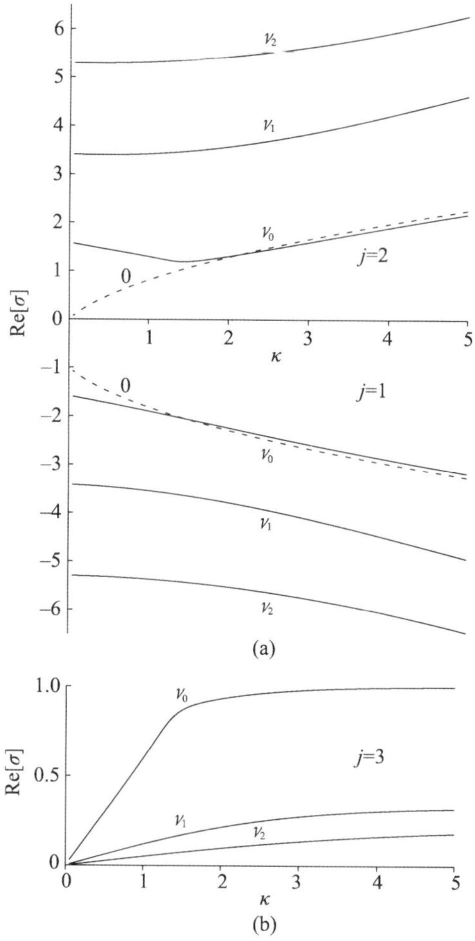

图 8　Re[σ]-κ曲线

$f=0.73\times10^{-4}\ \text{sec}^{-1}$；$\delta=0.679$；$\beta>0$；

---- 半无限宽陆架模型

3. 由图 6 还可看出，当$\mathcal{O}(\delta)<1$时，上述的 Coriolis 力对相对圆频率非对称性的影响小于 10%，而当$\mathcal{O}(\delta)=1$时，这一影响能高达 50%。其实这一性质也存在于更一般情形下的边缘波；事实上，Coriolis 力对边缘波的贡献为$\mathcal{O}(\kappa/\text{Re}[\sigma])=\mathcal{O}(\sqrt{\delta})$[1]。

4. 由图 5a 和图 8a 可见，当$\mathcal{O}(\delta)=1$时，对于一个在北半球上背陆面海的观察者而言，向右传播的零阶边缘波，由于 Coriolis 力的影响，其相对圆频率于$\kappa=1.5\sim1.75$附近呈现出一个极小值；因而，可以期望在这个相对波数区间中该零阶波存在零群速点。与半无限宽陆架上的零阶波不同[3]，事实上，两列互为反向传播的零阶边缘波的群速确是不对称的。特别，当$\beta=0$时，依据其相对群速的表达式$C_{gj}=-\text{d}\sigma_j/\text{d}\kappa=[(2\kappa\text{d}\ \nu_n/\text{d}\kappa+2\nu_n+1)\sigma_j-1]/[(2\nu_n+1)\kappa+(1-3\sigma_j^2)\delta]$，其群速为零的点应满足关系式$2\kappa\text{d}\ \nu_n/\text{d}\kappa+$

$2\nu_n+1=\sigma_j^{-1}$，依其计算，即可找到上述的这个零群速点$\kappa\approx1.75$。

5. 由图 3 和图 9 可见，湍阻尼效应减小边缘波的圆频率之大小，即使其周期增加；但，湍阻尼对边缘波的这种修正，在$\mathscr{O}(\delta)<1$的情形，近似 1‰或小于 1‰，而在$\mathscr{O}(\delta)=1$的情形，对零阶波为 10%和对高阶波至多为 1%的量阶。事实上，对于边缘波的一般情形，衡量湍阻尼效应大小的无因次参数为$\beta/|\mathrm{Re}[\omega]|$[1]，故有$\mathscr{O}(\beta/\mathrm{Re}[\omega])=\mathscr{O}(\sqrt{\delta}\mu/\kappa)$——表明了底坡小、多浅水域的宽陆架上传播的边缘波最易感受到湍阻尼的影响。

6. 由图 9 可见，在 f–平面上，湍阻尼对于向相反方向传播的两列边缘波圆频率的影响是不对称的，特别，对于$\mathscr{O}(\delta)=1$的情形，其影响显著不对称。对于向左传播的边缘波，湍阻尼对圆频率的影响相对地小，且随着相对波数的增加，这一影响单调递降；而对于向右传播的边缘波，湍阻尼对圆频率的影响相对地大，特别在零群速点附近，湍阻尼对零阶波的影响特别显著，但当远离这一点时，湍阻尼影响显著下降。图 9 也顺便指出，对于小波数波段，半无限宽陆架模型将夸大这种不对称性[1]。

7. 当$f=0$时，湍阻尼衰亡因子$\mathrm{Im}[\sigma_j]=\mu/2\,(j=1,2)$。但，在 f–平面上，湍阻尼衰亡因子如相对圆频率一样是不对称的；由图 10 可见，其不对称性随着相对波数κ的

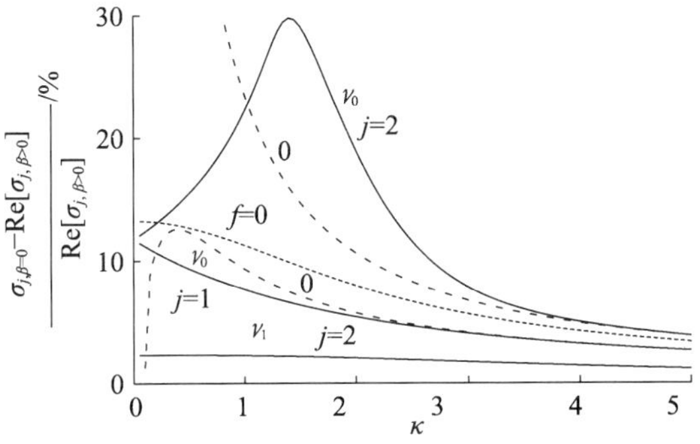

图 9　f–平面上湍阻尼影响的不对称性

$f=0.73\times10^{-4}\,\mathrm{sec}^{-1}$；$\delta=0.679$；　- - - -　半无限宽陆架模型

注：当$\delta=0.027$时这一不对称性≤1‰

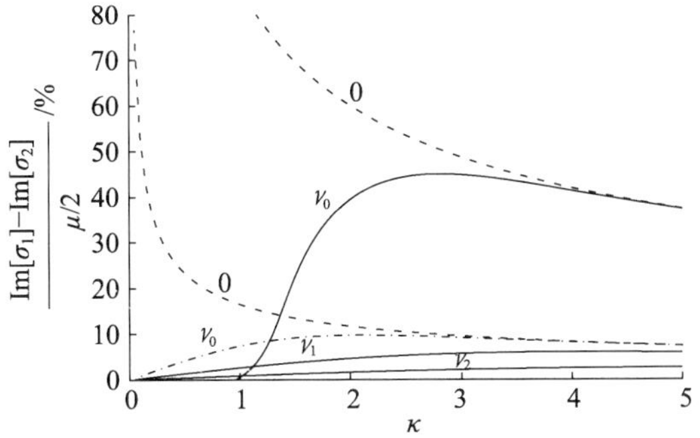

图 10　f–平面上湍阻尼衰亡因子的不对称性

$f=0.73\times10^{-4}\,\mathrm{sec}^{-1}$；$\beta>0$；—— $\delta=0.679$；　-·-·- $\delta=0.027$；　- - - -　半无限宽陆架模型

变化曲线类同于图 6 上所绘曲线，故其讨论同于 2 和 3，只要以湍阻尼衰亡因子代替相对圆频率即可。

必须指出，边缘波存在的最低波数点，即长波截断点，对于零阶、一阶和二阶波型分别近似取为 0.2，0.5 和 0.75[2]，这是因为我们假设了陆架外缘总是相接一个在 f–平面上、充满运动着的无黏性海水的 5000 米深平底大洋的缘故。

陆架波 ($|\mathrm{Re}[\sigma_3]|<1$)

1. 由图 11 可见，陆架波不同于边缘波，$\mathscr{O}(\delta)$ 的不同给予圆频率的影响不会产生量阶上的差异，对于零阶波也不过 10%，对于 1 阶波约为 1%，对于 2 阶波约为 1‰；并且二者相对圆频率的差异，特别是零阶波，主要显现于波数的某一区间之内。

2. 由图 12 可知，陆架波与边缘波相反，湍阻尼效应致使圆频率增加，这类似于惯

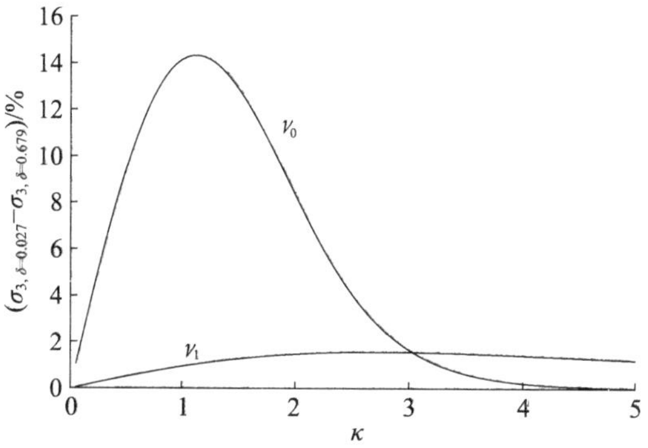

图 11　陆架波的陆架宽度效应

$f=0.73\times10^{-4}\ \mathrm{sec}^{-1}$；$\beta=0$；$\mathscr{O}(\nu_2)\leqslant1‰$

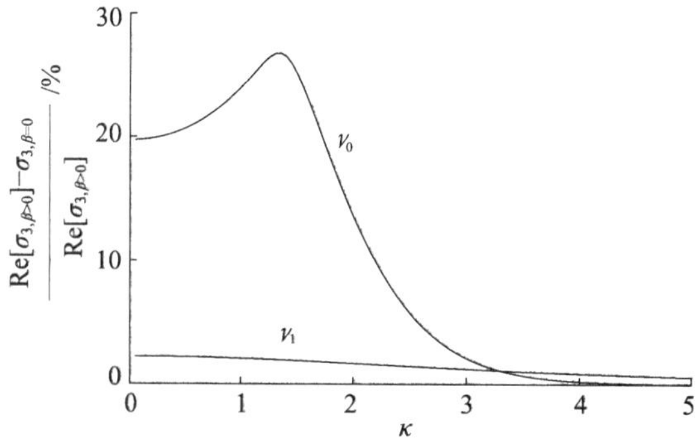

图 12　陆架波的湍阻尼效应

$f=0.73\times10^{-4}\ \mathrm{sec}^{-1}$；$\delta=0.679$

注：当 $\delta=0.027$ 时这一效应≤1‰

性波 $(\sigma=1)$；不过湍阻尼对相对圆频率的显著影响，特别对于 $\mathscr{O}(\delta)=1$ 情况下的零阶波，仅局限于某一波数区段之内，这又类似于在 f–平面上向右传播的零阶边缘波；对于 $\mathscr{O}(\delta)<1$ 的情形，陆架波与边缘波一样，湍阻尼对其影响也不过 1‰的量阶。

3. 由图 13 可见，零阶陆架波的湍阻尼衰亡因子与湍阻尼系数的比值随着波数的增加而接近于 1；特别，对于 $\mathscr{O}(\delta)<1$ 的情形，上述比值与 1 比较最大也不过 1%的量阶，也就是说，其湍阻尼衰亡因子与 Robinson 波的湍阻尼衰亡因子相差无几[1]。

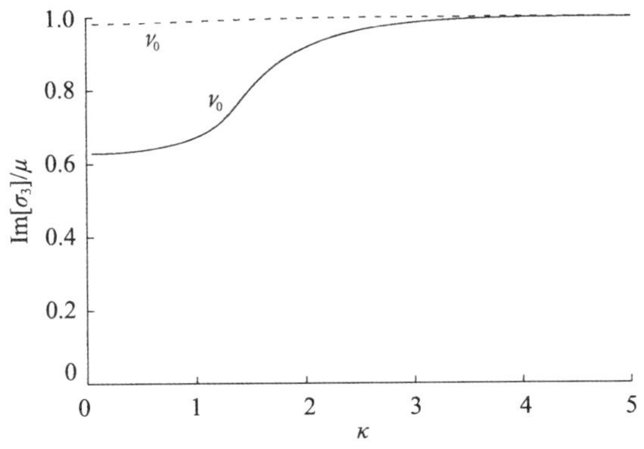

图 13　陆架波的湍阻尼衰亡因子

$f=0.73\times10^{-4}\,\text{sec}^{-1}$；$\beta>0$；—— δ=0.679；- - - - δ=0.027

模型与观测的比较

W. Munk，F. Snokgrass 和 G. Carrier[4]曾利用 Stokes 边缘波模型计算了沿美国东海岸传播的四次飓风潮所诱发的边缘波（余振）在 Atlantic City 和 Sandy Hook 两地所表现出来的振动周期；作者利用本文给出的弥散关系式，依据 Munk 等人的观点，也计算了上述的周期，并与实地观测值和依据 Stokes 波的计算结果作了比较，兹列于下页表。

表中的 T_0 为周期的观测值；T_∞ 代表依 Stokes 波计算的周期[4]；T 为依本模型但不考虑 Coriolis 力的无阻尼零阶边缘波计算的周期，T_β 和 ${}^fT_\beta$ 分别为经过湍阻尼修正后和又经过了 Coriolis 力修正后的零阶边缘波的计算周期。由表可见，f–平面上的阻尼边缘波模型构成了相对最优的方案。

最后应指出，本模型的湍阻尼效应是通过引进一个线化底摩擦表达式来实现的[1]；无疑，这在很大的程度上带有主观随意性，故这是本模型引入的物理假设中的一个最大的弱点。

理论周期和观测周期比较表

Atlantic City

α=5.0×10^{-4}, d=54.9 m, (l = 109.8 km), $\varphi = 40^\circ$N , (β_1 , β_2 , β_3)=(1,2,3,4)× 10^{-4} sec^{-1}

观测周期 T_∞/hr	飓风移行速度 v/(m/sec)	相对波数 κ	波长 λ/km	无阻尼波周期 T/hr	$\frac{T-T_0}{T_0}$/%	β_1 阻尼波周期 T_{β_1} /hr	$\frac{T_{\beta_1}-T_0}{T_0}$/%	β_2 阻尼波周期 T_{β_2} /hr	$\frac{T_{\beta_2}-T_0}{T_0}$/%	f-平面上 β_3 阻尼波周期 ${}^fT_{\beta_3}$ /hr	$\frac{{}^fT_{\beta_3}-T_0}{T_0}$/%	Stokes 波周期 T_∞ / hr	$\frac{T_\infty-T_0}{T_0}$/%
5.5	16.5~17.5	2.16~1.97	319.4~350.2	5.4~5.6	0	5.5~5.7	2	5.7~5.9	5	5.5~5.7	2	5.8~6.1	8
6.0	16.5	2.16	319.4	5.4	−10	5.5	−8	5.7	5	5.5	−8	5.8	−3
5.6	17.0	2.06	334.9	5.5	−2	5.6	0	5.8	4	5.6	0	6.0	7
?	20.6	1.57	439.4	5.9	?	6.0	?	6.3	?	6.2	?	7.3	?
					平均误差 4		平均误差 3		平均误差 5		平均误差 3		平均误差 6

Sandy Hook

α=4.2×10^{-4}, d=54.9 m, (l = 130.7 km), $\varphi = 40^\circ$N , (β_1 , β_2 , β_3)=(1,2,3,4)× 10^{-4} sec^{-1}

观测周期 T_∞/hr	飓风移行速度 v/(m/sec)	相对波数 κ	波长 λ/km	无阻尼波周期 T/hr	$\frac{T-T_0}{T_0}$/%	β_1 阻尼波周期 T_{β_1} /hr	$\frac{T_{\beta_1}-T_0}{T_0}$/%	β_2 阻尼波周期 T_{β_2} /hr	$\frac{T_{\beta_2}-T_0}{T_0}$/%	f-平面上 β_3 阻尼波周期 ${}^fT_{\beta_3}$ /hr	$\frac{{}^fT_{\beta_3}-T_0}{T_0}$/%	Stokes 波周期 T_∞ / hr	$\frac{T_\infty-T_0}{T_0}$/%
7.0	16.5~17.5	21.6~1.97	380.2~416.9	6.4~6.6	−7	6.5~6.7	−6	6.9~7.1	0	6.9~7.2	1	6.9~7.2	1
7.0	16.5	2.16	380.2	6.4	−9	6.5	−7	6.9	−1	6.9	−1	6.9	−1
7.2	17.0	2.06	398.6	6.5	−10	6.6	−8	7.0	−3	7.1	−1	7.1	−1
8.0	20.6	1.57	523.1	7.1	−11	7.2	−10	7.8	−3	8.3	3	8.6	8
					平均误差 9		平均误差 8		平均误差 2		平均误差 2		平均误差 3

参考文献

[1] 冯士筰, 1979. *f*–平面上的宽陆架诱导阻尼波. 海洋学报, 1(2): 177–192.

[2] Mysak, L. A., 1968. Edgewaves on a Gently Sloping Continental Shelf of Finite Width. *J. Mar. Res.*, 26(1): 24–33.

[3] Reid, R. O., 1958. Effect of Coriolis Force on Edge Waves (I) Investigation of the Normal Modes. *J. Mar. Res.*, 16(2): 109–144.

[4] Munk, W., F. Snodgrass & G. Carrier, 1956. Edge Waves on the Continental Shelf. *Science*, 123: 127–132.

论 f–和 β–坐标系*

冯士筰

旋转着的地球周围的大气和海洋中的行星尺度流场的描述，固然可以采用较精确的球坐标系统[1]，但是，更方便、更经常采用的却是所谓“f–平面”和“β–平面”坐标系[2]。譬如大气中 Rossby 波的引入[3]、大洋环流的西向强化的揭示[4]都是利用了 β–坐标系。研究这种坐标系的几何学和动力学具有相当的意义是毋庸置疑的，过去对这一问题有些学者也曾发表过专门的研究论文[5,6]，但具有相当的局限性，故作者试图在本文中进一步给出一个较系统的论证。因这一坐标系的探讨并非只涉及到一个纯粹的几何学问题而是紧密地联系着动力学，故首先应选择一个热力–动力学模型。为了使其具有一定的代表性和较为广泛的意义，作者选用了一种具有双参数热力学系统的 Newton-Boussinesq 流体模型作为论证的对象，为此，首先建议了一种导出的途径，特别提出了一种衡量这种旋转地球上的行星尺度流场的不可压缩性动力学准则。

不可压缩性动力学准则

如上所设，认为流体微团是一个双参数的热力学系统，并且取压强 p 和温度 T 作为两个独立参数。这样，原则上就给出了其状态方程如下

$$\rho = \rho(p,T) \tag{1}$$

其中，ρ 为流体密度。

本节，我们只考查压强改变 Δp 对密度改变 $\Delta\rho$ 的影响。显然，其量阶关系为[7]

$$O\left(\frac{\Delta\rho}{\rho}\right) = O\left(\frac{\Delta p}{\rho a^2}\right) \tag{2}$$

其中，a 为声速。

由于像大气和海洋中的这种行星尺度的薄层流场动力学具有准地转和准静力的特征，因此压强的水平改变 Δp_{H} 和铅垂改变 Δp_{V} 分别满足下述两个量阶等式

$$O(\Delta p_{\mathrm{H}}) = O(\rho f_0 UL) \tag{3}$$

$$O(\Delta p_{\mathrm{V}}) = O(\rho g D) \tag{4}$$

其中，f_0 为 Coriolis 参数之特征量，U 为水平流速特征量，g 为重力加速度，L 和 D 分

* 冯士筰. 1982. 论 f–和 β–坐标系. 山东海洋学院学报, 12(3): 1–10

别为流场的水平公尺度和铅垂尺度。

把式（3）和式（4）分别代入式（2），我们就得出了密度的水平改变 $\Delta\rho_{\mathrm{H}}$ 和铅垂改变 $\Delta\rho_{\mathrm{V}}$ 的量阶：

$$O\left(\frac{\Delta\rho_{\mathrm{H}}}{\rho}\right)=O\left(Ma^2/Ro\right) \tag{5}$$

$$O\left(\frac{\Delta\rho_{\mathrm{V}}}{\rho}\right)=O\left(c^2/a^2\right) \tag{6}$$

其中，$c=(gD)^{\frac{1}{2}}$ 为“长波相速”，$Ma=U/a$ 为 Mach 数，$Ro=U/f_0L$ 为 Rossby 数。

由此可见，无因次组合 Ma^2/Ro 和 c^2/a^2 即为压强变化所引起的密度变化之测度。从而建立了不可压缩性的必要条件：

$$Ma^2/Ro\ll 1 \tag{7}$$

$$c^2/a^2\ll 1 \tag{8}$$

条件（7）和（8）可称之为“不可压缩性动力学准则”。

强调指出，不可压缩性动力学准则（7）和（8）不同于流体力学教程中所导出的非旋转流体中的相应准则：$Ma^2\ll 1$。其物理原因是由于二者的基本力平衡各异。又由于准地转性，一般有 $O(Ro)\leqslant 1$，故与准则 $Ma^2\ll 1$ 比较，条件（7）—— $Ma^2/Ro\ll 1$ 是一个更强的条件。至于无因次组合 c^2/a^2，对于像海洋这种具有自由面的流场来讲，其物理意义显然是两种临界速度之平方比。

作为一例，引入大洋环流的典型数据，有 $O(U/a)=10^{-4}$，$O(Ro)=10^{-3}$，$O\left(\sqrt{gD}/a\right)=10^{-1}$，故 $O\left(Ma^2/Ro\right)=10^{-5}$ 和 $O\left(c^2/a^2\right)=10^{-2}$。可见，对于大洋环流问题，满足不可压缩性动力学准则（7）和（8）。

基本方程组

$$\frac{\partial\rho}{\partial t}+\nabla\cdot(\rho\boldsymbol{V})=0 \tag{9}$$

$$\rho\left\{\frac{\partial\boldsymbol{V}}{\partial t}+\boldsymbol{V}\cdot\nabla\boldsymbol{V}+2\boldsymbol{\Omega}\times\boldsymbol{V}\right\}=-\nabla p+\rho\boldsymbol{g}+\mu\left\{\nabla^2\boldsymbol{V}+\frac{1}{3}\nabla\nabla\cdot\boldsymbol{V}\right\} \tag{10}$$

$$\rho T\left\{\frac{\partial S}{\partial t}+\boldsymbol{V}\cdot\nabla S\right\}=\nabla\cdot(k\nabla T) \tag{11}$$

其中，t 表示时间变量，$\nabla\cdot$，∇ 和 ∇^2 分别表示散度、梯度和 Laplace 算子；$\boldsymbol{V}$ 表示速度向量，S 表示熵；$\boldsymbol{g}$ 表示重力，$\boldsymbol{\Omega}$ 表示地球自转角速度；μ 和 k 分别表示粘性系数和

导热系数。

热力学方程（11）的导出是基于热力学第一定律，不考虑流体内部的热源，忽略了粘性热耗散，仅计入热传导且引入了传导的 Fourier 定律。

进一步假定运动满足不可压缩性准则（7）和（8），即压强改变引起的密度改变可以略掉，则状态方程（1）可以近似写为

$$\rho=\rho(T) \tag{12}$$

同时，有 $T\dfrac{\mathrm{d}S}{\mathrm{d}t}=T\left(\dfrac{\partial S}{\partial T}\right)_{\mathrm{P}}\dfrac{\mathrm{d}T}{\mathrm{d}t}=C_{\mathrm{P}}\dfrac{\mathrm{d}T}{\mathrm{d}t}$；把后者代入方程（11），导出了温度扩散方程如下：

$$\rho C_{\mathrm{P}}\left\{\frac{\partial T}{\partial t}+\boldsymbol{V}\cdot\nabla T\right\}=\nabla\cdot(k\nabla T) \tag{13}$$

其中，C_{P} 为定压比热。

进一步按 Boussinesq 假设，$T=T_0+T'$，其中 T_0 为参考温度并为常数，而 T' 为小偏差，即设：$T'\ll T_0$；从而有 $\rho=\rho_0+\rho'$、$\rho_0\ll\rho(T_0)$ 且 $\rho'\ll\rho_0$，其中 ρ_0 为参考密度，亦设为常数。把 T 和 ρ 之上述表达式代入方程组（9）、（10）、（12）和（13）并略去小项，有 Boussinesq 近似方程组如下：

$$\nabla\cdot\boldsymbol{V}=0 \tag{14}$$

$$\frac{\partial\boldsymbol{V}}{\partial t}+\boldsymbol{V}\cdot\nabla\boldsymbol{V}+2\boldsymbol{\Omega}\times\boldsymbol{V}=-\nabla\phi+\nu\nabla^2\boldsymbol{V}-\alpha T'\boldsymbol{g} \tag{15}$$

$$\frac{\partial T'}{\partial t}+\boldsymbol{V}\cdot\nabla T'=\kappa\nabla^2 T' \tag{16}$$

其中，$\alpha=-\dfrac{1}{\rho_0}\left(\dfrac{\partial\rho}{\partial T}\right)_0$ 为热膨胀系数，$\nu=\mu/\rho_0$ 为运动粘性系数，$\kappa=k/\rho_0 C_{\mathrm{P}}$ 为导温系数即温度扩散系数；$\phi=p/\rho_0+\pi$，$\boldsymbol{g}=-\nabla\pi$。

在导出基本方程组（14）、（15）、（16）时应注意连续方程中 $O(\partial\rho'/\partial t)=U\rho'/L<O(\rho_0\nabla\cdot\boldsymbol{V})=U\rho_0/L$，以及已把状态方程代入了运动方程。

f–和 *β*–坐标系

在地球上，取经度为任意给定而纬度 θ 为 θ_0 的一点 o 作为参考原点，作局地正交曲线坐标系如下：ox 轴沿纬圈且指东为正，oy 轴沿经圈且指北为正，z 轴沿铅垂方向且向上为正，xoy 坐标曲面与未扰动的大气和海洋的交界面重合。

把基本方程组（14）~（16）投影于该正交曲线坐标系中，其分量形式如下：

$$\frac{\partial\left(uH_2H_3\right)}{\partial x}+\frac{\partial\left(vH_3H_1\right)}{\partial y}+\frac{\partial\left(wH_1H_2\right)}{\partial z}=0 \tag{17}$$

$$\begin{aligned}
&\frac{\partial u}{\partial t}+\frac{u}{H_1}\frac{\partial u}{\partial x}+\frac{v}{H_2}\frac{\partial u}{\partial y}+\frac{w}{H_3}\frac{\partial u}{\partial z}+\frac{uv}{H_1H_2}\frac{\partial H_1}{\partial y}+\frac{uw}{H_1H_3}\frac{\partial H_1}{\partial z}\\
&-\frac{v^2}{H_1H_2}\frac{\partial H_2}{\partial x}-\frac{w^2}{H_3H_1}\frac{\partial H_3}{\partial x}+2\Omega(-v\sin\theta+w\cos\theta)\\
=&-\frac{1}{H_1}\frac{\partial\phi}{\partial x}+\nu\left\{\frac{1}{H_1}\frac{\partial}{\partial x}\left\{\frac{1}{H_1H_2H_3}\left[\frac{\partial(uH_2H_3)}{\partial x}+\frac{\partial(vH_3H_1)}{\partial y}+\frac{\partial(wH_1H_2)}{\partial z}\right]\right\}\right.\\
&-\frac{1}{H_2H_3}\left\{\frac{\partial}{\partial y}\left[\frac{H_3}{H_1H_2}\frac{\partial(H_2v)}{\partial x}\right]-\frac{\partial}{\partial y}\left[\frac{H_3}{H_1H_2}\frac{\partial(H_1u)}{\partial y}\right]-\frac{\partial}{\partial z}\left[\frac{H_2}{H_3H_1}\frac{\partial(H_1u)}{\partial z}\right]\right.\\
&\left.\left.+\frac{\partial}{\partial z}\left[\frac{H_2}{H_3H_1}\frac{\partial(H_3w)}{\partial x}\right]\right\}\right\}
\end{aligned}\tag{18}$$

$$\begin{aligned}
&\frac{\partial v}{\partial t}+\frac{u}{H_1}\frac{\partial v}{\partial x}+\frac{v}{H_2}\frac{\partial v}{\partial y}+\frac{w}{H_3}\frac{\partial v}{\partial z}+\frac{vw}{H_2H_3}\frac{\partial H_2}{\partial z}+\frac{vu}{H_2H_1}\frac{\partial H_2}{\partial x}\\
&-\frac{w^2}{H_2H_3}\frac{\partial H_3}{\partial y}-\frac{u^2}{H_1H_2}\frac{\partial H_1}{\partial y}+2\Omega u\sin\theta\\
=&-\frac{1}{H_2}\frac{\partial\phi}{\partial y}+\nu\left\{\frac{1}{H_2}\frac{\partial}{\partial y}\left\{\frac{1}{H_1H_2H_3}\left[\frac{\partial(uH_2H_3)}{\partial x}+\frac{\partial(vH_3H_1)}{\partial y}+\frac{\partial(wH_1H_2)}{\partial z}\right]\right\}\right.\\
&-\frac{1}{H_3H_1}\left\{\frac{\partial}{\partial z}\left[\frac{H_1}{H_2H_3}\frac{\partial(H_3w)}{\partial y}\right]-\frac{\partial}{\partial z}\left[\frac{H_1}{H_2H_3}\frac{\partial(H_2v)}{\partial z}\right]-\frac{\partial}{\partial x}\left[\frac{H_3}{H_1H_2}\frac{\partial(H_2v)}{\partial x}\right]\right.\\
&\left.\left.+\frac{\partial}{\partial x}\left[\frac{H_3}{H_1H_2}\cdot\frac{\partial(H_1u)}{\partial y}\right]\right\}\right\}
\end{aligned}\tag{19}$$

$$\begin{aligned}
&\frac{\partial w}{\partial t}+\frac{u}{H_1}\frac{\partial w}{\partial x}+\frac{v}{H_2}\frac{\partial w}{\partial y}+\frac{w}{H_3}\frac{\partial w}{\partial z}+\frac{wu}{H_3H_1}\frac{\partial H_3}{\partial x}+\frac{wv}{H_3H_2}\frac{\partial H_3}{\partial y}\\
&-\frac{u^2}{H_3H_1}\frac{\partial H_1}{\partial z}-\frac{v^2}{H_2H_3}\frac{\partial H_2}{\partial z}-2\Omega u\cos\theta\\
=&-\frac{1}{H_3}\frac{\partial\phi}{\partial z}+\nu\left\{\frac{1}{H_3}\frac{\partial}{\partial z}\left\{\frac{1}{H_1H_2H_3}\left[\frac{\partial(uH_2H_3)}{\partial x}+\frac{\partial(vH_3H_1)}{\partial y}\right.\right.\right.\\
&\left.\left.+\frac{\partial(wH_1H_2)}{\partial z}\right]\right\}-\frac{1}{H_1H_2}\left\{\frac{\partial}{\partial x}\left[\frac{H_2}{H_3H_1}\frac{\partial(H_1u)}{\partial z}\right]-\frac{\partial}{\partial x}\left[\frac{H_2}{H_3H_1}\frac{\partial(H_3w)}{\partial x}\right]\right.\\
&\left.\left.-\frac{\partial}{\partial y}\left[\frac{H_1}{H_2H_3}\frac{\partial(H_2w)}{\partial y}\right]+\frac{\partial}{\partial y}\left[\frac{H_1}{H_2H_3}\frac{\partial(H_2v)}{\partial z}\right]\right\}\right\}+g\alpha T
\end{aligned}\tag{20}$$

$$\frac{\partial T}{\partial t}+\frac{1}{H_1H_2H_3}\left\{\frac{\partial(TuH_2H_3)}{\partial x}+\frac{\partial(TvH_3H_1)}{\partial y}+\frac{\partial(TwH_1H_2)}{\partial z}\right\}$$

$$=\frac{\kappa}{H_1H_2H_3}\left\{\frac{\partial}{\partial x}\left(\frac{H_2H_3}{H_1}\frac{\partial T}{\partial x}\right)+\frac{\partial}{\partial y}\left(\frac{H_3H_1}{H_2}\frac{\partial T}{\partial y}\right)+\frac{\partial}{\partial z}\left(\frac{H_1H_2}{H_3}\frac{\partial T}{\partial z}\right)\right\} \tag{21}$$

其中，u，v，w 为速度$\boldsymbol{V}$在 x，y，z 轴上的三个分量，为了书写简洁，T'之右上角之一撇被略掉了；H_1、H_2、H_3分别为相应于 x，y，z 的 Lamé 系数，其表达式为

$$\left.\begin{aligned}&H_1=(1+z/R)\cos(\theta_0+y/R)/\cos\theta_0\\&H_2=(1+z/R)\\&H_3=1\end{aligned}\right\} \tag{22}$$

R 为地球半径和未扰动时大洋深度之和。

引入 L、D、T、U、UD/L、$(2\Omega\sin\theta_0)UL$ 和 Θ 分别表示（x,y）、z、t、（u,v）、w、ϕ 和 T 的特征量，并用以无因次化方程组（17）~（22）。同时，我们引入 $D/R\ll 1$ 这一近似，这是因为无论对于大气流场、也无论对于海洋流场，其铅垂尺度远小于地球半径之故。于是，相应的无因次近似方程组如下：

$$\frac{\partial u}{\partial x}+H\frac{\partial v}{\partial y}+H\frac{\partial w}{\partial z}+\frac{\partial H}{\partial y}v=0 \tag{23}$$

$$\begin{aligned}&\varepsilon\frac{\partial u}{\partial t}+Ro\left\{\frac{u}{H}\frac{\partial u}{\partial x}+v\frac{\partial u}{\partial y}+w\frac{\partial u}{\partial z}+uv\frac{\partial H}{\partial y}\right\}-v(2\Omega\sin\theta)/f_0+\delta w(2\Omega\cos\theta)/f_0\\&=-\frac{1}{H}\frac{\partial\phi}{\partial x}+Ek_{\mathrm{H}}\left\{\frac{1}{H^2}\frac{\partial^2u}{\partial x^2}+\frac{\partial^2u}{\partial y^2}\right\}+Ek_{\mathrm{V}}\frac{\partial^2u}{\partial z^2}\\&+Ek_{\mathrm{H}}\left\{\frac{1}{H}\frac{\partial H}{\partial y}\frac{\partial u}{\partial y}+\frac{2}{H^2}\frac{\partial H}{\partial y}\frac{\partial v}{\partial x}+\frac{\partial}{\partial y}\left(\frac{1}{H}\frac{\partial H}{\partial y}\right)u\right\}\end{aligned} \tag{24}$$

$$\begin{aligned}&\varepsilon\frac{\partial v}{\partial t}+Ro\left\{\frac{u}{H}\frac{\partial v}{\partial x}+v\frac{\partial v}{\partial y}+w\frac{\partial v}{\partial z}-\frac{u^2}{H}\frac{\partial H}{\partial y}\right\}+u(2\Omega\sin\theta)/f_0\\&=-\frac{\partial\phi}{\partial y}+Ek_{\mathrm{H}}\left\{\frac{1}{H^2}\frac{\partial^2v}{\partial x^2}+\frac{\partial^2v}{\partial y^2}\right\}+Ek_{\mathrm{V}}\frac{\partial^2v}{\partial z^2}+Ek_{\mathrm{H}}\left\{-\frac{1}{H}\frac{\partial H}{\partial y}\frac{\partial v}{\partial y}\right.\\&\left.-\frac{2}{H^2}\frac{\partial H}{\partial y}\frac{\partial u}{\partial x}+\frac{\partial}{\partial y}\left(\frac{1}{H}\frac{\partial H}{\partial y}\right)v\right\}\end{aligned} \tag{25}$$

$$\begin{aligned}&\delta^2\left\{\varepsilon\frac{\partial w}{\partial t}+Ro\left[\frac{u}{H}\frac{\partial w}{\partial x}+v\frac{\partial w}{\partial y}+w\frac{\partial w}{\partial z}\right]\right\}-\delta u(2\Omega\cos\theta)/f_0\\&=-\frac{\partial\phi}{\partial z}+RiRoT+\delta^2\left\{Ek_{\mathrm{H}}\left[\frac{1}{H^2}\frac{\partial^2w}{\partial x^2}+\frac{\partial^2w}{\partial y^2}\right]+Ek_{\mathrm{V}}\frac{\partial^2w}{\partial z^2}+Ek_{\mathrm{H}}\frac{1}{H}\frac{\partial H}{\partial y}\frac{\partial w}{\partial y}\right\}\end{aligned} \tag{26}$$

$$\varepsilon\frac{\partial T}{\partial t}+Ro\left\{\frac{u}{H}\frac{\partial T}{\partial x}+v\frac{\partial T}{\partial y}+w\frac{\partial T}{\partial z}\right\}=\frac{Ek_{\mathrm{H}}}{Pr}\left\{\frac{1}{H^2}\frac{\partial^2T}{\partial x^2}+\frac{\partial^2T}{\partial y^2}\right\}+\frac{Ek_{\mathrm{V}}}{Pr}\frac{\partial^2T}{\partial z^2}+\frac{Ek_{\mathrm{H}}}{Pr}\cdot\frac{1}{H}\frac{\partial H}{\partial y}\frac{\partial T}{\partial y} \tag{27}$$

其中，$H=H_1=\cos\left[\theta_0+\left(\frac{L}{R}\right)y\right]/\cos\theta_0 f_0=2\Omega\sin\theta_0$，

$\varepsilon=(f_0T)^{-1}$，$Ri=\frac{g\alpha\Theta D}{U^2}$ 为 Richardson 数，

$Ek_{\mathrm{H}}=\nu/f_0L^2$ 为水平 Ekman 数，

$Ek_{\mathrm{V}}=\nu/f_0D^2$ 为铅垂 Ekman 数，

$Pr=\nu/\kappa$ 为 Prandtl 数，

$\delta=D/L$ 为空间尺度比。

应指出，本文中的无因次量与其相应的因次量利用了同一符号。

既然像大气和海洋中这样的行星尺度运动的铅垂尺度 D 远小于其水平尺度 L，故可以引入“薄层流场”的假定，即其空间尺度比 $\delta\ll1$。于是，在此近似下，无因次方程组（23）~（27）化简如下：

$$\frac{\partial u}{\partial x}+H\frac{\partial v}{\partial y}+H\frac{\partial w}{\partial z}+\frac{\partial H}{\partial y}v=0 \tag{28}$$

$$\begin{aligned}&\varepsilon\frac{\partial u}{\partial t}+Ro\left\{\frac{u}{H}\frac{\partial u}{\partial x}+v\frac{\partial u}{\partial y}+w\frac{\partial u}{\partial z}+uv\frac{\partial H}{\partial y}\right\}-(f/f_0)v\\&=-\frac{1}{H}\frac{\partial\phi}{\partial x}+Ek_{\mathrm{H}}\left\{\frac{1}{H^2}\frac{\partial^2u}{\partial x^2}+\frac{\partial^2u}{\partial y^2}\right\}+Ek_{\mathrm{V}}\frac{\partial^2u}{\partial z^2}\\&\quad+Ek_{\mathrm{H}}\left\{\frac{1}{H}\frac{\partial H}{\partial y}\frac{\partial u}{\partial y}+\frac{2}{H^2}\frac{\partial H}{\partial y}\frac{\partial v}{\partial x}+\frac{\partial}{\partial y}\left(\frac{1}{H}\frac{\partial H}{\partial y}\right)u\right\}\end{aligned} \tag{29}$$

$$\begin{aligned}&\varepsilon\frac{\partial v}{\partial t}+Ro\left\{\frac{u}{H}\frac{\partial v}{\partial x}+v\frac{\partial v}{\partial y}+w\frac{\partial u}{\partial z}-\frac{u^2}{H}\frac{\partial H}{\partial y}\right\}+(f/f_0)u\\&=-\frac{\partial\phi}{\partial y}+Ek_{\mathrm{H}}\left\{\frac{1}{H^2}\frac{\partial^2v}{\partial x^2}+\frac{\partial^2v}{\partial y^2}\right\}+Ek_{\mathrm{V}}\frac{\partial^2v}{\partial z^2}\\&\quad+Ek_{\mathrm{H}}\left\{-\frac{1}{H}\frac{\partial H}{\partial y}\frac{\partial v}{\partial y}-\frac{2}{H^2}\frac{\partial H}{\partial y}\frac{\partial u}{\partial x}+\frac{\partial}{\partial y}\left(\frac{1}{H}\frac{\partial H}{\partial y}\right)v\right\}\end{aligned} \tag{30}$$

$$\frac{\partial\phi}{\partial z}=(RiRo)T \tag{31}$$

$$\begin{aligned}&\varepsilon\frac{\partial T}{\partial t}+Ro\left\{\frac{u}{H}\frac{\partial T}{\partial x}+v\frac{\partial T}{\partial y}+w\frac{\partial T}{\partial z}\right\}\\&=\frac{Ek_{\mathrm{H}}}{Pr}\left\{\frac{1}{H^2}\frac{\partial^2T}{\partial x^2}+\frac{\partial^2T}{\partial y^2}\right\}+\frac{Ek_{\mathrm{V}}}{Pr}\frac{\partial^2T}{\partial z^2}+\frac{Ek_{\mathrm{H}}}{Pr}\frac{1}{H}\frac{\partial H}{\partial y}\frac{\partial T}{\partial y}\end{aligned} \tag{32}$$

其中，$f=2\Omega\sin\theta$ 为 Coriolis 参数。

由此立刻得出了一个重要结论：由方程（31）可知，准静力近似是薄层流场的一个直接后果——这完全类似于在 Prandtl 边界层中沿着边壁法线方向上其压强不变的结论。应指出，通常处理为 $RiRo=1$[2, 7, 8]，这只需适当选择 Θ 和 U 即可满足该等式；其物理意义是，温差所产生的浮力是不可忽视的因素，即包含了自由对流的情形。如果不把 $RiRo$ 简单地处理成 1，这意味着也可以包含另一个极端，即强迫对流的情况。

至于方程（32），一旦依据确定了对流项抑或传导项为主要因子以后，方程可以进一步化简。

把 Lamé 系数 H 和 Coriolis 参数展开如下：

$$H=1-\mathrm{tg}\theta_0\cdot\left(\frac{L}{R}\right)y+O(L^2/R^2) \tag{33}$$

$$f/f_0=1+\mathrm{ctg}\theta_0\cdot\left(\frac{L}{R}\right)y+O(L^2/R^2) \tag{34}$$

如果我们把兴趣限制在流场的水平尺度 L 不太大的情形，致使 $L^2/R^2\ll 1$，则 Lamé 系数和 Coriolis 参数表达式（33）和（34）可以线化如下：

$$H=1-\mathrm{tg}\theta_0\cdot\left(\frac{L}{R}\right)y \tag{35}$$

$$f/f_0=1+\mathrm{ctg}\theta_0\cdot\left(\frac{L}{R}\right)y \tag{36}$$

由此显然看出，只需满足下列不等式

$$\mathrm{tg}\theta_0\cdot(L/R)\ll 1 \tag{37}$$

则正交曲线坐标系中的基本方程组（28）~（32）可以化简成简单的直角坐标系中的形式：

$$\frac{\partial u}{\partial x}+\frac{\partial v}{\partial y}+\frac{\partial w}{\partial z}=0 \tag{38}$$

$$\varepsilon\frac{\partial u}{\partial t}+Ro\left\{u\frac{\partial u}{\partial x}+v\frac{\partial u}{\partial y}+w\frac{\partial u}{\partial z}\right\}-(f/f_0)v=-\frac{\partial\phi}{\partial x}+Ek_{\mathrm{H}}\left\{\frac{\partial^2 u}{\partial x^2}+\frac{\partial^2 u}{\partial y^2}\right\}+Ek_{\mathrm{V}}\frac{\partial^2 u}{\partial z^2} \tag{39}$$

$$\varepsilon\frac{\partial v}{\partial t}+Ro\left\{u\frac{\partial v}{\partial x}+v\frac{\partial v}{\partial y}+w\frac{\partial v}{\partial z}\right\}+(f/f_0)u=-\frac{\partial\phi}{\partial y}+Ek_{\mathrm{H}}\left\{\frac{\partial^2 v}{\partial x^2}+\frac{\partial^2 v}{\partial y^2}\right\}+Ek_{\mathrm{V}}\frac{\partial^2 v}{\partial z^2} \tag{40}$$

$$\frac{\partial\phi}{\partial z}=(RiRo)T \tag{41}$$

$$\varepsilon\frac{\partial T}{\partial t}+Ro\left\{u\frac{\partial T}{\partial x}+v\frac{\partial T}{\partial y}+w\frac{\partial T}{\partial z}\right\}=\frac{Ek_{\mathrm{H}}}{Pr}\left\{\frac{\partial^2 T}{\partial x^2}+\frac{\partial^2 T}{\partial y^2}\right\}+\frac{Ek_{\mathrm{V}}}{Pr}\frac{\partial^2 T}{\partial z^2} \tag{42}$$

顺便指出，注意到此时简化了的 Coriolis 力的表达式，可知该旋转球面上薄层流体的运动已等效于一个绕局地铅垂轴以角速度 $(f/2)\boldsymbol{k}$ 旋转的坐标系中的运动了（其中，$\boldsymbol{k}$ 为铅垂轴的单位向量）——这是均匀密度情形下所导出的同样结论的一个推广[9]。

显然，满足不等式 $\mathrm{tg}\theta_0 \cdot (L/R) \ll 1$，从而方程组（38）~（42）成立，可以在下列两种不同的情况下实现。

（Ⅰ）$L/R \ll 1$ 的情况

不等式 $L/R \ll 1$ 意味着此时该正交曲面已被切于参考点 0 上的一个“零阶曲面”——“切平面”所近似。在这种近似的“切平面”坐标系中，又可分为下面三种情形进一步讨论之。

（i）$\mathrm{tg}\theta_0 \sim 1$ 的情形——这大约相当于中纬度跨高纬度的地域。此时，当然有 $H=1$，从而方程组（38）~（42）成立；特别是由式（36）可见，其 Coriolis 参数 $f=f_0$。故此种情况可以不考虑 Coriolis 参数随纬度的改变——这相应于众所周知的“f–平面”的情形。

（ii）$\mathrm{tg}\theta_0 \ll 1$ 的情形——这大约相当于低纬度跨中纬度的地域。此时，当然亦有 $H=1$，从而方程组（38）~（42）成立。但由于此时 $\mathrm{ctg}\theta_0 \gg 1$，并由式（36）可见 $O(\mathrm{ctg}\theta_0 \cdot L/R)=1$，从而 Coriolis 参数为 y 的线函。即

$$f = f_0 + \beta L y \tag{43}$$

其中，$\beta = 2\Omega\cos\theta_0 / R$。

这相应于众所周知的“β–平面”的情形。

（iii）$\mathrm{tg}\theta_0 \gg 1$ 的情形——这大约是指在极地附近的高纬度处。由于此时 $\mathrm{tg}\theta_0 \cdot (L/R)$ 不能满足远小于 1 的条件，从而方程组（38）~（42）不成立；因而，这种情形在这里将被排除。事实上，在极地附近已利用了一种类似的“极平面”坐标系[10]。

（Ⅱ）$(L/R)^2 \ll 1$ 的情况

在这种情况下，为了满足不等式（37），即满足 $\mathrm{tg}\theta_0 \cdot (L/R) \leqslant 1$，从而方程组（38）~（42）成立，则参考纬度 θ_0 应被限制在如下的范围。即

$$O(\mathrm{tg}\theta_0) = O(L/R) \tag{44}$$

这大约相应于中、低纬度地域。

此时，由于 $(L/R)^2 \ll 1$ 但 L/R 不远小于 1，故该正交曲面是被一个切于参考点上的“1 阶曲面”所近似。又由式（36）知，既然 $\mathrm{ctg}\theta_0 \sim R/L$，从而 $O(\mathrm{ctg}\theta_0 \cdot L/R)=1$，故有式（43）成立，即应考虑 Coriolis 参数随纬度的改变——“β 效应”。显然，这相应于一种“β–曲面”的情形。

应强调指出，由 Lamé 系数的结构不难看出，对于以上两种情况——不论是“f–平面”、“β–平面”抑或是“β–曲面”近似，皆针对 y 向而言，而在 x 向上不被限制，也

勿须近似。实际上 x 与纬圈可以吻合，因而事实上皆为“曲面”。这一点只需把本文所使用的方法和导出的结果与曲壁上的 Prandtl 边界层中所使用的方法和结果相比较就十分清楚了[11]。

综上所述可以看出，通常习惯采用的术语“f–平面”和“β–平面”坐标至少是不完备的，看来采用术语“f–坐标系”和 β–坐标系”也许更为恰当。

为了给出一个数量上的概念，特举一算例如下：若对北半球而言，R 近似地取为地球半径6371千米，并假设0.2为可以忽略的小量；则计算表明——（Ⅰ）对于 $L/R=0.2\ll1$ 的情况，有 L=1274 千米~10^3 千米，且 $\theta_0\leqslant 20°$N 为“β–平面”情形，其流场最大北界大约为 30°N，而 20°N$\leqslant\theta_0\leqslant$50°N 为“$f$–平面”情形，其流场最大北界大致可达 60°N 处；（Ⅱ）对于 $(L/R)^2=0.2\ll1$ 的情况，有 L=2848 千米~3000 千米，且 $\max\theta_0\approx 30°$N，为“β–曲面”情形，其流场最大北界亦可达约为 60°N 的地方。

最后要强调指出，只要注意到下述几点，问题可以自然推广。

（1）大气和海洋中的行星尺度流场呈涡动状态，对于这种湍流运动，一个最简单的模型是依据 Boussinesq 假设，把湍流应力参数化，亦即可以用铅垂和水平湍粘性系数来替代上述方程组中的分子粘性系数即可[1,2]。如果不采用该假设，也可以完整地保留湍应力项而进行相似处理[12]。

（2）由于大气流场的热力学状态不同，要稍作改变[2,13]。对于大洋环流问题，可以把上述方程组直接进行推广和应用，这时只需把温度 T 理解为“表观温度”即可[14,15]。

（3）由于控制海洋环流的是一个由运动方程组导出的涡度方程[1,2,4]，看来在化简以前先保留惯性的变密度效应可能是有益的[15~17]；但显然这不会增加上述论证的根本困难，因而可以把它看作上述方程组的一个直接推广。

（4）把本文中采用的手段和方法去导出薄层流场中大尺度波的相应结论也不会有根本性的困难，只要注意到依据其动力学的基本假定以改变其相应的无因次参数即可[18]。

（5）假若流场的水平尺度，不管是 x 向，还是 y 向都不太大，以致满足条件 $L/R\ll1$，即问题可以立于一个“f–平面”坐标上，则此时 x 轴和 y 轴不必一定分别指东和指北，而可以指任何方向来构成直角坐标系[9,18]。

参考文献

[1] Kamenkovich, V. M., Fundamentals of Ocean Dynamics, Elsevier Oceanography series, 16, Elsevier Scientific Publishing Company, (1977).

[2] Pedlosky, J., Geophysical Fluid Dynamics, Springer-Verlag, New York, (1979).

[3] Rossby, C. G., Relation between variations in the intensity of the zonal circulation of the atmosphere and the displacements of the semi-permanent centers of the action, J. Mar. Res., Vol. 2, pp 38–55,

(1939).

[4] Stommel, H., Westward intensitication of wind-driven ocean currents, Trans. Am. Geophys. Union, Vol. 29, No. 2, pp 202–206, (1948).

[5] Veronis, G., On the approximations involved in transforming the equations of motion from a spherical surface to the β-plane, I. Barotropic systems, J. Mar. Res., Vol. 21, No. 2, pp 110–124, (1963).

[6] Veronis, G., On the approximations involved in transforming the equations of motion from a spherical surface to the β-plane, II. Baroclinic systems, J. Mar. Res., Vol. 21, No. 3, pp 199–204, (1963).

[7] Tritton, D. J., Physical Fluid Dynamics, VNR Com., New York, (1977).

[8] Friedlander, S., An Introduction to the Mathematical Theory of Geophysical Fluid Dynamics, N-H P Com., New-York, (1980).

[9] 冯士筰, 风暴潮导论, 科学出版社, (1982).

[10] Le Blond, P. H., Planetary waves in a symmetrical polar basin, Tellus, Vol. 16, No. 4, pp 503–512, (1964).

[11] Кочин, Н. Е., И. А. Кибедь, Н. В. Розе, Теоретическая Гибротехаиика часть II, ГИФ—МЛ, (1963).

[12] Stern, M. E., Ocean Circulation Physics, AP, New York, (1975).

[13] Dickinson, R. E., Rossby waves—long-period oscillations of oceans and atmospheres, Annual Review of Fluid Mechanics, Vol. 10, pp 159–195, (1978).

[14] Fofonoff, N. P., The Sea (Editor, M. N. Hill), Vol. 1, Section III, Interscience Pubs., New York, (1962).

[15] 冯士筰, 大洋风生–热盐环流模型, 山东海洋学院学报, 第 2 期, pp 1–14, (1979).

[16] 奚盘根, 张淑珍, 冯士筰, 东中国海环流的一种模型(Ⅰ), 山东海洋学院学报, 第 10 卷, 第 3 期, pp 13–25, (1980).

[17] 冯士筰, 张淑珍, 奚盘根, 东中国海环流的一种模型(Ⅱ), 山东海洋学院学报, 第11卷, 第2期, pp 8–26, (1981).

[18] Feng Shih-zao (冯士筰), A three-dimensional nonlinear model of tides, Scientia Sinica, Vol. XX, No. 4, pp 436–446, (1977).

论大洋环流的尺度分析及风旋度–热盐梯度方程式*

冯士筰

绪论

作者在文献[1]中曾简单地导出过一个描述大洋风生–热盐环流的“风旋度–热盐梯度方程式”，它是一个全流函数所满足的线性位涡方程，表达了海面上风应力旋度和界面上的热盐梯度在侧向湍摩擦所导致的涡度扩散下与行星涡度梯度的平衡。利用该方程，在一个矩形大洋中，作者提出了一个相当简单的准两层模型，并用以描述了大洋风生–热盐环流的主要特征。当不考虑热盐效应时，该模型即蜕化为 Munk 的大洋风生环流的经典模型[2]；因而，该模型可以看作经典的 Munk 风生大洋环流的一个推广。也就是说，该风生–热盐环流模型是由广义的 Sverdrup 流和粘性的行星边界层所组成的。所谓广义的 Sverdrup 流系指在风应力旋度场和热盐梯度场联合作用下产生的 Sverdrup 流动，并且当不考虑热盐效应时，它蜕化为经典的风生环流的 Sverdrup 流。这里所说的粘性行星边界层，准确地说，系指由侧向湍粘性所表征的广义 Munk 行星边界层；当不考虑热盐效应时，它蜕化为纯风生的 Munk 行星边界层。但是，由大洋风生洋流的研究中我们早已知道了大洋环流西向强化的 Munk 粘性理论不是唯一的西向强化理论。事实上，首次揭示洋流西向强化的理论是考虑底摩擦效应的 Stommel 模型[3]。特别是上层大洋环流的行星边界层具有显著的惯性性质，其本质为非线性的现象[4, 5]。为了更完善地描述和研讨大洋风生–热盐环流，应该建立一个比文献[1]中所导出的线性的风旋度–热盐梯度方程式更为完整的非线性的风旋度–热盐梯度方程式。这是本文试图完成的贡献之一。因此，本文可以看作文献[1]的一个推广。另外，在风旋度–热盐梯度方程中包含了一个界面上的热盐梯度因子，正是这个因子使得该模型有别于前人的模型。为了更深入地揭示这个热盐梯度因子，与文献[1]不同，本文采用了一个类似于文献[6]中所采用的系统尺度化的手段来导出风旋度–热盐梯度方程式。并且通过尺度分析泛泛地给出了现象的某些物理信息。这是本文试图做出的另一个贡献。因此，本文也可以看作文献[1]的进一步深化。

* 冯士筰. 1984. 论大洋环流的尺度分析及风旋度–热盐梯度方程式. 山东海洋学院学报, 14(1): 33–43

问题的提法

基于我们研究的有限目的，本文将简单地采用 β–坐标系作为描述大洋中行星尺度运动的参考系[7]。在其中描述海洋中行星尺度运动的热力–动力学方程组形如直角坐标系中方程的形式，其定常状态下的无因次方程组为

$$\nabla\cdot\boldsymbol{V}+\frac{\partial w}{\partial z}=0 \tag{1}$$

$$(1-\varepsilon\Theta)Ro\left[\boldsymbol{V}\cdot\nabla\boldsymbol{V}+w\frac{\partial\boldsymbol{V}}{\partial z}\right]+(1-\varepsilon\Theta)(1+\beta y)\boldsymbol{k}\times\boldsymbol{V}$$
$$=-\nabla\phi+Ek_{\mathrm{V}}\frac{\partial}{\partial z}\left(\nu_{\mathrm{V}}\frac{\partial\boldsymbol{V}}{\partial z}\right)+Ek_{\mathrm{H}}\nabla^2\boldsymbol{V} \tag{2}$$

$$\frac{\partial\phi}{\partial z}=(\varepsilon Ro/Fr^2)\Theta \tag{3}$$

$$Ro\left(\boldsymbol{V}\cdot\nabla\Theta+w\frac{\partial\Theta}{\partial z}\right)=\frac{Ek_{\mathrm{V}}}{Pr_{\mathrm{V}}}\frac{\partial}{\partial z}\left(k_{\mathrm{V}}\frac{\partial\Theta}{\partial z}\right)+\frac{Ek_{\mathrm{H}}}{Pr_{\mathrm{H}}}\nabla^2\Theta \tag{4}$$

其中诸无因次参数为

$\varepsilon=\alpha\Theta^*$， $Ro=\dfrac{V^*}{f_0L}$ （Rossby 数）， $\beta=\dfrac{\beta_0L}{f_0}$，

$Ek_{\mathrm{V}}=\dfrac{\nu}{f_0D^2}$ （铅垂 Ekman 数）， $Ek_{\mathrm{H}}=\dfrac{\nu_{\mathrm{H}}}{f_0L^2}$ （水平 Ekman 数），

$Pr_{\mathrm{V}}=\dfrac{\nu}{k}$ （铅垂 Prandtl 数）， $Pr_{\mathrm{H}}=\dfrac{\nu_{\mathrm{H}}}{k_{\mathrm{H}}}$ （水平 Prandtl 数），

$Fr=\dfrac{V^*}{\sqrt{gD}}$ （Froude 数）；

以上诸符号的意义说明如下：算子 $\nabla=\boldsymbol{i}\dfrac{\partial}{\partial x}+\boldsymbol{j}\dfrac{\partial}{\partial y}$， $x=\bar{x}/L$， $y=\bar{y}/L$， $\bar{x}$ 为纬圈坐标、朝东为正， $\bar{y}$ 为经圈坐标，朝北为正， L 为流场的水平尺度， $\boldsymbol{i}$， $\boldsymbol{j}$ 分别为 $\bar{x}$ 和 $\bar{y}$ 的单位向量； $z=\bar{z}/D$， z 为铅垂坐标，原点设在未扰动的海面上，朝上为正， $\boldsymbol{k}$ 为其单位向量，D 为流场的铅垂尺度； $\boldsymbol{V}=\bar{\boldsymbol{V}}/V^*, \bar{\boldsymbol{V}}=\boldsymbol{i}\bar{u}+\boldsymbol{j}\bar{v}$， $\bar{u}$ 和 $\bar{v}$ 分别为纬向和经向流分量，V^* 为流的尺度； $w=\bar{w}/(V^*D/L)$， $\bar{w}$ 为铅垂流速； $\Theta=\bar{\Theta}/\Theta^*$， $\Theta=\bar{T}-T_0$， $\bar{T}$ 为表观温度[1]， T_0 为参考表观温度， Θ^* 为 $\bar{\Theta}$ 之尺度， $\phi=\bar{\phi}/(f_0LV^*)$， $\bar{\phi}=\bar{P}/\rho_0+g\bar{z}$， $\bar{P}$ 为压强， ρ_0 为参考密度， g 为重力加速度。 $f_0=2\Omega\sin\theta_0$， Ω 为地转角速率， θ_0 为参考纬度； $\nu_{\mathrm{V}}=\bar{\nu}_{\mathrm{V}}/\nu$， $\bar{\nu}_{\mathrm{V}}$ 为铅垂湍粘性系数， ν 为其尺度； ν_{H} 为水平湍粘性系数； α 为

表观热膨胀系数，$\alpha = -\frac{1}{\rho_0}\left(\frac{\partial \bar{\rho}}{\partial \bar{T}}\right)_0$，其中 $\bar{\rho}$ 为密度，而导数的下标 0 表示参考点；$k_V = \bar{k}_V / k$，k_V 为铅垂湍扩散系数，k 为其尺度；k_H 为水平湍扩散系数；$\beta_0 = \frac{\partial f}{\partial \bar{y}}$，$f = f_0 + \beta_0 \bar{y}$ 为 Coriolis 参数。

无因次基本方程组（1）~（4）的导出过程可以参考文献[7]中方程组（38）~（42）的导出过程，事实上，前者是后者的一个推广了的形式，在水平运动方程（2）中包含了一个表达变密度的因子 $\varepsilon\Theta$，这个因子在经典的 Boussinesq 近似方程组（39）、（40）中是被略掉的。该因子保留的理由是基于直接用以描述大洋环流的不是运动方程本身而是位涡方程；二者之差异在于在前者中，地转平衡为基本的力平衡，在后者中，这个基本的力平衡项被消除了，从而，在基本运动方程中不显著的诸小项在位涡方程中可能将变得显著起来。再注意到在大洋的大尺度运动中，$O(Ro) = 10^{-4} \sim 10^{-3}$，$O(Ek_V, Ek_H) \leqslant 10^{-4} \sim 10^{-3}$ 而 $O(\varepsilon) = 10^{-4} \sim 10^{-3}$，这显然看出，按照经典的 Boussinesq 近似，在水平运动方程中略去 $\varepsilon\Theta$ 因子是值得商榷的。在水平运动方程（2）中保留变密度效应的物理意义是：该模型不仅如 Boussinesq 近似方程组那样考虑了变密度和重力的相互作用，而且还考虑了变密度和惯性力的相互作用——实质上指的是与 Coriolis 惯性力的相互作用。

众所周知，在洋流运动中变密度与重力的相互作用——浮力项是必须计及的，故有 $O(\varepsilon Ro / Fr^2)$ 为 1；并且为了全面地考查表观温度的湍扩散效应和对流效应，可令 $O\left(\frac{Ek_V}{Pr_V}, \frac{Ek_H}{Pr_H}\right) = O(Ro)$。再注意到，如果我们流场的典型数据采用 $L \sim 10^8$，$\beta_0 \sim 10^{-13}$，$f_0 \sim 10^{-4}$（本文中物理单位皆采用 cgs 制），则有 $O(\beta) = 10^{-1}$。因此，我们可以合理地假设

$$\max(\varepsilon, Ro, Ek_V, Ek_H) \leqslant O(\beta) < 1 \tag{5}$$

并且以 β 为小参数对场量展开如下：

$$(\boldsymbol{V}, w, \Theta, \phi) = (\boldsymbol{V}_0, w_0, \Theta_0, \phi_0) + \beta(\boldsymbol{V}_1, w_1, \Theta_1, \phi_1) + O(\beta^2) \tag{6}$$

把式（6）代入基本方程组（1）~（4），其零阶模型为

$$\nabla \cdot \boldsymbol{V}_0 + \frac{\partial w_0}{\partial z} = 0 \tag{7}$$

$$\boldsymbol{k} \times \boldsymbol{V}_0 = -\nabla \phi_0 \tag{8}$$

$$\frac{\partial \phi_0}{\partial z} = \left(\frac{\varepsilon Ro}{Fr^2}\right)\Theta_0 \tag{9}$$

$$Ro\left(\boldsymbol{V}_0 \cdot \nabla \Theta_0 + w_0 \frac{\partial \Theta_0}{\partial z}\right) = \frac{Ek_V}{Pr_V}\frac{\partial}{\partial z}\left(k_V \frac{\partial \Theta_0}{\partial z}\right) + \frac{Ek_H}{Pr_H}\nabla^2 \Theta_0 \tag{10}$$

显然，其零阶模型为地转运动。但由于浮力项的存在，该地转运动将不满足铅垂

刚性原则。不过若以 $\boldsymbol{k}$ 标乘方程（3）的旋度，将会发现运动仍保持水平无辐散场，即

$$\nabla \cdot \boldsymbol{V}_0 = 0 \tag{11}$$

因此，由方程（7）和（11）可见

$$\frac{\partial w_0}{\partial z} = 0 \tag{12}$$

可见，设若流场中存在一个水平界面 $z = z_0$ 上有 $w_0 = 0$，则整个流场中必有

$$w_0 = 0 \tag{13}$$

其物理意义是：该模型在零阶近似下实现了水平无辐散的地转运动的二维流场。应特别指出，我们采用这样一个相对简单的风生–热盐洋流模型是基于本文研究的有限目的：我们仅局限于去探讨稳定层化下的大尺度洋流运动，而不考虑铅垂对流和洋流的相互作用。

为了导出零阶模型的位涡方程，必须借助下列一阶扰动方程：

$$\begin{aligned}&\left(\frac{Ro}{\beta}\right)\boldsymbol{V}_0 \cdot \nabla \boldsymbol{V}_0 + \boldsymbol{k}\times\boldsymbol{V}_1 + y\boldsymbol{k}\times\boldsymbol{V}_0 - \left(\frac{\varepsilon}{\beta}\right)\Theta_0\boldsymbol{k}\times\boldsymbol{V}_0 \\ &= -\nabla\phi_1 + \left(\frac{Ek_{\rm V}}{\beta}\right)\frac{\partial}{\partial z}\left(\nu_{\rm V}\frac{\partial \boldsymbol{V}_0}{\partial z}\right) + \left(\frac{Ek_{\rm H}}{\beta}\right)\nabla^2\boldsymbol{V}_0\end{aligned} \tag{14}$$

假设我们感兴趣的流场是一个由上、下两个水平界面 $\bar{z} = \bar{z}_2$ 和 $\bar{z} = \bar{z}_1$ 所界的等厚度 D 的水层。在该水层中由下界至上界沿厚度积分方程（14）。并利用两个界面上湍应力的连续性条件，则导出全流满足的微分方程如下：

$$\begin{aligned}&\left(\frac{Ro}{\beta}\right)\boldsymbol{Q}_0 \cdot \nabla \boldsymbol{Q}_0 + \boldsymbol{k}\times\left[\boldsymbol{Q}_1 + y\boldsymbol{Q}_0 - \left(\frac{\varepsilon}{\beta}\right)'\int_{z_1}^{z_2}\boldsymbol{V}_0\Theta_0 \mathrm{d}z\right] \\ &= -\nabla\int_{z_1}^{z_2}\phi_1 \mathrm{d}z + \left(\frac{Ek_{\rm V}}{\beta}\right)\left[\frac{D\tau_{2*}}{V^*\nu_{2*}}\boldsymbol{\tau}_2 - \frac{D\tau_{1*}}{V^*\nu_{1*}}\boldsymbol{\tau}_1\right] + \left(\frac{Ek_{\rm H}}{\beta}\right)\nabla^2\boldsymbol{Q}_0\end{aligned} \tag{15}$$

其中全流为

$$(\boldsymbol{Q}_0, \boldsymbol{Q}_1) = \int_{z_1}^{z_2}(\boldsymbol{V}_0, \boldsymbol{V}_1)\mathrm{d}z \tag{16}$$

应力与密度之比为

$$(\boldsymbol{\tau}_2, \boldsymbol{\tau}_1) = \left(\left[\frac{D\tau_{2*}}{V^*\nu_{2*}}\right]^{-1}\left[\nu_{\rm V}\frac{\partial \boldsymbol{V}_0}{\partial z}\right]_{z=z_2}, \left[\frac{D\tau_{1*}}{V^*\nu_{1*}}\right]^{-1}\left[\nu_{\rm V}\frac{\partial \boldsymbol{V}_0}{\partial z}\right]_{z=z_1}\right) \tag{17}$$

而 τ_{1*}、ν_{1*} 和 τ_{2*}、ν_{2*}，分别为下界面和上界面上应力和湍粘系数之尺度。

把方程（15）施以旋度运算并标乘 $\boldsymbol{k}$，且利用流场中和界面上的质量连续性条件，

则导出下列全流函数所满足的方程：

$$\left(\frac{Ro}{\beta}\right)J(\psi,\nabla^2\psi)+\frac{\partial\psi}{\partial x}=\left(\frac{Ek_{\rm V}}{\beta}\right)\left\{\frac{D\tau_{2*}}{V^*\nu_{2*}}{\rm Curl}\,\tau_2-\frac{D\tau_{1*}}{V^*\nu_{1*}}{\rm Curl}\,\tau_1\right\}+\left(\frac{Ek_{\rm H}}{\beta}\right)\nabla^4\psi+\left(\frac{\varepsilon}{\beta}\right)\nabla\cdot\int_{z_1}^{z_2}\boldsymbol{V}_0\Theta_0{\rm d}z \quad (18)$$

其中，J（,）表示 Jacobian， ${\rm Curl}\,\tau=\boldsymbol{k}\cdot{\rm Curl}\,\boldsymbol{\tau}$ ； ψ 为全流函数，它满足下列方程

$$\boldsymbol{Q}_0=\boldsymbol{k}\cdot\nabla\psi \quad (19)$$

借助于热力学方程（10）的积分表达式并注意到质量守恒和界面上表观热流量的连续性条件，方程（18）归结为零阶模型的位涡方程如下：

$$\left(\frac{Ro}{\beta}\right)J(\psi,\nabla^2\psi)+\frac{\partial\psi}{\partial x}=\left(\frac{Ek_{\rm V}}{\beta}\right)\left\{\frac{D\tau_{2*}}{V^*\nu_{2*}}{\rm Curl}\,\tau_2-\frac{D\tau_{1*}}{V^*\nu_{1*}}{\rm Curl}\,\tau_1\right\}+\left(\frac{Ek_{\rm H}}{\beta}\right)\nabla^4\psi+\left(\frac{\varepsilon}{\beta}\right)\frac{Ek_{\rm V}}{RoPr_{\rm V}}\left\{\frac{\Lambda_{1*}}{Gk_{1*}}\Lambda_1-\frac{\Lambda_{2*}}{Gk_{2*}}\Lambda_2\right\}+\left(\frac{\varepsilon}{\beta}\right)\frac{Ek_{\rm H}}{R_0Pr_{\rm H}}\nabla^2\int_{z_1}^{z_2}\Theta_0{\rm d}z \quad (20)$$

其中，表观热流通量

$$(\Lambda_2,\Lambda_1)=\left(-\left[\frac{\Lambda_{2*}}{Gk_{2*}}\right]^{-1}\left[k_{\rm V}\frac{\partial\Theta_0}{\partial z}\right]_{z=z_2},\ -\left[\frac{\Lambda_{1*}}{Gk_{1*}}\right]^{-1}\left[k_{\rm V}\frac{\partial\Theta_0}{\partial z}\right]_{z=z_1}\right) \quad (21)$$

$G=\Theta^*/D$，而 Λ_{1*}、k_{1*} 和 Λ_{2*}、k_{2*} 分别为下界面和上界面上表观热流通量和扩散系数之尺度。

下面考查如文献[1]中作者所提出的一个大洋风生–热盐环流的准两层模型。对于上层大洋，位涡方程（20）化为

$$\left(\frac{Ro}{\beta}\right)J(\psi,\nabla^2\psi)+\frac{\partial\psi}{\partial x}=\left(\frac{Ek_{\rm V}}{\beta}\right)\left(\frac{D\tau_{{\rm a}*}}{V^*\nu_{{\rm a}*}}\right){\rm Curl}\,\tau_{\rm a}+\left(\frac{Ek_{\rm H}}{\beta}\right)\nabla^4\psi+\left(\frac{\varepsilon}{\beta}\right)\left(\frac{Ek_{\rm V}}{RoPr_{\rm V}}\right)\left(\frac{\Lambda_{{\rm i}*}}{k_{{\rm i}*}G}\right)\Lambda_{\rm i}\left\{1-\frac{\Lambda_{{\rm a}*}/k_{{\rm a}*}}{\Lambda_{{\rm i}*}/k_{{\rm i}*}}\cdot\frac{\Lambda_{\rm a}}{\Lambda_{\rm i}}\right\}+\left(\frac{\varepsilon}{\beta}\right)\left(\frac{Ek_{\rm H}}{RoPr_{\rm H}}\right)\nabla^2\int_{(D_{上})}\Theta_0{\rm d}z \quad (22)$$

其中，右下角标有 a 的量表示大洋表面处的量，如 $\boldsymbol{\tau}_{\rm a}$ 为作用于洋面上的风应力，$\Lambda_{\rm a}$ 为通过洋面的表观热流通量，右下角标为 i 的量表示准两层大洋中间界面处的量；而右下角标*则表示为相应量之尺度。

若进一步引入底摩擦的线性律

$$\bar{\boldsymbol{\tau}}_{\rm b}=R\bar{\boldsymbol{Q}}_0 \tag{23}$$

其中 R 为湍阻尼系数。

式（23）的无因次形式为

$$\boldsymbol{\tau}_{\rm b}=\frac{RDV^*}{\tau_{\rm b*}}\boldsymbol{Q}_0 \tag{24}$$

其中，右下角标 b 表示在洋底的量，而*表示其相应之尺度。

注意到式（24），则在下层大洋中，位涡方程（20）化为

$$\left(\frac{Ro}{\beta}\right)J(\psi,\nabla^2\psi)+\frac{\partial\psi}{\partial x}=-\left(\frac{Ek_{\rm V}}{\beta}\right)\frac{R}{\nu_{0*}/D^2}\nabla^2\psi+\left(\frac{Ek_{\rm H}}{\beta}\right)\nabla^4\psi$$
$$-\left(\frac{\varepsilon}{\beta}\right)\left(\frac{Ek_{\rm V}}{RoPr_{\rm V}}\right)\left(\frac{\Lambda_{\rm i*}}{k_{\rm i*}G}\right)\Lambda_{\rm i}+\left(\frac{\varepsilon}{\beta}\right)\left(\frac{Ek_{\rm H}}{RoPr_{\rm H}}\right)\nabla^2\int_{(D_{\text{下}})}\Theta_0{\rm d}z \tag{25}$$

尺度分析和风旋度–热盐梯度方程式

通过尺度分析，就一般情况将获得若干重要的物理信息，并且导出最终的风旋度–热盐梯度方程式。

对于上层大洋环流：

（i）上面已经提到了，对于大洋风生–热盐环流来说，浮力项是最基本的力项，从而，由方程（3）或方程（9）可知，有

$$\frac{\varepsilon Ro}{Fr^2}=1 \tag{26}$$

引入上层洋流的典型数据 $g\sim10^3$，$f_0\sim10^{-4}$，$L\sim10^8$，$D\sim10^5$，$\varepsilon\sim10^{-3}$，则由方程（26）得出 $V^*\sim10$（cm/sec），这是上层大洋中洋流的正确量阶——密度差是产生风生–热盐洋流的基本源动力[8]。

（ii）为了考查洋面上风应力对于上层洋流作用的大小，我们来估算方程（22）中 $\text{Curl}\,\tau_{\rm a}$ 前面的无因次参数 $\left(\frac{Ek_{\rm V}}{\beta}\right)\left(\frac{D\tau_{\rm a*}}{V^*\nu_{\rm a*}}\right)$ 的值。若除了在（i）中已引入的上层洋流的典型数据外再补充以 $\tau_{a*}\sim1$，$\beta_0\sim10^{-13}$，则无因次参数 $\left(\frac{Ek_{\rm V}}{\beta}\right)\left(\frac{D\tau_{\rm a*}}{V^*\nu_{\rm a*}}\right)\sim\frac{\nu/\nu_{\rm a*}}{V^*}$。由此显然可见，若假设 $\nu_{\rm a*}=\nu$，则洋面上的风应力只能产生流速为 1（cm/sec）量阶之环流，若我们认为 $V^*\sim10$（cm/sec）为正确量阶，则风应力将不是大洋环流的主要强迫力——这一推论正是 A. C. Саркисян 所坚持的论点[8]：形成上层大洋环流的大尺度流涡的基本作用力是热力而不是动力（指风应力）。不过若假设 $\nu_{\rm a*}/\nu\sim0.1$，则风应力就产生流速

为正确量阶之环流了。这表明了，在 $\nu_{a*}/\nu \sim 0.1$ 之假设下，风应力对于上层洋流也是基本的强迫力,即上层大洋环流为风生–热盐环流——这一推论与 H. Stommel 早已提出的论点是一致的[9]而作者也曾附和过这一论点[1]。显然，一个如通常被采用的常涡动系数模型是符合前一论点的充分条件；但这常铅垂涡动系数模型无疑是值得商榷的。按照 K. F. Bowden，离海面向下某个距离处将应存在一个铅垂涡动系数的最大值[10]。这样，或许后一论点更站得住脚。我们下面的讨论将沿着上层洋流为风生–热盐环流模型这一途径进行下去。此时有

$$\left(\frac{Ek_{\rm V}}{\beta}\right)\left(\frac{D\tau_{a*}}{V^{*}\nu_{a*}}\right)=1 \tag{27}$$

（iii）如前所述，为了全面考查表观温度的湍扩散效应和对流效应，应有方程（4）或方程（10）中的无因次参数 $O(Ro)=O\left(\dfrac{Ek_{\rm V}}{Pr_{\rm V}},\ \dfrac{Ek_{\rm H}}{Pr_{\rm H}}\right)$，于是可令

$$\frac{Ek_{\rm V}}{RoPr_{\rm V}}=\frac{Ek_{\rm H}}{RoPr_{\rm H}}=1 \tag{28}$$

应指出，为此并保证诸场量的正确量阶，必须设想一个相当大的表观温度湍扩散系数 $k_{\rm H}\sim 10^{9}$。

（iv）为了表明这样一个准两层模型中通过跃层的表观热流通量对洋流的作用，必须令方程（22）中 $\Lambda_{\rm i}$ 前面的无因次参数为 1，再注意引入式（28），则应有

$$\frac{\Lambda_{i*}/k_{i*}}{G}=\frac{\beta}{\varepsilon} \tag{29}$$

由于在这个准两层模型中，我们以一个准不连续面代替了跃层，故跨过这一压缩了的跃层之表观温度梯度 Λ_{i*}/k_{i*} 将远大于平均梯度 G，式（29）正表明了这一论断。于是我们可以合理地假定 $\Lambda_{i*}/k_{i*}\sim(\beta/\varepsilon)/G$，对于上层洋流，有 $\beta/\varepsilon\sim 10^{2}$。

联立（26）、（27）和（28）诸式，我们立刻得出上层大洋环流中的流速、表观温度和斜压深度的尺度 V^{*}、Θ^{*} 和 D 如下：

$$\Theta^{*}=\frac{f_0}{g\alpha k^{2}L}\left(\frac{\tau_{a*}}{\beta_0 L}\right)^{3}\left(\frac{\nu}{\nu_{a*}}\right)^{3} \tag{30}$$

$$D=\left(\frac{\tau_{a*}f_0}{g\alpha\Theta^{*}\beta_0}\right)^{1/2}\left(\frac{\nu}{\nu_{a*}}\right)^{1/2} \tag{31}$$

$$V^{*}=\frac{\tau_{a*}}{\beta_0 DL}\left(\frac{\nu}{\nu_{a*}}\right) \tag{32}$$

再注意到前已给出的 $G=\Theta^{*}/D$，则“跃层强度”Λ_{i*}/k_{i*} 也由（29）式确定了。

给定了上层洋流的典型数据如前，则由（30）、（31）和（32）分别估出 $\alpha\Theta^* = \varepsilon \sim 10^{-3}$，$D \sim 10^5$（cm），$V^* \sim 10$（cm/sec）。其物理意义表明了流场、表观温度场、跃层和斜压深度取决于外界参数的作用和内部湍流机制，并作相互调整。

把式（29）～（32）代入方程（22）和方程（25）就分别导出了描述上层大洋环流和下层大洋环流所满足的风旋度–热盐梯度方程式如下：

上层大洋

$$\left(\frac{Ro}{\beta}\right)J(\psi,\nabla^2\psi)+\frac{\partial\psi}{\partial x}=\mathrm{Curl}\,\tau_{\mathrm{a}}+\left(1-\frac{\Lambda_{\mathrm{a}*}/k_{\mathrm{a}*}}{\Lambda_{\mathrm{i}*}/k_{\mathrm{i}*}}\cdot\frac{\Lambda_{\mathrm{a}}}{\Lambda_{\mathrm{i}}}\right)\Lambda_{\mathrm{i}}+\left(\frac{Ek_{\mathrm{H}}}{\beta}\right)\nabla^4\psi+\left(\frac{\varepsilon}{\beta}\right)\nabla^2\int_{(D_{上})}\Theta_0\mathrm{d}z \tag{33}$$

下层大洋

$$\left(\frac{Ro}{\beta}\right)J(\psi,\nabla^2\psi)+\frac{\partial\psi}{\partial x}=-\Lambda_{\mathrm{i}}+\left(\frac{Ek_H}{\beta}\right)\nabla^4\psi-\left(\frac{\gamma}{\beta}\right)\nabla^2\psi+\left(\frac{\varepsilon}{\beta}\right)\nabla^2\int_{(D_{下})}\Theta_0\mathrm{d}z \tag{34}$$

其中 $\gamma=\left(\dfrac{\nu}{\nu_{\mathrm{b}*}}\right)\dfrac{R}{f_0}$。

应指出，当不考虑相对涡度，底摩擦和表观温度的水平扩散效应时，方程（33）和方程（34）分别蜕化为文献[4]中的方程（45）和方程（46）；因此，这里所导出的上述风旋度–热盐梯度方程当为文献[1]中所导出的线性风旋度–热盐梯度方程向非线性方程之推广。另外，也应指出，当不考虑热盐效应时，方程（33）或方程（34）蜕化成正压大洋中风生环流的位涡方程[6]；因此，这里导出的方程也可看作正压海洋中位涡方程向斜压海洋中位涡方程的推广。

对于下层大洋可以进行相似的尺度分析，从而获得一些进一步的信息：

（v）把式（28）应用于下层大洋，并假定对于上、下层大洋来说 k 和 L 皆一样，则推出

$$\frac{V_{下}^*}{V_{上}^*}=\left(\frac{D_{上}}{D_{下}}\right)^2 \tag{35}$$

其中，右下角标“上”和“下”分别表示该场量是属于上层大洋或下层大洋。

应指出，利用式（32）和（31）已给出了 $V_{上}^* \sim 10$（cm/sec）和 $D_{上} \sim 10^5$（cm），若进一步采用 4 km 作为整个大洋的平均深度——在本模型中即为 $D_{上}+D_{下}=4\times10^5$（cm），则依式（35）有下层洋流的尺度 $V_{下}^* \sim 1$（cm/sec）——正确量阶。

（vi）把式（26）应用于下层洋流，且假定上、下大洋中 g 和 α 相同，则有 $\varepsilon_{下}D_{下}/V_{下}^*=\varepsilon_{上}D_{上}/V_{上}^*$；把式（35）代入，则有

$$\frac{\Theta_{下}^{*}}{\Theta_{上}^{*}}=\frac{\varepsilon_{下}}{\varepsilon_{上}}=\left(\frac{D_{上}}{D_{下}}\right)^{3} \tag{36}$$

利用式（30）和式（36）可以给出下层大洋中密度的相对改变$\varepsilon_{下}\sim(10^{-5}\sim10^{-4})$——这是一个正确的量阶。

进一步的尺度分析和行星边界层

首先估计$(A_{a*}/k_{a*})/(A_{i*}/k_{i*})$的量阶。由于在大洋表面附近有一个混合相当均匀的水层，故可以合理假定$(A_{a*}/k_{a*})_{max}\sim G$，因而$\left[(A_{a*}/k_{a*})/(A_{i*}/k_{i*})\right]_{max}\sim\varepsilon/\beta$。至于涉及到了湍运动中的经验系数$\nu_H$和$R$的估值则是模型中最不能确定的参数了，不妨姑且采用某一经验数值范围：$\nu_H\sim(10^{6}\sim10^{9})$和$R\sim(10^{-9}\sim10^{-7})$。把它们以及上面引进的典型数据代入式（30）~（32）、（35）和（36）及其有关的无因次参数，再假定$\nu/\nu_{b*}\sim10$，则有方程（33）和（34）中诸无因次参数的估值如下表（表1）。

表 1

	上层大洋	下层大洋
Ro/β	10^{-2}	10^{-3}
Ek_H/β	$10^{-5}\sim10^{-2}$	$10^{-5}\sim10^{-2}$
γ/β	/	$10^{-3}\sim10^{-1}$
ε/β	10^{-2}	$10^{-4}\sim10^{-3}$

由表1可见，对于大洋的广大的内部区域，方程（33）和方程（34）被简化成

上层大洋

$$\frac{\partial\psi}{\partial x}=\mathrm{Curl}\,\tau_a+A_i \tag{37}$$

下层大洋

$$\frac{\partial\psi}{\partial x}=-A_i \tag{38}$$

当不考虑热盐效应时，即$A_i=0$时，方程（37）蜕化为正压大洋中著名的Sverdrup风旋度方程。由于一个热盐梯度因子A_i在这里被附加，故称为风旋度–热盐梯度方程式。由此可见，方程（33）和方程（34）也可称为广义的风旋度–热盐梯度方程式。

为了讨论方便且不影响其实质，我们假定风应力场仅为$\boldsymbol{\tau}=\boldsymbol{i}\tau_x(y)$且仅由中纬度的西风带及其以南的信风所组成的一个理想化模型而热盐梯度因子$A_i=A_i(y)$采用模型如文献[1]中所采用的稳定层化模型；仅讨论北半球。

在上面引入的风应力场和热盐梯度场的作用下，方程（37）和方程（38）的通解很容易得出，但基于本文的目的局限于一般的研究，故暂不给出其解而仅指出：由于它们完全相似于Sverdrup风旋度方程的结构，并且我们没有引入任何能的耗散项，故前者将如后者一样不能得到一个唯一的解，亦即不能确定大洋环流是东向强化抑或西向强化。为此必须至少引入一个能耗项，尽管它可能是小项；这样，我们如正压海洋中的风生洋流一样将会得出东边界为一流线而西边界附近将必然存在一股狭窄的强

流——形成了行星边界层。描述它们的方程为

上层大洋

$$\left(\frac{\sqrt{Ro/\beta}}{\delta}\right)^2\left\{\frac{\partial\psi}{\partial\zeta}\frac{\partial^3\psi}{\partial\zeta^2\partial y}-\frac{\partial\psi}{\partial y}\frac{\partial^3\psi}{\partial\zeta^3}\right\}+\frac{\partial\psi}{\partial\zeta}$$
$$=\left(\frac{\sqrt[3]{Ek_{\mathrm{H}}/\beta}}{\delta}\right)^3\frac{\partial^4\psi}{\partial\zeta^4}+\left(\frac{\varepsilon/\beta}{\delta}\right)\frac{\partial^2}{\partial\zeta^2}\int_{(D_{上})}\Theta_0\mathrm{d}z \tag{39}$$

和下层大洋

$$\left(\frac{\sqrt{Ro/\beta}}{\delta}\right)^2\left\{\frac{\partial\psi}{\partial\zeta}\frac{\partial^3\psi}{\partial\zeta^2\partial y}-\frac{\partial\psi}{\partial y}\frac{\partial^3\psi}{\partial\zeta^3}\right\}+\frac{\partial\psi}{\partial\zeta}$$
$$=\left(\frac{\sqrt[3]{Ek_{\mathrm{H}}/\beta}}{\delta}\right)^3\frac{\partial^4\psi}{\partial\zeta^4}-\left(\frac{\gamma/\beta}{\delta}\right)\frac{\partial^2\psi}{\partial\zeta^2}+\left(\frac{\varepsilon/\beta}{\delta}\right)\frac{\partial^2}{\partial\zeta^2}\int_{(D_{下})}\Theta_0\mathrm{d}z \tag{40}$$

其中：边界层坐标$\zeta=(x-x_{\mathrm{w}})/\delta$，$x_{\mathrm{w}}$和$\delta$分别为西边界$x$之坐标和行星边界层的无因次厚度（$O(\delta)<1$），$\psi=\psi(\zeta,y)$为边界层中的流函数。显然，当不考虑热盐效应时，上面给出的行星边界层方程将蜕化为正压海洋中风生洋流行星边界层的微分方程式（参见文献[6]）。

方程（39）和（40）中无因次参数的量阶列于表 2 如下。

表 2

	上层大洋	下层大洋
$\sqrt{Ro/\beta}$	10^{-1}	3×10^{-3}
$\sqrt[3]{Ek_{\mathrm{H}}/\beta}$	$2\times10^{-2}\sim2\times10^{-1}$	$2\times10^{-2}\sim2\times10^{-1}$
γ/β	/	$10^{-3}\sim10^{-1}$
ε/β	10^{-2}	$10^{-4}\sim10^{-3}$

综合方程（37）~（40）和表 2 并对比相应的正压大洋中风生环流的情形，我们立刻看出一些有趣的物理图象和有益的结论如下：（i）在上层大洋中形成了一个东西不对称的、顺时针的风生–热盐性质的大流涡，在中区南北半部之间有一广阔而缓慢地流向极地的水量迁移——由于热盐效应，该流量要比纯风生流量更加丰富；在下层大洋中形成了一个纯热盐性质的逆时针方向的大流涡，在中区南北半部之间有一广阔而缓慢地流向赤道的水量迁移——失去了热盐效应，深层水——即下层大洋的环流将不存在。（ii）注意到东边界为一流线，在西边界附近则形成了行星边界层：在上层大洋中的西边界层强流由南向北且比纯风生西边界流来得丰富，而在下层大洋中的西边界强流则为伏于上层行星边界层下方的一股热盐逆流。（iii）在行星边界层方程中包含了一个热盐项$[(\varepsilon/\beta)/\delta]\dfrac{\partial^2}{\partial\zeta^2}\int_{(D)}\Theta_0\mathrm{d}z$，其存在使得行星边界层呈斜压性质。不过从表 2 最后一栏ε/β的量阶可见，这一斜压性估计为一次要效应，特别对于深层热盐逆流更是如此。如果略去该项所表达的斜压效应，则方程（39）和（40）将与正压大洋中行星边界层方程

完全重合，故此时可称之为行星边界层的正压模型。（iv）由表 2 又可看出，上层大洋的行星边界层中的无因次参数 $O(\sqrt{Ro/\beta}) \geqslant O(\sqrt[3]{Ek_{\mathrm{H}}/\beta})$，因而该边界层将主要是惯性性质的、至多是惯性–粘性边界层；其无因次宽度 $\delta \sim \sqrt{Ro/\beta}$，即行星边界层的宽度 $1 \sim \delta L \sim 100$（km）——正确量阶。（v）至于下层大洋中行星边界层的性质及其宽度的量阶很难确定，这是因为粘性系数的量值、甚至量阶都带有相当大的经验性，而惯性又相对降低至上述湍粘系数之量值的经验范围以内了，但有一点可以肯定，即与上层海洋中的行星边界层相比较，下层大洋中的行星边界层的惯性效应确实下降了。

最后，对于风生–热盐环流，特别是对其行星边界层的问题，将在以后的论文中去探讨。

参考文献

[1] 冯士筰, 1979, 大洋风生–热盐环流模型, 山东海洋学院学报, 9(2): 1–14.

[2] Munk, W. H., 1950, J. Meteorol., 7: 79–93.

[3] Stommel, H., 1948, Trans. Amer. Geophys. Union, 29: 202–206.

[4] Morgan, G. W., 1956, Tellus, 8: 301–320.

[5] Charney, J. G., 1955, Proc. Nat. Acad. Sci. Wash., 41: 731–740.

[6] Pcdlosky, J., 1979, Geophysical Fluid Dynamics, Springer-verlag N. Y. Inc. pp 624.

[7] 冯士筰, 1982, 论 f–和 β–坐标系, 山东海洋学院学报, 12(3): 1–10.

[8] Саркисян, А. С., 1977, Чнсленный Аналиz и Прогноz Морских Течений, Гидрометеонzдат Ленинград, pp 261.

[9] Stommel, H., 1958, The Gulf Stream, Cambridge Uni. Press and Uni. of California Press, pp 156.

[10] Bowden, K. F., 1962, Turbulence, The Sea, vol. I, Section VI, 802–825.

The Baroclinic Residual Circulation in Shallow Seas I. The Hydrodynamic Models*

Feng Shizuo, Xi Pangen and Zhang Shuzhen

Circulation in continental shelf seas, semi-enclosed shallow seas, gulfs or bays and tidal estuaries is to be driven by tides, storm or wind, density and open boundary forces, with different spatial and temporal scales. Tides, sometimes storm surges, also dominate the circulation in shallow seas, such as the Bohai Sea and the Yellow Sea. However, the smaller longer-term residual currents, of which the order of magnitude is typically smaller than those of tides and storm surges, play a key role in the understanding of the transport phenomena of long-term processes, for example in an ecological system. Therefore, the studies on the dynamics of residual circulation are of considerable interest in both practice and theory. In recent years, many investigations on residual currents have been made e.g.[1–14]. Nevertheless, studies in this field are, for the most part, to treat the barotropic two-dimensional vertically-integrated or transport hydrodynamic models. It should be pointed out that the baroclinic effect on residual circulation, or the density-driven component of residual currents, might be necessary, at least not trivial, in such water areas as the northern reach of San Francisco Bay[2], the Bay of Fundy and Gulf of Maine[4] or the Bohai Sea and the Yellow Sea, or generally, the East China Sea[15–17]. On the other hand, there is a need to estimate the three-dimensional residual current in some physical processes. For example, the estimation of sediment fluxes requires that the vertical velocity profile at each location be known because the concentration of the sediment in the water column varies with depth. In this paper, both the baroclinic dynamic models of residual circulation are proposed, one of which is a two-dimensional transport model and the other is a three-dimensional model with variable eddy viscosity.

As pointed out by Cheng *et al.*①, there are two different modeling approaches that have been proposed and utilized to derive residual currents. The first approach is equivalent to deducing residual currents from current-meter records using filtering techniques or time averages of time-series records to remove tidal variations[6, 12]. In the second approach, filters or time averages over several tidal cycles are applied to the hydrodynamic equations to generate the governing equations for residual circulation[5, 7]. Both the baroclinic models for

---

* Feng S, Xi P, Zhang S. 1984. The baroclinic residual circulation in shallow seas I. The hydrodynamic models. Chinese Journal of Oceanology and Limnology, 2(1): 49–60

① 原文注：Cheng R T, V Casulli, 1982. On Lagrangian residual currents with application, in south San Francisco Bay, California. (personal communication) residual circulation proposed in the present paper are derived just based on the second approach. In this paper, we put emphasis on Eulerian residual currents.

Formulation

Based upon the nondimensional dynamic problem for barotropic shallow seas presented in the reference [18], a nondimensional dynamic problem for baroclinic shallow seas is proposed as follows:

$$\nabla \cdot \boldsymbol{V} + \frac{\partial w}{\partial z} = 0$$

$$\rho\left(k_{\mathrm{i}}\frac{\partial}{\partial t} + k_{\mathrm{c}}\boldsymbol{e}_3 \times\right)\boldsymbol{V} + \rho k_{\mathrm{i}} k_{\mathrm{n}}\left(\boldsymbol{V}\cdot\nabla + w\frac{\partial}{\partial z}\right)\boldsymbol{V}$$

$$= -\nabla(\zeta - k_{\mathrm{a}}\zeta_{\mathrm{a}} - k_{\mathrm{T}}\zeta_{\mathrm{T}}) - k_{\rho}\nabla\int_z^0 \rho' \mathrm{d}z + k_{\nu}\frac{\partial}{\partial z}\left(\nu\frac{\partial \boldsymbol{V}}{\partial z}\right)$$

$$Ro\left(\boldsymbol{V}\cdot\nabla + w\frac{\partial}{\partial z}\right)\rho' = \frac{Ek}{Pr}\frac{\partial}{\partial z}\left(\gamma\frac{\partial \rho'}{\partial z}\right) \tag{1}$$

$z = \chi\zeta$:

$$w = \left(k_{\mathrm{r}}\frac{\partial}{\partial t} + \chi\boldsymbol{V}\cdot\nabla\right)\zeta$$

$$\nu\frac{\partial \boldsymbol{V}}{\partial z} = k_{\tau}\boldsymbol{\tau}_{\mathrm{a}}$$

$$\nu\frac{\partial \rho'}{\partial z} = k_{\tau}\varGamma_{\mathrm{a}}$$

$z = -h$:

$$\boldsymbol{V} = w = 0$$

$$\left(\nu\frac{\partial \boldsymbol{V}}{\partial z} = k_{\mathrm{b}}\boldsymbol{\tau}_{\mathrm{b}}\right)$$

along the shore boundary C_1 :

$$\boldsymbol{n}_0 \cdot \int_{-h}^{\chi\zeta} \boldsymbol{V}\mathrm{d}z = 0$$

along the open boundary C_2 :

$$\boldsymbol{n}_0 \cdot \int_{-h}^{\chi\zeta} \boldsymbol{V}\mathrm{d}z = Q$$

or

$$\zeta = \mathscr{I}$$

where

$$\chi = \zeta_0 / D,\ k_i = SrEu^{-1},\ k_{\rm n} = Sr^{-1},\ k_{\rm c} = Ro^{-1}Eu^{-1},\ k_{\rm a} = \zeta_{\rm a0} / \zeta_0$$

$$k_{\rm T} = \zeta_{\rm T0} / \zeta_0,\ k_{\rm p} = \varepsilon / \chi,\ k_{\rm v} = EkRo^{-1}Eu^{-1},\ k_{\rm r} = \chi Sr,\ k_\tau = \frac{\tau_{\rm a0} D}{\rho_0 \nu_0 V_0}$$

$$\varepsilon = \rho_0' / \rho_0,\ k_{\rm b} = \frac{\tau_{\rm b0} D}{\rho_0 \nu_0 V_0},\ k_\gamma = \frac{\Gamma_{\rm a0} D}{\rho_0' \gamma_0}$$

the Rossby number $Ro = \dfrac{V_0}{fL}$, the Ekman number $Ek = \dfrac{\nu_0}{fD^2}$,

the Euler number $Eu = \dfrac{g\zeta_0}{V_0^2}$, the Strouhal number $Sr = \dfrac{L}{V_0 T}$,

the Prandtl number $Pr = \dfrac{\nu_0}{\gamma_0}$,

in the above, $\nabla = \boldsymbol{e}_1 \dfrac{\partial}{\partial x} + \boldsymbol{e}_2 \dfrac{\partial}{\partial y}$, (x,y,z) constitute an f-coordinates at the right-hand side, the plane (x,y) coincides with the undisturbed water surface, z is positive upward, $(\boldsymbol{e}_1, \boldsymbol{e}_2, \boldsymbol{e}_3)$ denote three unit vectors of coordinates; t denotes time; $\boldsymbol{V} = \boldsymbol{e}_1 u + \boldsymbol{e}_2 v$, u and v are the components of the current in x and y direction respectively and w denotes the vertical component of current; ζ is the elevation measured from the undisturbed sea surface; the density $\rho = 1 + \varepsilon\rho'$, ρ' is the variation of density, ρ_0 the reference constant density; f is the Coriolis parameter; g is the gravitational acceleration; $\zeta_{\rm a}$ and $\zeta_{\rm T}$ represent the effects of atmosphere pressure and tide-generating force, respectively; $\boldsymbol{\tau}_{\rm a} = \boldsymbol{e}_1 \tau_{\rm ax} + \boldsymbol{e}_2 \tau_{\rm ay}$, $\tau_{\rm ax}$, $\tau_{\rm ay}$ denote the components of wind stress at the sea surface in the x, y directions respectively; $\Gamma_{\rm a}$ denotes the density flux through the sea surface, h denotes the depth; $\boldsymbol{\tau}_{\rm b} = \boldsymbol{e}_1 \tau_{\rm bx} + \boldsymbol{e}_2 \tau_{\rm by}$ denotes the bottom friction; $\boldsymbol{n}_0 = \boldsymbol{e}_1 \cos\alpha_x + \boldsymbol{e}_2 \cos\alpha_y$, $\cos\alpha_x$ and $\cos\alpha_y$ are the direction-cosines of the boundary normal; Q and $\mathscr{S}$ denote the normal volume transport and the water elevation along the open boundary, respectively; ν and γ denote the eddy viscosity and diffusion coefficient respectively; D and L denote the vertical and horizontal scale respectively; T denotes the time scale; V_0 denotes the characteristic horizontal current velocity, ζ_0 denotes the characteristic amplitude of the elevation, ρ_0' is the characteristic quantity of ρ'; ν_0 and γ_0 denote the characteristic eddy viscosity and diffusion coefficient, respectively; $\tau_{\rm a0}$, $\tau_{\rm b0}$, $\zeta_{\rm a0}$, $\zeta_{\rm T0}$ and $\Gamma_{\rm a0}$ denote the characteristic quantities of $\boldsymbol{\tau}_{\rm a}$, $\boldsymbol{\tau}_{\rm b}$, $\zeta_{\rm a}$, $\zeta_{\rm T}$ and $\Gamma_{\rm a}$ respectively.

A scale analysis is made and we take the typical scales in Bohai Sea dynamics as the characteristic quantities: D~2×10^3, L~4×10^7, g~10^3, f~10^{-4}, χ ~10^{-1}, $\tau_{\rm a0}$ ~10 (for storm surges) or $O(\tau_{\rm a0}) \leqslant 1$ (for longer-term residual currents), ε~5×10^{-3} (in summer) (c.g.s.). $\partial\rho' / \partial t = 0$ will be explained later.

In view of the fact that the effect of the eddy viscosity on the motion in shallow seas are essential, we set $k_{\mathrm{v}}=1$; and as mentioned above, tides dominate the motion, so $k_r=1$ and $k_{\mathrm{i}}=1$. Thus, we have $k_{\mathrm{i}}k_{\mathrm{n}}=k_{\mathrm{n}}=Sr^{-1}=\chi$ and $EkRo^{-1}=Eu=Sr=\chi^{-1}$, and then the nondimensional problem (1) is reduced to the following

$$\nabla\cdot\boldsymbol{V}+\frac{\partial w}{\partial z}=0$$

$$\rho\left(\frac{\partial}{\partial t}+k_{\mathrm{c}}\boldsymbol{e}_3\times\right)\boldsymbol{V}+\chi\left(\boldsymbol{V}\cdot\nabla+w\frac{\partial}{\partial z}\right)\boldsymbol{V}$$

$$=-\nabla(\zeta-k_{\mathrm{a}}\zeta_{\mathrm{a}}-k_{\mathrm{T}}\zeta_{\mathrm{T}})-k_{\rho}\nabla\int_z^0\rho'\mathrm{d}z+\frac{\partial}{\partial z}\left(\nu\frac{\partial\boldsymbol{V}}{\partial z}\right)$$

$$\left(\boldsymbol{V}\cdot\nabla+w\frac{\partial}{\partial z}\right)\rho'=\frac{1}{\chi Pr}\frac{\partial}{\partial z}\left(\gamma\frac{\partial\rho'}{\partial z}\right)\tag{2}$$

$z=\chi\zeta$:

$$w=\left(\frac{\partial}{\partial t}+\chi\boldsymbol{V}\cdot\nabla\right)\zeta$$

$$\nu\frac{\partial\boldsymbol{V}}{\partial z}=k_{\tau}\boldsymbol{\tau}_{\mathrm{a}}$$

$$\gamma\frac{\partial\rho'}{\partial z}=k_{\gamma}\Gamma_{\mathrm{a}}$$

$z=-h$:

$$\boldsymbol{V}=w=0$$

$$\left(\nu\frac{\partial\boldsymbol{V}}{\partial z}=k_{\mathrm{b}}\boldsymbol{\tau}_{\mathrm{b}}\right)$$

along the shore boundary C_1:

$$\boldsymbol{n}_0\cdot\int_{-h}^{\chi\zeta}\boldsymbol{V}\mathrm{d}z=0$$

along the open boundary C_2:

$$\boldsymbol{n}_0\cdot\int_{-h}^{\chi\zeta}\boldsymbol{V}\mathrm{d}z=Q$$

or

$$\zeta=\mathscr{S}$$

where $T=\dfrac{L}{\sqrt{gD}}$, $V_0=\chi\sqrt{gD}$, $\nu_0=\dfrac{D^2}{T}$ (or $\nu_0/D^2=\dfrac{1}{T}$), $k_{\mathrm{c}}=fT$.

The scale analysis is continued as follows: the significant influence of Coriolis force on the dynamics in such shallow seas as the Bohai Sea is derived from $k_c = fT = f\frac{L}{\sqrt{gD}} = \frac{4}{\sqrt{2}} \sim 1$. Whereas $k_\tau = \frac{\tau_{a0}/\rho_0}{(\nu_0/D^2)V_0 D} = \frac{\tau_{a0}}{\rho_0}\frac{k_c}{fD\chi\sqrt{gD}} = 1$ for storm surges while $O(k_\tau) < 1$ as the seasonal mean winds, $O(\overline{\tau}_{a0}) \leqslant 1$ are taken, the tides are severely affected by storm surges but almost not by the seasonal mean winds. However, since the wind-driven current $V_0 = \frac{\overline{\tau}_{a0}}{\rho_0}\frac{1}{(\nu_0/D^2)D} \sim 10$ as the seasonal mean wind of order of magnitude of $\overline{\tau}_{a0} \sim 1$ is taken, it is expected that the residual circulation is influenced by the seasonal mean wind. It is derived from $O(k_\rho) = O\left(\frac{\varepsilon}{\chi}\right) = 5\times 10^{-2}$ that the baroclinic force has the trivial effect on tides and storm surges, namely, there is the barotropic dynamics for tides and storm surges. Therefore the solutions of tides and/or storm surges can be obtained by using the dynamic problem (2) with $\rho' = 0$. It should be emphasized to point out that the density effect on the residual circulation might be really essential in view of the density-driven current $O(V_0) = O(\chi\sqrt{gD}) = O(\varepsilon k_\rho^{-1}\sqrt{gD}) = O(\varepsilon\sqrt{gD}) = 10$. Thus it can be seen that the temporal scale of baroclinity, or ρ', is the same as that of residual currents and, of course, much larger than that of tides and storm surges. Since the very slowly time-varying residual current field can be treated as a steady-state one[7], we suppose reasonably ρ' to be independent of time and $\partial\rho'/\partial t$ dropped. The terms with ($\chi\varepsilon$) have been neglected for $O(\chi) < 1$ and $O(\varepsilon) < 1$.

As pointed out by N. Heaps[5], studies in this field have, for the most part, employed the two-dimensional vertically-integrated, or transport, hydrodynamic equations. An attempt will be made in the next section to generalize simply the two-dimensional problem residual currents from the barotropic to the baroclinic. Here, the nondimensional volume-transport problem is derived from (2) as follows:

$$\frac{\partial\zeta}{\partial t} + \frac{\partial U}{\partial x} + \frac{\partial V}{\partial y} = 0$$

$$\frac{\partial \mathscr{U}}{\partial t} + \chi\left\{\frac{\partial}{\partial x}\left(\frac{U^2}{h+\chi\zeta}\right) + \frac{\partial}{\partial y}\left(\frac{UV}{h+\chi\zeta}\right)\right\} - k_c\mathscr{V}$$

$$= -(h+\chi\zeta) + \frac{\partial}{\partial x}(\zeta - k_a\zeta_a - k_T\zeta_T) - k_\rho\int_{-h}^{\chi\zeta}\frac{\partial}{\partial x}\int_z^0 \rho'\mathrm{d}z'\mathrm{d}z + k_\tau\tau_{ax} - k_b\tau_{bx}$$

$$\frac{\partial \mathscr{V}}{\partial t} + \chi\left\{\frac{\partial}{\partial x}\left(\frac{UV}{h+\chi\zeta}\right) + \frac{\partial}{\partial y}\left(\frac{V^2}{h+\chi\zeta}\right)\right\} + k_c\mathscr{U}$$

$$= -(h+\chi\zeta)\frac{\partial}{\partial y}(\zeta - k_a\zeta_a - k_T\zeta_T) - k_\rho\int_{-h}^{\chi\zeta}\frac{\partial}{\partial y}\int_z^0 \rho'\mathrm{d}z'\mathrm{d}z + k_\tau\tau_{ay} - k_b\tau_{by}$$

$$\frac{\partial \zeta}{\partial t}+\frac{\partial \mathscr{U}}{\partial x}+\frac{\partial \mathscr{V}'}{\partial y}=\frac{k_{\mathrm{r}}\Gamma_{\mathrm{a}}}{\chi Pr} \tag{3}$$

where

$$\boldsymbol{e}_1 U+\boldsymbol{e}_2 V=\int_{-h}^{\chi\zeta}\boldsymbol{V}\mathrm{d}z\text{, the volume-transport,}$$

$$\boldsymbol{e}_1\mathscr{U}+\boldsymbol{e}_2\mathscr{V}=\int_{-h}^{\chi\zeta}\rho\boldsymbol{V}\mathrm{d}z=\int_{-h}^{\chi\zeta}(1+\varepsilon\rho')\boldsymbol{V}\mathrm{d}z=(\boldsymbol{e}_1 U+\boldsymbol{e}_2 V)+\varepsilon(\boldsymbol{e}_1\mathscr{U}'+\boldsymbol{e}_2\mathscr{V}')\text{, the mass-transport,}$$

$$\boldsymbol{e}_1\mathscr{U}'+\boldsymbol{e}_2\mathscr{V}'=\int_{-h}^{\chi\zeta}\rho'\boldsymbol{V}\mathrm{d}z$$

The scales of (U, V) and $(\mathscr{U},\mathscr{V})$ are $(V_0 D)$ and $(\rho_0 V_0 D)$ respectively; the scale of $(\mathscr{U}',\mathscr{V}')$ is $(\rho_0' V_0 D)$.

The bottom stress is generally assumed to be a quadratic law or a linear law; here we choose the former only;

$$\boldsymbol{\tau}_b=\mathscr{K}\frac{(U^2+V^2)^{\frac{1}{2}}}{(h+\chi\zeta)^2}(\boldsymbol{e}_1 U+\boldsymbol{e}_2 V) \tag{4}$$

where

$$\mathscr{K}=\frac{\rho_0 V_0^2}{\tau_{\mathrm{b}0}}k$$

k is a frictional coefficient of order 10^{-3}.

In the equations of motion, the horizontal dispersive stresses derived due to the velocity shear have been neglected because these terms are small compared with the bottom friction for shallow seas.

A Two-dimensional Transport Model

Noting that the residual motion is a nonlinear phenomenon and χ is a small nonlinear parameter, it is expected that the longer-term residual currents have the order of magnitude of χ [19]. We suppose that the total motion (U,V,ζ) is separated into a mainly tidal part $(\tilde{U},\tilde{V},\tilde{\zeta})$ and a residual part $(\bar{U},\bar{V},\bar{\zeta})$, where

$$(U,V,\zeta)=(\tilde{U},\tilde{V},\tilde{\zeta})+\chi(\bar{U},\bar{V},\bar{\zeta}) \tag{5}$$

$(\bar{U},\bar{V},\bar{\zeta})$ to be independent of t.

An operator "∧" may take the form of a time average:

$$(\hat{U},\hat{V},\hat{\zeta})=\frac{T}{T_0}\int_{-T_0/2T}^{T_0/2T}(U,V,\zeta)\mathrm{d}t \tag{6}$$

which filters out the tidal motion and storm surges to a greater or lesser extent according to the length of the averaging period, T_0, and its commensurability with tidal periods[5, 7]. Noting $(\hat{U},\hat{V},\hat{\zeta})=(\hat{\tilde{U}},\hat{\tilde{V}},\hat{\tilde{\zeta}})+\chi(\hat{\bar{U}},\hat{\bar{V}},\hat{\bar{\zeta}})=\chi(\bar{U},\bar{V},\bar{\zeta})$, the nondimensional problem (3) will be translated into the problem for the residual currents.

Noting $\boldsymbol{\tau}_\mathrm{a}=\tilde{\boldsymbol{\tau}}_\mathrm{a}+\delta_1\bar{\boldsymbol{\tau}}_\mathrm{a}\left(\delta_1=\dfrac{\bar{\tau}_{\mathrm{a}0}}{\tau_{\mathrm{a}0}}\text{ and }O(\delta_1)<1\right)$, $\zeta_\mathrm{T}=\tilde{\zeta}_\mathrm{T}$, $\zeta_\mathrm{a}=\tilde{\zeta}_\mathrm{a}+\delta_2\bar{\zeta}_\mathrm{a}$ ($\delta_2=\bar{\zeta}_{\mathrm{a}0}/\zeta_{\mathrm{a}0}$ and $O(\delta_2)<1$) and

$$\widehat{\chi\left\{\frac{\partial}{\partial x}\left(\frac{U^2}{h+\chi\zeta}\right)+\frac{\partial}{\partial y}\left(\frac{UV}{h+\chi\zeta}\right)\right\}}=\chi\left\{\frac{\partial}{\partial x}\left(\frac{\hat{\tilde{U}}^2}{h}\right)+\frac{\partial}{\partial y}\left(\frac{\widehat{\tilde{U}\tilde{V}}}{h}\right)\right\}+O(\chi^2)$$

$$\widehat{\chi\left\{\frac{\partial}{\partial x}\left(\frac{UV}{h+\chi\zeta}\right)+\frac{\partial}{\partial y}\left(\frac{V^2}{h+\chi\zeta}\right)\right\}}=\chi\left\{\frac{\partial}{\partial x}\left(\frac{\widehat{\tilde{U}\tilde{V}}}{h}\right)+\frac{\partial}{\partial y}\left(\frac{\hat{\tilde{V}}^2}{h}\right)\right\}+O(\chi^2)$$

$$\widehat{-\chi_\rho\int_{-h}^{\chi\zeta}\nabla\int_z^0\rho'\mathrm{d}z'\mathrm{d}z}=-\frac{\varepsilon}{\chi}\int_{-h}^0\nabla\int_z^0\rho'\mathrm{d}z'\mathrm{d}z+O(\chi\varepsilon)$$

$$\widehat{-(h+\chi\zeta)\nabla(\zeta-k_\mathrm{a}\zeta_\mathrm{a}-k_\mathrm{T}\zeta_\mathrm{T})}=-\chi h\nabla\left(\bar{\zeta}-k_\mathrm{a}\frac{\delta_2}{\chi}\bar{\zeta}_\mathrm{a}\right)$$
$$-\widehat{\chi\tilde{\zeta}\nabla\tilde{\zeta}}+\widehat{\chi\tilde{\zeta}\nabla k_\mathrm{a}\tilde{\zeta}_\mathrm{a}}+\widehat{\chi\tilde{\zeta}\nabla k_\mathrm{T}\tilde{\zeta}_\mathrm{T}}+O(\chi^2)$$

$$-k_\mathrm{b}\hat{\boldsymbol{\tau}}_\mathrm{b}\propto-\chi k_\mathrm{b}\mathscr{K}(\boldsymbol{e}_1\bar{U}+\boldsymbol{e}_2\bar{V})\frac{\widehat{(\tilde{U}^2+\tilde{V}^2)^{\frac{1}{2}}}}{h^2}\ \text{[1, 15]}$$

and neglecting $O(\chi^2)$, we have the problem for the residual currents

$$\frac{\partial\bar{U}}{\partial x}+\frac{\partial\bar{V}}{\partial y}=0$$

$$-k_\mathrm{c}\bar{\mathscr{V}}=-h\frac{\partial}{\partial x}\left(\bar{\zeta}-k_\mathrm{a}\frac{\delta_2}{\chi}\bar{\zeta}_\mathrm{a}\right)+\tau_x'+\left(k_\tau\frac{\delta_1}{\chi}\right)\bar{\tau}_{\mathrm{a}x}-(\varepsilon/\chi^2)\int_{-h}^0\frac{\partial}{\partial x}\int_z^0\rho'\mathrm{d}z'\mathrm{d}z-\mathscr{R}\bar{U}$$

$$k_\mathrm{c}\bar{\mathscr{U}}=-h\frac{\partial}{\partial y}\left(\bar{\zeta}-k_\mathrm{a}\frac{\delta_2}{\chi}\bar{\zeta}_\mathrm{a}\right)+\tau_y'+\left(k_\tau\frac{\delta_1}{\chi}\right)\bar{\tau}_{\mathrm{a}y}-(\varepsilon/\chi^2)\int_{-h}^0\frac{\partial}{\partial y}\int_z^0\rho'\mathrm{d}z'\mathrm{d}z-\mathscr{R}\bar{V}\tag{7}$$

$$\frac{\partial\bar{\mathscr{U}}'}{\partial x}+\frac{\partial\bar{\mathscr{V}}'}{\partial y}=\frac{k_\gamma}{\chi^2 Pr}\Gamma_\mathrm{a}$$

where $\mathscr{R}=\dfrac{\mathscr{C}}{h}$, $\mathscr{C}=k_\mathrm{b}\mathscr{K}\dfrac{\widehat{(\tilde{U}^2+\tilde{V}^2)^{\frac{1}{2}}}}{h}$

$$\tau_x'=-\left\{\widehat{\tilde{\zeta}\frac{\partial\tilde{\zeta}}{\partial x}}-\widehat{\tilde{\zeta}\frac{\partial}{\partial x}(k_\mathrm{a}\tilde{\zeta}_\mathrm{a})}-\widehat{\tilde{\zeta}\frac{\partial}{\partial x}(k_\mathrm{T}\tilde{\zeta}_\mathrm{T})}+\frac{\partial}{\partial x}\left(\frac{\hat{\tilde{U}}^2}{h}\right)+\frac{\partial}{\partial y}\left(\frac{\widehat{\tilde{U}\tilde{V}}}{h}\right)\right\}$$

$$\tau'_y = -\left\{\widehat{\tilde{\zeta}\frac{\partial\tilde{\zeta}}{\partial y}} - \overline{\tilde{\zeta}\frac{\partial}{\partial y}(k_a\tilde{\zeta}_a)} - \overline{\tilde{\zeta}\frac{\partial}{\partial y}(k_T\tilde{\zeta}_T)} + \frac{\partial}{\partial x}\left(\frac{\widehat{\tilde{U}\tilde{V}}}{h}\right) + \frac{\partial}{\partial y}\left(\frac{\hat{\tilde{V}}^2}{h}\right)\right\}$$

(τ'_x, τ'_y) have been called "tidal stresses" [7].

In shallow seas, the effect of bottom stress on the motion is essential, or $O(\mathscr{R}) = O(\mathscr{C})$, $O(k_b, \mathscr{K}) = 1$; thus, k~0.5 × 10^{-3}, which is a correct order of magnitude of k.

Introducing the stream function Ψ (8)

$$\frac{\partial\Psi}{\partial x} = \bar{V}, \quad \frac{\partial\Psi}{\partial y} = -\bar{U} \tag{8}$$

and eliminating $\bar{\zeta}$, the problem (7) reduces to the equation satisfied by Ψ as follows

$$\nabla\cdot\left(\frac{\mathscr{C}}{h}\nabla\Psi\right) - \frac{\mathscr{C}}{h^2}\nabla h\cdot\nabla\Psi - \frac{k_c}{h}\nabla h\cdot \boldsymbol{e}_3\times\nabla\Psi$$
$$= \boldsymbol{e}_3\cdot\nabla\times\left(\boldsymbol{\tau}' + k_\tau\frac{\delta_1}{\chi}\bar{\boldsymbol{\tau}}_a\right) + \frac{1}{h}\nabla h\cdot\boldsymbol{e}_3\times\left(\boldsymbol{\tau}' + k_\tau\frac{\delta_1}{\chi}\bar{\boldsymbol{\tau}}_a\right) - \mathscr{T}\Gamma_a + \mathscr{D} \tag{9}$$

where $\boldsymbol{\tau}' = \boldsymbol{e}_1\tau'_x + \boldsymbol{e}_2\tau'_y$, $\mathscr{T} = \left(\frac{k_c k_\gamma}{Pr}\right)\frac{\varepsilon}{\chi^2}$

$$\mathscr{D} = -\left(\frac{\varepsilon}{\chi^2}\right)h\left\{\frac{\partial}{\partial x}\left(\frac{1}{h}\int_{-h}^{0}\int_{z}^{0}\frac{\partial\rho'}{\partial y}\mathrm{d}z'\mathrm{d}z\right) - \frac{\partial}{\partial y}\left(\frac{1}{h}\int_{-h}^{0}\int_{z}^{0}\frac{\partial\rho'}{\partial x}\mathrm{d}z'\mathrm{d}z\right)\right\}$$

The lateral boundary conditions of the problem (2) reduces to (10) and (11):

$$C_1:\ \boldsymbol{n}_0\cdot(\boldsymbol{e}_1\bar{U} + \boldsymbol{e}_2\bar{V}) = 0 \tag{10}$$

$$C_2:\ \boldsymbol{n}_0\cdot(\boldsymbol{e}_1\bar{U} + \boldsymbol{e}_2\bar{V}) = \bar{Q} \tag{11}$$

or

$$\bar{\zeta} = \overline{\mathscr{S}}$$

here, we have assumed $(Q, \overline{\mathscr{S}}) = (\tilde{Q}, \tilde{\mathscr{S}}) + \chi(\bar{Q}, \overline{\mathscr{S}})$.

The equation (9) shows that (i) the "tidal stress" is an essential force of order of magnitude of 1, (ii) the wind stress has the order of magnitude of δ_1 / k in view of $O(k_\tau) = 1$, and (iii) both the thermohalinic forces are the same in order of magnitude, $O(\varepsilon/\kappa^2)$, since $Pr = 1$ and $O(k_\Gamma) = 1$, $O(k_c) = 1$.

The problem thus reduces to a single elliptic equation for the stream function of residual currents (9) and the boundary conditions (10) and (11).

A Three-dimensional Model

To study the three-dimensional residual circulation over seasonal time scales, the dynamic problem (2) must be directly time-averaged to remove variations of tidal period or shorter such as storm surges. As mentioned above, we may express each dependent variable in terms of a slowly-varying or steady-state residual part and a tidal part

$$(\boldsymbol{V},w,\zeta)=(\tilde{\boldsymbol{V}},\tilde{w},\tilde{\zeta})+\chi(\bar{\boldsymbol{V}},\bar{w},\bar{\zeta}) \tag{12}$$

$(\bar{\boldsymbol{V}},\bar{\zeta})$ assumed to be independent of t in this paper.

An operator "$\wedge$" may take the form of a time average:

$$(\hat{\boldsymbol{V}},\hat{w},\hat{\zeta})=\frac{T}{T_0}\int_{-T_0/2T}^{T_0/2T}(\boldsymbol{V},w,\zeta)\mathrm{d}t \tag{13}$$

with $(\hat{\tilde{\boldsymbol{V}}},\hat{\tilde{w}},\hat{\tilde{\zeta}})=0$ and $(\hat{\bar{\boldsymbol{V}}},\hat{\bar{w}},\hat{\bar{\zeta}})=(\boldsymbol{V},\bar{w},\hat{\bar{\zeta}})$. Thus, setting $\rho=1$ and dropping the diffusion equation, the time-averaged nondimensional problem (2) become

$$\nabla\cdot\bar{\boldsymbol{V}}+\frac{\partial\bar{w}}{\partial z}=0$$

$$\frac{\partial}{\partial z}\left(\nu\frac{\partial\bar{\boldsymbol{V}}}{\partial z}\right)-k_c\boldsymbol{e}_3\times\bar{\boldsymbol{V}}=\nabla\bar{\zeta}+\boldsymbol{\Pi}$$

$z=0$:

$$\bar{w}=\Theta \tag{14}$$

$$\nu\frac{\partial\bar{\boldsymbol{V}}}{\partial z}=\boldsymbol{\tau}$$

$z=-h$:

$$\bar{\boldsymbol{V}}=\bar{w}=0$$

along the shore boundary C_1:

$$\boldsymbol{n}_0\cdot\left(\int_{-h}^{0}\bar{\boldsymbol{V}}\mathrm{d}z+\boldsymbol{q}\right)=0$$

along the open boundary C_2:

$$\boldsymbol{n}_0\cdot\left(\int_{-h}^{0}\bar{\boldsymbol{V}}\mathrm{d}z+\boldsymbol{q}\right)=\bar{Q}$$

or

$$\bar{\zeta}=\bar{\mathscr{T}}$$

where

$$\boldsymbol{\Pi}=-\nabla\left(k_a\frac{\delta_2}{\chi}\bar{\zeta}_a\right)+\frac{\varepsilon}{\chi_2}\nabla\int_z^0\rho'\mathrm{d}z+\widehat{\left(\tilde{\boldsymbol{V}}\cdot\nabla\tilde{\boldsymbol{V}}+\tilde{w}\frac{\partial\tilde{\boldsymbol{V}}}{\partial z}\right)}$$

$$\Theta=\nabla\cdot(\widehat{\tilde{\boldsymbol{V}}\tilde{\zeta}}),\quad \boldsymbol{\tau}=k_\tau\frac{\delta_1}{\chi}\bar{\boldsymbol{\tau}}_a+\widehat{\left[-\frac{\partial}{\partial z}\left(\nu\frac{\partial\tilde{\boldsymbol{V}}}{\partial z}\right)\tilde{\zeta}\right]}$$

$$\boldsymbol{q}=\widehat{\tilde{\zeta}\tilde{\boldsymbol{V}}}$$

The nondimensional problem (14) is in the same form as the nondimensional problem (2) derived in the reference [18]. Thus, a Sturm-Liouville system proposed in the reference [18] can be used here and is written as follows

$$\left[\frac{\partial}{\partial z}\left(\nu\frac{\partial}{\partial z}\right)+\lambda\right]F(z)=0$$
$$\left(\frac{\partial F}{\partial z}\right)_{z=0}=0 \tag{15}$$
$$(F)_{z=-h}=0$$

Suppose $F = F(x,y,z)$ and λ being independent of z, and for each (x,y), let the ascending eigenvalues and corresponding eigenfunctions derived from (15) be denoted by

$$\lambda=\lambda_n,\quad F=F_n\qquad (n=1,2,\cdots)$$

Thus, we have

$$\bar{V}=\sum_{n=1}^{\infty}G_n\mathscr{U}_nF_n+\frac{\boldsymbol{\tau}+(h+z)}{\nu_0}\quad(\nu_0=\nu_{z=0}) \tag{16}$$

$$\bar{w}=-\sum_{n=1}^{\infty}\nabla\cdot(G_n\mathscr{U}_n\mathscr{H}_n)-\nabla\cdot\frac{\boldsymbol{\tau}(h+z)^2}{2\nu_0} \tag{17}$$

where

$$\mathscr{U}_n=(\beta_n^2+k_c^2)^{-1}\{(-H_n\beta_n+H_nk_c\boldsymbol{e}_3\times)\nabla\zeta+(\beta_n-k_c\boldsymbol{e}_3\times)\boldsymbol{\tau}_n-(\beta_n-k_c\boldsymbol{e}_3\times)\boldsymbol{\Phi}_n\}$$

$$G_n=\left(\int_{-h}^{0}F_n^2\mathrm{d}z\right)^{-1},\quad \mathscr{H}_n=\int_{-h}^{z}F_n\mathrm{d}z,\quad H_n=\int_{-h}^{0}F_n\mathrm{d}z$$

$$\beta_n=\lambda_n,\quad \boldsymbol{\Phi}_n=\int_{-h}^{0}\boldsymbol{\Pi}F_n\mathrm{d}z$$

$$\boldsymbol{\tau}_n=\frac{\boldsymbol{\tau}}{\nu_0}\left\{\int_{-h}^{0}F_n\frac{\partial\nu}{\partial z}\mathrm{d}z+(\times k_c\boldsymbol{e}_3)\left(hH_n-\int_{-h}^{0}\mathscr{H}_n\mathrm{d}z\right)\right\}$$

and

$$\left[\mathscr{A}\nabla^2+\left(\frac{\partial\mathscr{A}}{\partial x}-\frac{\partial\mathscr{B}}{\partial y}\right)\frac{\partial}{\partial x}+\left(\frac{\partial\mathscr{A}}{\partial y}+\frac{\partial\mathscr{B}}{\partial x}\right)\frac{\partial}{\partial y}\right]\zeta=\Theta+\mathscr{F}+\nabla\cdot\frac{h\boldsymbol{\tau}}{2\nu_0} \tag{18}$$

along the shore boundary C_1:

$$Q_{n0}=-q_{n0}-\frac{h^2}{2\nu_0}\tau_{n0}$$

along the open boundary C_2:

$$Q_{n0}=\bar{Q}-q_{n0}-\frac{h^2}{2\nu_0}\tau_{n0}$$

or

$$\bar{\zeta} = \mathscr{G}\bar{\mathscr{T}}$$

where

$$\mathscr{F} = -\sum_{n=1}^{\infty}\{\boldsymbol{e}_3 \cdot \nabla \times (B_n \boldsymbol{\Phi}_n) + \nabla \cdot (A_n \boldsymbol{\Phi}_n) - \boldsymbol{e}_3 \cdot \nabla \times (B_n \boldsymbol{\tau}_n) - \nabla \cdot (A_n \boldsymbol{\tau}_n)\}$$

$$Q_{n0} = -\mathscr{A}\frac{\partial \bar{\zeta}}{\partial n_0} + \mathscr{B}\frac{\partial \bar{\zeta}}{\partial s_0} - \sum_{n=1}^{\infty}\{A_n \Phi_{nn_0} - B_n \Phi_{ns_0} - A_n \tau_{nn_0} + B_n \tau_{ns_0}\}$$

$$A_n = \frac{H_n G_n \beta_n}{\beta_n^2 + k_c^2}, \quad B_n = \frac{H_n G_n k_c}{\beta_n^2 + k_c^2}, \quad \mathscr{A} = \sum_{n=1}^{\infty} A_n H_n, \quad \mathscr{B} = \sum_{n=1}^{\infty} B_n H_n$$

$$\frac{\partial \bar{\zeta}}{\partial n_0} = \frac{\partial \bar{\zeta}}{\partial x}\cos\alpha_x + \frac{\partial \bar{\zeta}}{\partial y}\cos\alpha_y, \quad \frac{\partial \bar{\zeta}}{\partial s_0} = -\frac{\partial \bar{\zeta}}{\partial y}\cos\alpha_x + \frac{\partial \bar{\zeta}}{\partial x}\cos\alpha_y$$

$$\Phi_{nn_0} = \Phi_{nx}\cos\alpha_x + \Phi_{ny}\cos\alpha_y, \quad \Phi_{ns_0} = -\Phi_{ny}\cos\alpha_x + \Phi_{nx}\cos\alpha_y$$

$$\tau_{nn_0} = \tau_{nx}\cos\alpha_x + \tau_{ny}\cos\alpha_y, \quad \tau_{ns_0} = -\tau_{ny}\cos\alpha_x + \tau_{nx}\cos\alpha_y$$

$$\tau_{n_0} = \tau_x \cos\alpha_x + \tau_y \cos\alpha_y, \quad q_{n_0} = q_x \cos\alpha_x + q_y \cos\alpha_y$$

here, s_0 denotes the tangent at the boundary which forms a right-hand Cartesian coordinate system with the normal n_0.

$$\nabla^2 = \frac{\partial^2}{\partial x^2} + \frac{\partial^2}{\partial y^2}$$

The nondimensional problem (14) now reduces to the problem of $\bar{\zeta}$ (18) and the expression for $\bar{V}$ (16), where the eigenfunction F_n is derived from the Sturm-Liouville system (15) on the finite closed interval $[-h, 0]$, as well as the expression for $\bar{w}$ (17).

Postscript

Based on a scale analysis for the currents in shallow seas, a nondimensional hydrodynamic problem for baroclinic shallow seas is proposed. A two-dimensional transport model and a three-dimensional model for baroclinic residual currents are developed respectively. The former is a direct generalization from the barotropic model of residual circulation presented by Nihoul and Randy (1975)[7] and Heaps (1978)[5] to the baroclinic one while in the latter the same Sturm-Liouville system as that proposed in the reference [18] is used, which can play a key role in the mathematical treatment for this model. Thus, no the mathematical difficulty will be added to the models proposed for computation.

References

[1] Cheng, R. T. & R. A. Walters, 1981. Modeling of Estuarine Hydrodynamics and Field Data Require-

ments, Finite Elements in Fluids, Vol. IV, Wiley
[2] Cheng, R. T., 1981. Modeling of Tidal and Residual Circulation in San Francisco Bay, California, Seminar on 2D-Flows, HEC Rept., U. S. Army Corps of Eng. Davis., Calif., pp. 1–14
[3] Chen Shijun *et al.*, 1982. Numerical modeling of circulation and pollutant dispersion in Jiaozhou Bay (II). J. of Shandong College of Oceanology 12(4): 1–12 (in Chinese with English abstract)
[4] Greenberg, D. A., 1983. Modelling the Mean Barotropic Circulation in the Bay of Fundy and Gulf of Maine. J. Phys. Oceanogr. 13(5): 886–906
[5] Heaps, N. S., 1978. Linearized Vertically-Integrated Equations for Residual Circulation in Coastal Seas. Deutsche Hydrographische Z. 31: 147–169
[6] Maier-Reimer, E., 1977. Residual circulation in the North Sea due to the M_2-tide and mean annual wind stress. Deutsche Hydrographische Z. 30: 69–80
[7] Nihoul, J. C. J., 1975. Modelling of Marine Systems, Part I, Ch. 2, ESPC N. Y., pp. 41–66
[8] Tee, K. T., 1976. Tide-induced residual current, a 2D nonlinear tidal model. J. Mar. Res. 34: 603-628
[9] Tee, K. T., 1977. Tide-induced residual current—varification of a numerical model. J. Phys. Oceanogr. 7: 396–402
[10] Tee, K. T., 1980. The structure of three-dimensional tide-induced current, Part II: Residual currents. J. Phys. Oceanogr. 10(12): 2035–2057
[11] Walters, R. A. & Cheng, R. T., 1979. A two-dimensional hydrodynamic model of a tidal estuary. Advances in Water Resources. 2: 177–184
[12] Wang Huatong *et al.*, 1980. Numerical Modeling of Circulation and Pollutant Dispersion in Jiaozhou Bay (I). J. of Shandong College of Oceanology. 10(1): 26–63 (in Chinese with English abstract)
[13] Yu Guangyao *et al.*, 1983. Numerical modeling of circulation and pollutant dispersion in Jiaozhuo Bay (III). J. of Shandong College of Oceanology. 13(1): 1–11 (in Chinese with English abstract)
[14] Zimmerman, J. T. F., 1979. On the Euler-Lagrange transformation and the stokes drift in the presence of oscillatory and residual currents. Deep Sea Research. 26A: 505–520
[15] Feng Shizuo, Zhang Shuzhen & Xi Pangen, 1981. Mathematical modeling of circulation in the East China Sea (II). J. of Shandong College of Oceanology. 11(2): 8–26 (in Chinese with English abstract)
[16] Xi Pangen, Zhang Shuzhen & Feng Shizuo, 1980. Mathematical modeling of circulation in the East China Sea (I). J. of Shandong College of Oceanology. 10(3): 13–25 (in Chinese with English abstract)
[17] Zhang Shuzhen, Xi Pangen & Feng Shizuo, 1984. Numerical modeling of the steady circulation in the Bohai Sea. J. of Shandong College of Oceanology. 14(2): 12–19 (in Chinese with English abstract)
[18] Feng Shizuo, 1984. A Three-dimensional Nonlinear Hydrodynamic Model with Variable Eddy Viscosity in Shallow Seas. Chin. J. Oceanol. Limnol. 2(2): 177–187 (English edition)
[19] Feng Shih-zao (Feng Shizuo), 1977. A Three-dimensional Nonlinear Model of Tides. Scientia Sinica. 20 (4): 436–446

A Three-dimensional Nonlinear Hydrodynamic Model with Variable Eddy Viscosity in Shallow Seas*

Feng Shizuo

Introduction

The investigation on shallow sea dynamics is of considerable interest not only in theory but also in practice. Particularly, since our country faces vast shallow seas, the study aiming at establishing some dynamical relations for our shallow seas becomes essential. We developed a three-dimensional nonlinear model of tides[1] and an ultra-shallow water storm surge model[2]. Based on the models, the preliminary numerical experiments on the shallow water tides, M_4 and MS_4[3] and on the wind surges in the Bohai Sea[4, 5] have been made, respectively. However, these models have physically a common weakpoint, i.e., the eddy viscosity contained in the models is assumed to be a constant, or, at most, a function of horizontal coordinates independent of the vertical coordinate. In order to improve these shallow water models, an ultra-shallow water storm surge model with variable eddy viscosity[6] and a tidal three-dimensional nonlinear model with variable eddy viscosity[7] have been proposed. It is a pity that both the improved models just mentioned are not satisfactory yet: the zeroth-order model for the former is unable to contain the Coriolis force while the eddy viscosity contained in the zeroth-order model of the latter is unable to be the function of vertical coordinate. In addition, the three-dimensional shallow sea circulation with variable eddy viscosity should be also investigated after the two-dimensional one[8, 9]. In this paper, an attempt is made to construct a mathematical model with variable eddy viscosity in shallow seas and to apply it to the boundary value problem of three-dimensional nonlinear tides and the ultra-shallow water storm surges as well as the steady shallow sea circulation on an *f*- or a *β*-coordinates. It should be emphasized that the eddy viscosity in the mathematical model mentioned above might be a physically acceptable arbitrary function of depth. Finally, a generalized linear law for bottom friction is proposed.

Formulation

The following hypotheses are made:

(i) The typical horizontal scale, *L*, of the field of flow is relatively small in comparison

---

* Feng S. 1984. A Three-dimensional nonlinear hydrodynamic model with variable eddy viscosity in shallow seas. Chinese Journal of Oceanology and Limnology, 2(2): 177–187

with the radius of the earth so that an f- or a β-coordinates can be adopted reasonably[10].

(ii) $D/L \ll 1$, where D denotes the vertical scale of the field of motion.

(iii) The density, ρ, of sea water is supposed to be a constant, namely, the fluid is considered as a homogeneous incompressible one.

(iv) The lateral eddy viscosity is ignored, only the vertical eddy viscosity ν in the model is presented.

Based upon these hypotheses a nondimensional dynamic problem for shallow seas is proposed as follows

$$\left.\begin{array}{l} \nabla \cdot \boldsymbol{u} + \dfrac{\partial w}{\partial z} = 0 \\ \left\{K_{\mathrm{i}}\left[\dfrac{\partial}{\partial t} + K_{\mathrm{n}}\left(\boldsymbol{u}\cdot\nabla + w\dfrac{\partial}{\partial z}\right)\right] + K_{\mathrm{c}} f\boldsymbol{e}_3 \times\right\}\boldsymbol{u} = -\nabla(\zeta - K_{\mathrm{a}}\zeta_{\mathrm{a}} - K_{\mathrm{T}}\zeta_{\mathrm{T}}) + K_{\mathrm{v}}\dfrac{\partial}{\partial z}\left(\nu\dfrac{\partial \boldsymbol{u}}{\partial z}\right) \\ \text{at the sea surface } z = \chi\zeta: \\ w = \left[K_{\mathrm{r}}\dfrac{\partial}{\partial t} + \chi(\boldsymbol{u}\cdot\nabla)\right]\zeta \\ \nu\dfrac{\partial \boldsymbol{u}}{\partial z} = K_{\tau}\boldsymbol{\tau}_{\mathrm{a}} \\ \text{at the sea bottom } z = -h: \\ \boldsymbol{u} = w = 0 \\ \text{along the shore boundary } \mathrm{C}_1: \\ \boldsymbol{N}\cdot\displaystyle\int_{-h}^{\chi\zeta}\boldsymbol{u}\mathrm{d}z = 0 \\ \text{along the open boundary } \mathrm{C}_2: \\ \boldsymbol{N}\cdot\displaystyle\int_{-h}^{\chi\zeta}\boldsymbol{u}\mathrm{d}z = \mathscr{Q} \\ \text{or} \\ \zeta = \mathscr{S} \end{array}\right\} \quad (1)$$

where

$$\chi = \mathscr{Z}/D,\ \ K_{\mathrm{i}} = SrEu^{-1},\ \ K_{\mathrm{n}} = Sr^{-1},\ \ K_{\mathrm{c}} = Ro^{-1}Eu^{-1},\ \ K_{\mathrm{a}} = \zeta_{\mathrm{a0}}/\mathscr{Z}$$

$$K_{\mathrm{T}} = \zeta_{\mathrm{T0}}/\mathscr{Z},\ \ K_{\mathrm{v}} = EkRo^{-1}Eu^{-1},\ \ K_{\mathrm{r}} = \chi Sr,\ \ K_{\tau} = \frac{\tau_{\mathrm{a0}}D}{\rho\mathscr{R}\mathscr{U}}$$

the Rossby number $Ro = \dfrac{\mathscr{U}}{f_0 L}$,

the Ekman number $Ek = \dfrac{\mathscr{R}}{f_0 D^2}$,

the Eular number $Eu = \dfrac{g\mathscr{G}}{\mathscr{U}^2}$,

the Strouhal number $Sr = \dfrac{L}{\mathscr{U}\mathscr{T}}$.

In the above, $\nabla = \boldsymbol{e}_1 \dfrac{\partial}{\partial x} + \boldsymbol{e}_2 \dfrac{\partial}{\partial y}$, (x,y,z) constitute an f- or a β-coordinates at the right-hand side, the plane (x,y) coincides with the undisturbed sea surface and x points to the east and y points to the north for the β-coordinates[11] while the restriction on (x,y) above is removed for the f-coordinates[12], z is positive upward, $(\boldsymbol{e}_1, \boldsymbol{e}_2, \boldsymbol{e}_3)$ denote the unit vectors of coordinates; t denotes time; $\boldsymbol{u} = \boldsymbol{e}_1 u + \boldsymbol{e}_2 v$, u and v are the components of the current in x, y directions respectively and w denotes the vertical component of current; ζ is the elevation measured from the undisturbed sea surface; f is the Coriolis parameter; g is the gravitational acceleration; ζ_a and ζ_T represent the effects of atmosphere pressure and tide-generating force, respectively; $\boldsymbol{\tau}_a = \boldsymbol{e}_1 \tau_{ax} + \boldsymbol{e}_2 \tau_{ay}$, τ_{ax}, τ_{ay} denote the components of wind stress in the x, y directions respectively; h denotes the depth; $N = \boldsymbol{e}_1 \cos\alpha_x + \boldsymbol{e}_2 \cos\alpha_y$, $\cos\alpha_x$ and $\cos\alpha_y$ are the direction-cosines of the boundary normal; $\mathscr{Q}$ and $\mathscr{S}$ denote the normal volume transport and the elevation along the open boundary, respectively; $\mathscr{T}$ denotes the characteristic time; $\mathscr{U}$, the characteristic horizontal current velocity; $\mathscr{G}$, the characteristic amplitude of the elevation; f_0 is the characteristic Coriolis parameter; $\mathscr{R}$ is the characteristic eddy viscosity; ζ_{ao}, ζ_{To} and τ_{ao} denote the characteristic quantities of ζ_a, ζ_T and $\boldsymbol{\tau}_a$ respectively.

In view of the fact that the effect of the rotation of the earth and that of the eddy viscosity on the motion in shallow seas are essential, we further put $K_c = 1$ $(Eu^{-1} = Ro)$ and $K_\nu = 1$ $(Ek = 1)$, and then $K_i = SrRo$ and $K_n K_i = Ro$, in the nondimensional problem (1). Thus, we shall formally reduce the nondimensional problem (1) to the following nondimensional one (2):

$$
\left.
\begin{aligned}
& \nabla \cdot \boldsymbol{u}' + \frac{\partial w'}{\partial z} = 0 \\
& \left[\frac{\partial}{\partial z}\left(\nu \frac{\partial}{\partial z} \right) + \Sigma(\sigma) - f\boldsymbol{e}_3 \times \right] \boldsymbol{u}' = \nabla \zeta' + \boldsymbol{\Pi} \\
z = 0: \quad & \\
& w' = Z(\zeta') + \Theta \\
& \nu \frac{\partial \boldsymbol{u}'}{\partial z} = \boldsymbol{\tau} \\
z = -h: \quad &
\end{aligned}
\right\} \qquad (2)
$$

$$u' = w' = 0$$

along the shore boundary C_1:

$$N \cdot \left(\int_{-h}^{0} u' \mathrm{d}z + q \right) = 0$$

along the open boundary C_2:

$$N \cdot \left(\int_{-h}^{0} u' \mathrm{d}z + q \right) = \mathscr{Q}'$$

or

$$\zeta' = \mathscr{S}' \tag{2}$$

where the variables $\boldsymbol{u}' = \boldsymbol{e}_1 u' + \boldsymbol{e}_2 v'$, u', v', w' and ζ' together with the parameters and the other quantities $\Sigma(\sigma)$, $\boldsymbol{\Pi} = \boldsymbol{e}_1 \pi_1 + \boldsymbol{e}_2 \pi_2$, $Z(\zeta')$, Θ, $\boldsymbol{\tau} = \boldsymbol{e}_1 \tau_1 + \boldsymbol{e}_2 \tau_2$, $\boldsymbol{q} = \boldsymbol{e}_1 q_1 + \boldsymbol{e}_2 q_2$, q_1, q_2, $\mathscr{Q}'$ and $\mathscr{S}'$, will be defined later according to the different cases.

A Sturm-liouville System

The Sturm-Liouville System proposed is similar to that presented by N. S. Heaps (1971)[13], is a Sturm-Liouville equation on a finite closed interval, $-h \leqslant z \leqslant 0$, together with two separate boundary conditions, of form

$$\left\{\begin{aligned} &\left[\frac{\partial}{\partial z} \left(\nu \frac{\partial}{\partial z} \right) + \lambda \right] F = 0 \\ &\left(\frac{\partial F}{\partial z} \right)_{z=0} = 0 \\ &(F)_{z=-h} = 0 \end{aligned}\right. \tag{3}$$

where λ is a parameter independent of z, while the eddy viscosity ν is a real-valued function of z and is positive, continuous and bounded in the interval given above.

This Sturm-Liouville System can be a regular or a singular one. When the physically acceptable profile of the eddy viscosity ν is selected and given, the solutions of the Sturm-Liouville System (3), as the eigenfunctions F_n, corresponding a set of eigenvalues of the Sturm-Liouville System, λ_n, will be obtained.

By using (3), it is shown that the eigenfunctions just mentioned, corresponding different eigenvalues, are orthogonal with weight function 1 on the interval $[-h, 0]$. Of course, it is necessary to assume that the eigenfunctions are square integrable for a singular Sturm-Liouville System. Thus, the orthogonality just mentioned and proved enables one to expand any reasonably smooth function $\mathscr{G}(z)$ as the infinite linear combinations of the eigenfunctions of the Sturm-Liouville System (3); namely,

$$\mathscr{G}(z) = \sum_{n=1}^{\infty} C_n F_n(z) \tag{4}$$

where

$$C_n = \int_{-h}^{0} F_n(z)\mathscr{G}(z)\mathrm{d}z \Big/ \int_{-h}^{0} F_n^2(z)\mathrm{d}z$$

Model

Setting

$$\boldsymbol{u}' = \boldsymbol{u} + \frac{\boldsymbol{\tau}(h+z)}{\nu_0}$$

where $\nu_0 = \nu_{z=0}$ and substituting this equation into (2), we have

$$\left.\begin{aligned}
&\nabla \cdot \boldsymbol{u} + \frac{\partial w'}{\partial z} = -\nabla \cdot \frac{\boldsymbol{\tau}(h+z)}{\nu_0} \\
&\left[\frac{\partial}{\partial z}\left(\nu \frac{\partial}{\partial z}\right) + \Sigma(\sigma) - f\boldsymbol{e}_3 \times\right]\boldsymbol{u} \\
&= \nabla \zeta' + \boldsymbol{\Pi} - \frac{\boldsymbol{\tau}}{\nu_0}\left[\times (h+z) f\boldsymbol{e}_3 + (h+z)\Sigma(\sigma) + \frac{\partial \nu}{\partial z}\right] \\
&z = 0: \\
&\qquad w' = Z(\zeta') + \Theta \\
&\qquad \nu \frac{\partial \boldsymbol{u}}{\partial z} = 0 \\
&z = -h: \\
&\qquad \boldsymbol{u} = w' = 0 \\
&\mathrm{C}_1: \\
&\qquad \boldsymbol{N} \cdot \left[\int_{-h}^{0} \boldsymbol{u}\mathrm{d}z + \boldsymbol{q} + \frac{h^2 \boldsymbol{\tau}}{2\nu_0}\right] = 0 \\
&\mathrm{C}_2: \\
&\qquad \boldsymbol{N} \cdot \left[\int_{-h}^{0} \boldsymbol{u}\mathrm{d}z + \boldsymbol{q} + \frac{h^2 \boldsymbol{\tau}}{2\nu_0}\right] = \mathscr{Q}'
\end{aligned}\right\} \tag{2$'$}$$

or

$$\zeta' = \mathscr{S}'$$

Suppose $F = F(x, \mathrm{y}, \mathrm{z})$ and λ being independent of z, and for each (x,y), let the ascending eigenvalues and corresponding eigenfunctions derived from the Sturm-Liouville System (3) be denoted by

$$\lambda=\lambda_n,\quad F=F_n \quad (n=1,2,\cdots)$$

Thus, the $\boldsymbol{u}$ as a sufficiently smooth function can be expanded into the following series by using the eigenfunctions and the expression (4)

$$\boldsymbol{u}=\sum_{n=1}^{\infty}G_n\mathscr{U}_nF_n \tag{5}$$

where

$$\mathscr{U}_n=\int_{-h}^{0}\boldsymbol{u}F_n\mathrm{d}z$$

$$G_n=\left[\int_{-h}^{0}\cdot F_n^2\mathrm{d}z\right]^{-1}$$

and we have

$$\boldsymbol{u}'=\sum_{n=1}^{\infty}G_n\mathscr{U}_nF_n+\frac{\boldsymbol{\tau}(h+z)}{\nu_0} \tag{6}$$

Multiplying the second equation of the problem (2)' by F_n, vertically integrating from $z=-h$ to $z=0$ and using the Sturm-Liouville System (3) and the expression of $\mathscr{U}_n$ contained in (5), yields

$$\left[(\Sigma(\sigma)-\lambda_n)-f\boldsymbol{e}_3\times\right]\mathscr{U}_n=H_n\nabla\zeta'+\boldsymbol{\Phi}_n-\boldsymbol{\tau}_n \tag{7}$$

where

$$\boldsymbol{\Phi}_n=\int_{-h}^{0}\boldsymbol{\Pi}F_n\mathrm{d}z,\quad H_n=\int_{-h}^{0}F_n\mathrm{d}z,\quad \mathscr{H}_n=\int_{-h}^{z}F_n\mathrm{d}z$$

$$\boldsymbol{\tau}_n=\frac{\boldsymbol{\tau}}{\nu_0}\left\{\int_{-h}^{0}F_n\frac{\partial\nu}{\partial z}\mathrm{d}z+\left[\Sigma(\sigma)+(\times f\boldsymbol{e}_3)\right]\left(hH_n-\int_{-h}^{0}\mathscr{H}_n\mathrm{d}z\right)\right\}$$

From the equation (7), we have the solution

$$\mathscr{U}_n=(\beta_n^2+f^2)^{-1}\{(-H_n\beta_n+H_nf\boldsymbol{e}_3\times)\nabla\zeta'+(\beta_n-f\boldsymbol{e}_3\times)\boldsymbol{\tau}_n-(\beta_n-f\boldsymbol{e}_3\times)\boldsymbol{\Phi}_n\} \tag{8}$$

where

$$\beta_n=\lambda_n-\Sigma(\sigma)$$

By using the first equation of the problem (2)′, the boundary condition on the bottom and the expression (5), the w' is derived, namely,

$$w'=-\sum_{n=1}^{\infty}\nabla\cdot(G_n\mathscr{U}_n\mathscr{H}_n)-\nabla\cdot\frac{\boldsymbol{\tau}(h+z)^2}{2\nu_0} \tag{9}$$

Substituting (8) into (9) and then (9) into the third equation of the problem (2)', the non-dimensional problem for ζ' is obtained as follows

$$\left.\begin{aligned}
&\left[\mathscr{A}\nabla^2+\left(\frac{\partial\mathscr{A}}{\partial x}-\frac{\partial\mathscr{B}}{\partial y}\right)\frac{\partial}{\partial x}+\left(\frac{\partial\mathscr{A}}{\partial y}-\frac{\partial\mathscr{B}}{\partial x}\right)\frac{\partial}{\partial y}\right]\zeta'-Z(\zeta')=\Theta+\mathscr{F}+\nabla\cdot\frac{h^2\boldsymbol{\tau}}{2\nu_0}\\
&\text{along the shore boundary } C_1:\\
&\qquad Q_{\mathrm{N}}=-q_{\mathrm{N}}-\frac{h^2}{2\nu_0}\tau_{\mathrm{N}}\\
&\text{along the open boundary } C_2:\\
&\qquad Q_{\mathrm{N}}=\mathscr{Q}'-q_{\mathrm{N}}-\frac{h^2}{2\nu_0}\tau_{\mathrm{N}}\\
&\text{or}\\
&\qquad \zeta'=\mathscr{S}'
\end{aligned}\right\}\quad(10)$$

where

$$\mathscr{F}=-\sum_{n=1}^{\infty}\{\boldsymbol{e}_3\cdot\nabla\times(B_n\boldsymbol{\Phi}_n)+\nabla\cdot(A_n\boldsymbol{\Phi}_n+\boldsymbol{e}_3\cdot\nabla\times(B_n\boldsymbol{\tau}_n)+\nabla\cdot(A_n\boldsymbol{\tau}_n)\}$$

$$Q_{\mathrm{N}}=-\mathscr{A}\frac{\partial\zeta'}{\partial N}+\mathscr{B}\frac{\partial\zeta'}{\partial T}-\sum_{n=1}^{\infty}\{A_n\Phi_{n\mathrm{N}}-B_n\Phi_{n\mathrm{T}}-A_n\tau_{n\mathrm{N}}+B_n\tau_{n\mathrm{T}}\}$$

$$A_n=\frac{H_nG_n\beta_n}{\beta_n^2+f^2},\quad B_n=\frac{H_nG_nf}{\beta_n^2+f^2},\quad \mathscr{A}=\sum_{n=1}^{\infty}A_nH_n,\quad \mathscr{B}=\sum_{n=1}^{\infty}B_nH_n$$

$$\frac{\partial\zeta'}{\partial N}=\frac{\partial\zeta'}{\partial x}\cos\alpha_x+\frac{\partial\zeta'}{\partial y}\cos\alpha_y,\quad \frac{\partial\zeta'}{\partial T}=-\frac{\partial\zeta'}{\partial y}\cos\alpha_x+\frac{\partial\zeta'}{\partial x}\cos\alpha_y$$

$$\Phi_{n\mathrm{N}}=\Phi_{n1}\cos\alpha_x+\Phi_{n2}\cos\alpha_y,\quad \Phi_{n\mathrm{T}}=-\Phi_{n2}\cos\alpha_x+\Phi_{n1}\cos\alpha_y$$

$$\tau_{n\mathrm{N}}=\tau_{n1}\cos\alpha_x+\tau_{n2}\cos\alpha_y,\quad \tau_{n\mathrm{T}}=-\tau_{n2}\cos\alpha_x+\tau_{n1}\cos\alpha_y$$

$$\tau_{\mathrm{N}}=\tau_1\cos\alpha_x+\tau_2\cos\alpha_y,\quad q_{\mathrm{N}}=q_1\cos\alpha_x+q_2\cos\alpha_y$$

here, T denotes the tangent at the boundary which forms a right-hand Cartesion coordinate system with the normal N, $\nabla^2=\frac{\partial^2}{\partial x^2}+\frac{\partial^2}{\partial y^2}$.

The nondimensional problem (2) is now reduced to the nondimensional problem of ζ', (10); and the expression for $\boldsymbol{u}'$, (6), and the vertical distribution for $\boldsymbol{u}'$ are given by (6) by means of the eigenfunction, F_n, derived from the Sturm-Liouville System (3) on the finite-closed interval $[-h,0]$. w' is derived by the expression (9).

Application Ⅰ: To Nonlinear Tides

The shallow sea model with variable eddy viscosity {(6), (10)} can be applied to developing a tidal three-dimensional nonlinear model with variable eddy viscosity.

In view of the fact that tidal waves are gravitational long-waves by nature, we put $K_{\mathrm{i}}=1$ and $K_{\mathrm{r}}=1$, i.e., $K_{\mathrm{i}}K_{\mathrm{n}}=Ro=Sr^{-1}=\chi$ introduced in the nondimensional problem (1). Further ignoring the wind stress and the effect of atmospheric pressure, i.e., putting $K_{\tau}=K_{\mathrm{a}}=0$, the nondimensional problem (1) is reduced to that of nonlinear tides proposed in the reference [1], where χ is a sole parameter. As treated in the reference [1], noting χ as a perturbation parameter and introducing a harmonic factor $\mathrm{e}^{-\mathrm{i}\sigma t}$, a tidal nondimensional problem of a certain constituent of jth order constituents is derived and this nondimensional problem can be expressed by using the nondimensional problem (2) with the definite variables and parameters as follows:

The Nondimensional Problem (2)

where

$$\begin{cases}(u',v',w',\zeta')=(\overline{u},\overline{v},\overline{w},\overline{\zeta});\ \Sigma(\sigma)=\mathrm{i}\sigma,\ Z(\zeta')=-\mathrm{i}\sigma\overline{\zeta}\\(\Pi_1,\Pi_2)=((\overline{\psi}_1-\overline{\xi}_1),(\overline{\psi}_2-\overline{\xi}_2)),\ \Theta=\overline{f}\\(\tau_1,\tau_2)=(\nu)_{z=0}(\overline{\gamma}_1,\overline{\gamma}_2),(q_1,q_2)=(\overline{\eta}_1,\overline{\eta}_2)\\\mathscr{Q}'=\mathscr{Q}',\ \mathscr{S}'=\mathscr{S}'\end{cases}\tag{11}$$

Here the symbols introduced and not defined are defined to be the same as those defined in the references [1] and [7].

Thus, substituting (11) into the expressions (5), (8) and the problem (10), the boundary-value problem of tidal elevation $\overline{\zeta}$ and the expressions for tidal currents $\overline{u}$, $\overline{v}$ are obtained; namely, the model {(6), (10)} with (11) describes the three-dimensional nonlinear problem of tides with variable eddy viscosity.

It should be pointed out that the model {(6), (10)} with (11) is not mathematically more complicated than those derived in the references [1] and [7] though the tidal model presented here should be essentially improved in comparison with those proposed in the references [1] and [7], In fact, following the models developed in the references [1] and [7], this is the latest effort made by author to extend the boundary-value problem in both a tidal two-dimensional linear model initiated by W. Hansen (1952)[14] and a three-dimensional linear model with a eddy viscosity being independent of vertical coordinate developed by L. Sgibnewa and A. Felsenbaum (1965)[15] to the one in a three-dimensional nonlinear model with variable eddy viscosity.

Application Ⅱ: To Ultra-shallow Water Storm Surges

The shallow sea model with variable eddy viscosity {(6), (10)} can be also applied to the problem for ultra-shallow water storm surges.

Putting $K_r = 1$, i.e., $Sr = \chi^{-1}$ and then $K_i K_n = \chi K_i$ and $K_i = \Xi$ where Ξ was defined in the reference [2], and $K_\tau = 1$ as well as $K_T = 0$, the nondimensional problem (1) is reduced to the one for shallow water storm surges established in the reference [2]. Furthermore, as pointed in the reference [2], $\mathscr{O}(\Xi) < 1$ for the ultra-shallow water storm surges, and then supposing $\mathscr{O}(\chi) < 1$, we have

$$(u, v, w, \zeta) = (u_0, v_0, w_0, \zeta_0) + \Xi (u_1, v_1, w_1, \zeta_1) \tag{12}$$

where (u_0, v_0, w_0, ζ_0) and (u_1, v_1, w_1, ζ_1) denote the zeroth-order solutions and the first-order perturbation solutions, respectively.

Introducing (12) into the nondimensional problem (1), we obtain the zeroth-order model and the first-order perturbation model for ultra-shallow water storm surges as follows:

(i) The zeroth-order model

The Nondimensional Problem (2)

where

$$\begin{cases} (u', v', w', \zeta') = (u_0, v_0, w_0, \zeta_0 - K_a \zeta_a);\ \Sigma(\sigma) = 0,\ Z(\zeta') = \partial \zeta_a / \partial t \\ \Pi_1 = \Pi_2 = \Theta = q_1 = q_2 = 0,\ (\tau_1, \tau_2) = (\tau_{ax}, \tau_{ay}) \\ \mathscr{Q}' = \mathscr{Q}, \mathscr{S}' = \mathscr{S}' - K_a \zeta_a \end{cases} \tag{13}$$

(ii) The first-order perturbation model

The Nondimensional Problem (2)

where

$$\begin{cases} (u', v', w', \zeta') = (u_1, v_1, w_1, \zeta_1);\ \Sigma(\sigma) = 0,\ Z(\zeta') = \partial \zeta_1 / \partial t \\ (\Pi_1, \Pi_2) = \dfrac{\partial}{\partial t}(u_{0},_{0}) + \chi \left\{ u_0 \dfrac{\partial}{\partial x} + v_0 \dfrac{\partial}{\partial y} + w_0 \dfrac{\partial}{\partial z} \right\} (u_0, v_0) \\ \Theta = \dfrac{\chi}{\Xi} \left\{ \dfrac{\partial (u_0 \zeta_0)}{\partial x} + \dfrac{\partial (v_0 \zeta_0)}{\partial y} \right\}_{z=0} \\ (\tau_1, \tau_2) = -\dfrac{\chi}{\Xi} \left\{ \zeta_0 \dfrac{\partial}{\partial z} \left[\nu \dfrac{\partial}{\partial z} (u_0, v_0) \right] \right\}_{z=0},\quad (q_1, q_2) = \dfrac{\chi}{\Xi} \left[\zeta_0 (u_0, v_0) \right]_{Cm} \quad (m = 1, 2) \\ \mathscr{Q}' = \mathscr{S}' = 0. \end{cases} \tag{14}$$

Thus, substituting (13) (or (14)) into the expressions (6), (8) and the problem (10), together with an additional initial condition, an initio-boundary-value problem for the elevation ζ_0 (or ζ_1) and the expression for the currents u_0, v_0 (or u_1, v_1) of the zeroth-order model (or the first-order perturbation model) of ultra-shallow water storm surges are obtained. In other words, the model {(6), (10)} with (13) (or (14)), together with the additional initial condition just mentioned, describes the zeroth-order problem (or the first-order perturbation problem) of ultra-shallow water storm surges with variable eddy viscosity.

To obtain the currents and the elevation from this model with variable eddy viscosity there is no additional mathematical difficulty compared with those developed in the references [2] and [6] though the former would be more complete than the latters.

Application Ⅲ: To the Steady Circulation

For a steady problem of general circulation there exists $Sr = 0$ and then $K_i = K_r = 0$. Putting $K_T = 0$ and supposing $\mathscr{O}(Ro) < 1$, the currents (u,v,w) and the elevation ζ can be expressed as follows

$$(u,v,w,\zeta) = (u_0,v_0,w_0,\zeta_0) + Ro(u_1,v_1,w_1,\zeta_1) \tag{15}$$

Further assuming $\mathscr{O}(\chi) < 1$, and substituting (15) into the nondimensional problem (1), both the zeroth-order model and the first-order perturbation model are obtained for the nonlinear steady circulation:

(i) The zeroth-order model

The Nondimensional Problem (2)

where

$$\begin{cases} (u',v',w',\zeta') = (u_0,v_0,w_0,\zeta_0 - K_a\zeta_a); \quad \Sigma(\sigma) = Z(\zeta') = \Pi_1 = \Pi_2 = \Theta = q_1 = q_2 = 0 \\ (\tau_1,\tau_2) = (\tau_{ax},\tau_{ay}); \quad \mathscr{Q}' = \mathscr{Q}, \ \mathscr{J}' = \mathscr{J} - K_a\zeta_a \end{cases} \tag{16}$$

(ii) The first-order perturbation model

The Nondimensional Problem (2)

where

$$\begin{cases} (u',v',w',\zeta') = (u_1,v_1,w_1,\zeta_1); \quad \Sigma(\sigma) = Z(\zeta') = 0 \\ (\Pi_1,\Pi_2) = \left\{ u_0\dfrac{\partial}{\partial x} + v_0\dfrac{\partial}{\partial y} + w_0\dfrac{\partial}{\partial z} \right\}(u_0,v_0); \quad \Theta = \dfrac{\chi}{Ro}\left\{ \dfrac{\partial(u_0\zeta_0)}{\partial x} + \dfrac{\partial(v_0\zeta_0)}{\partial y} \right\}_{z=0} \\ (\tau_1,\tau_2) = -\dfrac{\chi}{Ro}\left\{ \zeta_0\dfrac{\partial}{\partial z}\left[\nu\dfrac{\partial}{\partial z}(u_0,v_0) \right] \right\}_{z=0}, \quad (q_1,q_2) = \dfrac{\chi}{Ro}[\zeta_0(u_0,v_0)]_{Cm} \ (m = 1,2) \\ \mathscr{Q}' = \mathscr{S}' = 0 \end{cases} \tag{17}$$

It is quite evident to deduce that the model {(6), (10)} with (16) (or (17)) controls the zeroth-order problem (or the first-order perturbation problem) of steady circulation with variable eddy viscosity.

No doubt, this barotropic model, which is appropriate to the wind-driven circulation, cannot be applied to describe the wind-driven and the thermohalinic circulation. A baroclinic model with variable eddy viscosity will be developed in a other paper.

Bottom Friction

Introducing (6) into the expressions for bottom friction $\boldsymbol{\tau}_{\mathrm{b}} = (v\partial \boldsymbol{u}' / \partial z)_{z=-h}$ and for transport $\mathscr{Q} = \int_{-h}^{0} \boldsymbol{u}'\mathrm{d}z$, we have

$$\boldsymbol{\tau}_{\mathrm{b}} = \sum_{n=1}^{\infty} G_n \mathscr{U}_n \left(v \frac{\partial F_n}{\partial z} \right)_{z=-h} + \frac{v_{\mathrm{b}}}{v_0} \boldsymbol{\tau} \tag{18}$$

and

$$\mathscr{Q} = \sum_{n=1}^{\infty} G_n \mathscr{U}_n H_n + \frac{h^2}{2v_0} \boldsymbol{\tau} \tag{19}$$

where $$v_{\mathrm{b}} = v_{z=-h}$$

By using (8) and eliminating $\nabla \zeta'$, yields

$$\boldsymbol{\tau}_{\mathrm{b}} = (\beta - \bar{\beta}\boldsymbol{e}_3 \times)\mathscr{Q} + (\alpha - \bar{\alpha}\boldsymbol{e}_3 \times)\boldsymbol{\tau} - \boldsymbol{\Psi} + \boldsymbol{e}_3 \times \boldsymbol{\Phi} + (\beta - \bar{\beta}\boldsymbol{e}_3 \times)\mathscr{C} - (\beta + \bar{\beta}\boldsymbol{e}_3 \times)\mathscr{D} \tag{20}$$

where

$$\beta = \frac{\mathscr{U}\mathscr{I} + \mathscr{N}\mathscr{B}}{\mathscr{I}^2 + \mathscr{F}^2}, \quad \bar{\beta} = \frac{\mathscr{N}\mathscr{I} - \mathscr{U}\mathscr{B}}{\mathscr{I}^2 + \mathscr{B}^2}$$

$$\alpha = \frac{v_{\mathrm{b}}}{v_0} + \varepsilon - \beta\mathscr{F} + \bar{\beta}\mathscr{G}, \quad \bar{\alpha} = \mathscr{L} - \bar{\beta}\mathscr{F} - \beta\mathscr{G}$$

$$\mathscr{U} = \sum_{n=1}^{\infty} \frac{H_n G_n \beta_n}{\beta_n^2 + f^2} T_n, \quad \mathscr{N} = \sum_{n=1}^{\infty} \frac{H_n G_n f}{\beta_n^2 + f^2} T_n$$

$$\mathscr{E} = \sum_{n=1}^{\infty} \frac{G_n \beta_n}{\beta_n^2 + f^2} T_n \tau_n', \quad \mathscr{L} = \sum_{n=1}^{\infty} \frac{G_n f}{\beta_n^2 + f^2} T_n \tau_n'$$

$$\boldsymbol{\Psi} = \sum_{n-1}^{\infty} \frac{G_n \beta_n}{\beta_n^2 + f^2} T_n \boldsymbol{\Phi}_n, \quad \boldsymbol{\Phi} = \sum_{n=1}^{\infty} \frac{G_n f}{\beta_n^2 + f^2} T_n \boldsymbol{\Phi}_n$$

$$\mathscr{C} = \sum_{n=1}^{\infty} A_n \boldsymbol{\Phi}_n, \quad \mathscr{D} = \sum_{n=1}^{\infty} B_n \boldsymbol{\Phi}_n$$

$$T_n = \left(v \frac{\partial F_n}{\partial z} \right)_{z=-h}, \quad \mathscr{F} = \sum_{n=1}^{\infty} A_n \tau_n' + \frac{h^2}{2v_0}, \quad \mathscr{G} = \sum_{n=1}^{\infty} B_n \tau_n'$$

$$\tau_n' = \frac{1}{\nu_0}\left\{\int_{-h}^{0} F_n \frac{\partial \nu}{\partial z}\mathrm{d}z + \left[\Sigma(\sigma) - f\right]\left(hH_n - \int_{-h}^{0} \mathscr{H}_n \mathrm{d}z\right)\right\}$$

The expression (20) can be regarded as a generalized linear law between the bottom friction and the transport. For the linear model, the expression (20) reduces to

$$\boldsymbol{\tau}_{\mathrm{b}} = (\beta - \bar{\beta}\boldsymbol{e}_3\times)\boldsymbol{Q} + (\alpha - \bar{\alpha}\boldsymbol{e}_3\times)\boldsymbol{\tau} \tag{21}$$

which is frequently used in two-dimensional models with the transport in shallow sea dynamics and will be reduced to that derived by Feng and Shi[6] (1980) if the Coriolis force is omitted or to that developed by Qin (1981)[10] if the eddy viscosity is assumed to be a constant. The qualitative and quantitative analyses for the bottom friction and the linear law made in some detail in the references [6] and [10] will be not made here.

It should be pointed out with emphasis that, in view of the expression (20) derived from the relatively complete model proposed here, judging and extending the empirical linear relations between the bottom friction and the transport used frequently in two-dimensional models will become possible to some extent.

Postscript

The (initio-) boundary-value problem of free surface elevation (10), the expressions of currents (6) and (9) together with (8) and the expression for bottom friction (20) are derived. To solve these equations, a physically acceptable function of depth for eddy viscosity ν is first assumed, and then the eigenvalues λ_n, and the eigenfunctions F_n are obtained from the Sturm-Liouville System (3) with the additional condition of $F(x,y,0)=1$. Thus, a three-dimensional nonlinear hydrodynamic model with variable eddy viscosity in shallow seas is established in principle. Nevertheless, modifications and further development may be expected before satisfactory application to various concrete problems is achieved.

References

[1] Feng Shihzao (Feng Shizuo), 1977. A three-dimensional nonlinear model of tides. Scientia Sinica XX(4): 436–446.

[2] Chin Tsenghao and Feng Shihzao (Feng Shizuo), 1975. A preliminary study on the mechanism of shallow water storm surges. Scientia Sinica XVIII (2): 242–261.

[3] Sun Wenxin, Chen Zongyong and Feng Shizuo, 1981. Numerical simulation of the three-dimensional nonlinear tidal waves (I)—A numerical study for M_4 and MS_4 waves in the Bohai Sea. J. Shandong College Ocean. 11(1): 23–31. (in Chinese with English abstract)

[4] Sun Wenxin, Feng Shizuo and Qin Zenghao, 1979. Numerical modeling of an ultra-shallow water storm surge (I). Acta Oceanologia Sinica 1(2): 193–211. (in Chinese with English abstract)

[5] Sun Wenxin, Qin Zenghao and Feng Shizuo, 1980. Numerical modeling of an ultra-shallow water storm surge (II). J. Shandong College Ocean. 10(2): 7–19. (in Chinese with English abstract)

[6] Feng Shizuo and Shi Ping, 1980. An ultra-shallow water storm surge model with variable eddy viscosity. Acta Oceanologia Sinica 2(3): 1–11. (in Chinese with English abstract)

[7] Feng Shizuo and Sun Wenxin, 1983. A tidal three-dimensional nonlinear model with variable eddy viscosity (I), Chin. J, Ocean, Limn. 1(2): 166–170.

[8] Feng Shizuo, Zhang Shuzhen and Xi Pangen, 1981. Mathematical modeling of circulation in the East China Sea (II). J. Shandong College Ocean. 11(2): 8–26. (in Chinese with English abstract)

[9] Xi Pangen, Zhang Shuzhen and Feng Shizuo, 1980. Mathematical modeling of circulation in the East China Sea (I). J. Shandong College Ocean. 10(3): 13–25. (in Chinese with English abstract)

[10] Qin Zenghao, 1981. A simple air-sea planetary boundary layer model with special reference to Ekman currents. Ocean. Limn. Sin. 12(1): 9–24. (in Chinese with English abstract)

[11] Feng Shizuo, 1982. On the f- and β-coordinates, J. Shandong College Ocean. 12 (3): 1–10. (in Chinese with English abstract)

[12] Feng Shizuo, 1982. An Introduction to Storm Surges, Science Press, Beijing, pp. 241. (in Chinese)

[13] Heaps, N. S., 1971. On the numerical solution of the three-dimensional hydrodynamical equations for tides and storm surges. Mém. Soc. r. Sci. Liége 6(2): 143–180.

[14] Hansen, W., 1952. Gezeiten und Gezeitenströme der halbtaqiqen Hauptmondtide M_2 in der Nordsee. Deut. Hydrog. Z. Ergänzungsheft 1.

[15] Sgibnewa, L. and A. Felsenbaum, 1965. Theory on tides in water with friction. Doklady Akad. Nauk S. S. S. R. 164(2): 315–318. (in Russia)

On Lagrangian Residual Ellipse*

R.T. Cheng, Shizuo Feng, Pangen Xi

1. Introduction

The importance of residual current in studies of long-term transport of dissolved and suspended matter has long been recognized. Since progress in research of residual current has been relatively slow, confusion may still exist with regard to the definitions of various residual variables (Alfrink and Vreugdenhil, 1980). There is ample evidence that residual current has become the focus of research for many scientists who are concerned with the long-term transport processes in estuaries and in coastal seas. In the case of tide-induced residual current, the residual variables are an order of magnitude smaller than the tidal variables. Although the long-term transport of dissolved and suspended matter is of interest, the actual transport is characterized as a convection dominated process on a time scale on the order of the tidal period. Dominance of convection suggests that the actual transport is a Lagrangian process, the fate of dissolved solutes or suspended matter should be determined by a Lagrangian mean rather than by a Eulerian mean. In other words, the Eulerian residual current, the time averaged tidal current is not sufficient for the determination of the net mass transport. A mass transport velocity was introduced and shown to be the sum of the Eulerian residual current and the Stokes drift by Longuet-Higgins (1969). He suggested that the mass transport velocity should be used to compute the net mass transport correctly, although the concept of the Lagrangian residual current was not given.

Following a Lagrangian point of view, a concise definition of the Lagrangian residual current was first introduced by Zimmerman (1979). Using a numerical approach, the Lagrangian residual current has been calculated by Cheng and Casulli (1982) for South San Francisco Bay, California. Cheng and Casulli (1982) pointed out that the Lagrangian residual current is a function of the time (tidal current phase) when the water parcel is labelled and released. Further numerical experiments have been carried out by Cheng (1983) in an attempt to define the relation between the Lagrangian residual current and the tidal current phase. Unfortunately, because of the high degree of relative uncertainty in the computed residual currents, a definitive relation between the tidal and the residual currents was not attainable from the numerical solutions. Nevertheless, the computations have reinforced the fact that the Lagrangian residual current is a function of the tidal current phase.

---

* Cheng R T, Feng S, Xi P. 1986. On Lagrangian residual ellipse. In: J van de Kreeke (ed.) Physics of Shallow Estuaries and Bays. Berlin, New York: Springer-Verlag. 102–113

In this paper, an analytical approach is used to study the properties of Lagrangian residual circulation. In a weakly nonlinear tidal system, the zeroth order solution of the governing equations gives the astronomical tides and tidal currents. The higher order solutions include the Eulerian residual current, the Stokes drift and the Lagrangian residual current; they are the results of nonlinear interactions among the lower order solutions. Using a Lagrangian approach, the Lagrangian residual current in an M_2 system has been obtained up to the second order. The first order Lagrangian residual current is shown to be the sum of the Eulerian residual current and the Stokes drift, and the second order solution of the Lagrangian residual current is shown to be an ellipse on a hodograph plane.

2. Governing Equations

Circulation problems in well-mixed, tidal estuaries and shallow tidal embayments can be described by the familiar shallow water equations (Stoker, 1965). Let L_c and h_c denote the basin characteristic length and depth where L_c is on the order of $\sqrt{gh_c} / \omega_c$ with ω_c being the characteristic tidal frequency and g being the gravitational constant. Further, let u_c and ζ_c be the characteristic tidal current speed and tidal amplitude, then $l_c = u_c / \omega_c$ is essentially the characteristic value of tidal excursion. In tidal estuaries, the ratios of l_c / L_c, $u_c / \sqrt{gh_c}$, ζ_c / h_c are of the same order of magnitude and a parameter κ can be introduced so that

$$\kappa = l_c / L_c = u_c / \sqrt{gh_c} = \zeta_c / h_c \tag{1}$$

The value κ characterizes the relative magnitude of residual variables versus tidal variables. For many estuaries and shallow coastal seas around the world, κ has a value in the range of 0.0–0.2. Normalized by these characteristic values, the non-dimensional shallow water equations in two-dimensional Cartesian coordinates, (x, y), on the horizontal plane are the continuity equation

$$\frac{\partial \zeta}{\partial \theta} + \frac{\partial}{\partial x}\left[(h + \kappa\zeta)u\right] + \frac{\partial}{\partial y}\left[(h + \kappa\zeta)v\right] = 0 \tag{2}$$

the x-momentum equation

$$\frac{\partial u}{\partial \theta} + \kappa\left[u\frac{\partial u}{\partial x} + v\frac{\partial u}{\partial y}\right] - fv = \frac{-\partial}{\partial x}(\zeta + \zeta_a) + \tau_x^w - \tau_x^b \tag{3}$$

the y-momentum equation

$$\frac{\partial v}{\partial \theta} + \kappa\left[u\frac{\partial v}{\partial x} + v\frac{\partial v}{\partial y}\right] + fu = \frac{-\partial}{\partial y}(\zeta + \zeta_a) + \tau_y^w - \tau_y^b \tag{4}$$

Where (u, v) are the velocity components in the (x, y) directions;

θ is time;

ζ is the water surface elevation measured from mean sea level;

ζ_a is the water surface displacement due to atmospheric pressure;

(τ_x^w, τ_y^w) are the wind stress components in the (x, y) directions acting on the water surface;

(τ_x^b, τ_y^b) are the bottom stress components in the (x, y) directions;

f is the Coriolis parameter; and

h is the water depth measured from mean sea level.

In order to analyze the Lagrangian motions of labelled water parcels, the Lagrangian tidal velocity components (u_l, v_l) in the (x, y) directions are defined as the velocity of a labelled water parcel at time θ which is released from (x_0, y_0) at time θ_0. As time elapses, the position of the labelled water parcel can be given by

$$x = x_0 + \kappa\xi, \quad y = y_0 + \kappa\eta \tag{5}$$

with the initial conditions that $\xi = \eta = 0$ at $\theta = \theta_0$. The Lagrangian water parcel displacement, (ξ, η), or simply the Lagrangian displacement is normalized by l_c. Thus the Lagrangian tidal velocity, (u_l, v_l), can be expressed in terms of the Eulerian tidal velocity as

$$\begin{aligned} u_l(x_0, y_0, \theta) &= v(x_0 + \kappa\xi,\ y_0 + \kappa\eta, \theta) \\ v_l(x_0, y_0, \theta) &= v(x_0 + \kappa\xi,\ y_0 + \kappa\eta, \theta) \end{aligned} \tag{6}$$

where the initial position of the labelled water parcel is (x_0, y_0) at $\theta = \theta_0$. The Lagrangian displacement (ξ, η) can be obtained from integrating the differential equation of a streakline (parcel trajectory)

$$\frac{d\xi}{u_l} = \frac{d\eta}{v_l} = d\theta \tag{7}$$

or,

$$\begin{aligned} (\xi, \eta) &= \int_{\theta_0}^{\theta} \left[u_l(x_0, y_0, \theta'),\ v_l(x_0, y_0, \theta')\right] d\theta' \\ &= \int_{\theta_0}^{\theta} \left[u(x_0 + \kappa\xi, y_0 + \kappa\eta, \theta'), v(x_0 + \kappa\xi, y_0 + \kappa\eta, \theta')\right] d\theta' \end{aligned} \tag{8}$$

Note that Eq. (8) is an integral equation of (ξ, η). With appropriate initial and boundary conditions, Eqs. (2) to (8) describe completely the dependent variables $(\zeta, u, v, u_l, v_l, \xi, \eta)$.

3. Method of Solution

3.1 The first order Lagrangian residual current-mass transport velocity

Because the ratio of residual current to tidal current, κ, is normally a small parameter for most estuaries and coastal seas, the weakly nonlinear solution of the tide-induced residual currents can be obtained by means of a small perturbation technique (Feng, 1977). In fact, the small parameter κ also represents the ratio of the net Lagrangian displacement to tidal excursion, or the ratio of tidal current speed to tidal wave celerity. The dependent variables can be expanded in ascending orders of κ such that

$$(\zeta,u,v,u_1,v_1,\xi,\eta)=\sum(\zeta_j,u_j,v_j,[u_1]_j,[v_1]_j,\xi_j,\eta_j)\kappa^j \tag{9}$$

where the subscript j indicates the j-th order solution of the dependent variables.

By substituting Eq. (9) into Eqs. (2) to (8), they become a set of independent linear systems of equations. Each subsystem of equations contains some elements which stem from Eqs. (2) to (7). The zeroth order equations determine the tides and tidal currents, and the first order subsystem of equations determines the Eulerian residual current and residual water level, and the Stokes drifts as a consequence of the nonlinear coupling of the zeroth order solutions.

For clarity and without losing generality, the flows in an M_2 tidal system are considered. Systems including several astronomical tides have essentially the same properties except that the algebra is much more complex. As before, κ is assumed to be a small parameter but not negligible. The solutions of tides and tidal currents for an M_2 system can be written as

$$\begin{aligned}u_0&=u_0'\cos\theta+u_0''\sin\theta\\ v_0&=v_0'\cos\theta+v_0''\sin\theta\\ \zeta_0&=\zeta_0'\cos\theta+\zeta_0''\sin\theta\end{aligned} \tag{10}$$

where the amplitudes and the phases of M_2, (u_0,v_0,ζ_0), are indirectly given by (u_0',u_0''), (v_0',v_0''), and (ζ_0',ζ_0''). The first order harmonics in an M_2 system are due to (M_2+M_2) and (M_2-M_2) interactions, and they are the M_4 and the first order Eulerian residual water level and residual current. Or,

$$\begin{aligned}u_1&=u_1'\cos 2\theta+u_1''\sin 2\theta+u_{\text{er}}^{(1)}\\ v_1&=v_1'\cos 2\theta+v_1''\sin 2\theta+v_{\text{er}}^{(1)}\\ \zeta_1&=\zeta_1'\cos 2\theta+\zeta_1''\sin 2\theta+\zeta_{\text{er}}^{(1)}\end{aligned} \tag{11}$$

M_4 First order Eulerian Residual

Similarly, the second order solution include the shallow water harmonics with frequencies θ, and 3θ, and they are

$$\begin{aligned}u_2&=u_2'\cos 3\theta+u_2''\sin 3\theta+\Sigma_i(u_{2,i}'\cos\theta+u_{2,i}''\sin\theta)\\ v_2&=v_2'\cos 3\theta+v_2''\sin 3\theta+\Sigma_i(v_{2,i}'\cos\theta+v_{2,i}''\sin\theta)\\ \zeta_2&=\zeta_2'\cos 3\theta+\zeta_2''\sin 3\theta+\Sigma_i(\zeta_{2,i}'\cos\theta+\zeta_{2,i}''\sin\theta)\end{aligned} \tag{12}$$

M_6 other harmonics

where the summation over i includes all other harmonics of frequency θ. As these do not contribute to the Lagrangian residual current up to the second order, the specific values of these harmonics are not essential. The complete solution of the M_2 system up to $O(\kappa^2)$ is, of course, the sum of the zeroth, first, and second order solutions.

A time averaging operator is defined as the time average of a dependent variable over one tidal period. Then, the time averaged tidal current, the sum of Eqs. (10)–(12), becomes the Eulerian residual current, (u_{er}, v_{er}), i.e.

$$u_{er} = u_{er}^{(1)} + O(\kappa^2),\ v_{er} = v_{er}^{(1)} + O(\kappa^2) \tag{13}$$

where superscript indicates the order of approximation of the variable with respect to the κ expansion. The Eulerian residual current, which is accurate up to $O(\kappa^2)$, is normalized by κu_c.

By applying the time averaging operator to the Lagrangian tidal velocities, the results lead naturally to the Lagrangian residual current (u_{lr}, v_{lr}). By definition then,

$$\begin{aligned} u_{lr}(x_0, y_0, \theta_0) &= \frac{1}{2\pi\kappa}\int_{\theta_0}^{\theta_0+2\pi} u_l(x_0, y_0, \theta')\mathrm{d}\theta' \\ v_{lr}(x_0, y_0, \theta_0) &= \frac{1}{2\pi\kappa}\int_{\theta_0}^{\theta_0+2\pi} v_l(x_0, y_0, \theta')\mathrm{d}\theta' \end{aligned} \tag{14}$$

with the aid of Eq. (8), the Lagrangian residual current in Eq. (14) can be written as

$$\begin{aligned} u_{lr}(x_0, y_0, \theta_0) &= \frac{1}{2\pi\kappa}\left[x_0 + \kappa\xi(\theta_0 + 2\pi) - x_0\right] = \xi(\theta_0 + 2\pi)/2\pi \\ v_{lr}(x_0, y_0, \theta_0) &= \frac{1}{2\pi\kappa}\left[y_0 + \kappa\eta(\theta_0 + 2\pi) - y_0\right] = \eta(\theta_0 + 2\pi)/2\pi \end{aligned} \tag{15}$$

From a Lagrangian point of view, Eq. (8) describes the trajectories of labelled water parcels. By following the same water parcel to the end of a complete tidal cycle, one gets the net Lagrangian displacement of the labelled water parcel per tidal period. Thus, Eq. (15) becomes a concise definition of the Lagrangian residual current, and it also gives a lucid physical description of the properties of the Lagrangian residual current, (Zimmerman, 1979; Cheng and Casulli, 1982). It is worth noting that the Lagrangian residual current is a function of the time θ_0 when the labelled water parcel is released, Eq. (15).

Using the solutions of an M_2 tidal system given in Eqs. (10) to (12), and using the definition of the Lagrangian residual current, Eq. (15), the first order Lagrangian residual current can be shown to be

$$\begin{aligned} u_{lr} &= u_{er}^{(1)} + \left\langle (\partial u_0/\partial x)_0\, \xi_0 \right\rangle + \left\langle (\partial u_0/\partial y)_0\, \eta_0 \right\rangle + o(\kappa) \\ v_{lr} &= v_{er}^{(1)} + \left\langle (\partial v_0/\partial x)_0\, \xi_0 \right\rangle + \left\langle (\partial v_0/\partial y)_0\, \eta_0 \right\rangle + o(\kappa) \end{aligned} \tag{16}$$

where $< >$ is the notation for the time averaging operator, $(\ )_0$ denotes that the argument in $(\ )_0$ is evaluated at (x_0, y_0), and (ξ_0, η_0) is the zeroth order Lagrangian displacement. (ξ_0, η_0) can be calculated by

$$\xi_0 = \int_{\theta_0}^{\theta} u_0(x_0, y_0, \theta')\mathrm{d}\theta',\ \eta_0 = \int_{\theta_0}^{\theta} v_0(x_0, y_0, \theta')\mathrm{d}\theta' \tag{17}$$

The correlation terms in Eq. (16) are known as the Stokes drift, (u_{sd}, v_{sd}), (Longuet-Higgins, 1969; Feng *et al.*, 1986), and

$$u_{sd} = \left\langle (\partial u_0 / \partial x)_0 \xi_0 \right\rangle + \left\langle (\partial u_0 / \partial y)_0 \eta_0 \right\rangle \tag{18}$$
$$v_{sd} = \left\langle (\partial v_0 / \partial x)_0 \xi_0 \right\rangle + \left\langle (\partial v_0 / \partial y)_0 \eta_0 \right\rangle$$

The first order Lagrangian residual current, Eq. (16), can then be rewritten as

$$u_{lr} = u_{er}^{(1)} + u_{sd} + o(\kappa),\ v_{lr} = v_{er}^{(1)} + v_{sd} + o(\kappa) \tag{19}$$

At this order of the approximation, the Lagrangian residual current is shown to be the same as the mass transport velocity given by Longuet-Higgins (1969). In the past, the mass transport velocity (the sum of the Eulerian residual current and the Stokes drift) has been interpreted as the Lagrangian residual current. The present analysis ascertains that this usage of the mass transport velocity is an acceptable approximation to the Lagrangian residual current. However, it is important to recognize that this usage is valid only as the first order approximation in a weakly nonlinear system. For this reason, the present authors deliberately identify the mass transport velocity as the first order Lagrangian residual current as in the sense introduced by Longuet-Higgins (1969).

3.2 The dilemma in the Lagrangian residual current

The first order Lagrangian residual current presents a dilemma. From a physical point of view, the Lagrangian residual current is expected to be a function of the time when a water parcel is labelled and released, Eq. (15). When the labelled water parcels are released from the same position at different phases of the tidal current, each labelled water parcel inscribes a different trajectory in space over a tidal period. At the end of a tidal period the net Lagrangian displacements are not expected to be the same. Yet, the analytical results show that the first order Lagrangian residual current is not a function of time. Reported field experiments in the North Sea (Mulder, 1982) indicated that the measured net Lagrangian displacements cannot be adequately matched by the measured Eulerian residual current with the correction due to the Stokes drift.

To further illustrate this dilemma and the properties of the Lagrangian residual circulation, we resort to results from a numerical model of South San Francisco Bay, California (South Bay). The tidal current, the Eulerian residual and the Lagrangian residual currents in South Bay have been calculated by Cheng and Casulli (1982) by means of a numerical model. South Bay is a shallow, semi-enclosed embayment which, for the most part of the year, is isohaline. The circulation in the Bay is strongly affected by the basin topography. South Bay bathymetry is characterized by a deep relict channel (>10 m) which connects to a broad shoal east of the channel. The isobaths shown in Figure 1 through 4 are 3 and 6 m at mean sea level. The shallow-water equations, Eq. (2) to (4), have been solved by a standard alternating-direction-implicit (ADI), finite difference method (Leendertse, 1970; Leendertse and

Gritton, 1971). Consult Cheng and Casulli (1982) for a more detailed description of the numerical methods and other pertinent information relevant to the South Bay model. The numerical results presented here were obtained with the finite-difference grids of 500 m to give a more detailed spatial resolution of the circulation in South Bay than those reported in Cheng and Casulli (1982).

Shown in Fig. 1 is a typical tidal circulation pattern in South Bay. An important and unique property of the tidal current distribution is that the tidal current, both in magnitude and in direction, is strongly affected by the basin bathymetry. The magnitude of the tidal current is roughly proportional to the local water depth, and the direction of the tidal current is generally tangent to the local isobath (Cheng and Gartner, 1985). When we use this numerical model as the base-line flow field, the movements of labelled water parcels can be computed.

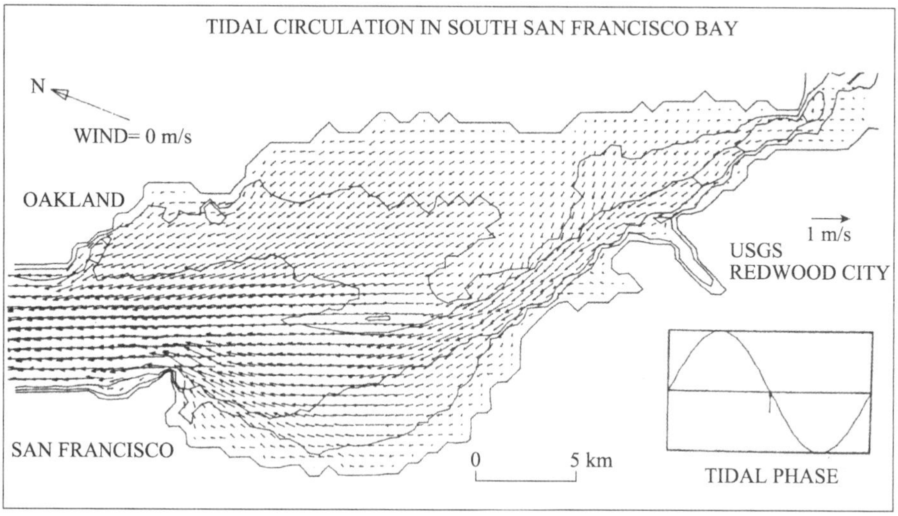

Fig. 1 Simulated tidal circulation in South San Francisco Bay, California using ADI finite difference method. A semi-diurnal tide, 1 m. amplitude was specified at open boundary. Shown in the figure is the tidal current distribution at near maximum ebb

Shown in Fig. 2 are the parcel trajectories computed in the Lagrangian sense over a period of idealized mixed diurnal and semi-diurnal tides. Note that the parcel trajectories follow the isobaths, because the local tide currents are generally tangent to the isobaths. The numerical results given in Fig. 2 are in agreement with the basic assumption that the net Lagrangian displacement, the distance between symbols + and Δ, is an order of magnitude smaller than the tidal excursion. Thus, the small κ approximation should be valid for South Bay.

When the net Lagrangian displacement is computed for every grid point in the model, the final results are the Lagrangian residual circulation, Fig. 3. The scale for the residual velocity vectors in Fig. 3 is an order of magnitude smaller than the scale for the tidal velocity vectors

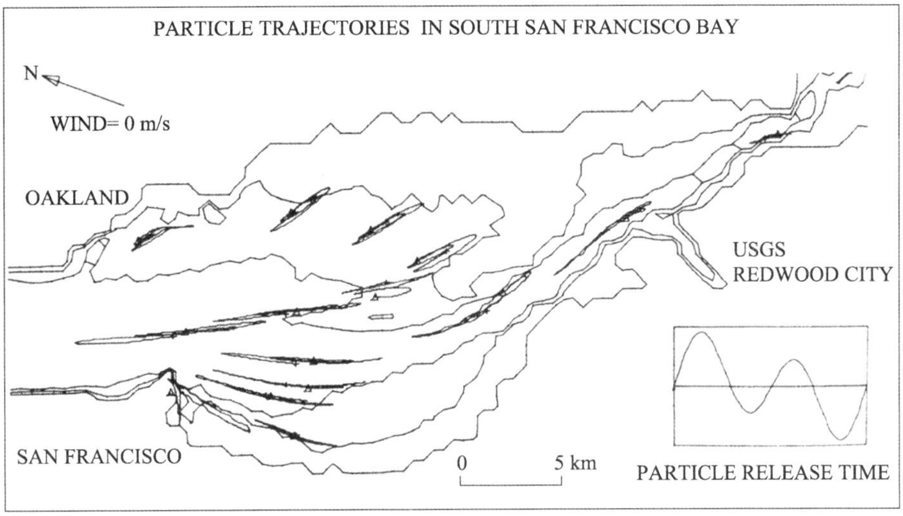

Fig. 2 Labelled water parcel trajectories over a diurnal period at a few selected locations. A mixed semi-diurnal and diurnal tide was specified at open boundary. The water parcels were labelled and released from positions marked + and were found at the end of 24 hours at positions marked by Δ

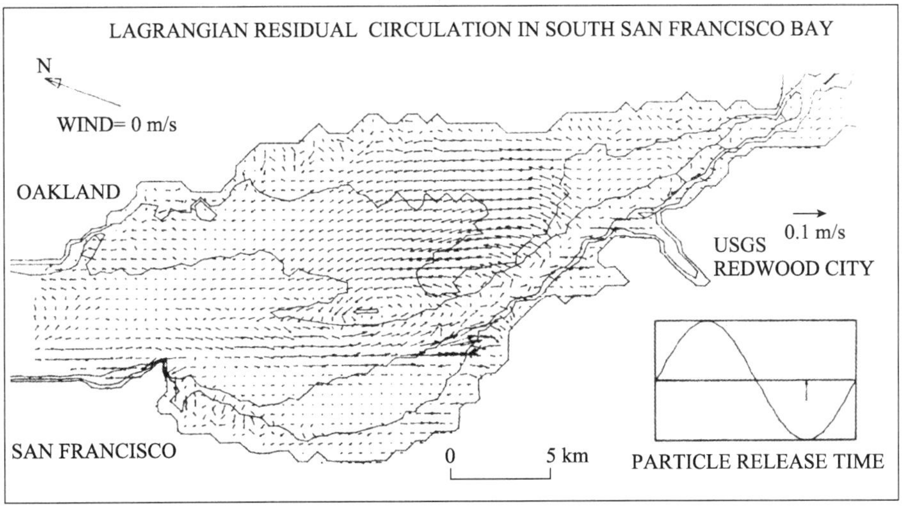

Fig. 3 Lagrangian residual circulation in South San Francisco Bay due to a semi-diurnal tide. The computations for the Lagrangian residual circulation were initiated when low water occurred at the open boundary. When there is no velocity vector plotted, the tracers have moved outside of the computation. The Lagrangian residual current there is unknown, not zero

in Fig. 2. Since the magnitudes for the velocity vectors in Figs. 2 and 3 are comparable, the actual ratio of residual current to tidal current, κ, is again shown to be small. More importantly, however, in separate computations when the labelled water parcels are released at one hour intervals throughout a 12-hour period, the computed Lagrangian residual currents at a

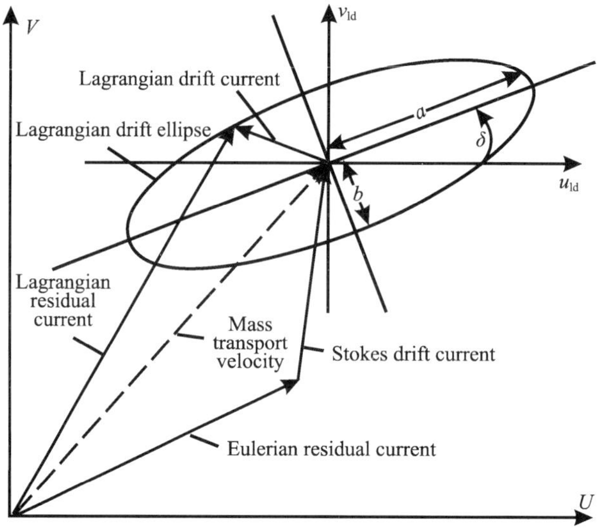

Fig. 4 The schematic diagram of the interrelations between the Eulerian residual current, the Stokes drift, the mass transport velocity, the Lagrangian drift, and the Lagrangian residual current

given position are not equal, but they are shown to be of the same order of magnitude (Cheng, 1983). Because the labelled water parcels are released at different phases of the tidal current, each water parcel inscribes a different trajectory. The net Lagrangian displacements are functions of the bathymetry within the region enclosed by the water parcel trajectory. Thus, there is no reason to expect the final net Lagrangian displacements to be the same after a complete tidal cycle.

Clearly, the first order Lagrangian residual current or the mass transport velocity is not adequate to capture the essential properties of the Lagrangian residual current in a weakly nonlinear system even when κ is small. To explore further the properties of the Lagrangian residual current, it is necessary to examine the next order of the approximation.

3.3 The Lagrangian residual ellipse

By following the definition, Eq. (15), the second order Lagrangian residual current is the second order Lagrangian displacement divided by 2π. It can be shown that

$$u_{lr} = u_{er} + u_{sd} + \kappa u_{ld} + O(\kappa^2),\ v_{lr} = v_{er} + v_{sd} + \kappa v_{ld} + O(\kappa^2) \tag{20}$$

where

$$u_{ld} = \left\langle (\partial u_1 / \partial x)_0 \xi_0 \right\rangle + \left\langle (\partial u_1 / \partial y)_0 \eta_0 \right\rangle + \left\langle (\partial u_0 / \partial x)_0 \xi_1 \right\rangle + \left\langle (\partial u_0 / \partial y)_0 \eta_1 \right\rangle + \left[\left\langle (\partial^2 u_0 / \partial x^2)_0 \xi_0^2 \right\rangle + 2\left\langle (\partial^2 u_0 / \partial x \partial y)_0 \xi_0 \eta_0 \right\rangle + \left\langle (\partial^2 u_0 / \partial y^2)_0 \eta_0^2 \right\rangle \right] / 2$$

$$v_{ld} = \left\langle (\partial v_1 / \partial x)_0 \xi_0 \right\rangle + \left\langle (\partial v_1 / \partial y)_0 \eta_0 \right\rangle + \left\langle (\partial v_0 / \partial x)_0 \xi_1 \right\rangle + \left\langle (\partial v_0 / \partial y)_0 \eta_1 \right\rangle + \left[\left\langle (\partial^2 v_0 / \partial x^2)_0 \xi_0^2 \right\rangle + 2\left\langle (\partial^2 v_0 / \partial x \partial y)_0 \xi_0 \eta_0 \right\rangle + \left\langle (\partial^2 v_0 / \partial y^2)_0 \eta_0^2 \right\rangle \right] / 2$$

In Eq. (20), the second order Lagrangian residual current depends on the first order Lagrangian displacement (x_{i1}, η_1) which is given as

$$\begin{aligned} \xi_1 &= \int_{\theta_0}^{\theta} \left[u_1(x_0, y_0, \theta') + (\partial u_0 / \partial x)_0 \xi_0 + (\partial u_0 / \partial y)_0 \eta_0 \right] \mathrm{d}\theta' \\ \eta_1 &= \int_{\theta_0}^{\theta} \left[v_1(x_0, y_0, \theta') + (\partial v_0 / \partial x)_0 \xi_0 + (\partial v_0 / \partial y)_0 \eta_0 \right] \mathrm{d}\theta' \end{aligned} \tag{21}$$

with the initial condition $\xi_1 = \eta_1 = 0$ at $\theta = \theta_0$. By substituting the solutions of M_2 into Eqs. (19), (20), and (21), and after carrying out the integration with time, the expression for $(u_{\mathrm{ld}}, v_{\mathrm{ld}})$ becomes

$$u_{\mathrm{ld}} = u'_{\mathrm{ld}} \cos\theta_0 + u''_{\mathrm{ld}} \sin\theta_0,\ v_{\mathrm{ld}} = v'_{\mathrm{ld}} \cos\theta_0 + v''_{\mathrm{ld}} \sin\theta_0 \tag{22}$$

where

$$\begin{aligned} u'_{\mathrm{ld}} &= u''_0 \partial(u_{\mathrm{er}} + u_{\mathrm{sd}}) / \partial x + v''_0 \partial(u_{\mathrm{er}} + u_{\mathrm{sd}}) / \partial y - (u_{\mathrm{er}} + u_{\mathrm{sd}}) \partial u''_0 / \partial x - (v_{\mathrm{er}} + v_{\mathrm{sd}}) \partial u''_0 / \partial y \\ u''_{\mathrm{ld}} &= -u'_0 \partial(u_{\mathrm{er}} + u_{\mathrm{sd}}) / \partial x - v'_0 \partial(u_{\mathrm{er}} + u_{\mathrm{sd}}) / \partial y + (u_{\mathrm{er}} + u_{\mathrm{sd}}) \partial u'_0 / \partial x + (v_{\mathrm{er}} + v_{\mathrm{sd}}) \partial u'_0 / \partial y \\ v'_{\mathrm{ld}} &= u''_0 \partial(v_{\mathrm{er}} + v_{\mathrm{sd}}) / \partial x + v''_0 \partial(v_{\mathrm{er}} + v_{\mathrm{sd}}) / \partial y - (u_{\mathrm{er}} + u_{\mathrm{sd}}) \partial v''_0 / \partial x - (v_{\mathrm{er}} + v_{\mathrm{sd}}) \partial v''_0 / \partial y \\ v''_{\mathrm{ld}} &= -u'_0 \partial(v_{\mathrm{er}} + v_{\mathrm{sd}}) / \partial x - v'_0 \partial(v_{\mathrm{er}} + v_{\mathrm{sd}}) / \partial y + (u_{\mathrm{er}} + u_{\mathrm{sd}}) \partial v'_0 / \partial x + (v_{\mathrm{er}} + v_{\mathrm{sd}}) \partial v'_0 / \partial y \end{aligned}$$

In Eq. (22), $(u_{\mathrm{er}}, v_{\mathrm{er}})$ is the Eulerian residual current, and $(u_{\mathrm{sd}}, v_{\mathrm{sd}})$ is the Stokes drift which has been given in Eq. (18). It is important to observe that the second order correction to the Lagrangian residual, $(u_{\mathrm{ld}}, v_{\mathrm{ld}})$, is the result of the nonlinear interactions between the astronomical constituents, i.e., the zeroth order solution, and the Eulerian residual current and the nonlinear interactions between the astronomical constituents and the Stokes drift. At this order, $(u_{\mathrm{ld}}, v_{\mathrm{ld}})$ is shown to be a function of θ_0, Eq. (22), which is the tidal current phase when the labelled water parcel is released from a fixed point (x_0, y_0). In the analysis of the second order dynamics, the distinct Lagrangian property is finally brought forth. It thus seems proper to name $(u_{\mathrm{ld}}, v_{\mathrm{ld}})$ as the Lagrangian residual drift velocity, or simply the Lagrangian drift. Expressing the above results in words, then

Lagrangian		Eulerian		Stokes'			Lagrangian	
Residual	=	Residual	+	Drift	+	κ	Drift	(23)
Current		Current		Velocity			Velocity	

The unique property of the Lagrangian drift is that as the initial phase angle θ_0 varies from 0 to 2π, the Lagrangian drift velocities trace out an ellipse on a hodograph plane, Fig. 4. The properties of the Lagrangian residual ellipse can be given explicitly. The semi-major (+ sign) and semi-minor (−sign) axes are

$$1/\sqrt{2}\left\{u'^2_{\mathrm{ld}} + u''^2_{\mathrm{ld}} + v'^2_{\mathrm{ld}} + v''^2_{\mathrm{ld}} \pm \left[(u'^2_{\mathrm{ld}} + u''^2_{\mathrm{ld}} + v'^2_{\mathrm{ld}} + v''^2_{\mathrm{ld}})^2 - 4(u'_{\mathrm{ld}} v''_{\mathrm{ld}} - u''_{\mathrm{ld}} v'_{\mathrm{ld}})^2\right]^{1/2}\right\}^{1/2} \tag{24}$$

The angle between the major axis of the residual ellipse and the x-axis is denoted by δ,

(Fig. 4), and

$$\delta = 1/2\tan^{-1}\left[2\frac{u'_{\mathrm{ld}}v'_{\mathrm{ld}} + u''_{\mathrm{ld}}v''_{\mathrm{ld}}}{(u'^2_{\mathrm{ld}} + u''^2_{\mathrm{ld}}) - (v'^2_{\mathrm{ld}} + v''^2_{\mathrm{ld}})}\right] \tag{25}$$

Further, the phase angle $\theta_{\max}$ which gives the maximum magnitude of the Lagrangian drift velocity is

$$\theta_{\max} = 1/2\tan^{-1}\left[2\frac{u'_{\mathrm{ld}}u''_{\mathrm{ld}} + v'_{\mathrm{ld}}v''_{\mathrm{ld}}}{(u'^2_{\mathrm{ld}} - u''^2_{\mathrm{ld}}) + (v'^2_{\mathrm{ld}} - v''^2_{\mathrm{ld}})}\right] \tag{26}$$

The properties of the Lagrangian residual ellipse are very similar to the parent tidal current ellipse. The second order solution reveals the generation mechanism of the Lagrangian residual current, and it explains the dilemma in the first order Lagrangian residual current. With the second order solution at hand, it is now natural to expect that the water parcels released at different phases of the tides give slightly different values of the Lagrangian residual current. The inter-relations between the Lagrangian residual current, the Stokes drift, the mass transport velocity (the sum of the Eulerian residual current and the Stokes drift), and the Lagrangian drift are depicted in Fig. 4. Furthermore, the second order Lagrangian residual current gives an error assessment for using the mass transport velocity as an approximation to the Lagrangian residual current.

4. The Sverdrup Wave

The properties of the Lagrangian displacement and the Lagrangian drift are further illustrated by means of an analytical example. Consider a two-dimensional Sverdrup wave, which is a harmonic wave in an infinite frictionless ocean of constant depth h_c (Defant, 1961). A typical form of the Sverdrup wave in dimensionless variables is

$$\begin{aligned}\zeta_0 &= \sin(\theta - x) = \zeta'_0\cos\theta + \zeta''_0\sin\theta \\ u_0 &= \sin(\theta - x) = u'_0\cos\theta + u''_0\sin\theta \\ v_0 &= f\cos(\theta - x) = v'_0\cos\theta + v''_0\sin\theta\end{aligned} \tag{27}$$

where $0 < f < 1$, and the wave speed is $C = (1 - f^2)^{-1/2}$, and where

$$\zeta'_0 = u'_0 = -\sin x,\quad \zeta''_0 = u''_0 = \cos x,\quad v'_0 = f\cos x, \text{ and } v''_0 = f\sin x.$$

The Sverdrup wave given in Eq. (27) is a progressive wave in the direction of the wave propagation, whereas the phase of the velocity component normal to the wave direction lags the water level by 90°. The tidal current ellipse rotates clockwise and has an aspect ratio of f. The Eulerian residual current for a Sverdrup wave is zero, and the Stokes and the Lagrangian drifts can be computed straightforwardly to give

$$u_{\mathrm{sd}} = 1/2,\quad v_{\mathrm{sd}} = 0 \tag{28}$$

and from Eq. (22),

$$u_{\text{ld}} = 1/2\sin x_0 \cos\theta_0 - 1/2\cos x_0 \sin\theta_0$$
$$v_{\text{ld}} = -f/2\cos x_0 \cos\theta_0 - f/2\sin x_0 \sin\theta_0$$

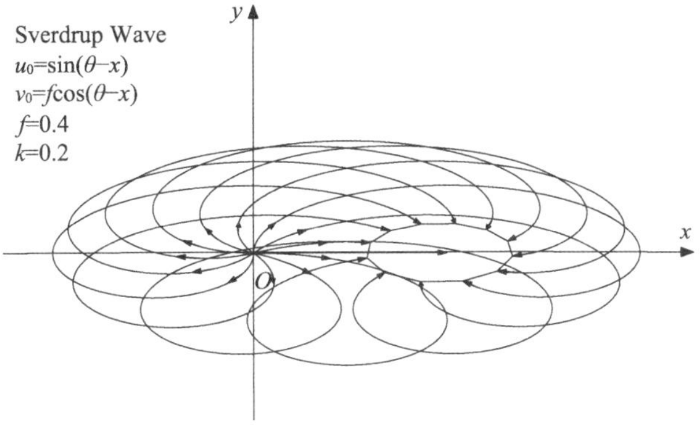

Fig. 5 The trajectories inscribed by labelled water parcels in a Sverdrup wave. The water parcels are released from the origin, and their termini at the end of a complete tidal period form an ellipse in space

The Lagrangian drift can be rewritten as

$$u_{\text{ld}} = 1/2\sin(\theta_0 - x_0 - \pi),\ v_{\text{ld}} = f/2\cos(\theta_0 - x_0 - \pi) \tag{29}$$

which is a typical form of an ellipse on a hodograph plane. The Lagrangian residual current becomes

$$\begin{array}{lccccc} u_{\text{lr}} = & 0 & + & 1/2 & +\kappa/2\sin(\theta_0 - x_0 - \pi) \\ v_{\text{lr}} = & 0 & + & 0 & +\kappa f/2\cos(\theta_0 - x_0 - \pi) \\ & \text{Eulerian} & & \text{Stokes} & \text{Lagrangian} \\ & \text{Residual} & & \text{Drift} & \text{Drift} \\ & \text{Current} & & \text{Velocity} & \text{Velocity} \end{array} \tag{30}$$

Shown in Fig. 5 are the trajectories inscribed by the water parcels released at $x_0 = 0$, and θ_0 at intervals of one twelfth of a period. The termini of labelled water parcels at the end of a tidal period form an ellipse in space. The eccentricities for the tidal current and for the Lagrangian residual ellipses are the same, and they are equal to f. Both ellipses are clockwise rotating, but they are 180° out of phase.

5. Summary and Conclusions

The dynamics of the tide-induced Lagrangian residual current has been discussed for a weakly nonlinear tidal system. The present analysis reveals the generation mechanism of the tide-induced Lagrangian residual current by means of a perturbation technique. Using a Lagrangian approach, the first order Lagrangian residual current has been shown to be the sum of the Eulerian residual current and the Stokes drift. Or, the mass transport velocity introduced by Longuet-Higgins (1969) has been proven to be the first order Lagrangian residual current in a weakly nonlinear tidal system. The dilemma that exists in the first order Lagrangian residual current can now be explained by the solution of the second order Lagrangian residual current. The second order correction to the Lagrangian residual current is named as the Lagrangian residual drift which is an ellipse on a hodograph plane. The second order solution of the Lagrangian residual current confirms the expectation that the Lagrangian residual current is a function of the time when the labelled water parcels are released. Moreover, with the second order solution at hand, an assessment can be made of the adequacy in the usage that the mass transport velocity as an approximation to the Lagrangian residual current.

References

Alfrink, B.J., & Vreugdenhil, C.B., 1981, Residual currents, Rept. R-1469-11, Delft Hydraul. Lab., Delft, The Netherlands.

Cheng, R.T., 1983, Euler-Lagrangian computations in estuarine hydrodynamics, Proc. of the Third Inter. Conf. on Num. Meth. in Laminar and Turbulent Flow, (Eds.) C. Taylor, J.A. Johnson, and R. Smith, p. 341–352, Pinderidge Press.

Cheng, R.T. & Casulli, V., 1982, On Lagrangian residual currents with applications in South San Francisco Bay, California, Water Resour. Res., v. 18, no. 6, p. 1652–1662.

Cheng, R.T., & Gartner, J.W., 1985, Harmonic analysis of tides and tidal currents in South San Francisco Bay, California, Estuarine Coastal Shelf Sci., V. 21, P.h 57–74.

Defant, A., 1961, Physical Oceanography, v. 1, Progamon Press, New York, 598 p.

Feng, S., 1977, A three-dimensional non-linear model of tides, Sci. Sin., v. 20, no. 4, p. 436–446.

Feng, S., Cheng, R.T. & Xi, P., 1986, On tide-induced Lagrangian residual current and residual transport: 1. Lagrangian residual current, Water Resur. Res., v. 22, no. 12, p. 1623–1634..

Leendertse, J.J., 1970, A water-quality simulation model for well-mixed estuaries and coastal seas, I. Principles of Computation, Resp. RM-6230-RC, Rand Corp., Santa Monica, California.

Leendertse, J.J. & Gritton, E.C., 1971, A water-quality simulation model for well-mixed estuaries and coastal seas, II. Computation Procedures, Rep. R-708-NYC, Rand Corp. Santa Monica, California.

Longuet-Higgins, M.S., 1969, On the transport of mass by time-varying ocean currents, Deep Sea Res., v. 16, 431–447.

Mudler, R., 1982, Eulerian and Lagrangian Analysis of Velocity of Field in the Southern North Sea, North Sea Dynamics, Ed. by Sündermann and Lenz, Springer-Verlag, Berlin, Heidelberg, p. 134–147.

Stoker, J.J., 1965, Water Waves, Interscience Publ., New York.

Zimmerman, J.T.F., 1979, On the Euler-Lagrangian transformation and the Stokes drift in the presence of oscillatory and residual currents, Deep Sea Res., 26A, p. 505–520.

On Tide-induced Lagrangian Residual Current and Residual Transport 1. Lagrangian Residual Current*

SHIZUO FENG, RALPH T. CHENG, PANGEN XI

Introduction

The long-term transports of suspended matter and dissolved solutes are transport phenomena which are dominated by convection due to tidal current within a tidal period but over many tidal cycles. Thus the nature of transport processes is strongly Lagrangian. As pointed out by numerous researchers (Nihoul and Ronday, 1975; Heaps, 1978), what determines the overall ecological balance or the long-term transport of dissolved or suspended matter in estuaries, coastal embayments, and shallow seas is residual current rather than tidal current. There is hardly any need for further elaboration on the importance of residual circulation.

For logistic reasons, the vast majority of tidal current data were taken in an Eulerian reference frame. Confusion often arises in studies of transport phenomena because the measured velocity (Eulerian) is not the velocity with which the same water mass (Lagrangian) is transported. Thus attention must be given to the adequacy of using Eulerian mean flow for estimates of long-term transport of dissolved and suspended matter. In this context, researchers have recently demonstrated the difference between an Eulerian mean velocity at a point and a Lagrangian mean velocity of a marked water parcel (Zimmerman, 1979; Cheng and Casulli, 1982; Yu and Chen, 1983). The latter has led to the concept of Lagrangian residual current, and research on Lagrangian residual current is a relatively recent topic.

Background

The concept of Lagrangian residual current as used in shallow-water oceanography stemmed from studies of finite amplitude water waves. The first use of the Lagrangian concept in large-scale circulation was given by Longuet-Higgins (1969), who showed that the mass transport velocity equals the sum of the Eulerian residual current and the Stokes drift (the Stokes formula), although a concise definition of the Lagrangian residual current was not given. The mass transport velocities for a double Kelvin wave were analyzed by Longuet-Higgins. In conclusion, Longuet-Higgins suggested that there are situations where the Stokes

* Feng S, Cheng R T, Xi P. 1986. On tide-induced Lagrangian residual current and residual transport 1. Lagrangian residual current. Water Resources Research, 22(12): 1623–1634

drift might even eclipse the Eulerian residual current and that the Lagrangian mean velocity should be used in determining the origin of water masses. The term "Lagrangian residual current" was used by Tee (1976) in his numerical model for residual currents, and perhaps this was the first modeling effort on the Lagrangian residual current. In Tee's model, the "Lagrangian residual current" was computed from the Eulerian mean volumetric transport divided by the mean-water depth. Strictly, the Stokes drift given by Tee (1976) is only valid for one-dimensional flows [see Feng *et al.* (1986) for detail].

Zimmerman (1979) and Cheng and Casulli (1982) have pointed out that the Stokes formula for mass transport is not sufficient to explain the Lagrangian nature of residual current which is quite fundamental in the dynamics of estuaries and coastal embayments. They have given the definition for the Lagrangian residual current as the net displacement of a marked water parcel divided by the elapsed time. The net Lagrangian displacement of a market water parcel, thus the Lagrangian residual current, is expected to be a function of the flow field in the neighborhood where the marked water parcel is released. A numerical model of South San Francisco Bay, California, was used in an attempt to define the relation between the Lagrangian residual current and the tidal current phase (Cheng, 1983), and similar calculations have been reported by Yu and Chen (1983). Because residual currents are an order of magnitude smaller than tidal currents, the computed residual currents contain a higher degree of uncertainty than the computed tidal currents. The numerical model results have only been able to confirm that the Lagrangian residual current is a function of tidal phase, and a quantitative relation between the Lagrangian residual current and tidal current phase could not be deduced (Cheng, 1983).

Discrepancies between the observed Lagrangian residual current and the mass transport velocity given by the Stokes formula have been reported (Dooley, 1974; Mulder, 1982; J. W. Ramster and J. A. Durance, Unpublished document, 1973). Substantial progress in understanding the nature of Lagrangian residual current was advanced by Zimmerman (1979); the Stokes formula for the mass transport velocity was shown to be an approximation of the Lagrangian residual current by means of an "Euler-Lagrangian" transformation.

The present approach

The principal objectives of the present study are aimed at clarification of the conceptual difference between the Eulerian and Lagrangian residual currents and at the generation mechanism of the Lagrangian residual current. In principle, a three-dimensional analysis should and can be carried out. In the interest of being able to present a lucid physical insight into the Lagrangian processes, the present analysis will be given in a two-dimensional space similar to that used in the relevant, previous investigations (Longuet-Higgins, 1969; Zimmerman, 1979; Cheng and Casulli, 1982). To avoid confusing the main issues, the effects due to surface wind stress, variations of barometric pressure, and baroclinic variations are not included. Thus the system is simplified to a two-dimensional barotropic flow in which all variables are depth-averaged values. If the two-dimensional barotropic flows obey the property of vertical rigidity (Stern, 1975; Pedlosky, 1979), the derived results for the Lagrangian residual current are definitive. The validity of the present analysis in a dissipative system depends upon the validity of the two-dimensional approximation for the tidal current field.

The present formulation and the method of solution follow closely the development of a three-dimensional nonlinear model given by Feng (1977). In this study a further extension in two dimensions of Feng's approach is made to include aspects of the Lagrangian processes (labeled particle dynamics). The ratio of residual current to tidal current is identified as a small parameter κ which characterizes the weak nonlinearity in the governing equations. The solutions of the problems are obtained by a perturbation technique. Specifically, a general relation between the Lagrangian residual current, Eulerian residual current, and Stokes drift have been obtained up to the second order. The first-order solution reproduces the Stokes formulation of the mass transport velocity, and the second-order solution shows that the Lagrangian residual is a periodic function of the tidal current phase. Further, the Lagrangian residual current is shown to be an ellipse on a hodograph plane. Several examples are given to demonstrate the unique Lagrangian properties of the residual current.

Governing Equations

For the stated purpose of the present study, circulation problems in shallow-water embayments are approximated by a two-dimensional barotropic flow system for which the governing equations are the vertically integrated conservation equations, or the shallow-water long-wave equations (Stoker, 1965). In order to write the governing equations in nondimensional form, the nondimensional variables will be denoted by a superscript "*", and all characteristic values will be denoted by a subscript "c". For tidal circulation there is only one obvious characteristic time scale, i.e., the tidal period, or the inverse of tidal frequency, ω_c. However there are several characteristic length scales present in such problems, namely, the basin horizontal length scale L_c, the basin characteristic depth h_c, the characteristic tidal amplitude ζ_c, and the characteristic values of tidal excursion tidal wave length, etc. Of course, not all of the characteristic lengths are independent. In Cartesian coordinates (x,y) and on the plane, the nondimensional variables are defined as follows:

$$\begin{aligned} &x = x^* / L_c \quad y = y^* / L_c \quad \theta = t^* / \omega_c \quad u = u^* / u_c \\ &v = v^* / u_c \quad \zeta = \zeta^* / \zeta_c \quad h = h^* / h_c \quad \gamma = \gamma^* / \omega_c \\ &S = S^* / S_c^* \quad f = f^* / \omega_c \end{aligned} \tag{1}$$

where

u, v	the vertically averaged velocity components in the x and y directions;
S	the specified incident tidal wave elevation;
ζ	the displacement of the free surface from a reference plane;
θ	time;
$h(x,y)$	the water depth measured from the reference plane;
f	the Coriolis parameter;
γ	the bottom friction coefficient.

The nondimensional shallow-water equations (Stoker, 1965) are the continuity equation,

$$\frac{\partial \zeta}{\partial \theta}+\frac{\partial}{\partial x}[(h+\kappa\zeta)u]+\frac{\partial}{\partial y}[(h+\kappa\zeta)v]=0 \tag{2}$$

the x-momentum equation,

$$\frac{\partial u}{\partial \theta}+\kappa\left[u\frac{\partial u}{\partial x}+v\frac{\partial u}{\partial y}\right]-fv=-\frac{\partial \zeta}{\partial x}-\gamma u \tag{3}$$

the y-momentum equation,

$$\frac{\partial v}{\partial \theta}+\kappa\left[u\frac{\partial v}{\partial x}+v\frac{\partial v}{\partial y}\right]+fu=-\frac{\partial \zeta}{\partial y}-\gamma v \tag{4}$$

The boundary condition at a shoreline boundary Γ_1 is

$$u\cos\alpha_x+v\cos\alpha_y=0 \tag{5}$$

where $\cos\alpha_x$ and $\cos\alpha_y$ are the direction cosines of the outward pointing normal of Γ_1 and at an open boundary Γ_2 the boundary is

$$\zeta=S \tag{6}$$

The parameter κ is defined directly from nondimensionalization of the continuity equation as

$$\kappa=\zeta_c/\eta_c=u_c/\omega_c L_c$$

In the momentum equations, $\omega_c L_c$ is required to be of the same order of magnitude as the tidal wave celerity $C=(gh_c)^{1/2}$. For a typical estuary or a shallow coastal embayment, and for semidiurnal tides, L_c and h_c are estimated to be of the order of 50 km and 10 m, respectively. Thus this required condition is approximately met. In (2)–(4) the surface wind stress and the horizontal dispersion terms have been neglected, and the bottom stresses have been assumed to be linearly proportional to the depth-averaged velocity.

Strictly, the bottom friction terms are nonlinear; they are proportional to the quadratic of tidal velocity and inversely proportional to the water depth. In addition to bottom stress terms, the convection terms in the momentum equations and the cross product terms in the continuity equation are also nonlinear. Pingree and Maddock (1978) estimated the relative importance of the bottom stress, convection, and the nonlinearity in the continuity equation as sources for generating higher-order harmonics to be in the ratios of

$$2:5:13$$

respectively. Recently, Parker (1984) investigated bottom friction effects in a one-dimensional tidal estuary. Although there might be significant effects that stemmed from the bottom friction terms, a complete analysis for other than a one-dimensional estuary is not known. In view of that, the bottom stresses play a relatively unimportant role in generating the higher-order harmonics, and for simplicity the bottom stress terms in (3) and (4) are assumed to be linear in the present analysis.

The notions of the vertically averaged Lagrangian velocity components u_l and v_l will be used to denote (the local velocity) at (x,y,θ) of a labeled water parcel which is released from (x_0,y_0) at the time $\theta=\theta_0$. Or

$$\begin{aligned} u_l(x_0,y_0,\theta)&=u(x,y,\theta) \\ v_l(x_0,y_0,\theta)&=v(x,y,\theta) \end{aligned} \tag{7}$$

where

$$\begin{aligned} x&=x(x_0,y_0,\theta_0,\theta) \\ y&=y(x_0,y_0,\theta_0,\theta) \end{aligned}$$

Thus an equation for a streakline (parcel trajectory) can be written as

$$\frac{\mathrm{d}\xi}{u_l}=\frac{\mathrm{d}\eta}{v_l}=\mathrm{d}\theta \tag{8}$$

where ξ and η are the Lagrangian displacements of a water parcel in the x and y directions at time θ. The Lagrangian tidal velocities and the Lagrangian displacements have been normalized by

$$\xi=\xi^*/l_c \quad \eta=\eta^*/l_c \quad u_l=u_l^*/u_c \quad v_l=v_l^*/u_c$$

where $l_c=u_c/\omega_c$ is essentially the characteristic value of tidal excursion. Following from the above discussion, κ can also be defined as $\kappa=l_c/L_c$ and thus

$$\kappa=l_c/L_c=u_c/C=\zeta_c/h_c \tag{9}$$

When the initial condition for (8) is taken as $(\xi,\eta)=0$ at $\theta=\theta_0$, the position of a labeled water parcel can be expressed as

$$\begin{aligned} x&=x_0+\kappa\xi \\ y&=y_0+\kappa\eta \end{aligned} \tag{10}$$

From (8) and (10) the Lagrangian displacements of the labeled parcel at time θ become

$$\begin{aligned} (\xi,\eta)&=\int_{\theta_0}^{\theta}\left[u_l(x_0,y_0,\theta'),\ v_l(x_0,y_0,\theta')\right]\mathrm{d}\theta' \\ &=\int_{\theta_0}^{\theta}\left[u(x_0+\kappa\xi,y_0+\kappa\eta,\theta'),\ v(x_0+\kappa\xi,y_0+\kappa\eta,\theta')\right]\mathrm{d}\theta' \end{aligned} \tag{11}$$

Because (u_1, v_1) are functions of (ξ, η), (11) is an integral equation.

The correct choice of the characteristic values is essential. The value of κ characterizes the ratios of residual variables to tidal variables. In many previous investigations, κ has been shown to be a small parameter in the range of 0–0.2 for most estuaries and coastal seas. Numerous examples of such estimates can be cited for estuaries, around the world. For example, Tee (1976) discussed the ratio of residual to tidal currents for Minas Basin at the head of the Bay of Fundy, Uncles and Jordan (1980) for the Seven estuary, and Cheng and Gartner (1984) for San Francisco Bay, California. Similar estimates were made for Chesapeake Bay by Elliott and Hendrix (1976), and for Narragansett Bay by Weisberg (1976). On the western boundary of the Pacific Ocean, Sun *et al.* (1981) considered Bohai Sea, and Yu and Chen (1983) investigated Jiaozhou Bay of China. For the North Sea of the Atlantic Ocean, Nihoul and Ronday (1975) and Heaps (1978) have independently made similar claims that the ratio of residual current to tidal current is of small order. It seems then that the condition that κ is of small order is valid for many estuaries and coastal seas, and this condition of small κ is the basic assumption for the present analysis.

Under small κ approximation the Lagrangian velocity components can be related to the Eulerian variables by Taylor series expansions about (x_0, y_0). Or,

$$u_1(x_0, y_0, \theta) = u(x_0, y_0, \theta) + \kappa[(\partial u / \partial x)_0 \xi + (\partial u / \partial y)_0 \eta] + \kappa^2[(\partial^2 u / \partial x^2)_0 \xi^2 + 2(\partial^2 u / \partial x \partial y)_0 \xi\eta + (\partial^2 u / \partial y^2)_0 \eta^2]/2 + O(\kappa^3) \quad (12)$$

and

$$v_1(x_0, y_0, \theta) = v(x_0, y_0, \theta) + \kappa[(\partial v / \partial x)_0 \xi + (\partial v / \partial y)_0 \eta] + \kappa^2[(\partial^2 v / \partial x^2)_0 \xi^2 + 2(\partial^2 v / \partial x \partial y)_0 \xi\eta + (\partial^2 v / \partial y^2)_0 \eta^2]/2 + O(\kappa^3) \quad (13)$$

where the notation $(\ )_0$ indicates that the term is evaluated at (x_0, y_0). The Lagrangian velocity components differ from the Eulerian velocities starting from terms of the order of κ, $\kappa^2, \cdots$. The system of equations (2) to (6), (8) and (10) to (13) forms a well-posed initial-boundary value problem whose solutions describe the tidal and residual circulation in coastal embayments.

Method of Solution

Because of the presence of a small parameter in the governing equations the solutions to this class of problems can be obtained by using a small perturbation technique (Van Dyke, 1964). All of the dependent variables are first expanded in ascending series of κ such that

$$(\zeta, u, v, u_1, v_1, \xi, \eta) = \sum_{j=0} (\zeta_j, u_j, v_j, [u_1]_j, [v_1]_j, \xi_j, \eta_j)\kappa^j \quad (14)$$

where the subscript j indicates the jth order solution. Substituting (14) into (2) to (6), the governing equations for the jth order solution become

$$\frac{\partial\zeta_j}{\partial\theta}+\frac{\partial(hu_j)}{\partial x}+\frac{\partial(hv_j)}{\partial y}=-\Phi_{j-1} \tag{15}$$

$$\frac{\partial u_j}{\partial\theta}-fv_j=-\frac{\partial\zeta_j}{\partial x}-\gamma u_j-X_{j-1} \tag{16}$$

$$\frac{\partial v_j}{\partial\theta}+fu_j=-\frac{\partial\zeta_j}{\partial y}-\gamma v_j-Y_{j-1} \tag{17}$$

On Γ_1,

$$u_j\cos\alpha_x+v_j\cos\alpha_y=0 \tag{18}$$

and on Γ_2,

$$\zeta_j=S_j \tag{19}$$

where

$$\Phi_{j-1}=\sum_{m=0}^{j-1}[\partial(\zeta_m u_{j-1-m})/\partial x+\partial(\zeta_m v_{j-1-m})/\partial y] \tag{20}$$

$$X_{j-1}=\sum_{m=0}^{j-1}[u_m\partial(u_{j-1-m})/\partial x+v_m\partial(u_{j-1-m})/\partial y] \tag{21}$$

$$Y_{j-1}=\sum_{m=0}^{j-1}[u_m\partial(v_{j-1-m})/\partial x+v_m\partial(v_{j-1-m})/\partial y] \tag{22}$$

$$\begin{aligned}S_j&=S \quad j=0\\ S_j&=0 \quad j>0\end{aligned} \tag{23}$$

When the subscript, the order of the solution, is less than zero, the variable is defined to be zero. By substituting the κ expansion, (14) into (12) and (13), the jth order Lagrangian velocity components in κ expansion become

$$\begin{aligned}[u_1]_j&=[u_1(x_0,y_0,\theta)]_j=\mathrm{d}\xi_j/\mathrm{d}\theta\\ &=u_j(x_0,y_0,\theta)+\sum_{m=0}^{j-1}[(\partial u_m/\partial x)_0\xi_{j-1-m}+(\partial u_m/\partial y)_0\eta_{j-1-m}]+\sum_{m=0}^{j-2}\sum_{n=0}^{m}[(\partial^2u_n/\partial x^2)_0\xi_{m-n}\xi_{j-2-m}\\ &\quad+2(\partial^2u_n/\partial x\partial y)_0\xi_{m-n}\eta_{j-2-m}+(\partial^2u_n/\partial y^2)_0\eta_{m-n}\eta_{j-2-m}]/2\end{aligned} \tag{24}$$

$$\begin{aligned}[v_1]_j&=[v_1(x_0,y_0,\theta)]_j=\mathrm{d}\eta_j/\mathrm{d}\theta\\ &=v_j(x_0,y_0,\theta)+\sum_{m=0}^{j-1}[(\partial v_m/\partial x)_0\xi_{j-1-m}+(\partial v_m/\partial y)_0\eta_{j-1-m}]+\sum_{m=0}^{j-2}\sum_{n=0}^{m}[(\partial^2v_n/\partial x^2)_0\xi_{m-n}\xi_{j-2-m}\\ &\quad+2(\partial^2v_n/\partial x\partial y)_0\xi_{m-n}\eta_{j-2-m}+(\partial^2v_n/\partial y^2)_0\eta_{m-n}\eta_{j-2-m}]/2\end{aligned} \tag{25}$$

The system of equations (15) to (25) are linear, and, in principle, the jth order solution can be solved consecutively in ascending order of j without much difficulty. The zeroth order solution is the linear approximation of the astronomical tides and tidal currents (Hansen, 1952). The higher-order solutions (j=1,2,$\cdots$) represent the higher-order harmonics which are generated from nonlinear interactions between astronomical tides and their associated harmonics. The residual water level and residual currents are embedded within these higher-order solutions. The integral equation for the nonlinear streakline trajectory, (11), can be linearized and simplified by using (24) and (25), and the jth order Lagrangian displacements become

$$\xi_j = \int_{\theta_0}^{\theta} [u_1(x_0, y_0, \theta')]_j \mathrm{d}\theta' \tag{26}$$

$$\eta_j = \int_{\theta_0}^{\theta} [v_1(x_0, y_0, \theta')]_j \mathrm{d}\theta' \tag{27}$$

Because the governing equations are linear, all the jth order harmonics satisfy the same set of the j th order equations. The j th order solution, in fact, is the sum of all jth order harmonics, or

$$(\zeta_j, u_j, v_j) = \sum_p (\zeta_{j,p}, u_{j,p}, v_{j,p}) \tag{28}$$

where the first index is the order of the solution, and the second index identities the tidal constituents of tidal harmonics. For example, when j=0, the zeroth order solutions are the astronomical tides. If O_1, K_1, P_1, M_2, N_2, and S_2 are considered, then p=1,2,$\cdots$,6 represent the six tidal constituents. The solutions for each tidal constituent and for each tidal harmonic can be solved independently in succeeding order of j. A systematic and detailed solution procedure for solving the jth order solution is given in the appendix.

Residual Currents Induced by an M_2 Tidal System

The Tide-induced residual currents

For clarity, in this section, only the higher-order shallow-water harmonics which are generated from nonlinear interactions of the astronomical M_2 tide and its harmonics are considered. Systems including interactions between several astronomical tides behave essentially in the same manner, except that the algebra is much more complex. As before, κ is assumed to be a small parameter but not negligible. Since all variables are periodic functions, they can be conveniently expressed in a complex notation

$$\Psi = \Psi' \cos\theta + \Psi'' \sin\theta = \mathrm{Re}[\tilde{\Psi} \mathrm{e}^{-\mathrm{i}\sigma\theta}]$$

where ψ stands for any dependent variable, i=$(-1)^{1/2}$, and σ denotes the frequency of the harmonics. The notation $\tilde{\psi}$ is related to ψ' and ψ'' by $\tilde{\Psi} = \Psi' + \mathrm{i}\Psi''$.

When only the M_2 tide is considered, the circular frequency of M_2 is naturally the characteristic frequency, and thus σ=1. The zeroth order solution, or the solution for M_2, is

$$\begin{aligned} u_0 &= u_0' \cos\theta + u_0'' \sin\theta \\ v_0 &= v_0' \cos\theta + v_0'' \sin\theta \\ \zeta_0 &= \zeta_0' \cos\theta + \zeta_0'' \sin\theta \end{aligned} \tag{29}$$

where the solution is written in an expanded form. An equivalent form of an amplitude and phase relation can be obtained easily from (29). The first-order harmonics of the M_2 system include the M_4 tide and the first-order Eulerian residual water level and residual currents, or

$$\begin{aligned} u_1 &= u_1'\cos2\theta + u_1''\sin2\theta + u_{\text{er}}^{(1)} \\ v_1 &= v_1'\cos2\theta + v_1''\sin2\theta + v_{\text{er}}^{(1)} \\ \zeta_1 &= \underbrace{\zeta_1'\cos2\theta + \zeta_1''\sin2\theta}_{M_4} + \underbrace{\zeta_{\text{er}}^{(1)}}_{\text{first-order Eulerian residual}} \end{aligned} \tag{30}$$

where M_4 is due to the (M_2+M_2) interaction and the Eulerian residual is due to (M_2–M_2) interaction. Similarly, the second-order solutions include the shallow-water harmonics with frequencies of σ and 3σ, and they are given as

$$\begin{aligned} u_2 &= u_2'\cos3\theta + u_2''\sin3\theta + \sum_i (u_{2,i}'\cos\theta + u_{2,i}''\sin\theta) \\ v_2 &= v_2'\cos3\theta + v_2''\sin3\theta + \sum_i (v_{2,i}'\cos\theta + v_{2,i}''\sin\theta) \\ \zeta_2 &= \underbrace{\zeta_2'\cos3\theta + \zeta_2''\sin3\theta}_{M_6} + \underbrace{\sum_i (\zeta_{2,i}'\cos\theta + \zeta_{2,i}''\sin\theta)}_{\text{other harmonics}} \end{aligned} \tag{31}$$

where the M_6 harmonic, ($2\sigma + \sigma$), or (M_4+M_2), is given explicitly. The other three higher-order harmonics are included in the summation over i; they are the results of interactions between M_4 and M_2, ($2\sigma - \sigma$), and the interactions of M_2 and the Eulerian residual currents ($\sigma \pm 0$). The complete solution of the M_2 system up to $O(\kappa^2)$ is the sum of (29), (30), and (31), and the harmonics of the order of $O(\kappa^3)$ and higher are not considered.

Introduce the notation of a time-averaging operator $<>$ for a variable ψ over a tidal period and in nondimensional form as

$$<\psi> = \frac{1}{2\pi}\int_{\theta_0}^{\theta_0+2\pi} \psi \, \mathrm{d}\theta' \tag{32}$$

where the integration is carried over the period of M_2. Thus, by time-averaging the sum of (29) through (31) the Eulerian residual current, $(u_{\text{er}}, v_{\text{er}})$, is shown to be

$$\begin{aligned} u_{\text{er}} &= u_{\text{er}}^{(1)}\kappa + O(\kappa^3) \\ v_{\text{er}} &= v_{\text{er}}^{(1)}\kappa + O(\kappa^3) \end{aligned} \tag{33}$$

The leading terms in (33) are of the order of κ, or the Eulerian residual current is an order smaller than the tidal current as expected (e.g., Tee, 1976; Dyke, 1980; Uncles and Jordan,

1980; Sun *et al.*, 1981; Cheng and Gartner, 1984, 1985), and thus the Eulerian residual currents should be properly normalized by $\kappa\mu_c$. With the correct scaling,

$$\begin{aligned} u_{er} &= u_{er}^{(1)} + O(\kappa^2) \\ v_{er} &= v_{er}^{(1)} + O(\kappa^2) \end{aligned} \tag{34}$$

the Eulerian residual current is shown to be of second-order accuracy.

By applying the same time-averaging operator on the Lagrangian velocities, and (13), the results lead naturally to the definition of the Lagrangian residual currents (u_{lr}, v_{lr}) as

$$\begin{aligned} u_{lr}(x_0, y_0, \theta_0) &= \frac{1}{2\pi}\int_{\theta_0}^{\theta_0+2\pi} u_l(x_0, y_0, \theta')\mathrm{d}\theta' = \frac{1}{2\pi\kappa}[x_0 + \kappa\xi(\theta_0 + 2\pi) - x_0] \\ &= \xi(\theta_0 + 2\pi)/2\pi \\ v_{lr}(x_0, y_0, \theta_0) &= \frac{1}{2\pi}\int_{\theta_0}^{\theta_0+2\pi} v_l(x_0, y_0, \theta')\mathrm{d}\theta' = \frac{1}{2\pi\kappa}[y_0 + \kappa\eta(\theta_0 + 2\pi) - y_0] \\ &= \eta(\theta_0 + 2\pi)/2\pi \end{aligned} \tag{35}$$

The Lagrangian residual current components, (35), are expressed as the net Lagrangian displacements divided by the tidal period. This result is consistent with the concise definition of the Lagrangian residual current given by Zimmerman (1979), and Cheng and Casulli (1982). The integrals in (35) can be evaluated by using the Lagrangian velocity components given in (24) and (25) with the aid of solutions of a M_2 system. The lowest-order term of (35) is shown to be of the same order of magnitudes as the Eulerian residual current, thus the Lagrangian residual current should also be normalized by κu_c to give

$$\begin{aligned} u_{lr} &= [u_{er}^{(1)} + <(\partial u_0/\partial x)_0\xi_0> + <(\partial u_0/\partial y)_0\eta_0>] + O(\kappa) \\ v_{lr} &= [v_{er}^{(1)} + <(\partial v_0/\partial x)_0\xi_0> + <(\partial v_0/\partial y)_0\eta_0>] + O(\kappa) \end{aligned} \tag{36}$$

where (ξ_0, η_0) is the zeroth order Lagrangian displacement defined in (26) and (27), or

$$\begin{aligned} \xi_0 &= \int_{\theta_0}^{\theta} [u_0(x_0, y_0, \theta')]\mathrm{d}\theta' \\ \eta_0 &= \int_{\theta_0}^{\theta} [v_0(x_0, y_0, \theta')]\mathrm{d}\theta' \end{aligned} \tag{37}$$

The Stokes' drift velocity components can be given in the form of

$$\begin{aligned} u_{sd} &= <(\partial u_0/\partial x)_0\xi_0> + <(\partial u_0/\partial y)_0\eta_0> \\ v_{sd} &= <(\partial v_0/\partial x)_0\xi_0> + <(\partial v_0/\partial y)_0\eta_0> \end{aligned} \tag{38}$$

At this order of the approximation the Lagrangian residual current turns out to be the sum of the Eulerian residual current and the Stokes drift. Thus the first-order Lagrangian residual current is the same as the mass transport velocity given by Longuet-Higgins (1969), i.e.,

$$\begin{aligned} u_{lr} &= u_{er} + u_{sd} + O(\kappa) \\ v_{lr} &= v_{er} + v_{sd} + O(\kappa) \end{aligned} \tag{39}$$

The Stokes drift was first used as a correction to the Eulerian residual current for estimating the net mass transport by Longuet-Higgins (1969) in the study of large-scale circulation. Quite often the mass transport velocity, the sum of the Eulerian residual current and Stokes drift, is referred to as the Lagrangian residual current. The result from the present analysis, (39), points out that this assessment of the Lagrangian residual current is correct only to the first order of approximation. The present analysis has arrived at (39) directly using a Lagrangian approach, and the main result ascertains that the first-order Lagrangian residual current is the same as the mass transport velocity given by Longuet-Higgins (1969). When κ is small but not negligibly small, the solution of the Lagrangian residual current at this order presents a dilemma. The Lagrangian residual current is expected to be dependent on the time (tidal phase) when a water parcel is labeled and released (Cheng and Casulli, 1982; Cheng, 1983), and yet (39) shows that the Lagrangian residual current is not a function of tidal phase. In order to bring out the unique Lagrangian characteristics of the Lagrangian residual current, it is necessary to examine the next higher order solution of the Lagrangian residual current. Although the inclusion of the next higher order solution gives only a higher-order correction to the quantitative values, it reveals important properties of the Lagrangian residual current.

The Lagrangian residual ellipse

The unique property of the Lagrangian residual velocity is its dependence on the trajectories that the labeled water parcels follow. The water parcels follow different trajectories depending upon the time (tidal phase) when the water parcels are labeled and released. Furthermore, the water parcel trajectories are functions of the spatial variations of tidal currents, which are strongly affected by basin bathymetry (Cheng and Casulli, 1982; Cheng, 1983; Cheng and Gartner, 1984). The first-order Lagrangian residual velocity given in (39) does not show this property. The second-order solution of the Lagrangian residual current has been obtained as

$$
\begin{aligned}
u_{\mathrm{lr}} &= u_{\mathrm{er}} + u_{\mathrm{sd}} + \kappa u_{\mathrm{ld}} + O(\kappa^2) \\
v_{\mathrm{lr}} &= v_{\mathrm{er}} + v_{\mathrm{sd}} + \kappa v_{\mathrm{ld}} + O(\kappa^2)
\end{aligned}
\tag{40}
$$

where

$$
\begin{aligned}
u_{\mathrm{ld}} =& <(\partial u_1/\partial x)_0 \xi_0> + <(\partial u_1/\partial y)_0 \eta_0> + <(\partial u_0/\partial x)_0 \xi_1> + <(\partial u_0/\partial y)_0 \eta_1> \\
&+ [<(\partial^2 u_0/\partial x^2)_0 \xi_0{}^2> + 2<(\partial^2 u_0/\partial x\partial y)_0 \xi_0 \eta_0> + <(\partial^2 u_0/\partial y^2)_0 \eta_0{}^2>]/2 \\
v_{\mathrm{ld}} =& <(\partial v_1/\partial x)_0 \xi_0> + <(\partial v_1/\partial y)_0 \eta_0> + <(\partial v_0/\partial x)_0 \xi_1> + <(\partial v_0/\partial y)_0 \eta_1> \\
&+ [<(\partial^2 v_0/\partial x^2)_0 \xi_0{}^2> + 2<(\partial^2 v_0/\partial x\partial y)_0 \xi_0 \eta_0> + <(\partial^2 v_0/\partial y^2)_0 \eta_0{}^2>]/2
\end{aligned}
\tag{41}
$$

The second-order solution of the Lagrangian residual velocity depends upon the first-order Lagrangian displacement (ξ_1, η_1) which is expressed as

$$\xi_1 = \int_{\theta_0}^{\theta} [u_1(x_0, y_0, \theta') + (\partial u_0 / \partial x)_0 \xi_0 + (\partial u_0 / \partial y)_0 \eta_0] \mathrm{d}\theta'$$
$$\eta_1 = \int_{\theta_0}^{\theta} [v_1(x_0, y_0, \theta') + (\partial v_0 / \partial x)_0 \xi_0 + (\partial v_0 / \partial y)_0 \eta_0] \mathrm{d}\theta' \tag{42}$$

Substituting the solution of an M_2 system into (38) and (41), after carrying out the integrations with time, the Stokes drift velocity components, (u_{sd}, v_{sd}), becomes

$$u_{sd} = \frac{1}{2}[u_0' \partial u_0'' / \partial x - u_0'' \partial u_0' / \partial x + v_0' \partial u_0'' / \partial y - v_0'' \partial u_0' / \partial y]$$
$$v_{sd} = \frac{1}{2}[u_0' \partial v_0'' / \partial x - u_0'' \partial v_0' / \partial x + v_0' \partial v_0'' / \partial y - v_0'' \partial v_0' / \partial y] \tag{43}$$

Similarly, the expressions for (u_{ld}, v_{ld}) are written as

$$u_{ld} = u_{ld}' \cos\theta_0 + u_{ld}'' \sin\theta_0$$
$$v_{ld} = v_{ld}' \cos\theta_0 + v_{ld}'' \sin\theta_0 \tag{44}$$

where

$$u_{ld}' = u_0'' \partial(u_{er} + u_{sd}) / \partial x + v_0'' \partial(u_{er} + u_{sd}) / \partial y - (u_{er} + u_{sd}) \partial u_0'' / \partial x - (v_{er} + v_{sd}) \partial u_0'' / \partial y$$
$$u_{ld}'' = -u_0' \partial(u_{er} + u_{sd}) / \partial x - v_0' \partial(u_{er} + u_{sd}) / \partial y + (u_{er} + u_{sd}) \partial u_0' / \partial x + (v_{er} + v_{sd}) \partial u_0' / \partial y$$
$$v_{ld}' = u_0'' \partial(v_{er} + v_{sd}) / \partial x + v_0'' \partial(v_{er} + v_{sd}) / \partial y - (u_{er} + u_{sd}) \partial v_0'' / \partial x - (v_{er} + v_{sd}) \partial v_0'' / \partial y$$
$$v_{ld}'' = -u_0' \partial(v_{er} + v_{sd}) / \partial x - v_0' \partial(v_{er} + v_{sd}) / \partial y + (u_{er} + u_{sd}) \partial v_0' / \partial x + (v_{er} + v_{sd}) \partial v_0' / \partial y \tag{45}$$

It is important to observe that the second-order correction to the Lagrangian residual, (u_{ld}, v_{ld}), is the combination of the nonlinear interactions between the astronomical constituents and the Eulerian residual and between the astronomical constituents and the Stokes drift, (45). In contrast, the Stokes drift, (43), is only a function of the nonlinear interactions among the astronomical tides. At this order, (u_{ld}, v_{ld}) in (44) is shown to be a function of θ_0, which is the tidal phase when the water parcel is labeled and released from a fixed point. When the second-order dynamics is considered, the distinct Lagrangian property has finally been brought forth. It thus seems proper to name (u_{ld}, v_{ld}) as the Lagrangian residual drift velocity or simply the Lagrangian drift. This main result, (40), can be expressed in words as

$$\begin{bmatrix}\text{Lagrangian}\\ \text{residual}\\ \text{velocity}\end{bmatrix} = \begin{bmatrix}\text{Eulerian}\\ \text{residual}\\ \text{velocity}\end{bmatrix} + \begin{bmatrix}\text{Stokes}\\ \text{drift}\\ \text{velocity}\end{bmatrix} + \kappa \begin{bmatrix}\text{Lagrangian}\\ \text{drift}\\ \text{velocity}\end{bmatrix} \tag{46}$$

There are some rather interesting properties associated with the Lagrangian drift. As the initial phase angle θ_0 varies from 0 to 2π, the Lagrangian drift velocities trace out an ellipse on a hodograph plane. Or, when the labeled water parcels are released from a fixed point continuously over a tidal period, the termini of the labeled parcels form an ellipse in space after a tidal cycle. The properties of the Lagrangian residual ellipse can be given explicitly.

The semimajor (plus sign) and semiminor (minus sign) axes are

$$1/\sqrt{2}\{u_{ld}'^2 + u_{ld}''^2 + v_{ld}'^2 + v_{ld}''^2 + [(u_{ld}'^2 + u_{ld}''^2 + v_{ld}'^2 + v_{ld}''^2)^2 - 4(u_{ld}' v_{ld}'' - u_{ld}'' v_{ld}')^2]^{1/2}\}^{1/2} \tag{47}$$

The angle between the major axis of the residual ellipse and the x axis is denoted by δ (Figure 1), and

$$\delta = \frac{1}{2}\tan^{-1}\left[2\frac{u'_{\mathrm{ld}}v'_{\mathrm{ld}} + u''_{\mathrm{ld}}v''_{\mathrm{ld}}}{(u'^2_{\mathrm{ld}} + u''^2_{\mathrm{ld}}) - (v'^2_{\mathrm{ld}} + v''^2_{\mathrm{ld}})}\right] \tag{48}$$

Further, the phase angle $\theta_{\max}$ which gives the Lagrangian residual drift velocity a maximum magnitude is

$$\theta_{\max} = \frac{1}{2}\tan^{-1}\left[2\frac{u'_{\mathrm{ld}}u''_{\mathrm{ld}} + v'_{\mathrm{ld}}v''_{\mathrm{ld}}}{(u'^2_{\mathrm{ld}} - u''^2_{\mathrm{ld}}) - (v'^2_{\mathrm{ld}} - v''^2_{\mathrm{ld}})}\right] \tag{49}$$

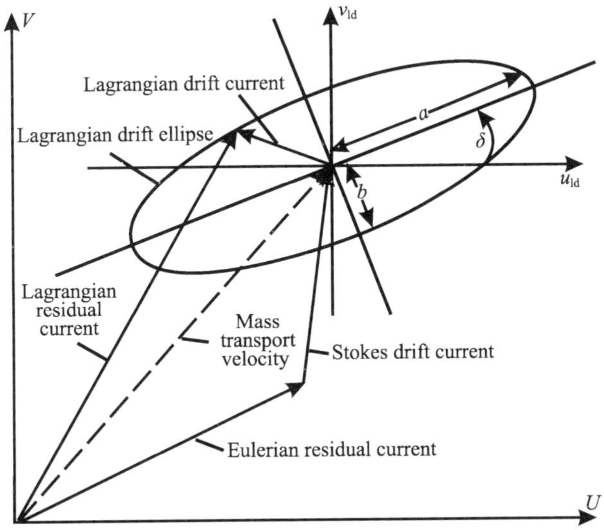

Fig. 1 The Lagrangian residual ellipse and the interrelations between the Lagrangian residual current, Eulerian residual current, Stokes drift, and the Lagrangian residual drift

The properties of the Lagrangian residual ellipse are very similar to the properties of the parent tidal current ellipse, as the Lagrangian drift is induced by the tidal current field. Figure 1 summarizes, graphically, the interrelations between the Lagrangian drift, Stokes drift, and the Eulerian residual velocity, and the properties of the Lagrangian residual ellipse.

A Few Classes of Flows

In this section, several examples are given to further illustrate the properties of the net Lagrangian displacement and the Lagrangian residual drift. In principle, there should not be any difficulty in obtaining the solutions for astronomical tides and tidal currents (i.e., the zeroth order solution) and for the Eulerian residual current. Since the emphasis here is on the Lagrangian processes of the fluid flows, applications of the complete solution procedure in realistic basins will be reported separately. Instead, several known solutions are used as the starting point for further scrutiny of the Lagrangian drift.

Velocity field without shear

It has been suggested that the net Lagrangian displacement is generated by velocity shear due either to nonuniform amplitude or nonuniform phase distributions of tidal current in space. To demonstrate this point, Awaji *et al.* (1980) have shown in a few cases that the velocity shear is responsible for the net Lagrangian displacement. Although the examples given by Awaji *et al.* (1980) are correct, they did not include the class of flows without shear. An interesting question can be posed: Is velocity shear a necessary condition for the existence of Lagrangian residual drift? In order to answer this question, consider a unidirectional flow without shear for which v=0 and $\partial u / \partial y = 0$. Since all velocity components in the y direction are zero, the Stokes and Lagrangian drifts, (43) and (44), can be simplified to

$$u_{\mathrm{sd}} = \left[u_0' \partial u_0'' / \partial x - u_0'' \partial u_0' / \partial x\right] / 2 \tag{50}$$

and

$$\begin{aligned} u_{\mathrm{ld}}' &= u_0'' \partial(u_{\mathrm{er}} + u_{\mathrm{sd}}) / \partial x - (u_{\mathrm{er}} + u_{\mathrm{sd}}) \partial u_0'' / \partial x \\ u_{\mathrm{ld}}'' &= -u_0' \partial(u_{\mathrm{er}} + u_{\mathrm{sd}}) / \partial x + (u_{\mathrm{er}} + u_{\mathrm{sd}}) \partial u_0' / \partial x \end{aligned} \tag{51}$$

It becomes immediately obvious that when the velocity shear is zero, the Stokes drift and Lagrangian drift are not necessarily zero. Even when the Eulerian residual velocity is also zero, the Lagrangian drift could still be nonzero; see (51).

Consider the solution of an ideal progressive wave in an infinite domain and in one dimension (Lamb, 1932); the zeroth order solution, in nondimensional form, is

$$\begin{aligned} \zeta_0 &= \cos(\theta - x) \\ u_0 &= \cos(\theta - x) \end{aligned} \tag{52}$$

and the first-order solution is

$$\begin{aligned} \zeta_1 &= -3x / 4\cos(2\theta - 2x) \\ u_1 &= -\frac{1}{8}\cos(2\theta - 2x) - 3x / 4\sin(2\theta - 2x) \end{aligned} \tag{53}$$

From inspection of (52) and (53) the Eulerian residual velocity is shown to be zero. Recall that the Stokes drift is induced from the nonlinear interaction of the zeroth-order solution, (50), and the Lagrangian drift is induced from the nonlinear interactions between the zeroth-order solution and the Eulerian residual current and the Stokes drift, (51). In the expanded form, (52) becomes

$$\begin{aligned} \zeta_0 &= \zeta_0' \cos\theta + \zeta_0'' \sin\theta \\ u_0 &= u_0' \cos\theta + u_0'' \sin\theta \end{aligned} \tag{54}$$

where $\zeta_0' = u_0' = \cos x$ and $\zeta_0'' = u_0'' = \sin x$. By direct substitutions the Lagrangian residual current can be calculated to give

$$u_{\mathrm{lr}}=u_{\mathrm{er}}+u_{\mathrm{sd}}+\kappa u_{\mathrm{ld}} \tag{55}$$

where $u_{\mathrm{er}}=0$, $u_{\mathrm{sd}}=\frac{1}{2}$, and $u_{\mathrm{ld}}=-\frac{1}{2}\cos(\theta_0-x_0)$.

The Stokes drift is a constant and is independent of the tidal current phase. The Lagrangian drift velocity is obviously an ellipse on a hodograph plane, although the Lagrangian drift ellipse is a degenerate ellipse, or a straight line.

Consider a subclass of flows for which the phase of the tidal current is further required to be constant, or $u_0''/u_0'=\text{constant}$. A typical standing wave belongs to this class, and (50) shows that the Stokes drift is identically zero. Consequently, the Lagrangian drift is also zero. Comparing these two examples, the Lagrangian drift is nonzero for a progressive wave, and it is zero for a standing wave when the phase is constant. These results seem to suggest that the phase variation of the tidal current plays a more influential role in the mechanism which generates Lagrangian residual current than the amplitude variation of the tidal current.

The tidal velocity field is uniform in the direction of the wave front: Sverdrup wave

Next, consider the class of two-dimensional tidal waves in an infinite domain. Assuming that the tidal wave propagates in the x direction, and all the dependent variables are uniform in y, or $\partial(u,v,\zeta)/\partial y=0$. Consequently, the Stokes and the Lagrangian drifts are invariant in y and can be simplified to

$$\begin{aligned}u_{\mathrm{sd}}&=\left[u_0'\partial u_0''/\partial x-u_0''\partial u_0'/\partial x\right]/2\\ v_{\mathrm{sd}}&=\left[u_0'\partial v_0''/\partial x-u_0''\partial v_0'/\partial x\right]/2\end{aligned} \tag{56}$$

and

$$\begin{aligned}u_{\mathrm{ld}}'&=u_0''\partial(u_{\mathrm{er}}+u_{\mathrm{sd}})/\partial x-(u_{\mathrm{er}}+u_{\mathrm{sd}})\partial u_0''/\partial x\\ u_{\mathrm{ld}}''&=-u_0'\partial(u_{\mathrm{er}}+u_{\mathrm{sd}})/\partial x+(u_{\mathrm{er}}+u_{\mathrm{sd}})\partial u_0'/\partial x\\ v_{\mathrm{ld}}'&=u_0''\partial(v_{\mathrm{er}}+v_{\mathrm{sd}})/\partial x-(u_{\mathrm{er}}+u_{\mathrm{sd}})\partial v_0''/\partial x\\ v_{\mathrm{ld}}''&=-u_0'\partial(v_{\mathrm{er}}+v_{\mathrm{sd}})/\partial x+(u_{\mathrm{er}}+u_{\mathrm{sd}})\partial v_0'/\partial x\end{aligned} \tag{57}$$

A typical example in this class is the Sverdrup wave. The Sverdrup wave is a two-dimensional harmonic wave in an infinite, constant depth, and frictionless domain (Defant, 1961); it is a progressive wave in nature and its frequency must be smaller than the local inertia frequency. Semidiurnal tides are in this class of waves, and some phenomena of tidal wave propagation can be represented by the Sverdrup wave. A typical form of the Sverdrup wave in dimensionless form is given by Defant (1961) as

$$\begin{aligned}\zeta_0&=\sin(\theta-x)=\zeta_0'\cos\theta+\zeta_0''\sin\theta\\ u_0&=\sin(\theta-x)=u_0'\cos\theta+u_0''\sin\theta\\ v_0&=f\cos(\theta-x)=v_0'\cos\theta+v_0''\sin\theta\end{aligned} \tag{58}$$

where $0 < f < 1$, the wave speed is $C=(1-f^2)^{-1/2}$, and

$$\zeta_0' = u_0' = -\sin x \quad \zeta_0'' = u_0'' = \cos x$$
$$v_0' = f\cos x \qquad v_0'' = f\sin x$$

The characteristics of the Sverdrup wave are that the tidal velocity component in the direction of wave propagation is in phase with the water level and the phase of the velocity component normal to the wave direction lags the water level by 90°. The major axis of the tidal ellipse parallels the direction of wave propagation, and the maximum tidal current speeds are reached at high and low waters. The tidal current ellipse whose aspect ratio is f rotates clockwise in the northern hemisphere.

The solution for the Eulerian residual currents, (u_{er}, v_{er}), is trivial for a Sverdrup wave in an infinite domain. Substituting the solution for the Sverdrup wave, (58), into (56) and (57) gives the Stokes and the Lagrangian drifts as

$$u_{sd} = 1/2 \quad v_{sd} = 0 \tag{59}$$

and

$$u_{ld} = u_{ld}'\cos\theta + u_{ld}''\sin\theta$$
$$v_{ld} = v_{ld}'\cos\theta + v_{ld}''\sin\theta$$

where

$$u_{ld}' = \frac{1}{2}\sin x_0 \qquad u_{ld}'' = -\frac{1}{2}\cos x_0$$
$$v_{ld}' = -f/2\cos x_0 \qquad v_{ld}'' = -f/2\sin x_0$$

Rewriting the Lagrangian drift in the form of amplitude and phase, the Lagrangian residual drift takes the typical form of an ellipse,

$$\begin{aligned} u_{ld} &= \frac{1}{2}\sin(\theta_0 - x_0 - \pi) \\ v_{ld} &= f/2\cos(\theta_0 - x_0 - \pi) \end{aligned} \tag{60}$$

on a hodograph plane. The Lagrangian residual current can be written in a simplified form as

$$\begin{aligned} u_{lr} &= \frac{1}{2} + \kappa\left[\frac{1}{2}\sin(\theta_0 - x_0 - \pi)\right] \\ v_{lr} &= \kappa[f/2\cos(\theta_0 - x_0 - \pi)] \end{aligned} \tag{61}$$

Of particular interest are the properties of the Lagrangian drift ellipse. In this case, the major axis of the Lagrangian residual ellipse coincides with that of the tidal ellipse, and the lengths of the semimajor and semiminor axes are equal to 1/2 and f/2, respectively. The eccentricities for these ellipses are the same and are equal to f. Both ellipses are clockwise rotating, but they are 180° out of phase (Figure 2a). Several trajectories of labeled water parcels which are released from the same point but at different phases of the tidal current are plotted in Figure 2b. The water parcels end at different positions after a complete tidal cycle, their termini (net Lagrangian displacement) forming an ellipse in space.

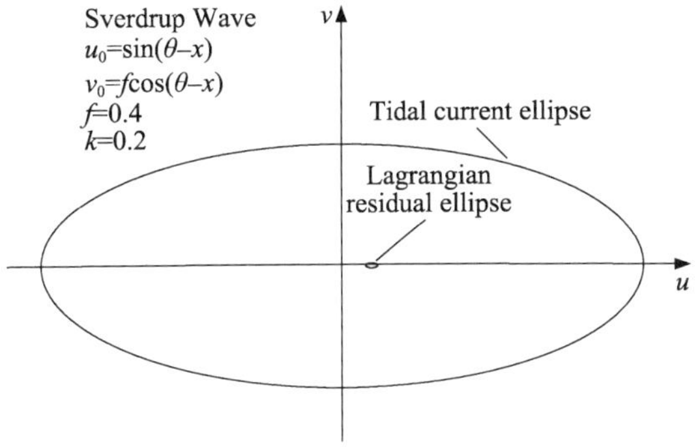

Fig. 2a The tidal current ellipse and the Lagrangian residual ellipse of a Sverdrup wave

The maximum Lagrangian displacement is obtained when the water parcel is released at low water (phase=3π/2) when the tidal current is pointing in the opposite direction to the wave propagation. On the contrary, if the water parcel is released at high water, when the tidal current is pointing in the same direction as the wave propagation, the Lagrangian displacement is a minimum. The Stokes drift is always positive or is in the direction of wave propagation. Since the Lagrangian drift is smaller than the Stokes drift, the mean Lagrangian residual current is also always positive, or in the direction of wave propagation.

The tidal velocity field is unidirectional but not uniform in the direction of the wave front

The class of flows just discussed can be further extended by removing the constraint that the dependent variables are constant in the direction of the wave front. Instead, the velocity normal to the wave direction is assumed to be zero, or v=0. Again, for this lass of flows, the Eulerian residual velocity can be shown to be identically zero. Under these conditions, the Stokes and the Lagrangian drifts are simplified to

$$\begin{aligned} u_{\mathrm{sd}} &= \frac{1}{2}\left[u_0'\partial u_0'' / \partial x - u_0''\partial u_0' / \partial x\right] \\ v_{\mathrm{sd}} &= 0 \end{aligned} \tag{62}$$

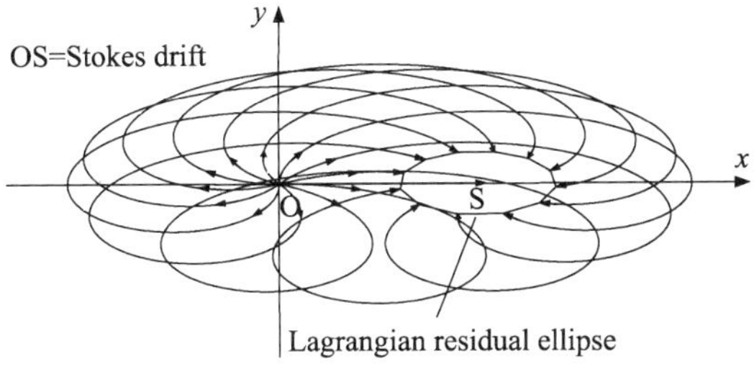

Fig. 2b The water parcel trajectories due to a Sverdrup wave

The labeled water parcels are released from "0", and their termini form an ellipse in space after a complete tidal cycle and

$$
\begin{aligned}
u'_{\mathrm{ld}} &= u''_0 \partial(u_{\mathrm{er}} + u_{\mathrm{sd}}) / \partial x - (u_{\mathrm{er}} + u_{\mathrm{sd}}) \partial u''_0 / \partial x \\
u''_{\mathrm{ld}} &= -u'_0 \partial(u_{\mathrm{er}} + u_{\mathrm{sd}}) / \partial x + (u_{\mathrm{er}} + u_{\mathrm{sd}}) \partial u'_0 / \partial x \\
v'_{\mathrm{ld}} &= v''_{\mathrm{ld}} = 0
\end{aligned} \tag{63}
$$

Note that all residual velocities in the y direction are zero.

The Kelvin wave

The Kelvin wave in an infinitely long, constant depth channel is a typical flow in this class (LeBlond and Mysak, 1978). The Kelvin wave can also be visualized as the propagation of tidal waves in a narrow strait. The solution for a Kelvin wave in dimensionless form is

$$
\begin{aligned}
\zeta_0 &= \mathrm{e}^{-fy} \cos(\theta - x) \\
u_0 &= \mathrm{e}^{-fy} \cos(\theta - x)
\end{aligned} \tag{64}
$$

and note that $u_{\mathrm{er}} = 0$. In the expanded form, (64) becomes

$$
\begin{aligned}
\zeta_0 &= \zeta'_0 \cos\theta + \zeta''_0 \sin\theta \\
u_0 &= u'_0 \cos\theta + u''_0 \sin\theta
\end{aligned} \tag{65}
$$

where

$$
u'_0 = \zeta'_0 = \mathrm{e}^{-fy} \cos x \quad \text{and} \quad u''_0 = \zeta''_0 = \mathrm{e}^{-fy} \sin x .
$$

From (62) and (63), the Stokes and the Lagrangian drifts are found to be

$$
\begin{aligned}
u_{\mathrm{sd}} &= \frac{1}{2} \mathrm{e}^{-2fy} \\
u'_{\mathrm{ld}} &= -\frac{1}{2} \mathrm{e}^{-3fy} \cos x_0 \\
u''_{\mathrm{ld}} &= -\frac{1}{2} \mathrm{e}^{-3fy} \sin x_0
\end{aligned} \tag{66}
$$

Or,

$$
u_{\mathrm{ld}} = \frac{1}{2} \mathrm{e}^{-3fy} \cos(\theta_0 - x_0 - \pi)
$$

Some observations can be made for the Kelvin wave: (1) The Stokes drift is always positive, in the same direction of the wave propagation; (2) the Lagrangian drift is 180° out of phase with the tidal current; (3) the rates at which u_0, u_{sd}, and u_{ld} decrease across the channel are proportional to e^{-fy} : e^{-2fy}: e^{-3fy}. In the northern hemisphere, (f>0); when the observer looks in the direction of the wave propagation, all dependent variables decrease away to his/her left.

An amphidromic system

A more interesting flow is the superposition of two opposite traveling Kelvin waves which form an amphidromic system (Figure 3). This type of amphidromic system is common in coastal seas, coastal bays, and semienclosed embayments. The Kelvin wave pair is, of course, a simplest model representing such a system. The solution of a Kelvin wave pair is

$$\begin{aligned}\zeta_0 &= \mathrm{e}^{-fy}\cos(\theta - x) - \mathrm{e}^{-fy}\cos(\theta + x) \\ u_0 &= \mathrm{e}^{-fy}\cos(\theta - x) + \mathrm{e}^{fy}\cos(\theta + x)\end{aligned} \tag{67}$$

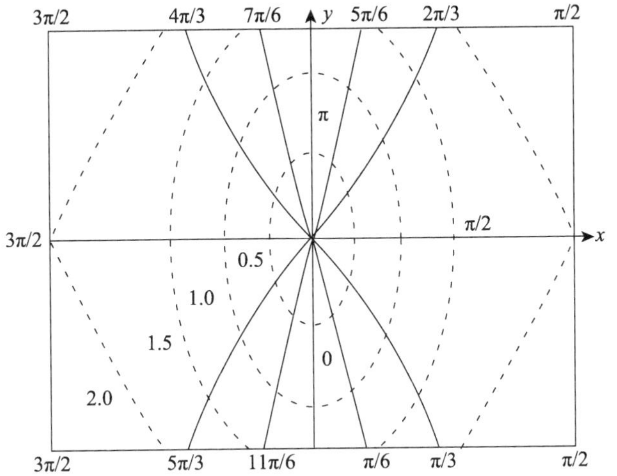

Fig. 3a The cophase lines (solid) and the coamplitude lines (dotted) of the water level in an amphidromic system

The coordinates of the solution are chosen so that the origin of the coordinates is the amphidromic point. Rewriting (67) in the expanded form, the solution of the Kelvin wave pair becomes

$$\begin{aligned}\zeta_0 &= \zeta_0' \cos\theta + \zeta_0'' \sin\theta \\ u_0 &= u_0' \cos\theta + u_0'' \sin\theta\end{aligned} \tag{68}$$

where

$$\zeta_0' = -2\sinh(fy)\cos x \quad \zeta_0'' = 2\cosh(fy)\sin x$$
$$u_0' = 2\cosh(fy)\cos x \quad u_0'' = -2\sinh(fy)\sin x$$

Once the tidal current is known, (68), the Stokes and the Lagrangian drifts can be calculated straightforwardly to give

$$\begin{aligned}u_{\mathrm{sd}} &= -\sinh(2fy) \\ u_{\mathrm{ld}} &= u_{\mathrm{ld}}' \cos\theta + u_{\mathrm{ld}}'' \sin\theta\end{aligned} \tag{69}$$

with

$$u_{\mathrm{ld}}' = -2\sinh(2fy)\sinh(fy)\cos x$$
$$u_{\mathrm{ld}}'' = 2\sinh(2fy)\cosh(fy)\sin x$$

The Lagrangian drift ellipse is

$$u_{\mathrm{ld}} = |u_{\mathrm{ld}}|\cos(\theta_0 - \alpha) \tag{70}$$

where $|u_{\rm ld}| = (u'^2_{\rm ld} + u''^2_{\rm ld})^{1/2}$ and $\alpha = \tan^{-1}(u''_{\rm ld} / u'_{\rm ld})$

For the convenience of discussion and without losing generality, let $f=0.5$, $x=[-\pi/2, \pi/2]$, and $y=[-1.2, 1.2]$. The cophase lines and coamplitude lines of the water level are shown in Figure 3a. The origin of the coordinates is the amphidromic point. Because of the condition that v=0 all tidal ellipses take a degenerate form of a straight line, or the tidal flows are bidirectional oscillatory flows. By taking advantage of this property, the distribution of the tidal current in space and in time can be exhibited in a similar amplitude-phase diagram. Depicted in Figure 3b are the cophase and coamplitude lines of u_0. The phase different between the water level and tidal current is a variable which is neither near zero nor 90° (Figures 3a and 3b), and the wave characteristics in this amphidromic system are neither those of progressive waves nor those of standing waves.

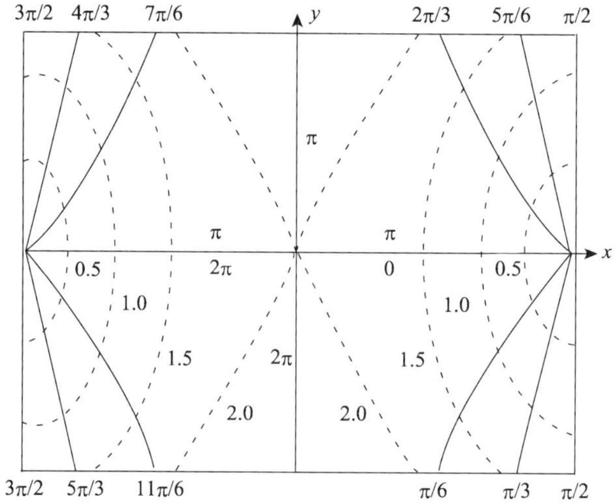

Fig. 3b The cophase lines (solid) and the coamplitude lines (dotted) of the tidal current in an amphidromic system

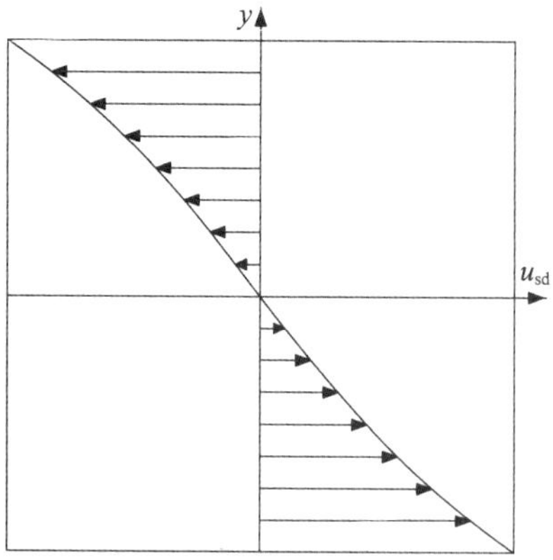

Fig. 3c The Stokes drift which is an antisymmetric function of y in an amphidromic system

The Stokes drift is an antisymmetric function of y (Figure 3c), which induces an anticlockwise drift gyre in the northern hemisphere. The Stokes drift is positive (in the direction of the x axis) in the lower half of the channel and negative in the upper half of the channel. The net drift across the channel, however, is zero.

Because v=0, $v_{\rm er}$, and $v_{\rm lr}$ are also zero, the Lagrangian drift ellipses are degenerated to lines parallel to the x axis. The temporal and spatial distributions of the Lagrangian drift ellipse are quite complex; however, their properties can be completely depicted by plotting the cophase and coamplitude lines (Figure 3d). Note that the cophase lines of the water elevation (Figure 3a) and the cophase lines of the Lagrangian drift ellipses (Figure 3d) are identical. This means that the maximum Lagrangian drift can be obtained by releasing the labeled water parcels at high water. The maximum Lagrangian residual currents are found near the four corners of the amphidromic system, while the maximum tidal current is located near the amphidromic point. The Lagrangian residual current is small near the x axis; in fact, it is exactly zero on the x axis.

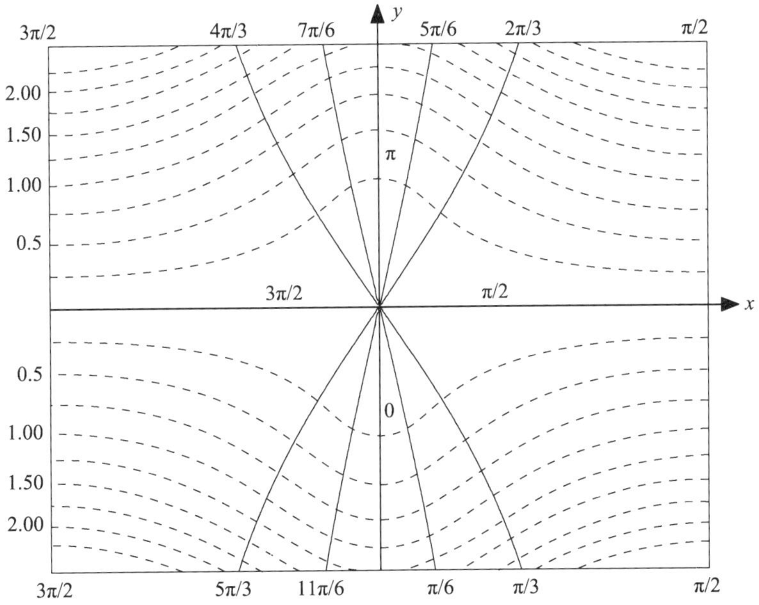

Fig. 3d The cophase lines (solid) and the coamplitude lines (dotted) of the Lagrangian drift in an amphidromic system

Conclusion

The dynamics of the tide-induced Lagrangian residual current has been discussed for shallow-water estuaries and coastal embayments. The analysis is based on a two-dimensional barotropic model on a horizontal plane. The depth-averaged barotropic shallow-water equations are used to describe the tides and tidal currents in the system. Under the weakly nonlinear approximation for the tidal system the present analysis has systematically revealed that the Lagrangian residual current is generated from nonlinear coupling among astronomical tidal constituents and their higher-order harmonics. Using a Lagrangian approach, the first-order Lagrangian residual current has been shown to be the sum of the Eulerian residual current and the Stokes drift, or the first-order approximation of the Lagrangian residual current has been shown to be the same as the mass transport velocity introduced by Longuet-Higgins (1969).

The second-order Lagrangian residual current has the characteristics of tidal period variations, and the second-order correction terms are referred to as the Lagrangian residual drift. On a hodograph plane the Lagrangian residual current describes an ellipse over a complete tidal cycle. Although the quantitative second-order correction to the Lagrangian residual current may appear to be unimportant, it is only when the second-order dynamics is examined that the generating mechanism of the Lagrangian drift can be fully revealed and understood. It has been shown that the Lagrangian drifts is generated from the nonlinear interactions between the astronomical tides with both the Eulerian residual current and Stokes drift. The property that the Lagrangian residual current depends on the phase of tidal current has led to the notion of a Lagrangian residual ellipse. The properties of the Lagrangian residual ellipse have been further examined through several examples. The implication from the present study is rather significant. Results of the present study suggest that it is the Lagrangian residual current, not the Eulerian residual current, which determines long-term transport phenomena. Further demonstrations of the role of the Lagrangian residual current in the long-term (intertidal) transport equation have been discussed by Cheng *et al.* (1985), and the discussions of the Lagrangian residual transport are the subject of Feng *et al.* (1986).

Appendix

Some details for the method of the jth order solutions are given in this appendix for completeness. The governing system of equations, (15) to (25), is linear for any order of the solution, and thus it can be solved systematically in ascending order of j. Since each solution $(\zeta_{j,p}, u_{j,p}, v_{j,p})$ is periodic and has a fixed frequency $\sigma_{j,p}$, the time dependency in the governing equations can be removed by introducing the complex notation

$$\psi = \psi' \cos\sigma\theta + \psi'' \sin\sigma\theta = \mathrm{Re}\left[\tilde{\psi} \mathrm{e}^{-\mathrm{i}\sigma\theta}\right] \tag{A1}$$

where ψ stands for any dependent variable, i equals $(-1)^{1/2}$, and σ stands for the frequency of any harmonic. The notation $\tilde{\psi}$ is defined as $\tilde{\psi} = \psi' + \mathrm{i}\psi''$, and Re[ ]= real part of [ ].

In the new notation, the right-hand sides of (20), (21), and (22), Φ_{j-1}, X_{j-1} and Y_{j-1}, can be written as

$$\Phi_{j-1} = \sum_{m=0}^{j-1} \sum_{p,q} \mathrm{Re}[\phi_{p+q} \exp\{-\mathrm{i}(\sigma_{m,p} + \sigma_{j-1-m,q})\theta\} + \phi_{p-q} \exp\{-\mathrm{i}(\sigma_{m,p} - \sigma_{j-1-m,q})\theta\}] \tag{A2}$$

$$X_{j-1} = \sum_{m=0}^{j-1} \sum_{p,q} \mathrm{Re}[\chi_{p+q} \exp\{-\mathrm{i}(\sigma_{m,p} + \sigma_{j-1-m,q})\theta\} + \chi_{p-q} \exp\{-\mathrm{i}(\sigma_{m,p} - \sigma_{j-1-m,q})\theta\}] \tag{A3}$$

$$Y_{j-1} = \sum_{m=0}^{j-1} \sum_{p,q} \mathrm{Re}[\Upsilon_{p+q} \exp\{-\mathrm{i}(\sigma_{m,p} + \sigma_{j-1-m,q})\theta\} + \Upsilon_{p-q} \exp\{-\mathrm{i}(\sigma_{m,p} - \sigma_{j-1-m,q})\theta\}] \quad \text{(A4)}$$

where

$$\phi_{p+q} = [\partial(\tilde{\zeta}_{m,p}\tilde{u}_{j-1-m,q}) / \partial x + \partial(\tilde{\zeta}_{m,p}\tilde{v}_{j-1-m,q}) / \partial y] / 2$$

$$\phi_{p-q} = [\partial(\tilde{\zeta}_{m,p}\tilde{u}_{j-1-m,q}{}^{*}) / \partial x + \partial(\tilde{\zeta}_{m,p}\tilde{v}_{j-1-m,q}{}^{*}) / \partial y] / 2$$

$$\chi_{p+q} = [\tilde{u}_{m,p}\partial\tilde{u}_{j-1-m,q} / \partial x + \tilde{v}_{m,p}\partial\tilde{u}_{j-1-m,q} / \partial y] / 2$$

$$\chi_{p-q} = [\tilde{u}_{m,p}\partial\tilde{u}_{j-1-m,q}{}^{*} / \partial x + \tilde{v}_{m,p}\partial\tilde{u}_{j-1-m,q}{}^{*} / \partial y] / 2$$

$$\Upsilon_{p+q} = [\tilde{u}_{m,p}\partial\tilde{v}_{j-1-m,q} / \partial x + \tilde{v}_{m,p}\partial\tilde{v}_{j-1-m,q} / \partial y] / 2$$

$$\Upsilon_{p-q} = [\tilde{u}_{m,p}\partial\tilde{v}_{j-1-m,q}{}^{*} / \partial x + \tilde{v}_{m,p}\partial\tilde{v}_{j-1-m,q}{}^{*} / \partial y] / 2$$

in which the subscripts $p+q$ and $p-q$ indicate that the modulations are due to the sum and the difference of the p th and q th harmonics, and the superscript $*$ denotes complex conjugates. The notation $\sum_{p,q}$ stands for summation over all possible harmonics within the $(j-1)$ th order solutions. Without causing any confusion, from this point on, the symbol $\sim$ and subscripts for all variables are dropped. The governing equations, (15) and (17), for the j th order harmonics have been transformed to a steady state boundary value problem of complex variables,

$$-\mathrm{i}\sigma\zeta + \partial(hu) / \partial x + \partial(hv) / \partial y = -\phi \quad \text{(A5)}$$

$$-\mathrm{i}\sigma u - fv + \partial\zeta / \partial x + \gamma u = -\chi \quad \text{(A6)}$$

$$-\mathrm{i}\sigma v + fu + \partial\zeta / \partial y + \gamma v = -\Upsilon \quad \text{(A7)}$$

On Γ_1

$$u\cos\alpha_x + v\cos\alpha_y = 0 \quad \text{(A8)}$$

and ζ is given on Γ_2,

$$\zeta = S \quad \text{(A9)}$$

Substituting (u,v), which can be found explicitly from (A6) and (A7), into (A5) gives rise to a boundary value problem of ζ as

$$\begin{aligned} & hA\nabla^2\zeta + [A(\partial h / \partial x) - B(\partial h / \partial y) + h(\partial A / \partial x - \partial B / \partial y)]\partial\zeta / \partial x + \\ & [A(\partial h / \partial y) + B(\partial h / \partial x) + h(\partial A / \partial y + \partial B / \partial x)]\partial\zeta / \partial y + \mathrm{i}\sigma\zeta = (\phi - F) \end{aligned} \quad \text{(A10)}$$

The boundary condition on Γ_1 is

$$A\partial\zeta / \partial n + B\partial\zeta / \partial n = -(A\psi_n - B\psi_\tau) \quad \text{(A11)}$$

in which

$$\psi_n = \chi\cos\alpha_x + \Upsilon\cos\alpha_y \quad \text{and} \quad \psi_\tau = \chi\cos\alpha_y - \Upsilon\cos\alpha_x$$

where n and τ are the outward pointing normal and tangent directions of Γ_1. The boundary condition on Γ_2 becomes

$$\zeta = S \tag{A12}$$

The symbols in (A10) are defined as

$$\begin{aligned} A &= (\gamma - \mathrm{i}\sigma)/[(\gamma - \mathrm{i}\sigma)^2 + f^2] \\ B &= f/[(\gamma - \mathrm{i}\sigma)^2 + f^2] \\ F &= \partial(Ah\chi + Bh\Upsilon)/\partial x + \partial(Ah\Upsilon - Bh\chi)/\partial y \end{aligned} \tag{A13}$$

Once the solution of ζ is obtained, the velocity field can be calculated from

$$u = -[A\partial\zeta/\partial x + B\partial\zeta/\partial y + A\chi + B\Upsilon] \tag{A14}$$

$$v = -[A\partial\zeta/\partial y - B\partial\zeta/\partial x + A\Upsilon - B\chi] \tag{A15}$$

This formulation and the method of solution were first proposed and used by Hansen (1952) for the linear approximation of tides and tidal currents. This approach was further extended to nonlinear three-dimensional problems by Feng (1977) with applications in Bohai Sea, China (Sun *et al.*, 1981; Feng and Sun, 1983). Parenthetically, this formulation is based on the assumptions that κ is a small parameter and the assumption that the bottom stress is linear.

References

Awaji, T., N. Imasato, and H. Kunishi, Tidal exchange through a strait: A numerical experiment using a simple model basin, *J. Phys. Oceanogr.*, 10(11), 1499–1508, 1980.

Cheng, R. T., Euler-Lagrangian computations in estuarine hydrodynamics, in *Proceedings of Third International Conference on Numerical Methods in Laminar and Turbulent Flow*, edited by C. Taylor, J. A. Johnson, and R. Smith, pp. 341–352, Pineridge Press, Swansea, United Kingdom, 1983.

Cheng, R. T., and V. Casulli, On Lagrangian residual currents with applications in South San Francisco Bay, California, *Water Resour. Res.*, 18, 1652–1662, 1982.

Cheng, R. T., and J. W. Gartner. Tides, tidal, and residual currents in San Francisco Bay, California: Results of measurements, 1979–1980, Parts I to V, *U.S. Geol. Surv. Water Resour. Invest. Rep.*, 84–4339, 1–1747, 1984.

Cheng, R. T., and J. W. Gartner. Harmonics analysis of tides and tidal currents in South San Francisco Bay, California, *Estuarine Coastal Shelf Sci.*, 21, 57–74, 1985.

Cheng, R. T., S. Feng, and P. Xi, On inter-tidal transport equation, paper presented at William and Mary Charter Day Symposium on Circulation in Estuaries, January 1985, Va. Inst. of Mar. Sci., Gloucester Point, Va., 1985.

Defant, A., *Physical Oceanography*, vol.1, 598 pp., Pergamon, New York, 1961.

Dooley, H. D., A comparison of drogue and current meter measurement in shallow water, Rapports on Procés-Verbaux des Réunions, *Rep.*, 167, pp. 225–230, Conseil Permanent International pourl' Exploration de la Mer, 1974.

Dyke, P. P. G., On the Stokes' drift induced by tidal motions in a wide estuary, *Estuarine Coastal Mar. Sci.*, 11, 17–25, 1980.

Elliott, A. J., and T. E. Hendrix, Intensive observations of the circulation in the Potomac estuary, *Spec. Rep.*, 55, 35 pp., Chesapeake Bay Inst., Baltimore, Md., 1976.

Feng, S., A three-dimensional non-linear model of tides, *Sci. Sin.*, 20, 436–446, 1977.

Feng, S., and W. Sun, A tidal three-dimensional nonlinear model with variable eddy viscosity, *Chin. J. Oceanol. Limnol.*, 1, 166–170, 1983.

Feng, S., R. T. Cheng, and P. Xi, On tide-induced Lagrangian residual current and residual transport, 2, Residual transport with applications in South San Francisco Bay, California, *Water Resour. Res.*, this issue, 1986.

Hansen, W., Gezeiten und Gezeitenstrome der halbtagigen Hauptmondtide M_2 in der Nordsee, *Dtsch. Hydrogr. Z.*, 1, erganzungsh., 1952.

Heaps, N. S., Linearized vertically-integrated equations for residual circulation in coastal seas, *Deut. Hydrogr. Z.*, 31, 147–169, 1978.

Lamb, H., *Hydrodynamics*, 6th ed., Cambridge University Press, New York, 1932.

LeBlond, P. H., and L. Mysak. *Waves in the Ocean*, Elsevier, New York, 1978.

Longuet-Higgins, M. S., On the transport of mass by time-varying ocean currents, *Deep Sea Res.*, 16, 431–447, 1969.

Mulder, R., Eulerian and Lagrangian analysis of velocity fields in the southern North Sea, in *North Sea Dynamics*, edited by Sundermann/Lenz, pp. 134–147, Springer-Verlag, New York, 1982.

Nihoul, J. C. J., and R. C. Ronday, The influence of the tidal stress on the residual circulation, *Tellus*, 27, 484–489, 1975.

Parker, B. B., Frictional effects on the tidal dynamics of a shallow estuary, P.h.D. dissertation, Johns Hopkins Univ., Baltimore, Md., 1984.

Pedlosky, J., *Geophysical Fluid Dynamics*, 624 pp., Spinger-Verlag, New York, 1979.

Pingree, R. D., and L. Maddock, The M_4 tide in the English channel derived from a nonlinear numerical model, *Deep Sea Res.*, 25, 53–63, 1978.

Stern, M. E., *Ocean circulation Physics*, Academic, Orlando, Fla., 1975.

Stoker, J. J., *Water Waves*, Wiley-Interscience, New York, 1965.

Sun, W., S. Chen, and S. Feng, Numerical simulation of three-dimensional nonlinear tidal waves, I, A numerical study for M_4 and MS_4 waves in the Bohai Sea, *J. Shandong Coll. Oceanol.*, 11, 23–31, 1981.

Tee, T. K., Tide-induced residual current: A 2-D nonlinear numerical model, *J. Mar. Res.*, 31, 603–628, 1976.

Uncles, R. J., and M. B. Jordan, A one-dimensional representation of residual currents in the Severn estuary and associated observations, *Estuarine Coastal Mar. Sci.*, 10, 39–60, 1980.

Van Dyke, M., *Perturbation Methods in Fluid Mechanics*, Academic, Orlando, Fla., 1964.

Weisberg, R. H., The nontidal flow in the Providence River of Narragansett Bay: A stochastic approach to estuarine circulation, *J. Phys. Oceanogr.*, 6, 721–734, 1976.

Yu, G., and S. Chen. Numerical modeling of the circulation and the pollutant dispersion in Jiaozhou Bay, Ⅲ, The Lagrangian residual current and the pollutant dispersion, *J. Shandong Coll. Oceanol.*, 13, 1–14, 1983.

Zimmerman, J. T. F., On the Euler-Lagrangian transformation and the Stokes drift in the presence of oscillatory and residual currents, *Deep Sea Res.*, 26A, 505–520, 1979.

On Tide-induced Lagrangian Residual Current and Residual Transport
2. Residual Transport with Application in South San Francisco Bay, California *

SHIZUO FENG, RALPH T. CHENG, PANGEN XI

1. Introduction

In recent years, studies of hydrodynamics in oceanographic and environmentally related problems have focused on transport processes that determine the long-term fate of dissolved and suspended matter. Transports of solutes, sediment, salinity, nutrients, and other tracers are fundamental to the interactive physical, chemical, and biological processes in an ecological system. Although the apparent dominating transport mechanism appears to be convection due to tidal currents in coastal estuaries and shallow embayments, it has been generally agreed upon that long-term transport processes are controlled and determined by residual circulation and not by tidal circulation. The residual circulation is generated owing to the nonlinear interactions of the tides and tidal currents and owing to the rectification of basin bathymetry, atmospheric forcing, thermal and salinity gradients, and fresh water inflows at the head of an estuary. The tidal currents can be viewed as the sum of a periodic part and an aperiodic part, where the aperiodic component (the residual) is an order of magnitude smaller than the periodic component. Effects in transport processes due to the periodic tidal current components have intrinsically cancelled themselves, and the net transport of mass and dissolved matter results mainly from the aperiodic tidal current components or the residual circulation. Thus, for transport processes whose characteristic time scale is much greater than a tidal period, the residual circulation plays a more important role than the tidal currents. In addition to convection, the tracers are mixed with adjacent water mass by shear effect dispersion, turbulent mixing, and other mixing mechanisms which are directly or indirectly related to large-scale circulation. The impacts of long-term transport processes often have significant implications for an ecological system. The motivation of the present study is prompted by a need for a further understanding of the long-term behavior of transport process, which is of great value in the development and protection of marine resources in estuaries, coastal embayments, and shallow seas.

---

* Feng S, Cheng R T, Xi P. 1986. On tide-induced Lagrangian residual current and residual transport 2. Residual transport with application in South San Francisco Bay, California. Water Resources Research, 22(12): 1635–1646

Owing to a lack of complete knowledge of circulation in a three-dimensional system, a necessary simplification is the use of a two-dimensional depth-averaged system to approximate the dynamics in estuaries or shallow seas. Because a two-dimensional flow field is used, the transport equation is also necessarily to be approximated by a depth-averaged, two-dimensional equation in the horizontal plane. Commonly, two contrasting approaches are used in dealing with long-term transport problems. In the first, the conservation equation for an intratidal time scale is solved with a time step much shorter than a tidal period using a very accurate numerical solution scheme. Simulations are carried out to cover a sufficiently long period of time (many tidal cycles), and the long-term properties of solutes (intratidal properties) are obtained by averaging the results from the intratidal simulations. At present, this approach is not considered to be practical because of the relatively high computing time requirements and the lack of a suitable numerical algorithm for the simulations. Often the accumulated numerical errors in a long-term simulation can render the final results meaningless. Nevertheless, this may still be an attractive approach when practical computer models, which can provide better accuracy, and higher spatial and temporal resolutions with reasonable computing cost become available.

The second approach is to reformulate the problem by applying a Eulerian time average over several tidal periods to the intratidal conservation equation. This approach is ad hoc for the purpose of arriving at a new governing equation for which the time-averaged properties are considered as the dependent variables and are solved directly. In this approach the intertidal conservation equation takes the form of

$$\boldsymbol{u}_{\mathrm{er}}^{*} \cdot \nabla^{*} < C^{*} > = \frac{1}{H^{*}} \nabla^{*} \cdot (H^{*} D^{*} \nabla^{*} < C^{*} >) \tag{1}$$

which is the time-averaged, two-dimensional, convection-dispersion equation. In (1), $< >$ is a Eulerian time-average operator over one or a few tidal cycles, and C^{*}, H^{*}, and D^{*} are the concentration, the total water depth, and the dispersion coefficient tensor, respectively. The convective velocity, $\boldsymbol{u}_{\mathrm{er}}^{*}$, is the Eulerian residual velocity. Included in the dispersion tensor are the horizontal turbulent diffusion, the dispersion induced by the shear effects of tidal current in the vertical plane (Bowden, 1965), and the phase effects dispersion or the so-called "tidal dispersion". The phase effects refer to the non-zero correlation between the fluctuation of tidal current and the fluctuation of concentrations as a result of time averaging (Officer, 1976; Fischer *et al.*, 1979). Quite often (1) is used to describe long-term transport of conservative tracers (or solutes) without sources and sinks (e.g., Fischer *et al.*, 1979). Equation (1) can be written in a transient form by including a term for the local time rate of change on its left-hand side. In the transient case the Eulerian residual velocity and the dispersion tensor are generally time dependent. Since the time dependent properties of the Eulerian residual velocity and the dispersion tensor are generally not known, the slow-varying transient term is not considered.

Because the intertidal transport equation, (1), is strictly the result of a mathematical averaging (over time) of the intratidal conservation equation, from a physical point of view, (1) has two basic weaknesses. First, it is questionable whether the Eulerian residual velocity gives the proper representation for convection in the long-term transport of solutes as indicated in (1). Recently, researchers have recognized that the long-term transport process are clearly of a Lagrangian nature, and the transport of water mass or tracers (solutes) should be determined by the Lagrangian residual current rather than by the Eulerian residual current (Longuet-Higgins, 1969; Zimmerman, 1979; Awaji *et al.*, 1980; Cheng and Casulli, 1982; Cheng *et al.*, 1984a; Feng *et al.*, 1986). Therefore the use of the Eulerian residual velocity to represent convection needs further examination.

Second, the dispersion tensor has arrived at its present form owing to a fundamental hypothesis that the correlation between the tidal current and the concentration fluctuation can be expressed in a Fickian form. Values for the dispersion coefficients have been estimated using data from concurrent measurements of tidal velocity and tracer concentration over an extended period of time. The estimated coefficients of the dispersion tensor are at least 2 orders of magnitude larger than the horizontal turbulent mixing coefficient and at least an order of magnitude greater than the dispersion coefficients due to shear effect (Dyer, 1973, 1974; Fischer, 1976; Uncles and Jordan, 1979; Uncles and Radford, 1980; Winterwerp, 1983; Lewis and Lewis, 1983). Therefore, when considering the tidal dispersion (or the dispersion due to phase effect), the turbulent mixing and the shear effect dispersion become relatively unimportant. The tidal dispersion coefficients have also been estimated from arguments of dimensional analysis (Stommel and Farmer, 1952; Officer, 1976), from computing the actual Lagrangian water mass movements in a tidal current field by means of a numerical model (Awaji, 1982), and from estimates made using a statistical approach (Zimmerman, 1978). For lack of a concise estimate of the tidal dispersion coefficients, some studies simply neglected the phase effect terms (Pritchard, 1954; Bowden, 1965; Fischer *et al.*, 1979). The tidal dispersion terms have also been explained as being caused by branchings and shoalings in shallow estuaries: such mechanisms have been referred to as "tidal trapping", and /or "tidal pumping" (Fischer, 1976; Fischer *et al.*, 1979). However, some other researchers may hold different views (e.g., Winterwerp, 1983). In any case, the so-called tidal dispersion terms in (1) are the results of a Fickian hypothesis in a mathematical averaging of the conservation equation. The physical meaning and the generation mechanisms of the tidal dispersion terms are not well understood. Furthermore, most previous studies have been concerned with estuaries which are principally one-dimensional; a general consideration of mixing induced by tidal currents in a wide two- or three-dimensional embayment is not known. Yet, it has been suggested that basin-wide residual gyres might have greater effects on dispersion than all of the known mixing mechanisms (Fischer, 1976; Cheng, 1983).

The main purpose of this paper is to derive a long-term transport equation, the intertidal transport equation, which correctly describes the Lagrangian nature of transport process without introducing the Fickian hypothesis for the phase effect dispersion. The dependent variables are the tracer or solute concentrations averaged over a period of time substantially longer than one or a few tidal periods. In the formulation of transport problems, it seems logical to use the volumetric transports (the depth-averaged velocity multiplied by the water depth). In this paper the volumetric residual transport variables are introduced and examined. In part 1 of this paper we used a perturbation method to give solutions to the Lagrangian residual currents up to the second order of κ, which is the ratio of residual current to tidal current. We have shown some interrelations between the Lagrangian residual current, the Eulerian residual current, the Stokes drift, and the Lagrangian drift. The interrelations between the Lagrangian residual transport, the Eulerian residual transport, the Stokes drift transport, and the Lagrangian drift transport are derived and further examined in this paper.

Without confusing the main issue, and for clarity, the effects due to surface wind forcing, thermal stratification, and salinity gradients are not considered in the present analysis. Further, we assume that tracers (including salinity) are dilute and passive or that the nonlinear interaction between the tracer concentration and the dynamics of the tidal flow system is small and can be neglected.

In the following sections the order of magnitude of terms in the conservation equation is first examined. A small parameter κ is again identified, and a formal solution to the conservation equation can be given by a perturbation technique as used by Feng *et al.* (1986). With the aid of a formal solution to the intratidal conservation equation, the intertidal conservation equation is then derived. The long-term transport equation (intertidal governing equation) should be valid for the interior of a basin where the tidal dispersion terms never appear. The tide-induced dispersion becomes important only in a boundary region where the mixing process intensities owing to the presence of irregular shoreline boundaries and owing to radical variations of local bathymetry (change of characteristic length scale). An analytic solution is given to support the present theoretical model, and finally, a salinity transport problem in South San Francisco Bay, California, is used as a practical test case to assess the applicability of the present approach.

2. Governing Equations

The tides and the tidal and residual currents in a weakly nonlinear system have been the subject of discussion by Cheng *et al.* (1984a) and Feng *et al.* (1986). Through an order of magnitude argument, a parameter κ has been identified. The parameter κ is shown to be either the ratio of tidal excursion to the horizontal characteristic length, the ratio of tidal current to the tidal wave celerity, the ratio of residual current to tidal current, or the ratio of tidal amplitude to the mean water depth. In a coastal estuary or a shallow embayment when

κ is indeed small, the weakly nonlinear approximation is valid, and the solutions for the dynamics of such a flow system can be obtained systematically by a perturbation method as discussed in detail in part 1 of this paper.

An intratidal transport equation which takes the form of a convection dispersion equation can be used to describe adequately the transport of dilute, passive tracers over a time period much shorter than a tidal period, i.e.

$$\frac{\partial C^*}{\partial t^*}+u^*\frac{\partial C^*}{\partial x^*}+v^*\frac{\partial C^*}{\partial y^*}=\frac{1}{(h^*+\zeta^*)}\nabla^*\cdot[(h^*+\zeta^*)D^*\nabla^*C^*] \tag{2}$$

where

$C^*(x^*,y^*,t^*)$ concentration of solute;

t^* time;

(x^*,y^*) Cartesian coordinates on a horizontal plane;

(u^*,v^*) velocity components in the (x^*,y^*) directions;

$\zeta^*(x^*,y^*,t^*)$ free-surface displacement;

$h^*(x^*,y^*)$ basin bathymetry;

D^* combined turbulent mixing coefficient and the dispersion coefficient due to shear effects;

∇^* either a divergence or a gradient operator.

Under the same assumptions and approximations as Feng *et al.* (1986), let the characteristic values for the following parameters be denoted as: concentration C_c; tidal velocity, u_c; water surface displacement, ζ_c; water depth, h_c; time, ω_c^{-1}; horizontal length, L_c; ω_c the characteristic circular frequency of the tides, thus a characteristic length scale for tidal excursion can be given as $l_c=u_c/\omega_c$. Let D_c denote the characteristic dispersion coefficient due to shear effects; the turbulent mixing is negligible when compared to shear effect induced dispersion (Bowden, 1965). After substitution, the nondimensional intratidal conservation equation becomes

$$\frac{\partial C}{\partial \theta}+\kappa\left(u\frac{\partial C}{\partial x}+v\frac{\partial C}{\partial y}\right)=\frac{\kappa/Pe}{h+\kappa\zeta}\nabla\cdot[(h+\kappa\zeta)D\nabla C] \tag{3}$$

where variables without superscript are nondimensional and θ is the nondimensional time. There are two nondimensional parameters in (3), namely κ and Pe. The physical meaning of the small parameter κ is the same as used in Feng *et al.* (1986); κ characterizes the relative importance of nonlinearity in a tidal dynamic system. The significance of a Peclet number, Pe, is, of course, the characterization of the relative importance of convection versus dispersion. Specifically, κ and Pe are defined as

$$\kappa = l_c / L_c \quad \text{and} \quad Pe = u_c L_c / D_c \tag{4}$$

In a tidal dynamics system where either semidiurnal or diurnal tides dominate, ω_c is of the order of 10^{-4} (s^{-1}). Assuming that u_c is of the order of 50 to 100 cm/s, then l_c is of the order of (0.5 to 1.0)$\times10^6$ cm. Further, if we consider only the class of large and broad estuaries or shallow seas for which L_c is assumed to be of the order of (0.5 to 1.0)$\times10^7$ cm, then the nondimensional parameter κ is in the range of 0.05 and 0.2, and Pe is of the order of (0.5 to 1.0)$\times10^9/D_c$, where D_c is in cgs units. If we estimate the characteristic dispersion coefficient D_c to be of the order of 10^6 cm^2/s (Okubu, 1967; Bowden, 1967), then Pe is of the order of (0.5 to 1.0)$\times10^3$. From these estimations, Pe is clearly shown to be much larger than order one. The large value of Pe is an indication that solute transport mechanism in the interior of a tidal system is strongly convection dominated and the dispersion effects are probably negligible. The dispersion terms are included in some numerical solutions only for controlling numerical properties of the computations. Usually there is no significant difference in simulation results between cases with and without dispersion in a simulation of one or several tidal cycles (Leendertse, 1970; Leendertse and Gritton, 1971; Fischer *et al.*, 1979). In summary, then, from the order of magnitude estimates of κ and Pe, we have concluded that κ is of small order and Pe is of the order of κ^{-2} or κ^{-3}.

3. Residual Transport in an M_2 System

In a two-dimensional analysis, the tidal velocity distribution in the vertical is assumed to be nearly uniform, and the velocity components on the horizontal plane represent the depth-averaged values of the tidal velocity components. Alternatively, the volumetric transport (or simply transport), which is defined as the depth-averaged velocity times the water depth, can be used in the analysis. In studies of long-term transport of solutes in estuaries and shallow bays, sometimes the problems can be more meaningfully formulated using the volumetric transports instead of the depth-averaged velocities. When the transports are used, the continuity equation takes a simple linear form (Cheng and Walters, 1982).

By definition, the volumetric transports, (U,V), in the x and y directions and in dimensionless form are given as

$$(U,V) = (u,v)(h + \kappa\zeta) \tag{5}$$

where the volumetric transports are normalized by $u_c h_c$. The nondimensional intratidal conservations equations, (3), can be written in terms of the transports as

$$(h+\kappa\zeta)\frac{\partial C}{\partial \theta} + \kappa\left[U\frac{\partial C}{\partial x} + V\frac{\partial C}{\partial y}\right] = \frac{\kappa}{Pe}\nabla\cdot[(h+\kappa\zeta)D\nabla C] \tag{6}$$

Once the transport variables have been introduced, it is interesting to ask: (1) Is there any difference between the Eulerian and the Lagrangian residual transports? and (2) What is the interrelation between the Eulerian and the Lagrangian residual transports? In fact, the term "Lagrangian residual transport" has not been clearly defined in the literature. Only the residual mass transport (referred to as the Lagrangian residual transport) has been previously defined as the mass transport velocity multiplied by the mean water depth, where the mass transport velocity is the sum of the Eulerian residual velocity and the Stokes drift. As pointed out by Zimmerman (1979) and Cheng and Casulli (1982) and will be further clarified below, this usage is valid only for some special cases as the first order of approximation. Conceptually, the Lagrangian residual transport and the residual mass transport are not the same.

By definition, the Eulerian residual transport $(U_{\mathrm{er}}, V_{\mathrm{er}})$ are simply the time-averaged Eulerian volumetric transports, i.e.,

$$(U_{\mathrm{er}}, V_{\mathrm{er}}) = (<U>, <V>) \tag{7}$$

where $< >$ is the Eulerian time-averaging operator. In a weakly nonlinear system for which κ is of small order, the tidal currents (u,v), the water level ζ, and the volumetric transports (U,V) can be expanded in an ascending series of κ as

$$(u,v,\zeta) = \sum_{j=0,1,2,\cdots} (u_j, v_j, \zeta_j)\kappa^j \tag{8}$$

and

$$(U,V) = \sum_{j=0,1,2,\cdots} (U_j, V_j)\kappa^j \tag{9}$$

where the subscript j indicates the jth order solution. Substituting (8) and (9) into (5) and equating coefficients of corresponding powers of κ, the jth order transport become

$$\begin{aligned} U_j &= hu_j + \sum_{m=0}^{j-1} u_m \zeta_{j-1-m} \\ V_j &= hv_j + \sum_{m=0}^{j-1} v_m \zeta_{j-1-m} \end{aligned} \tag{10}$$

Since (10) is linear, the first few orders of the Eulerian residual transport can be obtained straightforwardly. The zeroth-order Eulerian residual transports are

$$\begin{aligned} <U_0> &= <u_0> h \\ <V_0> &= <v_0> h \end{aligned} \tag{11}$$

the first-order Eulerian residual transports are

$$\begin{aligned} <U_1> &= <u_1> h + <u_0\zeta_0> \\ <V_1> &= <v_1> h + <v_0\zeta_0> \end{aligned} \tag{12}$$

and the second order Eulerian residual transports are

$$\begin{aligned} <U_2> &= <u_2>h + <u_1\zeta_0> + <u_0\zeta_1> \\ <V_2> &= <v_2>h + <v_1\zeta_0> + <v_0\zeta_1> \end{aligned} \tag{13}$$

and the like.

For clarity and without losing generality, the flows in an M_2 tidal system are considered, and the solutions for the M_2 tidal system can be written as

$$\begin{aligned} u_0 &= u_0' \cos\theta + u_0'' \sin\theta \\ v_0 &= v_0' \cos\theta + v_0'' \sin\theta \\ \zeta_0 &= \zeta_0' \cos\theta + \zeta_0'' \sin\theta \end{aligned} \tag{14}$$

where the amplitudes and the phases of M_2 are indirectly given by (u_0', u_0''), (v_0', v_0''), and (ζ_0', ζ_0'). The first-order harmonics in the M_2 system are the M_4 and the first-order Eulerian residual water level and residual currents, or

$$\begin{aligned} u_1 &= u_1' \cos 2\theta + u_1'' \sin 2\theta + u_{\text{er}}{}^{(1)} \\ v_1 &= v_1' \cos 2\theta + v_1'' \sin 2\theta + v_{\text{er}}{}^{(1)} \\ \zeta_1 &= \underbrace{\zeta_1' \cos 2\theta + \zeta_1'' \sin 2\theta}_{M_4} + \underbrace{\zeta_{\text{er}}{}^{(1)}}_{\text{Eulerian residual}} \end{aligned} \tag{15}$$

where $(u_{\text{er}}{}^{(1)}, v_{\text{er}}{}^{(1)}, \zeta_{\text{er}}{}^{(1)})$ are the first-order Eulerian residuals, Feng *et al.* (1986). The second-order harmonics include components with frequencies of θ and 3θ. Substituting the solutions for an M_2 system into (11) and (13), it can be shown that

$$<U_0> = <V_0> = <U_2> = <V_2> = 0 \tag{16}$$

The Eulerian residual transport components become

$$\begin{aligned} U_{\text{er}} &= hu_{\text{er}} + <u_0\zeta_0> + O(\kappa^2) \\ V_{\text{er}} &= hv_{\text{er}} + <v_0\zeta_0> + O(\kappa^2) \end{aligned} \tag{17}$$

where

$$\begin{aligned} <u_0\zeta_0> &= (u_0'\zeta_0' + u_0''\zeta_0'')/2 \\ <v_0\zeta_0> &= (v_0'\zeta_0' + v_0''\zeta_0'')/2 \end{aligned}$$

U_{er} and V_{er} in (17) have been normalized by $\kappa u_c h_c$. This relation between the Eulerian residual current, $(u_{\text{er}}, v_{\text{er}})$, and the Eulerian residual transport, $(U_{\text{er}}, V_{\text{er}})$, has been given by Tee (1976), Zimmerman (1979), Alfrink and Vreugdenhil (1981), and Cheng and Casulli (1982). In the past the expression in (17) was thought of as the first-order approximation for $(U_{\text{er}}, V_{\text{er}})$. Because $<U_2>$ and $<V_2>$ are identically zero, (17) is accurate to the second order.

To relate the Eulerian residual transport to the Lagrangian residual current, ζ_0 must be eliminated from (17). By substituting the zeroth order solution, (14), into the zeroth order continuity equation,

$$\partial\zeta_0/\partial\theta + \partial(hu_0)/\partial x + \partial(hv_0)/\partial y = 0 \tag{18}$$

the continuity equation can be written in an equivalent form of

$$
\begin{aligned}
\zeta_0' &= \partial(hu_0'')/\partial x + \partial(hv_0'')/\partial y \\
-\zeta_0'' &= \partial(hu_0')/\partial x + \partial(hv_0')/\partial y
\end{aligned} \tag{19}
$$

The Lagrangian displacements (ξ,η) have been introduced by Feng *et al.* (1986), the zero order Lagrangian displacements are defined as

$$
\begin{aligned}
\xi_0 &= \int_{\theta_0}^{\theta} u_0(x_0, y_0, \theta')\mathrm{d}\theta' \\
\eta_0 &= \int_{\theta_0}^{\theta} v_0(x_0, y_0, \theta')\mathrm{d}\theta'
\end{aligned} \tag{20}
$$

with the initial conditions, $\xi_0 = 0$, $\eta_0 = 0$ at $\theta = \theta_0$. In terms of (ξ_0, η_0), the Stokes' drift $(u_{\mathrm{sd}}, v_{\mathrm{sd}})$ can be written as

$$
\begin{aligned}
u_{\mathrm{sd}} &= <(\partial u_0/\partial x)\xi_0> + <(\partial u_0/\partial y)\eta_0> \\
v_{\mathrm{sd}} &= <(\partial v_0/\partial x)\xi_0> + <(\partial v_0/\partial y)\eta_0>
\end{aligned} \tag{21}
$$

Substitute (19) into (17), and then add to the following identities

$$
\begin{aligned}
&[hv_0'\partial u_0''/\partial y - hv_0''\partial u_0'/\partial y + u_0''\partial(hv_0')/\partial y - u_0'\partial(hv_0'')/\partial y]/2 = \partial(h<u_0\eta_0>)/\partial y \\
&[hu_0'\partial v_0''/\partial x - hu_0''\partial v_0'/\partial x + v_0''\partial(hu_0')/\partial x - v_0'\partial(hu_0'')/\partial x]/2 = \partial(h<v_0\xi_0>)/\partial x
\end{aligned} \tag{22}
$$

When the final results are compared with the expression for the Stokes drift given in (21), the Stokes drift multiplied by h can be written as

$$
\begin{aligned}
hu_{\mathrm{sd}} &= <u_0\zeta_0> + \partial(h<u_0\eta_0>)/\partial y \\
hv_{\mathrm{sd}} &= <v_0\zeta_0> + \partial(h<v_0\xi_0>)/\partial x
\end{aligned} \tag{23}
$$

Equation (23) should be equivalent to (28) and (29) given by Longuet-Higgins (1969) (We believe there is a misprint in (29) of Longuet-Higgins (1969)). One of the main results obtained by Feng *et al.* (1986) is that the Lagrangian residual current has been shown to be the sum of the Eulerian residual current and the Stokes drift with a high-order correction term known as the Lagrangian drift $(u_{\mathrm{ld}}, v_{\mathrm{ld}})$, i.e.,

$$
\begin{aligned}
u_{\mathrm{lr}} &= u_{\mathrm{er}} + u_{\mathrm{sd}} + \kappa u_{\mathrm{ld}} \\
v_{\mathrm{lr}} &= v_{\mathrm{er}} + v_{\mathrm{sd}} + \kappa u_{\mathrm{ld}}
\end{aligned} \tag{24}
$$

Multiply (24) by h, and with the use of (17) and (23), we have

$$
\begin{aligned}
hu_{\mathrm{lr}} &= U_{\mathrm{er}} + \partial(h<u_0\eta_0>)/\partial y + \kappa u_{\mathrm{ld}}h \\
hv_{\mathrm{lr}} &= V_{\mathrm{er}} + \partial(h<v_0\xi_0>)/\partial \mathrm{x} + \kappa v_{\mathrm{ld}}h
\end{aligned} \tag{25}
$$

The second term on the right-hand side of (25) is associated with the Stokes drift, and by analogy the term can be defined as the Stokes drift transport $(U_{\mathrm{sd}}, V_{\mathrm{sd}})$, or

$$
\begin{aligned}
U_{\mathrm{sd}} &= \partial(h<u_0\eta_0>)/\partial y \\
V_{\mathrm{sd}} &= \partial(h<v_0\xi_0>)/\partial x
\end{aligned} \tag{26}
$$

where

$$< u_0\eta_0 > = (u_0''v_0' - v_0''u_0')/2$$
$$< v_0\xi_0 > = (v_0''u_0' - u_0''v_0')/2$$

Similarly, the Lagrangian residual transports can be introduced and defined as

$$\begin{aligned} U_{\mathrm{lr}} &= hu_{\mathrm{lr}} \\ V_{\mathrm{lr}} &= hv_{\mathrm{lr}} \end{aligned} \tag{27}$$

and the Lagrangian drift transports as

$$\begin{aligned} U_{\mathrm{ld}} &= hu_{\mathrm{ld}} \\ V_{\mathrm{ld}} &= hv_{\mathrm{ld}} \end{aligned} \tag{28}$$

With these definitions, (25) can be written as

$$\begin{aligned} U_{\mathrm{lr}} &= U_{\mathrm{er}} + U_{\mathrm{sd}} + \kappa U_{\mathrm{ld}} + O(\kappa^2) \\ V_{\mathrm{lr}} &= V_{\mathrm{er}} + V_{\mathrm{sd}} + \kappa V_{\mathrm{ld}} + O(\kappa^2) \end{aligned} \tag{29}$$

The first-order Lagrangian residual transport is then the sum of the Eulerian residual transport and the Stokes drift transport. The next-order correction is the Lagrangian drift transport which is due to the Lagrangian drift. In an M_2 system, the Eulerian residual transport, (17), is accurate to second order. Historically, the Eulerian residual transport has been referred to as the "Lagrangian" residual transport (net mass transport). This usage is clearly incorrect, (29), and several authors have attempted to clarify this confusion (Zimmerman, 1979; Alfrink and Vreugdenhil, 1981; Cheng and Casulli, 1982). As given in (26), the Stokes drift transport is zero only for one-dimensional flows. In one-dimensional flows the Eulerian residual transport and the Lagrangian residual transport are thus equivalent at the first order of approximation, (29). For two-dimensional problems, however, the Lagrangian residual transport differs from the Eulerian residual transport starting from the first order of approximation because the Stokes drift transport is not zero in general. In studies of long-term transport of suspended and dissolved matter, it is important to recognize the conceptual difference between the various residual transport variables: erroneous findings may result if the physical meanings of these variables are misinterpreted and misused.

4. Intratidal Transport Processes

Consider the transport processes in coastal estuaries or in shallow bays in which the tidal circulation can be described by a weakly nonlinear dynamic system as given by Feng *et al.* (1986). Since κ is assumed to be a small parameter, the concentration C can be expanded in an ascending power series of κ as

$$C = \sum_{j=0,1,2,\cdots} C_j \kappa^j \tag{30}$$

where C_j is the solution of the jth order perturbation.

We further assume that $1/Pe$ is not greater than κ^2, or $1/Pe << \kappa$; (3) suggests that the solutions for this class of transport problems can be accurately described by convection alone, and the solutions so obtained should be valid up to second order of κ, or $O(\kappa^2)$. Because of the change in the local characteristic length scales in the narrow region adjacent to shoreline boundaries, the dispersion effects may not be negligible within the dispersion boundary layer. Similar to other boundary layer properties in hydrodynamics (Schlichting, 1962), the thickness of the dispersion boundary layer can be estimated to be on the order of $(\kappa/Pe)^{1/2}$. It can be shown that the thickness of a momentum boundary layer within which the momentum dispersion is important is of the same order of magnitude as the dispersion boundary layer. Outside of the momentum and dispersion boundary layers the small parameter expansions for the dynamic system and for the transport of solutes are consistent and compatible. The transport processes within the boundary layer must be treated separately.

By substituting (30) into (3), the zeroth order equation becomes simply

$$\partial C_0 / \partial \theta = 0 \tag{31}$$

or the zeroth order solution C_0 is independent of time, and C_0 is a function of (x,y) only. When the Eulerian time-averaging operator, $< >$, is applied to the expression given in (30), it becomes

$$< C > = C_0 + \kappa \sum_{j=1,2,\cdots} < C_j > \kappa^{j-1} \tag{32}$$

From a different point of view, the concentration C can be expressed as the summation of a time-averaged value, $<C>$, and a tidal fluctuating component $\tilde{C}$, or $C=<C>+\tilde{C}$. The tidal fluctuation component $\tilde{C}$ is of the order of κ, and $\tilde{C}$ can be written as

$$\tilde{C} = \kappa \sum_{j=1,2,\cdots} \tilde{C}_j \kappa^{j-1} \tag{33}$$

where $\tilde{C}_j = C_j - < C_j >$. The fact that the fluctuating component of C is of smaller order has been postulated (Fischer, 1972) and observed in the field (e.g., Pritchard, 1945).

The governing equation for the first-order solution is

$$\partial C_1 / \partial \theta + u_0 \partial C_0 / \partial x + v_0 \partial C_0 / \partial y = 0 \tag{34}$$

where (u_0, v_0) are the zeroth order tidal current components due to astronomical tides. Quite often the one-dimensional form of (34) has been used in analyzing the salt balance in long and narrow estuaries. The results of the analysis are accepted by researchers as good approximations to field observations (Officer, 1976).

Because $\partial C_1 / \partial \theta = \partial \tilde{C}_1 / \partial \theta$ and because C_0 is independent of time, the solution for C_1 can be obtained by integrating (34) to give

$$C_1 = < C_1 > + \tilde{C}_1 = < C_1 > - \nabla C_0 \cdot \int V_0 \mathrm{d}\theta \tag{35}$$

where V_0 is a vertical notation of (u_0, v_0). Because C_0 is not a function of time, ∇C_0 has been written outside of the time integration. When V_0 is a harmonic function and is essentially one-dimensional, then C_1 is 90° out of phase from the tidal current V_0. This phase relation between concentration and tidal current has been shown to be valid near the mouth of some estuaries where the so-called tidal pumping and trapping mechanisms do not exist (Pritchard, 1954; Dyer, 1973; Officer, 1976; Fischer *et al.*, 1979). Clearly, this phase relation is only valid when the transport mechanism is convection dominated. In such a situation, the salinity at the mouth of an estuary reaches a maximum at the end of a flood, and the salinity is a minimum at the end of an ebb.

The governing equation for the second order solution, C_2, is

$$\partial C_2 / \partial \theta + u_1 \partial C_0 / \partial x + v_1 \partial C_0 / \partial y + u_0 \partial C_1 / \partial x + v_0 \partial C_1 / \partial y = 0 \tag{36}$$

where (u_1, v_1) are the first-order solution of the tidal current. A solution to (36) will not be considered here. To conclude the discussion of the intratidal transport process, we assume $Pe = \kappa^{-2}$. Then the governing equation for the third-order solution C_3 can be obtained as

$$\begin{aligned} &\partial C_3 / \partial \theta + u_0 \partial C_2 / \partial x + v_0 \partial C_2 / \partial y + u_1 \partial C_1 / \partial x + v_1 \partial C_1 / \partial y \\ &+ u_2 \partial C_0 / \partial x + v_2 \partial C_0 / \partial y = (1/h) \nabla \cdot (hD \nabla C_0) \end{aligned} \tag{37}$$

where (u_2, v_2) are the second-order solution of the dynamic system. It should be pointed out that if $1/Pe$ is substantially different form κ^{2}, then (37) is no longer valid.

5. Intertidal Transport Processes: Long-term Transport Processes

In order to derive the inter-tidal transport equation, the governing equation for C_2 is examined first. When the Eulerian time-averaging operator, $< >$, is applied to (36), and noting that C_0 is independent of time. (36) becomes

$$u_{er}{}^{(1)} \partial C_0 / \partial x + v_{er}{}^{(1)} \partial C_0 / \partial y + < u_0 \partial C_1 / \partial x > + < v_0 \partial C_1 / \partial y > = 0 \tag{38}$$

where the Eulerian time-averaging of (u_1, v_1) gives rise to the first-order Eulerian residual current $(u_{er}{}^{(1)}, v_{er}{}^{(1)})$. At the zeroth order of approximation, the zeroth-order Lagrangian displacements (ξ_0, η_0) can be calculated from integrating the Eulerian tidal velocities (Feng *et al.*, 1986). Suppose we consider only one major tidal constituent, say M_2, whose solutions (u_0, v_0, ζ_0) are known and given in (14); then the following identities hold true:

$$\begin{aligned} &< u_0 \xi_0 > = < v_0 \eta_0 > = < u_0 \eta_0 > + < v_0 \xi_0 > = 0 \\ &< u_0 \partial \eta_0 / \partial x > + < \xi_0 \partial v_0 / \partial x > = 0 \\ &< v_0 \partial \xi_0 / \partial y > + < \eta_0 \partial u_0 / \partial y > = 0 \cdots \end{aligned} \tag{39}$$

etc. Recall also that the Stokes drift velocity components have been given in (21). By substituting the solutions for C_1 given in (35) into (38), it can be shown that

$$<u_0\partial C_1/\partial x+v_0\partial C_1/\partial y>=u_{sd}\partial C_0/\partial x+v_{sd}\partial C_0/\partial y \tag{40}$$

Equation (38) can be rewritten as the intertidal conservation equation for C_0, or

$$u_{lr}^{(1)}\partial C_0/\partial x+v_{lr}^{(1)}\partial C_0/\partial y=0 \tag{41}$$

where

$$(u_{lr}^{(1)},v_{lr}^{(1)})=(u_{er}^{(1)},v_{er}^{(1)})+(u_{sd},v_{sd})$$

are the first-order Lagrangian residual currents (Feng *et al.*, 1986). At this order of approximation the first-order Lagrangian residual current is the same as the mass transport velocity defined by Longuet-Higgins (1969).

In order to demonstrate further the Lagrangian properties in long-term transport of solutes the governing equation for $<C_1>$ can be obtained by applying the Eulerian time-averaging operator to (37). After substitution of some identities into (37) it becomes

$$u_{lr}^{(1)}\partial<C_1>/\partial x+v_{lr}^{(1)}\partial<C_1>/\partial y=(1/h)\nabla\cdot(h<D>\nabla C_0) \tag{42}$$

or

$$u_{lr}^{(1)}\partial<C>/\partial x+v_{lr}^{(1)}\partial<C>/\partial y=(\kappa/h)\nabla\cdot(h<D>\nabla C_0) \tag{43}$$

where $<C>=C_0+\kappa<C_1>$. The derivation of (42) is quite tedious (given in the appendix), but the implications from the resultant equation are quite significant. Note that (43) takes the familiar form of the convection dispersion equation, however the convection velocity is the first-order Lagrangian residual velocity, and is not the Eulerian residual current. In fact, if we assume that $Pe=1/\kappa$, then a similar analysis can be carried out to give results similar to the results given above. The final result is nearly the same as (43) with $<C>$ replacing C_0 on the right-hand side of (43). Even in this case, the dispersion terms on the right-hand side of the equation are higher-order corrections to the basically convective transport.

In section 3 the residual transports have been introduced and defined, and at the first order of approximation the Lagrangian residual transport $\left(U_{lr}^{(1)},V_{lr}^{(1)}\right)$ equals the sum of the Eulerian residual transport (U_{er},V_{er}) and the Stokes drift transport (U_{sd},V_{sd}), (29). It can be shown easily that all the residual transport variables are solenoid. In terms of the transport variables the governing intertidal transport equations of the zeroth and the first orders can be shown to be

$$U_{lr}^{(1)}\partial C_0/\partial x+V_{lr}^{(1)}\partial C_0/\partial y=0 \tag{44}$$

and

$$U_{lr}^{(1)}\partial<C_0+\kappa C_1>/\partial x+V_{lr}^{(1)}\partial<C_0+\kappa C_1>/\partial y=\kappa\nabla\cdot(h<D>\nabla C_0) \tag{45}$$

where the first-order Lagrangian residual transport are defined as

$$(U_{lr}^{(1)},V_{lr}^{(1)})=h(u_{lr}^{(1)},v_{lr}^{(1)}) \tag{46}$$

There are several points of significance in the resultant intertidal conservation equations,

(41), (43), (44) and (45). Within the interior of a weakly nonlinear coastal estuary or shallow sea the transport of passive solutes in an intertidal time scale is a convection-dominated process. The convection can be correctly represented by either the first-order Lagrangian residual velocity or the first-order Lagrangian residual transport. At the lowest order of approximation (i.e., $<C>=C_0$), the transport process is shown to be determined purely by convection; see (41) or (44). The isoconcentration contours for C_0 coincide with the streamlines which represent the first-order Lagrangian residual velocity or the first-order Lagrangian residual transport.

Now, let us compare the more commonly used approaches for describing the long-term transport processes in estuaries and shallow seas with the present formulation. As mentioned previously, the direct integration of the intratidal conservation equation for solutes, (2), is possible, but it is considered impractical at this time. The second approach is the solution of a time-averaged conservation equation which takes a form similar to (1). The present approach, in fact, is similar to the latter in concept; an intertidal conservation equation has been obtained in a form of convection dispersion equation; see (41) through (45). Instead of the Eulerian residual velocity as it appeared in (1) the convective terms in the intertidal transport equation have been shown to be the first-order Lagrangian residual velocity or the Lagrangian residual transport. In retrospect, the results of the present formulation are not particularly surprising. This conclusion had been hinted at by Longuet-Higgins (1969), who stated, "In determining the origin of water masses, it is the Lagrangian mean which is most relevant".

There is a subtle but import difference between the present and the conventional formulations. In deriving (1) the mathematical problem is not closed without the introduction of an approximation to the correlations between tidal current and concentration. In the absence of a rigorous means to relate these correlations to other physical parameters a Fickian hypothesis was introduced with which the correlation terms were assumed to be given in the form of dispersion. Since these phase effect correlation terms have been forced to behave as dispersion, they have been referred to as tidal dispersion (Bowden, 1967; Dyer, 1973; Officer, 1976; Fischer *et al.*, 1979). The correlation terms have been neglected in some analysis of the salt balance near the mouth of a one-dimensional estuary (Bowden, 1967; Fischer *et al.*, 1979). What has been shown above is that the concentration and tidal current correlation terms can be precisely expressed in a convective form with the Stokes drift as the convective velocity. Thus the introduction of the Fickian hypothesis is no longer necessary. This approximation is valid, at least in the interior of a tidal basin (under the small κ approximation) where the so-called tidal trapping and tidal pumping do not exist.

6. Coastal Dispersion Boundary Layer

The concentration gradient at a shoreline boundary should be tangent to the shore because, from a physical point of view, no mass transfer is allowed in the direction normal to the boundary. Up to the second-order solution, the long-term transport processes in the interior of a basin have been shown to be essentially convection dominated, and dispersion is of secondary importance. Specifically, the dispersion terms are absent in (41) and (44). The dispersion terms in (43) and (45) are not the higher-order derivatives for C_1; rather, they are the high-order corrections to C_1 due to C_0. Thus (41), (43), (44) and (45) are first-order hyperbolic partial differential equations, and a no-flux condition at the shoreline boundary can not be satisfied.

The mixing process at or near a shoreline boundary are rather complex and are not well understood (Fischer *et al.*, 1979). Because of the possible presence of shoaling or branching channels near shore, The so-called tidal pumping and tidal trapping are additional mechanisms for mixing in a narrow coastal boundary layer. Since tidal pumping or tidal trapping in unique to each specific situation, each case requires a special treatment. Thus tidal pumping and tidal trapping are excluded in the following considerations; instead, we assume that the shoreline is generally smooth and without large curvature. The presence of a shoreline changes the local length scale normal to the shore: with the introduction of the new length scale, dispersion becomes significant. A coastal dispersion boundary layer exists within which the shear effect dispersion and turbulent diffusion are expected to be equally as important as convection. Within the dispersion boundary layer, there is a continuous transition of the concentration from the interior of a basin to the shoreline boundary where a no-flux condition is required and satisfied.

The thickness of the dispersion boundary layer can be estimated to be of the order of $\kappa^{1/2}$ from inspection of (43) (see Schlichting, 1962). Introducing the local boundary layer coordinates along (x_b) and normal (y_b) to shore (Figure 1) and in dimensionless form, x_b and y_b are related locally as

$$x_b \sim y_b / \kappa^{1/2} \tag{47}$$

Using (47) in the continuity equation, the order of magnitude of the Lagrangian residual current within the dispersion boundary layer can be estimated to be

$$(u_{lr})_b \sim (v_{lr})_b / \kappa^{1/2} \tag{48}$$

where $[(u_{lr})_b, (v_{lr})_b]$ are the components of the Lagrangian residual current in the (x_b, y_b) directions. Substituting (47) and (48) into (43) gives

$$(u_{lr})_b \partial < C_b > / \partial x_b + (v_{lr})_b \partial < C_b > / \partial y_b = (D / h) \partial (h \partial (C_0)_b / \partial y_b) / \partial y_b + O(\kappa) \tag{49}$$

where variables with subscript b are defined within the dispersion boundary layer and $< C_b > = (C_0)_b + \kappa < (C_1)_b >$. For simplicity, the dispersion coefficient D has been assumed to be a constant. Within the small κ approximation, when all terms less than order 1 are neglected, (49) becomes

$$(u_{lr})_b \partial < (C_0)_b > / \partial x_b + (v_{lr})_b \partial < (C_0)_b > / \partial y_b = (D / h) \partial (h \partial (C_0)_b / \partial y_b) / \partial y_b \tag{50}$$

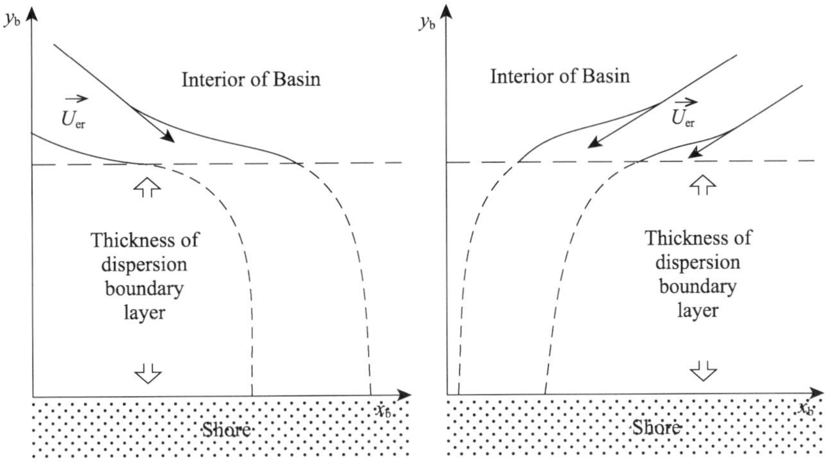

Fig. 1 Schematic diagram of the coastal boundary layer. The solid lines are the "streamlines" of the Lagrangian residual current. These streamlines coincide with isoconcentration contours with in the interior of a tidal basin. The dotted lines are the isoconcentration contours within the dispersion boundary layer. The concentration distribution must satisfy the boundary conditions at shore and at the edge of the dispersion boundary layer

which is a classic form of a boundary layer equation (Schlichting, 1962). The boundary conditions for (50) are

$$\partial(C_0)_b / \partial y_b = 0 \qquad y_b = 0 \tag{51}$$

and

$$(C_0)_b \to C_0 \qquad as\ y_b \to \infty$$

The problem defined by (50) and (51) is well posed for a solution of $(C_0)_b$. To carry the solution of $(C_0)_b$ further without detailed scrutiny of $[(u_{lr})_b, (v_{lr})_b]$ seems meaningless, and a precise method of solution for $[(u_{lr})_b, (v_{lr})_b]$ is not known. Nevertheless, the present discussion points out an area of research which may benefit the understanding of long-term transport problems. Once the solution for $< C_b >$ is obtained, the concentration distribution is consistent with the correct physics; namely, a continuous transition from no-flux at shore through a dispersion boundary layer to the interior region of a basin where the dispersion is negligible. The general forms of $< C_b >$ are qualitatively sketched and given in Figure 1.

7. An Analytical Example: An Amphidromic System

In this section a simple analytical example is used to further examine the present long-term transport model before it is to be tested in a practical situation such as San Francisco Bay, California. Keeping simplicity in mind, we consider the case of two Kelvin waves propagating in opposite directions in an infinitely long, constant depth, frictionless channel. The resultant flow field is a series of amphidromic system. The Lagrangian residual current in a typical amphidromic system has been considered by Feng *et al.* (1986). Because the flow is assumed to be frictionless, the vertical distributions of flow properties are uniform and are consistent with the condition of vertical rigidity (Stern, 1975). Thus a two-dimensional representation of the flow field on the horizontal plane is quite proper, and consequently, the definitions and properties for the Lagrangian residual current and the Lagrangian residual transport are rigorous. The flow properties for this system have been discussed in detail by Feng *et al.* (1986), the zeroth order tidal velocity can be given as

$$u_0 = u_0' \cos\theta + u_0'' \sin\theta \tag{52}$$

where

$$u_0' = 2\cosh(fy)\cos x$$
$$u_0'' = -2\sinh(fy)\sin x$$

and f is the Coriolis parameter. All cross-channel (y direction) velocity components are zero. Since the Eulerian residual velocity is zero, the mass transport velocity, the first-order Lagrangian residual current, and the Stokes drift are identical, and they can be written as

$$u_{\mathrm{lr}}^{(1)} = u_{\mathrm{sd}} = -\sinh(2fy) \tag{53}$$

Consistent with the assumption of vertical rigidity, the shear induced dispersion is neglected. The turbulent mixing is also neglected; thus, in the interior of the system, only convection is considered. Using the properties of the flow field given above, the intertidal transport equation, (41), is reduced to

$$\partial C_0 / \partial x = 0 \tag{54}$$

or C_0 is a function of y only and is independent of x. This general property is consistent with the fact that all streamlines representing the Lagrangian residual current are parallel to the x axis as shown in Figure 2. The concentration C_0 remains a constant along the streamline, and it only varies across the channel including the cross section at the open boundary. Since all velocity components in the y direction are zero, there is no water mass transported in the y-direction. Thus the solution of C_0 is completely determined by the boundary condition specified at the upstream side of the open boundary.

As pointed out previously, in the interior of the amphidromic system, the long-term transport is determined by the Lagrangian residual current rather than by the Eulerian residual current. In this case, the Eulerian residual current is zero! The net mass transport is due to the Stokes drift which is the second component of the first-order Lagrangian residual current (mass transport velocity). This situation is similar to the case investigated by Longuet-Higgins (1969). Further, the long-term transport equation is essentially a pure convection equation, the boundary condition specified at the upstream open boundary completely determining the solution along the streamline of the Lagrangian residual current. In this simple analytical example it can be shown that the concentration is constant within a coastal boundary layer adjacent to the shore. The value of the concentration in the boundary layer equals the value of concentration just outside the coastal dispersion boundary layer.

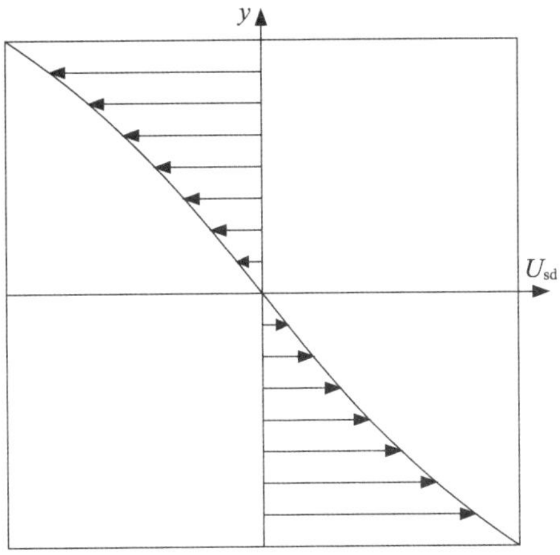

Fig. 2 The Stokes drift as a function of y only in an amphidromic system in which all y-direction velocity components are zero and the Eulerian residual current is identically zero. Thus the Stokes drift becomes the same as the first-order Lagrangian residual current (or the mass transport velocity)

8. An Application to South San Francisco Bay, California

The final verification of the present approach is a test of the theory against field data in a realistic situation. Often in the analysis of a realistic system a number of factors might not have been included in the consideration. Often, too, the available field data are insufficient to adequately verify a long-term transport model. Therefore a quantitative comparison of the present approach with field observations of salt transport in South San Francisco Bay (South Bay) is not feasible, but a qualitative comparison is plausible and of interest. A general description of the San Francisco Bay system and South San Francisco Bay can be found in the work by Cheng and Gartner (1984, 1985), and a map of the study area is shown in Figure 3a . There is no major river or tributary leading into South Bay, and South Bay is vertically well mixed during most of summer and fall. In early spring, when large volumes of fresh water enter South Bay from the north, the system becomes stratified. As the fresh water inflows from the north are reduced, the system restores to a nearly isohaline condition over the next 2 to 3 months. During this period, salinity becomes a convenient conservative tracer for studies of long-term transport properties. The restoration of salinity to an isohaline condition in South Bay has been computed numerically using a Eulerian-Lagrangian method on a tidal time scale by Cheng *et al.* (1984b). Their approach follows the developments of transport process on a tidal time scale: qualitative results have been obtained. Alternatively, a direct estimate of salinity distribution will be made using the present long-term transport equation.

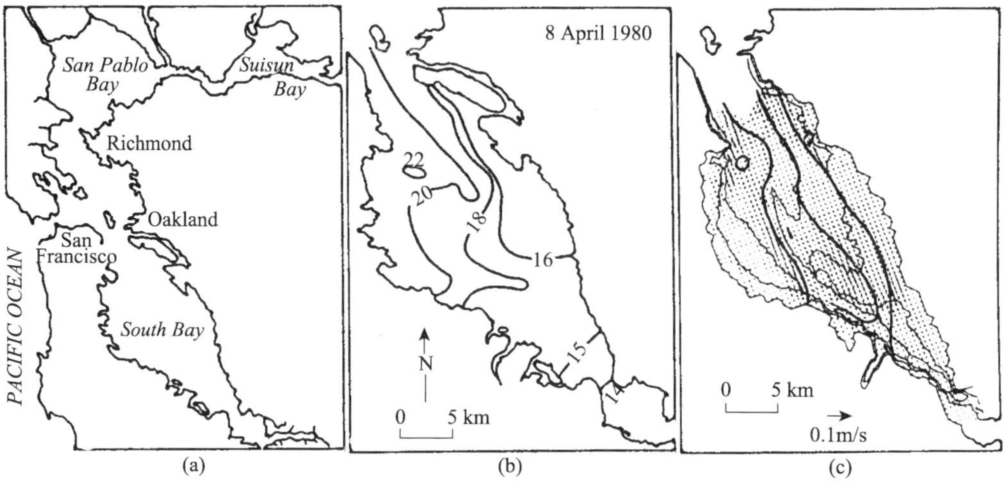

Fig. 3 (a) San Francisco Bay estuarine system and South San Francisco Bay, (b) surface salinity measured on April 8, 1980, (c) computed Lagrangian residual circulation in South San Francisco Bay along with the streamlines representing the Lagrangian residual current field

According to the present theory, (41) or (44), the isohaline contours of salinity should be identical to the contours of the "streamline" representing the mass transport velocity or the first-order Lagrangian residual current. There is no known direct measurement of the Lagrangian residual current in South San Francisco Bay; in the following discussion, the Lagrangian residual currents are inferred from the results of a two-dimensional vertically averaged numerical model. Extensive hydrographic measurements of water quality constituents were made by the U.S. Geological Survey in 1980 (Dedini *et al.*, 1981). In early 1980 there were some episodic storm events which caused the introduction of a large volume of fresh water into South Bay from the northern reach of the Bay system. After mid-March, no substantial additional freshwater inflows were introduced from the north. During the subsequent weeks, winds were generally calm and variable, and the wind-driven component of the residual currents could be neglected.

Based on a 500-m finite-difference grid, the tidal and residual circulation in South Bay have been calculated and reported by Cheng (1983). The Lagrangian residual current was calculated by following the movement of a labeled water mass as described by Cheng and Casulli (1982). The salinity distribution in South Bay for April 8, 1980, is shown in Figure 3b, and the computed Lagrangian residual current is depicted in Figure 3c. Superimposed on the residual current vectors are the interpreted streamlines for the Lagrangian residual current field. Without including the other factors such as the coastal dispersion boundary layer and the residual currents due to stratification, the patterns of isohaline contours for salinity in Figure 3b and the streamlines for the computed Lagrangian residual currents (Figure 3c) are in surprisingly good qualitative agreement. The present theory for predicating a long-term transport process in an estuary seems to be satisfactory, at least qualitatively.

9. Conclusion and Discussion

Under the weakly nonlinear approximation an intertidal transport equation has been obtained for a vertically well-mixed tidal estuary. The new governing equation differs from the traditional formulation in two fundamental ways. First, the classical approach uses the Eulerian residual velocity as convection in the intertidal transport equation, whereas in the present analysis the convection is shown to be the first-order Lagrangian residual current (Cheng *et al.*, 1984a) or the mass transport velocity (Longuet-Higgins, 1969). The present approach reveals more of the actual mechanism with which solutes are transport over an intertidal time scale. In other words, the present approach is more descriptive of the physical mechanism which is responsible for the long-term transport of solutes. Second, the mathematical problem in the classical formulation (the governing differential equation) is actually not well posed or is not closed without the introduction of an ad hoc hypothesis which relates the correlations between flow properties and solutes or the phase effects to the time-averaged variables. The introduction of an ad hoc Fickian hypothesis has led to the so-called tidal dispersion. Although researchers have attempted to quantify values of the tidal dispersion coefficients, past attempts seem to have been inconclusive. In contrast, no such ad hoc hypothesis is needed in the present formulation, the phase effect correlation terms (or the tidal dispersion terms) having been shown to behave like convection, with the convection velocity identified as the Stokes drift. The only assumption involved in the present analysis is that the estuarine system is weakly nonlinear, and the results should be valid in the interior of the system.

The present results can now be used to explain the large variability in the estimates of tidal dispersion coefficients. When the two fundamentally different formulations for the long-term transport equations are compared, it simply implies that the coefficients of the tidal dispersion tensor should be a function of the Stokes drift and a function of the dependent variable itself. Accordingly, the values for tidal dispersion coefficients are determined by the dynamics of the system. As a result of the spring-neap tidal variations, the residual currents and the Stokes drift and thus the mixing characteristics would also vary throughout the spring-neap cycle (Uncles and Radford, 1980). Since the tidal dispersion coefficients are functions of the Stokes drift, they are not expected to be constants.

The dispersion boundary layer near shore, which we have barely touched upon, is an important area of research that requires further scrutiny. Transport properties near shore where shoaling and branching play an important part in the mixing are much more complicated, and they might ultimately determine the overall solute distribution. Derivation of a general theory for tidal pumping or tidal trapping might be quite difficult, and it is likely that the near-shore mixing properties for different basins need to be treated individually on a case by case basis.

Appendix: Derivation of (42)

In an M_2 tidal system the solution for C_2 can be obtained from (36) through direct substitution of the known lower-order solutions give

$$C_2 = < C_2 > - \xi_0 \frac{\partial < C_1 >}{\partial x} - \eta_0 \frac{\partial < C_1 >}{\partial y} + C'(x,y)\cos 2\theta + C''(x,y)\sin 2\theta \tag{A1}$$

where $C'(x,y)$ and $C''(x,y)$ represent the terms stemming from the nonlinear interactions between the gradients of C_0 and the higher-order harmonics of the tidal flows. The detailed expressions for $C'(x,y)$ and $C''(x,y)$ are unimportant: it is only necessary to identify that these terms are associated with $\cos 2\theta$ and $\sin 2\theta$. The known expression for (u_0, v_0), (u_1, v_1), C_0, C_1, and C_2 are substituted into (37), and then the Eulerian time average operator is applied over the resultant equation. Note that the right-hand side of (37) is independent of time, and also note the following identities:

$$< \frac{\partial C_3}{\partial \theta} > = 0$$

$$< u_2 \frac{\partial C_0}{\partial x} > = < v_2 \frac{\partial C_0}{\partial y} > = 0$$

$$< u_1 \frac{\partial C_1}{\partial x} > = u_{\mathrm{er}} \frac{\partial < C_1 >}{\partial x}$$

and

$$< v_1 \frac{\partial C_1}{\partial y} > = v_{\mathrm{er}} \frac{\partial < C_1 >}{\partial y}$$

After time averaging, (37) becomes

$$u_{\mathrm{er}} \frac{\partial < C_1 >}{\partial x} + v_{\mathrm{er}} \frac{\partial < C_1 >}{\partial y} + < u_0 \frac{\partial C_2}{\partial x} > + < v_0 \frac{\partial C_2}{\partial y} > = \frac{(\kappa Pe^{1/2})^{-2}}{h} \nabla \cdot hD\nabla C_0 \tag{A2}$$

With the solutions for C_2 and (u_0, v_0) known, then

$$\begin{aligned} < u_0 \frac{\partial C_2}{\partial x} > &= - < u_0 \frac{\partial \xi_0}{\partial x} \frac{\partial < C_1 >}{\partial x} + u_0 \frac{\partial \eta_0}{\partial x} \frac{\partial < C_1 >}{\partial y} + u_0 \eta_0 \frac{\partial^2 < C_1 >}{\partial x \partial y} > \\ < v_0 \frac{\partial C_2}{\partial y} > &= - < v_0 \frac{\partial \xi_0}{\partial y} \frac{\partial < C_1 >}{\partial x} + v_0 \xi_0 \frac{\partial^2 < C_1 >}{\partial y \partial x} > + v_0 \frac{\partial \eta_0}{\partial y} \frac{\partial < C_1 >}{\partial y} > \end{aligned} \tag{A3}$$

Recall that the Stokes drift is defined as

$$u_{\mathrm{sd}} = < \xi_0 \frac{\partial u_0}{\partial x} + \eta_0 \frac{\partial u_0}{\partial y} >$$

with the aid of the identities $<\xi_0(\partial u_0/\partial x)>+<u_0(\partial\xi_0/\partial x)>=0$, the first term on the right-hand side of (A3) can be rewritten as

$$-<u_0\frac{\partial\xi_0}{\partial x}\frac{\partial<C_1>}{\partial\alpha}>=-<-\xi_0\frac{\partial u_0}{\partial x}\frac{\partial<C_1>}{\partial x}>=u_{sd}\frac{\partial<C_1>}{\partial x}-<\eta_0\frac{\partial u_0}{\partial y}\frac{\partial<C_1>}{\partial x}>$$

and similarly,

$$-<v_0\frac{\partial\eta_0}{\partial y}\frac{\partial<C_1>}{\partial y}>=v_{sd}\frac{\partial<C_1>}{\partial y}-<\xi_0\frac{\partial v_0}{\partial x}\frac{\partial<C_1>}{\partial y}>$$

Finally, with the use of the identities

$$<u_0\frac{\partial\eta_0}{\partial x}+\xi_0\frac{\partial v_0}{\partial x}>=<\eta_0\frac{\partial u_0}{\partial y}+v_0\frac{\partial\xi_0}{\partial y}>=<u_0\eta_0+v_0\xi_0>=0$$

the sum of the correlation terms in (A3) becomes

$$<u_0\frac{\partial C_2}{\partial x}>+<v_0\frac{\partial C_2}{\partial y}>=u_{sd}\frac{\partial<C_1>}{\partial x}+v_{sd}\frac{\partial<C_1>}{\partial y} \tag{A4}$$

and (A2) becomes the intertidal transport equation of $<C_1>$ as given in (42).

References

Alfrink, B. J., and C. B. Vreugdenhil, Residual currents, Rep. R-1469-11, Delft Hydraulic Laboratory, Delft, The Netherlands, 1981.

Awaji, T., Water mixing in a tidal current and the effect of turbulence on tidal exchange through a strait, J. Phys. Oceanogr., 12(6), 501–514, 1982.

Awaji, T., N. Imasato, and H. Kunishi, Tidal exchange through a strait: A numerical experiment using a simple model basin, J. Phys. Oceanogr., 10, 1499–1508, 1980.

Bowden, K. F., Horizontal mixing in the sea due to a shearing current, J. Fluid Mech., 21, 83–95, 1965.

Bowden, K. F., Circulation and diffusion, in Estuaries, edited by G. H. Lauff, pp. 15–36, AAAS, Washington, D. C., 1967.

Cheng, R. T., Euler-Lagrangian computations in estuarine hydrodynamics, in Proceedings of the Third International Conference on Numerical Methods in Laminar and Turbulent Flow, edited by C. Taylor, J. A. Johnson, and R. Smith, pp. 341–352, Pineridge Press, Swansea, United Kingdom, 1983.

Cheng, R. T., and V. Casulli, On Lagrangian residual currents with applications in South San Francisco Bay, California, Water Resour. Res., 18(6), 1652–1662, 1982.

Cheng, R. T., and J. W. Gartner, Tides, tidal and residual circulation in San Francisco Bay, California. Results of measurements—1979–1980, parts Ⅰ to Ⅴ, U.S. Geol. Surv. Water Resour. Invest. Rep., 84-4339, 1–1747, 1984.

Cheng, R. T., and J. W. Gartner, Harmonic analysis of tides and tidal currents in south San Francisco Bay, California, Estuarine, Coastal Shelf Sci., 21, 57–74, 1985.

Cheng, R. T., S. Feng, and P. Xi. On Lagrangian residual ellipse, paper presented at international Conference on Physics of Shallow Estuaries and Bays, University of Miami. Miami, Fla., August 1984, 1984a.

Cheng, R. T., V. Casulli., and S. N. Milford. Eulerian-Lagrangian solution of the convection-dispersion equation in natural coordinates, Water Resour. Res., 20(7), 944–952, 1984b.

Dedini, L. A., L. A. Schemel, and M. A. Tembreull, Salinity and temperature measurements in San Francisco Bay waters, U.S. Geol. Surv. Open File Rep., 82–125, 1–130, 1981.

Dyer. K. R., Estuaries: A Physical Introduction, 140pp., John Wiley, New York, 1973.

Dyer, K. R., The salt balance in stratified estuaries, Estuarine, Coastal and Shelf Science, 2, 273–281, 1974.

Feng, S., R. T. Cheng, and P. Xi, On tide-induced residual current and residual transport, 1, Lagrangian residual current, Water Resour. Res., 22(12), 1623–1634, 1986.

Fischer, H. B., Mass transport mechanisms in partially stratified estuaries, J. Fluid Mech., 53, 671–687, 1972.

Fischer, H. B., Mixing and dispersion in estuaries, Annu. Rev. Fluid Mech., 8, 107–133, 1976.

Fischer, H. B., E. J. List, R. C. Y. Koh, J. Imberger, and N. H. Brooks, Mixing in Inland and Coastal Waters, 483pp., Academic, Orlando, Fla., 1979.

Leendertse, J. J., A water-quality simulation model for well-mixed estuarines and coastal seas, Principles of computation, vol. I. Rep. RM-6230-R6, Rand Corp., Santa Monica, Calif., 1970.

Leendertse, J. J., and E. C. Gritton, Computation procedures, vol. II, Rep. R-708-NUC, Rand Corp., Santa Monica, Calif., 1971.

Lewis, R. E., and J. O. Lewis, The principal factors contributing to the flux of salt in a narrow, partially stratified estuary, Estuarine Coastal and Mar. Sci., 16, 599–626, 1983.

Longuet-Higgins, M. S., On the transport of mass by time varying ocean currents, J. Deep Sea Res., 16, 431–447, 1969.

Officer, C. B., Physical Oceanography of Estuaries, 465 pp., John Wiley, New York, 1976.

Okubu, A., The effect of shear in an oscillatory current on horizontal diffusion from an instantaneous source, Int. J. Oceanogr. Limnol., 1, 194–204, 1967.

Pritchard, D. W., A study on the salt balance in a coastal plain estuary, J, Mar. Res., 13, 133–144, 1954.

Schlichting, H., Boundary Layer Theory, 6th Ed., McGraw Hill, 747 pp., 1962.

Stern, M. E., Ocean Circulation Physics, Academic, Orlando, Fla, 1975.

Stommel, H., and H. G. Farmer, On the nature of estuarine circulation, I, vol. 52–88, Woods Hole Oceanographic Institution, Woods Hole, Mass., 1952.

Tee, T. K., Tide-induced residual current: A 2-D nonlinear numerical model, J. Mar. Res., 31, 603–628, 1976.

Uncles, R. J., and B. Jordan, Residual fluxes of water and salt at two stations in the Severn estuary, Estuarine Coastal Mar. Sci., 9, 287–302, 1979.

Uncles, R. J., and P. J. Radford, Seasonal and spring-neap tidal dependence of axial dispersion coefficients in the Severn—A wide, vertically mixed estuary, J. Fluid Mech., 98(4), 703–726, 1980.

Winterwerp, J. C., Decomposition of the mass transport in narrow estuaries, Estuarine Coastal Shelf Sci., 16, 627–638, 1983.

Zimmerman, J. T. F., Dispersion by tide-induced residual current vortices, in Hydrodynamics of Estuaries and Fjords, edited by J. C. J. Nihoul, pp. 207–216, Elsevier, New York, 1978.

Zimmerman, J. T. F., On the Euler-Lagrangian transformation and the Stokes' drift in the presence of oscillatory and residual currents, Deep Sea Res., 26A, 505–520, 1979.

A Three-dimensional Weakly Nonlinear Dynamics on Tide-induced Lagrangian Residual Current and Mass-transport*

Feng Shizuo (冯士筰)

Introduction

As well known, studies of the environmental hydrodynamics have put focus on the longer-term transport processes of suspended matter and dissolved substances in estuaries, coastal embayments, shallow seas and continental shelf seas. In fact, transports of solutes, salinity, nutrients, sediments and other tracers are really fundamental to the interactive physical, chemical, biological processes in an ecological system. We also know that tides dominate the circulation in the coastal seas and the apparent dominating transport mechanism is tidal convection. Thus the nature of the longer-term transport processes mentioned above is strongly Lagrangian and it has been generally agreed on that these longer-term transport processes are determined by the Lagrangian mean velocity of a marked water parcel and not by the Eulerian mean velocity at a point, or the Eulerian residual current. The Lagrangian mean velocity of a marked parcel may lead to a concept of Lagrangian residual current. It should be pointed out that the study on Lagrangian residual currents is a relatively recent undertaking, and thus any further investigation of the Lagrangian residual current may be a valuable contribution.

It has been revealed recently that, unlike the Eulerian residual current, the Lagrangian residual current depends not only upon the point in space where the marked water parcel is released but also upon the tidal phase at the time when the marked water parcel is released, and the Lagrangian residual velocity describes an ellipse over a complete tidal cycle on a hodograph plane (Feng *et al.*, 1986a). In fact, a numerical simulation of South San Francisco Bay, California, was used in an attempt to define the relation between the Lagrangian residual current and the tidal phase (Cheng, 1983), and similar modeling of the Lagrangian residual circulation in the Jiaozhou Bay have been made by Yu and Chen (1983). However, these proposed models have certain weakpoint as they are two-dimensional and depth-averaged. The Lagrangian residual current should be rigorously treated in a three-dimensional space from the point of view of dynamics (Alfrink and Vreugdenhil, 1981; Feng *et al.*, 1986a),

* Feng S. 1986. A three-dimensional weakly nonlinear dynamics on tide-induced Lagrangian residual current and mass-transport. Chinese Journal of Oceanology and Limnology, 4(2): 139–158

although the Lagrangian residual current can be defined in a horizontal, two-dimensional space when the two-dimensional barotropic flow has the property of vertical rigidity (Stern, 1975). A study on the three-dimensional Lagrangian residual circulation is of much importance from both the theoretical and the practical points of view. In the present paper, we propose a three-dimensional model for the Lagrangian residual current and investigate the three-dimensional structure of the Lagrangian residual circulation.

A longer-term transport equation, namely, a convection-diffusion equation for the tidal cycle, averaged concentration of any conservative and passive tracer, describes a balance of convection and diffusion or dispersion. In the classical longer-term transport equation, the convection velocity is the Eulerian residual velocity. As stated above, however, the longer-term transport processes are Lagrangian, so the transport of any tracer should be determined by the Lagrangian residual current rather than by the Eulerian residual current. Therefore, the use of the Eulerian residual velocity to represent convection needs further examination. On the other hand, in the classical longer-term transport equation an assumption of "tidal dispersion" had to be made, the coefficients of which have been estimated based upon data from concurrent measurements of the tidal velocity and the concentration over an extended period of time (Dyer, 1973, 1974; Fischer, 1976; Uncles and Jordan, 1979; Uncles and Radford, 1980; Winterwerp, 1983; Lewis and Lewis, 1983), upon arguments of dimensional analysis (Stommel and Famer, 1952), upon computations of the actual Lagrangian water mass movements in a tidal current field by means of a numerical model (Awaji, 1982), or upon statistical approach (Zimmerman, 1978). Sometimes, the so called "tidal dispersion" terms were simply neglected (Pritchard, 1954; Bowden, 1965; Fischer *et al.*, 1979). Of course, the "tidal dispersion" terms in the longer-term transport equation are the results of a hypothesis in a mathematical average of the governing equation, but their physics and dynamics are not well understood. In fact, the alternative to the classical longer-term transport equation has been proposed recently (Feng *et al.*, 1986b), but it is a two-dimensional, depth-averaged equation. Obviously, a corresponding three-dimensional equation is expected, and in the present paper, we have proposed such an equation, which describes the Lagrangian nature of convection transport without introducing the "tidal dispersion" terms.

The Lagrangian residual circulation and the longer-term transport processes are to be driven by tides, storm or wind, and density and open boundary forces. To avoid confusing the main issues, however, the effects of surface wind stress, variations of barometric pressure and baroclinic variations are not included. Thus the present study is confined to the tide-induced Lagrangian residual circulation and longer-term transport processes.

Formulation

Based on a nonlinear three-dimensional tidal model (Feng, 1977) with an additional equation for a streakline and a convection-diffusion equation for the concentration of any conservative and passive indicator substances in the water, a nondimensional dynamic problem is presented as follows:

$$\nabla \cdot \boldsymbol{u} = 0 \tag{1}$$

$$\frac{\partial u}{\partial \theta} + \kappa \boldsymbol{u} \cdot \nabla u - fv = -\frac{\partial \mathscr{G}}{\partial x} + \frac{\partial}{\partial z}\left(\nu \frac{\partial u}{\partial z}\right) \tag{2}$$

$$\frac{\partial v}{\partial \theta} + \kappa \boldsymbol{u} \cdot \nabla v + fu = -\frac{\partial \mathscr{G}}{\partial y} + \frac{\partial}{\partial z}\left(\nu \frac{\partial v}{\partial z}\right) \tag{3}$$

$$\frac{\mathrm{d}(N\boldsymbol{\xi})}{\mathrm{d}\theta} = \boldsymbol{u} \tag{4}$$

$$\frac{\partial s}{\partial \theta} + \kappa \boldsymbol{u} \cdot \nabla S = \varepsilon \frac{\partial}{\partial z}\left(k \frac{\partial s}{\partial z}\right) \tag{5}$$

$z = \kappa \mathcal{Z}$:

$$w = \frac{\partial \mathscr{G}}{\partial \theta} + \kappa\left(u \frac{\partial \mathscr{G}}{\partial x} + v \frac{\partial \mathcal{Z}}{\partial y}\right) \tag{6}$$

$$\frac{\partial u}{\partial z} = \frac{\partial v}{\partial z} = \frac{\partial s}{\partial z} = 0 \tag{7}$$

$z = -h$:

$$\boldsymbol{u} = 0 \tag{8}$$

where

$$\frac{\partial s}{\partial z} = 0 \tag{9}$$

$$\boldsymbol{x} = (x, y, z) = \left(\frac{x_*}{L}, \frac{y_*}{L}, \frac{z_*}{h_c}\right), \quad \theta = t_* \omega; \quad h = \frac{h_*}{h_c}$$

$$\boldsymbol{u} = (u, v, w) = \left(\frac{u_*}{u_c}, \frac{v_*}{u_c}, \frac{w_*}{w_c}\right), \quad \mathscr{G} = \frac{\mathscr{G}_*}{\mathscr{G}_c}$$

$$\boldsymbol{\xi} = (\xi, \eta, \zeta) = \left(\frac{\xi_*}{\xi_c}, \frac{\eta_*}{\xi_c}, \frac{\zeta_*}{\mathscr{G}_c}\right), \quad S = \frac{S_*}{S_c}$$

$$\nabla = \boldsymbol{e}_1 \frac{\partial}{\partial x} + \boldsymbol{e}_2 \frac{\partial}{\partial y} + \boldsymbol{e}_3 \frac{\partial}{\partial z}$$

$$u_c = \kappa \sqrt{g h_c}, \quad w_c = \frac{h_c}{L} u_c, \quad \xi_c = N \kappa L$$

$$f = f_* / \omega, \quad \nu = \frac{\nu_*}{\nu_c}, \quad k = \frac{k_*}{k_c}$$

$$\nu_c = \omega h_c^2; \quad \omega = \left(\frac{L}{\sqrt{g h_c}}\right)^{-1}$$

and

$$\kappa = \mathscr{Z}_c / h_c \tag{10}$$

$$N = 1 + (n-1)\kappa \tag{11}$$

$$\varepsilon = \frac{k_c / h_c^2}{\omega} \tag{12}$$

t_* denotes time; (x_*, y_*, z_*) form Cartesian coordinates on an f-plane with corresponding unit vectors $(\boldsymbol{e}_1, \boldsymbol{e}_2, \boldsymbol{e}_3)$; (u_*, v_*, w_*) are the velocity components in the (x_*, y_*, z_*) -directions; $\mathscr{Z}_*$ is the displacement of the free surface from the undisturbed sea surface; h_* is the water depth measured from the undisturbed sea surface, g is the gravitational acceleration; f_* is the Coriolis parameter; v_* is the eddy viscosity, k_* is the eddy diffusion coefficient; S_* is the concentration of solute or any conservative and passive tracer; (ξ_*, η_*, ζ_*) are the Lagrangian displacements in the (x_*, y_*, z_*) directions of a water parcel at time t_*, of which the initial condition is $(\xi_*, \eta_*, \zeta_*) = 0$ when $t_* = t_{*0}$, and thus the position of the marked water parcel can be expressed as $(x_*, y_*, z_*) = (x_{*0}, y_{*0}, z_{*0}) + (\xi_*, \eta_*, \zeta_*)$; L and h_c denote the horizontal and vertical scales, respectively; ω^{-1} is the time scale; the quantities with the subscript "c" indicate the characteristic values of corresponding dimensional quantities, the meaning of N with n will be explained later in the section "Discussion on the Lagrangian Residual Circulation" (see Fig. 1), and n will be given in the operator (27); and then, N is assumed to be the order of 1 here, namely

$$\mathscr{O}(N) = 1 \tag{13}$$

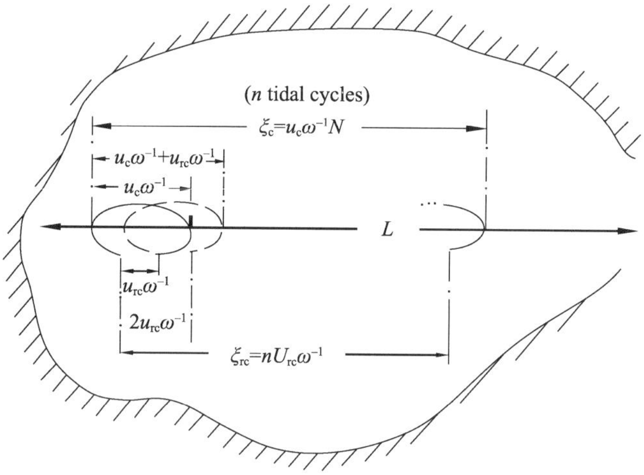

Fig. 1 $N = 1 + (n-1)\kappa$, $\kappa = u_{rc}/u_c$

In fact, the nonlinear three-dimensional dynamic problem, (1)–(9), is a further extension of the nonlinear two-dimensional dynamic problem proposed by Feng *et al.* (1986a, 1986b).

It should be pointed out with emphasis that the extension of the two-dimensional dynamics for the Lagrangian residual current to the three-dimensional one is of much importance from both theoretical and practical points of view.

As might be expected, the nondimensional parameter x is small for most coastal seas (Tee, 1976; Elliott and Hendrix, 1976; Weisberg, 1976; Heaps, 1978; Uncles *et al.* 1980, Sun *et al.*, 1981; Cheng and Gartner, 1984). For example, for Bohai Sea, $\mathscr{O}(\kappa)=10^{-1}$ ($\mathscr{G}_c \sim 2\ \mathrm{m}$, $h_c \sim 20\ \mathrm{m}$). Thus, we shall reasonably suppose

$$\mathscr{O}(\kappa)<1. \tag{14}$$

In view of the nondimensional parameter κ being a measure of the nonlinearity of the dynamic problem (1)–(9), the condition (14) implies that the dynamic system is weakly nonlinear, so a weakly nonlinear theory on the Lagrangian residual circulation will be treated in the present paper. Here, of course, the nonlinearity due to the turbulent viscosity has been excluded, with a hypothesis of the linearized form of the eddy viscosity coefficient, or $\nu=\nu(x,y,z;t)$.

The parameter ε is a measure of the relative importance for eddy diffusion. If ω is taken to be $1.5\times10^{-4}(\mathrm{sec}^{-1})$ (Corresponding to the circular frequency of M_2), h_c is 20 m, and k_c is taken to be 10 ($\mathrm{cm}^2/\mathrm{sec}$) (K. B. Bowden, 1965), then the nondimensional parameter ε is the order of 10^{-2}; thus ε will be assumed to be a parameter smaller than κ, or

$$\mathscr{O}(\varepsilon)=\kappa^2 \tag{15}$$

The equality (15) implies that the transport mechanism of solutes in the interior of a tidal system is strongly convection dominated and the diffusion effects are relatively small (Leendertse, 1970; Leendertse and Gritton, 1971; Fischer *et al.*, 1979).

Noting the scale of tidal excursion, $\xi_c=\kappa NL$, and the conditions (13) and (14), the velocity of a marked water parcel can be expanded in Taylor series expansions about $\boldsymbol{x}_0$, or

$$\begin{aligned}\boldsymbol{u}(\boldsymbol{\kappa}_0+xN\boldsymbol{\xi},\theta)&=\boldsymbol{u}(\boldsymbol{x}_0,\theta)+\kappa N\boldsymbol{\xi}\cdot(\nabla\boldsymbol{u})_0+\kappa^2\frac{1}{2}\left\{(N\xi)^2\left(\frac{\partial^2\boldsymbol{u}}{\partial x^2}\right)_0+(N\eta)^2\left(\frac{\partial^2\boldsymbol{u}}{\partial y^2}\right)_0\right.\\&\quad\left.+(N\zeta)^2\left(\frac{\partial^2\boldsymbol{u}}{\partial z^2}\right)_0+2(N^2\xi\eta)\left(\frac{\partial^2\boldsymbol{u}}{\partial x\partial y}\right)_0+2(N\zeta)\left(\frac{\partial^2\boldsymbol{u}}{\partial x\partial z}\right)_0+2(N^2\eta\zeta)\left(\frac{\partial^2\boldsymbol{u}}{\partial y\partial z}\right)_0\right\}\\&\quad+\mathscr{O}(\kappa^3)\end{aligned} \tag{16}$$

where the notation $(\ )_0$ indicates that the term is evaluated at $\boldsymbol{x}_0$.

Of course, the displacement of the marked water parcel can be expressed as

$$N\boldsymbol{\xi}=\int_{\theta_0}^{\theta}\boldsymbol{u}(\boldsymbol{x}_0+\kappa N\boldsymbol{\xi},\theta')\mathrm{d}\theta' \tag{17}$$

where $\theta_0=t_0\omega$.

As pointed out by Feng (1977) and by Feng *et al.* (1986a), using a perturbation technique, all of the dependent variables can be expanded in ascending series of the small parameter, κ, as follows

$$\mathscr{V} = \sum_{j=0,1,\cdots} \kappa^j \mathscr{V}_j ; \tag{18}$$

where $\mathscr{V}_j$ are the j-th order perturbation solutions of $\mathscr{V}$ and $\mathscr{V} = (\boldsymbol{u}, \mathscr{G}, \xi, S)$.

A substitution of (18) into (1)–(9) with (16) and (17) yields the j-th order model, where the tedious and complicated expressions and equations derived from the equations (1)–(3) and the conditions (6)–(9) can be found in the previous paper (Feng, 1977). Here we shall show the additional expressions for the streakline and the convection-diffusion equations only. The former is

$$N\boldsymbol{\zeta}_j = \int_{\theta_0}^{\theta} \boldsymbol{u}_j(\boldsymbol{x}_0 + \kappa N\boldsymbol{\xi}_j, \theta')\mathrm{d}\theta' \tag{19}$$

where

$$\begin{aligned}\boldsymbol{u}_j(\boldsymbol{x}_0 + \kappa N\boldsymbol{\xi}_j, \theta) = \boldsymbol{u}_j(\boldsymbol{x}_0, \theta) + \sum_{m=0}^{j-1} N\boldsymbol{\xi}_{j-1-m} \cdot (\nabla \boldsymbol{u}_m)_0 + \\ + \sum_{m=0}^{j-2}\sum_{n=0}^{m} \frac{N^2}{2}\Bigg\{\left(\frac{\partial^2 \boldsymbol{u}_n}{\partial x^2}\right)_0 \xi_{m-n}\xi_{j-2-m} + \left(\frac{\partial^2 \boldsymbol{u}_n}{\partial y^2}\right)_0 \eta_{m-n}\eta_{j-2-m} \\ + \left(\frac{\partial^2 \boldsymbol{u}_n}{\partial z^2}\right)_0 \zeta_{m-n}\zeta_{j-2-m} + 2\left(\frac{\partial^2 \boldsymbol{u}_n}{\partial x \partial y}\right)_0 \zeta_{m-n}\eta_{j-2-m} \\ +2\left(\frac{\partial^2 \boldsymbol{u}_n}{\partial x \partial z}\right)_0 \xi_{m-n}\zeta_{j-2-m} + 2\left(\frac{\partial^2 \boldsymbol{u}_n}{\partial y \partial z}\right)_0 \eta_{m-n}\zeta_{j-2-m}\Bigg\} + \sum_{m=0}^{j-3} \cdots\cdots\end{aligned} \tag{20}$$

and the latter

$$\frac{\partial S_j}{\partial \theta} + \sum_{m=0}^{j-1} \boldsymbol{u}_m \cdot \nabla S_{j-1-m} = \boldsymbol{\varepsilon}\frac{\partial}{\partial z}\left(\kappa \frac{\partial S_{j-2}}{\partial z}\right) \tag{21}$$

where

$$\boldsymbol{\varepsilon} = \varepsilon / \kappa^2$$

The subscripts indicate the order of the perturbation solution; when the subscript is less than zero, the variable is defined to be zero. Noting (15), there is $\mathscr{O}(\boldsymbol{\varepsilon}) = 1$.

If the higher order tides coming from the external ocean have been excluded, the zeroth order model represents the astronomical tides and the higher order models represent the higher order constituents which are generated from the nonlinear coupling between the astronomical tides and their associated higher order constituents. It is noticed that the Eulerian residual current is embedded within these higher order models.

The linearity of the j-th order model implies that the solutions for each tidal constituent of the *j*-th order model can be solved independently in terms of the related lower order tidal constituents. In fact, the dynamic problem of any tidal constituent that we are interested in can be reduced to a boundary-value problem of the elliptic differential equation for tidal elevation and an expression for the vertical distribution of tidal current, particularly, of Eulerian residual current, the details of which can be found in the previous papers (Feng, 1977; Sun *et al.*, 1981; Feng and Sun, 1983; Feng, 1984; Sung, 1986, 1987). In the present paper, we suppose that the problems of tidal elevation and tidal currents, particularly that of Eulerian residual current, have been solved, and the Lagrangian displacements of the labelled water parcels and the concentration of solutes are obtained respectively using the equations (19)–(21). Thus the basis has been laid for the solution to solve the Lagrangian residual current and the longer-term mass-transport.

Lagrangian Residual Current

For clarity, a nonlinear M_2 tidal system is used (instead of a complicated tidal system including several astronomical tides and associated shallow water constituents) to examine the Lagrangian residual current. As well known, the first order constituents of the M_2 tidal system contain the M_4 tide and the first order Eulerian residual current, and the second order constituents of the M_2 system include the M_6 tide and the others with frequencies being equal to the frequency of the M_2 tide. The harmonics of the order of $\mathcal{O}(\kappa^j)$ $(j=3,4,\cdots)$ are not considered. It is natural to select the circular frequency of M_2 as the characteristic circular frequency, and thus the nondimensional circular frequency and the period of M_2 are 1 and 2π respectively. The *j*-th order perturbation solutions have been supposed to be solved as mentioned above and are written as follows.

The zeroth order model, or M_2 tide:

$$\begin{cases} \boldsymbol{u}_0 = \boldsymbol{u}_0' \cos\theta + \boldsymbol{u}_0'' \sin\theta \\ \mathcal{G}_0 = \mathcal{G}_0' \cos\theta + \mathcal{G}_0'' \sin\theta \end{cases} \tag{22}$$

the first order model, or M_4 + first order Eulerian residual:

$$\begin{cases} \boldsymbol{u}_1 = \boldsymbol{u}_1' \cos(2\theta) + \boldsymbol{u}_1'' \sin(2\theta) + \boldsymbol{u}_{\text{er},1} \\ \mathcal{G}_1 = \mathcal{G}' \cos(2\theta) + \mathcal{G}'' \sin(2\theta) + \mathcal{G}_{\text{er},1} \end{cases} \tag{23}$$

the second order model, or M_6 + the other harmonics

$$\begin{cases} \boldsymbol{u}_2 = \boldsymbol{u}_2' \cos(3\theta) + \boldsymbol{u}_2'' \sin(3\theta) + \sum_{i=1}^{3} (\boldsymbol{u}_{2,i}' \cos\theta + \boldsymbol{u}_{2,i}'' \sin\theta) \\ \mathcal{G}_2 = \mathcal{G}_2' \cos(3\theta) + \mathcal{G}_2'' \sin(3\theta) + \sum_{i=1}^{3} (\mathcal{G}_{2,i}' \cos\theta + \mathcal{G}_{2,i}'' \sin\theta) \end{cases} \tag{24}$$

where the superscripts " ′ " and " ″ " indicate the harmonic coefficients and the summation

$\sum_{i=1}^{3}$ contains the other constituents of the second order model. Substituting (22)–(24) into (18), the solutions are obtained to correct to the second order, or $\mathscr{O}(\kappa^2)$ approximation:

$$\boldsymbol{u} = \sum_{j=0}^{2} \kappa^j \boldsymbol{u}_j + O(\kappa^3) \tag{25}$$

$$\mathscr{G} = \sum_{j=0}^{2} \kappa^j \mathscr{G}_j + O(\kappa^3) \tag{26}$$

where $\boldsymbol{u}_j$ and $\mathscr{G}_j$ are expressed by (22)–(24).

Introduce a time-averaging operator of a variable $\mathscr{V}$ over one or more tidal periods, namely n tidal cycles, as

$$\langle \mathscr{V} \rangle = \frac{1}{2\pi n}\int_{\theta_0}^{\theta_0+2\pi n} \mathscr{V} \,\mathrm{d}\theta' \tag{27}$$

where $n = 1,2,\cdots$ which means that n is introduced into (11).

By putting the time-averaging operator (27) on the Lagrangian velocity, $\boldsymbol{u}(\boldsymbol{x}(\boldsymbol{x}_0,\theta),\theta)$, the result is called the Lagrangian residual current, $\boldsymbol{u}_{\mathrm{lr}} = \langle \boldsymbol{u}(\boldsymbol{x}(\boldsymbol{x}_0,\theta),\theta)\rangle$. The Lagrangian residual current is different from the Eulerian residual current $\boldsymbol{u}_{\mathrm{er}}$ in that the time-averaging is to be evaluated by following the water parcel. This averaging procedure leads naturally to an equivalent definition of the Lagrangian residual current as

$$\boldsymbol{u}_{\mathrm{lr}} = \frac{1}{2\pi n}\int_{\theta_0}^{\theta_0+2\pi n} \boldsymbol{u}(\boldsymbol{x}(\boldsymbol{x}_0,\theta'),\theta')\mathrm{d}\theta' = \frac{1}{2\pi n} N\boldsymbol{\xi}(\theta_0 + 2\pi n) \tag{28}$$

where the equation (4) is used to derive (28).

The Lagrangian residual current is also expressed as the net Lagrangian displacement over n tidal cycles $(N\boldsymbol{\xi})$ $(n = 1,2,\cdots)$ divided by the n tidal periods $(2\pi n)$, as shown in (28).

Differing from the Eulerian residual current which is a function of the spatial coordinate, say $\boldsymbol{x}_0$ only, the Lagrangian residual current is the function of not only spatial coordinate $\boldsymbol{x}_0$ but also the temporal coordinate of tidal phase θ_0 when the marked water parcel is released (Feng *et al.*, 1986a; Cheng and Casulli, 1983; Zimmerman, 1979). In fact, the Lagrangian residual current should depend also on the number of tidal cycles, n. This matter needs further examination and discussion in the next section.

A substitution of (19)–(20) and (22)–(25) into (28) yields the Lagrangian residual velocity induced by the M_2-tidal system to be correct to the second order harmonics, $\mathscr{O}(\kappa^2)$, as follows

$$\boldsymbol{u}_{\mathrm{lr}} = \kappa(\boldsymbol{u}_{\mathrm{er}} + \boldsymbol{u}_{\mathrm{sd}}) + \kappa^2 \boldsymbol{u}_{\mathrm{ld}} + \mathscr{O}(\kappa^3) \tag{29}$$

Of course, the Lagrangian residual velocity should be properly normalized by $u_{\mathrm{rc}} = \kappa u_{\mathrm{c}}$, and with this correct scaling, the Lagrangian residual current becomes

$$\boldsymbol{u}_{\mathrm{lr}} = \boldsymbol{u}_{\mathrm{er}} + \boldsymbol{u}_{\mathrm{sd}} + \kappa \boldsymbol{u}_{\mathrm{ld}} + \mathscr{O}(\kappa^2) \tag{30}$$

where the Eulerian residual velocity $\boldsymbol{u}_{\mathrm{er}}$, is generated by the nonlinear coupling of M_2 with M_2 and is correct to the order of $\mathscr{O}(\kappa^2)$ since the next order of nonzero Eulerian residual is on the order of $\mathscr{O}(\kappa^3)$. The Stokes' drift velocity $\boldsymbol{u}_{\mathrm{sd}}$ is given as

$$\boldsymbol{u}_{\mathrm{sd}} = \langle N\xi_0 \cdot (\nabla \boldsymbol{u}_0)_0 \rangle \tag{31}$$

and the Lagrangian (residual) drift velocity $\boldsymbol{u}_{\mathrm{ld}}$ is expressed as

$$\boldsymbol{u}_{\mathrm{ld}} = \boldsymbol{u}'_{\mathrm{ld}} \cos\theta_0 + \boldsymbol{u}''_{\mathrm{ld}} \sin\theta_0 \tag{32}$$

$$\boldsymbol{u}'_{\mathrm{ld}} = \boldsymbol{u}''_0 \cdot \nabla(\boldsymbol{u}_{\mathrm{er}} + \boldsymbol{u}_{\mathrm{sd}}) - (\boldsymbol{u}_{\mathrm{er}} + \boldsymbol{u}_{\mathrm{sd}}) \cdot \nabla \boldsymbol{u}''_0$$

$$\boldsymbol{u}''_{\mathrm{ld}} = -\boldsymbol{u}'_0 \cdot \nabla(\boldsymbol{u}_{\mathrm{er}} + \boldsymbol{u}_{\mathrm{sd}}) + (\boldsymbol{u}_{\mathrm{er}} + \boldsymbol{u}_{\mathrm{sd}}) \cdot \nabla \boldsymbol{u}'_0$$

The first order Lagrangian residual velocity expressed as the sum of the Eulerian residual and Stokes' drift velocities, $\boldsymbol{u}_{\mathrm{er}} + \boldsymbol{u}_{\mathrm{sd}}$, refers to the mass transport velocity, which was first introduced by Longuet-Higgins (1969) and called Stokes' formula. The Lagrangian (residual)drift velocity, $\boldsymbol{u}_{\mathrm{ld}}$, was first revealed and named in two-dimensional space, or in the problem of a vertically integrated model (Feng *et al*., 1986a). In the present paper, a generalization of the Lagrangian (residual) drift velocity from the two-dimensional to the three-dimensional problem has been made and expressed in the formula (30) with (32). It is important to note that the Lagrangian (residual) drift velocity shows really the distinct Lagrangian property because it is the function of θ_0, the tidal phase when the marked water parcel is released from a fixed point, $\boldsymbol{x}_0$. In the previous two-dimensional problem of a vertically integrated model, the two-dimensional Lagrangian (residual) drift velocity traces out an ellipse on a hodograph plane as the initial phase angle θ_0 varies from 0 to 2π; or, when the marked water parcels are released from a fixed point $\boldsymbol{x}_0$ continuously over a tidal period, the terminus of the marked water parcels after a tidal cycle form an ellipse in space (Feng *et al*., 1986a). The Lagrangian residual velocity derived in three-dimensional space (30), or the Lagrangian (residual) drift velocity (32), has a similar behavior. The two horizontal components of Lagrangian residual velocity can be expressed as follows

$$u_{\mathrm{lr}} = u_{\mathrm{er}} + u_{\mathrm{sd}} + \kappa u_{\mathrm{ed}} \tag{33}$$

$$v_{\mathrm{lr}} = v_{\mathrm{er}} + v_{\mathrm{sd}} + \kappa v_{\mathrm{ed}} \tag{34}$$

where u_{er}, v_{er} and u_{sd}, v_{sd} and u_{ld}, v_{ld} are the horizontal components of the Eulerian residual velocity, Stokes' drift velocity, and the Lagrangian (residual) drift velocity, respectively, and

$$u_{\mathrm{sd}} = \langle N\xi_0 \cdot (\nabla u_0)_0 \rangle \tag{35}$$

$$v_{\mathrm{sd}} = \langle N\xi_0 \cdot (\nabla v_0)_0 \rangle \tag{36}$$

$$u_{\mathrm{ld}} = u'_{\mathrm{ld}} \cos\theta_0 + u''_{\mathrm{ld}} \sin\theta_0 \tag{37}$$

$$v_{\rm ld} = v'_{\rm ld}\cos\theta_0 + v''_{\rm ld}\sin\theta_0 \tag{38}$$

where

$$u'_{\rm ld} = \boldsymbol{u}''_0\cdot\nabla(u_{\rm er}+u_{\rm sd}) - (\boldsymbol{u}_{\rm er}+\boldsymbol{u}_{\rm sd})\cdot\nabla u''_0$$
$$u''_{\rm ld} = -\boldsymbol{u}'_0\cdot\nabla(u_{\rm er}+u_{\rm sd}) + (\boldsymbol{u}_{\rm er}+\boldsymbol{u}_{\rm sd})\cdot\nabla u'_0$$
$$v''_{\rm ld} = \boldsymbol{u}''_0\cdot\nabla(v_{\rm er}+v_{\rm sd}) - (\boldsymbol{u}_{\rm er}+\boldsymbol{u}_{\rm sd})\cdot\nabla v''_0$$
$$v'_{\rm ld} = -\boldsymbol{u}'_0\cdot\nabla(v_{\rm er}+v_{\rm sd}) + (\boldsymbol{u}_{\rm er}+\boldsymbol{u}_{\rm sd})\cdot\nabla v'_0$$

The expressions (37) and (38) say that the horizontal components of Lagrangian (residual) drift velocity, $u_{\rm ed}$ and $v_{\rm ed}$, trace out an ellipse on a hodograph plane as the tidal (current) phase θ_0 when the marked water parcels are released from the point $\boldsymbol{x}_0$ continuously varies from 0 to 2π. The properties of the Lagrangian residual ellipse can be given explicitly. The semi-major (+ sign) and semiminor (− sign) axes are indicated as a and b in the expression

$$\left.\begin{matrix} a \\ b \end{matrix}\right\} = \frac{1}{\sqrt{2}}\left\{u'^2_{\rm ld} + u''^2_{\rm ld} + v'^2_{\rm ld} + v''^2_{\rm ld} \pm \left[\left(u'^2_{\rm ld} + u''^2_{\rm ld} + v'^2_{\rm ld} + v''^2_{\rm ld}\right)^2 - 4\left(u'_{\rm ld}v''_{\rm ld} - u''_{\rm ld}v'_{\rm ld}\right)^2\right]^{1/2}\right\}^{1/2} \tag{39}$$

where the angle between the major axis of the residual ellipse and the x-axis is denoted by δ, and

$$\delta = \frac{1}{2}{\rm tg}^{-1}\left[2\frac{u'_{\rm ld}v'_{\rm ld} + u''_{\rm ld}v''_{\rm ld}}{(u'^2_{\rm ld} + u''^2_{\rm ld}) - (v'^2_{\rm ld} + v''^2_{\rm ld})}\right] \tag{40}$$

and the phase angle $\theta_{\max}$ which gives the Lagrangian (residual) drift velocity a maximum magnitude is

$$\theta_{\max} = \frac{1}{2}{\rm tg}^{-1}\left[2\frac{u'_{\rm ld}u''_{\rm ld} + v'_{\rm ld}v''_{\rm ld}}{(u'^2_{\rm ld} - u''^2_{\rm ld}) + (v'^2_{\rm ld} - v''^2_{\rm ld})}\right] \tag{41}$$

The Lagrangian residual ellipse in the three-dimensional model differs from that in the horizontally two-dimensional model because the former is the function of not only the horizontal coordinates (x_0,y_0) but also the vertical coordinate z_0, and thus is a three-dimensional structure.

In addition to the horizontal components of the Lagrangian residual current, the vertical component of the Lagrangian residual current $w_{\rm lr}$ is given, or

$$w_{\rm lr} = w_{\rm er} + w_{\rm sd} + \kappa w_{\rm ed} \tag{42}$$

where $w_{\rm er}$, $w_{\rm sd}$ and $w_{\rm ed}$ are the vertical components of the Eulerian residual, Stokes' drift, and Lagrangian (residual) drift velocities, respectively, and

$$w_{\rm sd} = \left\langle N\boldsymbol{\xi}_0\cdot(\nabla w_0)\right\rangle \tag{43}$$

$$w_{\rm ld} = w'_{\rm ld}\cos\theta_0 + w''_{\rm ld}\sin\theta_0 \tag{44}$$

where

$$w_{\rm ld} = \boldsymbol{u}''_0\cdot\nabla(w_{\rm er}+w_{\rm sd}) - (\boldsymbol{u}_{\rm er}+\boldsymbol{u}_{\rm sd})\cdot\nabla w''_0$$

$$w''_{\mathrm{ld}} = -\boldsymbol{u}'_0 \cdot \nabla(w_{\mathrm{er}} + w_{\mathrm{sd}}) + (\boldsymbol{u}_{\mathrm{er}} + \boldsymbol{u}_{\mathrm{sd}}) \cdot \nabla w'_0$$

Finally, the Lagrangian residual velocity (to be correct to the order of $\mathcal{O}(\kappa^2)$) can be summed up in words to read

$$\begin{pmatrix} \text{Lagrangian} \\ \\ \text{Residual} \\ \text{Velocity} \end{pmatrix} = \begin{pmatrix} \text{Eulerian} \\ \\ \text{Residual} \\ \text{Velocity} \end{pmatrix} + \begin{pmatrix} \text{Stokes'} \\ \\ \text{Drift} \\ \text{Velocity} \end{pmatrix} + \kappa \begin{pmatrix} \text{Lagrangian} \\ \text{(residual)} \\ \text{Drift} \\ \text{Velocity} \end{pmatrix} \tag{45}$$

In 3-D (Longuet-Higgins, 1969)
In 2-D (Feng *et al.*, 1986a)
In 3-D (The present paper)

Discussion on the Lagrangian Residual Circulation

The concept of Lagrangian residual current was first introduced into the large scale currents and waves in the ocean from the theory of surface waves by Longuet-Higgins (1969), who showed that the mass transport velocity equals the sum of the Eulerian residual velocity and Stokes' drift velocity (the Stokes' formula). The term "Lagrangian residual current" was used by Tee (1976) in his numerical model for tidal and residual circulation based on the Stokes' formula. The Lagrangian residual current has been defined as the mean velocity of a marked water parcel, or as the net displacement of a marked water parcel divided by the averaging period over one or more tidal cycles as pointed out by Zimmerman (1979) and Cheng *et al.* (1982). The tide-induced Lagrangian residual current is expected to be a function of the tidal phase when the marked water parcel is released since the net displacement of a marked water parcel is evidently expected to be a function of the flow field in the neighborhood where the marked water parcel is released, which has been illustrated by Cheng (1983) and by Yu and Chen (1983) in their numerical models for tidal and residual currents respectively. Specifically, a quantitative relation between the tide-induced Lagrangian residual current and the tidal phase is further derived based upon a weakly nonlinear tidal and residual current theory (Feng *et al.*, 1986a and the formula (30) in the present paper). In accordance with the definition of Lagrangian residual current expressed by the formula (28), however, the Lagrangian residual current should be pointed out to be usually also dependent on the number of averaging tidal cycles, n. An attempt will be made to discuss it below.

A sketch (Fig. 1) shows the relations between n and N and the scales of tided excursion, ξ_{c}, and net Lagrangian displacement, ξ_{rc}, over n tidal cycles (n = 1,2,$\cdots$). If we select a typical value of χ to be 0.1, the corresponding values of n, N and several ratios of scales for lengths are exhibited in the following table.

Averaging Period of Time	Day	Week	Month	Season	Year
n	1	10	30	100	300
$N=1+(n-1)\kappa$	1	1.9	3.9	10.9	30.9
$\frac{n}{N}$	1	5.3	7.7	9.2	9.7
$\frac{\xi_c}{L}=\kappa N$	0.1	0.19	0.39	1.09	3.09
$\frac{\xi_{rc}}{\xi_c}=\frac{n}{N}\kappa$	0.1	0.53	0.77	0.92	0.97
$\frac{\xi_{rc}}{L}=\kappa^2 n$	0.01	0.10	0.30	1	3

It is naturally shown that the net Lagrangian displacements of a marked water parcel for a month or a season can be greater than one that over a tidal cycle in the order of magnitude, and thus the Lagrangian residual drift which traces out an ellipse over a tidal cycle plays a more considerable role in the dispersion, for example, of pollutants, for the former than for the latter though the Lagrangian (residual) drift velocities are in the same order of magnitude in both cases. However, we should point out that, if the assumption that $\mathcal{O}(N)=1$ is valid in the cases of such averaging period of time as a day, a week or a month, but is not valid for longer periods then the theory on the Lagrangian residual current proposed in the present paper seems to be false and the averaging period of time in the problem on the Lagrangian residual current should be extended to about a season or a year. Unfortunately, in reality, there is of course always some residual motion, which adds up cycle after cycle and produces water parcel displacements over such longer terms as a season or a year that are much larger than the diameter of the tidal ellipse. It is worth while to use the following approach to solve these problems on the longer-term processes just mentioned. In fact,

$$\begin{aligned}\boldsymbol{u}_{\mathrm{lr}} &= \boldsymbol{u}_{\mathrm{er}}(\boldsymbol{x}_0,\theta_0;n)=\frac{1}{2\pi n}\int_{\theta_0}^{\theta_0+2\pi n}\boldsymbol{u}(\overline{x}(\boldsymbol{x}_0,\theta'),\theta')\mathrm{d}\theta' \\ &=\frac{1}{n}\sum_{j=1}^{n}\frac{1}{2\pi}\int_{\theta_0+(j-1)2\pi}^{\theta_0+2\pi j}\boldsymbol{u}(\boldsymbol{x}(\boldsymbol{x}_0,\theta'),\theta')\mathrm{d}\theta' \\ &=\frac{1}{n}\sum_{j=1}^{n}\boldsymbol{u}_{\mathrm{lr}}(\boldsymbol{x}_{j-1},\theta_0+(j-1)2\pi;1)\end{aligned}\tag{46}$$

where

$$\boldsymbol{x}_{j-1}=\begin{cases}\boldsymbol{x}_0+\kappa^2 2\pi\sum_{i=1}^{j-1}\boldsymbol{u}_{\mathrm{lr}}(\boldsymbol{x}_{i-1},\theta_0+(i-1)2\pi;1) & (j=2,3,\cdots)\\ \boldsymbol{x}_0 & (j=1)\end{cases}$$

It is worthy of note that we come to the conclusion that the Lagrangian residual velocity generated by averaging the marked water parcel over such long period of time as a season or a year, $\boldsymbol{u}_{\mathrm{lr}}(\boldsymbol{x}_0,\theta_0;n)$, where $n\gg 1$ and $\mathcal{O}(N)>1$, can be constructed as an arithmetic mean of the Lagrangian residual velocities, $\boldsymbol{u}_{\mathrm{lr}}(\boldsymbol{x}_{j-1},\theta_0+(j-1)2\pi;1)$ $(j=1,2,\cdots n)$.

Lagrangian Residual Velocity as an Eulerian Field Variable

The Lagrangian residual velocity derived above, and expressed by formula (30), has really shown the distinct Lagrangian property because it is the function of θ_0, the tidal phase when the marked water parcel is released from a fixed point, $\boldsymbol{x}_0$. Noting that $\boldsymbol{x}_0$ and θ_0 are to be selected arbitrarily, the Lagrangian residual velocity could be reasonably described as an Eulerian Field variable and the aggregate of such local velocities may be specified as an Eulerian field of flow. Using $(\boldsymbol{x},\theta)$ instead of $(\boldsymbol{x}_0,\theta_0)$ in the flow field of Lagrangian residual circulation, the Lagrangian residual velocity is described as the function of position in space $(\boldsymbol{x})$ and time (θ),

$$\boldsymbol{u}_{\mathrm{lr}}=\boldsymbol{u}_{\mathrm{er}}(\boldsymbol{x})+\boldsymbol{u}_{\mathrm{sd}}(\boldsymbol{x})+\kappa\boldsymbol{u}_{\mathrm{ld}}(\boldsymbol{x},\theta) \tag{47}$$

where

$$\boldsymbol{u}_{\mathrm{ld}}=\boldsymbol{u}'_{\mathrm{ld}}(\boldsymbol{x})\cos\theta+\boldsymbol{u}''_{\mathrm{ld}}(\boldsymbol{x})\sin\theta \tag{48}$$

particularly, the horizontal components of Lagrangian residual velocity can be expressed as

$$u_{\mathrm{lr}}=u_{\mathrm{er}}(\boldsymbol{x})+u_{\mathrm{sd}}(\boldsymbol{x})+\kappa u_{\mathrm{ld}}(\boldsymbol{x},\theta) \tag{49}$$

$$v_{\mathrm{lr}}=v_{\mathrm{er}}(\boldsymbol{x})+v_{\mathrm{sd}}(\boldsymbol{x})+\kappa v_{\mathrm{ld}}(\boldsymbol{x},\theta) \tag{50}$$

where

$$u_{\mathrm{ld}}=u'_{\mathrm{ld}}(\boldsymbol{x})\cos\theta+u''_{\mathrm{ld}}(\boldsymbol{x})\sin\theta \tag{51}$$

$$v_{\mathrm{ld}}=v'_{\mathrm{ld}}(\boldsymbol{x})\cos\theta+v''_{\mathrm{ld}}(\boldsymbol{x})\sin\theta \tag{52}$$

It should be pointed out that the Lagrangian residual velocity of Eulerian type can be really constructed as an incompressible flow field because it satisfies the continuity equation for the incompressible flow. In fact, (Ⅰ) a direct substitution of the notation for the time-averaging operator (27) to the continuity equation (1) yields $\nabla\cdot\boldsymbol{u}_{\mathrm{er}}=0$; (Ⅱ) noting that $\nabla\cdot\boldsymbol{u}_0=0$ and introducing (19) $(j=0)$ and the first expression of (22) into the formula (31), we obtain $\nabla\cdot\boldsymbol{u}_{\mathrm{sd}}=0$ in terms of taking the divergence of the Stokes' drift velocity; (Ⅲ) by taking the divergence of (32) and using $\nabla\cdot\boldsymbol{u}_{\mathrm{er}}=\nabla\cdot\boldsymbol{u}_{\mathrm{sd}}=0$ just derived, then $\nabla\cdot\boldsymbol{u}_{\mathrm{ld}}=0$ is shown; and thus it is demonstrated that $\nabla\cdot\boldsymbol{u}_{\mathrm{er}}=\nabla\cdot\boldsymbol{u}_{\mathrm{sd}}=\nabla\cdot\boldsymbol{u}_{\mathrm{ed}}=0$ or

$$\nabla\cdot\boldsymbol{u}_{\mathrm{lr}}=0 \tag{53}$$

By applying the time-averaging operator (27) on the Lagrangian residual velocity expressed by the formula (47), the mass-transport velocity, $\boldsymbol{u}_{\mathrm{lM}}$, is derived to be as

$$\boldsymbol{u}_{\mathrm{lM}}=\langle\boldsymbol{u}_{\mathrm{lr}}\rangle \tag{54}$$

where

$$\boldsymbol{u}_{\mathrm{lM}}=\boldsymbol{u}_{\mathrm{er}}(\boldsymbol{x})+\boldsymbol{u}_{\mathrm{sd}}(\boldsymbol{x})$$

This reveals that it is the mass-transport velocity which is the Eulerian mean of the Lagrangian residual velocity over one or a few tidal cycles and the mass-transport velocity is correct to the second order of approximation rather than to the first order. Thus (47) can be rewritten as

$$\boldsymbol{u}_{\mathrm{lr}}=\boldsymbol{u}_{\mathrm{lM}}(\boldsymbol{x})+\kappa\boldsymbol{u}_{\mathrm{ld}}(\boldsymbol{x},\theta) \tag{55}$$

where

$$\begin{aligned}\boldsymbol{u}_{\mathrm{lM}}(\boldsymbol{x})&=\langle\boldsymbol{u}_{\mathrm{lr}}(\boldsymbol{x},\theta)\rangle=\boldsymbol{u}_{\mathrm{er}}(\boldsymbol{x})+\boldsymbol{u}_{\mathrm{sd}}(\boldsymbol{x})\\ \boldsymbol{u}_{\mathrm{ld}}&=\boldsymbol{u}'_{\mathrm{ld}}(\boldsymbol{x})\cos\theta+\boldsymbol{u}''_{\mathrm{ld}}(\boldsymbol{x})\sin\theta\end{aligned} \tag{56}$$

The formula (55) shows that the Lagrangian residual velocity is similar to the tidal currentvelocity as a sum of the tidally periodic fluctuation part and the tidal cycle average, but the tidally periodic part, $\boldsymbol{u}_{\mathrm{ld}}$, is smaller than the tidal cycle mean, $\boldsymbol{u}_{\mathrm{lM}}$, in the order of magnitude for the Lagrangian residual current. As well known, however, the tidally periodic part of the tidal current is typically greater than the residual part in the order of magnitude. It should he emphasized that the Lagrangian residual velocity field is different from the Eulerian residual velocity field which is a steady field because the Lagrangian residual velocity field is a time-dependent field as mentioned above.

In particular, the horizontal components of Lagrangian residual velocity become

$$u_{\mathrm{lr}}=u_{\mathrm{lM}}(\boldsymbol{x})+\kappa u_{\mathrm{ld}}(\boldsymbol{x},\theta) \tag{57}$$

$$v_{\mathrm{lr}}=v_{\mathrm{lM}}(\boldsymbol{x})+\kappa v_{\mathrm{ld}}(\boldsymbol{x},\theta) \tag{58}$$

where

$$u_{\mathrm{lM}}(\boldsymbol{x})=\langle u_{\mathrm{lr}}(\boldsymbol{x},\theta)\rangle=u_{\mathrm{er}}(\boldsymbol{x})+u_{\mathrm{sd}}(\boldsymbol{x}) \tag{59}$$

$$v_{\mathrm{lM}}(\boldsymbol{x})=\langle v_{\mathrm{lr}}(\boldsymbol{x},\theta)\rangle=v_{\mathrm{er}}(\boldsymbol{x})+u_{\mathrm{sd}}(\boldsymbol{x}) \tag{60}$$

By integrating the equations (57), (58) the horizontal mass-transports are easily obtained, or

$$U_{\mathrm{lr}}=U_{\mathrm{er}}(x,y)+U_{\mathrm{sd}}(x,y)+\kappa U_{\mathrm{ld}}(x,y,\theta) \tag{61}$$

$$V_{\mathrm{lr}}=V_{\mathrm{er}}(x,y)+V_{\mathrm{sd}}(x,y)+\kappa V_{\mathrm{ld}}(x,y,\theta) \tag{62}$$

where

$$(U_{\mathrm{lr}},V_{\mathrm{lr}})=\int_{-h}^{0}(u_{\mathrm{lr}},v_{\mathrm{lr}})\mathrm{d}z$$

$$(U_{\mathrm{er}},V_{\mathrm{er}})=\int_{-h}^{0}(u_{\mathrm{er}},v_{\mathrm{er}})\mathrm{d}z$$

$$(U_{\mathrm{sd}},V_{\mathrm{sd}})=\int_{-h}^{0}(u_{\mathrm{sd}},v_{\mathrm{sd}})\mathrm{d}z$$

$$(U_{\mathrm{ld}},V_{\mathrm{ld}})=\int_{-h}^{0}(u'_{\mathrm{ld}},v'_{\mathrm{ld}})\mathrm{d}z\cdot\cos\theta+\int_{-h}^{0}(u''_{\mathrm{ld}},v''_{\mathrm{ld}})\mathrm{d}z\cdot\sin\theta$$

and their tidal cycle averages are

$$\langle U_{\mathrm{lr}}\rangle=U_{\mathrm{er}}(x,y)+U_{\mathrm{sd}}(x,y) \tag{63}$$

$$\langle V_{\text{lr}} \rangle = V_{\text{er}}(x,y) + V_{\text{sd}}(x,y) \tag{64}$$

Longer-term Transport Equation

A tidally averaged convection-diffusion equation for the concentration of any passive solute is also called a longer-term transport equation and it can be derived from the equation (21). The zeroth-order, the first-order and the second-order equations are respectively obtained as follows

$$\frac{\partial S_0}{\partial \theta} = 0 \tag{65}$$

$$\frac{\partial S_1}{\partial \theta} + \boldsymbol{u}_0 \cdot \nabla S_0 = 0 \tag{66}$$

$$\frac{\partial S_2}{\partial \theta} + \boldsymbol{u}_0 \cdot \nabla S_1 + \boldsymbol{u}_1 \cdot \nabla S_0 = \boldsymbol{\varepsilon} \frac{\partial}{\partial z}\left(k \frac{\partial S_0}{\partial z} \right) \tag{67}$$

The equation (65) indicates that the tidal cycle average of the concentration S can be approximately evaluated by S_0, and thus it is enough to derive the convection-diffusion equation which is satisfied by S_0 instead of the tidally averaged concentration $\langle S \rangle$.

Substituting the equation (66) into the equation (67) and noting the equation (65), a tidal cycle average of the equation (67) yields the longer-term transport equation

$$\boldsymbol{u}_{\text{IM}} \cdot \nabla S_0 = \boldsymbol{\varepsilon} \frac{\partial}{\partial z}\left(\langle k \rangle \frac{\partial S_0}{\partial z} \right) \tag{68}$$

where $\boldsymbol{u}_{\text{IM}}$ is expressed by (56).

The equation derived here, (68), is different from the classical longer-term transport equation (Fischer *et al.*, 1979). On the one hand, in the latter, the convection has been unreasonably represented by the Eulerian residual velocity, but in the equation (68) the convection is reasonably expressed by the Eulerian mean of the Lagrangian residual velocity, namely, by the mass-transport velocity. On the other hand, an assumption on the so called "tidal dispersion" has to be introduced into the classical longer-term transport equation (Fischer *et al.*, 1979), the equation (68), however, may describe correctly the Lagrangian nature of longer-term transport processes without introducing the Fickian hypothesis for tided dispersion.

A longer-term transport equation satisfied by the depth-averaged quantity of tidal cycle mean of the concentration, or by $\overline{S}_0$ approximately, where $\overline{S}_0 = \frac{1}{h}\int_{-h}^{0} S_0 \mathrm{d}z$, can be derived, by integrating the equation (68) over the depth and using the continuity equations and the boundary conditions (6)–(9), to be as

$$U_{\text{IM}} \frac{\partial \overline{S}_0}{\partial x} + V_{\text{IM}} \frac{\partial \overline{S}_0}{\partial y} = \frac{1}{\kappa Pe} \text{ (Shear Effect)} \tag{69}$$

where

$$U_{\mathrm{lM}} = U_{\mathrm{er}} + U_{\mathrm{sd}}, \quad V_{\mathrm{lM}} = V_{\mathrm{er}} + V_{\mathrm{sd}}$$

$$Pe = \frac{u_{\mathrm{c}} L}{\mathscr{D}_{\mathrm{c}}}, \text{ the Peclet number,}$$

$\mathscr{D}_{\mathrm{c}}$ is the scale of the dispersion coefficient due to the shear effect (Bowden, 1965).

Noting $(\kappa Pe)^{-1}$ to be a smaller order quantity (Bowden, 1965; Feng *et al.*, 1986b) and further neglecting this term, the equation (69) is reduced to the form of

$$U_{\mathrm{lM}} \frac{\partial \overline{S}_0}{\partial x} + V_{\mathrm{lM}} \frac{\partial \overline{S}_0}{\partial y} = 0 \tag{70}$$

The equation (70) has been validly derived if the condition on a horizontally two-dimensional problem of tides has been satisfied as pointed out in the previous paper (Feng *et al.*, 1974b). Of course, the equation (70) derived here in a three-dimensional space behaves as the depth average of a three-dimensional flow field. It should be pointed out that, however, this equation is valid for "the interior" of a basin because the diffusion or dispersion becomes important in a "boundary region" (Feng *et al.*, 1974b).

Conclusion

In view of the three-dimensional behaviour in space of the Lagrangian motion of a water parcel, and based on the fact that $\mathscr{O}(\kappa) < 1$, a three-dimensional weakly-nonlinear theory on tide-induced Lagrangian residual circulation and longer-term transport processes in tidal estuaries and coastal seas is formed. Differing from the Eulerian residual circulation, which is steady-state, the Lagrangian residual circulation might be expressed as a sum of the tidally periodic fluctuation part and the tidal cycle mean part, which is similar to a tidal circulation. This does not surprise us since the net Lagrangian displacement of a marked water parcel in a tidal current field depends not only on the position where the marked water parcel is released but also on the tidal phase when the marked water parcel is released. The mass-transport velocity is the Eulerian mean of the Lagrangian residual velocity and is correct to the second order, $\mathscr{O}(\kappa^2)$, rather than to the first order, $\mathscr{O}(\kappa)$. And further, a formula of the Lagrangian residual current is proposed for such long-term processes as a season or a year, or $n \gg 1$ and $\mathscr{O}(N) > 1$. And finally, differing from the classical equation, a new, longer-term transport equation for any conservative and passive tracer is derived. This equation is briefly characterized by the Lagrangian convection without introducing the so-called "tidal dispersion". The convection velocity is but the Eulerian mean of the Lagrangian residual velocity, or the mass-transport velocity.

References

Alfrink, B. J. and C. B. Vreugdenhil, 1981. Residual Currents. Rep. R 1469-11, Delft Hydraul. Lab., Delft,

The Netherlands, 32 pp.

Awaji, T., 1982. Water mixing in a tidal current and the effect of turbulence on tidal exchange through a strait. J. Physical Oceanogr. 12(6): 501–514.

Bowden, K. F., 1965. Horizontal mixing in the sea due to a shearing current, J. Fluid Mech. 21: 83–95.

Cheng, R. T., 1983. Euler-Lagrangian computation in estuarine hydrodynamics. Proc. Third Inter. Conf. on Num. Meth. in Laminar and Turbulent Flow, (Eds.), C. Taylor, J. A. Johnson and R. Smith, Pineridge Press, New York, pp. 341–352.

Cheng, R. T. and Casulli, V., 1982. On Lagrangian residual currents with applications in South San Francisco Bay, California. Water Resour. Res. 18(6): 1652–1662.

Cheng, R. T. and Garther, J. W., 1984. Tides, tidal and residual currents in San Francisco Bay, California: results of measurements, 1979–1980, USGS Open-file Report, Repl. 84–xx, 120 pp.

Dyer, K. R., 1973. Estuaries: A Physical Introduction. Wiley, London, pp. 140.

Dyer, K. R., 1974. The salt balance in stratified estuaries. Estuarine and Coastal Marine Science 2: 273–281.

Elliott, A. J. and T. E. Hendrix, 1976. Intensive observations of the circulation in the Potomac estuary. Chesapeake Bay Institute, Sepc. Rept., No. 55, 35 pp.

Feng, S., 1982. Numerical models adopted in the East China Sea studies and the preliminary results of their application to general circulation, tides, storm surges and pollutant dispersion. Proceedings of J. O. A., Abstracts of the Poster Sessions (microfiche), Dalhousie Univ., Halifax, Nova Scotia, Canada.

Feng, S., 1984. A three-dimensional nonlinear hydrodynamic model with variable eddy viscosity in shallow seas. Chinese Journal of Oceanology and Limnology, 2 (2): 177–187.

Feng, S. (Feng Shizuo), 1977. A three-dimensional non-linear model of tides. Sci. Sin. 20 (4): 436–446.

Feng, S., R. T. Cheng and P. Xi, 1986a. On tide-induced Lagrangian residual current and residual transport: 1. Lagrangian residual current. Water Resour. Res., 22 (12): 1623–1634.

Feng, S., R. T. Cheng and P. Xi, 1986b. On tide-induced Lagrangian residual current and residual transport: 2. Residual transport with application in south San Francisco Bay, California. Water Resour. Res., 22(12): 1635–1646.

Feng, S. and W. Sun, 1983. A tidal three-dimensional non-linear model with variable eddy viscosity. Chinese J. of Oceanology & Limnology 1(2): 166–170.

Fischer, H. B., 1976. Mixing and dispersion in estuaries, Ann. Rev. Fluid Mech. 8: 107–133.

Fischer, H. B. *et al.*, 1979. Mixing in inland and coastal waters. Academic Press, N. Y., 483 pp.

Heaps, N. S., 1978. Linearized vertically-integrated equations for residual circulation in coastal seas. Deut. Hydrog. Z. 31: 147–169.

Leendertse, J. J., 1970. A water-quality simulation model for well-mixed estuaries and coastal seas. Vol. 1, Principles of Computation, Rep. RM-6230-R6, Rand Corp., Santa Monica, California.

Leendertse, J. J. and E. C. Gritton, 1971. Vol. II, Computation Procedures; Vol. III, Jamaica Bay Simulation, Rep. R-708-NUC and R-709-NYC, Rand Corp. Santa Monica, California.

Lewis, R. E. and J. O. Lewis, 1983. The principal factors contributing to the flux of salt in a narrow, partially stratified estuary. Estuarine Coastal Mar. Sci. 16: 599–626.

Longuet-Higgins, M. S., 1969. On the transport of mass by time-varying ocean currents. Deep Sea Res. 16:

431–447.

Pritchard, D. W., 1954. A study on the salt balance in a coastal plain estuary. J. Mar. Res. 13: 133–144.

Stern, M. E., 1975. Ocean Circulation Physics. Academic Press, New York, 246 pp.

Stommel, H. and H. G. Farmer, 1952. On the nature of estuarine circulation, Part I. Reference 52–88, Woods Hole Oceanographic Institution, Woods Hole, Massachusetts.

Sun, W., Z. Chen and S. Feng, 1981. Numerical simulation of three-dimensional non-linear tidal waves (I) —A numerical study for M_4 and MS_4 waves in the Bohai Sea. J. Shandong Coll. Oceanol. 11(1): 23–31. (In Chinese with English abstract)

Song, L., 1986. A Hydrodynamic Velocity-splitting Model with a Depth-varying Eddy viscosity in Shallow Seas (I)—The Velocity-splitting Model. J. Shandong Coll. of Oceanol. 16(3): 15–27. (In Chinese with English Abstract)

Song, L., 1987. A Hydrodynamic Velocity-splitting Model with a Depth-varying Eddy viscosity in Shallow Seas (II)—An application of the velocity-splitting Model to the Zeroth-order Model of the Ultra-shallow water storm surges. J. Shandong Coll. of Oceanol. 17(2): 8–19. (In Chinese with English abstract)

Tee, T. K., 1976. Tide-induced residual current: A 2-D non-linear numerical model. J. Mar. Res. 31: 603–628.

Uncles, R. J. and M. B. Jordan, 1979. Residual fluxes of water and salt at two stations in the Severn estuary. Estuarine Coastal Mar. Sci. 9: 287–302.

Uncles, R. J. and M. B. Jordan, 1980. A one-dimensional representation of residual currents in the Severn estuary and associated observations. Estuarine Coastal Mar. Sci. 10: 39–60.

Uncles, R. J. and P. J. Radford, 1980. Seasonal and spring-neap tidal dependence of axial dispersion coefficients in the Severn—a wide, vertically mixed estuary. J. Fluid Mech. 98(4): 703–726.

Weisberg. R. H., 1976. The nontidal flow in the Providence River of Narraganseh Bay: A stochastic approach to estuarine circulation. J. Phys. Oceanogr. 6: 721–734.

Winterwerp, J. C., 1983. Decomposition of the mass transport in narrow estuaries. Estuarine Coastal Shelf Sci. 16: 627–638.

Yu, G. and S. Chen. 1983. Numerical modeling of the circulation and pollutant dispersion in Jiaozhou Bay (III)—the Lagrangian residual current and the pollutant dispersion. J. Shandong Coll. Oceanol. 13(1): 1–14. (In Chinese with English abstract)

Zimmerman, J. T. F., 1978. Dispersion by tide-induced residual current vortices. Hydrodynamics of Estuaries and Fjords, (Ed.) J. C. J. Nihoul, Elsevier Scientific Publishing Co., N. Y., pp. 207–216.

Zimmerman, J. T. F., 1979. On the Euler-Lagrangian transformation and the Stokes' drift in the presence of oscillatory and residual currents. Deep Sea Res. 26A: 505–520.

A Three-dimensional Weakly Nonlinear Model of Tide-induced Lagrangian Residual Current and Mass-transport, with an Application to the Bohai Sea *

Shizuo Feng

1. Introduction

In recent years, research on the hydrodynamics of coastal seas and tidal estuaries has focused on the inter-tidal processes which determine the long-term transport and distribution of important water properties such as salinity and temperature, and the concentration of pollutants or any tracer. While the dominant observable motion in coastal seas, such as the Bohai Sea and the East China Sea, is the rotary circulation associated with tides, the inter-tidal transport processes are mainly dependent on the residual circulation, but not on the tidal currents themselves, at least not in a direct way. It is becoming increasingly clear that the residual circulation should be described and studied by means of the Lagrangian residual velocity, since it is more relevant to use a Lagrangian mean velocity rather than a Eulerian mean velocity to determine the origin of water masses. The Lagrangian mean velocity of a marked water parcel leads to the concept of Lagrangian residual velocity, or Lagrangian residual current. It should be pointed out that the study of Lagrangian residual currents is a relatively recent topic, and thus any further investigation of Lagrangian residual currents may be a valuable contribution both from a theoretical and from a practical point of view.

As a first approximation of the Lagrangian residual velocity, the mass-transport velocity has been shown to be the sum of the Eulerian residual velocity and of the Stokes' drift velocity (Longuet-Higgins, 1969). As is well known, the Eulerian residual velocity has been defined as the Eulerian mean velocity, i.e., as the velocity averaged at a fixed spatial point over one or several tidal cycles. The Lagrangian residual velocity, however, is related to the Lagrangian mean velocity of a marked water parcel, i.e., to the velocity averaged by following the marked water parcel over the tidal cycles. It is almost self-evident that, in contrast to the Eulerian residual velocity, the Lagrangian residual velocity is not only a function of the location where the parcel is released but also of the tidal phase at which that parcel is released. Indeed, the Lagrangian mean velocity depends upon the trajectory that the marked water parcel follows in the tidal field (Zimmerman, 1979; Cheng and Casulli, 1982; Cheng,

---

* Feng S. 1987. A three-dimensional weakly nonlinear model of tide-induced Lagrangian residual current and mass-transport, with an application to the Bohai Sea. In: J C J Nihoul, B M Jamart (eds.) Three Dimensional Models of Marine and Estuarine Dynamics, Elsevier Oceanography Series. Amsterdam, Oxford, New York, Tokyo: Elsewier. 471–488

1983). By considering second order dynamics, this dependence of the Lagrangian residual velocity on the tidal phase has been revealed in a depth-averaged two-dimensional model (Feng *et al.*, 1986a). It should be pointed out that a depth-averaged two-dimensional analysis of the Lagrangian residual velocity is, in general, questionable from the viewpoint of hydrodynamics because the interpretation of averaging a Lagrangian velocity over depth is not very clear unless the vertical water column is assumed to move in the tidal field as a rigid body (Stern, 1975; Alfrink and Vreugdenhil, 1981; Feng, 1986). The dynamics of the Lagrangian residual current should be rigorously treated in a three-dimensional space. Therefore, a study of a three-dimensional theory of Lagrangian residual circulation in coastal seas is of much importance not only for practical applications, but also to satisfy the requirements of Lagrangian dynamics. In this paper, a weakly nonlinear three-dimensional model of Lagrangian residual current is introduced. The residual circulation in continental shelf seas, semi-enclosed shallow seas, gulfs or bays and tidal estuaries is, generally speaking, driven by tides, winds acting on the sea surface, variations of water density, and open boundary forces. While much effort has been applied to the study of residual circulation driven by winds and density variations, only recently has attention been drawn to the tide-induced residual circulation (Nihoul and Ronday, 1975; Heaps, 1978; Tee, 1976). In this paper, to avoid confusing the main issues, the effects due to surface wind and baroclinic variations are not included; only the dynamics of tide-induced Lagrangian residual circulation is discussed. The effects of wind and density variations on the Lagrangian residual circulation have been examined in previous papers (Feng, 1990; Feng et al., 1990).

Here, the concept and the definition of Lagrangian residual velocity are reexamined. As stated above, the Eulerian residual velocity has been defined as the Eulerian mean velocity. Nevertheless, it would not seem proper to maintain that the Lagrangian residual velocity is synonymous with the Lagrangian mean velocity of a marked water parcel. Undoubtedly, the Lagrangian mean velocity of a marked water parcel in the tidal field is a Lagrangian velocity as far as inter-tidal processes are concerned, since it is proportional to the net displacement of the marked water parcel over the averaging period (several tidal cycles). On the other hand, it is natural to expect that the Lagrangian residual velocity could be used as a Eulerian field variable, and the aggregate of such local velocities may be specified as an Eulerian field of flow for the inter-tidal processes. Thus it can be seen that the Lagrangian residual velocity cannot be unconditionally defined as the Lagrangian mean velocity of a marked water parcel unless the latter satisfies the continuity equation for an incompressible fluid. These conditions are discussed in sections 3 and 6 of this paper.

Field equations governing the Eulerian residual circulation have been presented by Nihoul and Ronday (1975), who show that the introduction of nonlinear coupling between tides into the equations leads to the generation of a "tidal stress". Furthermore, a three-dimensional baroclinic model for the Eulerian residual circulation has also been proposed (Feng *et al.*, 1984), in which there exists not only a "tidal stress" but also a "tidal surface source" due to the nonlinear interaction between tides. However, it is more relevant to the present problem to derive a set of field equations governing the mean circulation of the Lagrangian residual current. In this paper, a set of field equations for the tide-induced mean circulation is introduced. These equations govern the mean Lagrangian residual circulation and are the most appropriate for the description of inter-tidal processes.

A long-term transport equation, namely, a convection-diffusion equation for the inter-tidal processes describes the balance between convective and dispersive transports for the tidally averaged concentration of any conservative and passive tracer. As is well known, in the conventional long-term transport equation, the convection velocity is the Eulerian residual velocity and an assumption of "tidal dispersion" has to be made (Fischer *et al.*, 1979). As stated above, however, for the inter-tidal transport processes the convective transport should be characterized by the Lagrangian residual velocity. Furthermore, the physics and dynamics of the "tidal dispersion" are not well understood (Pritchard, 1954; Bowden, 1967; Fischer *et al.*, 1979). A new inter-tidal transport equation has already been proposed, but it is a two-dimensional depth-averaged equation (Feng *et al.*, 1986b). In this paper, we derive a three-dimensional inter-tidal transport equation which describes correctly the Lagrangian nature of convective transport without introducing the "tidal dispersion" terms.

In the last two sections of this paper, the Lagrangian dynamics of tracer spreading on long-term scales and a preliminary application of the proposed model to the Bohai Sea, China, are briefly described.

2. Formulation of the Model

Based on a nonlinear three-dimensional tidal model (Feng, 1977), the generalized nondimensional dynamic problem proposed by Feng (1986) is as follows:

i) Field equations:

$$\nabla \cdot \boldsymbol{u} = 0 \tag{1}$$

$$\frac{\partial u}{\partial \theta} + \kappa \boldsymbol{u} \cdot \nabla u - fv = -\frac{\partial Z}{\partial x} + \frac{\partial}{\partial z}\left(\nu \frac{\partial u}{\partial z}\right) \tag{2}$$

$$\frac{\partial v}{\partial \theta} + \kappa \boldsymbol{u} \cdot \nabla v + fu = -\frac{\partial Z}{\partial y} + \frac{\partial}{\partial z}\left(\nu \frac{\partial v}{\partial z}\right) \tag{3}$$

$$\frac{\mathrm{d}}{\mathrm{d}\theta}(N\xi) = \boldsymbol{u} \tag{4}$$

$$\frac{\partial S}{\partial \theta} + \kappa \boldsymbol{u} \cdot \nabla S = \varepsilon \frac{\partial}{\partial z}\left(k \frac{\partial S}{\partial z}\right) \tag{5}$$

ii) Boundary conditions:

at $z = \kappa Z$,

$$w = \frac{\partial Z}{\partial \theta} + \boldsymbol{\kappa}\left(u\frac{\partial Z}{\partial x} + v\frac{\partial Z}{\partial y} \right) \tag{6}$$

and

$$\frac{\partial u}{\partial z} = \frac{\partial v}{\partial z} = \frac{\partial S}{\partial z} = 0 \tag{7}$$

at $z = -h$,

$$\boldsymbol{u} = 0 \tag{8}$$

and

$$\frac{\partial S}{\partial z} = 0 \tag{9}$$

where

$$\boldsymbol{x} = (x, y, z) = \left(\frac{x_*}{L}, \frac{y_*}{L}, \frac{z_*}{h_c} \right)$$

$$\theta = t_* \omega$$

$$h = \frac{h_*}{h_c}$$

$$S = \frac{S_*}{S_c}$$

$$Z = \frac{Z_*}{Z_c}$$

$$\boldsymbol{u} = (u, v, w) = \left(\frac{u_*}{u_c}, \frac{v_*}{u_c}, \frac{w_*}{w_c} \right)$$

$$\boldsymbol{\xi} = (\xi, \eta, \zeta) = \left(\frac{\xi_*}{\xi_c}, \frac{\eta_*}{\xi_c}, \frac{\zeta_*}{\zeta_c} \right)$$

$$\nabla = \boldsymbol{e}_1 \frac{\partial}{\partial x} + \boldsymbol{e}_2 \frac{\partial}{\partial y} + \boldsymbol{e}_3 \frac{\partial}{\partial z}$$

$$u_c = \kappa \sqrt{g h_c}$$

$$w_c = \frac{h_c}{L} u_c$$

$$\xi_c = \kappa N L$$

$$\zeta_c = \kappa N h_c$$

$$f = \frac{f_*}{\omega}$$

$$\nu = \frac{\nu_*}{\nu_c}$$

$$k = \frac{k_*}{k_c}$$

$$\nu_c = \omega h_c^2$$

$$\omega = \frac{\sqrt{gh_c}}{L}$$

Three nondimensional parameters, κ, N, and ε have been introduced. They are defined by

$$\kappa = \frac{Z_c}{h_c} \tag{10}$$

$$N = 1 + (n-1)\kappa \tag{11}$$

$$\varepsilon = \frac{k_c}{\omega h_c^2} \tag{12}$$

In these equations, t_* denotes time; (x_*, y_*, z_*) form a Cartesian coordinate system on an f-plane with corresponding unit vectors $(\boldsymbol{e}_1, \boldsymbol{e}_2, \boldsymbol{e}_3)$; (u_*, v_*, w_*) are the velocity components in the $(\boldsymbol{e}_1, \boldsymbol{e}_2, \boldsymbol{e}_3)$ directions; Z_* is the displacement of the free surface from the undisturbed sea surface; h_* is the water depth measured from the undisturbed sea surface; g is the gravitational acceleration; f_* is the Coriolis parameter; ν_* is the eddy viscosity, k_* is the eddy diffusivity; S_* is the concentration of any conservative and passive tracer; (ξ_*, η_*, ζ_*) are the Lagrangian displacements in the $(\boldsymbol{e}_1, \boldsymbol{e}_2, \boldsymbol{e}_3)$ directions of a marked water parcel at time t_*, for which the initial condition is $(\xi_*, \eta_*, \zeta_*) = 0$ when $t_* = t_{0*}$, and thus the position of the marked water parcel can be expressed as $(x_*, y_*, z_*) = (x_{0*}, y_{0*}, z_{0*}) + (\xi_*, \eta_*, \zeta_*)$; L and h_c denote the horizontal and vertical scales, respectively; ω^{-1} is the time scale; the quantities with the subscript "c" indicate the characteristic value of the corresponding dimensional quantities; n is the number of tidal cycles over which the tidal current velocity is averaged to filter the periodic current velocity and to derive the residual velocity; κN is a measure of the relative scale of tidal excursion to L; ε / κ is a measure of the relative effect on a concentration of the eddy diffusion compared to convection.

Let us assume (Feng, 1986):

$$O(\kappa N) = O(\kappa) \tag{13}$$

$$O(\kappa) < 1 \tag{14}$$

$$O(\varepsilon / \kappa) = O(\kappa) \tag{15}$$

These conditions imply weakly nonlinear dynamics: the nonlinearity due to the eddy viscosity and diffusion are excluded, and the transport mechanism is dominated by convection.

Given the scale of tidal excursion, $\xi_c = \kappa NL$, and conditions (13) and (14), the velocity of a marked water parcel can be expanded in a Taylor series expansion about $\boldsymbol{x}_0$, or

$$u(x_0+\kappa N\xi,\theta)=u(x_0,\theta)+\kappa N\xi\cdot(\nabla u)_0+\frac{\kappa^2N^2}{2}\left\{\xi^2\left(\frac{\partial^2 u}{\partial x^2}\right)_0+\eta^2\left(\frac{\partial^2 u}{\partial y^2}\right)_0+\zeta^2\left(\frac{\partial^2 u}{\partial z^2}\right)_0+2\xi\eta\left(\frac{\partial^2 u}{\partial x\partial y}\right)_0+2\xi\zeta\left(\frac{\partial^2 u}{\partial x\partial z}\right)_0+2\eta\zeta\left(\frac{\partial^2 u}{\partial y\partial z}\right)_0\right\}+O(\kappa^3)\quad(16)$$

where the notation $(\ )_0$ indicates that the term is evaluated at x_0.

The displacement of the marked water parcel can be expressed by

$$N\xi=\int_{\theta_0}^{\theta}u(x_0+\kappa N\xi,\ \theta')\mathrm{d}\theta'\quad(17)$$

where $\theta_0=\omega t_0$.

A perturbation technique with the small parameter κ has been used to derive and solve the jth-order model (j = 0,1,···) of the dynamic problem proposed, (1)–(9) (Feng, 1977, 1986). In the present paper, we suppose that the solution for the tidal currents, and in particular the Eulerian residual currents and the Lagrangian displacements of marked water parcels, have been obtained. Thus, the basis to calculate the Lagrangian residual velocity and the concentration of any conservative and passive tracer has been laid.

For clarity, a nonlinear M_2 tidal system is used, instead of a complicated tidal system including several astronomical tides and associated shallow water constituents, to examine the Lagrangian residual current. It is natural to select the circular frequency of M_2 as the characteristic circular frequency, and thus the nondimensional circular frequency and period of M_2 are 1 and 2π respectively. The zeroth order model, or M_2 tide, can be expressed; in a nondimensional form, as $u_0=u_0'\cos\theta+u_0''\sin\theta$, where the superscripts $'$ and $''$ indicate the harmonic coefficients. As is well known, the first order model contains the M_4 tide and the (first order) Eulerian residual velocity. The second order model contains the M_6 tide and other harmonics of frequency equal to that of the M_2 tide. Here, the harmonics of the order of $O(\kappa^j)$ ($j=3,4,\cdots$) are not considered.

3. Lagrangian Residual Current Induced by an M2-tidal System

Let us use angular brackets < > to denote a tidal cycle average of any quantity A, or $\langle A\rangle=\frac{1}{2\pi n}\int_{\theta_0}^{\theta_0+2\pi n}A\mathrm{d}\theta$, where n = 1,2,···. The Eulerian residual velocity, u_{ER}, defined as the Eulerian mean velocity (i.e., the velocity averaged at a fixed spatial point, say x_0, over n tidal cycles) is expressed as $u_{\mathrm{ER}}=\langle u(x_0,\theta)\rangle$. The Lagrangian mean velocity, u_{L}, of a marked water parcel released from the point x_0 at time θ_0 and moving over the n tidal cycles is defined as $u_{\mathrm{L}}=\langle u(x(x_0,\theta),\theta)\rangle$. However, as pointed out in section 1, the Lagrangian residual velocity u_{LR} cannot be unconditionally defined as the Lagrangian mean velocity u_{L} unless u_{L} satisfies the continuity equation for an incompressible fluid. In fact, in the proposed weakly nonlinear model, assumption (13) ensures that $\boldsymbol{u}_{\mathrm{L}}$ satisfies the continuity equation for an incompressible fluid. We shall prove that point at the end of this section. Thus, here the Lagrangian residual velocity can be expressed as $\boldsymbol{u}_{\mathrm{L}}$, or

$$u_{\mathrm{LR}} = \frac{1}{2\pi n}\int_{\theta_0}^{\theta_0+2\pi n} \boldsymbol{u}(\boldsymbol{x}(\boldsymbol{x}_0,\theta),\theta)\,\mathrm{d}\theta \tag{18}$$

The substitution of (16) and (17) into (18) yields the Lagrangian residual velocity induced by an M_2-tidal system, to be correct to the second order harmonics and normalized by $u_{\mathrm{rc}} = \kappa u_{\mathrm{c}}$, as follows

$$\boldsymbol{u}_{\mathrm{LR}} = \boldsymbol{u}_{\mathrm{LM}} + \kappa \boldsymbol{u}_{\mathrm{LD}} \tag{19}$$

where

$$\boldsymbol{u}_{\mathrm{LM}} = \boldsymbol{u}_{\mathrm{ER}} + \boldsymbol{u}_{\mathrm{SD}} \tag{20}$$

and

$$\begin{aligned} \boldsymbol{u}_{\mathrm{LD}} &= \boldsymbol{u}'_{\mathrm{LD}}\cos\theta_0 + \boldsymbol{u}''_{\mathrm{LD}}\sin\theta_0 \\ \boldsymbol{u}'_{\mathrm{LD}} &= \boldsymbol{u}''_0 \cdot \nabla \boldsymbol{u}_{\mathrm{LM}} - \boldsymbol{u}_{\mathrm{LM}} \cdot \nabla \boldsymbol{u}''_0 \\ \boldsymbol{u}''_{\mathrm{LD}} &= -\boldsymbol{u}'_0 \cdot \nabla \boldsymbol{u}_{\mathrm{LM}} + \boldsymbol{u}_{\mathrm{LM}} \cdot \nabla \boldsymbol{u}'_0 \end{aligned} \tag{21}$$

The first order Lagrangian residual velocity expressed as the sum of the Eulerian residual velocity $\boldsymbol{u}_{\mathrm{ER}}$ and the Stokes' drift velocity $\boldsymbol{u}_{\mathrm{SD}} = \langle N\xi_0 \cdot (\nabla \boldsymbol{u}_0)_0 \rangle$ is the mass- transport velocity $\boldsymbol{u}_{\mathrm{LM}}$, which was first introduced by Longuet-Higgins (1969). Equation (20) is called Stokes' formula. The Lagrangian drift velocity, $\boldsymbol{u}_{\mathrm{LD}}$, was only recently revealed and named (Feng *et al.*, 1986a; Feng, 1986). The second order dynamics shows the dependence of the Lagrangian residual velocity on the tidal phase θ_0 at which the marked water parcel is released from the fixed point $\boldsymbol{x}_0$.

In the previous depth-averaged model, the "two-dimensional" Lagrangian drift velocity traces out an ellipse on a hodograph plane as the initial tidal phase θ_0 varies from 0 to 2π; in other words, when the "marked water columns" are released from a fixed position continuously over a tidal period, the ensemble of the terminal positions of the "marked water columns" after a tidal cycle form an ellipse in the "two-dimensional space", the Lagrangian residual ellipse (Cheng *et al.*, 1986; Feng *et al.*, 1986a). The Lagrangian residual velocity derived in a three-dimensional space, equation (19), or the Lagrangian drift velocity, (21), has a similar but three-dimensional structure (Feng, 1986). Here it should be pointed out that this unique property of the Lagrangian residual velocity reflects its Lagrangian nature since the Lagrangian residual velocity is born of the Lagrangian mean velocity of a marked water parcel in the tidal field but the latter depends on the trajectory that such a parcel follows and the parcels follow different trajectories depending upon the time, θ_0, of their release at $\boldsymbol{x}_0$.

Noting that $(\boldsymbol{x}_0,\theta_0)$ is to be selected arbitrarily, and then using $(\boldsymbol{x},\theta)$ instead of $(\boldsymbol{x}_0,\theta_0)$, the Lagrangian residual velocity can be viewed as a Eulerian field variable and the aggregate of such local velocities may be specified as a Eulerian field of flow provided that the Lagrangian residual velocity expressed by (19)–(21) satisfies the continuity equation for an incompressible fluid. As a matter of fact, by taking the divergence of (19)–(21) and going through some algebraic manipulations, we have

$$\nabla \cdot \boldsymbol{u}_{\mathrm{LR}} = 0 \tag{22}$$

Hence, the definition used here of the Lagrangian residual velocity as the Lagrangian mean velocity of a marked water parcel is valid. The case in which $\boldsymbol{u}_{\mathrm{LR}}$ cannot be defined by (18) will be examined and discussed in section 6.

Thus, (19)–(21) show that the Lagrangian residual velocity field is similar to the tidal current velocity field in the sense that it is a sum of a tidally periodic fluctuation plus the tidal cycle average since $\boldsymbol{u}_{\mathrm{LM}} = \langle \boldsymbol{u}_{\mathrm{LR}} \rangle$. However, given that $O(|\kappa \boldsymbol{u}_{\mathrm{LD}}| / |\boldsymbol{u}_{\mathrm{LM}}|) = \kappa$, they are different because the tidally periodic part of the tidal current velocity field is typically greater than the residual part by one or more orders of magnitude. In contrast to the Eulerian residual velocity field which is steady, the Lagrangian residual velocity field is obviously a time-dependent field of flow.

4. A Set of Field Equations for the Mean Lagrangian Residual Circulation Induced by an M_2-tidal System

As stated in section 1, while much effort has been applied to the study of residual circulation driven by the wind on the sea surface and the variation of water density, only recently has attention been drawn to the tide-induced residual circulation. However, in coastal seas, where the dominant observable motions are tides, the residual circulation is induced not only by the wind on the sea surface and the horizontal gradient of water density but also by the nonlinear coupling of tides, as pointed out by Nihoul and Ronday (1975). A scale analysis on the general circulation in the Bohai Sea and the East China Sea has also shown that the tide-induced residual circulation is, in general, a component of the general residual circulation (Feng *et al.*, 1994). The residual circulation is conventionally derived from current-meter records using filter techniques or time averages of time series records to remove tidal variations, i.e., the residual circulation is conventionally defined as the Eulerian residual circulation. However, it is becoming increasingly clear that the residual circulation should be related to the Lagrangian residual velocity since the problem of residual circulation is to describe and understand the inter-tidal transport processes (Csanady, 1982; Feng, 1990). Thus, it might be appropriate to define the (tide-induced) residual circulation as the (tide-induced) mean Lagrangian residual circulation. To describe the tide-induced mean Lagrangian residual circulation and to study some of its characteristics, a set of field equations governing the tidal cycle average of the Lagrangian residual velocity, or the mass-transport velocity, $\boldsymbol{u}_{\mathrm{LM}}$, is derived as follows:

i) Field equations:

$$\nabla \cdot \boldsymbol{u}_{\mathrm{LM}} = 0 \tag{23}$$

$$-fv_{\mathrm{LM}} = -\frac{\partial \langle Z_1 \rangle}{\partial x} + \frac{\partial}{\partial z}\left(\nu \frac{\partial u_{\mathrm{LM}}}{\partial z}\right) + \pi_1 \tag{24}$$

$$fu_{\mathrm{LM}} = -\frac{\partial \langle Z_1 \rangle}{\partial y} + \frac{\partial}{\partial z}\left(\nu \frac{\partial v_{\mathrm{LM}}}{\partial z}\right) + \pi_2 \tag{25}$$

ii) Boundary conditions:

at $z = 0$,

$$w_{\mathrm{LM}} = 0 \tag{26}$$

and

$$\frac{\partial u_{\mathrm{LM}}}{\partial z} = \frac{\partial v_{\mathrm{LM}}}{\partial z} = 0 \tag{27}$$

at $z = -h$,

$$\boldsymbol{u}_{\mathrm{LM}} = 0 \tag{28}$$

where

$$\begin{aligned}\pi_1 = &-\frac{\partial}{\partial x}\left\langle \frac{1}{2}\xi_0 \cdot \nabla Z_0 \right\rangle + \frac{\partial}{\partial z}\left\langle \left(\frac{5}{2}\frac{\partial \xi_0}{\partial x} + \frac{\partial \eta_0}{\partial y}\right)\nu\frac{\partial u_0}{\partial z} + \left(\frac{\partial \xi_0}{\partial y} + \frac{1}{2}\frac{\partial \eta_0}{\partial x}\right)\nu\frac{\partial v_0}{\partial z} \right\rangle \\ &+ \frac{\partial}{\partial z}\left\langle \nabla\ \nu \cdot \xi_0 \frac{\partial u_0}{\partial z} \right\rangle - \left\langle \frac{\partial \boldsymbol{\xi}_0}{\partial z} \cdot \nabla\left(\nu\frac{\partial u_0}{\partial z}\right) \right\rangle - \frac{1}{2}\left\langle \frac{\partial^2 \xi_0}{\partial z \partial x}\left(\nu\frac{\partial u_0}{\partial z}\right) \right. \\ &\left. + \frac{\partial^2 \eta_0}{\partial z \partial x}\left(\nu\frac{\partial v_0}{\partial z}\right) + \xi_0 \frac{\partial^2}{\partial z \partial x}\left(\nu\frac{\partial u_0}{\partial z}\right) + \eta_0 \frac{\partial^2}{\partial z \partial x}\left(\nu\frac{\partial v_0}{\partial z}\right) \right\rangle\end{aligned}$$

$$\begin{aligned}\pi_2 = &-\frac{\partial}{\partial y}\left\langle \frac{1}{2}\xi_0 \cdot \nabla Z_0 \right\rangle + \frac{\partial}{\partial z}\left\langle \left(\frac{5}{2}\frac{\partial \eta_0}{\partial y} + \frac{\partial \xi_0}{\partial x}\right)\nu\frac{\partial v_0}{\partial z} + \left(\frac{\partial \eta_0}{\partial x} + \frac{1}{2}\frac{\partial \xi_0}{\partial y}\right)\nu\frac{\partial u_0}{\partial z} \right\rangle \\ &+ \frac{\partial}{\partial z}\left\langle \nabla\ \nu \cdot \xi_0 \frac{\partial v_0}{\partial z} \right\rangle - \left\langle \frac{\partial \boldsymbol{\xi}_0}{\partial z} \cdot \nabla\left(\nu\frac{\partial v_0}{\partial z}\right) \right\rangle - \frac{1}{2}\left\langle \frac{\partial^2 \xi_0}{\partial z \partial y}\left(\nu\frac{\partial u_0}{\partial z}\right) \right. \\ &\left. + \frac{\partial^2 \eta_0}{\partial z \partial y}\left(\nu\frac{\partial v_0}{\partial z}\right) + \xi_0 \frac{\partial^2}{\partial z \partial y}\left(\nu\frac{\partial u_0}{\partial z}\right) + \eta_0 \frac{\partial^2}{\partial z \partial y}\left(\nu\frac{\partial v_0}{\partial z}\right) \right\rangle\end{aligned}$$

In these equations, (π_1, π_2) represent the nonlinear coupling of astronomical tides and can be naturally named "tide-induced body force". The tidal force contains two parts. The first term characterizes the nonlinear interaction between the tidal displacement and the tidal elevation, and it is horizontally irrotational. The second part represents the effect of eddy viscosity; this term is rotational. $(u_{\mathrm{LM}}, v_{\mathrm{LM}})$ and w_{LM} are horizontal and vertical components of $\boldsymbol{u}_{\mathrm{LM}}$ respectively, and $\langle Z_1 \rangle$ is the residual elevation.

The conceptual difference between the set of field equations for the Lagrangian residual circulation derived here, (23)–(28), and that for the Eulerian residual circulation (Nihoul and Ronday, 1975; Feng *et al.*, 1984) is revealed by the differences in the kinematic and kinetic boundary conditions at $z = 0$. Equation (26) shows that there is no "tidal surface source", or $\frac{\partial}{\partial x}\langle u_0 Z_0\rangle + \frac{\partial}{\partial y}\langle v_0 Z_0\rangle$, at $z = 0$ and equation (27) shows that there is no "tidal stress", or $-\left\langle Z_0 \frac{\partial^2}{\partial z^2}(u_0, v_0)\right\rangle$, at $z = 0$. Thus, the handling of the continuity equation is simplified when the mean Lagrangian residual circulation is used to describe inter-tidal processes. There, are other attractive features to this formulation.

If a material surface in the water is specified geometrically by the equation $F(\boldsymbol{x},\theta) =$ const., F is a quantity which is invariant for a water parcel on the surface, so that:

$$\frac{\mathrm{D}F}{\mathrm{D}\theta} = \frac{\partial F}{\partial \theta} + \kappa \boldsymbol{u} \cdot \nabla F = 0 \tag{29}$$

at all points on the surface. In particular, the equation of any surface bounding the sea water must satisfy (29). $F(\boldsymbol{x},\theta)$ can also be written as $F_0 + \kappa F_1 + \kappa^2 F_2 + O(\kappa^3)$, like the other variables. Substituting this expansion into equation (29), and taking a tidal cycle average of the latter, we have

$$u_{\mathrm{LM}} \cdot \nabla \langle F \rangle = 0 \tag{30}$$

where F_0 has been approximately expressed by $\langle F \rangle$. The kinematic boundary condition at the sea surface (26) can be derived as a special case of equation (30) (Feng, 1990).

Equation (30) shows that in the individual derivative of the tidal cycle average of a material surface, $\frac{\mathrm{D}}{\mathrm{D}\theta}\langle F \rangle$, the convective velocity is the mean Lagrangian residual velocity rather than the Eulerian residual velocity. This further confirms that the mean Lagrangian residual circulation is appropriate to the description of inter-tidal flow processes. In fact, equation (30) is also valid when F is any conservative quantity in the flow field.

The set of field equations governing the mass-transport velocity, (23)–(28), can in principle be solved if the forcing function (π_1, π_2), i.e., the tide-induced body force, is given and if the eddy viscosity $\nu = \nu(\boldsymbol{x})$ is assumed to be known. A numerical model to solve the set of equations (23)–(28) was proposed by Song (1987). If the sea water is assumed to be an inviscid fluid, the tide-induced body force (π_1, π_2) is reduced to an irrotational force and the set of equations (23)–(28) describes some geostrophic motions. This implies that the careful and correct selection of the eddy viscosity, $\nu(\boldsymbol{x})$, is of much importance for the numerical simulation of the mean Lagrangian residual circulation in coastal seas.

5. Inter-tidal Transport Equation

The convection-diffusion equation for the tidal cycle average of the concentration of any conservative and passive tracer is also called the long-term transport equation, or the inter-tidal transport equation (Feng *et al.*, 1986b). It can be derived by means of equation (5). Substituting $S = S_0 + \kappa S_1 + \kappa^2 S_2 + O(\kappa^3)$ into equation (5), the zeroth order equation is $\partial S_0 / \partial\theta = 0$, which says that the tidal cycle average of the concentration, $\langle S \rangle$, can be approximately evaluated by S_0. The first order equation is $\partial S_1 / \partial\theta + \boldsymbol{u}_0 \cdot \nabla S_0 = 0$, and the second order equation is as follows

$$\frac{\partial S_2}{\partial\theta} + \boldsymbol{u}_0 \cdot \nabla S_1 + \boldsymbol{u}_1 \cdot \nabla S_0 = E\frac{\partial}{\partial z}\left(k\frac{\partial S_0}{\partial z}\right) \tag{31}$$

where $E = \varepsilon / \kappa^2$, and, in view of (15), we have $O(E) = 1$.

Substituting the first order equation into equation (31) and using the zeroth order equation, a tidal cycle average of equation (31) yields the inter-tidal transport equation, or

$$u_{\mathrm{LM}} \cdot \nabla\langle S \rangle = E\frac{\partial}{\partial z}\left(\langle k \rangle\frac{\partial\langle S \rangle}{\partial z}\right) \tag{32}$$

to the zeroth order approximation, in which S_0 has been approximated by $\langle S \rangle$.

The equation derived here, (32), is different from the traditional long-term transport equation (Fischer *et al.*, 1979). On the one hand, in the latter, the convection has been unreasonably represented by the Eulerian residual velocity, but in equation (32) the convective transport is reasonably expressed by the Eulerian mean of the Lagrangian residual velocity, i.e., by the mass-transport velocity. On the other hand, an assumption on the so-called "tidal dispersion" has to be introduced into the conventional long-term transport equation (Fischer *et al.*, 1979), whereas equation (32) can describe correctly the Lagrangian nature of the convection affecting inter-tidal transport processes without introducing any hypothesis for tidal dispersion.

Let $\overline{A}$ denote the depth-averaged quantity of *A*, or $\overline{A} = \frac{1}{h}\int_{-h}^{0} A\mathrm{d}z$. An inter-tidal transport equation satisfied by $\overline{\langle S \rangle}$ is derived as follows:

$$\overline{u}_{\mathrm{LM}}\frac{\partial\overline{\langle S \rangle}}{\partial x} + \overline{v}_{\mathrm{LM}}\frac{\partial\overline{\langle S \rangle}}{\partial y} = Pe^{-1}\frac{1}{h}\nabla \cdot (h\underline{D} \cdot \nabla\overline{\langle S \rangle})$$

where Pe denotes the Peclet number for the residual motion, $Pe = \kappa u_c L / D_c$; $\underline{D}$ denotes the dispersion coefficient tensor due to the vertical shear of the horizontal component of the mass-transport velocity (Bowden, 1965), and D_c represents its scale. Since Pe is usually a large parameter (Bowden, 1965; Feng *et al.*, 1986b), the convective transport effect is greater than the dispersive one. Thus the equation just derived is reduced to

$$\overline{u}_{\mathrm{LM}}\frac{\partial\overline{\langle S \rangle}}{\partial x} + \overline{v}_{\mathrm{LM}}\frac{\partial\overline{\langle S \rangle}}{\partial y} = 0 \tag{33}$$

Equation (33) suggests that, at the zeroth order approximation, the transport process is determined purely by the convection, so that the concentration isolines for $\overline{\langle S\rangle}$ coincide with the streamlines of the depth-averaged mass-transport velocity. As a matter of fact, an equation similar to (33) can be derived under similar conditions for the horizontally two-dimensional tidal problem (Feng *et al.*, 1986b). Of course, equation (33), derived in a three-dimensional space, behaves as the depth-average of a three-dimensional mass-transport velocity. It should be pointed out that this equation is valid in the "interior" of a basin because the diffusion or dispersion becomes more pronounced in the "boundary region" (Feng *et al.*, 1986b).

6. Long-term Transport Processes

The conclusions derived above concerning the Lagrangian residual velocity and the inter-tidal transport processes are based upon a weakly nonlinear dynamical model of tidal flow and intra-tidal transport processes. Several hypotheses have been made, including that $O(\kappa N)=O(\kappa)$, or $O(N)=1$, in which $O(\kappa)<1$. Given that $N=1+(n-1)\kappa$, the condition $O(N)=1$ is valid in the cases of $O(n)<\kappa^{-2}$. If we assume $O(\kappa)=10^{-1}$, then $O(n)=\kappa^{-2}\sim 100$, which implies that the theory and model given above could not be directly applied to the cases of long averaging period of time such as a season or a year. It is worth giving the following approach to these problems of long-term transport processes.

Instead of the Lagrangian residual velocity, the original Lagrangian mean velocity of a marked water parcel has to be used here, or

$$u_{\mathrm{L}}=\frac{1}{2\pi n}\int_{\theta_0}^{\theta_0+2\pi n}\boldsymbol{u}(\boldsymbol{x}(\boldsymbol{x}_0,\theta),\theta)\mathrm{d}\theta \tag{34}$$

The definition (34) can be reduced to (18) only when $O(n)<\kappa^{-2}$, of course, including $n=1$. Nevertheless, the Lagrangian mean velocity expressed by (34) can be formulated by means of the Lagrangian residual velocity. As a matter of fact, noting that

$$\begin{aligned}\boldsymbol{u}_{\mathrm{L}}\equiv\boldsymbol{u}_{\mathrm{L}}(\boldsymbol{x}_0,\theta_0;n)&=\frac{1}{2\pi n}\int_{\theta_0}^{\theta_0+2\pi n}\boldsymbol{u}(\boldsymbol{x}(\boldsymbol{x}_0,\theta),\theta)\mathrm{d}\theta\\&=\frac{1}{n}\sum_{j=1}^{n}\frac{1}{2\pi}\int_{\theta_0+(j-1)2\pi}^{\theta_0+2\pi j}\boldsymbol{u}(\boldsymbol{x}(\boldsymbol{x}_0,\theta),\theta)\mathrm{d}\theta=\frac{1}{n}\sum_{j=1}^{n}\boldsymbol{u}_{\mathrm{L}}(\boldsymbol{x}_{j-1},\theta_0+(j-1)2\pi;1)\end{aligned}$$

where $\boldsymbol{x}_j=\boldsymbol{x}_0+\kappa^2 2\pi\sum_{i=0}^{j-1}\boldsymbol{u}_{\mathrm{L}}(\boldsymbol{x}_i,\theta_0+2\pi i;1)\ (j=1,2,\cdots n)$, using (18), and substituting (19)–(21) into the expression just derived, we have

$$\boldsymbol{u}_{\mathrm{L}}(\boldsymbol{x}_0,\theta_0;n)={}^{\mathrm{M}}\boldsymbol{u}_{\mathrm{L}}(\boldsymbol{x}_0;n)+\kappa\,{}^{\mathrm{D}}\boldsymbol{u}_{\mathrm{L}}(\boldsymbol{x}_0,\theta_0;n) \tag{35}$$

where

$$ {}^{\mathrm{M}}\boldsymbol{u}_{\mathrm{L}}(\boldsymbol{x}_0;n)=\frac{1}{n}\sum_{j=1}^{n}\boldsymbol{u}_{\mathrm{LM}}(\boldsymbol{x}_{\mathrm{M},j-1}) \tag{36} $$

$$ \begin{aligned} {}^{\mathrm{D}}\boldsymbol{u}_{\mathrm{L}}(\boldsymbol{x}_0,\theta_0;n) &= {}^{\mathrm{D}}\boldsymbol{u}'_{\mathrm{L}}(\boldsymbol{x}_0;n)\cos\theta_0+{}^{\mathrm{D}}\boldsymbol{u}''_{\mathrm{L}}(\boldsymbol{x}_0;n)\sin\theta_0 \\ {}^{\mathrm{D}}\boldsymbol{u}'_{\mathrm{L}}(\boldsymbol{x}_0;n) &=\frac{1}{n}\left\{\sum_{j=1}^{n}\boldsymbol{u}'_{\mathrm{LD}}(\boldsymbol{x}_{\mathrm{M},j-1})+\kappa^2 2\pi\sum_{j=1}^{n-1}\left[\boldsymbol{u}'_{\mathrm{LD}}(\boldsymbol{x}_{\mathrm{M},j-1})\cdot\nabla\sum_{i=j}^{n-1}\boldsymbol{u}_{\mathrm{LM}}(\boldsymbol{x}_{\mathrm{M},i})\right]\right\} \\ {}^{\mathrm{D}}\boldsymbol{u}''_{\mathrm{L}}(\boldsymbol{x}_0;n) &=\frac{1}{n}\left\{\sum_{j=1}^{n}\boldsymbol{u}''_{\mathrm{LD}}(\boldsymbol{x}_{\mathrm{M},j-1})+\kappa^2 2\pi\sum_{j=1}^{n-1}\left[\boldsymbol{u}''_{\mathrm{LD}}(\boldsymbol{x}_{\mathrm{M},j-1})\cdot\nabla\sum_{i=j}^{n-1}\boldsymbol{u}_{\mathrm{LM}}(\boldsymbol{x}_{\mathrm{M},i})\right]\right\} \end{aligned} \tag{37} $$

$$ \begin{aligned} &\boldsymbol{x}_{\mathrm{M},j}=\boldsymbol{x}_0+\kappa^2 2\pi\sum_{i=0}^{j-1}\boldsymbol{u}_{\mathrm{LM}}(\boldsymbol{x}_{\mathrm{M},i}) \\ &\boldsymbol{x}_{\mathrm{M},0}=\boldsymbol{x}_0 \text{ and } \sum_{a}^{b}=0 \text{ if } b<a \end{aligned} \tag{38} $$

Obviously, the formulae (35)–(38) will be exactly reduced to the formulae (19)–(21), using the Lagrangian residual velocity $\boldsymbol{u}_{\mathrm{LR}}$ instead of the Lagrangian mean velocity $\boldsymbol{u}_{\mathrm{L}}(\boldsymbol{x}_0,\theta_0;1)$, if $n = 1$. It is certainly expected that the formulae (35)–(38) can be approximately reduced to the formulae (19)–(21), using the Lagrangian residual velocity $\boldsymbol{u}_{\mathrm{LR}}$ instead of the Lagrangian mean velocity $\boldsymbol{u}_{\mathrm{L}}(\boldsymbol{x}_0,\theta_0;n)$, if $O(n)<\kappa^{-2}$ even though $n > 1$. In fact, we have $\boldsymbol{x}_{\mathrm{M},j}=\boldsymbol{x}_0$ as $O(n)<\kappa^{-2}$ in view of the terms $\left(\kappa^2 2\pi\sum_{j}^{n}\right)=O(\kappa^2 n)<1$ contained in the formulae (35)–(38); and then,

$$ {}^{\mathrm{M}}\boldsymbol{u}_{\mathrm{L}}(\boldsymbol{x}_0;n)=\frac{1}{n}\sum_{j=1}^{n}\boldsymbol{u}_{\mathrm{LM}}(\boldsymbol{x}_0)=u_{\mathrm{LM}}(\boldsymbol{x}_0) $$

and $({}^{\mathrm{D}}\boldsymbol{u}'_{\mathrm{L}}(\boldsymbol{x}_0;n),{}^{\mathrm{D}}\boldsymbol{u}''_{\mathrm{L}}(\boldsymbol{x}_0;n))=(\boldsymbol{u}'_{\mathrm{LD}}(\boldsymbol{x}_0),\boldsymbol{u}''_{\mathrm{LD}}(\boldsymbol{x}_0))$; and finally, $\boldsymbol{u}_{\mathrm{L}}(\boldsymbol{x}_0,\theta_0;n)=\boldsymbol{u}_{\mathrm{LR}}(\boldsymbol{x}_0,\theta_0)$, approximately.

When n is of an order of magnitude equal to or greater than κ^{-2}, the Lagrangian mean velocity $\boldsymbol{u}_{\mathrm{L}}(\boldsymbol{x}_0,\theta_0;n)$ cannot be proven to satisfy the continuity equation for incompressible fluids. Thus, the Lagrangian residual velocity cannot be defined as the Lagrangian mean velocity, or $\boldsymbol{u}_{\mathrm{L}}(\boldsymbol{x}_0,\theta_0;n)$, when $O(n)\geqslant\kappa^{-2}$. It is evident that the Lagrangian mean velocity $\boldsymbol{u}_{\mathrm{L}}(\boldsymbol{x}_0,\theta_0;n)$ is, in general, not only a function of $(\boldsymbol{x}_0,\theta_0)$ but also dependent on n. Explicitly containing n reflects rationally the experience of the Lagrangian motion of the marked water parcel moving from the initial position $\boldsymbol{x}_0$ at time θ_0 to the terminal position $\boldsymbol{x}_n$ at time $\theta_0+2\pi n$. Because the Lagrangian mean velocity is directly proportional to the net Lagrangian displacement of the marked water parcel, with a proportionality coefficient $(2\pi n)^{-1}$, in view of (35)–(38), the latter behaves well like the former, and the horizontal projection of the net Lagrangian displacement will trace out an ellipse over a tidal cycle. The equations also show that the net Lagrangian displacements, or $2\pi n\boldsymbol{u}_{\mathrm{L}}(\boldsymbol{x}_0,\theta_0;n)$, of marked water parcels released from the position $\boldsymbol{x}_0$ at tidal phase θ_0 over such a long time as a season can be of an order of magnitude greater than those over one or a few of tidal cycles. Thus, the Lagrangian residual drifts, or $2\pi n\kappa{}^{\mathrm{D}}\boldsymbol{u}_{\mathrm{L}}(\boldsymbol{x}_0,\theta_0;n)$, of which the horizontal components trace out an ellipse over a tidal cycle, play a more pronounced role in the dispersion, for example, of pollutants on "long" time scales than on "short" scales, although the Lagrangian drift velocities, or $\kappa{}^{\mathrm{D}}\boldsymbol{u}_{\mathrm{L}}(\boldsymbol{x}_0,\theta_0;n)$, are of the same order of magnitude in both cases. For example, the ratio of the Lagrangian residual drift for $O(n)=\kappa^{-2}$ to that for $O(n)=1$ is $O(\kappa^{-2})$, which suggests the importance, at least the potential importance, of the Lagrangian residual drifts for long-term dispersion phenomena though $O(\kappa\left|{}^{\mathrm{D}}\boldsymbol{u}_{\mathrm{L}}(\boldsymbol{x}_0,\theta_0;n)\right|/\left|{}^{\mathrm{M}}u_{\mathrm{L}}(\boldsymbol{x}_0;n)\right|)=O(\kappa)<1$.

7. An Application to the Bohai sea, China

The dynamic model proposed in the present paper should be verified through a test of the theory against field data in a realistic situation. This three-dimensional model can be conveniently applied to the summer residual circulation and the inter-tidal transport processes in the Bohai Sea, China (Fig. 1). A scale analysis has shown that the tide-induced nonlinear effect and the Huanghai Sea Warm Current which enters the Bohai Sea through the Bohai Strait may be the principal factors contributing to the formation of the summer circulation in the Bohai Sea. The wind stress on the sea surface and the baroclinic effects are negligible and less important, respectively (Feng *et al.*, 1994). Thus, the total transport through the Bohai Strait has to be prescribed; that transport can be obtained from the calculation of the current speed based upon field data. For the tides, we can use an existing three-dimensional nonlinear numerical model of the M_2-tide in the Bohai Sea. The calculated depth-averaged mass- transport velocity field as the mean residual circulation of the summer in the Bohai Sea has recently been obtained (Sun *et al.*, 1989), and is exhibited in Fig. 2. According to equation (33), the concentration isolines for the depth-averaged tidal cycle mean of the concentration of any conservative and passive tracer must approximately coincide with the streamlines of the depth-averaged mass-transport velocity. The salinity can be conveniently used as such a tracer in the Bohai Sea for the summer. The salinity distribution at the depth of 10 m in the Bohai Sea for June, 1958, as derived from observations, is shown in Fig. 3. This picture can be used qualitatively to represent a typical summer distribution of the depth-averaged salinity in the Bohai Sea. Even without including other factors likely to affect the distribution of salinity, such as stratification, the Huanghe River runoff and the coastal dispersion boundary layer, the patterns of isohaline contours of Fig. 3 and the mass-transport velocity field of Fig. 2 seem to be in surprisingly good qualitative agreement.

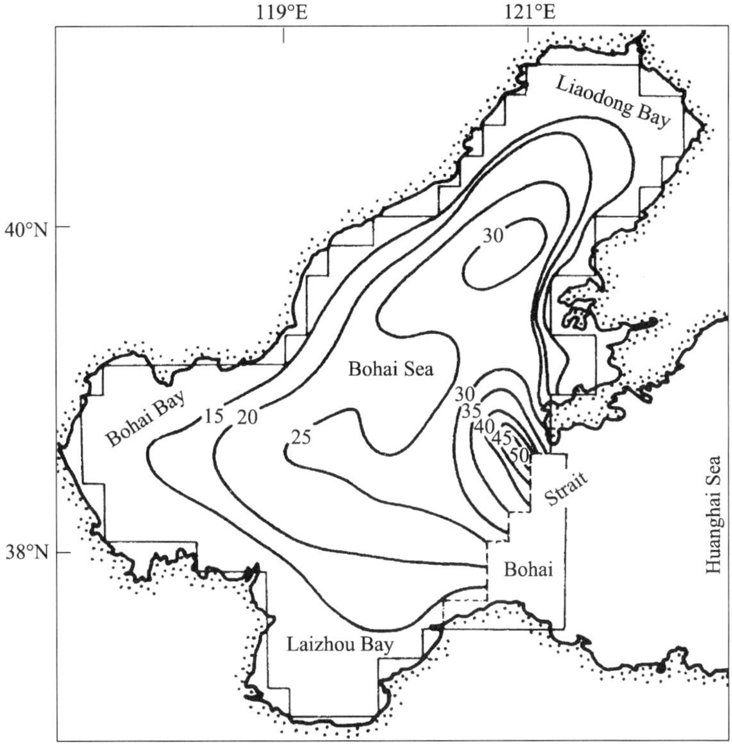

Fig. 1 Depth distribution of the Bohai Sea, China, (m)

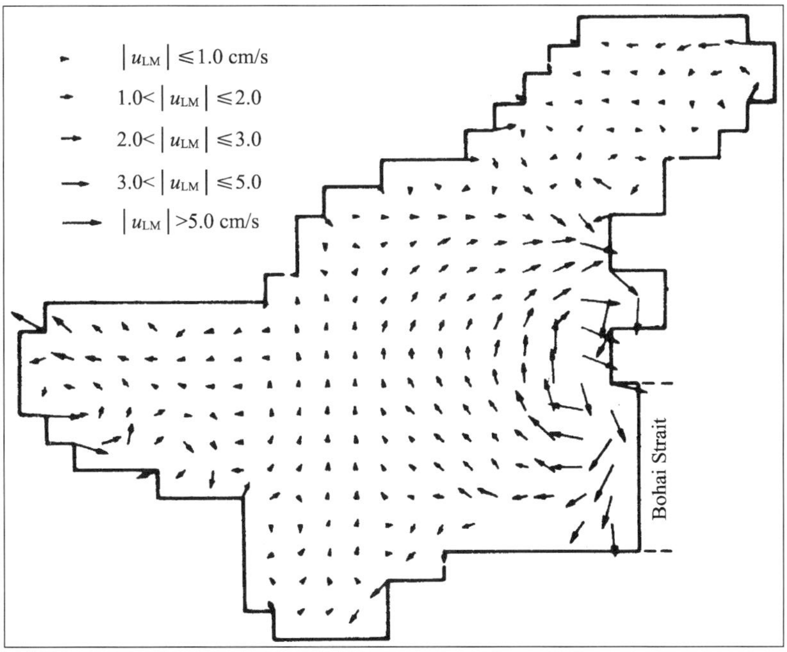

Fig. 2 Computed depth-averaged mass-transport velocity field (cm/sec)

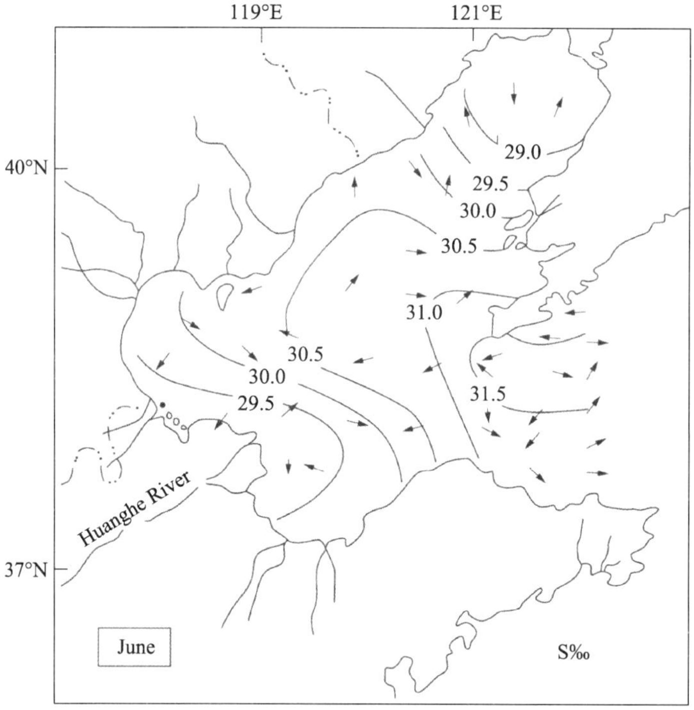

Fig. 3 Salinity distribution (‰) measured at the depth of 10 m in the Bohai Sea in June1958
Arrows denote observed Eulerian residual currents

Furthermore, the pattern of the mean Lagrangian residual circulation in the Bohai Sea for the summer can be used to explain dynamically the existence of an area, situated in the north-eastern part of Laizhou Bay, where the water is of relatively high salinity and transparency and of relatively low temperature and concentration of suspended sediments. These conditions explain the presence, in that area, of such bottom fauna as spatangia (Su *et al.*, 1989). It is of interest to note that the pattern of the summer circulation in the Bohai Sea is only one big counterclockwise gyre if the coupled mean tide-induced Lagrangian residual circulation is not considered (Feng *et al.*, 1994). Such a circulation cannot be used to interpret the existence of the area stated above. The present theory for predicting long-term transport processes and studying their dynamics in coastal seas seems to be satisfactory, at least qualitatively.

8. Conclusion

Based upon a three-dimensional weakly nonlinear theory, the Lagrangian residual velocity is approximately expressed as a sum of a tidally periodic fluctuation, which we call the Lagrangian drift velocity, and of the tidal cycle average, i.e., the mass-transport velocity. The Lagrangian residual velocity is different from the Eulerian residual velocity, which is steady, and it is similar to the tidal current velocity. A set of field equations governing the mass-transport velocity is derived. These equations show that the mean Lagrangian residual circulation is more relevant than the Eulerian one to the description of inter-tidal flow processes. In particular, a new inter-tidal transport equation for the tidal cycle average of the concentration of any conservative and passive tracer is proposed, in which the convective transport is characterized by the mass-transport velocity and there is no need for an ad hoc hypothesis on "tidal dispersion". It is also shown that the Lagrangian residual drift, which is related to the Lagrangian drift velocity, is potentially the most important factor controlling the dispersion or spreading of tracers in the long-term transport processes. And finally, when the theory is preliminarily applied to the Bohai Sea, China, a good qualitative agreement between the present theory and field data is revealed.

References

Alfrink, B.J. and Vreugdenhil, C.B., 1981. Residual currents. Delft Hydraul. Lab., Delft, The Netherlands, Rep. R 1469–11, 42 pp.

Bowden, K.F., 1965. Horizontal mixing in the sea due to a shearing current. J. Fluid Mech., 21: 83–95.

Bowden, K.F., 1967. Circulation and diffusion. In: G.H. Lauff (Editor), Estuaries. Publ., No. 83, AAAS, Washington, D.C., pp. 15–36.

Cheng, R.T., 1983. Euler-Lagrangian computations in estuarine hydrodynamics. In: C. Taylor, J.A. Johnson and R. Smith (Editors), Proc. of the Third Intern. Conf. on Num. Meth. in Laminar and Turbulent Row. Pineridge Press, pp. 341–352.

Cheng, R.T. and Casulli, V., 1982. On Lagrangian residual currents with applications in South San Francisco Bay, California. Water Resour. Res., vol. 18, 6: 1652–1662.

Cheng, R.T., Feng, S. and Xi, P., 1986. On Lagrangian residual ellipse. In: J. van de Kreeke (Editor), Intern. Conf. on Physics of Shallow Estuaries and Bays, Lecture Notes on Coastal and Estuarine Studies. Springer-Verlag, pp. 102–113.

Csanady, G.T., 1982. Circulation in the Coastal Ocean. D. Reidel Publ. Comp., Dordrecht/Boston/London, 279 pp.

Feng, S., 1977. A three-dimensional nonlinear model of tides. Scientia Sinica, vol. 20, 4: 436–446.

Feng, S., 1986. A three-dimensional weakly nonlinear dynamics on tide-induced Lagrangian residual current and mass-transport. Chinese J. of Oceanology and Limnology, vol. 4, 2: 139–158.

Feng, S., 1990. On the fundamental dynamics of barotropic circulation in shallow seas. Acta Oceanologica Sinica, Vol. 9, 3: 315–329.

Feng, S., Cheng, R.T., Sun, W., Xi, P., and Song, L., 1990. Lagrangian residual current and longterm transport processes in a weakly nonlinear bavoclinic system. In: H. Wang, J. Wang, H. Dai (Editors), Physics of Shallow Seas, Beijing: China Ocean Press, pp. 1–20.

Feng, S., Cheng, R.T. and Xi, P., 1986a. On tide-induced Lagrangian residual current and residual transport, Part I: Residual current. Water Resour. Res., vol. 22, 12: 1623–1634.

Feng, S., Cheng, R.T. and Xi, P., 1986b. On tide-induced Lagrangian residual current and residual transport,

Part II: Residual transport with application in South San Francisco Bay, California. Water Resour. Res., vol. 22, 12: 1635–1646.

Feng, S., Xi, P. and Zhang, S., 1984. The baroclinic residual circulation in shallow seas. Chinese J. of Oceanology and Limnology, vol. 2, 1: 49–60.

Feng, S., Zhang, S., and Xi, P., 1994. A Langrangiun model of circulation in Bohai Sea. In: D. Zhou, Y. Liang, C. Zeng (Editors), Oceanography of China Seas, Dordrecht, Boston, London: Kluwer Academic Publishers, pp. 83–90.

Fischer, H.B., List, E.J., Koh, R.C.Y., Imberger, J. and Brooks, N.H., 1979. Mixing in Inland and Coastal Waters. Academic Press, New York, 483 pp.

Heaps, N.S., 1978. Linearized vertically-integrated equations for residual circulation in coastal seas. Deut. Hydrog. Z., 31: 147–169.

Longuet-Higgins, M.S., 1969. On the transport of mass by time-varying ocean currents. Deep Sea Res., 16: 431–447.

Nihoul, J.C.J. and Ronday, F.C., 1975. The influence of the "tidal stress" on the residual circulation. Tellus, 27: 484–489.

Pritchard, D.W., 1954. A study on salt balance in a coastal plain estuary. J. Mar. Res., 13: 133–144.

Song, L., 1987. A hydrodynamic velocity-splitting model with a depth-varying eddy viscosity in shallow seas. Acta Oceanologica Sinica, vol. 6, S1: 135–151.

Stern, M.E., 1975. Ocean Circulation Physics. Academic Press, New York, 246 pp.

Su, Z., Wiseman, W.T., Fan, Y., Gao, S., Qian, Q. and Yang, Z., 1986. Analyses of hydrologic characteristics of the area adjacent to the Huanghe Estuary (personal communication).

Sun, W., Xi, P. and Song, L., 1989. Numerical calculation of the three-dimensional tide-induced Lagrangian residual circulation in the Bohai Sea. Journal of Ocean University of Qingdao, vol. 19, 2: 27–36.

Tee, T.K., 1976. Tide-induced residual current: A 2-D nonlinear numerical model. J. Mar. Res., 31: 603–628.

Zimmerman, J.T.F., 1979. On the Euler-Lagrangian transformation and the Stokes' drift in the presence of oscillatory and residual currents. Deep Sea Res., 26A: 505–520.

Lagrangian Residual Current and Long-term Transport Processes in a Weakly Nonlinear Baroclinic System *

Feng Shizuo, Ralph T. Cheng, Sun Wenxin, Xi Pangen and Song Lina

Introduction

One of the most important problems in shallow water oceanography is the study of circulation and water properties associated with long-term transport of mass and dissolved solutes in estuaries, bays and shallow seas. In such shallow water systems, the dynamics is generally nonlinear, and the directly observable variables are the tides, tidal currents and instantaneous solute concentrations. It is clear that the long-term fate of dissolved solutes is determined by the residual circulation resulting from nonlinear interactions between the tides and tidal currents amongst others. As pointed out by Longuet-Higgins (1969), it is more meaningful to use the Lagrangian mean velocity of a water parcel than the Eulerian mean velocity to determine the origin of water masses. The Lagrangian mean velocity can be derived from the net displacement of a water parcel over one or a few tidal cycles (Zimmerman, 1979; Cheng and Casulli, 1982). In general, the Lagrangian mean velocity is a function of the spatial position where the water parcel is released, the time (tidal phase) of release, and the number of the tidal cycles over which the tidal velocity is averaged (Feng, 1986). Furthermore, the Lagrangian residual velocity can be treated as an Eulerian field variable, it is solenoidal and satisfies an incompressible continuity equation (Feng, 1987). The dynamics and the mechanisms which control the coupling between the intertidal transport processes of mass and dissolved solutes, and the Lagrangian residual circulation, are extremely complex, and their inter-relations are not yet very clear.

Prediction or deduction of residual circulation is not an easy task because the residual variables are not directly observable or directly computable, they are deduced from measurements or from computed tidal velocity field. Consequently, the residual current is often masked by a slight inaccuracy in the measurements or in the predications of the tidal current (Cheng, 1983). Alternatively, analytical approach has been used in research of the Lagrangian residual circulation and the associated intertidal transport processes. For example, the tide-induced Lagrangian residual currents and mass-transport in a weakly nonlinear, baro-

* Feng S, Cheng R T, Sun W, Xi P, Song L. 1990. Lagrangian residual current and long-term transport processes in a weakly nonlinear baroclinic system. In: H Wang, J Wang, H Dai (eds.) Physics of Shallow Seas. Beijing: China Ocean Press. 1–20

tropic system with passive tracers have been considered previously (Cheng *et al.*, 1986; Feng *et al.*, 1986a, b; Feng, 1986, 1987). The residual circulation in estuaries, bays and shallow seas is also affected by winds and baroclinic forcings. In this paper, we extend previous analyses to include nonlinear couplings due to the presence of salinity gradients and their effects on the properties of tidal currents, residual currents, and long-term transport processes. An application of the present theory to summer condition in the Bohai Sea, China is given and discussed.

Formulation and Scale Analysis

Assuming the Boussinesq approximation for density variation is valid, the nondimensional governing equations for a baroclinic tidal dynamic system are the continuity equation

$$\nabla \boldsymbol{u} = 0 \tag{1}$$

the x- and y-momentum equations,

$$(\partial / \theta + \kappa \boldsymbol{u} \cdot \nabla)u - fv = -\partial Z / \partial x + \partial / \partial z(\nu \partial u / \partial z) - \delta \partial / \partial x \int_z^{\kappa Z} S\mathrm{d}z \tag{2}$$

$$(\partial / \theta + \kappa \boldsymbol{u} \cdot \nabla)v + fu = -\partial Z / \partial y + \partial / \partial z(\nu \partial v / \partial z) - \delta \partial / \partial y \int_z^{\kappa Z} S\mathrm{d}z \tag{3}$$

the equation of particle path (streakline equation),

$$\mathrm{D}\boldsymbol{\xi} / \mathrm{D}\theta = \boldsymbol{u} \tag{4}$$

and the conservation equation of salt,

$$(\partial / \partial \theta + \kappa \boldsymbol{u} \cdot \nabla)S = \epsilon \partial / \partial z(k \partial S / \partial z) \tag{5}$$

The boundary conditions at $z = \kappa Z$ are

$$w = [\partial / \partial \theta + \kappa(u \partial / \partial x + v \partial / \partial y)]Z \tag{6}$$

$$\partial u / \partial z = \partial v / \partial z = 0 \tag{7}$$

$$\partial S / \partial z = 0 \tag{8}$$

and the boundary condition at $z = -h$ are

$$\boldsymbol{u} = 0 \quad \text{and} \quad \partial S / \partial z = 0 \tag{9}$$

The above system of equations is given in a non-dimensional form, and the notations and the characteristic values used in the non-dimensional parameters are given in Table 1.

Furthermore, the following non-dimensional parameters are defined as:

$$\kappa = \zeta_c / h_c \tag{10}$$

$$\delta = \alpha_c / \kappa \tag{11}$$

$$\epsilon = k_c / \omega h_c^2 \tag{12}$$

where α_c is a scale factor of density variations. Additionally, the following notations

$$\nabla = \boldsymbol{e}_1 \partial / \partial x + \boldsymbol{e}_2 \partial / \partial y + \boldsymbol{e}_3 \partial / \partial z$$

Table 1 Definitions of Notations

Variable Name	Non-dimensional Symbol	Definitions	Characteristic value
t^*	θ	time	ω^{-1}, tidal frequency
(x^*, y^*)	(x, y)	Cartesian coordinates on a horizontal *f*-plane	L, length
z^*	z	vertical coordinate	h_c, typical depth
(u^*, v^*)	(u, v)	velocity components in (x, y) directions	u_c, tidal velocity
w^*	w	*z*-velocity component	$w_c = (u_c h_c / L)$
Z^*	Z	water elevation measured from mean sea level	Z_c, tidal amplitude
(ξ^*, η^*)	(ξ, η)	the Lagrangian displacements of a marked water parcel in (x, y) directions	$\xi_c (= \kappa L)$
ζ^*	ζ	the Lagrangian displacements of a marked water parcel in *z*-direction	$\zeta_c (= \kappa h_c)$
h^*	h	the water depth measured from mean sea level	h_c
S^*	S	concentration of a conservative tracer	S_c, Eg, temperature salinity, pollutant, etc.
f^*	f	the coriolis parameter	ω, tidal frequency
ν^*	ν	the eddy viscosity	$\nu_c (= \omega h_c)$
k^*	k	the eddy diffusivity	k_c
g		gravitational acceleration	
$(\boldsymbol{e}_1, \boldsymbol{e}_2, \boldsymbol{e}_3)$		unit vectors in (x, y, z)	

$$\boldsymbol{u} = (u, v, w)$$

$$\boldsymbol{\xi} = (\xi, \eta, \zeta)$$

are used with the initial condition for the Lagrangian displacement $(\xi, \eta, \zeta) = 0$ when $\theta = \theta_0$. Thus the position of a marked water parcel can be written as

$$(x, y, z) = (x_0, y_0, z_0) + \kappa(\xi, \eta, \zeta) \tag{13}$$

The subscript "0" denotes values at the initial time θ_0. The boundary condition, Eq. (7), implies that the wind stress on the sea surface is zero. The non-dimensional equation for streaklines of marked water parcels, Eq. (4), has been derived by taking $\xi_c = u_c \omega^{-1} = \kappa L$, which implies that the tidal excursion is of the order of κ when compared to the basin length (Feng *et al.*, 1986a). We further assume that $O(\kappa) \simeq o(1)$ and $O(\varepsilon / \kappa) \simeq \kappa$ which are valid for numerous shallow estuaries and bays (Feng *et al.*, 1986a, b). These assumptions are consistent with the fact that convection dominates the transport processes in a weakly nonlinear system.

In the coupled baroclinic system of equations, Eqs. (1)–(9), the maximum value of α_c for most estuaries is estimated to be 0.05, and an order of magnitude smaller for coastal seas. Solutions for such a weakly nonlinear system can be obtained by using a small parameter perturbation method (Feng, 1977; Feng *et al.*, 1986a, b). In order not to create a second small parameter, we assume that

$$O(\delta) \simeq \kappa \tag{14}$$

which gives the correct order of magnitude of each terms. When all the dependent variables are expanded in an ascending power series of κ, and the solutions of the nonlinear system can be obtained systematically.

For simplicity and clarity, the flows in an M_2-tidal system which include salinity are considered in the following sections. Using the small perturbation method, governing equations for the zeroth-, first- and second-order solutions can be obtained. Denoting the order of the solutions by a subscript on the dependent variables, the zeroth order governing equations consist of the continuity equation

$$\nabla \cdot \boldsymbol{u}_0 = 0 \tag{15}$$

the x- and y-momentum equations,

$$\partial u_0 / \partial \theta - f v_0 = -\partial Z_0 / \partial x + \partial / \partial z(\nu \partial u_0 / \partial z) \tag{16}$$

$$\partial v_0 / \partial \theta + f u_0 = -\partial Z_0 / \partial y + \partial / \partial z(\nu \partial v_0 / \partial z) \tag{17}$$

the equation of particle path (streakline equation),

$$\xi_0 = \int_{\theta_0}^{\theta} \boldsymbol{u}_0(\boldsymbol{x}_0, \theta') \mathrm{d}\theta' \tag{18}$$

and the conservation equation of salt,

$$\partial S_0 / \partial \theta = 0 \tag{19}$$

The boundary conditions at $z = 0$ are

$$w_0 = \partial Z_0 / \partial \theta \tag{20}$$

$$\partial u_0 / \partial z = \partial v_0 / \partial z = 0 \tag{21}$$

$$\partial S_0 / \partial z = 0 \tag{22}$$

and the boundary condition at $z = -h$ are

$$\boldsymbol{u}_0 = 0 \quad \text{and} \quad \partial S_0 / \partial z = 0 \tag{23}$$

The first order governing equations are the continuity equation

$$\Delta \boldsymbol{u}_1 = 0 \tag{24}$$

the x- and y-momentum equations,

$$\partial u_1 / \partial \theta - f v_1 = -\partial Z_1 / \partial x + \partial / \partial z(\nu \partial u_1 / \partial z) - \delta_1 \partial / \partial x \int_z^0 S_0 \mathrm{d}z - \boldsymbol{u}_0 \cdot \nabla u_0 \tag{25}$$

$$\partial u_1 / \partial \theta + f u_1 = -\partial Z_1 / \partial y + \partial / \partial z(\nu \partial v_1 / \partial z) - \delta_1 \partial / \partial y \int_z^0 S_0 \mathrm{d}z - \boldsymbol{u}_0 \cdot \nabla v_0 \tag{26}$$

the equation of particle path (streakline equation),

$$\xi_1 = \int_{\theta_0}^{0} [\boldsymbol{u}_1(\boldsymbol{x}_0, \theta') + \xi_0 \cdot (\nabla \boldsymbol{u}_0)_0] \mathrm{d}\theta' \tag{27}$$

and the conservation equation of salt,

$$\partial S_1 / \partial \theta + \boldsymbol{u}_0 \nabla S_0 = 0 \tag{28}$$

The boundary conditions at $z=0$ are

$$w_1=\partial Z_1/\partial\theta+\partial(u_0Z_0)/\partial x+\partial(v_0Z_0)/\partial y \tag{29}$$

$$\partial(u_1,v_1)/\partial z=-Z_0\partial^2(u_0,v_0)/\partial z^2 \tag{30}$$

$$\partial S_1/\partial z=-Z_0\partial^2 S_0/\partial z^2 \tag{31}$$

and the boundary conditions at $z=-h$ are

$$\boldsymbol{u}_1=0 \quad \text{and} \quad \partial S_1/\partial z=0 \tag{32}$$

Similarly, the second order governing equations are the continuity equation

$$\Delta\boldsymbol{u}_2=0 \tag{33}$$

the x- and y-momentum equations,

$$\begin{aligned}\partial u_2/\partial\theta-fv_2&=-\partial Z_2/\partial x+\partial/\partial z(\nu\partial u_2/\partial z)\\&-\delta_1\partial/\partial x\left[\int_z^0 S_1\mathrm{d}z+(S_0)_S Z_0\right]-\left[\boldsymbol{u}_0\cdot\nabla u_1+\boldsymbol{u}_1\cdot\nabla u_0\right]\end{aligned} \tag{34}$$

$$\begin{aligned}\partial v_2/\partial\theta+fu_2&=-\partial Z_2/\partial y+\partial/\partial z(\nu\partial v_2/\partial z)\\&-\delta_1\partial/\partial y\left[\int_z^0 S_1\mathrm{d}z+(S_0)_S Z_0\right]-\left[\boldsymbol{u}_0\cdot\nabla v_1+\boldsymbol{u}_1\cdot\nabla v_0\right]\end{aligned} \tag{35}$$

the equation of a particle path (streakline equation),

$$\begin{aligned}\boldsymbol{\xi}_2=\int_{\theta_0}^{\theta}\Bigg\{&\boldsymbol{u}_2(\boldsymbol{x}_0,\theta')+\boldsymbol{\xi}_0\cdot(\nabla\boldsymbol{u}_1)_0+\boldsymbol{\xi}_1\cdot(\nabla\boldsymbol{u}_0)_0+1/2\Bigg[\left(\frac{\partial^2\boldsymbol{u}_0}{\partial x^2}\right)_0\xi_0^2+\left(\frac{\partial^2\boldsymbol{u}_0}{\partial y^2}\right)_0\eta_0^2+\left(\frac{\partial^2\boldsymbol{u}_0}{\partial z^2}\right)_0\zeta_0^2\\&+\left[\left(\frac{\partial^2\boldsymbol{u}_0}{\partial x\partial y}\right)_0\xi_0\eta_0+\left(\frac{\partial^2\boldsymbol{u}_0}{\partial x\partial z}\right)_0\xi_0\zeta_0+\left(\frac{\partial^2\boldsymbol{u}_0}{\partial y\partial z}\right)_0\eta_0\zeta_0\right]\Bigg]\Bigg\}\mathrm{d}\theta'\end{aligned} \tag{36}$$

and the conservation equation of salt,

$$\partial S_2/\partial\theta+\boldsymbol{u}_0\cdot\nabla S_1+\boldsymbol{u}_1\cdot\nabla S_0=E\partial/\partial z(k\partial S_0/\partial z) \tag{37}$$

The boundary conditions at $z=0$ are

$$w_2=\partial Z_2/\partial\theta+\partial/\partial x\left[u_0Z_1+u_1Z_0+\frac{1}{2}Z_0^2\partial u_0/\partial z\right]+\partial/\partial y\left[v_0Z_1+v_1Z_0+\frac{1}{2}Z_0^2\partial v_0/\partial z\right] \tag{38}$$

$$\partial u_2/\partial z=-\left[Z_1\partial^2u_0/\partial z^2+Z_0\partial^2u_1/\partial z^2+\frac{1}{2}Z_0^2\partial^3u_0/\partial z^3\right] \tag{39a}$$

$$\partial v_2/\partial z=-\left[Z_1\partial^2v_0/\partial z^2+Z_0\partial^2v_1/\partial z^2+\frac{1}{2}Z_0^2\partial^3v_0/\partial z^3\right] \tag{39b}$$

$$\partial S_2/\partial z=-\left[Z_1\partial^2S_0/\partial z^2+Z_0\partial^2S_1/\partial z^2+\frac{1}{2}Z_0^2\partial^3S_0/\partial z^3\right] \tag{40}$$

and the boundary condition at $z=-h$ are

$$\boldsymbol{u}_2=0 \quad \text{and} \quad \partial S_2/\partial z=0 \tag{41}$$

where $E=\epsilon/\kappa^2$, $\delta_1=\delta/\kappa$, $O(\delta_1)\simeq O(E)\simeq 1$, and the notations $(\ )_0$ or $(\ )_s$ indicate that the term is evaluated at $\boldsymbol{x}_0$ or on the water surface, $z=0$, respectively. Some important and qualitative properties of the dynamic system can be revealed form these equations without actually solving them.

The zeroth order equations, Eqs. (15)–(23), reveal that the astronomical tides and tidal currents are independent of the effects due to the presence of a salt field; that is, in the zeroth-order approximation the thermodynamics and the hydrodynamics are uncoupled. Thus, the solutions for a three dimensional, linear, barotropic, astronomical tidal system can be solved independently. The zeroth-order approximation of the water parcel displacements can be calculated from Eq. (18). Eq. (19) shows that the zeroth-order solution of salt, S_0, is independent of time, i.e. S_0 is a function of space only.

In the first-order equations, Eqs. (24)–(32), the role of the salt field is in the form of forcing functions in the momentum equations. Note that only the first-order Eulerian residual, which is the time independent component of the first order solution, is affected by the salt field S_0. Moreover, the first order solution of salt is governed by a pure convection equation, and tidal current, $\boldsymbol{u}_0$, and S_1, are 90° out of phase, Eq. (28). This property is valid and has been observed and confirmed near the mouth of estuaries (Pritchard, 1954; Dyer, 1973; Officer, 1976; Fischer *et al.*, 1979). From integrating Eq. (28), S_1 can be expressed as

$$S_1=\langle S_1\rangle-\nabla S_0\cdot\int \boldsymbol{u}_0 \mathrm{d}\theta' \tag{42}$$

where the notation $\langle\ \rangle$ denotes the tidal cycle average.

The second-order governing equations include additional nonlinear interactions induced by the horizontal gradient of S_1, although the second-order Eulerian residual in a barotropic M_2-tidal system is zero (Feng *et al.*, 1986a; Feng, 1986a). In fact, there exists a nonlinear coupling between S_0 and the first-order Lagrangian residual velocity. Thus far, higher order solutions for the coupled baroclinic system are not known.

First-order Dynamics

First-order Lagrangian residual velocity

Because both $\boldsymbol{u}_0$ and ξ_0 are independent of the presence of salt, the Stokes' drift velocity, $\boldsymbol{u}_{\mathrm{sd}}=\langle \xi_0\nabla \boldsymbol{u}_0\rangle$, is also independent of the baroclinic effect. According to the definition of the Lagrangian residual velocity and following the approach adopted in previous papers (Feng *et al.*, 1986a; Feng, 1986), the first-order Lagrangian residual velocity can be shown and expressed as

$$[\boldsymbol{u}_{\mathrm{lr}}]_1=[\boldsymbol{u}_{\mathrm{er}}]_1+\boldsymbol{u}_{\mathrm{sd}} \tag{43}$$

where the first-order Eulerian residual velocity, $[\boldsymbol{u}_{\mathrm{er}}]_1$, is due to both the nonlinear couplings between M_2 and itself, and due to the horizontal gradient of S_0. The first-order Lagrangian residual velocity, $[\boldsymbol{u}_{\mathrm{lr}}]_1$, contains both the barotropic and baroclinic contributions, and the latter is introduced from the Eulerian residual velocity not the Stokes' drift velocity. As shown in previous papers (Feng *et al.*, 1986b; Feng, 1987), $\nabla \boldsymbol{u}_{\mathrm{lr}} = 0$, so that the Lagrangian residual velocity in a weakly nonlinear, baroclinic system can be treated as an Eulerian field variable. Furthermore, the Lagrangian residual current satisfies the following equation,

$$\boldsymbol{u}_{\mathrm{lr}} \cdot \nabla \langle F \rangle = 0 \tag{44}$$

where $F(\boldsymbol{x},\theta)$ is a conservative property associated with a fixed water parcel on the surface, so that

$$\mathrm{D}F / \mathrm{D}\theta = \partial F / \partial \theta + \kappa \boldsymbol{u} \nabla F = 0 \tag{45}$$

for any point on the free water surface.

It should be pointed out that the Eulerian residual velocity does not satisfy the convection equation, Eq. (44) though it satisfies the continuity equation $\nabla \boldsymbol{u}_{\mathrm{er}} = 0$. These properties suggest that the Lagrangian residual velocity rather than the Eulerian residual velocity controls the inter-tidal transport properties.

Governing equations for $\boldsymbol{u}_{\mathrm{lr}}$ and S_0

From a practical standpoint, the long-term, directly observable variables are the tides and tidal currents, and the vast majority of field data are collected from fixed stations, or the data are Eulerian in nature. The Eulerian residual circulation in coastal seas can be deduced, as a common practice, from long-term observations (such as current-meter records) as defined in the Eulerian sense. However, as pointed out by Feng *et al.* (1986a, b), the usefulness of the Eulerian residual circulation in studies of intertidal processes is increasingly doubtful. When the focus of the study is to describe and understand intertidal transport processes, the more relevant variable is the Lagrangian residual current. It is then of interest to show how $\boldsymbol{u}_{\mathrm{lr}}$ and S_0 are inter-related. The set of coupled governing equations for $\boldsymbol{u}_{\mathrm{lr}}$ and S_0 are the continuity equation,

$$\nabla \boldsymbol{u}_{\mathrm{lr}} = 0 \tag{46}$$

the x- and y-momentum equations,

$$-f v_{\mathrm{lr}} = -\partial \langle Z_1 \rangle / \partial x + \partial / \partial z(\nu \partial u_{\mathrm{lr}} / \partial z) + \pi_1 - \delta_1 \partial / \partial x \int_z^0 S_0 \mathrm{d}z' \tag{47}$$

$$f u_{\mathrm{lr}} = -\partial \langle Z_1 \rangle / \partial y + \partial / \partial z(\nu \partial v_{\mathrm{lr}} / \partial z) + \pi_2 - \delta_1 \partial / \partial y \int_z^0 S_0 \mathrm{d}z' \tag{48}$$

and the conservation equation of salt,

$$\boldsymbol{u}_{\mathrm{lr}} \nabla S_0 = E\partial / \partial z(\langle k \rangle \partial S_0 / \partial z) \tag{49}$$

The boundary conditions are at $z = 0$,

$$\begin{aligned} & w_{\mathrm{lr}} = 0 \\ & \partial(u_{\mathrm{lr}}, v_{\mathrm{lr}}) / \partial z = 0 \\ & \partial S_0 / \partial z = 0 \end{aligned} \tag{50}$$

and at $z = -h$,

$$\boldsymbol{u}_{\mathrm{lr}} = 0 \quad \text{and} \quad \partial S_0 / \partial z = 0 \tag{51}$$

where

$$\begin{aligned} \pi_1 = & -\frac{\partial}{\partial x}\langle \boldsymbol{\xi}_0 \nabla Z_0 / 2 \rangle + \frac{\partial}{\partial z}\left\langle \left(\frac{5}{2}\partial \xi_0 / \partial x + \partial \eta_0 / \partial y \right) \nu \partial u_0 / \partial z \right. \\ & + \left. \left(\partial \xi_0 / \partial y + \frac{1}{2}\partial \eta_0 / \partial x \right) \nu \partial v_0 / \partial z \right\rangle + \frac{\partial}{\partial z}\langle \nabla \nu \boldsymbol{\xi}_0 \partial u_0 / \partial z \rangle \\ & - \langle (\partial \boldsymbol{\xi}_0 / \partial z) \nabla (\nu \partial u_0 / \partial z) \rangle + \frac{1}{2}\langle \partial^2 \xi_0 / \partial z \partial x (\nu \partial u_0 / \partial z) \\ & + \partial^2 \eta_0 / \partial z \partial x (\nu \partial v_0 / \partial z) + \xi_0 \partial^2 / \partial z \partial x (\nu \partial u_0 / \partial z) + \eta_0 \partial^2 / \partial z \partial x (\nu \partial v_0 / \partial z) \\ \pi_2 = & -\frac{\partial}{\partial y}\langle \boldsymbol{\xi}_0 \nabla Z_0 / 2 \rangle + \frac{\partial}{\partial z}\left\langle \left(\frac{5}{2}\partial \eta_0 / \partial y + \partial \xi_0 / \partial x \right) \nu \partial v_0 / \partial z \right. \\ & + \left. \left(\partial \eta_0 / \partial x + \frac{1}{2}\partial \xi_0 / \partial y \right) \nu \partial u_0 / \partial z \right\rangle + \frac{\partial}{\partial z}\langle \nabla \nu \boldsymbol{\xi}_0 \partial \ v_0 / \partial z \rangle \\ & - \langle (\partial \boldsymbol{\xi}_0 / \partial z) \nabla (\nu \partial v_0 / \partial z) \rangle + \frac{1}{2}\langle \partial^2 \xi_0 / \partial z \partial y (\nu \partial u_0 / \partial z) \\ & + \partial^2 \eta_0 / \partial z \partial y (\nu \partial v_0 / \partial z) + \xi_0 \partial^2 / \partial z \partial y (\nu \partial u_0 / \partial z) + \eta_0 \partial^2 / \partial z \partial y (\nu \partial v_0 / \partial z) \end{aligned}$$

where (π_1, π_2) are the tide-induced 'body forces', and represent the nonlinear couplings of astronomical tides and tidal currents; and $(u_{\mathrm{lr}}, v_{\mathrm{lr}}, w_{\mathrm{lr}})$ are the components of $\boldsymbol{u}_{\mathrm{lr}}$, and $\langle Z_1 \rangle$ is the first order residual elevation.

It is noted that $\boldsymbol{u}_{\mathrm{lr}}$ and S_0 are coupled and nonlinear, Eqs. (46)–(51), although in intra-tidal processes the salt field and the tidal current are uncoupled in a weakly nonlinear baroclinic system, Eqs. (15)–(23). By vertically averaging Eq. (49), the resultant is the depth averaged governing equation,

$$\bar{u}_{\mathrm{lr}} \partial \bar{S}_0 / \partial x + \bar{v}_{\mathrm{lr}} \partial \bar{S}_0 / \partial y = 0 \tag{52}$$

where the over bar indicates the depth averaged values. Eq. (52) is formally indentical to the corresponding results given by Feng *et al.* (1986b) and Feng (1987). It is interesting to note the similarity between Eq. (52) and the results given previously, and to recognize that the long-term transport process in a weakly nonlinear tidal system is nearly pure convection, for which the convective velocity is the Lagrangian residual current.

Eqs. (46)–(51) have been given in a general form. In many applications, $O(\delta_1)$ can be quite small and neglected, the salinity is then uncoupled, and can be treated as a passive tracer. As a matter of fact, the summer circulation in the Bohai Sea, China, is just such an example (Feng *et al.*, 1994).

Second-order Dynamics

In barotropic systems, the solutions of tide-induced Lagrangian residual current up to the second order have been derived by Cheng *et al.* (1984) and Feng *et al.* (1986a) for a two-dimensional system, and Feng (1986, 1987) for a three-dimensional system. The first-order Lagrangian residual velocity is the same as the mass transport velocity given by Longuet-Higgins (1969), and the second-order solution shows that the Lagrangian residual current is not only a function of space, but is also a function of the tidal phase. Thus, it is interesting to examine further the baroclinic effects in the second-order Lagrangian residual velocity.

As pointed out in Section II, the Eulerian residual current is affected by baroclinic forcing, and there is a baroclinic component in the first-order and the second-order solutions of the Eulerian residual current. Because the Lagrangian displacements, and thus the Lagrangian residual currents are functions of the Eulerian residual, Eqs. (18), (27) and (36), the Lagrangian residual current (correct to the second-order approximation) can be formally written as

$$[\boldsymbol{u}_{\mathrm{lr}}]=[\boldsymbol{u}_{\mathrm{lr}}]_1+\kappa[\boldsymbol{u}_{\mathrm{lr}}]_2 \tag{53}$$

where the second-order perturbation $[\boldsymbol{u}_{\mathrm{lr}}]_2$ is

$$[\boldsymbol{u}_{\mathrm{lr}}]_2=[\boldsymbol{u}_{\mathrm{er}}]_2+\boldsymbol{u}_{\mathrm{ld}} \tag{54}$$

in which $[\boldsymbol{u}_{\mathrm{er}}]_2$ is the additional baroclinic component of the second-order Eulerian residual current. The Lagrangian drift velocity $\boldsymbol{u}_{\mathrm{ld}}$ has a time dependent property, and it can be expressed as

$$\boldsymbol{u}_{\mathrm{ld}}=\boldsymbol{u}'_{\mathrm{ld}}\cos\theta_0+\boldsymbol{u}''_{\mathrm{ld}}\sin\theta_0 \tag{55}$$

where

$$\begin{aligned}\boldsymbol{u}'_{\mathrm{ld}}&=\boldsymbol{u}''_0\nabla\boldsymbol{u}_{\mathrm{lr}}-\boldsymbol{u}_{\mathrm{lr}}\nabla\boldsymbol{u}''_0\\ \boldsymbol{u}''_{\mathrm{ld}}&=-\boldsymbol{u}'_0\nabla\boldsymbol{u}_{\mathrm{lr}}+\boldsymbol{u}_{\mathrm{lr}}\nabla\boldsymbol{u}'_0\end{aligned} \tag{56}$$

and where

$$\boldsymbol{u}_0=\boldsymbol{u}'_0\cos\theta+\boldsymbol{u}''_0\sin\theta$$

The second-order solution takes on formally the same form as a Lagrangian drift velocity, $\boldsymbol{u}_{\mathrm{ld}}$, (Cheng *et al.*, 1984; Feng *et al.*, 1986a; Feng, 1986, 1987), while the baroclinic effects are implicitly included in the first-order Eulerian residual velocity. In general, the second-order Lagrangian residual velocity is also solenoid, or it satisfies an incompressible continuity equation. Figure 1 summarizes graphically the interrelations between these residual variables.

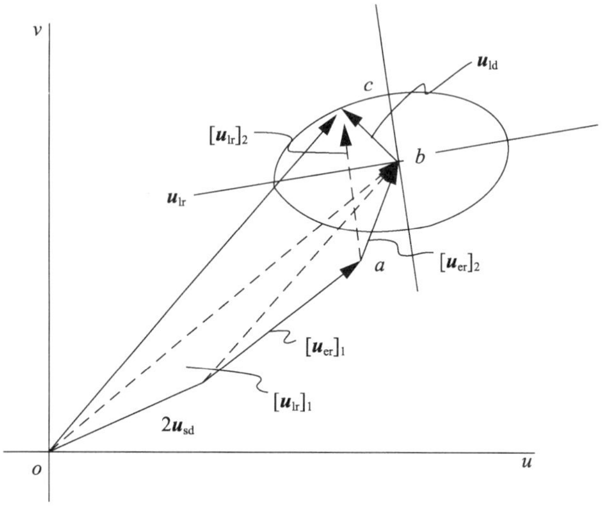

Fig. 1 Schematic diagram of inter-relations between the Eulerian residual current, Stokes' drift, and the Lagrangian residual current. The vectors oa and ob are the first and second order mass transport velocities

Application-summer Circulation and Transport in the Bohai Sea, China

To test the present theory, the summer residual circulation and associated transport processes of water mass, in the Bohai Sea, China is examined using combined analytical and numerical techniques. The Bohai Sea is a semi-closed coastal sea which is located at the north-eastern corner of China. The Bohai Sea can be considered as a sub-embayment of the Huanghai Sea, and is connected to the Huanghai Sea by the Bohai Strait (Fig. 2). The Huanghai Sea warm water which enters the Bohai Sea through Bohai Strait is one of the main contributing factors to the formation of residual circulation in the Bohai Sea. In order to model the residual circulation in the summer, the values of the volumetric transport of water through the Bohai Strait are prescribed, which are deduced from calculation of the current speed based on the field data collected in the summer (Zhang *et al.*, 1984). The East Asia monsoon dominates the meteorology of this region, and the prevailing winds are northeasterly in the winter and southwesterly in the summer. A typical wind stress on the sea surface is 0.005 dyne/cm^2 for the summer conditions. Therefore, effects of wind stress on circulation is negligible for the Bohai Sea, the characteristic tidal elevation and mean water depth are 2 m and 20 m, and the tidal current and the residual current speeds are typically 100 cm/sec and less than 10 cm/sec, so that κ is on the order of 10^{-1}. The relative deviation of salinity in the Bohai Sea is typically 1‰ for summer conditions, so that $O(\delta) \simeq 0.01$ and $O(\delta_1) = 0.1$. Therefore, the

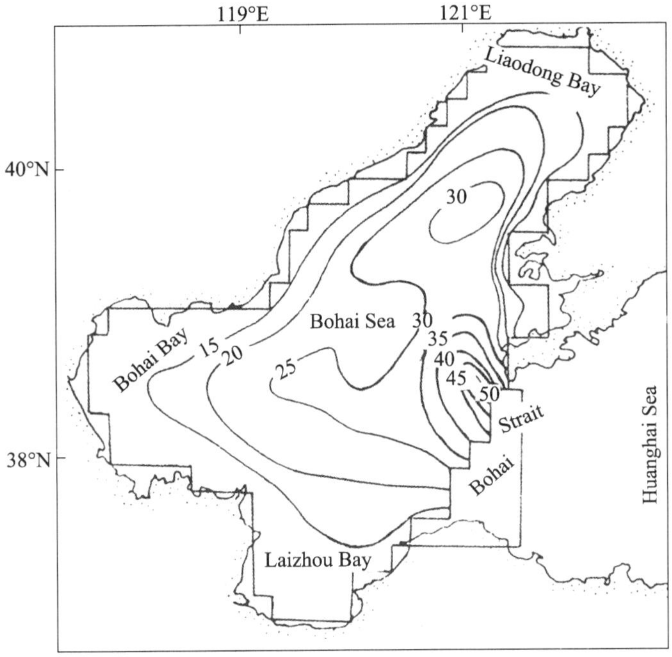

Fig. 2 Bathymetric distribution in the Bohai Sea, China. Units are in meters.

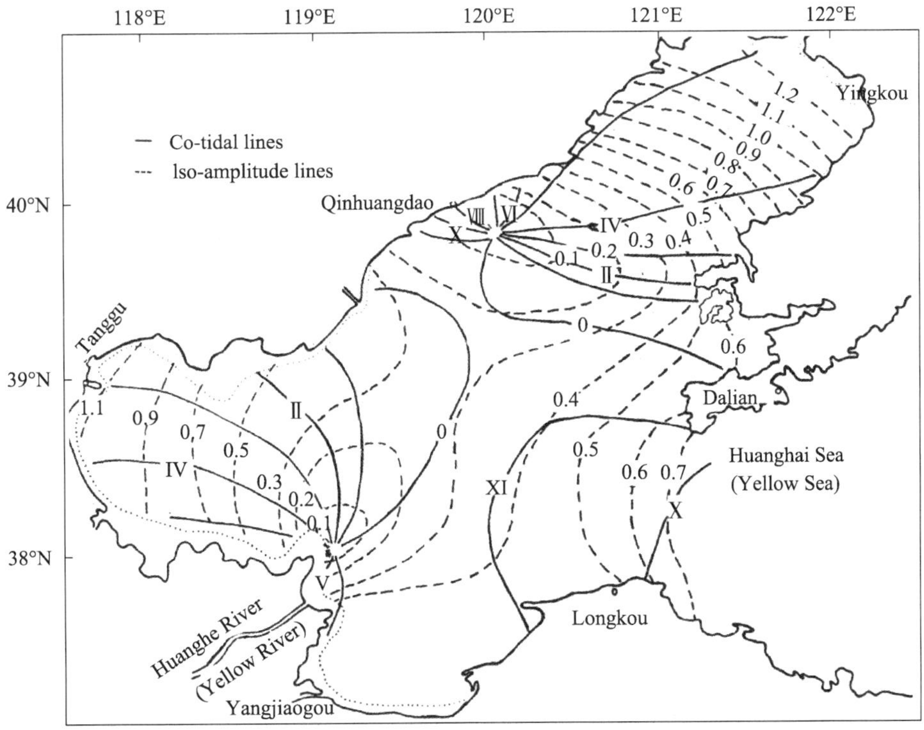

Fig. 3 Co-tidal and co-range lines for M_2 tidal constituent in the Bohai Sea

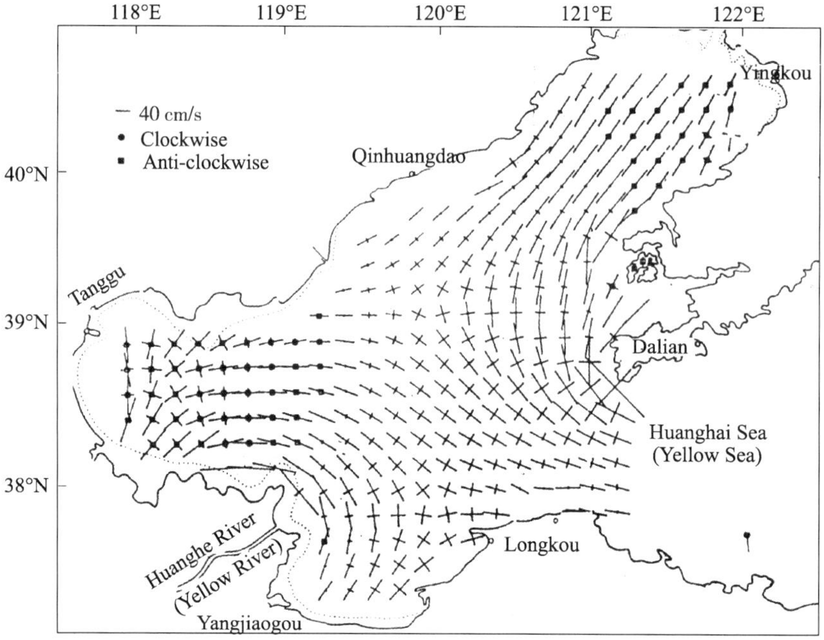

Fig. 4 Tidal current ellipses for M_2 at the surface in the Bohai Sea

baroclinic effect on summer circulation is of secondary importance, and can be omitted. The M_2-tide is the principal tidal constituent. Some of the numerical results for the zeroth order solution are taken from a previous numerical modelling study; the solutions were obtained from a velocity-splitting model, using a depth-varying eddy viscosity (Song, 1987; Feng, 1977). The cotidal and corange lines of the M_2-tide and exhibited in Fig. 3a, b, and the tidal current ellipses of the M_2-tide are depicted in Fig. 4. These results are valid when compared to field data and to the results of previous studies.

The present theory is scrutinized with the aid of this numerical mode. Additional computations are carried out to obtain the first-order and second-order Lagrangian residual circulation in the Bohai Sea for summer conditions. The first-order Lagrangian residual currents have been obtained in a three-dimensional space, and the vertical profiles at some locations are depicted in Fig. 5. The spatial distribution of the first and second order Lagrangian residual currents are plotted and given in Figs. 6 and 7. These results are compared, directly or indirectly, with field observations (Figs. 8 and 9).

For our purpose, a qualitative comparison between the theory and observations will be made. A depth-averaged first-order Lagrangian residual current field (mass-transport velocity) is obtained by integrating the three-dimensional solutions as the mean summer circulation in the Bohai Sea, Fig. 6. According to Eq. (52), the concentration (salinity) gradient is perpendicular to the streamline of the depth-averaged first order Lagrangian residual current or

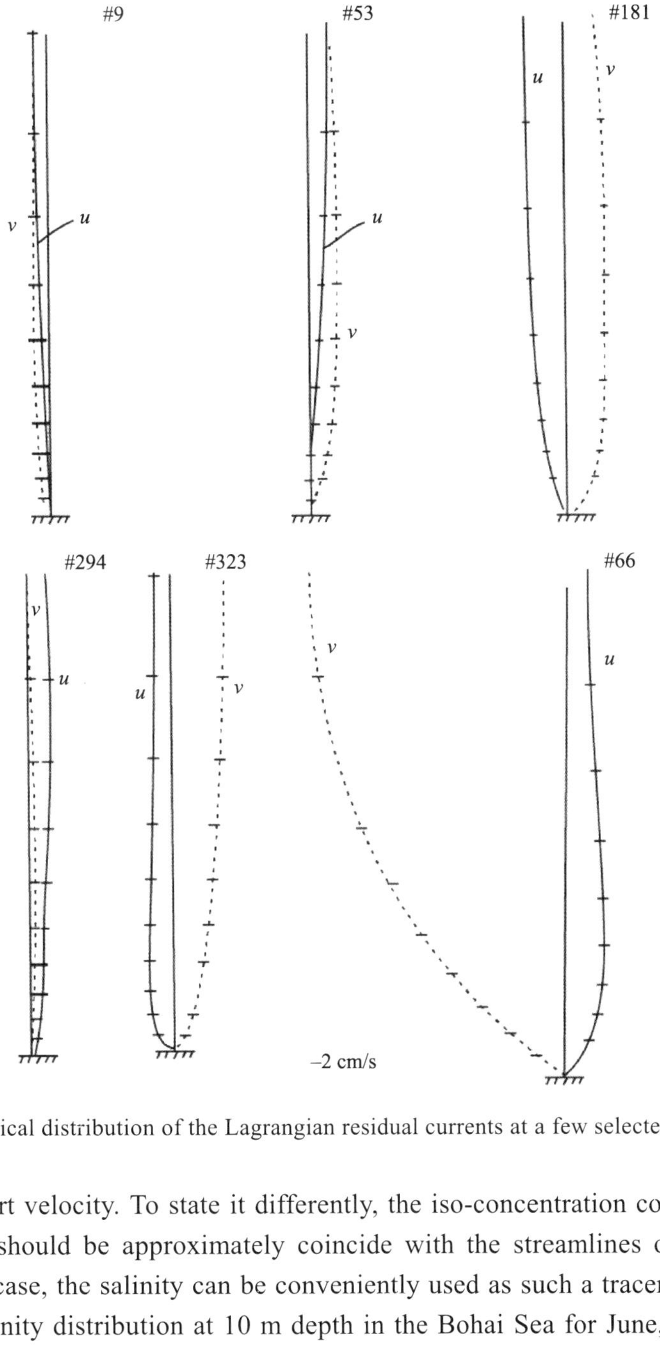

Fig. 5 The vertical distribution of the Lagrangian residual currents at a few selected locations

the mass transport velocity. To state it differently, the iso-concentration contours for a conservative tracer should be approximately coincide with the streamlines of mass-transport velocity. In this case, the salinity can be conveniently used as such a tracer. Figure 8 shows the observed salinity distribution at 10 m depth in the Bohai Sea for June, 1958. There are some minor factors, such as baroclinity, the Huanghe River runoff, and the coastal dispersion boundary layers, that are not included in the numerical simulation. The patterns of measured isohaline contours for salinity in Fig. 8, and the streamlines for the mass-transport velo- city in Fig. 6, show strong similarities, and they are in surprisingly good qualitative agreement. More interestingly, the flow pattern of the mass-transport can be used to explain some

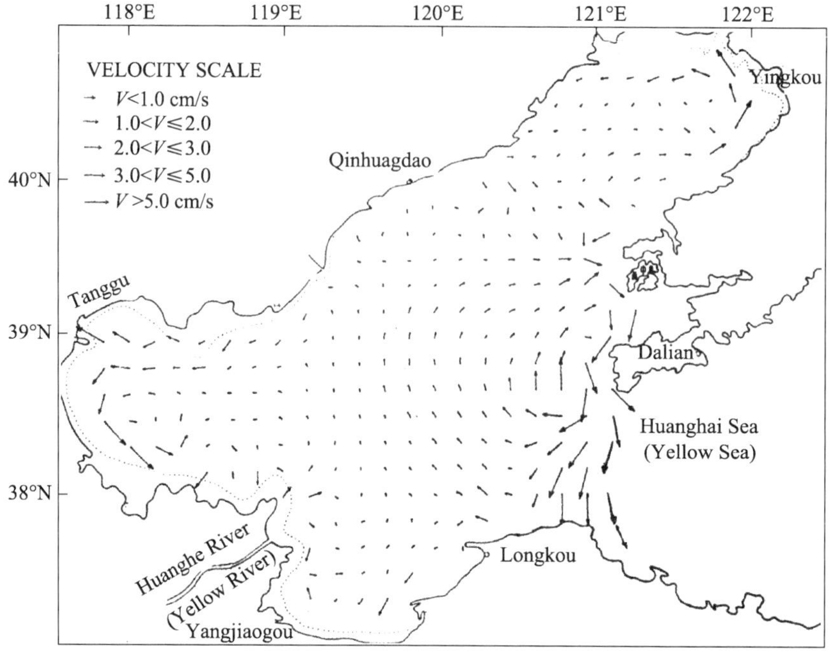

Fig. 6 The Lagrangian residual currents at the surface in the Bohai Sea

features that are apparent in the satellite photograph near the northern boundary of the Laizhou Bay (Fig. 9). To the north of the Laizhou Bay is the mouth of Huanghe River (Yellow River) where the water mass is expected to be heavily laden with sediments. However, from the satellite photograph, there exists a region near the north-east corner of Laizhou Bay where the water is of relatively high salinity, transparency, relatively low temperature, and low suspended sediments. The pattern of the Lagrangian residual currents, Fig. 6, shows that this relatively transparent water mass is originated from the Huanghai Sea. According to the flow pattern of the Lagrangian residual current, this water mass enters the Laizhou Bay and follows the southern coastline of the Bohai sea to reach the north-eastern region of Laizhou Bay. The velocity vectors shown in Fig. 8 are the observed Eulerian residual velocities. Based on the Eulerian residual velocity field, there is no explanation for the presence of this distinct water property in this region can be given.

The second order Lagrangian residual currents have also been computed according to Eq. (55). Shown in Fig. 7 are the horizontal components of the Lagrangian drift velocity components, u_{ld} and v_{ld}. Their time-dependent properties are displayed in Fig. 7 in terms of the Lagrangian residual ellipses (at the sea surface) in the Bohai Sea for the summer. Note the overall magnitude of the Lagrangian residual drifts; evidently, the Lagrangian residual drifts are relatively large in shallow water near the coasts, where nonlinear effects are strong. Although the Lagrangian drifts represent the second order corrections, their importance (at least conceptually) in assisting understanding the dynamics of a weakly nonlinear system should not be overlooked.

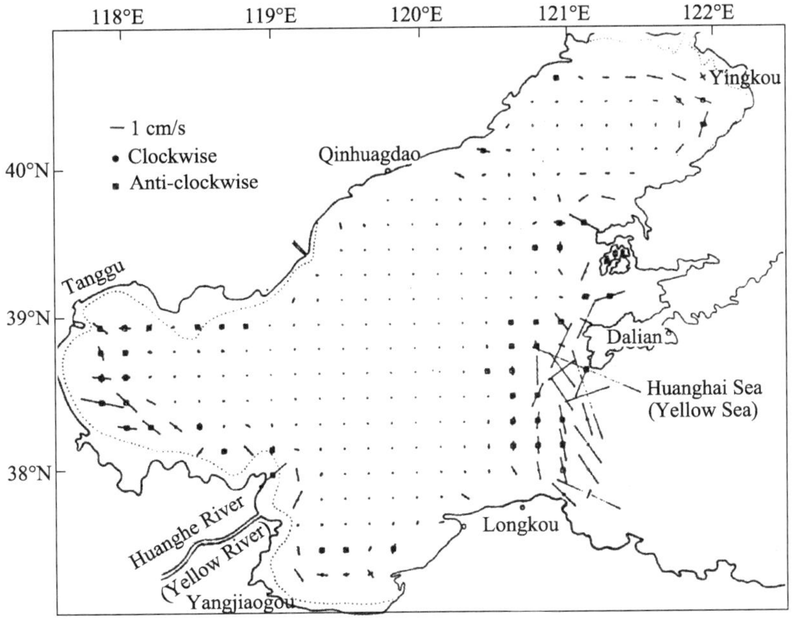

Fig. 7 The Lagrangian residual ellipses in the Bohai Sea

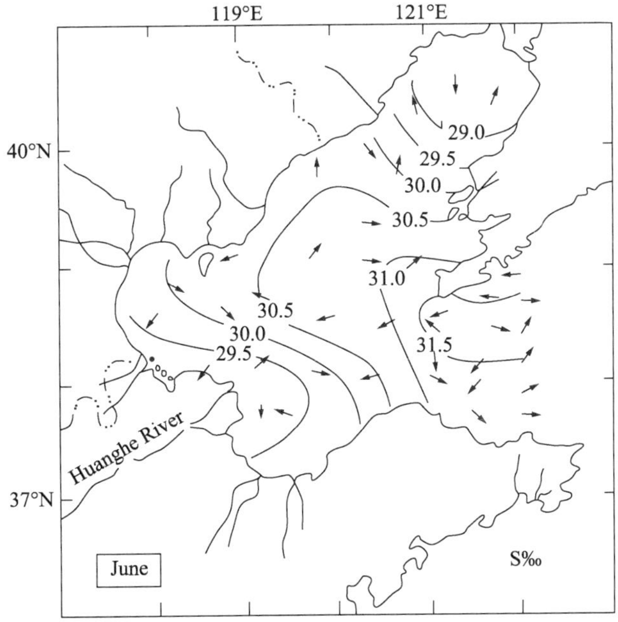

Fig. 8 The isohaline contours for the Bohai Sea at 10 m depth. Data were collected in June 1958

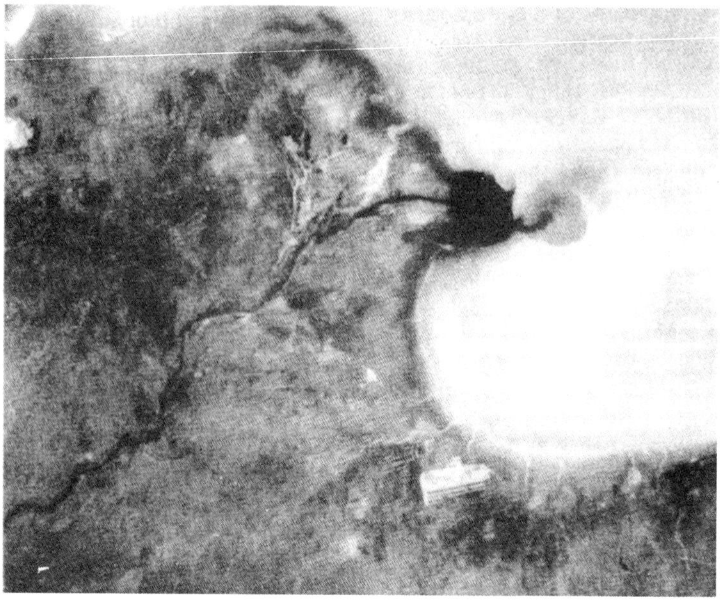

Fig. 9 Satellite photograph near the northern boundary of Laizhou Bay. The Huanghe River is clearly visible. A distinct region of low turbidity water near the mouth of the Huanghe River can be easily identified

Conclusion

Under the weakly nonliner approximation, it has been shown that the astronomical tides are independent of baroclinic effects due to the presence of a salt field. At the first order of approximation, the Stokes' drift is also not affected by the baroclinity; however, the Eulerian residual current is coupled to the salt field. A set of coupled governing equations for the first-order Lagrangian residual velocity and the zeroth-order salinity has been is derived. The second-order solution of the Lagrangian residual current is a Lagrangian drift velocity which is formally the same as the tide-induced Lagrangian residual drift. In this case, the values of the Lagrangian drift would have included the baroclinic effects through the first order Eulerian residual currents. The present theory has been applied to the Bohai Sea, China, with the aid of numerical modeling techniques. These results permit a qualitative explanation of the summer circulation pattern and its associated transport processes. The field observations in the Bohai Sea support, qualitatively, the trends of long-term transport properties which has been predicted by the present theoretical results.

References

Cheng, R.T., and Casulli, V., 1982, On Lagrangian residual currents with application in South San Francisco Bay, California, Wat. Resour. Res., 18(6), pp. 1652–1662.

Cheng, R.T., 1983, Euler-Lagrangian computations in estuarine hydro-dynamics, In Proceedings of the

Third International Conference on Numerical Methods in Laminar and Turbulent Flows, edited by C. Taylor *et al.*, pp. 341–352, Pineridge Press.

Cheng, R.T., Feng, S., and Xi, P., 1984, On Lagrangian residual ellipse, presented at Inter. Symposium on Physics of shallow Estuaries and Bays Univ. of Miami, Miami, FL., also In: J. van de Kreeke (Editor), Inter. Conf. on Physics of Shallow Estuaries and Bays, Lecture Notes on Coastal and Estuarine Studies, Springer-Verlag, pp. 102–113.

Dyer, K.R., 1973, Estuaries: A Physical Introduction, John Wiley Publishers, New York, pp. 140.

Feng, S., 1977, A three-dimensional nonlinear model of tides, Scientia Sinica, 20(4), pp. 436–446.

Feng, S., 1986, A Three-dimensional weakly nonlinear dynamics on tide-induced Lagrangian residual current and mass-transport, Chinese J. Oceanology and Limnology, 4(2), pp. 139–158.

Feng, S., 1987, A three-dimensional weakly nonlinear model of tide-induced Lagrangian residual current and mass-transport with an application to the Bohai Sea, In Proceedings of the 18th Liege Colloquium on Ocean Hydrodynamics, Three-dimensional Models of Marine and Estuarine Dynamics, J.C.J. Nihoul and B.M. Jamart Eds., Elsevier, pp. 471–488.

Feng, S., Cheng, R.T., and Xi, P., 1986a, On tide-induced Lagrangian residual current and residual transport, Part I: Residual current, Wat. Resour. Res., Vol. 22, No. 12, pp. 1623–1634.

Feng, S., Cheng, R.T., and Xi, P., 1986b, On tide-induced Lagrangian residual current and residual transport, Part II: Residual transport with applications in South San Francisco Bay, California, Wat. Resour. Res., Vol. 22, No. 12, pp. 1635–1646.

Feng, S., Zhang, S., and Xi, P., 1994, A langrangian model of circulation in Bohai Sea, D. Zhou, Y. Liang and C. Zeng (Editors), Oceanology of China Seas, Dordrecht, Boston, London: Kluwer Academic Publishers, pp. 83–90.

Fischer, H. B., List, E. J., Koh, R. C. Y., Imberger, J., and Brooks, N. H., 1979, Mixing in Inland and Coastal Waters, Academic Press, New York, 483 pp.

Longuet-Higgins, M. S., 1969, On the transport of mass by time-varying ocean current, Deep Sea Res., 16, pp. 431–447.

Officer, C. B., 1976, Physical Oceanography of Estuaries, John Wiley Publishers, New York, 465 pp.

Pritchard, D. W., 1954, A study on the salt balance in a coastal plain estuary, J. Mar. Res., 13, pp. 133–144.

Song, L., 1986, A hydrodynamic velocity-splitting model with a depth-varying eddy viscosity in shallow seas, Acta Oceanologica Sinica, 6(S1), pp. 135–151.

Zhang, S., Xi, P., and Feng, S., 1984, Numerical modeling of the steady circulation in the Bohai Sea, J. Shandong College of Oceanology, 14(2), pp. 12–19. (In Chinese, with English abstract).

Zimmerman, J.T.F., 1979, On the Euler-Lagrangian transformation and the Stokes' drift in the presence of oscillatory and residual currents, Deep Sea Res., 26A, pp. 505–520.

On the Lagrangian Residual Velocity and the Mass-transport in a Multi-frequency Oscillatory System *

Shizuo Feng

I. Introduction

In recent years, studies of environmental hydrodynamics of tidal estuaries and coastal seas have focused on residual circulation and long-term transport processes. Intertidal transports of temperature, salinity, solutes, pollution, sediment, nutrients, fish eggs, and other tracers are fundamental to the interactive physical, chemical, and biological processes in an ecological system. In shallow, coastal seas, such as the Bohai Sea, China, the dominant observable motion is tidal currents and the apparent dominating transport mechanism appears to be convection due to tidal currents. Over a seasonal time scale, however, the overall ecological balance and long-term transport processes are mainly dependent on the residual circulation, rather than tidal currents. It is becoming increasingly clear that the residual circulation should be described by means of the net displacements of water parcels, or the Lagrangian mean velocity, but not by an Eulerian mean velocity. The Lagrangian mean velocity has led to the concept of a Lagrangian residual velocity, and the research on the Lagrangian residual velocity and the coupled intertidal transport processes is a relatively recent and difficult topic. Numerous investigators, including Longuet-Higgins (1969), Moore (1970), Tee (1976), Zimmerman (1979), Ianniello (1979), Cheng and Casulli (1982), Backhaus (1985) and Hamrick (1987), have made important contributions to this topic. A weakly nonlinear theory of Lagrangian residual currents and intertidal transports has been investigated by the author and his collaborators (Feng *et al.*, 1986a, b, c; Cheng *et al.*, 1986; Feng, 1986, 1987, 1988a, b; Zheng, 1988; Wei, 1988; Sun *et al.*, 1989; Cheng *et al.*, 1989).

To the author's knowledge, most studies of the Lagrangian residual flow and long-term transports refer to a single-frequency tidal system, such as an M_2-nonlinear system in coastal seas and tidal estuaries because the most important tidal harmonics is, normally, the lunar semi-diurnal constituent, M_2. Other principal astronomical tidal harmonics are given in Table 1 (Hansen, 1962). The tidal motions in coastal seas or tidal estuaries are generally caused by the ocean tides, and dynamically characterized by the shallow water constituents

---

* Feng S. 1990. On the Lagrangian residual velocity and the mass-transport in a multi-frequency oscillatory system. In: R T Cheng (ed.) Residual Currents and Long-term Transport. New York: Springer-Verlag. 34–48

and residual motions resulting from the nonlinear couplings of the astronomical tides or, more generally, of the lower-order constituents (Feng, 1977).

In this paper, a weakly nonlinear baroclinic shallow water system is analyzed using a perturbation method. The Lagrangian residual velocity induced by a multi-frequency tidal system is derived and examined, up to the second order, and compared with that induced by a single-frequency tidal system. A generalized set of field equations for the first-order Lagrangian residual velocity, and the zeroth-order apparent concentration is proposed and used to describe a time-averaged water circulation coupled with the intertidal transport processes in coastal seas or tidal estuaries. The system is driven by the wind stress over the sea surface, the heat flux across the sea surface, the horizontal gradient of water density, and the tidal body force due to the nonlinear interactions of the constituents of the multi-frequency astronomical tidal system. This set of equations is solved numerically to model the summer circulation in the Bohai Sea, China, using a two-frequency representation. The calculated circulation pattern is compared with that induced by an M_2 system, and verified by field data. For a more thorough understanding of the second-order dynamics, the Lagrangian drift velocity induced by an M_2-S_2 tidal system is discussed in detail.

Table 1 The principal terms of astronomical tide

Name	Frequency/(°/hour)	Relative Coefficient
M_2	28.98	0.9085
S_2	30.00	0.4227
K_1	15.04	0.5305
O_1	13.94	0.3771

II. A Baroclinic Shallow Water System

The governing equations describing a baroclinic shallow water system are the continuity equation

$$\nabla \cdot \boldsymbol{U} = 0 \tag{1}$$

the x-momentum equation

$$\frac{\partial u}{\partial t} + \boldsymbol{U} \cdot \nabla u - fv = -g\frac{\partial \zeta}{\partial x} + \frac{\partial}{\partial z}\left(\nu \frac{\partial u}{\partial z}\right) - g\beta \frac{\partial}{\partial x}\int_z^{\zeta} c\mathrm{d}z' \tag{2}$$

the y-momentum equation

$$\frac{\partial v}{\partial t} + \boldsymbol{U} \cdot \nabla v + fu = -g\frac{\partial \zeta}{\partial y} + \frac{\partial}{\partial z}\left(\nu \frac{\partial v}{\partial z}\right) - g\beta \frac{\partial}{\partial y}\int_z^{\zeta} c\mathrm{d}z' \tag{3}$$

the conservation equation

$$\frac{\partial c}{\partial t} + \boldsymbol{U} \cdot \nabla c = \frac{\partial}{\partial z}\left(K \frac{\partial c}{\partial z}\right) \tag{4}$$

and the associated boundary conditions at $z = \zeta$

$$w=\frac{\partial\zeta}{\partial t}+u\frac{\partial\zeta}{\partial x}+v\frac{\partial\zeta}{\partial y} \tag{5}$$

$$(\tau_x,\tau_y)=\nu\frac{\partial(u,v)}{\partial z} \tag{6}$$

$$K\frac{\partial c}{\partial z}=\Gamma \tag{7}$$

and at $z=-h$,

$$\boldsymbol{U}=0 \tag{8}$$

$$\partial c/\partial z=0 \tag{9}$$

and the expression of the "apparent concentration"

$$c=(S-S_*)+\sum_{j=1}^{m}(\beta_j/\beta)(C_j-C_{j*})-(\alpha/\beta)(\theta-\theta_*) \tag{10}$$

where t is time;

(x,y,z) are the Cartesian coordinates on an f-plane;

$\nabla=\boldsymbol{i}\partial/\partial x+\boldsymbol{j}\partial/\partial y+\boldsymbol{k}\partial/\partial z$ is a vector operator;

$(\boldsymbol{i},\boldsymbol{j},\boldsymbol{k})$ are unit vectors;

$\boldsymbol{U}=\boldsymbol{i}u+\boldsymbol{j}v+\boldsymbol{k}w$ is the velocity vector;

ζ and h are the water surface elevation and the water depth measured from mean sea level, respectively;

f is the Coriolis parameter;

ρ is the water density;

ν and K are the eddy viscosity and the eddy diffusivity, respectively;

g is the gravitational acceleration;

(τ_x,τ_y) are the wind stress components in the (x,y) directions at the water surface divided by the water density;

$\Gamma=-(\alpha/\beta)\ Q/\rho C_{\mathrm{P}}$, where α is the coefficient of thermal expansion;

β is the fractional increase in density per unit increase in salinity;

Q is the heat flux across the sea surface;

C_{P} is the specific heat;

c is the "apparent concentration", which is formally similar to the "apparent temperature" defined by Fofonoff (1962). C_j is the concentration of j-tracer, β_j is the corresponding constant; the notation "$*$" indicates the corresponding reference constant. A generalized linearized equation of state containing salinity, S, and temperature, θ, and $C_j\,(j=1,2,\cdots,m)$, $(\rho-\rho_*)=\beta c$ is used, and θ, S and C_j, satisfy respectively the same conservation equation as given in Eq. (4) (Nihoul, 1975).

In order to close this basic set of equations, we suppose that the eddy coefficients, such as ν and K, are the function of spatial and temporal coordinates only. The other nonlinear terms of this set of equations are in the order of magnitude characterized by a nondimensional parameter $\kappa = \zeta_c / h_c = U_c / \sqrt{gh_c}$, where the subscript c indicates the characteristic value of the corresponding quantity.

Let us assume that (i) $O(\kappa) < 1$; (ii) $O\left[\frac{\partial}{\partial z}\left(k\frac{\partial c}{\partial z}\right)\right] = \kappa^2$; and (iii) $O(g\beta\frac{\partial}{\partial x}\int_z^\zeta c\mathrm{d}z'$, $g\beta\frac{\partial}{\partial y}\int_z^\zeta c\mathrm{d}z') = O(\tau_x, \tau_y) = \kappa$; (Feng, 1987; Feng *et al.*, 1990; Feng *et al.*, 1984). These assumptions imply that (i) the dynamic system is weakly nonlinear, and a perturbation technique with the small parameter κ can be used to solve the jth-order solution of this system ($j = 0, 1, \cdots$) (Feng, 1977); (ii) the transport mechanism is dominated by convection, and thus the zeroth-order apparent concentration, C_0, is independent of time (Feng, 1987); (iii) the astronomical tides, such as M_2, S_2, K_1, O_1, are independent of the effects due to baroclinic forcing and due to the wind stress forcing (Feng *et al.*, 1990; Feng, 1988a).

The streakline of a marked water parcel can be described by a displacement vector $(\xi = \boldsymbol{X} - \boldsymbol{X}_0)$, or

$$\xi = \int_{t_0}^{t} \boldsymbol{U}(\boldsymbol{X}_0 + \xi, t')\mathrm{d}t' \tag{11}$$

where $(\boldsymbol{X}_0, t_0)$ indicates that the marked water parcel is released from the spatial point $\boldsymbol{X}_0$ at instant t_0.

In the studies of long-term transport processes, one of the most important quantities is the net displacement of the marked water parcel released from $\boldsymbol{X}_0$ at t_0 and moving over several tidal periods T.

$$\xi_r = \int_{t_0}^{t_0+T} \boldsymbol{U}(\boldsymbol{X}_0 + \xi, t')\mathrm{d}t'$$

or equivalently, the Lagrangian mean velocity

$$\boldsymbol{U}_{LR} = (1/T)\int_{t_0}^{t_0+T} \boldsymbol{U}(\boldsymbol{X}_0 + \xi, t)\mathrm{d}t \tag{12}$$

Introducing the tidal excursion ξ_c as the scale of ξ and L as the horizontal scale of the flow field, and supposing $(\xi_c / L) = \kappa$, the Lagrangian residual velocity can be defined by the Lagrangian mean velocity, $\boldsymbol{U}_{LR}$ as given in Eq. (12) (Zimmerman, 1979; Cheng and Casulli, 1982; Feng, 1987).

III. First-order Dynamics

III. 1 Mass-transport velocity

Taking an appropriate averaging time for removing, at least approximately, all dominant tidal oscillations in an m-frequency tidal system, and substituting $\boldsymbol{U}(\boldsymbol{X}_0 + \xi, t)$ and Eq. (11)

into Eq. (12) yields the first-order Lagrangian residual velocity, or the mass-transport velocity, $\boldsymbol{U}_{\mathrm{L}}$ as

$$\boldsymbol{U}_{\mathrm{L}}=\boldsymbol{U}_{\mathrm{E}}+\boldsymbol{U}_{\mathrm{S}}=\sum_{k=0}^{m}{}^{k}\boldsymbol{U}_{\mathrm{L}} \tag{13}$$

where $(\boldsymbol{U}_{\mathrm{E}},\boldsymbol{U}_{\mathrm{S}})=\sum_{k=0}^{m}({}^{k}\boldsymbol{U}_{\mathrm{E}},{}^{k}\boldsymbol{U}_{\mathrm{S}})$ and ${}^{k}\boldsymbol{U}_{\mathrm{L}}={}^{k}\boldsymbol{U}_{\mathrm{E}}+{}^{k}\boldsymbol{U}_{\mathrm{S}}$;

${}^{k}\boldsymbol{U}_{\mathrm{E}}$ and ${}^{k}\boldsymbol{U}_{\mathrm{S}}$ are the Eulerian residual velocity and the Stokes' drift velocity, respectively. K (K= 1,2,$\cdots$,m), indicates the corresponding quantities induced by the k-constituent of astronomical tides; except ${}^{0}\boldsymbol{U}_{\mathrm{E}}$ contains a wind-driven barotropic component and a density-driven baroclinic component. It is noted that ${}^{0}\boldsymbol{U}_{\mathrm{S}}=0$, because the Stokes' drift velocity is independent of the wind stress and the baroclinic effect (Feng, 1988a; Feng *et al.*, 1990).

Eq. (13) reveals that the mass-transport velocity $\boldsymbol{U}_{\mathrm{L}}$ in an m-frequency baroclinic tidal system can be written as the sum of all k-constituent induced mass-transport velocities and including both the wind-driven barotropic and the density-driven baroclinic components.

III. 2 A set of field equations governing U_{L} and C_0

By tidally time-averaging the Eqs. (1)–(9), and noting

$$\langle\psi_0\ \phi_0\rangle=\left\langle\sum_{k=1}^{m}{}^{k}\psi_0\sum_{k=1}^{m}{}^{k}\phi_0\right\rangle=\sum_{k=1}^{m}\langle{}^{k}\psi_0\,{}^{k}\phi_0\rangle$$

where ψ_0 and ϕ_0 represent the zeroth-order variables, and < > is a tidally time-averaging operator, the set of field equations governing $\boldsymbol{U}_{\mathrm{L}}$ and C_0 is derived as follows:

$$\nabla\cdot\boldsymbol{U}_{\mathrm{L}}=0 \tag{14}$$

$$-fv_{\mathrm{L}}=-g\frac{\partial\langle\zeta_1\rangle}{\partial x}+\frac{\partial}{\partial z}\left(\nu\frac{\partial u_{\mathrm{L}}}{\partial z}\right)-g\beta\frac{\partial}{\partial x}\int_z^0 C_0\mathrm{d}z'+\Pi_1 \tag{15}$$

$$fu_{\mathrm{L}}=-g\frac{\partial\langle\zeta_1\rangle}{\partial y}+\frac{\partial}{\partial z}\left(\nu\frac{\partial v_{\mathrm{L}}}{\partial z}\right)-g\beta\frac{\partial}{\partial y}\int_z^0 C_0\mathrm{d}z'+\Pi_2 \tag{16}$$

$$\boldsymbol{U}_{\mathrm{L}}\cdot\nabla C_0=\frac{\partial}{\partial z}\left(\langle K\rangle\frac{\partial C_0}{\partial z}\right) \tag{17}$$

at $z=0$

$$w_{\mathrm{L}}=0 \tag{18}$$

$$\nu\frac{\partial(u_{\mathrm{L}},v_{\mathrm{L}})}{\partial z}=\left(\langle\tau_x\rangle,\langle\tau_y\rangle\right) \tag{19}$$

$$K(\partial C_0/\partial z)=\langle\Gamma\rangle \tag{20}$$

and at $z=-h$

$$\boldsymbol{U}_{\mathrm{L}} = 0 \tag{21}$$

$$\partial C_0 / \partial z = 0 \tag{22}$$

where

$$(\Pi_1, \Pi_2) = \sum_{k=1}^{m} ({}^k\Pi_1, {}^k\Pi_2) \tag{23}$$

$$\begin{aligned}
{}^k\Pi_1 = & -g\frac{\partial}{\partial x}\left\langle \frac{1}{2}{}^k\boldsymbol{\xi}_0 \cdot \nabla {}^k\zeta_0 \right\rangle + \frac{\partial}{\partial z}\left\langle \left(\frac{5}{2}\frac{\partial {}^k\xi_0}{\partial x} + \frac{\partial {}^k\eta_0}{\partial y}\right)\nu\frac{\partial {}^k u_0}{\partial z} + \left(\frac{\partial {}^k\xi_0}{\partial y} + \frac{1}{2}\frac{\partial {}^k\eta_0}{\partial x}\right)\nu\frac{\partial {}^k v_0}{\partial z} \right\rangle \\
& + \frac{\partial}{\partial z}\left\langle \nabla\, \nu \cdot {}^k\boldsymbol{\xi}_0 \frac{\partial {}^k u_0}{\partial z} \right\rangle - \left\langle \frac{\partial {}^k\boldsymbol{\xi}_0}{\partial z} \cdot \nabla\left(\nu\frac{\partial {}^k u_0}{\partial z}\right) \right\rangle - \frac{1}{2}\left\langle \frac{\partial^2\, {}^k\xi_0}{\partial z \partial x}\left(\nu\frac{\partial {}^k u_0}{\partial z}\right) \right. \\
& \left. + \frac{\partial^2\, {}^k\eta_0}{\partial z \partial x}\left(\nu\frac{\partial {}^k v_0}{\partial z}\right) + {}^k\xi_0 \frac{\partial^2}{\partial z \partial x}\left(\nu\frac{\partial {}^k u_0}{\partial z}\right) + {}^k\eta_0 \frac{\partial^2}{\partial z \partial x}\left(\nu\frac{\partial {}^k v_0}{\partial z}\right) \right\rangle
\end{aligned}$$

$$\begin{aligned}
{}^k\Pi_2 = & -g\frac{\partial}{\partial y}\left\langle \frac{1}{2}{}^k\boldsymbol{\xi}_0 \cdot \nabla {}^k\zeta_0 \right\rangle + \frac{\partial}{\partial z}\left\langle \left(\frac{5}{2}\frac{\partial {}^k\eta_0}{\partial y} + \frac{\partial {}^k\xi_0}{\partial x}\right)\nu\frac{\partial {}^k v_0}{\partial z} + \left(\frac{\partial {}^k\eta_0}{\partial x} + \frac{1}{2}\frac{\partial {}^k\xi_0}{\partial y}\right)\nu\frac{\partial {}^k u_0}{\partial z} \right\rangle \\
& + \frac{\partial}{\partial z}\left\langle \nabla\, \nu \cdot {}^k\boldsymbol{\xi}_0 \frac{\partial {}^k v_0}{\partial z} \right\rangle - \left\langle \frac{\partial {}^k\boldsymbol{\xi}_0}{\partial z} \cdot \nabla\left(\nu\frac{\partial {}^k v_0}{\partial z}\right) \right\rangle - \frac{1}{2}\left\langle \frac{\partial^2\, {}^k\xi_0}{\partial z \partial y}\left(\nu\frac{\partial {}^k u_0}{\partial z}\right) \right. \\
& \left. + \frac{\partial^2\, {}^k\eta_0}{\partial z \partial y}\left(\nu\frac{\partial {}^k v_0}{\partial z}\right) + {}^k\xi_0 \frac{\partial^2}{\partial z \partial y}\left(\nu\frac{\partial {}^k u_0}{\partial z}\right) + {}^k\eta_0 \frac{\partial^2}{\partial z \partial y}\left(\nu\frac{\partial {}^k v_0}{\partial z}\right) \right\rangle
\end{aligned}$$

where $\boldsymbol{U}_{\mathrm{L}} = \boldsymbol{i}u_{\mathrm{L}} + \boldsymbol{j}v_{\mathrm{L}} + \boldsymbol{k}w_{\mathrm{L}}$, and ζ_0 and ζ_1 are respectively the zeroth- and first-order solutions of ζ.

This set of field equations, [Eqs.(14)–(23)], is an extension of the corresponding set of field equations proposed in the previous papers (Feng, 1987; Feng *et al.*, 1990; Feng, 1988). It can be used to describe the (residual) circulation and long-term transport processes in coastal seas and tidal estuaries, and the nonlinear coupling between them. If $({}^k\Pi_1, {}^k\Pi_2) = C_j = 0$ the set of equations derived here will be reduced to the classical set of equations for the wind-driven and thermohaline circulation in coastal seas (Ramming and Kowalik, 1980). In addition to the wind stress, and the baroclinic forcing, (Π_1, Π_2) is a tide-induced body force, termed "tidal body force" (Feng, 1987), and represents the nonlinear couplings between the Lagrangian displacements of astronomical tides and tidal currents or tidal elevation. Using the "velocity-splitting method", the three-dimensional velocity field in an m-frequency tidal system can easily be solved for the barotrophic flows, or for the diagnosis mode of the baroclinic flows (Song, 1987).

III. 3 An application to the Bohai Sea, China

To test the dynamic model formulation, the summer residual circulation and associated transport processes of water mass in the Bohai Sea, China are examined using numerical solutions. The Bohai Sea is a semi-enclosed coastal shallow sea, which is located at the north-eastern corner of China and connected to the Huanghai Sea by the Bohai Strait (Fig. 1). The summer residual circulation in the Bohai Sea is driven mainly by the warm water branch of the Huanghai Sea and the nonlinear M_2-tide, which enter the Bohai Sea from the Huanghai Sea through the Bohai Strait (Feng *et al.*, 1990). The depth-averaged field of the mass-transport velocity calculated by Feng *et al.* (1990) and Zheng (1988) has revealed some interesting features and explained some observations. Here a supplement to these is that a clockwise gyre in the Bohai Bay might explain the existence of calcite and particularly muscovite in the Bohai Bay since the calcite and muscovite are the characteristic minerals of the Huanghe River [Figs. 2, 3, and 4, copied from Zheng (1988)]. It should be noted the Zheng's numerical modeling was based on the set of field equations governing the M_2-induced mass-transport velocity (Feng, 1987), i.e., the set of field equations Eqs. (14)–(23) with wind stress and baroclinic terms set to zero, and $\Gamma = 0$ and $(\Pi_1, \Pi_2) = ({}^{M_2}\Pi_1, {}^{M_2}\Pi_2)$, where the notation "$M_2$" inidcates that the tidal body force is generated by M_2 alone.

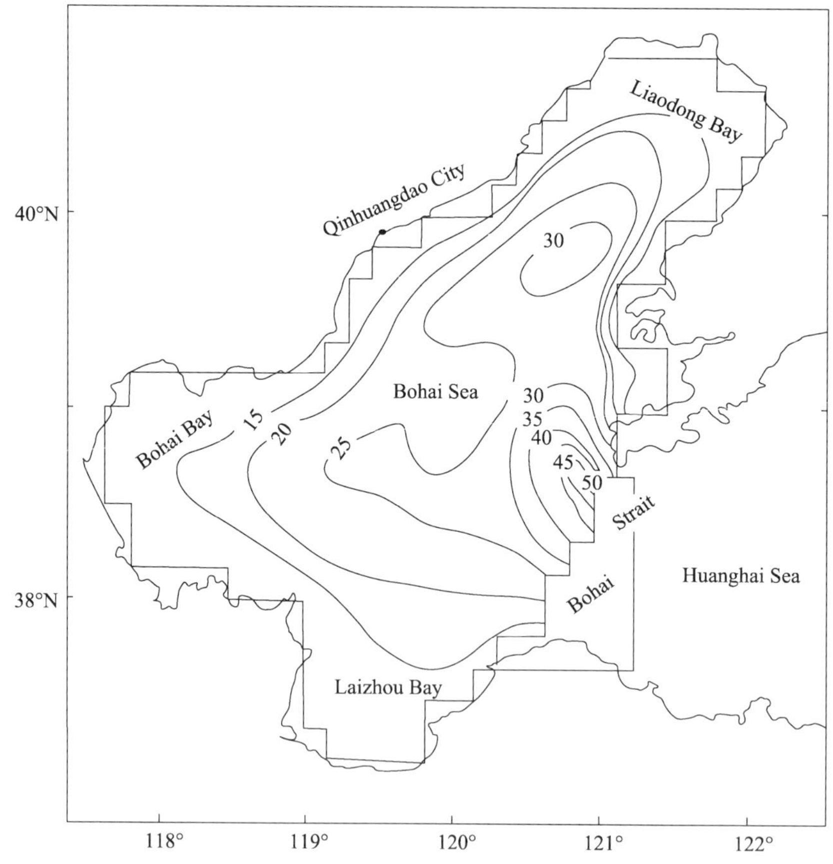

Figure 1 Depth distribution of the Bohai Sea, China

Besides M_2, as a representative of the diurnal constituent, K_1 is also important for the Bohai Sea, particularly in some regions, such as the region near Qinhuangdao city where there exists an amphidromic point of M_2-tide (Feng *et al.*, 1990). In order to model the summer residual circulation in the Bohai Sea, an M_2-"K_1" system has been used where "K_1" is an approximation to K_1 constituent having the same amplitude as K_1 and the frequency to be half that of M_2 frequency (Wei, 1988). Thus, the tidal body force, (Π_1, Π_2), expressed by Eq. (23) is greatly simplified. The depth-averaged field of the mass-transport velocity has been obtained by depth-averaging the three-dimensional solution of the mass-transport velocity, and is depicted in Fig. 5. The three-dimensional mass-transport velocity was calculated based on Eqs. (14)–(16), (18), (19) and (21), and omitting the baroclinic force and the wind stress in Eqs. (15), (16) and (19), and by using the "velocity-splitting method" (for further details, see Zheng, 1988, and Wei, 1988). In the region near Qinhuangdao City, where there is an amphidromic point of M_2-tide, a principal difference between the two calculated residual circulation patterns appears (Cf. Fig. 5 and Fig. 2). The longshore currents are opposite in direction, which is just what we expected. In fact, a north-eastward Eulerian residual flow along the coast has been pointed out by Prof. Shi (1983), Fig. 6, through field observations. In order to make a comparison between computed and observed results, the three-dimensional Eulerian residual velocity has been calculated by means of the formula $\boldsymbol{U}_{\mathrm{E}} = \boldsymbol{U}_{\mathrm{L}} - \boldsymbol{U}_{\mathrm{S}}$ and then the depth-averaged Eulerian residual velocity is derived by depth-averaging $\boldsymbol{U}_{\mathrm{E}}$ as shown in Fig. 6, and they are in good qualitative agreement. Thus, while "K_1" is an approximate constituent, it is heuristic and interesting to note that the tide-induced component of the (summer) residual circulation in the Bohai Sea should more appropriately be generated by an M_2-K_1 system than by an M_2-system alone.

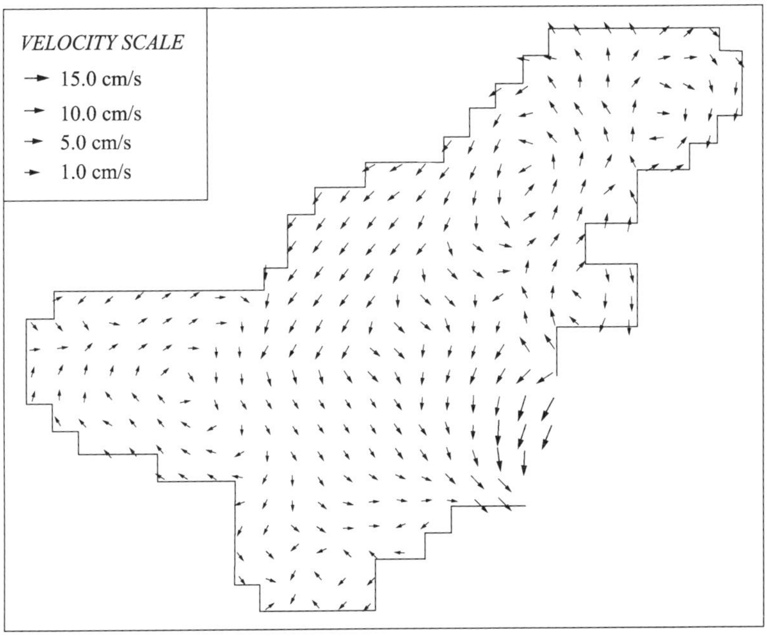

Figure 2 M_2-induced depth-averaged mass-transport velocity in the Bohai Sea, associated with the part of the Huanghai warm water coming from the Huanghai Sea through the Bohai Strait

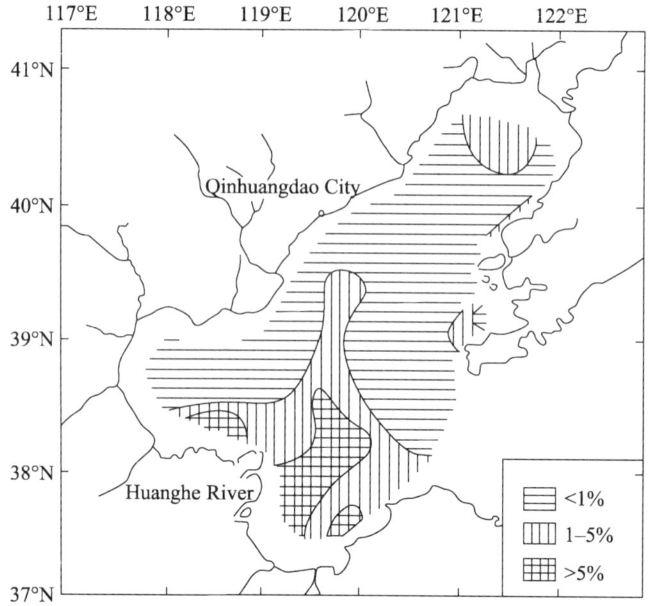

Figure 3 Distribution of muscovite (white mica) in the Bohai Sea. The muscovite is one of the characteristic minerals of the Huanghe River

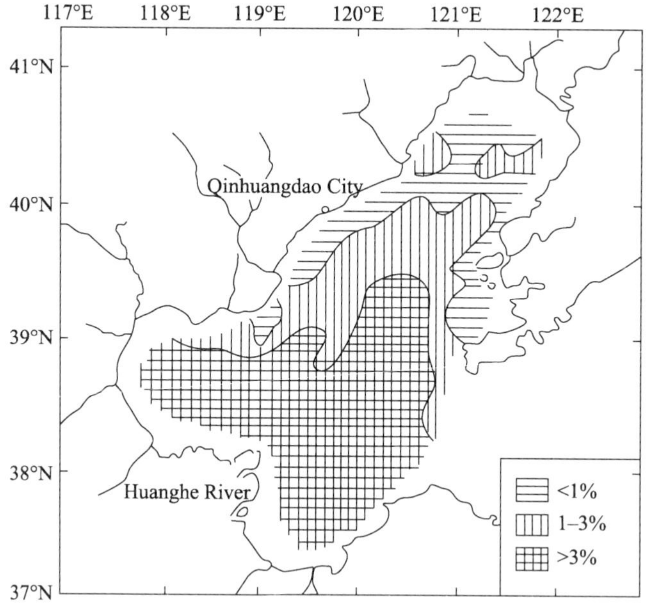

Figure 4 Distribution of calcite in the Bohai Sea. The calcite is one of the characteristic minerals of the Huanghe River

IV. The Lagrangian Drift Velocity

In a single-frequency tidal system, the Lagrangian drift velocity, $\boldsymbol{U}_{\mathrm{LD}}$, a second-order perturbation term of the Lagrangian residual velocity has been derived by Cheng *et al.* (1986)

and Feng *et al.* (1986a) for a two-dimensional system, by Feng (1986, 1987) for a three-dimensional barotropic system, and by Feng *et al.* (1990) for a three-dimensional baroclinic system. In this paper, the Lagrangian drift velocity induced by a multi-frequency tidal system will be examined. To avoid confusion with the main issues, the effects due to the sea surface wind stress and baroclinic forcing are not included.

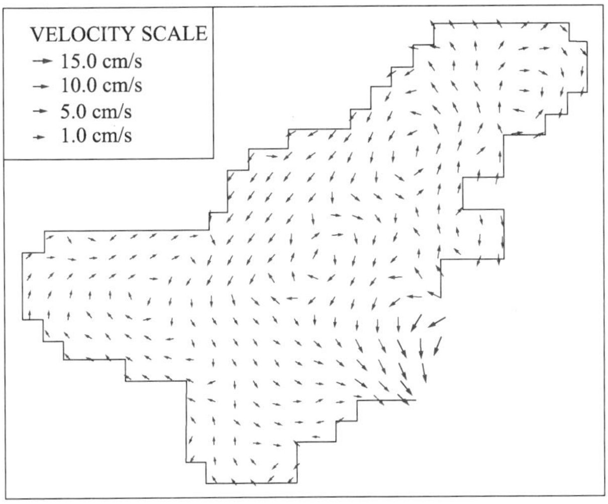

Figure 5 (M_2, "K_1")-induced depth-averaged mass-transport velocity in the Bohai Sea, associated with the part of the Huanghai warm water coming from the Huanghai Sea through the Bohai Strait

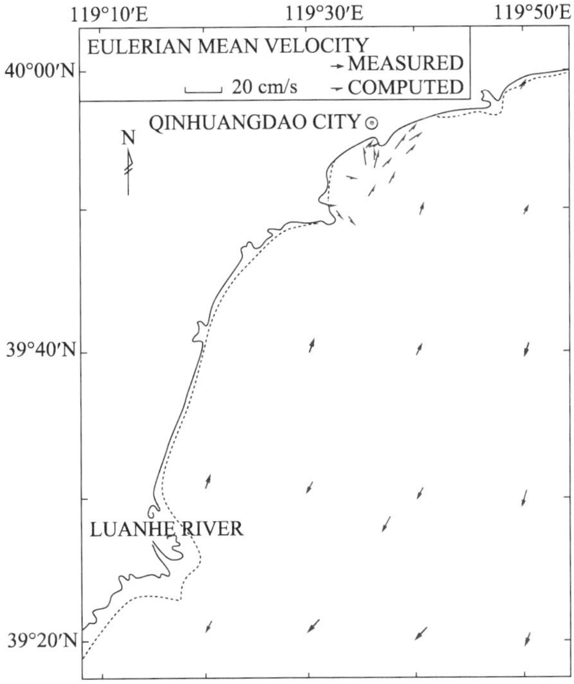

Figure 6 A comparison between the computed (M_2, "K_1")-induced depth-averaged Eulerian mean velocity and field observations

IV.1 An M_2-S_2 nonlinear system

For simplicity, the derivation of the Lagrangian drift velocity induced by an M_2-S_2 nonlinear system is summarized as follows

$$\boldsymbol{U}_{\mathrm{LD}}=-\xi_0(t_0)\cdot\nabla\boldsymbol{U}_{\mathrm{L}}+\boldsymbol{U}_{\mathrm{L}}\cdot\nabla\xi_0(t_0) \tag{24}$$

where

$$\boldsymbol{U}_{\mathrm{L}}={}^{\mathrm{M}}\boldsymbol{U}_{\mathrm{L}}+{}^{\mathrm{S}}\boldsymbol{U}_{\mathrm{L}} \tag{25}$$

$$\xi_0(t_0)={}^{\mathrm{M}}\xi_0(t_0)+{}^{\mathrm{S}}\xi_0(t_0)$$

$${}^{k}\xi_0(t_0)=-\frac{1}{\sigma_k}\,{}^{k}\boldsymbol{U}_0''\cos(\sigma_k t_0)+\frac{1}{\sigma_k}\,{}^{k}\boldsymbol{U}_0'\sin(\sigma_k t_0)$$

where $({}^{k}\boldsymbol{U}_0',{}^{k}\boldsymbol{U}_0'')$ are the harmonic coefficients of the velocity of the k-constituent of astronomical tide, which are inlayed in the expression for the zeroth-order tidal velocity—${}^{k}\boldsymbol{U}_0={}^{k}\boldsymbol{U}_0'\cos(\sigma_k t)+{}^{k}\boldsymbol{U}_0''\sin(\sigma_k t)$; σ is the frequency; and $k=\mathrm{M}$ or $k=\mathrm{S}$ refer to M_2 or S_2 quantities, respectively.

Substituting Eq. (25) into Eq. (24), we have

$$\boldsymbol{U}_{\mathrm{LD}}={}^{\mathrm{M}}\boldsymbol{U}_{\mathrm{LD}}+{}^{\mathrm{S}}\boldsymbol{U}_{\mathrm{LD}}+{}^{\mathrm{I}}\boldsymbol{U}_{\mathrm{LD}} \tag{26}$$

where

$${}^{k}\boldsymbol{U}_{\mathrm{LD}}=-{}^{k}\xi_0(t_0)\cdot\nabla{}^{k}\boldsymbol{U}_{\mathrm{L}}+{}^{k}\boldsymbol{U}_{\mathrm{L}}\cdot\nabla{}^{k}\xi_0(t_0)\;(k=\mathrm{M}\text{ or }k=\mathrm{S})$$

$${}^{\mathrm{I}}\boldsymbol{U}_{\mathrm{LD}}={}^{\mathrm{MS}}\boldsymbol{U}_{\mathrm{LD}}+{}^{\mathrm{SM}}\boldsymbol{U}_{\mathrm{LD}}$$

$${}^{\mathrm{MS}}\boldsymbol{U}_{\mathrm{LD}}=-{}^{\mathrm{M}}\xi_0(t_0)\cdot\nabla{}^{\mathrm{S}}\boldsymbol{U}_{\mathrm{L}}+{}^{\mathrm{S}}\boldsymbol{U}_{\mathrm{L}}\cdot\nabla{}^{\mathrm{M}}\xi_0(t_0)$$

$${}^{\mathrm{SM}}\boldsymbol{U}_{\mathrm{LD}}=-{}^{\mathrm{S}}\xi_0(t_0)\cdot\nabla{}^{\mathrm{M}}\boldsymbol{U}_{\mathrm{L}}+{}^{\mathrm{M}}\boldsymbol{U}_{\mathrm{L}}\cdot\nabla{}^{\mathrm{S}}\xi_0(t_0)$$

Eq. (26) shows that the Lagrangian drift velocity induced by an M_2-S_2 system, $\boldsymbol{U}_{\mathrm{LD}}$, contains not only the sum of the M_2- and S_2-induced Lagrangian drift velocities, ${}^{\mathrm{M}}\boldsymbol{U}_{\mathrm{LD}}+{}^{\mathrm{S}}\boldsymbol{U}_{\mathrm{LD}}$, but also ${}^{\mathrm{I}}\boldsymbol{U}_{\mathrm{LD}}$—a product of nonlinear interaction between M_2 and S_2 terms. Thus, while the mass-transport velocity—the first-order Lagrangian residual velocity expressed in Eq. (13), or Eq. (25), has the additive property, the Lagrangian drift velocity, $\boldsymbol{U}_{\mathrm{LD}}$, is a second-order perturbation of the Lagrangian residual velocity, it cannot be expressed as the sum of the Lagrangian drift velocities due to M_2 and S_2.

Substituting the expression for ${}^{k}\xi_0(t_0)$ into Eq. (24), the Lagrangian drift velocity can be also written as

$$\boldsymbol{U}_{\mathrm{LD}}={}^{\mathrm{M}\sigma}\boldsymbol{U}_{\mathrm{LD}}+{}^{\mathrm{S}\sigma}\boldsymbol{U}_{\mathrm{LD}} \tag{27}$$

where

$${}^{k\sigma}\boldsymbol{U}_{\mathrm{LD}}={}^{k\sigma}\boldsymbol{U}_{\mathrm{LD}}'\cos(\sigma_k t_0)+{}^{k\sigma}\boldsymbol{U}_{\mathrm{LD}}''\sin(\sigma_k t_0),\;(k=\mathrm{M}\text{ or }k=\mathrm{S})$$

$$^{k\sigma}\boldsymbol{U}_{\mathrm{LD}} = {}^{k\sigma}\boldsymbol{U}'_{\mathrm{LD}} + {}^{\mathrm{I}}\boldsymbol{U}'_{\mathrm{LD}},\ {}^{k\sigma}\boldsymbol{U}''_{\mathrm{LD}} = {}^{k\sigma}\boldsymbol{U}''_{\mathrm{LD}} + {}^{\mathrm{I}}\boldsymbol{U}''_{\mathrm{LD}},$$

$$^{k}\boldsymbol{U}'_{\mathrm{LD}} = (1/\sigma_k)\left[{}^{k}\boldsymbol{U}''_0 \cdot \nabla {}^{k}\boldsymbol{U}_{\mathrm{L}} - {}^{k}\boldsymbol{U}_{\mathrm{L}} \cdot \nabla {}^{k}\boldsymbol{U}''_0\right]$$

$$^{k}\boldsymbol{U}''_{\mathrm{LD}} = (1/\sigma_k)\left[-{}^{k}\boldsymbol{U}'_0 \cdot \nabla {}^{k}\boldsymbol{U}_{\mathrm{L}} + {}^{k}\boldsymbol{U}_{\mathrm{L}} \cdot \nabla {}^{k}\boldsymbol{U}'_0\right]$$

$$^{\mathrm{I}}\boldsymbol{U}'_{\mathrm{LD}} = (1/\sigma_k)\left[{}^{k}\boldsymbol{U}''_0 \cdot \nabla {}^{q}\boldsymbol{U}_{\mathrm{L}} - {}^{q}\boldsymbol{U}_{\mathrm{L}} \cdot \nabla {}^{k}\boldsymbol{U}''_0\right]$$

$$^{\mathrm{I}}\boldsymbol{U}''_{\mathrm{LD}} = (1/\sigma_k)\left[-{}^{k}\boldsymbol{U}'_0 \cdot \nabla {}^{q}\boldsymbol{U}_{\mathrm{L}} + {}^{q}\boldsymbol{U}_{\mathrm{L}} \cdot \nabla {}^{k}\boldsymbol{U}'_0\right]$$

in which $q = \mathrm{S}$ if $k = \mathrm{M}$, or $q = \mathrm{M}$ if $k = \mathrm{S}$.

Eq. (27) shows the dependence of the Lagrangian drift velocity on the phases of both M_2 and S_2 when the water parcel is labelled and released. As the initial phases vary from 0 to 2π, the horizontal projections of ${}^{\mathrm{M}\sigma}\boldsymbol{U}_{\mathrm{LD}}$ and ${}^{\mathrm{S}\sigma}\boldsymbol{U}_{\mathrm{LD}}$ trace out a respective ellipse on a hodograph plane. A comparison between Eq. (27) and Eq. (26) shows the difference between ${}^{k\sigma}\boldsymbol{U}_{\mathrm{LD}}$ and ${}^{k}\boldsymbol{U}_{\mathrm{LD}}$ (k = M, S) due to the nonlinear interaction.

IV.2 Generalized theory

Substituting $\boldsymbol{U}(\boldsymbol{X}_0 + \boldsymbol{\xi}, t)$ and Eq. (11) into Eq. (12), the Lagrangian drift velocity, or simply the Lagrangian drift, $\boldsymbol{U}_{\mathrm{LD}}$, induced by an m-frequency tidal system can be derived as

$$\boldsymbol{U}_{\mathrm{LD}} = -\boldsymbol{\xi}_0(t_0) \cdot \nabla \boldsymbol{U}_{\mathrm{L}} + \boldsymbol{U}_{\mathrm{L}} \cdot \nabla \boldsymbol{\xi}_0(t_0) + \boldsymbol{U}'_{\mathrm{L}} \tag{28}$$

where $\boldsymbol{U}'_{\mathrm{L}} = \boldsymbol{U}'_{\mathrm{E}} + \boldsymbol{U}'_{\mathrm{S}}$, and $\boldsymbol{U}'_{\mathrm{L}}$ is independent of t_0 and is the additional term when compared to Eq. (24). $\boldsymbol{U}'_{\mathrm{E}} = \langle \boldsymbol{U}_2(\boldsymbol{X}_0, t)\rangle$, a second-order velocity, is the time average of $\boldsymbol{U}_2$, which contains in general not only the second-order constituents but also the residual motions. Thus there is normally an additional Eulerian Residual velocity, $\boldsymbol{U}'_{\mathrm{E}} = \langle \boldsymbol{U}_2(\boldsymbol{X}_0, t)\rangle$, which is due to the nonlinear couplings between the zeroth- and first-order constituents, and itself is of second order. $\boldsymbol{U}_{\mathrm{S}}$ is composed of three parts, which are respectively extracted from $\langle \boldsymbol{\xi}_0(t) \cdot (\nabla \boldsymbol{U}_1)_0 \rangle$, $\langle \boldsymbol{\xi}_1(t) \cdot (\nabla \boldsymbol{U}_0)_0 \rangle$ and

$$\frac{1}{2}\left\langle \xi_0^2 \left(\frac{\partial^2 \boldsymbol{u}_0}{\partial x^2}\right)_0 + \eta_0^2 \left(\frac{\partial^2 \boldsymbol{u}_0}{\partial y^2}\right)_0 + \lambda_0^2 \left(\frac{\partial^2 \boldsymbol{u}_0}{\partial z^2}\right)_0 + 2\xi_0\eta_0 \left(\frac{\partial^2 \boldsymbol{u}_0}{\partial x \partial y}\right)_0 + 2\xi_0\lambda_0 \left(\frac{\partial^2 \boldsymbol{u}_0}{\partial x \partial z}\right)_0 + 2\eta_0\lambda_0 \left(\frac{\partial^2 \boldsymbol{u}_0}{\partial y \partial z}\right)_0 \right\rangle$$

where $\boldsymbol{\xi}_0 = \boldsymbol{i}\xi_0 + \boldsymbol{j}\eta_0 + \boldsymbol{k}\lambda_0$, and the notation $(\ )_0$ indicates the term is evaluated at $\boldsymbol{X}_0$.

An expanded form of Eq. (28) can be obtained as

$$\boldsymbol{U}_{\mathrm{LD}} = \sum_{k=1}^{m} {}^{k\sigma}\boldsymbol{U}_{\mathrm{LD}} + \boldsymbol{U}'_{\mathrm{K}} + \boldsymbol{U}'_{\mathrm{S}} \tag{29}$$

where

$$^{k\sigma}\boldsymbol{U}_{\mathrm{LD}} = {}^{k\sigma}\boldsymbol{U}_{\mathrm{LD}} \cos(\sigma_k t_0) + {}^{k\sigma}\boldsymbol{U}_{\mathrm{LD}} \sin(\sigma_k t_0)$$

$$^{k\sigma}\boldsymbol{U}_{\mathrm{LD}} = (1/\sigma_k)\left[{}^{k}\boldsymbol{U}_0 \cdot \nabla \boldsymbol{U}_{\mathrm{L}} - \boldsymbol{U}_{\mathrm{L}} \cdot \nabla {}^{k}\boldsymbol{U}''_0\right]$$

$$^{k\sigma}\boldsymbol{U}_{\mathrm{LD}} = (1/\sigma_k)\left[-{}^{k}\boldsymbol{U}_0 \cdot \nabla \boldsymbol{U}_{\mathrm{L}} - \boldsymbol{U}_{\mathrm{L}} \cdot \nabla {}^{k}\boldsymbol{U}_0'\right]$$

Eq. (29) shows the dependence of the Lagrangian drift velocity induced by an m-frequency tidal system on all phases of k-constituent ($k = 1,2,\cdots,m$) when the water parcel is released, and as these phases vary from 0 to 2π, the horizontal projections of $^{k\sigma}\boldsymbol{U}_{\mathrm{LD}}$ ($k= 1,2,\cdots,m$) trace out a respective ellipse on a hodograph plane. It is noted that the Lagrangian drift velocity does not have the additive property, because it involves the nonlinear interactions among the k-constituent ($k = 1,2,\cdots,m$).

IV.3 An example

It has been recognized that the Stokes' drift velocity induced by an M_2-tidal current velocity field without shear and with a constant phase is identically zero, and thus the Lagrangian drift velocity is also zero if the Eulerian residual velocity is supposed to be zero (Feng *et al.*, 1986a). This result seems to suggest that the phase variation of tidal current plays an influential role in the mechanism which generates Lagrangian residual flow. This statement is only valid for a single-frequency oscillatory system.

Consider a coupled M_2-S_2 induced unidirectional ideal flow without shear for which $v = 0$ and $\partial u / \partial y = 0$, and the phase of the M_2-tidal current is further required to be constant, or $^{\mathrm{M}}u_0'' / {}^{\mathrm{M}}u_0' = \tan(\alpha_{\mathrm{M}})$, where α_{M} is the constant phase angle. A direct substitution yields

$$^{\mathrm{M}}u_{\mathrm{L}} = {}^{\mathrm{M}}u_{\mathrm{S}} = 0$$

and

$$^{\mathrm{M}\sigma}U_{\mathrm{LD}} = \frac{\sec\alpha_{\mathrm{M}}}{\sigma_{\mathrm{M}}}\left[{}^{\mathrm{M}}u_0' \frac{\partial\, {}^{\mathrm{S}}u_{\mathrm{L}}}{\partial x} - {}^{\mathrm{S}}u_{\mathrm{L}} \frac{\partial\, {}^{\mathrm{M}}u_0'}{\partial x}\right]\cos(\sigma_{\mathrm{M}} t_0 - \beta_{\mathrm{M}})$$

where $\beta_{\mathrm{M}} = -\cot a_{\mathrm{M}}$, and to avoid confusing the main issues, $^{\mathrm{M}}U_{\mathrm{E}}$ has been supposed to be zero.

It becomes immediately obvious that although the tidal current velocity shear is zero and the phase of M_2-tidal current is constant, $^{\mathrm{M}\sigma}U_{\mathrm{LD}}$ is an ellipse on a hodograph plane, or more precisely, a straight line, or a degenerate ellipse. This Lagrangian residual ellipse and M_2-tidal current ellipse are 90° out of phase.

V. Conclusions

The Lagrangian residual velocity and associated intertidal transport processes in a three-dimensional, weakly nonlinear, multi-frequency oscillatory, baroclinic shallow water system have been examined and analyzed using a perturbation method. The first-order dynamics show that the mass-transport velocity is composed of a tide-induced component, a wind-driven barotropic component and a density-driven baroclinic component, and the tide-induced mass-transport velocity is the sum of the mass-transport velocities by the respective constituents of astronomical tide contained in the multi-frequency tidal system. A set of generalized field equations governing the mass-transport velocity and the zeroth-order apparent concentration is presented and used to describe the long-term mean circulation and transport processes in coastal seas and tidal estuaries. The present theory has been applied to the Bohai Sea, China with the aid of numerical solutions of the model equations. A comparison between the model and the field observations implies that the tide-induced residual circulation in the Bohai Sea resembles more closely to an M_2-K_1 tidal system than to an M_2-tidal system. The second-order dynamics show that the Lagrangian drift velocity induced by a multi-frequency tidal system involves nonlinear coupling terms between the products of the respective constitutents of tides. The dependence on the periodicities of all the astronomical tidal constituents contained in the multi-frequency oscillatory system show up in the Lagrangian residual ellipses on a hodograph plane.

References

Backhaus, J. O., 1985: A Three-dimensional model for the simulation of shelf sea dynamics. Dt. Hydrogr. Z., 38, 185–187.

Cheng, R. T., and V. Casulli, 1982: On Lagrangian residual currents with applications in South San Francisco Bay, California. Water Resour. Res., 18, 1652–1662.

Cheng, R. T., S. Feng, and P. Xi, 1986: On Lagrangian residual ellipse, In: Intern. Conf. on Physics of Shallow Estuaries and Bays, J. van de Kreeke (Editor), Lecture Notes on Coastal and Estuarine Studies, 16, 102–113, Springer-Verlag.

Cheng, R. T., S. Feng, and P. Xi, 1989: On inter-tidal transport equation, In: Estuarine Circulation, B. J. Neilson, J. Brubaker, and A. Kuo (Editors), Humana Press, Inc., 133–156.

Feng, S., 1977: A three-dimensional nonlinear model of tides. Scientia Sinica, Vol. 20, 4, 436–446.

Feng, S., 1986: A three-dimensional weakly nonlinear dynamics on tide-induced Lagrangian residual current and mass-transport. Chinese J. of Oceanology and Limnology, Vol. 4, 2, 139–158.

Feng, S., 1987: A three-dimensional weakly nonlinear model of tide-induced Lagrangian residual current and mass-transport, with an application to the Bohai Sea, In: Three-dimensional Models of Marine and Estuarine Dynamics, J. C. J. Nihoul and B. M. Jamart (Editors), Elsevier Oceanography Series, 45, 471–488, Elsevier.

Feng, S., 1988a: Fundamentals of hydrodynamics of shallow water circulation and transport, In: Proc. of Conf. on Engineering Mechanics and Termophysics, 199–208, Qinghua Univeristy Press (in Chinese).

Feng, S., 1988b: The focus of marine environment hydrodynamic research, J. of Shandong College of Oceanology, Vol. 18, 2 (2), 1–7 (in Chinese).

Feng, S., R. T. Cheng, and P. Xi, 1986a: On tide-induced Lagrangian residual current and residual transport, Part I, Residual current, Water Resour. Res., Vol. 22, 12, 1623–1634.

Feng, S., R. T. Cheng, and P. Xi, 1986b: On tide-induced Lagrangian residual current and residual transport, Part II, Residual transport with application in South San Francisco Bay, California, Water Resour. Res.,

Vol. 22, 12, 1635–1646.

Feng, S., Cheng R. T., Sun, W., Xi, P. and L. Song, 1990: Lagrangian residual current and longterm transport processes in a weakly nonlinear Baroclinic system, In: Physics of Shallow Seas, H. Wang, J Wang and H. Dai(Editors), Beijing: China Ocean Press, 1–20.

Feng, S., P. Xi, and S. Zhang, 1984: The baroclinic residual circulation in shallow seas. Chinese J. of Oceanology and Limnology. Vol. 2, 1, 49–60.

Fofonoff, N. P., 1962: Dynamics of ocean currents, In: M. N. Hill (Editor), The Sea, Vol. I, Physical Oceanography, Section III, Interscience Publishers, 323–396.

Hamrick, J. M., 1987: Time averaged estuarine mass transport equations, Hydraulic Engineering, In: R. M. Ragan (Editor), Proc. of the 1987 National Conf. on Hydraulic Engineering, 624–629, ASCE.

Hansen, W., 1962: Tides, In: M. N. Hill (Editor), The Sea, vol. I, Physical Oceanography. Interscience Publishers, CH. 23, Interscience, 764–801.

Ianniello, J. P., 1979: Tidally induced residual currents in estuaries of variable breadth and depth. J. Phys. Oceanogr., 9, 963–974.

Longuet-Higgins, M. S., 1969: On the transport of mass by Time-Varying Ocean currents. Deep Sea Res., 16, 431–447.

Moore, D., 1970: The mass transport velocity induced by free oscillations at a single frequency. Geophysical Fluid Dynamics, 1, 237–247.

Nihoul, J. C. J., 1975: Modeling of Marine System, Elsevier, 272 pp.

Ramming, H. G., and Z. Kowalik, 1980: Numerical Modeling of Marine Hydrodynamics, Elsevier Oceanography Series, Vol. 26, Elsevier.

Song, L., 1987: A hydrodynamic velocity-splitting model with depth-varying eddy viscosity in shallow seas. Acta Oceanologica Sinica, vol. 6, Supp. I, 135–151.

Sun, W., P. Xi, and L. Song, 1989: Numerical calculation of the three-dimensional tide-induced Lagrangian residual circulation in the Bohai Sea. J. of Ocean University of Qingdao, vol. 19, 2(1), 27–36.

Tee, K. T., 1976: Tide-induced residual current, a 2-d numerical nonlinear tidal model. J. Mar. Res., 34, 603–628.

Wei, G., 1988: On Lagrangian Residual Currents and Long-term Transports by Tides at Both Frequency, M. S. thesis, Ocean Univ. Press (in Chinese), 54 pp.

Zheng, L., 1988: A Numerical Study on the Three-dimensional Hydrodynamic Equations for the Mass-transport Velocity, with an Application to the Bohai Sea, M. S. thesis, Ocean Univ. Press (in Chinese), 74 pp.

Zimmerman, J. T. F., 1979: On the Euler-Lagrangian transformation and the Stokes drift in the presence of oscillatory and residual currents. Deep Sea Res., 26A, 505–520.

The Dynamics on Tidal Generation of Residual Vorticity*

Feng Shi-zuo (冯士筰)

The vorticity analysis is fundamental to a more thorough understanding of the ocean circulation dynamics.

The problem of circulation in a tidally dominant shallow sea, bay, or estuary should be described by the Lagrangian mean velocity of a water parcel, i.e. by the velocity averaged by following the water parcel over the tidal cycles, since it is closely related to the description of long-term transport processes. In a weakly nonlinear baroclinic shallow sea, the mass-transport velocity can appropriately be used to embody the lowest order velocity field of shallow sea circulation[1, 2]. The study on the vorticity dynamics for the mass-transport velocity is of evident interest.

A nonlinearly coupled thermo-hydrodynamic set of field equations governing the mass-transport velocity, residual elevation and the apparent concentration in a weakly nonlinear, baroclinic, three-dimensional system dominated by the multi-frequency oscillations, has recently been derived[3]. In the present note, based on this set of field equations, a vorticity equation for the depth-averaged mass-transport velocity is further derived, and then used to show a residual vorticity budget, particularly to reveal the dynamics on tidal generation of residual vorticity.

The vorticity equation is

$$
\begin{aligned}
r\omega = f\nabla h\cdot\boldsymbol{u}+\boldsymbol{k}\cdot\Bigg\{&\frac{r}{h}\nabla h\times\boldsymbol{u}-\frac{1}{2}g\beta h\nabla h\times\nabla C+\frac{g\beta}{h}\nabla h\times\nabla\int_{-h}^{0}\int_{z}^{0}c'\mathrm{d}z'\mathrm{d}z \\
&+\nabla\times\boldsymbol{\tau}_a-\frac{1}{h}\nabla h\times\boldsymbol{\tau}_a+\nabla\times\boldsymbol{\Pi}-\frac{1}{h}\nabla h\times\boldsymbol{\Pi}\Bigg\}
\end{aligned}
\tag{1}
$$

where

$$
\boldsymbol{\Pi}=\boldsymbol{\Pi}_{\mathrm{N}}+\boldsymbol{\Pi}_{\mathrm{b}}+\boldsymbol{\Pi}_{v} \tag{2}
$$

$$
\boldsymbol{\Pi}_i=\sum_{k=1}^{m}{}^{k}\boldsymbol{\Pi}_i \quad (i=\mathrm{N}, \mathrm{b}, \text{ or } v)
$$

$$
{}^{k}\boldsymbol{\Pi}_{\mathrm{N}}=-g\nabla\int_{-h}^{0}\left\langle\frac{1}{2}{}^{k}\xi_0\cdot\nabla{}^{k}Z_0\right\rangle\mathrm{d}z
$$

$$
{}^{k}\boldsymbol{\Pi}_{\mathrm{b}}=\boldsymbol{i}\,{}^{k}\Pi_{\mathrm{b}x}+\boldsymbol{j}\,{}^{k}\Pi_{\mathrm{b}y}
$$

---

* Feng S. 1991. The dynamics on tidal generation of residual vorticity. Chinese Science Bulletin, 36(24): 2043–2046

$$ {}^{k}\Pi_{bx} = \left\langle \left(\frac{5}{2}\frac{\partial {}^{k}\xi_0}{\partial x} + \frac{\partial {}^{k}\eta_0}{\partial y} \right) \nu \frac{\partial {}^{k}u_0}{\partial z} + \left(\frac{\partial {}^{k}\xi_0}{\partial y} + \frac{1}{2}\frac{\partial {}^{k}\eta_0}{\partial x} \right) \nu \frac{\partial {}^{k}v_0}{\partial z} \right\rangle_{z=-h} $$

$$ {}^{k}\Pi_{by} = \left\langle \left(\frac{5}{2}\frac{\partial {}^{k}\eta_0}{\partial y} + \frac{\partial {}^{k}\xi_0}{\partial x} \right) \nu \frac{\partial {}^{k}v_0}{\partial z} + \left(\frac{\partial {}^{k}\eta_0}{\partial x} + \frac{1}{2}\frac{\partial {}^{k}\xi_0}{\partial y} \right) \nu \frac{\partial {}^{k}u_0}{\partial z} \right\rangle_{z=-h} $$

$$ {}^{k}\boldsymbol{\Pi}_{\text{v}} = \boldsymbol{i}\,{}^{k}\Pi_{\text{v}x} + \boldsymbol{j}\,{}^{k}\Pi_{\text{v}y} $$

$$ {}^{k}\Pi_{\text{v}x} = -\int_{-h}^{0} \left\langle \frac{\partial}{\partial z} {}^{k}\xi_0 \cdot \nabla \left(\nu \frac{\partial {}^{k}u_0}{\partial z} \right) + \frac{\partial}{\partial \zeta} {}^{k}\zeta_0 \frac{\partial}{\partial z} {}^{k}\zeta_0 \frac{\partial}{\partial \zeta} \left(\nu \frac{\partial {}^{k}u_0}{\partial z} \right) + \frac{1}{2} \left[\frac{\partial^2 {}^{k}\xi_0}{\partial z \partial x} \left(\nu \frac{\partial {}^{k}u_0}{\partial z} \right) \right.\right. $$
$$ \left.\left. + \frac{\partial^2 {}^{k}\eta_0}{\partial z \partial x} \left(\nu \frac{\partial {}^{k}v_0}{\partial z} \right) + {}^{k}\xi_0 \frac{\partial^2}{\partial z \partial x} \left(\nu \frac{\partial {}^{k}u_0}{\partial z} \right) + {}^{k}\eta_0 \frac{\partial^2}{\partial z \partial x} \left(\nu \frac{\partial {}^{k}v_0}{\partial z} \right) \right] \right\rangle \mathrm{d}z $$

$$ {}^{k}\Pi_{\text{v}y} = -\int_{-h}^{0} \left\langle \frac{\partial}{\partial z} {}^{k}\boldsymbol{\xi} \cdot \nabla \left(\nu \frac{\partial {}^{k}v_0}{\partial z} \right) + \frac{\partial}{\partial z} {}^{k}\zeta_0 \frac{\partial}{\partial z} \left(\nu \frac{\partial {}^{k}v_0}{\partial z} \right) + \frac{1}{2} \left[\frac{\partial^2 {}^{k}\xi_0}{\partial z \partial y} \left(\nu \frac{\partial {}^{k}u_0}{\partial z} \right) \right.\right. $$
$$ \left.\left. + \frac{\partial^2 {}^{k}\eta_0}{\partial z \partial y} \left(\nu \frac{\partial {}^{k}v_0}{\partial z} \right) + {}^{k}\xi_0 \frac{\partial^2}{\partial z \partial y} \left(\nu \frac{\partial {}^{k}u_0}{\partial z} \right) + {}^{k}\eta_0 \frac{\partial^2}{\partial z \partial y} \left(\nu \frac{\partial {}^{k}v_0}{\partial z} \right) \right] \right\rangle \mathrm{d}z $$

$\boldsymbol{u} = \boldsymbol{i}u + \boldsymbol{j}v$ denotes the depth-averaged mass-transport velocity; $\omega = \boldsymbol{k} \cdot \nabla \times \boldsymbol{u}$ is the vertical component of relative vorticity of $\boldsymbol{u}$, or the residual vorticity; $\nabla = \boldsymbol{i}\partial/\partial x + \boldsymbol{j}\partial/\partial y$ denotes a horizontal vector operator; (x,y,z) constitute Cartesian coordinates on an f-plane, $(\boldsymbol{i}, \boldsymbol{j}, \boldsymbol{k})$ are coordinate unit vectors, $\boldsymbol{k}$ points upwards, and the origin of coordinates is located at the undisturbed sea level; $\times$ and $\cdot$ denote respectively the vector and scalar multiplication; r is a damping coefficient in the linear law of bottom stress; f is the Coriolis parameter, and is supposed to be a constant for estuaries, bays, or coastal seas; h is the water depth measured from the undisturbed sea level; g is the gravitational acceleration; β is the fractional increase in density per unit increase in salinity in a generalized linear equation of state; $(c + c')$ is the time-averaged apparent concentration, where c is its depth-averaged quantity and c' is a deviation produced due to the vertical shear of the concentration; $\boldsymbol{\tau}_a$ is the wind stress at the sea surface; $\boldsymbol{\Pi}$ is an integral along the water depth of the "tidal body force" [3], and can relevantly be termed as the "tidal stress" (note : this "tidal stress" introduced here is entirely different from the "tidal stress" derived and termed by Nihoul and Ronday (1975)[4]); k $(1,2,\cdots,m)$ indicate the corresponding quantities induced by the k-constituent of astronomical tide containing m constituents; $({}^{k}u_0, {}^{k}v_0)$ are tidal currents. ${}^{k}Z_0$ is the tidal elevation, and ${}^{k}\xi_0 = \boldsymbol{i}\,{}^{k}\xi_0 + \boldsymbol{j}\,{}^{k}\eta_0$ and ${}^{k}\zeta_0$ are respectively the eddy viscosities, and supposed to be the function of spacial coordinated; $< >$ denotes a time-mean operator.

Eq. (l) expresses a vorticity budget of shallow sea circulation. The meaning of every term in Eq. (l), except the terms containing the tidal stress $\boldsymbol{\Pi}$, has been explained by means of the traditional analysis on the vorticity based on the vorticity equation for wind-driven and thermohaline circulation in shallow seas. The tidal stress is a link of transferring the vorticity from the tidal field to the residual circulation. The expression of $\boldsymbol{\Pi}$, (2), implies the three kinds of mechanism of tidal-generation of residual vorticity, which are discussed as follows.

$\boldsymbol{\Pi}_{\mathrm{N}}$ represents the cumulative effect of the nonlinear coupling between the horizontal gradient of elevation of astronomical tide and the zeroth order displacements of water parcels in the tidal field, and is an irrotational force, i.e. $\nabla\times\boldsymbol{\Pi}_{\mathrm{N}}=0$. Thus, this component of tidal stress can generate the residual vorticity only through the interaction between itself and the bottom fluctuation, in other words, through the term $(-h^{-1}\nabla h\times\boldsymbol{\Pi}_{\mathrm{N}})\cdot\boldsymbol{k}$, which is similar to the mechanism of residual vorticity generation by the baroclinic force—the "bottom-baroclinic effect".

$\boldsymbol{\Pi}_{\mathrm{b}}$ represents the cumulative effect of the nonlinear coupling between the bottom fricton due to tidal currents and the spatial inhomogeneity of the zeroth order displacements of water parcels near the sea bottom, and is a rotational component of tidal stress. Its contribution to the residual vorticity, in general, is expressed through the last two terms at the right-hand side of Eq. (1), $(\nabla\times\boldsymbol{\Pi}_{\mathrm{b}})\cdot\boldsymbol{k}$ and $(-h^{-4}\nabla h\times\boldsymbol{\Pi}_{\mathrm{b}})\cdot\boldsymbol{k}$. However, it should be stressed that, in a flat-bottomed sea, not only $(-h^{-1}\nabla h\times\boldsymbol{\Pi}_{\mathrm{b}})=0$, but also $\nabla\times\boldsymbol{\Pi}_{\mathrm{b}}=0$ since $\Pi_{\mathrm{b}}=0$ as $h=\mathrm{const}$. In fact, from the viewpoint of dynamics, water parcels on a sea bottom cannot move along the sea bottom due to the no-slip boundary condition, or the displacements of those water parcels are identically zero; thus, $\xi_0=\partial\xi_0/\partial x=\partial\xi_0/\partial y=0$ at $z=-h$ as $h=\mathrm{const}$. Therefore, the contribution of $\boldsymbol{\Pi}_{\mathrm{b}}$ to the generation of the residual vorticity is essentially through the interaction between itself and the bottom fluctuation.

$\boldsymbol{\Pi}_{\mathrm{v}}$ is another rotational component of tidal stress, and is closely related to the vertical shear of tidal current and the eddy viscosity. It represents the cumulative effect of the nonlinear coupling between the spacially inhomogeneous turbulent stress due to tidal currents and the zeroth order displacements of water parcels. Differing from the irrotational component of tidal stress $\boldsymbol{\Pi}_{\mathrm{N}}$, $\boldsymbol{\Pi}_{\mathrm{v}}$ gives the contribution to the generation of residual vorticity through both of $(\nabla\times\boldsymbol{\Pi}_{\mathrm{v}})\cdot\boldsymbol{k}$ and $(-h^{-1}\nabla h\times\boldsymbol{\Pi}_{\mathrm{v}})\cdot\boldsymbol{k}$. Also differing from the rotational component of tidal stress $\boldsymbol{\Pi}_{\mathrm{b}}$, $\boldsymbol{\Pi}_{\mathrm{v}}$ generates the residual vorticity both formally and essentially through the last two terms at the right-hand side of Eq. (1), which is similar to the mechanism of residual vorticity generation by the wind stress (note the corresponding terms containing the wind stress in Eq. (1), or the fifth and sixth terms at the right-hand side). It is important to point out that the tide-induced residual vorticity in a flat bottomed sea can be generated and then maintained by the vertical inhomogeneity of tidal currents through $\nabla\times\boldsymbol{\Pi}_{\mathrm{v}}\neq 0$. This is a typical conclusion about the residual vorticity generation by a three-dimensional tidal model.

It is of much interest to note that, all the terms in $\boldsymbol{\Pi}$ do involve the displacements of water parcels in the tidal field, which implies that the generation mechanisms of tide- induced residual vorticity are really dependent on the tidal Lagrangian dynamics.

References

[1] 冯士筰, 清华大学工程力学与工程热物理学术会议文集, 清华大学出版社, 北京, 1988, pp. 199–208.

[2] Feng, Shizuo, Acta Oceanologica Sinica, 9(1990), 3: 315.

[3] Feng, Shizuo, Lecture Notes on Coastal and Estuarine Studies, 38, Springer-Verlag, New York, 1990, pp. 34–48.

[4] Nihoul, J. C. J. & Ronday, F. C., Tellus, 27(1975), 484.

第九章 浅海环流物理及数值模拟*

冯士筰

海洋环流问题是海洋科学中最重要的基础课题之一，是物理海洋学研究的核心。

众所周知，全球大洋环流的数值模拟及其相伴的气候问题已构成研究的一条热线。与此同时，随着近岸海洋开发的日益增长，近海环境问题越来越受到人们的关注，因此，有关浅海环流及其输运的问题已成为当前浅海物理海洋学特别是浅海环境动力学研究的焦点。

事实上，一个河口或一个海湾，都可以视为一个生态系统，对其中生物的、化学的、地质的和地球物理的过程及其与环流的相互作用进行统一探讨。在这样一个正在逐渐形成中的交叉学科的发展中，浅海环流问题的探讨是最基础的课题。由此可见，本文所涉及的研究内容，不仅具有现实的应用意义和动力学的理论意义，而且也应对环境科学这一综合学科的发展作出贡献。

9.1 研究背景

海洋环流通常系指海洋中平均的“气候式”（准）定常流动。像在渤海或东中国海这样一些半封闭水域或陆架浅海及其附属的海湾或河口中，最显著的经久不息的运动是伴随着天文潮波的周期性潮流。为了基本上消除潮流的主要振动分量从而获得一种（准）定常流，通常借助于对流场空间任一选定点上的流速资料进行低通滤波，或简单地作一个潮周期上的时间平均。这种潮周期平均速度称作 Euler 平均速度，或 Euler 余流，它常被用来作为在这样一些边缘浅海中海洋环流的一种体现。以 Euler 余流来体现的这种浅海环流，也可称为 Euler 余环流。

在浅海环流动力学的研究史上，最原始的观点认为：浅海环流能够简单地被一个风生–热盐环流的定常模型所描述。这实际上意味着，对流速资料的潮平均已把潮周期运动及其影响全部消除，剩余流动仅被平均的风场和密度场的水平梯度所驱动。其控制方程组可参看文献[1]，基本特点即为一组斜压浅海定常方程。直至 70 年代中期，Csanady（1975[2], 1976[3]）与 Nihoul 和 Ronday（1975[4]）分别指出了上述经典方程组的不完备性。因为在边缘浅海中，显著的潮运动表现了一定程度的非线性，故天文潮

* 冯士筰. 1992. 第九章 浅海环流物理及数值模拟. 见：冯士筰，孙文心 主编. 科学与工程计算丛书——物理海洋数值计算. 郑州：河南科学技术出版社. 543–610

运动必将与环流耦合，占优势的天文潮自身的非线性耦合的潮周期平均效应可以产生所谓的“潮应力”，它像海面平均风应力那样作用于 Euler 余环流[4]。Heaps（1978）就深度积分的全流模型给出了这种方程组的一个系统推导和论述[5]。若把这种观点用于三维浅海 Euler 余环流的动力学模型，就会发现，不仅存在海面上的“潮应力”，还存在“海面潮源”和“潮致体力”[6]。结论是：由于浅海系统的非线性，其环流的强迫力，不仅有海面风应力和热盐因素，还应包含潮汐导致的非线性耦合效应。但遗憾的是，以这种 Euler 余流来体现浅海环流的观点已受到质疑[7~9]。

事实上，环流问题与水温、盐度这样一些重要的海水特性，与营养盐、沉积物或污染物质等浓度的长期输运过程以及浮游生物、鱼卵等的迁移和分布规律的描述和确定都有着密切的联系。而且人们越来越认识到，这些物质、动量和能量的输运过程主要并不取决于该 Euler 余流，而是依赖于在流场中运动着的流体微团经过一个或多个潮周期后所导致的净位移，或（等价地）依赖于正比于该净位移的 Lagrange 平均速度——即在上述潮周期时段内对该流体微团的随体速度进行的平均[10~14]。由此可见，浅海环流的概念及其输运过程的描述紧密地依赖于上述的 Lagrange 净位移或 Lagrange 平均速度，而不是 Euler 余流——Euler 平均速度。

由 Lagrange 平均速度可以产生所谓“Lagrange 余流”的概念，在这方面，真正形成一股研究势头不过是近 10 年的事情，事实上，直至 80 年代中期，还需要阐明两种余流概念间的差异[13, 14]。当然，与这种 Lagrange 平均速度有关的研究可以上溯到 60 年代末 Longuet-Higgins（1969）的一篇文章[10]，在文中，他把有限振幅表面波中的物质输运速度（mass transport velocity）和 Stokes 公式推广到海洋中大尺度运动的问题中，并以双 Kelvin 波为例作了详细分析。70 年代中期，Tee（1976）的工作[15]可能是在文献中所见到的试图依据 Stokes 公式作浅海余流数值模拟的首次努力，同时，也正是在这篇论文中使用了 Lagrange 余流这一术语。利用 Stokes 公式数值计算和解析研究感潮河口或海湾中的物质输运速度的论文，还可以举出若干例子[16~20]。应该指出，真正开始系统论述 Lagrange 余流概念和计算方法的论文当推 Zimmerman（1979）[11]与 Cheng 和 Casulli（1982）[12]。Zimmerman 在指出了物质输运速度仅为一个 Euler-Lagrange 双重变换的一阶近似以后，进一步在原则上给出了一个计算 Lagrange 平均速度的方法[11]。Cheng 和 Casulli 利用“（标识）微团追踪法”（Parcel-tracking method）对旧金山南湾的 Lagrange 余流进行了数值计算[12]。与其相类似的计算还见于 Awaji（1982）[21]和 Awaji 等人（1980）[22]以及俞光耀和陈时俊（1983）[23]的论文中。其实，从流体动力学的观点看，“在一个潮流场中求出任一（标识）流体微团的轨迹，从而给出经过一个潮周期或多个潮周期后该流体微团的净位移或等价的 Lagrange 平均速度”，即为上述微团追踪法的原理[24]。应当指出，利用 Stokes 公式计算物质输运速度与以微团追踪方法计算

Lagrange 平均速度这两种途径一直并行至今[24]。

笔者及其合作者自 80 年代中期开始，发表了一系列论文，建立了有关 Lagrange 余流及其输运过程的一种弱非线性理论[13, 14, 25, 26, 8, 24, 9, 27~33]。其有关贡献叙述如下。

“Lagrange 余流”一直与“Lagrange 平均速度”等价使用，也就是说，文献中的英文术语“Lagrange residual velocity (Lagrangian residual current)”一直与“Lagrangian mean velocity”在等价的意义上使用。其实，如前所述，Lagrange 平均速度正比于在流场中释放的标识流体微团经过一个或多个潮周期后所导致的净位移，因此它是一个典型的 Lagrange 量。一般说，它不仅是标识流体微团被释放时所占据的空间点的函数，也是该流体微团释放时潮位相的函数[34]，并且它还依赖于用以平均的潮周期数[26]。一般说，Lagrange 平均速度不能构成一个不可压缩流场，亦即不满足其散度为零的条件。因此，笔者建议区分“Lagrange 平均速度”与“Lagrange 余流”这两个术语，且将 Lagrange 余流定义为满足散度为零的 Lagrange 平均速度[8]。本章将在这一规定下使用这两个术语。应当指出，在我们所提出的弱非线性系统中导出的 Lagrange 平均速度能够用来定义 Lagrange 余流，因为它为一管量场。

物质输运速度被导出为最低阶的 Lagrange 余流，并且为一定常场[13]。一个有趣之点是：Euler 余流虽为管量场，但并不满足流场中流体微团组成的物质面守恒的条件，如在海面上存在所谓“海面潮源”以及在侧面固体边界上出现了“源”和“汇”[6, 35]。而物质输运速度不仅为一管量场，而且可在最一般的意义上证明满足上述流场中的物质面守恒方程[8]。由此可见，物质输运速度与 Euler 余流相较，更有资格构成一“流场”，也就是说，即使只以此连续性质之优劣来衡量，以物质输运速度而不是以传统的 Euler 余流来体现环流的流场也更为合理[9]。事实上，也许正像 Csanady 所指出的那样，Euler 平均速度除了表示一个平均数以外，不再有更丰富的内涵[7]。

物质输运速度的这一良好性质在伴随环流问题的长期物质浓度输运过程研究中得到了进一步的确证。物质浓度的潮周期平均场满足的对流–扩散方程也称为“长期输运方程”（long-term transport equation），这一方程是海洋环境问题中关键性的方程[36, 37]。经典的长期输运方程存在两个问题：对流速度为 Euler 余流；存在一个所谓的“潮弥散”（tidal dispersion）。第一，既然如上所述，决定水微团长期输运的是 Lagrange 平均速度而非 Euler 平均速度，那么，在长期输运方程中对流速度为何不是 Lagrange 平均速度却是 Euler 平均速度？第二，“潮弥散”项的物理意义是什么？不同的研究者曾设想了不同的机制[38, 39]，而某些研究对此则干脆略而不计[40, 41, 36]。80 年代中期，一个全新的长期输运方程被导出[14]，随后被推广[26, 8, 42, 9]，该方程的特征为：对流速度为物质输运速度，并且不存在附加的“潮弥散”项。应指出，这一长期输运方程，不仅控制了平均浓度，也控制了像温度或盐度这样一些量的平均场——它们在斜压浅海或河口中往往与环流的速度场构成一个非线性耦合的系统[9]。

综上所述，以最低阶的 Lagrange 余流——物质输运速度构成的速度场来描述浅海环流问题，至少在最低阶的水平上有实际意义，可称之为“Lagrange 余环流”。本章即着重介绍这一内容，并加以展开和讨论。

应强调指出，涉及利用 Lagrange 余环流（或更一般地，导出 Lagrange 平均速度）来描述浅海环流问题，其原始的浅海动力学定解问题原则上应严格在三维空间来处理[8]。众所周知，在浅海或河口中处理潮和风暴潮这样一些流体动力学问题或物质输运问题时，经常采用深度平均、宽度平均或截面平均等模型[参看 Heaps（1978）[5]，Gerritsen（1986）[43]，Prandle（1986）[20]，Huang 等（1986）[16]]。只要我们所关心的是场量的空间平均值或积分值，上述空间平均模型就是相对简单而有效的，原则上不存在动力学上的疑难问题。但是，对于涉及 Lagrange 余流的原始问题，上面这种空间平均模型，一般说来，则是违反 Lagrange 动力学原则的。譬如，在一个深度平均的二维潮流场中，被释放的标识流体微团应意味着一个自海底至海面的铅垂水柱微元；一般地说，在释放该水柱时刻组成该水柱的流体微团，在释放时刻以后，由于这些微团的速度大小和方向皆可能不同，从而该水柱微元将可能“四分五裂”、溃不成体，故此时的 Lagrange 位移的动力学意义就不存在了！何来 Lagrange 余流？也就是说，除非把潮波问题处理成理想流体中的长波运动[44]，才能保持运动着的流体柱元的铅垂“刚性”[45]，从而在一个流速与深度无关的二维潮流场中释放一个铅垂流体柱元并形成其 Lagrange 余流，这才有确定的动力学意义。但在河口或陆架浅海动力学中，由于海水层化、内部湍流以及底摩擦等效应的存在，致使运动水柱微元的铅垂“刚性”难以满足。因此，一般说，对于涉及 Lagrange 余流或更普遍地涉及 Lagrange 平均速度的问题作三维空间模型的研究，就不只是在通常的意义下为了更广泛的应用上的需要，而首先是由于动力学分析所必需[26, 8]。显然，问题的这种“空间三维性”给解析研究和数值模拟都带来了困难。

前面已经提到，可以利用 Stokes 公式得到物质输运速度，通常是先求解 Euler 平均速度，再附加上一个 Stokes 漂移速度[10]。因为只要求得潮流问题的解，Stokes 漂移速度的获得便迎刃而解，故这一方法的关键是求解 Euler 平均速度。对于充分混合的浅海或感潮河口，问题往往处理为正压的深度平均的二维模型，其求解 Euler 平均速度的原始方法是先解一个潮的非线性问题，再对潮流作一潮周期平均[46, 15, 47~54, 12, 4]。但 Nihoul 和 Ronday（1975）对此方法的精度提出了质疑，并给出了前面提到的先导出 Euler 余流满足的动力学方程组，然后再求解该方程组以得到 Euler 平均速度的途径[4]。两种方法的系统比较可参看文献[43]。尽管在精度比较上没得到什么肯定的结论，但对于弱非线性系统而言，后一种途径具有可将潮致余流与风生余流线性叠加的明显优点。因为物质输运速度导源于一个浅海的弱非线性系统，故下面涉及以物质输运速度来体现浅海环流的讨论将仅限于弱非线性问题，且只沿着上述后一种途径进行下去。我们更感兴趣的控制浅海 Euler 余流的三维方程组已被导出，它对于正压或斜压诊断模型是有效的[6]，虽然并不是理想的[24]。特别应指出，对于斜压非线性耦合问题，这一途径将是无效的，至少是苍白无力的[55]。这里指的是：一个控制 Euler 余流的运动方程如何与被非线性耦合的控制物质输运速度和温、盐潮周期平均场的长期输运方程联立求解呢？事实上，为了探讨以最低阶的 Lagrange 余流——物质输运速度来描述浅海或感潮河口的环流问题，建立其自身的封闭方程组和完整的定解问题是最基础的工作。笔者及其合作者，基于笔者 1987 年首次提出的控制潮致浅海 Lagrange 余环流的运动方程组[8]，建立了斜压浅海系统 Lagrange 余环流的定解问题[9, 31]，并推广至普遍情形下的广义方程组[33]。这就奠定了这项研究的动力学基础。

作为这一环流理论的最初步的试验和应用，笔者及其合作者针对渤海环流，依据上述 Lagrange 余环流方程组，利用“速度分解法”[56~58]作了数值实验，获得了若干具有启发性且令人鼓舞的结果[27, 28, 30, 32]。

但也应该指出，以物质输运速度构成的速度场来描述浅海环流问题，仅是一种最低阶的近似。事实上，以 Lagrange 余流来描述浅海或感潮河口中的环流问题具有更广泛和丰富的内涵[33]，何况在更一般的意义上，它应该以 Lagrange 平均速度这一更普遍的概念来描述。这些内容本章也作了简单讨论。

9.2 斜压浅海系统

海水运动具有多种时空尺度，其中何种尺度是重要的且为人们所感兴趣，则取决于我们所提出的特定问题。而正确选择解决这一特定问题的物理模型和适当地搜集海上观测资料，又与上述这些我们感兴趣的时空尺度紧密相关。伴随长期输运过程的浅海环流问题的时间尺度，应包括半日潮或日潮所具有的中时间尺度和长期生态变化过程所对应的大时间尺度，即相应于时间频谱上的惯性频率及其低频段。应强调指出，与经典的环流问题不同，正如 9.1 节所述，我们必须考虑的不仅是长期效应的环流尺度本身，还应考虑它与中时间尺度运动潮汐的非线性耦合效应。因为在一个海湾或边缘浅海中，环流流场波及整个海域并深达海底，故其空间尺度在水平方向上可相当于海域的幅员，而在铅垂方向上可相当于平均水深。具有这种时空尺度的浅海环流动力学，显然必须考虑地球旋转所产生的柯氏力和海盆地形对运动的影响。一般说来，运动的时空尺度取决于相应的外力和（或）系统的固有性质，在一个斜压浅海系统中，在我们所感兴趣的上述时空尺度范围内，一般说来，潮汐运动是经久不息的占优势的运动，且半日太阴分潮 M_2 可以作为其代表。尽管风暴潮或海啸具有不低于天文潮波的量级，但由于它们的偶发性，对环流问题将没有多少贡献。海面上的持续风场，或其月、季平均效应，虽然在量级上小于风暴，但一般对浅海中的长期环流却有贡献。海–气界面热交换的日、季变化有时也影响环流。在感潮河口或邻近河口的海域中，由于淡水侵入形成的盐楔或层化对局部环流将有重要影响。尽管在浅海中，上述的时间中尺度和大尺度运动是我们直接感兴趣的描述对象，但海水运动是处于连续不断的高频湍流状态的。事实上，我们所处理的变量和场量都是其平均值，并且湍效应已被参数化；对于三维空间问题，Boussinesq 湍假设常被采用，即在广义 Newton 分子摩擦律中以一个湍粘性系数替代分子粘性系数，与此相似，在热、盐或浓度输运方程中亦如此处理。因此湍尺度是我们所关心的现象中最小的尺度，湍不仅产生了湍粘性效应，而且影响海水微团的局地混合特征。

流场的铅垂尺度远小于其水平尺度，这种空间尺度的畸形导致了两个动力学后果：① 在铅垂方向上准静力近似成立；② 因流场的水平尺度远小于地球的半径，则旋转地球坐标系中的问题可以化简为一个“f 平面”上的问题[59, 60]。若进一步引入有关海水密度，或更一般的海水热力学变量的 Boussinesq 近似[59]，则在 f 平面上的右手直角坐标系中，有如下描述斜压浅海系统的方程组。

连续方程：

$$\nabla\cdot\boldsymbol{u}=0 \tag{9.2.1}$$

运动方程：

$$\frac{\partial u}{\partial t}+\boldsymbol{u}\cdot\nabla u-fv=-g\frac{\partial\tilde{\zeta}}{\partial x}+\frac{\partial}{\partial z}\left(\nu\frac{\partial u}{\partial z}\right)-g\beta\frac{\partial}{\partial x}\int_z^{\tilde{\zeta}}c\mathrm{d}z' \tag{9.2.2}$$

$$\frac{\partial v}{\partial t}+\boldsymbol{u}\cdot\nabla v+fu=-g\frac{\partial\tilde{\zeta}}{\partial y}+\frac{\partial}{\partial z}\left(\nu\frac{\partial v}{\partial z}\right)-g\beta\frac{\partial}{\partial y}\int_z^{\tilde{\zeta}}c\mathrm{d}z' \tag{9.2.3}$$

浓度输运方程：

$$\frac{\partial c}{\partial t}+\boldsymbol{u}\cdot\nabla c=\frac{\partial}{\partial z}\left(k\frac{\partial c}{\partial z}\right) \tag{9.2.4}$$

在海面（$z=\tilde{\zeta}$），

运动学边界条件：

$$w=\frac{\partial\tilde{\zeta}}{\partial t}+u\frac{\partial\tilde{\zeta}}{\partial x}+v\frac{\partial\tilde{\zeta}}{\partial y} \tag{9.2.5}$$

动力学边界条件：

$$(\tau_x,\tau_y)=\nu\frac{\partial(u,v)}{\partial z} \tag{9.2.6}$$

通量条件：

$$k\frac{\partial c}{\partial z}=\Gamma \tag{9.2.7}$$

在海底（ $z=-h$ ），

粘性条件：

$$\boldsymbol{u}=0 \tag{9.2.8}$$

通量条件：

$$\frac{\partial c}{\partial z}=0 \tag{9.2.9}$$

式中，t 为时间；(x,y,z) 构成 f 平面上的右手直角坐标系，(x,y) 含于未扰动的海平面中，z 向上为正；$\nabla=\boldsymbol{i}\frac{\partial}{\partial x}+\boldsymbol{j}\frac{\partial}{\partial y}+\boldsymbol{k}\frac{\partial}{\partial z}$ 为向量算子；$(\boldsymbol{i},\boldsymbol{j},\boldsymbol{k})$ 乃相应于 (x,y,z) 坐标轴的单位向量；$\boldsymbol{u}=\boldsymbol{i}u+\boldsymbol{j}v+\boldsymbol{k}w$ 为速度向量；$\tilde{\zeta}$ 为从未扰动海平面起算的海面坐标；h 为从未扰动海平面下量的水深；f 为 Coriolis 参数；ρ 为海水密度；ν 为湍粘性系数；k 为扩散系数；g 为重力加速度；(τ_x,τ_y) 为海面风应力与水密度之比；$\varGamma=-\left(\frac{\alpha}{\beta}\right)\frac{Q}{\rho C_p}$ 为海面通量(α 为热涨系数，β 为单位盐度下密度的增量，Q 为海面热通量，C_p 为比热)；而

$$c=(s-s_*)+\sum_{j=1}^{m}\left(\frac{\beta_j}{\beta}\right)(c_j-c_{j*})-\left(\frac{\alpha}{\beta}\right)(\varTheta-\varTheta_*) \tag{9.2.10}$$

其中，c 为“表观浓度”（apparent concentration），类似于“表观温度”（apparent temperature），后者由 Fofonoff（1962）[61]引进，前者见文献[33]；c_j 为第 j 个示踪物或分量的浓度；β_j 为相应的常数；s 为盐度；$\varTheta$ 为温度；*指出相应的常数参考量。

应指出，盐度和温度以及 $c_j\,(j=1,2,\cdots,m)$ 满足形如式（9.2.4）的输运方程；并且如下的一个广义的状态方程已被利用[62]：

$$\frac{\rho-\rho_*}{\rho_*}=\beta c \tag{9.2.11}$$

这一线化方程中的 c 由式（9.2.10）表示。由式（9.2.11）也可看出，以密度替代表现浓度 c，一个形如式（9.2.4）的密度输运方程可被导出和利用[63]，这也是这一想法和处理的最初来源。

在运动方程（9.2.2）和（9.2.3）中忽略了海面大气压强梯度力和引潮力，但这并不影响其普遍性，因为它们作为水平无旋力附加到水位水平梯度力上并无任何困难。不过应当指出，引潮力直接在边缘浅海引起的潮汐与外海或大洋传入的潮汐相比较，是小到可以忽略的。这样，在陆架、海湾或河口等边缘浅海中的潮波运动，主要是通过开边界传入外海或大洋潮波的反应。此外，在方程（9.2.2）、（9.2.3）及（9.2.4）中分别忽略了水平湍摩擦项及水平湍扩散项，这意味着在问题中不考虑侧边界层，略去侧湍摩擦边界层的一个直接后果是侧向边界条件可采用理想流体的边界条件，即流体

沿岸界流动。当然，在开边界上，将依不同模型采取不同的自然边界条件。海底的粘性边界条件（9.2.8）和海面的运动学边界条件（9.2.5）以及上述流体仅能沿海岸流动的侧边界条件，皆为下面表达场中流体面 $F(x,y,z;t)$ 守恒条件方程（9.2.12）的特例[44]：

$$\frac{\partial F}{\partial t}+\boldsymbol{u}\cdot\nabla F=0 \tag{9.2.12}$$

如 9.1 节中提到的，既然环流问题依赖于在流场中运动着的流体微团经过一个或多个潮周期后所产生的净位移，那么写出其轨迹方程就是必需的。若把在任意时刻 t_0 占据任意空间位置 (x_0,y_0,z_0) 的流体微团释放，并为了方便而称之为“标识流体微团”，则描述其轨迹的方程可用其位移向量 $\boldsymbol{z}=\boldsymbol{x}-\boldsymbol{x}_0$ 来表达：

$$\frac{\mathrm{d}\xi}{u}=\frac{\mathrm{d}\eta}{v}=\frac{\mathrm{d}\zeta}{w}=\mathrm{d}t\ , \tag{9.2.13}$$

其中 $(\boldsymbol{x},\boldsymbol{x}_0,\boldsymbol{\xi})=\boldsymbol{i}(x,x_0,\xi)+\boldsymbol{j}(y,y_0,\eta)+\boldsymbol{k}(z,z_0,\zeta)$。

为了对斜压浅海中由方程组（9.2.1）~（9.2.13）所描述的这一热力–动力学非线性耦合系统作统一的尺度分析，较为方便的是导出其相应的无因次方程组。

如上所述，在像渤海或东中国海这样的边缘浅海中，一般说，本质上为惯性重力长波的潮汐及相伴的潮流是最显著的经久不息的运动，故相应于式（9.2.1）~（9.2.13）的无因次动力学问题可导出如下：

$$\nabla\cdot\boldsymbol{u}=0 \tag{9.2.14}$$

$$\frac{\partial u}{\partial\theta}+\kappa\boldsymbol{u}\cdot\nabla u-fv=-\frac{\partial\tilde{\zeta}}{\partial x}+\frac{\partial}{\partial z}\left(\nu\frac{\partial u}{\partial z}\right)-\delta\frac{\partial}{\partial x}\int_z^{\kappa\tilde{\zeta}}c\mathrm{d}z' \tag{9.2.15}$$

$$\frac{\partial v}{\partial\theta}+\kappa\boldsymbol{u}\cdot\nabla v+fu=-\frac{\partial\tilde{\zeta}}{\partial y}+\frac{\partial}{\partial z}\left(\nu\frac{\partial v}{\partial z}\right)-\delta\frac{\partial}{\partial y}\int_z^{\kappa\tilde{\zeta}}c\mathrm{d}z' \tag{9.2.16}$$

$$\frac{\partial c}{\partial\theta}+\kappa\boldsymbol{u}\cdot\nabla c=\varepsilon\frac{\partial}{\partial z}\left(k\frac{\partial c}{\partial z}\right) \tag{9.2.17}$$

$\mathrm{z}=\kappa\tilde{\zeta}$：

$$w=\frac{\partial\tilde{\zeta}}{\partial\theta}+\kappa\left(u\frac{\partial\tilde{\zeta}}{\partial x}+v\frac{\partial\tilde{\zeta}}{\partial y}\right) \tag{9.2.18}$$

$$k_\tau(\tau_x,\tau_y)=\nu\frac{\partial(u,v)}{\partial z} \tag{9.2.19}$$

$$\frac{\partial c}{\partial z}=k_\Gamma\Gamma \tag{9.2.20}$$

$z=-h$：

$$\boldsymbol{u}=0 \tag{9.2.21}$$

$$\frac{\partial c}{\partial z}=0 \tag{9.2.22}$$

和

$$\frac{\partial F}{\partial \theta}+\kappa \boldsymbol{u}\cdot\nabla F=0 \tag{9.2.23}$$

以及

$$\frac{\mathrm{d}\xi}{u}=\frac{\mathrm{d}\eta}{v}=\frac{\mathrm{d}\zeta}{w}=\frac{\mathrm{d}\theta}{N} \tag{9.2.24}$$

其中的无因次场量与相应的因次方程中的因次场量采用了相同的符号，其特征量采用具有下角标“c”的同样符号表示（譬如流速的 x 轴投影，其因次量和无因次量皆为 u，而其特征量为 u_c）；时间坐标特征量为 ω^{-1}，空间坐标水平特征量和铅垂特征量分别为 L 和 h_c；f 特征量本应为其自身，这是因为 f 为常数，但这里的 f 等于因次 Coriolis 参数与时间尺度 ω^{-1} 之积；而 $u_c=\kappa\sqrt{gh_c}$，$w_c=\left(\frac{h_c}{L}\right)u_c$，$\xi_c=\kappa NL$，$\zeta_c=\kappa Nh_c$，$\nu_c=\omega {h_c}^2$，$\omega=\frac{\sqrt{gh_c}}{L}$；以及我们导出的下列 6 个无因次参数：

$$\kappa=\frac{\tilde{\zeta}_c}{h_c}=\frac{u_c}{\sqrt{gh_c}} \tag{9.2.25}$$

$$\delta=\frac{\Delta}{\kappa} \tag{9.2.26}$$

$$\varepsilon=\frac{k_c}{\omega h_c^2} \tag{9.2.27}$$

$$k_\tau=\frac{\tau_c h_c}{\nu_c u_c} \tag{9.2.28}$$

$$k_\varGamma=\frac{\varGamma_c h_c}{k_c c_c} \tag{9.2.29}$$

$$N=1+(n-1)\kappa \tag{9.2.30}$$

其中 Δ 表示相对密度改变的尺度，且方程（9.2.10）与（9.2.11）相应的无因次方程皆未表明；n 表示将来为取得余流用以平均的潮周期的个数[26]；$\varGamma_c=\frac{\alpha}{\beta}\frac{1}{\rho C_p}Q_c$。

由方程（9.2.14）~（9.2.24）可见，这一边缘浅海的热力–动力学非线性耦合系统的基本特征取决于由式（9.2.25）~（9.2.30）表示的 6 个无因次参数。

非线性参数 κ 乃水面起伏与水深之比，或流速与长波波速之比，其实它也是余流与潮流之比的测度[13, 15, 64~67, 5, 35]。一般说，κ 为一小参数。如中国渤海和美国旧金山南湾（South San Francisco Bay）的有关特征量分别为：水深 20 m 和 10 m，潮差 2 m 和 1 m，潮流为 1 m/s，故它们的无因次参数 κ 取 10^{-1}。因此，我们可引入假设：

$$O(\kappa) < 1 \qquad (9.2.31)$$

无因次参数ε与κ之比$\left(\dfrac{\varepsilon}{\kappa}\right)$乃表观浓度的湍扩散输运与对流输运之比的测度。依据像渤海这样的浅海中的典型数据，ω取 M_2 潮圆频率$1.5\times10^{-4}\ \mathrm{s}^{-1}$，$k_c$依据 Bowden（1965）[41]，可以取 10 cm^2/s，则有$O\left(\dfrac{\varepsilon}{\kappa}\right)=0.25$，故作假设如下[26]：

$$O\left(\frac{\varepsilon}{\kappa}\right)=\kappa \qquad (9.2.32)$$

上式表明，我们感兴趣的是对流输运占优势的浓度输运过程。

无因次参数δ是以表观浓度的水平梯度表示的斜压效应大小的测度，若以浅海和河口的典型值来估算，$O(\Delta)=$（5‰~5%），这是一个相当保守的估值，故我们可以假设

$$O(\delta)=\kappa \qquad (9.2.33)$$

无因次参数k_τ表达了海面风效应大小的测度。若排除偶发的风暴潮事件，可取$O(\tau_c)=1\ \mathrm{cm}^2/\mathrm{s}$和$O(\nu_c)=100\ \mathrm{cm}^2/\mathrm{s}$，则对于像渤海这样的浅海有$O(k_\tau)=10^{-1}$，故我们合理假定[6]：

$$O(k_\tau)=\kappa \qquad (9.2.34)$$

无因次参数k_Γ正比于海面热通量的尺度Q_c，因而是海面热通量效应大小的测度，为了保持普遍性，可暂令其为 1 的量级。

无因次参数N表示在t_0时刻由$\boldsymbol{x}_0$空间点释放的流体微团，在经过n个潮周期后的位移尺度ξ_c与经过 1 个潮周期后的位移尺度$u_c\omega^{-1}$之比，显然，N最小为 1，而当n无限增长时，N也将无限增长[26]。在这里，我们假设

$$O(N)=1 \qquad (9.2.35)$$

这等价于假设

$$O(n)<\kappa^{-2} \qquad (9.2.36)$$

这一假设限制了标识流体微团被释放后经历的时间长度$\omega^{-1}n$，即$O(\omega^{-1}n)<\omega^{-1}\kappa^{-2}$。例如，取$\kappa$为$10^{-1}$，则标识流体微团经历的时间长度应小于 100 个潮周期，即不能超过季平均问题。

若进一步排除湍粘性和湍扩散的非线性效应，即排除了在湍粘性和湍扩散系数中显含流速等场变量及其梯度的诸模型，而使该系数仅限于时空坐标的函数，则上述问题将蜕化为一弱非线性系统。

注意到$\xi_c=\kappa NL$，以及假设（9.2.31）和（9.2.35），则流体微团的随体速度在$\boldsymbol{x}_0$点的 Taylor 展式为

$$\boldsymbol{u}(\boldsymbol{x}_0+\kappa N\boldsymbol{\xi},\theta)=\boldsymbol{u}(\boldsymbol{x}_0,\theta)+\kappa \mathrm{N}\boldsymbol{\xi}\cdot(\nabla\boldsymbol{u})_0+\frac{1}{2}\kappa^2N^2\left\{\xi^2\left(\frac{\partial^2\boldsymbol{u}}{\partial x^2}\right)_0+\eta^2\left(\frac{\partial^2\boldsymbol{u}}{\partial y^2}\right)_0\right.$$
$$\left.+\zeta^2\left(\frac{\partial^2\boldsymbol{u}}{\partial z^2}\right)_0+2\xi\eta\left(\frac{\partial^2\boldsymbol{u}}{\partial x\partial y}\right)_0+2\xi\zeta\left(\frac{\partial^2\boldsymbol{u}}{\partial x\partial z}\right)_0+2\eta\zeta\left(\frac{\partial^2\boldsymbol{u}}{\partial y\partial z}\right)_0\right\}+O(\kappa^3N^3) \quad (9.2.37)$$

其中，$(\quad)_0$ 表示在 $\boldsymbol{x}_0$ 处取值。该流体微团相应的位移借助积分轨迹方程（9.2.24）可表示如下：

$$N\boldsymbol{\xi}=\int_{\theta_0}^{\theta}(\boldsymbol{x}_0+\kappa N\boldsymbol{\xi},\theta')\mathrm{d}\theta' \quad (9.2.38)$$

其中 $\theta_0=\omega t_0$。

在这样一个以非线性小参数 κ 为表征的弱非线性斜压浅海系统中，场变量可展成 κ 的幂级数：

$$A=\sum_{j=0}^{1,2,\cdots}A_j\kappa^j \quad (9.2.39)$$

式中，A 表示任何一个场变量，A_j 表示其 j 阶模型的解，并且皆以无因次形式表示。

把解（9.2.39）代入无因次方程组（9.2.14）~（9.2.22），可得到第 j 阶解满足的描述第 j 阶模型的方程组，其零阶模型为：

$$\nabla\cdot\boldsymbol{u}_0=0 \quad (9.2.40)$$

$$\frac{\partial u_0}{\partial\theta}-fv_0=-\frac{\partial\tilde{\zeta}_0}{\partial x}+\frac{\partial}{\partial z}\left(\nu\frac{\partial u_0}{\partial z}\right) \quad (9.2.41)$$

$$\frac{\partial v_0}{\partial\theta}+fu_0=-\frac{\partial\tilde{\zeta}_0}{\partial y}+\frac{\partial}{\partial z}\left(\nu\frac{\partial v_0}{\partial z}\right) \quad (9.2.42)$$

$$\frac{\partial c_0}{\partial\theta}=0 \quad (9.2.43)$$

$z=0$：

$$w_0=\frac{\partial\tilde{\zeta}_0}{\partial\theta} \quad (9.2.44)$$

$$\frac{\partial(u_0,v_0)}{\partial z}=0 \quad (9.2.45)$$

$$\frac{\partial c_0}{\partial z}=k_\Gamma\Gamma \quad (9.2.46)$$

$z=-h$：

$$\boldsymbol{u}_0=0 \quad (9.2.47)$$

$$\frac{\partial c_0}{\partial z}=0 \quad (9.2.48)$$

特别是若将式(9.2.39)代入式(9.2.37)、(9.2.38),则可得到在零阶模型中 Lagrange 速度及位移的表达式:

$$\boldsymbol{u}_0(\boldsymbol{x}_0+\kappa N\boldsymbol{\xi},\theta)=\boldsymbol{u}_0(\boldsymbol{x}_0,\theta) \tag{9.2.49}$$

$$N\boldsymbol{\xi}_0=\int_{\theta_0}^{\theta}\boldsymbol{u}_0(\boldsymbol{x}_0,\theta')\mathrm{d}\theta' \tag{9.2.50}$$

方程组(9.2.40)~(9.2.48)表明了零阶模型的线性性质,即在动力学变量$(\boldsymbol{u}_0,\tilde{\zeta}_0)$与热力学变量$c_0$之间不存在耦合。事实上,在边界条件(9.2.44)、(9.2.45)和(9.2.47)下,方程组(9.2.40)~(9.2.42)描述了正压天文潮运动,海面风和斜压效应的影响不存在。如果我们进一步补充上下面两个侧边界条件:在开边界B_0上给出已知的潮位$\tilde{Z}$并在岸界B_c给出流体不可流入条件

$$\tilde{\zeta}_0\Big|_{(x,y)\in B_0}=\tilde{Z} \tag{9.2.51}$$

和

$$\int_{-h}^{0}\boldsymbol{u}_0\cdot\boldsymbol{n}\mathrm{d}z\Big|_{(x,y)\in B_\mathrm{c}}=0 \tag{9.2.52}$$

则方程组(9.2.40)~(9.2.42)和边界条件(9.2.44)、(9.2.45)、(9.2.47)及(9.2.51)、(9.2.52)组成了零阶模型——天文潮的定解问题[35]。

表观浓度的零阶场c_0满足的方程(9.2.43)和边界条件(9.2.46)、(9.2.48)仅给出了c_0与时间无关而仅为空间坐标函数的信息,但并未给出其解或求解的途径;为求得c_0,必须进一步作高阶摄动。

其一阶模型如下:

$$\nabla\cdot\boldsymbol{u}_1=0 \tag{9.2.53}$$

$$\frac{\partial u_1}{\partial\theta}-fv_1=-\frac{\partial\tilde{\zeta}_1}{\partial x}+\frac{\partial}{\partial z}\left(\nu\frac{\partial u_1}{\partial z}\right)-\frac{\delta}{\kappa}\frac{\partial}{\partial x}\int_z^0 c_0\mathrm{d}z'-\boldsymbol{u}_0\cdot\nabla u_0 \tag{9.2.54}$$

$$\frac{\partial v_1}{\partial\theta}+fu_1=-\frac{\partial\tilde{\zeta}_1}{\partial y}+\frac{\partial}{\partial z}\left(\nu\frac{\partial v_1}{\partial z}\right)-\frac{\delta}{\kappa}\frac{\partial}{\partial y}\int_z^0 c_0\mathrm{d}z'-\boldsymbol{u}_0\cdot\nabla v_0 \tag{9.2.55}$$

$$\frac{\partial c_1}{\partial\theta}=-\boldsymbol{u}_0\cdot\nabla c_0 \tag{9.2.56}$$

$z=0$:

$$w_1=\frac{\partial\tilde{\zeta}_1}{\partial\theta}+\frac{\partial(u_0\tilde{\zeta}_0)}{\partial x}+\frac{\partial(v_0\tilde{\zeta}_0)}{\partial y} \tag{9.2.57}$$

$$\nu\frac{\partial(u_1,v_1)}{\partial z}=\frac{k\tau}{\kappa}(\tau_x,\tau_y)-\nu\tilde{\zeta}_0\frac{\partial^2(u_0,v_0)}{\partial z^2} \tag{9.2.58}$$

$$\frac{\partial c_1}{\partial z}=-\tilde{\zeta}_0\frac{\partial^2 c_0}{\partial z^2} \tag{9.2.59}$$

$z=-h$：

$$\boldsymbol{u}_1=0 \tag{9.2.60}$$

$$\frac{\partial c_1}{\partial z}=0 \tag{9.2.61}$$

Lagrange 速度及位移为

$$\boldsymbol{u}_1(\boldsymbol{x}_0+\kappa N\xi,\theta)=\boldsymbol{u}_1(\boldsymbol{x}_0,\theta)+N\boldsymbol{\xi}_0\cdot(\nabla\boldsymbol{u}_0)_0 \tag{9.2.62}$$

$$N\boldsymbol{\xi}_1=\int_{\theta_0}^{\theta}\boldsymbol{u}_1(\boldsymbol{x}_0,\theta')\mathrm{d}\theta'+\int_{\theta_0}^{\theta}N\boldsymbol{\xi}_0\cdot(\nabla\boldsymbol{u}_0)_0\mathrm{d}\theta' \tag{9.2.63}$$

方程（9.2.54）和（9.2.55）表明，一阶流动包括风海流、热盐流和伴随浅海分潮的浅海分潮流以及 Euler 余流。

方程（9.2.56）表明：表观浓度的一阶解 c_1 为潮周期振荡形式；进一步与 c_0 为一定常场的结论相对照，可见，表观浓度的周期振动部分与其定常部分相比要小一个量级；且对于一维空间的河口模型问题，该振荡与潮流位相差为 π/2 ——在接近外海的河口处，这点已被观测所证实[40, 68, 69, 36]。

表达式（9.2.50）和（9.2.63）已表明：在这一弱非线性系统中，满足非线性轨迹方程（9.2.24）的位移向量可简单地由对时间的积分得到。

如果需要，可以再摄动而获得高阶模型。

9.3　浅海环流概念的基本构想

鉴于海洋环流问题是紧密地联系着海水特性、溶解物和悬浮质等浓度的长期输运过程、迁移规律及其最终分布的描述和确定，则显然海洋环流问题可由海水微团经过 1 个或多个潮周期这样足够长的时段后的净位移来描述。在流场中 $\boldsymbol{x}_0$ 处，由 t_0 时刻开始释放的标识流体微团经 n 个潮周期 T 后所产生的净位移向量可借助于积分轨迹方程（9.2.13）而得到：

$$\delta\xi=\int_{t_0}^{t_0+nT}\boldsymbol{u}(\boldsymbol{x}_0+\xi(t),t)\mathrm{d}t \tag{9.3.1}$$

其中 $\delta\xi$ 表示净位移。

完全等价地，我们可以借助于净位移定义一个平均速度 $\dfrac{\delta\xi}{nT}$，并用它来描述海洋环流问题。若以 $\boldsymbol{u}_{\mathrm{LM}}$ 表示该平均速度，则代入式（9.3.1）有

$$\boldsymbol{u}_{\mathrm{LM}}=\frac{1}{nT}\int_{t_0}^{t_0+nT}\boldsymbol{u}(\boldsymbol{x}_0+\boldsymbol{\xi}(t),t)\mathrm{d}t \tag{9.3.2}$$

显然，由式（9.3.2）可见，$\boldsymbol{u}_{\mathrm{LM}}$ 表达了该标识流体微团随体速度 $\boldsymbol{u}(\boldsymbol{x}_0+\boldsymbol{\xi}(t),t)$ 在 n 个潮周期时段内的平均值；故 $\boldsymbol{u}_{\mathrm{LM}}$ 可称为 Lagrange 平均速度。

作为对照，我们给出在流场空间 $\boldsymbol{x}_0$ 处经 n 个潮周期时段内流速向量的平均值 $\boldsymbol{u}_{\mathrm{E}}$：

$$\boldsymbol{u}_{\mathrm{E}}=\frac{1}{nT}\int_{t_0}^{t_0+nT}\boldsymbol{u}(\boldsymbol{x}_0,t)\mathrm{d}t \tag{9.3.3}$$

鉴于 $\boldsymbol{u}_{\mathrm{E}}$ 乃定点 $\boldsymbol{x}_0$ 处流速的平均值，故称之为 Euler 平均速度，或 Euler 余流。

在一个周期潮运动占优势的浅海或感潮河口中，Lagrange 净位移和 Euler 余流的出现，除了有像海面风应力和密度水平梯度力等非潮强迫力的作用外，由于浅海流体动力学为一非线性系统，故还取决于潮的非线性耦合作用。由此可见，浅海环流问题本质上为一非线性现象，其中包括了中时间尺度的潮运动和大时间尺度的环流运动之间的非线性相互作用。

应该指出，Lagrange 净位移 $\delta\xi$，或等价意义上的 Lagrange 平均速度 $\boldsymbol{u}_{\mathrm{LM}}$ 与 Euler 平均速度之间，在概念上是完全不同的。为了清晰地阐明这一区别，首先讨论一个不考虑任何非潮强迫力的纯潮汐的非线性浅海系统的问题，即潮致余环流问题。

Euler 余流 $\boldsymbol{u}_{\mathrm{E}}$ 与平均时段 nT 或用以平均的潮周期的数目 n 无关。也与平均开始时刻 t_0 无关，一般说，它仅为空间坐标的函数[8]。

Lagrange 净位移，或等价的 Lagrange 平均速度 $\boldsymbol{u}_{\mathrm{LM}}$，一般说，不仅取决于标识流体微团释放时的空间坐标，也取决于释放时刻本身，还取决于用以平均的潮周期个数[8]，也就是说，正比于 Lagrange 净位移的 Lagrange 平均速度是一个典型的 Lagrange 量。

我们进一步考查一个 M_2 的弱非线性系统，并在无因次化后再作讨论。若选取 ω 为 M_2 的圆频率，则其无因次圆频率为 1，而无因次周期为 2π。引入潮平均算子

$$\langle\ \rangle=\frac{1}{2\pi n}\int_{\theta_0}^{\theta_0+2\pi n}\mathrm{d}\theta \tag{9.3.4}$$

则无因次净位移、Lagrange 平均速度和 Euler 余流可分别表示为 $N\delta\xi=2\pi n\left\langle\boldsymbol{u}(\boldsymbol{x}_0+\kappa N\xi,\theta)\right\rangle=2\pi n\boldsymbol{u}_{\mathrm{LM}}$ 和 $\boldsymbol{u}_{\mathrm{E}}=\left\langle\boldsymbol{u}(\boldsymbol{x}_0,\theta)\right\rangle$。注意，同前面一样，因次量和无因次量用同一符号表示。

把式（9.2.37）代入（9.3.4），可得 Lagrange 平均速度，其最低阶表达式为物质输运速度 $\boldsymbol{u}_{\mathrm{L}}$：

$$\boldsymbol{u}_{\mathrm{LM}}=\boldsymbol{u}_{\mathrm{L}}+O(\kappa) \tag{9.3.5}$$

其中，

$$\boldsymbol{u}_{\mathrm{L}}=\boldsymbol{u}_{\mathrm{E}}+\boldsymbol{u}_{\mathrm{S}} \tag{9.3.6}$$

而 Stokes 漂移速度为

$$\boldsymbol{u}_{\mathrm{S}}=\left\langle N\xi_0\cdot(\nabla\boldsymbol{u}_0)\right\rangle \tag{9.3.7}$$

注意，上三式中平均速度已再次被 $\kappa\boldsymbol{u}_{\mathrm{c}}$ 尺度化。这表明 Euler 余流和 Lagrange 平均速度皆为 κ 量级，即比潮流小一个数量级[13]。

式（9.3.6）表明，即使在最低阶的意义上，Lagrange 平均速度与 Euler 平均速度也相差一个同阶的量——Stokes 漂移速度[10]，而后者是由于潮流的速度梯度和潮场中水微团位移非线性耦合的潮周期平均效应产生的，其中的水微团位移的出现表明了 Lagrange 动力学的性质。$\boldsymbol{u}_{\mathrm{S}}$ 只是空间坐标的函数，故在一个弱非线性系统中，Lagrange 平均速度在最低阶的意义上仅为空间坐标的函数。

把式（9.2.37）代入（9.3.4），且取二阶近似，则有

$$\boldsymbol{u}_{\mathrm{LM}} = \boldsymbol{u}_{\mathrm{L}} + \kappa\boldsymbol{u}_{\mathrm{LD}} + O(\kappa^2) \tag{9.3.8}$$

其中 $\boldsymbol{u}_{\mathrm{LD}}$ 为 Lagrange 漂移速度：

$$\boldsymbol{u}_{\mathrm{LD}} = -N\xi_0(\theta_0)\cdot\nabla\boldsymbol{u}_{\mathrm{L}} + \boldsymbol{u}_{\mathrm{L}}\cdot\nabla[N\boldsymbol{\xi}_0(\theta_0)] \tag{9.3.9}$$

式中

$$N\xi_0(\theta_0) = [\int \boldsymbol{u}_0(\boldsymbol{x}_0,\theta)\mathrm{d}\theta]_{\theta=\theta_0} \quad (\theta_0 = \omega t_0) \tag{9.3.10}$$

式（9.3.8）表明，以式（9.3.9）表示的 Lagrange 漂移速度 $\boldsymbol{u}_{\mathrm{LD}}$——一个二阶摄动项表达了 Lagrange 平均速度对于流体微团释放时刻潮位相 θ_0 的依赖关系。其中 $\boldsymbol{\xi}_0(\theta_0)$ 表明，这种依赖关系是由于在潮流场中某空间点处于不同潮位相 θ_0 时释放的流体微团所描述的轨迹不同而产生的。事实上，展开式（9.3.9），有

$$\boldsymbol{u}_{\mathrm{LD}} = \boldsymbol{u}'_{\mathrm{LD}}\cos\theta_0 + \boldsymbol{u}''_{\mathrm{LD}}\sin\theta_0 \tag{9.3.11}$$

其中

$$\boldsymbol{u}'_{\mathrm{LD}} = \boldsymbol{u}''_0\cdot\nabla\boldsymbol{u}_{\mathrm{L}} - \boldsymbol{u}_{\mathrm{L}}\cdot\nabla\boldsymbol{u}''_0$$
$$\boldsymbol{u}''_{\mathrm{LD}} = -\boldsymbol{u}'_0\cdot\nabla\boldsymbol{u}_L + \boldsymbol{u}_L\cdot\nabla\boldsymbol{u}'_0$$

$\boldsymbol{u}'$ 和 $\boldsymbol{u}''_0$ 为天文潮 M_2 的流速的调和系数[70]，嵌于下式中：$\boldsymbol{u}_0 = \boldsymbol{u}'_0\cos\theta + \boldsymbol{u}''_0\sin\theta$。

式（9.3.11）表明，随着流体微团释放时刻潮位相 θ_0 由 0 变化至 2π，Lagrange 漂移速度 $\boldsymbol{u}_{\mathrm{LD}}$ 的水平投影将在速度图上描出一个椭圆[25]；而式（9.3.1）又表明，Lagrange 净位移 $\delta\xi$ 的水平投影将在流场空间描出一个椭圆。它们称为 Lagrange 漂移椭圆。

表达式（9.3.8）的导出基于一个弱非线性潮系统，在此系统中，海水微团运移的时段不包括太长的时间，亦即用以平均的潮周期个数不是太大，以保持假设 $O(N)=1$ 的成立。但显然，$O(N)>1$ 的情形完全可能是我们感兴趣的问题。此时，若除了 $O(N)>1$ 被肯定外，以上的弱非线性系统的假设全部保留，则 Lagrange 平均速度 $\boldsymbol{u}_{\mathrm{LM}}$ 能够在二阶近似的意义下表示出来[8]：

$$\begin{aligned}\boldsymbol{u}_{\mathrm{LM}} = \boldsymbol{u}_{\mathrm{LM}}(\boldsymbol{x}_0,\theta_0;n) &= \frac{1}{2\pi n}\int_{\theta_0}^{\theta_0+2\pi n}\boldsymbol{u}(\boldsymbol{x}(\boldsymbol{x}_0,\theta),\theta)\mathrm{d}\theta\\ &= \frac{1}{n}\sum_{j=1}^{n}\frac{1}{2\pi}\int_{\theta_0+(j-1)2\pi}^{\theta_0+2\pi j}\boldsymbol{u}(\boldsymbol{x}(\boldsymbol{x}_0,\theta),\theta)\mathrm{d}\theta\\ &= \frac{1}{n}\sum_{j=1}^{n}\boldsymbol{u}_{\mathrm{LM}}(\boldsymbol{x}_{j-1},\theta_0+(j-1)2\pi;1)\end{aligned}$$

其中

$$\boldsymbol{x}_j=\boldsymbol{x}_0+\kappa^2 2\pi\sum_{i=0}^{j-1}\boldsymbol{u}_{\rm LM}(\boldsymbol{x}_j,\theta_0+2\pi i;1)\ (j=1,2,\cdots,n)$$

将式（9.3.2）代入式（9.3.6）、（9.3.8）、（9.3.11），上式变为

$$\boldsymbol{u}_{\rm LM}=\boldsymbol{u}_{\rm LM}(\boldsymbol{x}_0,\boldsymbol{\theta}_0;n)={}^{\rm L}\boldsymbol{u}_{\rm LM}(\boldsymbol{x}_0;n)+\kappa\,{}^{\rm D}\boldsymbol{u}_{\rm LM}(\boldsymbol{x}_0,\theta_0;n) \tag{9.3.12}$$

式中，

$${}^{\rm L}\boldsymbol{u}_{\rm LM}(\boldsymbol{x}_0;n)=\frac{1}{n}\sum_{j=1}^{n}\boldsymbol{u}_{\rm L}(\boldsymbol{x}_{{\rm L},j-1}) \tag{9.3.13}$$

$${}^{\rm D}\boldsymbol{u}_{\rm LM}(\boldsymbol{x}_0,\theta_0;n)={}^{\rm D}\boldsymbol{u}'_{\rm LM}(\boldsymbol{x}_0;n)\cos\theta_0+{}^{\rm D}\boldsymbol{u}''_{\rm LM}(\boldsymbol{x}_0;n)\sin\theta_0 \tag{9.3.14}$$

$${}^{\rm D}\boldsymbol{u}'_{\rm LM}(\boldsymbol{x}_0;n)=\frac{1}{n}\left\{\sum_{j=1}^{n}\boldsymbol{u}'_{\rm LD}(\boldsymbol{x}_{{\rm L},j-1})+\kappa^2 2\pi\sum_{j=1}^{n-1}\left[\boldsymbol{u}'_{\rm LD}(\boldsymbol{x}_{{\rm L},j-1})\cdot\nabla\sum_{i=j}^{n-1}\boldsymbol{u}_{\rm L}(\boldsymbol{x}_{{\rm L},i})\right]\right\}$$

$${}^{\rm D}\boldsymbol{u}''_{\rm LM}(\boldsymbol{x}_0;n)=\frac{1}{n}\left\{\sum_{j=1}^{n}\boldsymbol{u}''_{\rm LD}(\boldsymbol{x}_{{\rm L},j-1})+\kappa^2 2\pi\sum_{j=1}^{n-1}\left[\boldsymbol{u}''_{\rm LD}(\boldsymbol{x}_{{\rm L},j-1})\cdot\nabla\sum_{i=j}^{n-1}\boldsymbol{u}_{\rm L}(\boldsymbol{x}_{{\rm L},i})\right]\right\}$$

$$\boldsymbol{x}_{{\rm L},j}=\boldsymbol{x}_0+\kappa^2 2\pi\sum_{i=0}^{j-1}\boldsymbol{u}_{\rm L}(\boldsymbol{x}_{{\rm L},i}) \tag{9.3.15}$$

$$\boldsymbol{x}_{{\rm L},0}=\boldsymbol{x}_0$$

$$\sum_{a}^{b}=0\ （当 b<a\ 时）$$

显然，当$n=1$时，式（9.3.12）~（9.3.15）将严格蜕化为式（9.3.8）~（9.3.11）；或虽然$n>1$，但当$O(n)<\kappa^{-2}$时，式（9.3.12）~（9.3.15）将近似化为式（9.3.8）~（9.3.11）。事实上，鉴于式（9.3.12）~（9.3.15）中诸项$\left(\kappa^2 2\pi\sum_{j}^{n}\right)=O(\kappa^2 n)<1$，故当$O(n)<\kappa^{-2}$时，有$\boldsymbol{x}_{{\rm L},j}=\boldsymbol{x}_0$，从而

$${}^{\rm L}\boldsymbol{u}_{\rm LM}(\boldsymbol{x}_0;n)=\frac{1}{n}\sum_{j=1}^{n}\boldsymbol{u}_{\rm L}(\boldsymbol{x}_0)=\boldsymbol{u}_{\rm L}(x_0)$$

$$\left[{}^{\rm D}\boldsymbol{u}'_{\rm LM}(\boldsymbol{x}_0;n),{}^{\rm D}\boldsymbol{u}''_{\rm LM}(\boldsymbol{x}_0;n)\right]=\left[\boldsymbol{u}'_{\rm LD}(\boldsymbol{x}_0),\boldsymbol{u}''_{\rm LD}(\boldsymbol{x}_0)\right]$$

最终有

$$\boldsymbol{u}_{\rm LM}(\boldsymbol{x}_0,\theta_0;n)=\boldsymbol{u}_{\rm LM}(\boldsymbol{x}_0,\theta_0)$$

被式（9.3.8）~（9.3.11）所表达。

不过，当$O(n)<\kappa^{-2}$时，即当$O(N)>1$时，表达式（9.3.12）~（9.3.15）不能蜕化或近似化为式（9.3.8）~（9.3.11）；此时，（9.3.12）~（9.3.15）所表示的 Lagrange 平均速度则不仅是(x_0,θ_0)的函数，而且一般说也取决于$n\cdot\boldsymbol{u}_{\rm LM}(\boldsymbol{x}_0,\theta_0;n)$或等价的

$\delta\xi = 2\pi n \times \boldsymbol{u}_{\mathrm{LM}}(\boldsymbol{x}_0, \boldsymbol{\theta}_0; n)$ 显含用以平均的潮周期个数 n，这反映了 θ_0 时刻由 $\boldsymbol{x}_0$ 点释放的标识流体微团在到达终点 $\boldsymbol{x}_n$ 前在时段 $2\pi n$ 内迁移的过程，从而更清楚地表明了其 Lagrange 轨迹的性质。

由此可见，Lagrange 平均速度与 Euler 平均速度的概念何其不同！因此，以 Euler 余环流来体现浅海环流问题这一经典做法，与用 Lagrange 平均速度所描述的浅海环流问题自然会有很大的差异。

由上面讨论的（仅为 M_2 潮致）Lagrange 平均速度的问题已可看出，以其来描述浅海环流问题将十分繁复，故应找到一种相对简捷而有效的途径。

显然，在一个弱非线性浅海系统中，Lagrange 平均速度的最低阶解——物质输运速度自然是我们最感兴趣的，也就是说，在最低阶的意义上，以物质输运速度来体现浅海环流的速度场是我们所期望的[9]。

下面我们来讨论物质输运速度 $\boldsymbol{u}_{\mathrm{L}}$ 的性质。

一般情况下，正比于标识流体微团的净位移的 Lagrange 平均速度是一个典型的 Lagrange 动力学量，但在本文所建立的这一斜压浅海弱非线性系统中，Lagrange 平均速度却能构成大时间尺度环流问题中的不可压缩流速场，这至少在二阶近似的意义上是真确的。事实上，形式上取式（9.3.8）之空间散度，有 $\nabla \cdot \boldsymbol{u}_{\mathrm{LM}} = \nabla \cdot \boldsymbol{u}_{\mathrm{L}} = \nabla \cdot \boldsymbol{u}_{\mathrm{LD}} = 0$ [9]，因此，在这一弱非线性系统中导出的 Lagrange 平均速度可定义为 Lagrange 余流[8]。物质输运速度 $\boldsymbol{u}_{\mathrm{L}}$ 即为最低阶 Lagrange 余流，且满足构成不可压缩流场的必要条件，即为一管量场：

$$\nabla \cdot \boldsymbol{u}_{\mathrm{L}} = 0 \tag{9.3.16}$$

9.1 节中已提到，物质输运速度不仅为一管量场，而且在最一般的意义上被证明满足环流场中的物质面守恒方程。事实上，方程（9.2.23）中的 F 之零阶量 F_0 可作为潮周期平均 $\langle F \rangle$ 的近似，且满足以下方程：

$$\boldsymbol{u}_{\mathrm{L}} \cdot \nabla \langle F \rangle = 0 \tag{9.3.17}$$

作为对照，我们指出：Euler 余流虽亦为一管量场，但却不满足物质面守恒方程，从而在海面和侧向岸界皆出现了潮致“源”和“汇”[6]。

方程（9.3.16）和（9.3.17）表明了物质输运速度具有良好的流体连续性。由此亦可看出，以物质输运速度来体现环流的流场将比 Euler 余流更为合理。

以物质输运速度来体现浅海环流的对流性质这一点，可通过讨论和分析表观浓度所满足的长期输运方程来加以证明，这是最有效的[14]。

为了导出长期输运方程，也就是说，为了导出表观浓度的潮周期平均所满足的对流–扩散方程，除了表观浓度的零阶方程（9.2.43）和一阶方程（9.2.56）外，还必须导出其二阶方程：

$$\frac{\partial c_2}{\partial \theta}+\boldsymbol{u}_1\cdot\nabla c_0+\boldsymbol{u}_0\cdot\nabla c_1=E\frac{\partial}{\partial z}\left(k\frac{\partial c_0}{\partial z}\right) \tag{9.3.18}$$

其中 $E=\dfrac{\varepsilon}{\kappa^2}$。由假设（9.2.32）可知 E 为 1 的量级。

式（9.2.43）和（9.2.56）表明，表观浓度的平均场<c>可以用其零阶场 c_0 来近似。这样，导出 c_0 满足的对流–扩散方程即为表观浓度潮周期平均场<c>所满足的长期输运方程。事实上，由潮周期平均方程（9.3.18）可得

$$\boldsymbol{u}_{\mathrm{E}}\cdot\nabla c_0=E\frac{\partial}{\partial z}\left(\langle k\rangle\frac{\partial c_0}{\partial z}\right)+\langle-\boldsymbol{u}_0\cdot\nabla c_1\rangle \tag{9.3.19}$$

或以 $\langle c\rangle$ 近似代替 c_0，上述方程变为

$$\boldsymbol{u}_{\mathrm{E}}\cdot\nabla\langle c\rangle=E\frac{\partial}{\partial z}\left(\langle k\rangle\frac{\partial\langle c\rangle}{\partial z}\right)+\langle-\boldsymbol{u}_0\cdot\nabla c_1\rangle \tag{9.3.20}$$

其中 $\langle-\boldsymbol{u}_0\cdot\nabla c_1\rangle$ 表示潮流周期振荡部分与表观浓度周期变化部分的梯度之间非线性耦合的潮周期平均效应[36]。

如果我们坚持经典的概念，即以 Euler 余流 $\boldsymbol{u}_{\mathrm{E}}$ 来体现浅海环流速度场，亦即以 Euler 余流表征表观浓度平均场<c>的对流输运：$\boldsymbol{u}_{\mathrm{E}}\cdot\nabla\langle c\rangle$，则必须对方程（9.3.20）右端第二项 $\langle-\boldsymbol{u}_0\cdot\nabla c_1\rangle$ 作出解释。正如 9.1 节中所提到的，这是通常所用的长期物质浓度输运方程中引起集中争论的项。除了有人认为应干脆把它置零以外，通常把它看作是与潮振荡和浓度变化相关而产生的一种弥散效应，称为“潮弥散”，并为使方程封闭而采用了 Fick 假设[36]。

其实，如果我们放弃经典概念，即放弃以 Euler 余流来体现浅海环流速度场，从而表征表观浓度平均场的对流输运的观点，则问题就迎刃而解了[14]。事实上，把方程（9.2.56）和 $\boldsymbol{u}_0$ 代入式（9.3.20）或（9.3.19），有

$$\langle-\boldsymbol{u}_0\cdot\nabla c_1\rangle=-\boldsymbol{u}_{\mathrm{s}}\cdot\nabla\langle c\rangle$$

从而，长期输运方程为

$$\boldsymbol{u}_{\mathrm{L}}\cdot\nabla\langle c\rangle=E\frac{\partial}{\partial z}\left(\langle k\rangle\frac{\partial\langle c\rangle}{\partial z}\right) \tag{9.3.21}$$

方程（9.3.21）表明，环流场中表观浓度时均场的对流输运应该以物质输运速度 $\boldsymbol{u}_{\mathrm{L}}$ 来表征，而不是以 Euler 余流来表征，所谓“潮弥散”项并不存在。

由此可见，在最低阶的意义上，以物质输运速度来体现浅海环流的速度场，应该说是恰如其分的，并且进而表明，浅海环流基本上为一定常场。

不过应该强调指出，此结论的真确性应取决于这一弱非线性系统主要假设的真确性。特别应考虑到假设（9.2.31）和（9.2.35）中 $O(\kappa)<1$ 和 $O(N)=1$ 的局限性，譬如，

仅就上面在 M_2 非线性系统中给出的当 $O(N)>1$ 时的 Lagrange 平均速度表达式（9.3.12）来说，此时，Lagrange 平均速度已不能构成一流场了[8, 26, 9]。

尽管理论模型有局限性，本章仍将着力探讨以物质输运速度作为浅海环流速度场这一途径，并介绍其基础动力学及其初步应用。这是因为，与以 Euler 余流来体现浅海环流的经典概念相比较，物质输运速度毕竟在环流尺度上具有明确的输运意义，循着这条路线，可能为环流的研究开拓新的前景。

9.4　基本方程组

在上一节，为了把注意力集中于浅海环流概念的基本构想，我们简单地假设了一个 M_2 非线性潮系统作为分析的对象。实际上，众所周知，天文潮波系乃是一系列分潮波的叠加。其中的主要四大分潮为[70]：太阴半日分潮 M_2，圆频率为 28.98°/h，相对系数为 0.9085；太阳半日分潮 S_2，圆频率为 30.00°/h，相对系数为 0.4227；太阴—太阳赤纬全日分潮 K_1，圆频率为 15.04°/h，相对系数为 0.5305；太阴赤纬全日分潮 O_1，圆频率为 13.94°/h，相对系数为 0.3771。由相对系数的大小可知，太阴半日分潮 M_2 为最主要的分潮波。

边缘海、海湾或感潮河口中的潮汐，一般说来，是由外海或大洋传入的天文潮波及其非线性耦合所产生的浅水分潮和余流所组成的[35]。每一个天文分潮波自身的非线性耦合将首先同时产生一个圆频率为其 2 倍的倍潮和一个圆频率为零的余流。例如，M_2 将产生倍潮 M_4 和一余流分量。每两个天文分潮波之间的非线性耦合将首先同时产生一个圆频率为二者之和的复合潮和一个圆频率为二者之差的长周期复合潮。例如，M_2–S_2 将产生复合潮 MS_4 和一长周期潮 MS_f——一个周期为半个月的复合潮波。

显然，当我们作日潮周期平均时，像 MS_f 这样的长周期振动将不可能被滤掉而必然表现出来，即为长周期流研究的课题。但是当我们以足够长的时段作平均时（譬如月平均甚至季平均等等），这些半月、月周期振动将被消除，或至少近似被滤掉。若再进一步假设，对更长周期的分潮（如年或多年分潮）不感兴趣，则显露出来的剩余运动——Euler 余流必包含像 M_2，S_2，K_1 和 O_1 这样一些主要天文潮自我耦合产生的效应，至少在一般情况下必须作这样的考虑。

在一个潮汐的弱非线性系统中，现已证明：像 M_2，S_2，K_1 和 O_1 组成的天文潮系统产生的一阶 Euler 余流为其中每一个天文分潮自我耦合产生的 Euler 余流分量之和，即 Euler 余流具有线性叠加的性质[35]。我们进一步证明，在一个弱非线性的斜压浅海系统中，Euler 余流可以作为潮致 Euler 余流分量、风生和斜压分量之和[6]。

对应于式（9.3.7）的 Stokes 漂移速度因次表达式为

$$\boldsymbol{u}_{\mathrm{s}}=\left\langle\xi_0\cdot(\nabla\boldsymbol{u}_0)_0\right\rangle \tag{9.4.1}$$

其中因次潮平均算子< >为

$$\langle\ \rangle=\frac{1}{T}\int_{t_0}^{t_0+T}\mathrm{d}t \tag{9.4.2}$$

用以平均的时段 T 是如此之长，以致能滤掉主要潮振动，或至少在近似意义上如此；但它又不太长，以保持 $O(N)=1$ 之假设。

把对应于（9.2.31）的因次表达式代入上式并取最低阶近似，即得弱非线性的斜压浅海系统中的物质输运速度：

$$\boldsymbol{u}_{\mathrm{L}}=\boldsymbol{u}_{\mathrm{E}}+\boldsymbol{u}_{\mathrm{S}}=\sum_{k=0}^{m}{}^{k}\boldsymbol{u}_{\mathrm{L}} \tag{9.4.3}$$

式中，

$$(\boldsymbol{u}_{\mathrm{E}},\boldsymbol{u}_{\mathrm{S}})=\sum_{k=0}^{m}({}^{k}\boldsymbol{u}_{\mathrm{E}},{}^{k}\boldsymbol{u}_{\mathrm{S}})$$

$${}^{k}\boldsymbol{u}_{\mathrm{L}}={}^{k}\boldsymbol{u}_{\mathrm{E}}+{}^{k}\boldsymbol{u}_{\mathrm{S}}$$

注意，左上角标 $k=1,2,\cdots,m$ 指出该变量为相应于第 k 个天文分潮非线性耦合产生的量，而 ${}^{0}\boldsymbol{u}_{\mathrm{E}}$ 中包括一个风生正压分量和一个密度水平梯度力产生的斜压分量； ${}^{0}\boldsymbol{u}_{\mathrm{S}}=0$, 这是因为式（9.4.1）表明 Stokes 漂移速度仅取决于零阶模型中的天文潮变量之故。

式（9.4.3）表明：在一个由 m 个自外海或大洋传入的天文潮波组成的弱非线性斜压浅海系统中，物质输运速度乃 m 个天文潮波各自导致的物质输运速度和风生以及密度水平梯度力导致的正压和斜压分量之和[33]。

下面将导出这个在最一般意义下的物质输运速度所满足的动力学方程组，作为在最低阶近似下浅海环流的基本方程组。

将式（9.4.3）、（9.4.1）及潮周期平均算子（9.4.2）代入对应于方程（9.2.53）~（9.2.55）、（9.2.57）、（9.2.58）、（9.2.60）的因次方程组和对应于方程（9.3.18）的因次方程，并且鉴于式

$$\begin{aligned}\left\langle\psi_0\cdot\varphi_0\right\rangle&=\left\langle\sum_{k=1}^{m}{}^{k}\psi_0\cdot\sum_{k=1}^{m}{}^{k}\varphi_0\right\rangle\\&=\sum_{k=1}^{m}\left\langle{}^{k}\psi_0\cdot{}^{k}\varphi_0\right\rangle\end{aligned} \tag{9.4.4}$$

其中 ψ_0 ， φ_0 表示任意零阶变量，我们有场方程：

$$\nabla\cdot\boldsymbol{u}_{\mathrm{L}}=0 \tag{9.4.5}$$

$$-fv_{\mathrm{L}}=-g\frac{\partial\left\langle\tilde{\zeta}_1\right\rangle}{\partial x}+\frac{\partial}{\partial z}\left(\nu\frac{\partial u_{\mathrm{L}}}{\partial z}\right)-g\beta\frac{\partial}{\partial x}\int_z^0 c_0\mathrm{d}z'+\pi_1 \tag{9.4.6}$$

$$fu_{\mathrm{L}} = -g\frac{\partial\left\langle\tilde{\zeta}_1\right\rangle}{\partial y} + \frac{\partial}{\partial z}\left(\nu\frac{\partial v_{\mathrm{L}}}{\partial z}\right) - g\beta\frac{\partial}{\partial y}\int_z^0 c_0 \mathrm{d}z' + \pi_2 \tag{9.4.7}$$

$$\boldsymbol{u}_{\mathrm{L}} \cdot \nabla c_0 = \frac{\partial}{\partial z}\left(\langle k\rangle\frac{\partial c_0}{\partial z}\right) \tag{9.4.8}$$

$z = 0$：

$$w_{\mathrm{L}} = 0 \tag{9.4.9}$$

$$\nu\frac{\partial(u_{\mathrm{L}}, v_{\mathrm{L}})}{\partial z} = \left(\left\langle\tau_x, \tau_y\right\rangle\right) \tag{9.4.10}$$

$$k\frac{\partial c_0}{\partial z} = \langle\Gamma\rangle \tag{9.4.11}$$

$z = -h$：

$$\boldsymbol{u}_{\mathrm{L}} = 0 \tag{9.4.12}$$

$$\frac{\partial c_0}{\partial z} = 0 \tag{9.4.13}$$

式中，

$$(\pi_1, \pi_2) = \sum_{k=1}^{m}({}^k\pi_1, {}^k\pi_2) \tag{9.4.14}$$

$$\left.\begin{aligned} {}^k\pi_1 &= {}^k\pi_{\mathrm{N1}} + {}^k\pi_{\mathrm{b1}} + {}^k\pi_{\nu 1} \\ {}^k\pi_2 &= {}^k\pi_{\mathrm{N2}} + {}^k\pi_{\mathrm{b2}} + {}^k\pi_{\nu 2} \end{aligned}\right\} \tag{9.4.15}$$

$${}^k\pi_{\mathrm{N1}} = -g\frac{\partial}{\partial x}\left\langle\frac{1}{2}\,{}^k\boldsymbol{\xi}_0 \cdot \nabla\,{}^k\tilde{\zeta}_0\right\rangle$$

$${}^k\pi_{\mathrm{N2}} = -g\frac{\partial}{\partial y}\left\langle\frac{1}{2}\,{}^k\boldsymbol{\xi}_0 \cdot \nabla\,{}^k\tilde{\zeta}_0\right\rangle$$

$${}^k\pi_{b1} = \frac{\partial}{\partial z}\left\langle\left(\frac{5}{2}\frac{\partial\,{}^k\xi_0}{\partial x} + \frac{\partial\,{}^k\eta_0}{\partial y}\right)\nu\frac{\partial\,{}^k u_0}{\partial z} + \left(\frac{\partial\,{}^k\xi_0}{\partial y} + \frac{1}{2}\frac{\partial\,{}^k\eta_0}{\partial x}\right)\nu\frac{\partial\,{}^k v_0}{\partial z}\right\rangle + \frac{\partial}{\partial z}\left\langle\nabla\nu\cdot{}^k\boldsymbol{\xi}_0\frac{\partial\,{}^k u_0}{\partial z}\right\rangle$$

$${}^k\pi_{b2} = \frac{\partial}{\partial z}\left\langle\left(\frac{5}{2}\frac{\partial\,{}^k\eta_0}{\partial y} + \frac{\partial\,{}^k\xi_0}{\partial x}\right)\nu\frac{\partial\,{}^k v_0}{\partial z} + \left(\frac{\partial\,{}^k\eta_0}{\partial x} + \frac{1}{2}\frac{\partial\,{}^k\xi_0}{\partial y}\right)\nu\frac{\partial\,{}^k u_0}{\partial z}\right\rangle + \frac{\partial}{\partial z}\left\langle\nabla\nu\cdot{}^k\boldsymbol{\xi}_0\frac{\partial\,{}^k v_0}{\partial z}\right\rangle$$

$$\begin{aligned} {}^k\pi_{\nu 1} = -\Bigg\langle & \frac{\partial\,{}^k\boldsymbol{\xi}_0}{\partial z}\cdot\nabla\left(\nu\frac{\partial^2\,{}^k u_0}{\partial z}\right) + \frac{1}{2}\Bigg[\frac{\partial^2\,{}^k\xi_0}{\partial z\partial x}\left(\nu\frac{\partial\,{}^k u_0}{\partial z}\right) + \frac{\partial^2\,{}^k\eta_0}{\partial z\partial x}\left(\nu\frac{\partial\,{}^k v_0}{\partial z}\right) \\ & + {}^k\xi_0\frac{\partial^2}{\partial z\partial x}\left(\nu\frac{\partial\,{}^k u_0}{\partial z}\right) + {}^k\eta_0\frac{\partial^2}{\partial z\partial x}\left(\nu\frac{\partial\,{}^k v_0}{\partial z}\right)\Bigg]\Bigg\rangle \end{aligned}$$

$${}^k\pi_{\nu 2} = -\Bigg\langle\frac{\partial\,{}^k\boldsymbol{\xi}_0}{\partial z}\cdot\nabla\left(\nu\frac{\partial\,{}^k v_0}{\partial z}\right) + \frac{1}{2}\Bigg[\frac{\partial^2\,{}^k\xi_0}{\partial z\partial y}\left(\nu\frac{\partial\,{}^k u_0}{\partial z}\right) + \frac{\partial^2\,{}^k\eta_0}{\partial z\partial y}\left(\nu\frac{\partial\,{}^k v_0}{\partial z}\right)$$

$$+{}^k\xi_0\frac{\partial^2}{\partial z\partial y}\left(\nu\frac{\partial^k u_0}{\partial z}\right)+{}^k\eta_0\frac{\partial^2}{\partial z\partial y}\left(\nu\frac{\partial^k v_0}{\partial z}\right)\Bigg]\Bigg]\Bigg\rangle$$

其中 $\boldsymbol{u}_{\mathrm{L}}=\boldsymbol{i}u_{\mathrm{L}}+\boldsymbol{j}v_{\mathrm{L}}+\boldsymbol{k}w_{\mathrm{L}}$。

式（9.4.5）~（9.4.15）即为在最低阶意义上的斜压浅海环流的基本方程组，用以描述最低阶 Lagrange 余流——物质输运速度 $\boldsymbol{u}_{\mathrm{L}}$、余水位 $\langle\tilde{\zeta}_1\rangle$ 和零阶表观浓度 c_0 的分布及其相互作用。

在解释并讨论该基本方程组以前，为了对照，先给出 Euler 余环流 $\boldsymbol{u}_{\mathrm{E}}$ 满足的控制方程组[6]：

$$\nabla\cdot\boldsymbol{u}_{\mathrm{E}}=0 \tag{9.4.16}$$

$$-fv_{\mathrm{E}}=-g\frac{\partial\langle\tilde{\zeta}_1\rangle}{\partial x}+\frac{\partial}{\partial z}\left(\nu\frac{\partial u_{\mathrm{E}}}{\partial z}\right)-g\beta\frac{\partial}{\partial x}\int_z^0 c_0\mathrm{d}z'-\langle\boldsymbol{u}_0\cdot\nabla u_0\rangle \tag{9.4.17}$$

$$fu_{\mathrm{E}}=-g\frac{\partial\langle\tilde{\zeta}_1\rangle}{\partial y}+\frac{\partial}{\partial z}\left(\nu\frac{\partial v_E}{\partial z}\right)-g\beta\frac{\partial}{\partial y}\int_z^0 c_0\mathrm{d}z'-\langle\boldsymbol{u}_0\cdot\nabla v_0\rangle \tag{9.4.18}$$

$$\boldsymbol{u}_{\mathrm{E}}\cdot\nabla c_0=\frac{\partial}{\partial z}\left(\langle k\rangle\frac{\partial c_0}{\partial z}\right)-\langle\boldsymbol{u}_0\cdot\nabla c_1\rangle \tag{9.4.19}$$

$z=0$：

$$w_{\mathrm{E}}=\frac{\partial\langle u_0\tilde{\zeta}_0\rangle}{\partial x}+\frac{\partial\langle v_0\tilde{\zeta}_0\rangle}{\partial y} \tag{9.4.20}$$

$$\nu\frac{\partial(u_{\mathrm{E}},v_{\mathrm{E}})}{\partial z}=\left(\langle\tau_x\rangle,\langle\tau_y\rangle\right)-\nu\left\langle\tilde{\zeta}_0\frac{\partial^2(u_0,v_0)}{\partial z^2}\right\rangle \tag{9.4.21}$$

$$k\frac{\partial c_0}{\partial z}=\langle\Gamma\rangle \tag{9.4.22}$$

$z=-h$：

$$\boldsymbol{u}_{\mathrm{E}}=0 \tag{9.4.23}$$

$$\frac{\partial c_0}{\partial z}=0 \tag{9.4.24}$$

式中，$\boldsymbol{u}_{\mathrm{E}}=\boldsymbol{i}u_{\mathrm{E}}+\boldsymbol{j}v_{\mathrm{E}}+\boldsymbol{k}w_{\mathrm{E}}$；$-(\langle\boldsymbol{u}_0\cdot\nabla u_0\rangle,\langle\boldsymbol{u}_0\cdot\nabla v_0\rangle)$ 为“潮体力”；$-\langle\boldsymbol{u}_0\cdot\nabla c_1\rangle$ 为“潮弥散”；$\dfrac{\partial\langle u_0\tilde{\zeta}_0\rangle}{\partial x}+\dfrac{\partial\langle v_0\tilde{\zeta}_0\rangle}{\partial y}$ 为“海面潮源”；$\nu\left\langle\tilde{\zeta}_0\dfrac{\partial^2(u_0,v_0)}{\partial z^2}\right\rangle$ 为“潮应力”。

将方程组（9.4.16）~（9.4.24）与方程组（9.4.5）~（9.4.15）相比，我们发现，前者中如“潮弥散”、“海面潮源”和“潮应力”诸项皆为多余项。“潮弥散”假设的不确实性前面已作了讨论。注意到方程（9.4.10），“潮应力”根本不存在，表明了良好的在界面上的应力连续性质。式（9.4.9）表明：在式（9.4.20）中表现出来的虚假的“海面潮源”根本不存在，表明以物质输运速度来体现浅海环流流场确有良好的流场连续性质。事实上，式（9.4.9）表明的 $w_L=0$ 乃是流体物质面守恒 $\boldsymbol{u}_L\cdot\nabla\langle F\rangle=0$［式（9.3.17）］的特例。由此可见，与 Euler 余流相比，以 Lagrange 余流的最低阶近似——物质输运速度来体现浅海环流流场，有着无可争辩的良好的流场连续性质。

“潮体力” (π_1,π_2) 表明了中时间尺度的潮汐对大时间尺度环流的耦合作用，该体积力做的功表明了潮振荡能量向定常环流的转移，从而表明，在一个风作用下的以潮周期运动占优势的斜压浅海中，环流的驱动力，除了惯常认为的海面风应力和热盐产生的斜压力外，尚有潮周期运动非线性耦合所产生的效应——“潮体力”（tidal body forces）(π_1,π_2)。也就是说，在浅海系统中，其环流不仅包括风生正压和热盐斜压分量，而且也耦合了潮生环流分量。

由“潮体力” (π_1,π_2) 表达式（9.4.14）可见，一个多频天文潮系统产生的“潮体力” (π_1,π_2) 乃组成该天文潮系统的每一个天文分潮波各自产生的“潮体力” $({}^k\pi_1,{}^k\pi_2)$ $(k=1,2,\cdots,m)$ 之和。

由 $({}^k\pi_1,{}^k\pi_2)$ 的表达式（9.4.15）可见，其中每一项皆正比于天文潮场中标识流体微团的位移、其分量或它们的导数，这表明“潮体力”的产生紧密联系于 Lagrange 轨迹运动与潮流场的非线性耦合效应。同时，“潮体力”原则上由两部分组成：水平无旋力 $({}^k\pi_{\mathrm{N1}},{}^k\pi_{\mathrm{N2}})$ 和湍粘性导致的有旋力 $({}^k\pi_{\mathrm{b1}}+{}^k\pi_{v1},{}^k\pi_{\mathrm{b2}}+{}^k\pi_{v2})$。

当我们不考虑粘性效应时，也就是说，若流体为理想流体，潮流及场中标识流体微团的位移可看作与铅垂坐标无关时，潮体力中 $({}^k\pi_{\mathrm{b1}},{}^k\pi_{\mathrm{b2}})=({}^k\pi_{v1},{}^k\pi_{v2})=0$，亦即 $({}^k\pi_1,{}^k\pi_2)=({}^k\pi_{N1},{}^k\pi_{N2})$，且仅为 (x,y,t) 的函数。运动方程（9.4.6）和（9.4.7）化简为

$$-fv_{\mathrm{L}}=-g\frac{\partial\varphi}{\partial x} \tag{9.4.25}$$

$$fu_{\mathrm{L}}=-g\frac{\partial\varphi}{\partial y} \tag{9.4.26}$$

其中 $\varphi=\left\langle\tilde{\zeta}_1\right\rangle+\sum_{k=1}^{m}\left\langle\frac{1}{2}{}^k\xi_0\cdot\nabla{}^k\tilde{\zeta}_0\right\rangle+\beta\int_z^0 c_0\mathrm{d}z'$。

如果不考虑潮体力的影响，即如果将 φ 表达式中的第二项置零，则方程（9.4.25）和（9.4.26）表达了经典的地转运动。考虑潮体力表达的天文潮非线性耦合效应 $({}^k\pi_{N1},{}^k\pi_{N2})$，方程（9.4.25）和（9.4.26）可认为表述了一个“广义的地转运动”，因为此时它们保留了地转运动的主要特征。事实上，对上述两方程交叉求导且作和，代入连续方程（9.4.5），并考虑到运动学边界条件（9.4.9），则有

$$\frac{\partial u_{\mathrm{L}}}{\partial x}+\frac{\partial v_{\mathrm{L}}}{\partial y}=w_{\mathrm{L}}=0 \tag{9.4.27}$$

从而表明了它为水平无辐散运动，并且水平成层状。若进一步忽略斜压影响，则该运动将蜕化为纯二维的，满足 Taylor-Proudman 定理，形成 Taylor 柱[71]，特别应指出，“广义地转运动”如地转运动一样，不能从动力学观点完整而唯一地确定问题的流场和余水位场，从而启示人们注意湍流粘性对确定运动的重要意义，应小心而准确地选择湍粘性系数的铅垂剖面形式等等[8]。

当忽略中时间尺度的潮运动与大尺度环流之间的非线性能量转移时，亦即当令 $(\pi_1,\pi_2)=0$ 时，方程组（9.4.5）~（9.4.13）蜕化为描述浅海中风生热盐环流的经典方程组[1]。

应注意，尽管在一个斜压浅海的弱非线性系统中，“动力学”和“热力学”已被解开了耦合[见零阶模型（9.2.40）~（9.2.50）和一阶模型（9.2.53）~（9.2.63）]，但描述浅海环流和长期输运的方程组（9.4.5）~（9.4.15）却仍为一个典型的动力学–热力学非线性耦合系统。

当然，对于某些浅海湾来说，作为正压浅海模型近似，式（9.4.6）和（9.4.7）中的斜压力可以忽略，则流（以及余水位）和浓度方程可以先后求解。

9.5　全流及其输运模型

虽然 9.1 节中已指出，涉及利用 Lagrange 余环流来描述浅海环流问题，其原始方程组原则上应为三维空间方程组（9.2.1）~（9.2.9），方有确定的动力学意义，但是，这种空间三维问题，经潮周期平均后构成的描述浅海环流及其长期输运的场方程组（9.4.5）~（9.4.15），其深度平均或全流模型已有明确的动力学意义。当我们并不关心环流场的铅垂结构而仅对水平输运过程感兴趣时，构成浅海环流基本方程组（9.4.5）~（9.4.15）的全流方程或深度平均变量的方程组就显得特别方便而有效，不仅对数值模拟和预测问题是如此，对动力学某些方面（如涡动力学）的研究亦是如此。

引入全流 $\boldsymbol{U}_\mathrm{L}=\boldsymbol{i}U_\mathrm{L}+\boldsymbol{j}V_\mathrm{L}$：

$$(U_\mathrm{L},V_\mathrm{L})=\int_{-h}^{0}(u_\mathrm{L},v_\mathrm{L})\mathrm{d}z \tag{9.5.1}$$

同时引入深度平均速度 $\bar{\boldsymbol{u}}_\mathrm{L}=\boldsymbol{i}\bar{u}_\mathrm{L}+\boldsymbol{j}\bar{v}_\mathrm{L}$：

$$(\bar{u}_\mathrm{L},\bar{v}_\mathrm{L})=\frac{1}{h}(U_\mathrm{L},V_\mathrm{L}) \tag{9.5.2}$$

依深度积分方程组（9.4.5）~（9.4.8），利用海面和海底诸边界条件（9.4.9）~（9.4.13），可导出全流方程组：

$$\nabla\cdot\boldsymbol{U}_\mathrm{L}=0 \tag{9.5.3}$$

$$fV_\mathrm{L}=-gh\frac{\partial\langle\tilde{\xi}_1\rangle}{\partial x}+\tau_x-\tau_{bx}-g\beta\int_{-h}^{0}\int_{z}^{0}\frac{\partial c_0}{\partial x}\mathrm{d}z'\mathrm{d}z+\Pi_1 \tag{9.5.4}$$

$$fU_{\rm L} = -gh\frac{\partial\left\langle\tilde{\zeta}_1\right\rangle}{\partial y} + \tau_y - \tau_{\rm by} - g\beta\int_{-h}^{0}\int_{z}^{0}\frac{\partial c_0}{\partial y}{\rm d}z'{\rm d}z + \Pi_2 \tag{9.5.5}$$

$$\boldsymbol{U}_{\rm L}\cdot\nabla\overline{c}_0 = \left\langle\Gamma\right\rangle + \nabla\cdot\left[h(-\overline{\boldsymbol{u}_{\rm L}'c_0'})\right] \tag{9.5.6}$$

式中，$\boldsymbol{u}_{\rm L}' = \boldsymbol{u}_{\rm L} - \overline{\boldsymbol{u}}_{\rm L}$；$c_0' = c_0 - \overline{c}_0$；$(\Pi_1, \Pi_2) = \int_{-h}^{0}(\pi_1, \pi_2){\rm d}z$; $(\tau_{\rm bx}, \tau_{\rm by})$ 为底应力。

方程组（9.5.3）~（9.5.6）是不封闭的，必须给出某些补充的假设和规律。

尽管一个广义的底应力的线性律已经由湍粘性的线化假设而导出[57]，但是这里为了简便且不影响主题，只简单地假定底应力正比于全流向量：

$$\boldsymbol{\tau}_{\rm b} = \frac{r}{h}\boldsymbol{U}_{\rm L} \tag{9.5.7}$$

其中 r 为阻尼系数，应正比于速度本身的幅度，但此处仅设为常数。

长期输运的深度积分方程（9.5.6）中右端第二项表示流速和浓度铅垂分布不均匀导致的弥散效应，常称之为“切变效应”（shear effect），并且通常被参数化为[41]

$$-\overline{\boldsymbol{u}_{\rm L}'c_0'} = \boldsymbol{D}\cdot\nabla c_0 \tag{9.5.8}$$

其中张量 $\boldsymbol{D}$ 称作由切变效应产生的弥散系数。但当这种切变效应与对流输运相比时至多为 $O(\kappa)$，故可略而不计。因此，方程（9.5.6）可化简为

$$\boldsymbol{U}_{\rm L}\cdot\nabla\overline{c}_0 = \left\langle\Gamma\right\rangle \tag{9.5.9}$$

运动方程（9.5.4）和（9.5.5）中右端的斜压项可分解为两部分：

$$-g\beta\int_{-h}^{0}\int_{z}^{0}\begin{bmatrix}\dfrac{\partial c_0}{\partial x}\\ \dfrac{\partial c_0}{\partial y}\end{bmatrix}{\rm d}z'{\rm d}z = -\frac{h^2}{2}g\beta\begin{bmatrix}\dfrac{\partial\overline{c}_0}{\partial x}\\ \dfrac{\partial\overline{c}_0}{\partial y}\end{bmatrix} - g\beta\int_{-h}^{0}\int_{z}^{0}\begin{bmatrix}\dfrac{\partial c_0'}{\partial x}\\ \dfrac{\partial c_0'}{\partial y}\end{bmatrix}{\rm d}z'{\rm d}z \tag{9.5.10}$$

其中包括 c_0' 的项为空间三维函数。

潮体力 (π_1, π_2) 的深度积分 (Π_1, Π_2) 可分解为三个分量：

$$\begin{bmatrix}\Pi_1,\\ \Pi_2\end{bmatrix} = \sum_{k=1}^{m}\begin{bmatrix}{}^k\Pi_{\rm N1} + {}^k\Pi_{\rm b1} + {}^k\Pi_{\nu1},\\ {}^k\Pi_{\rm N2} + {}^k\Pi_{\rm b2} + {}^k\Pi_{\nu2}\end{bmatrix} \tag{9.5.11}$$

式中，

$${}^k\Pi_{\rm N1} = \int_{-h}^{0}{}^k\pi_{\rm N1}\,{\rm d}z = -g\frac{\partial}{\partial x}\int_{-h}^{0}\left\langle\frac{1}{2}{}^k\boldsymbol{\xi}_0\cdot\nabla{}^k\tilde{\zeta}_0\right\rangle{\rm d}z$$

$${}^k\Pi_{\rm N2} = \int_{-h}^{0}{}^k\pi_{\rm N2}\,{\rm d}z = -g\frac{\partial}{\partial y}\int_{-h}^{0}\left\langle\frac{1}{2}{}^k\boldsymbol{\xi}_0\cdot\nabla{}^k\tilde{\zeta}_0\right\rangle{\rm d}z$$

$${}^k\Pi_{\rm b1} = \int_{-h}^{0}{}^k\pi_{\rm b1}\,{\rm d}z = -\left\langle\left(\frac{5}{2}\frac{\partial\,{}^k\xi_0}{\partial x} + \frac{\partial\,{}^k\eta_0}{\partial y}\right)\nu\frac{\partial\,{}^ku_0}{\partial z} + \left(\frac{\partial\,{}^k\xi_0}{\partial y} + \frac{1}{2}\frac{\partial\,{}^k\eta_0}{\partial x}\right)\nu\frac{\partial\,{}^kv_0}{\partial z}\right\rangle_{z=-h}$$

$$ {}^k\Pi_{b2}=\int_{-h}^{0}{}^k\pi_{b2}\mathrm{d}z=-\left\langle\left(\frac{5}{2}\frac{\partial\,{}^k\eta_0}{\partial y}+\frac{\partial\,{}^k\xi_0}{\partial x}\right)\nu\frac{\partial\,{}^k v_0}{\partial z}+\left(\frac{\partial\,{}^k\eta_0}{\partial x}+\frac{1}{2}\frac{\partial\,{}^k\xi_0}{\partial y}\right)\nu\frac{\partial\,{}^k u_0}{\partial z}\right\rangle_{z=-h} $$

$$ {}^k\Pi_{v1}=\int_{-h}^{0}{}^k\pi_{v1}\mathrm{d}z,\quad {}^k\Pi_{v2}=\int_{-h}^{0}{}^k\pi_{v2}\mathrm{d}z $$

$({}^k\pi_{v1}, {}^k\pi_{v2})$ 见式（9.4.15）。

引入满足连续方程（9.5.3）的流函数 ψ_{L}：

$$ \frac{\partial\psi_{\mathrm{L}}}{\partial x}=V_{\mathrm{L}},\quad \frac{\partial\psi_{\mathrm{L}}}{\partial y}=-U_{\mathrm{L}} \tag{9.5.12} $$

并代入由运动方程（9.5.4）和（9.5.5）交叉微商作和产生的涡度方程，得到 ψ_{L} 满足的方程：

$$ \begin{aligned} &\nabla\cdot\left(\frac{r}{h}\nabla\psi_{\mathrm{L}}\right)-\frac{r}{h^2}\nabla h\cdot\nabla\psi_{\mathrm{L}}+\frac{f}{h}J(h,\psi_{\mathrm{L}})\\ &=-\frac{h}{2}g\beta J(h,\bar{c}_0)+\frac{1}{h}g\beta J\left(h,\int_{-h}^{0}\int_{z}^{0}c_0'\mathrm{d}z'\mathrm{d}z\right)\\ &\quad+\boldsymbol{k}\cdot\left\{\nabla\times\langle\boldsymbol{\tau}_{\mathrm{a}}\rangle-\frac{1}{h}\nabla h\times\langle\boldsymbol{\tau}_{\mathrm{a}}\rangle+\nabla\times\boldsymbol{\Pi}-\frac{1}{h}\nabla h\times\boldsymbol{\Pi}\right\} \end{aligned} \tag{9.5.13} $$

式中，$J(a,b)$ 表示 Jacobian 算子，$J(a,b)=\dfrac{\partial a}{\partial x}\dfrac{\partial b}{\partial y}-\dfrac{\partial a}{\partial y}\dfrac{\partial b}{\partial x}$；$\boldsymbol{\tau}_{\mathrm{a}}=\boldsymbol{i}\tau_x+\boldsymbol{j}\tau_y$；潮体力深度积分 $\boldsymbol{\Pi}=\boldsymbol{i}\Pi_1+\boldsymbol{j}\Pi_2$。

物质输运速度的全流流函数 ψ_{L} 满足的涡度方程（9.5.13）表明，环流的全流场或深度平均流场不仅为风生–热盐性质的，而且亦为潮生性质的。并且，潮体力 (π_1,π_2) 的深度积分 $\boldsymbol{\Pi}$ 在方程（9.5.13）中表现为与风应力 $\boldsymbol{\tau}_{\mathrm{a}}$ 同样的形式，即可称为“潮应力”[注意，这里的潮应力 $\boldsymbol{\Pi}$ 不同于 Nihoul 等（1975）[4]在 Euler 余流中命名的“潮应力”]。

当略去“潮应力”后，方程（9.5.13）蜕化为经典的风生–热盐环流的涡度方程[1]。

作为一个诊断方程，利用（9.5.13）求解全流函数，与经典的风生–热盐环流涡度方程求解相比，不增加任何实质上的困难。其数值解法参见文献[1]。

事实上，当表观浓度 c 表示的是像热、盐这样一些斜压变量时，一般说，涡度方程（9.5.13）必须与全流型的长期输运方程（9.5.9）联立求解，后者以 ψ_{L} 表示的形式为

$$ J(\psi_{\mathrm{L}},\bar{c}_0)=\langle\Gamma\rangle \tag{9.5.14} $$

也就是说，（9.5.13）、（9.5.14）组成一联立方程组以求解 ψ_{L} 和 $\bar{c}_0$。

不过应指出，方程组（9.5.13）和（9.5.14）并不封闭，这是因为涡度方程（9.5.13）中包含了表示表观浓度 c 的铅垂分布的项 c_0'，从物理角度看，这实质上是反映了热盐效应的空间三维性质。只有当 c_0' 远小于 $\bar{c}_0$ 时，从而忽略包含 c_0' 的项，才能确实利用联

立方程组（9.5.13）、（9.5.14）求解 ψ_{L} 和 $\overline{c}_0$ 。

应该指出，热盐环流固有的空间三维性质可能表明：一个斜压浅海的环流问题采用全流模型也许是不可取的。

也应指出，若我们仅局限于全流和水平输运的“诊断”，上面已表明，利用这一全流函数满足的涡度方程（9.5.13）还是简便而有效的。

一个有趣的事实是，即使我们的兴趣已集中于全流所表达的水平二维深度平均环流和输运的问题，也必须首先完成一个天文潮三维空间问题的计算，以便预先给出“潮体力”(π_1,π_2)并进而给出其深度积分“潮应力”$\boldsymbol{\Pi}$ 函数，这是 Lagrange 余流固有的三维空间性质的再次显现。事实上，全流$(U_{\mathrm{L}},V_{\mathrm{L}})$或其流函数 ψ_{L} 乃物质输运速度$(u_{\mathrm{L}},v_{\mathrm{L}})$的深度积分，后者是最低阶或一阶的 Lagrange 余流，是由天文潮流梯度点乘潮流场中标识流体微团 Lagrange 位移后潮周期平均的效应——显然，一般说，Lagrange 位移是一个典型的空间三维变量。

全流问题的侧边界条件为

$$J(\psi_{\mathrm{L}},B_{\mathrm{c}})=0 \tag{9.5.15}$$

其中，B_{c} 如前所述，乃侧边岸界方程。方程（9.5.15）可由环流场中物质面守恒方程（9.3.17）推出：令 $F_0=B_{\mathrm{c}}$，并依深度积分（9.3.17），然后代入式（9.5.12），即导出方程（9.5.15），这就是固体边界上流函数为一常数的条件：

$$\psi_{\mathrm{L}}\big|_{(x,y)\in B_{\mathrm{c}}}=\mathrm{const} \tag{9.5.16}$$

在开边界 B_0 上给出流量分布，即流函数为已知函数 $\varPsi(x,y\in B_0)$：

$$\psi_{\mathrm{L}}\big|_{(x,y)\in B_{\mathrm{c}}}=\varPsi(x,y) \tag{9.5.17}$$

描述表观浓度零阶近似的深度平均 $\overline{c}_0$ 的长期输运方程（9.5.9）或（9.5.14）表明：在最低阶意义上，深度平均环流场中浓度的水平输运为纯对流性质的，并取决于海面通量的潮周期平均场 $\langle\varGamma\rangle$。

如果海面外源 $\varGamma$ 或至少其潮周期平均值 $\langle\varGamma\rangle$ 为零，则方程（9.5.9）或（9.5.14）化简为

$$J(\psi_{\mathrm{L}},\overline{c}_0)=0 \tag{9.5.18}$$

方程（9.5.18）表明，当不考虑海面外源时，深度平均环流将沿深度平均零阶表观浓度等值线流动。

流线与浓度等值线相重合这一特性表明：像盐度、温度或浓度的场观测值所连成的等值线，既可作为深度平均环流理论结果的验证资料，也可用以推断该环流流型。

应指出，方程（9.5.18）及其推论仅在海域广大的内区是正确的，也就是说，理论的确实区域不包括侧边界层[14]。事实上，沿海岸边界的浓度梯度法向投影应为零。从

物理观点看，这表明在海岸与海水之间无物质或能量的转移。但方程(9.5.18)和(9.5.16)却表明，沿海岸为浓度的等值线，也就是说，在岸界，无物质和能量转移的上述条件不能被满足。

沿岸界或在岸界附近的海水混合过程是相当复杂的，人们对其了解甚少。由于浅滩或河口中分叉的存在，所谓“潮泵”（tidal pumping）和“潮陷”（tidal trapping）都属于近岸狭窄区域中附加的混合机制[36]，但由于这些机制因地而异，故下面的讨论中将排除它们，而假定海岸足够光滑且无大的曲率。

事实上，在海岸附近一相对狭窄区域内，像切变效应导致的弥散乃至湍混合效应与对流输运同等重要。并且，应该期望，在这样一个弥散边界层中，浓度将由海域的广大内区的值连续过渡至边界层岸界处，且满足浓度梯度法向分量为零的条件。

对应于长期输运方程（9.5.6）和参数化条件（9.5.8）的无因次方程已表明，弥散与对流相比不大于 $O(\kappa)$ [14]。这样，依据边界层技术来估值，该侧向弥散边界层（dispersion boundary layer）的无因次宽度 δ 应不大于 $O(\sqrt{\kappa})$，即

$$O(\delta) \leqslant \kappa^{1/2} \qquad (9.5.19)$$

如果假设 $(x_{\mathrm{b}}, y_{\mathrm{b}})$ 为局部边界层坐标中的切线和法线坐标，则描述弥散边界层中浓度潮周期时均的深度平均场 $\overline{c}_0$ 的长期输运方程为

$$(\boldsymbol{U}_{\mathrm{L}})_{\mathrm{b}} \cdot \nabla_{\mathrm{b}} (\overline{c}_0)_{\mathrm{b}} = (\Gamma)_{\mathrm{b}} + \mathrm{D} \frac{\partial}{\partial y_{\mathrm{b}}} \left[h \frac{\partial (\overline{c}_0)_{\mathrm{b}}}{\partial y_{\mathrm{b}}} \right] \qquad (9.5.20)$$

式中，带有下标 b 的字母皆为弥散边界层中的变量和算子，且为了简单已把张量 $\boldsymbol{D}$ 假设为一常值标量 D。岸界 $(y_{\mathrm{b}} = 0)$ 和开边界 $(y_{\mathrm{b}} \to \infty)$ 处的边界条件为

$y_{\mathrm{b}} = 0$：

$$\frac{\partial (\overline{c}_0)_{\mathrm{b}}}{\partial y_{\mathrm{b}}} = 0 \qquad (9.5.21)$$

$y_{\mathrm{b}} \to \infty$：

$$(\overline{c}_0)_{\mathrm{b}} \to c_0 \qquad (9.5.22)$$

方程（9.5.20）~（9.5.22）构成的定解问题表达了在海岸附近的弥散边界层中长期输运过程的物理机制。除去对流、海面通量和弥散三者平衡外，应特别指出：浓度横过边界层的改变所产生的弥散效应才是重要的。

9.6 涡度分析

环流的流型和动力机制，就其本质而言，与其说取决于运动方程本身，还不如说取决于其涡度方程更为恰当。对于风生–热盐环流问题涡度的分析，在物理海洋学或海

洋动力学的教科书上通常都可以找到，也就是说，这是一个相当经典的问题，而潮致浅海或河口 Euler 余环流的涡度分析仅有 10 年或不超过 15 年的历史。Euler 余环流流涡（vortices）已在不同的浅海中通过各异的方法先后发现了：通过潮流的数值实验，对于陆架海，有 Nihoul 和 Ronday（1975）[4]，对于内陆海，有 Tee（1976）[15]；通过水利学模拟实验，有 Sugimoto（1975）[72]和 Yanagi（1976）[73]；通过海上观测对流速的分析也发现了这些流涡[74~76]。关于海洋底形不均匀致使余涡产生的动力学理论和统计研究，可参看文献[77]。但正如本章所断言的那样，平均环流或余环流，在浅海或感潮河口中，不应以 Euler 余流来描述，而应以 Lagrange 平均速度来描述，特别是在本文所提出的弱非线性系统中，可以用物质输运速度来体现最低阶近似下的浅海平均环流的速度场。由此可见，直接研究物质输运速度所满足的涡度方程，显然是令人感兴趣的。应当指出，间接探讨涡度由潮运动向余运动转移的工作，已由 Robinson 做了[78]，但由于他只依据一个潮流所满足的深度平均涡度方程来推断余涡机制，故只能是定性的描述，而不能形成一个系统的数学理论。本文推出的全流或深度平均环流的涡度方程（9.5.13），是探讨浅海或感潮河口中以物质输运速度来体现平均环流场的涡度机制的有力工具。事实上，方程（9.5.13）可改写为显含涡度的形式：

$$
\begin{aligned}
r\omega_{\mathrm{L}} = f\nabla h\cdot\bar{\boldsymbol{v}}_{\mathrm{L}} + \boldsymbol{k}\cdot\Bigg\{ & \frac{r}{h}\nabla h\times\bar{\boldsymbol{v}}_{\mathrm{L}} - \frac{1}{2}g\,\beta h\nabla h\times\nabla\bar{c}_0 + \frac{g\beta}{h}\nabla h\times\nabla\int_{-h}^{0}\int_{z}^{0}c_0'\mathrm{d}z'\mathrm{d}z \\
& + \nabla\times\langle\boldsymbol{\tau}_a\rangle - \frac{\nabla h}{h}\times\langle\boldsymbol{\tau}_a\rangle + \nabla\times\boldsymbol{\Pi} - \frac{\nabla h}{h}\times\boldsymbol{\Pi}\Bigg\}
\end{aligned}
\qquad (9.6.1)
$$

式中，$\bar{\boldsymbol{v}}_{\mathrm{L}} = \boldsymbol{i}\bar{u}_{\mathrm{L}} + \boldsymbol{j}\bar{v}_{\mathrm{L}}$；$\omega_{\mathrm{L}}$ 为深度平均的物质输运速度之相对涡度的铅垂分量：

$$
\omega_{\mathrm{L}} = \frac{\partial\bar{v}_{\mathrm{L}}}{\partial x} - \frac{\partial\bar{u}_{\mathrm{L}}}{\partial y} \qquad (9.6.2)
$$

可简称为涡度或余涡度。

上式表示一个涡度平衡方程，表明了涡度收支。左端项表示由于底摩擦效应导致的涡度的耗散，而右端诸项则表示涡度产生的不同效应：第一、二两项分别表示地转与海底地形的相互作用，以及底应力和底形的相互作用；三、四两项表示斜压与底形之间的相互作用，有时称为“底斜效应”（the baroclinity-topography effect）；第六项和第八项表示海面风应力与底形之间的相互作用以及潮应力与底形之间的相互作用；第五项和第七项分别表示风应力旋度和潮应力旋度效应。

应当指出，涉及海面风应力和斜压效应的诸涡度产生项的分析完全相似于经典的风生–热盐环流问题，此处不作详细讨论。我们的兴趣是详细探讨和分析潮致余流的涡度，即潮致余涡问题。

涡度方程（9.6.1）的最后两项，$\nabla\times\boldsymbol{\Pi}$ 和 $-\dfrac{\nabla h}{h}\times\boldsymbol{\Pi}$，描述了潮流场中相对涡度向余

流场中余涡度的非线性转移。由 $\boldsymbol{\Pi}$ 的表达式（9.5.11）可以看出，这种涡度的非线性转移，即潮致余涡度的产生，具有三种机制：

其一，潮应力的无旋分量($^k\Pi_{N1}, {}^k\Pi_{N2}$)表达了标识流体微团的 Lagrange 位移与天文分潮波之间的非线性耦合效应，并且由于 $\nabla\times{}^k\boldsymbol{\Pi}_N=0$ ($^k\boldsymbol{\Pi}_N=\boldsymbol{i}\,{}^k\Pi_{N1}+\boldsymbol{j}\,{}^k\Pi_{N2}$)，故仅有 $-\dfrac{\nabla h}{h}\times{}^k\boldsymbol{\Pi}_N$ 起作用，即潮应力的无旋分量仅通过与海盆深度的改变的耦合才起作用。进一步可以看出，这一潮应力的无旋分量必须有偏离海盆深度等值线的梯度方向的分力，方能起到产生潮致余涡的作用。

其二，$^k\boldsymbol{\Pi}_b=\boldsymbol{i}\,{}^k\Pi_{b1}+\boldsymbol{j}\,{}^k\Pi_{b2}$ 表达了潮流场中的底应力与标识流体微团的 Lagrange 位移变化之间的非线性耦合效应，它为其旋度部分与深度梯度耦合部分之和：$\nabla\times{}^k\boldsymbol{\Pi}_b+\left(-\dfrac{\nabla h}{h}\times{}^k\boldsymbol{\Pi}_b\right)$，且第二项的特性显然与无旋分量 $^k\boldsymbol{\Pi}_N$ 相同，而首项的作用则类似于风应力旋度项。

最后，$^k\boldsymbol{\Pi}_v=\boldsymbol{i}\,{}^k\Pi_{v1}+\boldsymbol{j}\,{}^k\Pi_{v2}$，乃潮应力的另一有旋分量，它来源于潮流场中标识流体微团的位移与湍应力相互非线性耦合的效应，其特性类似于 $^k\boldsymbol{\Pi}_b$ 组成的有旋分量。

应强调指出：潮应力中包含潮流场中水微团的位移向量这一点充分表明，潮致余涡来源于并取决于潮的 Lagrange 动力学的机制，亦即后者决定了涡度由潮环流场向余环流场的转移。

海盆底形和侧界的起伏变化常常是产生涡度的重要条件。由方程（9.6.1）可见，尽管海深的不均匀并非产生涡度和维持涡度的必要条件，但显然是相当重要的条件。众所周知，斜压力产生涡度总是取决于深度不均匀性的，故如上所述，称之为底斜效应，即平底海洋中斜压力对涡度的改变无任何贡献。相反，风应力对涡度产生的贡献，除去耦合了深度变化的分量外，尚有风应力旋度的分量，不考虑斜压力和潮应力的平底海洋的环流动力学满足著名的风旋度方程。这样，我们所考虑的深度不均匀的浅海中潮应力产生的潮余涡问题，将是研究海底地形对潮致余涡的产生和改变的贡献的典型问题[79]。其中的两种特殊情形是令人感兴趣的，现分述如下。

就平底海洋中的潮致余涡问题而论，其控制方程可由（9.6.1）蜕化为

$$r\omega_L=\boldsymbol{k}\cdot\nabla\times\boldsymbol{\Pi}_v \tag{9.6.3}$$

其中，$\boldsymbol{\Pi}_v=\sum_{k=1}^{m}{}^k\boldsymbol{\Pi}_v$。显然，相对涡度是由潮流场内部湍粘性与微团位移的耦合产生并维持的。

就非平底海洋但潮流场铅垂均匀的情形而言，其控制方程由（9.6.1）蜕化为

$$r\omega_L=f\nabla h\cdot\bar{\boldsymbol{v}}_L+\boldsymbol{k}\cdot\frac{r}{h}\nabla h\times\bar{\boldsymbol{v}}_L-\boldsymbol{k}\cdot\frac{\nabla h}{h}\times\boldsymbol{\Pi}+\boldsymbol{k}\cdot\nabla\times\boldsymbol{\Pi}_b \tag{9.6.4}$$

式中，$\Pi = \Pi_{\mathrm{b}} + \Pi_{\mathrm{N}}$，$\Pi_{\mathrm{b}} = \sum_{k=1}^{m} {}^{k}\Pi_{\mathrm{b}}$，$\Pi_{\mathrm{N}} = \sum_{k=1}^{m} {}^{k}\Pi_{\mathrm{N}}$。

在这一情形下，潮流场内部湍粘性由于流场的铅垂均匀性已不起作用；但底应力却仍然存在，并且通过 ${}^{k}\Pi_{\mathrm{b}}$ 表达了对潮余涡诞生和维持的作用。

当一个平底海洋中存在一铅垂充分混合的潮流运动时，由于其潮流的铅垂分布相对均匀，故 $\Pi_{v} = 0$，从而方程（9.6.3）简化为

$$r\omega_{\mathrm{L}} = 0 \tag{9.6.5}$$

此方程由方程（9.6.4）亦可推出，因为此时有 $\nabla h = 0$ 和 $\Pi_{\mathrm{b}} = 0$。

对应于涡度方程（9.6.5）的以流函数表示的方程为

$$\nabla^2 \psi_{\mathrm{L}} = 0 \tag{9.6.6}$$

事实上，方程（9.6.6）亦可由以流函数表示的涡度方程（9.5.13）蜕化而来。

由方程（9.6.5）和（9.6.6）可见，在一个平底的浅海中，铅垂方向相对均匀的潮流运动所导致的深度平均的余涡度为零，因此其潮致积分环流不存在。此时，相对余涡的诞生和维持，除依赖于外海进入流的影响外，仅取决于海域上空海面风应力旋度的作用。

注意，以上所得有关余涡度的种种结论，仅为深度平均环流问题的结果[84]。

9.7　渤海环流数值实验

渤海环流的数值实验可作为本章介绍的浅海环流理论模型的一个初步应用，它提供了某些令人感兴趣的结果。

渤海是嵌入中国北部大陆的一个内海，北依辽宁，西靠河北，南面山东，仅在其东向，通过渤海海峡与黄海沟通（图 9.1）。

渤海乃一典型浅海，平均水深仅 20 m；南北长约 500 km，东西宽约 300 km;含辽东、渤海、莱州三湾，其水深分布趋势是由海峡向沿岸逐渐变浅；渤海湾和莱州湾水深平均约 10 m，辽东湾较深。以水深分布表示的渤海海底地形见图 9.1。

渤海上空基本上被亚细亚季风场所控制，冬季多东北风，夏季为西南风，春秋为过渡季节。冬季海水由于充分混合，密度几乎呈均匀状态；夏季海水有层化现象。通过渤海海峡的水交换问题尚无定论，但通常认为，黄海暖流余脉由海峡北部进入渤海，而通过海峡南端流出的海水沿岸南下，汇入黄海沿岸水。此外，沿岸尚有以黄河为主的几大河的淡水入海，在夏季丰水期，应考虑它们对河口附近水域海水混合和局部环流的影响。

关于渤海环流问题，管秉贤教授等已依据历史调查资料和分析结果绘出了月平均表面环流图[80]。它表明，典型的渤海夏季环流图式为一简单的逆时针流环（图 9.2）。

其后，有人以一个浅海风生–热盐环流数值模型模拟了渤海夏季环流，同样表明它确为一逆时针环流[81]。同时，后者还揭示了这一渤海夏季风生–热盐环流的动力学机制：在渤海实际底形条件下，渤海夏季环流的型式和流速大小主要取决于通过海峡的出入海流形成的“边界力”；风应力和热盐力对流场形式几乎无影响，仅对流速大小略有作用——风应力引起的增值一般不超过 1%，而热盐力引起的增值不超过 10%；黄河径流仅对其邻近海域有影响。顺便指出，由于渤海冬季风效应增强，海峡水交换已趋复杂，甚至要考查海冰动力学，故为了突出本文主旨，这里将仅涉及渤海夏季环流。

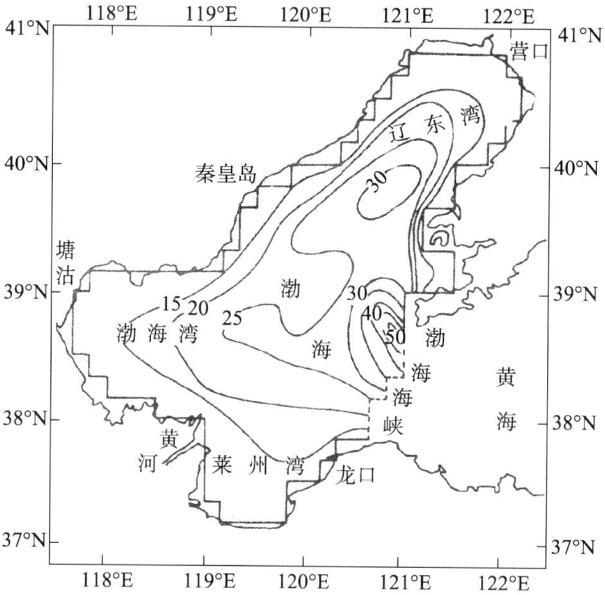

图 9.1　渤海水深图

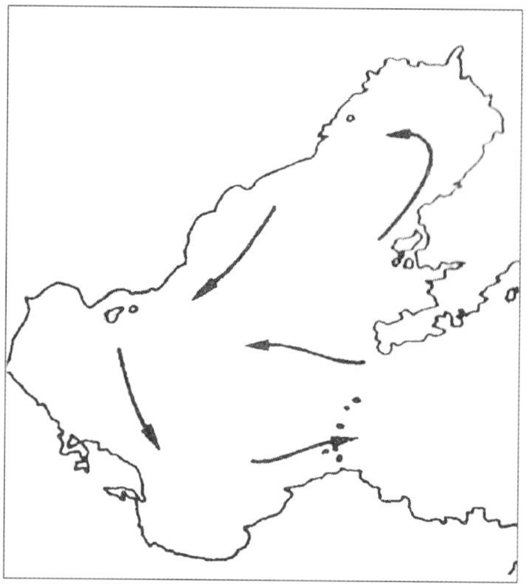

图 9.2　渤海夏季表面环流图

以上风生–热盐环流模型未计入潮汐非线性耦合效应产生的“潮体力”导致的潮汐余流分量，因此是一个不完整的模型。对于像渤海这样的大潮差浅水域中的环流问题，必须考虑潮体力效应。本章介绍的浅海环流理论和动力学模型提供了这种可能。

由太平洋通过东中国海大陆架进入渤海的潮波及其相伴的潮流形成了渤海流场中占优势的运动，其量级分别为 2 m/s 和 100 cm/s。M_2 分潮可以作为这一海域潮波系统的代表，故下面介绍的渤海夏季环流问题中的潮效应主要是由 M_2 分潮体现的，也就是说，其潮体力主要来源于 M_2 的非线性耦合。K_1 分潮可以作为日潮的代表，但为了简单，这里仅以一个假想的“K_1 天文分潮”来考察 K_1 导致的潮体力对渤海环流的影响。所谓假想的 K_1 分潮，其中的“K_1”系指其振幅与 K_1 的振幅相同，而其圆频率为 M_2 圆频率的 1/2。

在选择渤海夏季环流模型以前，除了上述观测分析和数值计算经验之外，对渤海动力学作一量级分析是有益的。

首先，既然潮致余流分量为一非线性效应，故其相对大小的测度应为非线性的无因次参数 κ 。把上述渤海有关参数代入式（9.2.25），得到渤海动力学非线性效应 $O(\kappa)=10^{-1}$ 。这表明潮致余流分量为潮流的10^{-1}量级。注意，潮流量级为 100 cm/s，故潮致余流量级为 10 cm/s。这与通过渤海海峡进入的黄海暖流余脉形成的渤海夏季环流的量级是一致的[81]。这进一步说明了考虑潮体力对于渤海夏季环流的重要意义。

测度密度水平梯度大小的无因次参数为 δ ；渤海夏季密度相对变化的典型数据 $O(\Delta)=10^{-3}$ [81]，代入式（9.2.26），有 $O(\delta)=10^{-2}$ 。这表明热盐环流分量仅为渤海潮流的 1%；也即仅为潮致余流的10^{-1}。热盐力引起的流速增值不超过 10%这一估值与数值模拟的结果是一致的[81]。

渤海风生环流分量的大小应由式（9.2.28）表示的无因次参数 k_τ 测度。渤海夏季海面风应力场的典型数据为 $\tau_c=0.005\ \ \mathrm{cm^2/s}$。若合理选择渤海动力学的湍粘性系数 $\nu_c=100\ \ \mathrm{cm^2/s}$，则 $O(k_\tau)=10^{-3}$ 。这意味着渤海夏季环流的风海流分量仅为整个环流速度的 1%左右，这一论断与数值模拟的结果也是一致的[81]。

由此可见，在黄海暖流余脉形成的渤海海峡“边界力”的作用下，以一个潮致余环流来体现渤海夏季环流是一种良好的近似。尽管本文描述斜压浅海环流的场方程组（9.4.5）~（9.4.15）用以数值模拟渤海夏季环流在原则上可行，但实现这一非线性耦合方程组在数值方法上的困难，渤海海面平均风场制造的精度问题，以及由此引出的数值误差，可能会掩没物理因素考虑全面所产生的优点。看来，至少作为开端，也许以一个潮致余环流模型来近似数值实现渤海夏季环流的模拟，不仅有效，而且方便。

当不考虑海面风应力和斜压效应时，式（9.4.5）~（9.4.15）就蜕化为描述潮致余环流的方程组[8]。其中“潮体力”(π_1,π_2)可由描述天文潮运动的零阶模型（9.2.40）~

（9.2.52）数值求解后进行潮平均而获得。郑连远（1988）依据这一理论模型及上述相应的方程组，利用求解浅海三维空间动力学问题的“流速分解法”[56]，实现了 M_2 潮致物质输运速度流场的计算，给出了渤海夏季环流的最低阶近似结果[28]，图 9.3 表示其深度平均流场。

从图 9.3 中可以看出，黄海暖流余脉从渤海海峡北侧老铁山水道进入渤海后沿岸北上，到达长兴岛之前有部分海水转西再向南，几乎沿着 30 m 等深线流动，最后由渤海海峡南端流出渤海。另一部分海水沿辽东湾东岸继续北上，到达湾顶后分为两支：一支向东，形成一顺时针流环；另一支向西，沿辽东湾西岸南下，沿渤海湾口南下，经莱州湾从海峡南侧流出渤海，它形成渤海中的逆时针大流环。应当指出，后一支流在南下途中有部分海水向东北回流，形成了辽东湾中又一支逆时针流环。另外，在渤海湾中存在一顺时针流环，形成了渤海湾环流。

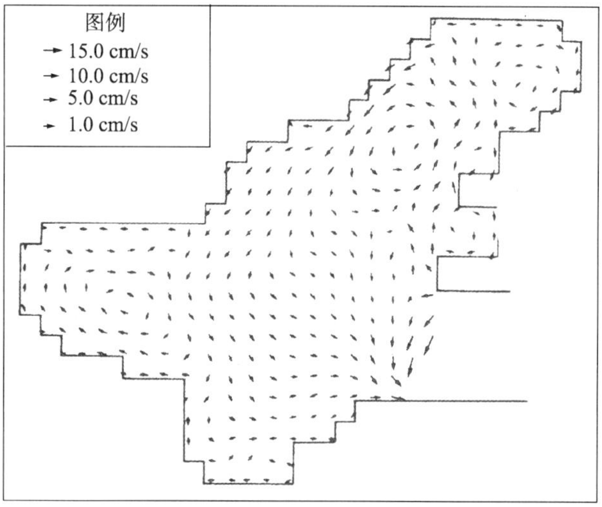

图 9.3　渤海夏季 M_2 导致的深度平均物质输运速度环流图

注意，当不考虑潮体力作用时，渤海夏季环流蜕化为一简单的逆时针大流环[28]，这与以一个风生–热盐环流模型模拟的渤海夏季环流的简单图景是一致的[81]。

再注意到上述渤海夏季环流中的热盐环流分量和风生环流分量分别仅为总环流的 10%和 1%，可见这里引进的郑连远数值实现的渤海夏季环流的复杂流型[28]（图 9.3）是潮致余环流分量导致的结果。特别应指出，渤海湾顺时针流环乃潮致效应的结果。

由此可见，渤海夏季环流主要为潮致余环流，其中海峡附近深水区主要被外海水所占据，热盐和风生环流分量是相对微弱的。顺便指出，进一步计入风与斜压效应后，除了流场中的环流数值稍有增加和中央流型局部稍有改变外，宏观上则与潮致余环流基本一致[32]。还应指出，对渤海夏季环流的这一认识完全改变了历史上的观点：认为渤海夏季环流是风生–热盐性质的。

渤海中直接作 Lagrange 余流观测的现场实验很少，有效的资料就更少，这给我们直接验证计算结果带来了不便。但间接、定性的验证还是可行的，尤其是当我们侧重于环流概貌及其物理机制研究的时候，后者显得特别有效。其实上述对渤海夏季环流的数值模拟，虽然依据的都是实际资料，但由于许多细节上的不确定因素，所以其意义与其说在于追求精确的定量模拟，倒不如说主要在于定性机制的探讨。

公式（9.5.18）告诉我们：当不考虑海面外源时，深度平均环流将沿时均表观浓度的深度平均场的等值线流动。这一近似规律，为以温度、盐度或其他示踪物的浓度分布来验证浅海环流的流型提供了可能。

苏志清提供了一组渤海夏季盐度的同步观测资料，经王辉作了深度平均和整理并给于图 9.4 上[32]。由图 9.4 不难看出，在渤海内区，深度平均的等盐线分布与深度平均环流流型大体一致。

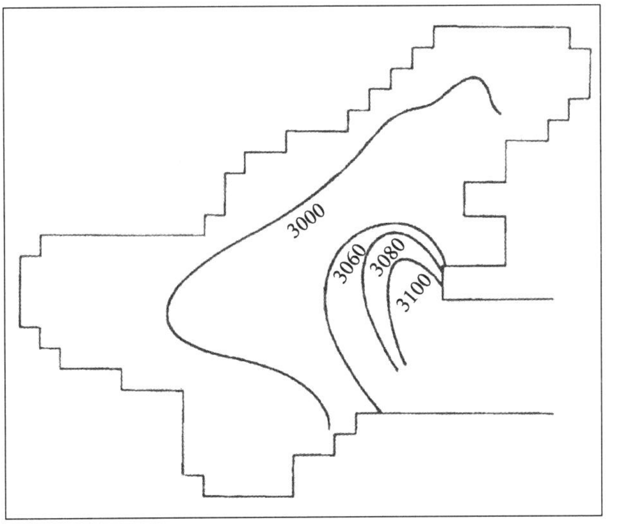

图 9.4　深度平均盐度分布图

黄河水以含沙量高而举世闻名，其入海后，除部分泥沙在黄河口门附近沉积外，悬浮沙粒将被环流携带而继续迁移和散布，因而构成了流场中很好的示踪物。平仲良根据地球资源卫星发回的渤海湾图像对渤海湾中混浊水羽的散布途径作了解译，进而推演出了渤海湾表层环流（图 9.5）[82]。比较图 9.5 和图 9.3（计算得到的渤海夏季表面环流图像基本上与图 9.3 所表明的环流流型一致，故未绘出），发现二者大体一致，这对于“渤海湾环流为一顺时针流环”的判断是一个有力的支持。事实上，携带泥沙、高浓度悬浮物以及黄河水中特有的白云母和方解石等矿物的黄河水入海后，有一部分沿渤海湾南岸的西向流动这一理论计算结果，还可以解释为什么渤海湾南侧和中区沉积有上述物质的动力学机制（图 9.6~图 9.8）[28, 32]。

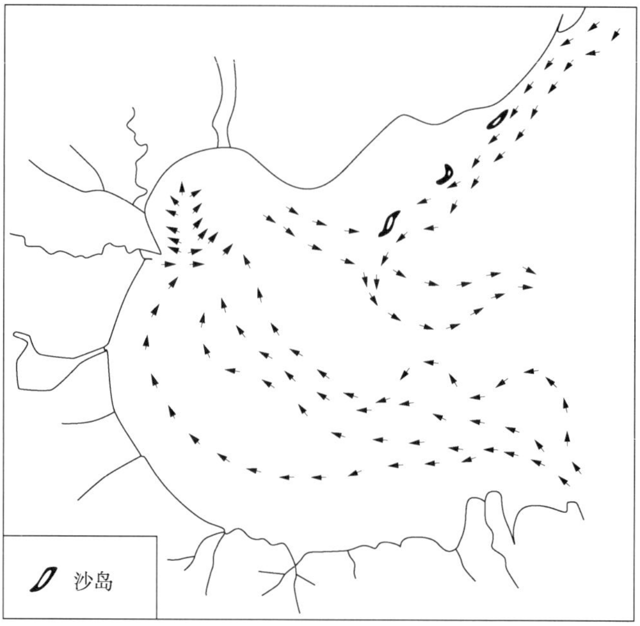

图 9.5　黄河的浑浊水羽和流向的解译图

1978 年 6 月 20 日，N38°15′/E118°45′黄河的浑浊水羽绕过山东钓口，西行至马家堡、歧口、塘沽沿海一带北上，海河水入海后随海流北上，沿湾顶拐弯，从西方向下，在南堡、曹妃甸南面海域与东北方向下来的滦河水汇合，流入渤海中部

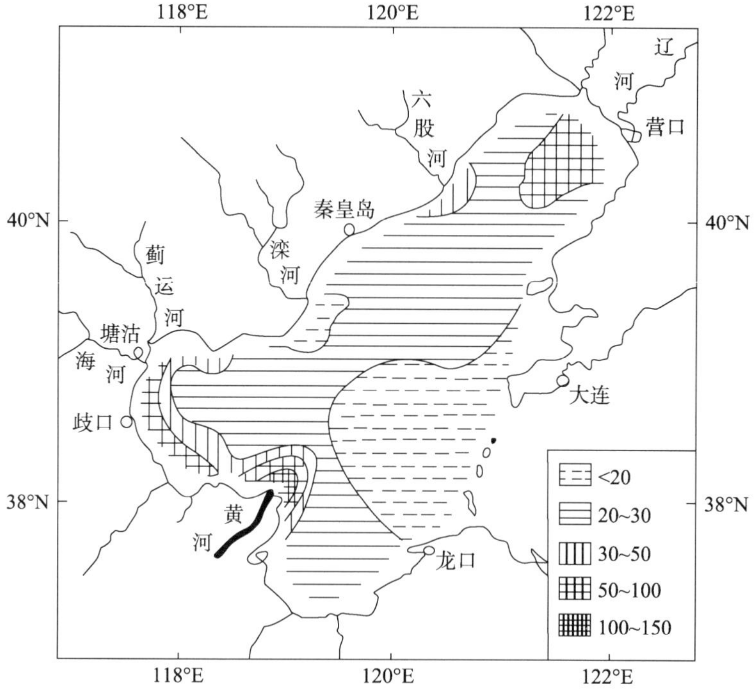

图 9.6　7 月份表层海水悬浮体含量（mg/L）分布

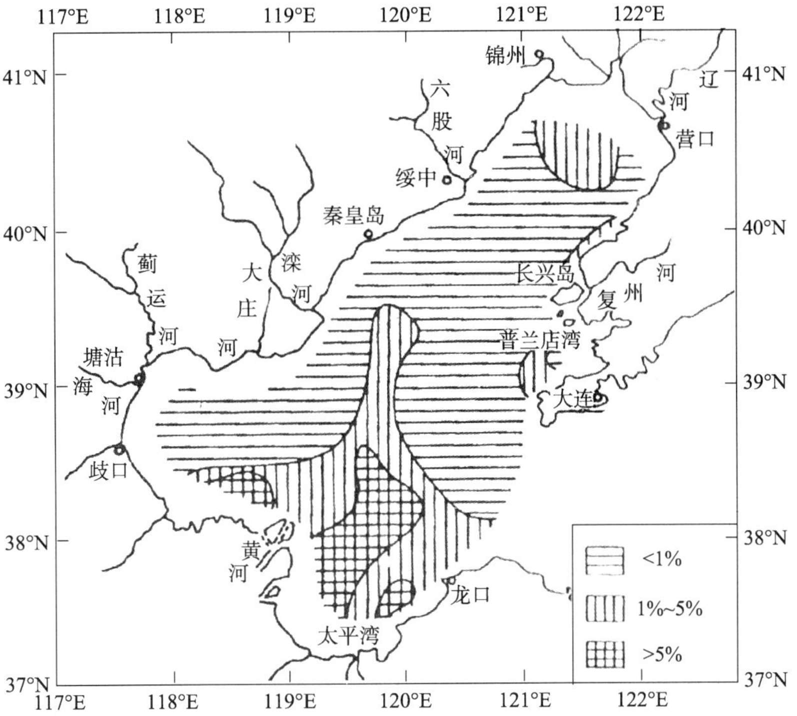

图 9.7　白云母百分含量分布

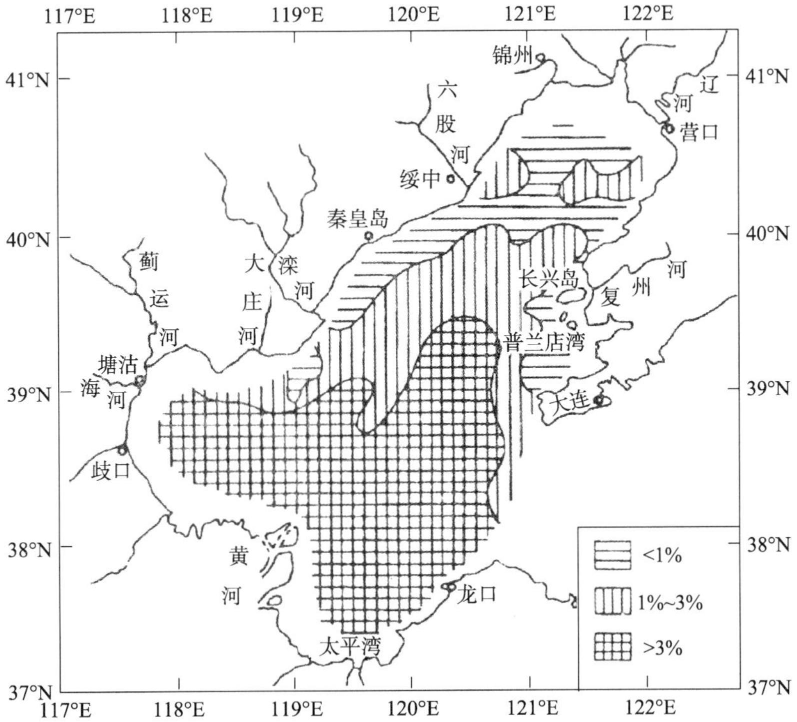

图 9.8　方解石百分含量分布

应该指出，在秦皇岛邻近海域，理论计算结果与观测值是不符的。由图 9.3 可见，该区海流沿岸南下，即流动呈西南向。但是，侍茂崇教授于 1983 年夏季对此海区所作的现场观测发现：海流沿岸北上，也就是说，流动呈东北向（图 9.9）。鉴于秦皇岛附近海区存在 M_2 潮波系统的一个无潮点，秦皇岛验潮资料表明其有日潮性质，所以，人们自然期待利用一个 M_2-K_1 潮波系统导致的 Lagrange 余环流来解开这一局部死结，魏更生（1988）[27]利用一个上面提到的 M_2-“K_1” 潮致 Lagrange 余环流模型，得到了物质输运速度，绘制了体现渤海夏季环流的深度平均流场[27]（图 9.10）。由图 9.10 与图 9.3

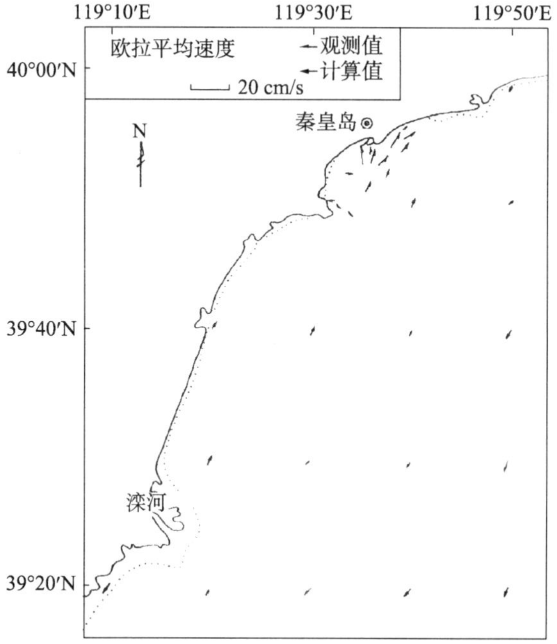

图 9.9　M_2-“K_1” 潮系统导致的深度平均的 Euler 余流计算值与实测值的比较

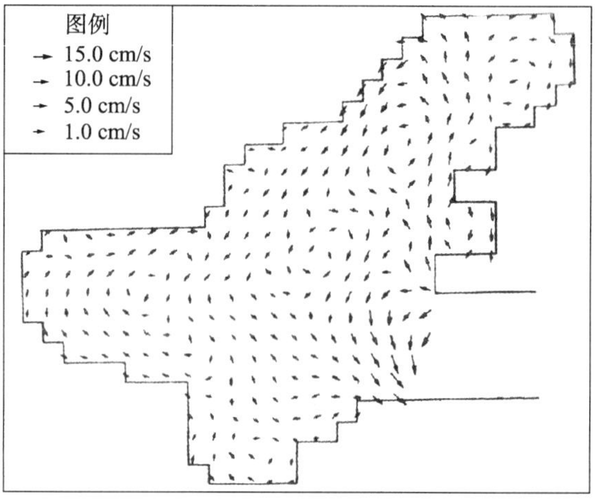

图 9.10　M_2-“K_1” 潮系统导致的深度平均物质输运速度环流图

的比较可以发现：二者流型基本相似，仅在流速上前者比后者略有增加；但在秦皇岛附近海区二者之间显示了明显的差异——流动反向，这就解释了侍茂崇教授的上述观测。图 9.9 中 Euler 平均速度的计算值是为了与观测的 Euler 平均速度相比较而依据公式（9.4.3）求出来的。结论是：以一个 M_2 潮致余环流来近似体现渤海夏季环流，这对绝大部分海区来说是符合实际的；考虑到秦皇岛附近海区的特点，一个相应的 M_2-K_1 潮致模型看来更为全面。

最后应该指出，像对渤海这样的浅海环流及其输运问题的描述，并不止于本节所介绍的以物质输运速度来体现环流这一最低阶近似。如前所述，浅海环流问题应具有更丰富的内涵。Lagrange 余流的二阶近似已表明了它对于流体微团释放时潮位相的依赖，且当不考虑海面上的非定常风应力时，这种依赖关系可以用一个潮周期的函数来表示。Lagrange 漂移的这一性质，对于长期输运过程具有潜在的重要意义。这是因为，尽管 Lagrange 漂移速度与潮流相比恒为 $O(\kappa^2)$，但 Lagrange 漂移与潮位移相比时，当后者与波长同阶的时候，可以达到 $O(\kappa)$，而微团的净位移可以达到与波长同阶[26]。也就是说，用流体微团的净位移（或等价地，用 Lagrange 平均速度）来描述浅海环流及其输运问题时，当过程时间长到使得包括用以平均的潮周期个数 n 不再满足不等式 $O(n)<\kappa^2$ 的时候，$O(N)=1$ 的假设被破坏，Lagrange 平均速度将不能构成一个不可压缩流场，此时，浅海环流问题中“场”的概念也消失了。不过，在仍局限于本文提出的弱非线性动力学的前提下，当考察 $O(n)\geqslant\kappa^2$ 的情形时，可以利用公式（9.3.12），通过简单的计算给出 Lagrange 净位移向量[83]。

最后应强调指出，这方面涉及的其他问题还很多，必须一一解决，正如笔者在 1988 年所说的那样：“如果本文中关于浅海环流及其输运问题的论述尚有可取之处的话，也只不过是一个开端而已”[9]。

参考文献

[1] Ramming, H. G., Kowalik, Z., Numerical Modelling of Marine Hydrodynamics, Elsevier Science Pub. Com., Amsterdam-Oxford-New York, 1980.

[2] Csanady, G. T., Lateral momentum flux in boundary currents, J. Phys. Oceanogr., 5 (1975), 705.

[3] Csanady, G. T., Mean circulation in shallow seas, J. Geophys. Res., 81 (1976), 5389.

[4] Nihoul, J. C. J., Ronday, F. C., The influence of the “tidal stress” on the residual circulation, Tellus, 27 (1975), 484.

[5] Heaps, N. S., Linearized vertically-integrated equations for residual circulation in coastal seas, Deut. Hydrog. Z., 31 (1978), 147.

[6] Feng, S. *et al.*, The baroclinic residual circulation in shallow seas, Chinese J. of Oceanol. and Limmol., 2 (1984), No.1, 49.

[7] Csanady, G. T., Circulation in the Coastal Ocean, D. Reidel Pub. Com., Dordrecht-Boston-London, 1982.

[8] Feng, S., A three-dimensional weakly nonlinear model of tide-induced Lagrangian residual current and mass-transport, with an application to the Bohai Sea, Three-dimensional Models of Marine and Estuarine Dynamics, ed. by J. C. J. Nihoul and B. M. Jamart, Elsevier Science Pub. Com., Amsterdam-Oxford-New York, 1987, pp. 471–488.

[9] 冯士筰，论浅海环流及其输运的流体动力学基础，清华大学工程力学与工程热物理学术会议论文集，清华大学出版社，1988, pp. 199–208.

[10] Longuet-Higgins, M. S., On the transport of mass by time-varying ocean currents, Deep Sea Res., 16 (1969), 431.

[11] Zimmerman, J. T. F., On the Euler-Lagrangian transformation and the Stokes' drift in the presence of oscillatory and residual currents, Deep Sea Res., 26A (1979), 505.

[12] Cheng, R. T., Casulli, V., On Lagrangian residual currents with applications in South San Francisco Bay, California, Water Resour. Res., 18 (1982), No. 6, 1652.

[13] Feng, S. *et al.*, On tide-induced Lagrangian residual current and residual transport, Part I: Residual current, Water Resour. Res., 22 (1986), No. 12, 1623.

[14] Feng, S. *et al.*, On tide-induced Lagrangian residual current and residual transport, Part II: Residual transport with application in South San Francisco Bay, California, Water Resour. Res., 22 (1986), No. 12, 1635.

[15] Tee, T. K., Tide-induced residual current: A 2-D nonlinear numerical model, J. Mar. Res., 31 (1976), 603.

[16] Huang, P. S. *et al.*, Analysis of residual currents using a two-dimensional model, Physics of Shallow Estuaries and Bays, ed. by J. ven de Kreeke, Springer-Verlag, Berlin-Heidelberg-New York-Tokyo, 1986, pp. 71–80.

[17] Dyke, P. P. G., On the Stokes' drift induced by tidal motions in a wide estuary, Estuarine, Coastal, and Marine Science, 11 (1980), 17.

[18] Ianniello, J. P., Tidally-induced residual currents in estuaries of constant breadth and depth, J. Mar. Res., 35 (1977), 1988, No. 4, 755.

[19] Ianniello, J. P., Tidally-induced residual currents in estuaries of variable breadth and depth, J. Phys. Oceanogr., 9 (1979), No. 5, 962.

[20] Prandle, D., Generalised theory of estuarine dynamics, Physics of Shallow Estuaries and Bays, ed. by J. ven de Kreeke, Springer-Verlag. Berlin-Heidelbery-New York-Tokyo, 1986, pp. 42–57.

[21] Awaji, T., Water mixing in a tidal current and the effect of turbulence on tidal exchange through a strait, J. Phys. Oceanogr., 12 (1982), No. 6, 501.

[22] Awaji, T. *et al.*, Tidal exchange through a strait: A numerical experiment using a simple model basin, J. Phys. Oceanogr., 10 (1980), 1499.

[23] 俞光耀，陈时俊，胶州湾环流和污染扩散数值模拟 (III)，拉格朗日余流和污染扩散，山东海洋学院学报, 13 (1983), No. 1, 1.

[24] 冯士筰，海洋环境流体动力学研究的焦点，山东海洋学院学报, 18 (1988), No. 2, 1.

[25] Cheng, R. T. *et al.*, On Lagrangian residual ellipse, Physics of Shallow Estuaries and Bays, ed. by J.

ven de Kreeke, Springer-Verlag, Berlin-Heidelbery-New York-Tokyo, 1986, pp. 102–113.

[26] Feng, S., A three-dimensional weakly nonlinear dynamics on tide-induced Lagrangian residual current and mass-transport, Chinese J. of Oceanol., and Limnol., 4 (1986), No. 2, 139.

[27] 魏更生, 论双频潮致拉格朗日余流和长期输运, 硕士论文, 青岛海洋大学, 1988.

[28] 郑连远, 三维潮致拉格朗日余流的数值计算及其在渤海中的应用, 硕士论文, 青岛海洋大学, 1988.

[29] Cheng, R. T. *et al.*, On inter-tidal transport equation, Estuarine Circulation, ed. by B. J. Neilson, J. Brubaker, and A Kuo, The Humana Press Inc., U. S., 1989, pp. 133–156.

[30] Sun, W. *et al.*, Numerical calculation of the three-dimensional tide-induced Lagrangian residual circulation in the Bohai Sea, J. of Ocean University of Qingdao, 19 (1989), No. 2(1), 27.

[31] Feng, S. *et al.*, Lagrangian residual current and long-term transport processes in a weakly nonlinear baroclinic system, Physics of Shallow Seas, ed. by H. Wang, J. Wang and H. Dai, China Ocean Press, Beijing, 1990, pp. 3–20.

[32] 王辉, 一种三维风生–热盐–潮致拉格朗日余流的数值计算及其在渤海中的应用, 硕士论文, 青岛海洋大学, 1989.

[33] Feng, S., On the Lagrangian residual velocity and mass-transport in a multi-frequency oscillatory system, Physics of Shallow Estuaries and Bays, ed. by R. T. Cheng, Springer-Verlag, Berlin-Heidelbery-New York-Tokyo, 1990, pp. 18–34.

[34] Cheng, R. T., Euler-Lagrangian computations in estuarine hydrodynamics, Proc. of the Third Intern. Conf. on Numerical Method in Laminar and Turbulent Flows, ed. by C. Taylor, Pineridge Press, 1983, pp. 341–352.

[35] Feng, S., A three-dimensional non-linear model of tides, Scientia Sinica, 20 (1977), No. 4, 436.

[36] Fischer, H. B. *et al.*, Mixing in Inland and Coastal Waters, Academic Press, New York, 1979.

[37] Hamrick, J., Subtidal circulation and transport in estuaries, Advancements in Aerodynamics, Fluid Mechanics and Hydrolics, ASCE, (1986), 1.

[38] Fischer, H. B., Mixing and dispersion in estuaries, Annu. Rev. Fluid Mech., 8 (1976), 107.

[39] Winterwerp, J. C., Decomposition or the mass transport in narrow estuaries, Estuarine Coastal Shelf Sci., 16 (1983), 627.

[40] Pritchard, D. W., A study on the salt balance in a coastal plain estuary, J. Mar. Res., 13 (1954), 133.

[41] Bowden, K. F., Horizontal mixing in the sea due to a shearing current, J. Fluid Mech., 21 (1965), 83.

[42] Hamrick, J., Time-averaged estuarine mass transport equations, Proc. of the 1987 National Conf. on Hydraulic Engineering, ed. by R. M. Ragan, ASCE, 1987, pp. 624–629.

[43] Gerritsen, H., Residual currents, a comparison of two modeling approaches, Physics of Shallow Estuaries and Bays, ed. by J. ven de Kreeke, Springer-Verlag, Berlin-Heidelberg-New York-Tokyo, 1986, pp. 81–101.

[44] Кочин, Н. Е. и тд, Теоретическая Гидромеханика, Госуд. Изд. Хизико-Мат. Литер., Москва, 1963.

[45] Stern, M. E., Ocean Circulation Physics, Academic, Orlando, Fla., 1975.

[46] Flather, R. A., A tidal model of the North-West European continental shelf, Mém. Soc. R. Sc. Liége, 6° Série, 10 (1976), 141.

[47] Ramming, H. G., A Nested North Sea model with fine resolution in shallow coastal seas, Mèm. Soc. R. Sc. Liège, 6° Sèrie, 10 (1976), 9.

[48] Maier-Reimer, E., Residual circulation in the North Sea due to the M_2-tide and mean annual wind stress, Dt. Hydrogr. Z., 30 (1977), 69.

[49] Pingree, R. D., Maddock, L., Tidal residuals in the English Channel. J. Mar. Biol. Ass. U. K., 57 (1977), 339.

[50] Prandle, D., Residual flows and elevation in the Southern North Sea, Proc. R. Soc. Lond., A 359 (1978), 189.

[51] 王化桐等, 胶州湾环流和污染扩散数值模拟 (I), 山东海洋学院学报, 10 (1980), No. 1, 26.

[52] Backhaus, J. O., A three-dimensional model for the simulation of shelf sea dynamics, Dt. Hydrogr. Z., 38 (1985), 165.

[53] Leendertse, J. J. *et al.*, Two-dimensional tidal models for the Delta Works, Transport Models for Inland and Coastal Waters, ed. by H. B. Fischer, Academic Press. New York, 1981, pp. 408–450.

[54] 李龙章等, 渤海海峡拉格朗日余流的数值模拟, 海洋学报, 10 (1988), 31.

[55] Hamrick, J., The dynamics of long-term advective transport in estuaries, Physics of Shallow Estuaries and Bays, ed. by R. T. Cheng, Springer-Verlag, Berlin-Heidelberg-New York-Tokyo, 1990, pp. 1–17.

[56] 孙文心, 超浅海风暴潮的进一步研究, 山东海洋学院学报, 17 (1987), No. 1, 34,

[57] Song, L., A hydrodynamic velocity-splitting model with a depth-varying eddy viscosity in shallow seas, Acta Oceanol. Sinica, 6 (1987), Supp 1, 135.

[58] 唐永明, 一种三维浅海流体动力学模型及其在渤海风潮中的应用, 青岛海洋大学硕士学位论文, 1987.

[59] 冯士筰, 论 f–和 β–坐标系, 山东海洋学院学报, 12 (1982), No. 3, 1.

[60] 冯士筰, 风暴潮导论, 科学出版社, 1982.

[61] Fofonoff, N. P., Dynamics of ocean currents, The Sea (I), Physical Oceanography, ed. by M. N. Hill, Interscience Publishers, New York-London, 1962, pp. 323–396.

[62] Nihoul, J. C. J., Modeling of Marine System, Elsevier Science Pub. Com., Amsterdam-Oxford-New York, 1975.

[63] Линенкин, П. С., Об определении толщины бароклииного слоя моря, ДАН СССР, 10 (1955), No. 3, 461.

[64] Uncles, R. J., Jordan, M. B., A one-dimensional representation of residual currents in the severn estuary and associate observations, Estuarine Coastal Marine Sci., (1980), No. 10, 39.

[65] Cheng, R, T., Gartner, J. W., Tides, tidal, and residual currents in San Francisco Bay, California: Results of measurements, 1979–1980, Parts I–V, U. S. Geol. Surv. Water Resour, Inv. Rep., 84–4339, 1–1747, 1984.

[66] Elliott, A. J., Hendrix, T. E., Intensive observations of the circulation in the Potomac estuary, Spec. Rep. 55, 35 pp., Chesapeake Bay Inst., Baltimore, Md., 1976.

[67] Weisberg, R. H., The nontidal flow in the Providence River of Narragansett Bay: A stochastic approach to estuarine circulation, J. Phys. Oceanogr., 6 (1976), 721.

[68] Dyer, K. R., Estuaries: A Physical Introduction, John Wiley, New York, 1973.

[69] Officer, C. B., Physical Oceanography of Estuaries, John Wiley, New York, 1976.
[70] Hansen, W., Tides, The Sea (I), Physical Oceanography, ed. by M. N. Hill, Interscience Publishers, New York-London, 1962, pp. 764–801.
[71] Pedlosky, J., Geophysical Fluid Dynamics, Springer-Verlag, New York, 1979.
[72] Sugimoto, T., Effect of boundary geometries on tidal currents and tidal mixing, J. Oceanogr. Soc. Japan, (1975), No. 3l, 1.
[73] Yanagi, T., Fundamental study on the tidal residual circulation, I, J. Oceanogr. Soc. Japan, 32 (1976), 199.
[74] Zimmerman, J. T. F., Mixing and flushing of tidal embayments in the western Dutch Wadden Sea- II, Neth. J. Sea Res., 10 (1976), 397.
[75] Riepma, H., Spatial variability of residual currents in an area of the southern North Sea, I. C. E. S., CM / 1977 / C: 43, 1977.
[76] Tee, K. T., Tide-induced residual current-verification of a numerical model, J. Phys. Oceanogr., 7 (1977), 396.
[77] Zimmerman, J. T. F., Dispersion by tide-induced residual current vortices, Hydrodynamics of Estuaries and Fjords, ed. by J. C. J. Nihoul, Elsevier Oceanogr. Ser., Amsterdam, 1978, pp. 207–216.
[78] Robinson, I. S., Tidal vorticity and residual circulation, Deep Sea Res., 28A (1981), No. 3, 195.
[79] Zimmerman, J. T. F., Topographic generation of residual circulation by oscillatory (tidal) currents, Geophys. Astrophys. Fluid Dynamics, 11 (1978), 35.
[80] 管秉贤等，渤黄东海表层海流图，中国科学院海洋研究所, 1977.
[81] 张淑珍等，渤海环流数值模拟，山东海洋学院学报, 14 (1984), No. 2, 12.
[82] 平仲良，从 ERTS 图像上观测渤海湾表层流，海洋与湖沼, 14 (1983), No. 3, 297.
[83] Feng, S. *et al.*, Coastal circulation physics, a report submitted to the 5th International Conf. on Physics of Estuaries and Coastal Seas, Gregynog, Wales, U. K., 1990, pp. 9–13.
[84] 冯士筰，潮致余涡的动力机制，科学通报, 14 (1991), 1059.

A Turbulent Closure Model of Coastal Circulation*

FENG Shi-Zuo (冯士筰) and LU You-Yu (鹿有余)

In a tidally dominant coastal ocean, bay, or estuary it has been proposed that the problem of circulation should be described by the Lagrangian mean velocity, not the Eulerian mean velocity used traditionally[1, 2]. The first-order Lagrangian mean velocity of a weakly nonlinear coastal ocean system, or the mass-transport velocity has been used to body the velocity field of steady circulation as the lowest order coastal circulation, and a set of field equations for it has been derived[2-4]. It should be pointed out that in the set of field equations just mentioned the eddy viscosity has been supposed to be a known empirical function of spatial coordinates. In fact, this hypothesis on the eddy viscosity cancels the nonlinear effect of turbulent stresses. In order to determine in an objective manner the eddy viscosity and understand more thoroughly the dynamics of coastal circulation, in the present note, an attempt is made to introduce the k - ε model of turbulence closure, where k is the turbulent kinetic energy and ε is the rate of dissipation of k, and to derive a generalized set of field equations for the problem of coastal circulation.

According to Колмогоров-Prandtl[5], the eddy viscosity ν is expressed as

$$\nu = c_\mu \frac{k^2}{\varepsilon} \tag{1}$$

where k and ε are respectively described by the following transport equations

$$\mathscr{D}(k) = P + G - \varepsilon \tag{2}$$

and

$$\mathscr{D}(\varepsilon) = c_{1\varepsilon}\frac{\varepsilon}{k}\left[(P+G)(1+c_{3\varepsilon}Ri_{\mathrm{f}}) - \frac{c_{2\varepsilon}}{c_{1\varepsilon}}\varepsilon\right] \tag{3}$$

in which the operator $\mathscr{D}(p) = \left[\frac{\partial}{\partial t} + \boldsymbol{u}\cdot\nabla - \frac{\partial}{\partial z}\left(\lambda_p \frac{\partial}{\partial z}\right)\right](p)$; the production of k, $P = \nu\left[\left(\frac{\partial u}{\partial z}\right)^2 + \left(\frac{\partial v}{\partial z}\right)^2\right]$ and the buoyancy production/destruction of k, $G = g\beta\lambda_{\mathrm{c}}\frac{\partial c}{\partial z}$; Ri_{f} is the flux Richardson number; c_μ, $c_{1\varepsilon}$, $c_{2\varepsilon}$ and $c_{3\varepsilon}$ are nondimensional empirical coefficients; the gradient operator $\nabla = \boldsymbol{i}\frac{\partial}{\partial x} + \boldsymbol{j}\frac{\partial}{\partial y} + \boldsymbol{k}\frac{\partial}{\partial z}$; $(\boldsymbol{i}x, \boldsymbol{j}y, \boldsymbol{k}z)$ constitute Cartesian coordi-

* Feng S, Lu Y. 1993. A turbulent closure model of coastal circulation. Chinese Science Bulletin, 38(20): 1737–1741

nates on an f-plane; t is time; $\boldsymbol{u}=\boldsymbol{i}u+\boldsymbol{j}u+\boldsymbol{k}w$ is a velocity vector; c is the apparent concentration[4]; β is the coefficient of (thermal) expansion; g is the gravitational acceleration; $\lambda_p=\nu/\sigma_p$ is the eddy diffusivity, where σ_p is the turbulent Prandtl/Schmidt number, and $p=c,k,\varepsilon$.

Equations (1)—(3) with the original equations (cf. Eqs. (1)—(10) in Ref. [4]) constitute the k - ε turbulent closure problem of a three-dimensional baroclinic shallow sea.

The principal hypotheses proposed in Ref. [4] are introduced here as follows: (i) the non-dimensional nonlinear parameter, $\kappa=\mathscr{O}(|\boldsymbol{u}\cdot\nabla|/|\partial/\partial t|)$, is a small parameter; (ii) the transport mechanism is dominated by convection, or $\mathscr{O}\left(\left|\frac{\partial}{\partial z}\left(\lambda_p\frac{\partial}{\partial z}\right)\right|\Big/|\boldsymbol{u}\cdot\nabla|\right)=\kappa$; (iii) the wind stress on the sea surface $\boldsymbol{\tau}=(\boldsymbol{i}\tau_x+\boldsymbol{j}\tau_y)$ and the baroclinic force are small, or $\mathscr{O}(|\tau|)=\mathscr{O}\left(\left|g\beta\nabla_{\mathrm{H}}\int_z^{\zeta}c\mathrm{d}z'\right|\right)=\kappa$ (∇_{H} is a horizontal gradient operator); thus, the zeroth-order transport quantity p_0 (the subscript "0" denotes the zeroth-order approximation) is independent of time, and $\langle p\rangle\approx p_0$ (the operator < > denotes a tidal cycle average).

In view of an application of a "local-equilibrium" turbulence model to the numerical simulation of tides[6], here we suppose formally

$$\mathscr{O}(P+G-\varepsilon)=\kappa^2 \tag{4}$$

which implies a "quasi-local-equilibrium" turbulence model, since the tide-oscillatory motion is dominant in the coastal ocean. Noting $c_{1\varepsilon}$ ~1.44, $c_{2\varepsilon}$ ~1.92, $c_{3\varepsilon}$ ~0.8 and the uncertainty of the definition of Ri_{f} , the sum of terms on the right-hand side of Eq. (3) is also assumed to be the order of magnitude of κ^2 , which is a weak hypothesis.

By means of a perturbation approach with the small parameter κ , expanding all variables, including k, ε and ν , and through some mathematical treatments similar to those made in Ref. [4], a generalized set of field equations for the k - ε turbulent closure problem of coastal circulation is derived as follows:

$$\nabla\cdot\boldsymbol{u}_{\mathrm{L}}=0 \tag{5}$$

$$-fv_{\mathrm{L}}=-g\frac{\partial\langle\zeta_1\rangle}{\partial x}+\frac{\partial}{\partial z}\left(\langle\nu\rangle\frac{\partial u_{\mathrm{L}}}{\partial z}\right)-g\beta\frac{\partial}{\partial x}\int_z^0\langle c\rangle\mathrm{d}z'+\pi_1 \tag{6}$$

$$fu_{\mathrm{L}}=-g\frac{\partial\langle\zeta_1\rangle}{\partial y}+\frac{\partial}{\partial z}\left(\langle\nu\rangle\frac{\partial v_{\mathrm{L}}}{\partial z}\right)-g\beta\frac{\partial}{\partial y}\int_z^0\langle c\rangle\mathrm{d}z'+\pi_2 \tag{7}$$

$$\boldsymbol{u}_{\mathrm{L}}\cdot\nabla\langle c\rangle=\frac{\partial}{\partial z}\left(\langle\lambda_{\mathrm{c}}\rangle\frac{\partial\langle c\rangle}{\partial z}\right) \tag{8}$$

$$\boldsymbol{u}_{\mathrm{L}}\cdot\nabla\langle k\rangle=\frac{\partial}{\partial z}\left(\langle\lambda_k\rangle\frac{\partial\langle k\rangle}{\partial z}\right)+\langle\nu\rangle\left\langle\left(\frac{\partial u_0}{\partial z}\right)^2+\left(\frac{\partial v_0}{\partial z}\right)^2\right\rangle+g\beta\langle\lambda_{\mathrm{c}}\rangle\frac{\partial\langle c\rangle}{\partial z}-\langle\varepsilon\rangle \tag{9}$$

$$\boldsymbol{u}_{\mathrm{L}}\cdot\nabla\langle\varepsilon\rangle=\frac{\partial}{\partial z}\left(\langle\lambda_{\varepsilon}\rangle\frac{\partial\langle\varepsilon\rangle}{\partial z}\right)+c_{1\varepsilon}\frac{\langle\varepsilon\rangle}{\langle k\rangle}\left\{\left(\langle\nu\rangle\left\langle\left(\frac{\partial u_0}{\partial z}\right)^2+\left(\frac{\partial v_0}{\partial z}\right)^2\right\rangle\right.\right.$$
$$\left.\left.+\beta g\langle\lambda_{\mathrm{c}}\rangle\frac{\partial\langle c\rangle}{\partial z}\right)\left(1+c_{3\varepsilon}Ri_{\mathrm{f}}\right)-\frac{c_{2\varepsilon}}{c_{1\varepsilon}}\langle\varepsilon\rangle\right\} \tag{10}$$

$$\langle\nu\rangle=c_{\mu}\frac{\langle k\rangle^2}{\langle\varepsilon\rangle},\ \ (c_{\mu}\sim 0.09) \tag{11}$$

$$\langle\lambda_p\rangle=\frac{\langle\nu\rangle}{\sigma_p},\ (p=c,k,\varepsilon\ \text{and}\ \ \sigma_k\sim 1,\sigma_{\varepsilon}\sim 1.3) \tag{12}$$

$z=0$:

$$w_{\mathrm{L}}=0 \tag{13}$$

$$\langle\nu\rangle\frac{\partial}{\partial z}(u_{\mathrm{L}},v_{\mathrm{L}})=\left(\langle\tau_x\rangle,\langle\tau_y\rangle\right) \tag{14}$$

$$\langle\lambda_{\mathrm{c}}\rangle\frac{\partial\langle c\rangle}{\partial z}=\langle\Gamma\rangle \tag{15}$$

$$\langle k\rangle=\frac{\langle|\tau|\rangle^{[5]}}{\sqrt{c_{\mu}}} \tag{16}$$

$$\langle\varepsilon\rangle=\frac{\langle|\tau|\rangle^{3/2}}{\mathscr{K}\Delta z} \tag{17}$$

$z=-h$:

$$\boldsymbol{u}_{\mathrm{L}}=0 \tag{18}$$

$$\langle k\rangle=\langle\varepsilon\rangle=0^{[7]} \tag{19}$$

$$\frac{\partial\langle c\rangle}{\partial z}=0 \tag{20}$$

where

$$\pi_1=-g\frac{\partial}{\partial x}\left\langle\frac{1}{2}\xi_0\cdot\nabla\zeta_0\right\rangle+\frac{\partial}{\partial z}\left\langle\left(\frac{5}{2}\frac{\partial\xi_0}{\partial x}+\frac{\partial\eta_0}{\partial y}\right)\langle\nu\rangle\frac{\partial u_0}{\partial z}+\left(\frac{\partial\xi_0}{\partial y}+\frac{1}{2}\frac{\partial\eta_0}{\partial x}\right)\langle\nu\rangle\frac{\partial v_0}{\partial z}\right\rangle$$
$$-\left\langle\frac{\partial\xi_0}{\partial z}\cdot\nabla\left(\langle\nu\rangle\frac{\partial u_0}{\partial z}\right)\right\rangle-\frac{1}{2}\left\langle\frac{\partial^2\xi_0}{\partial z\partial x}\left(\langle\nu\rangle\frac{\partial u_0}{\partial z}\right)+\frac{\partial^2\eta_0}{\partial z\partial x}\left(\langle\nu\rangle\frac{\partial v_0}{\partial z}\right)\right.$$
$$\left.+\xi_0\frac{\partial^2}{\partial z\partial x}\left(\langle\nu\rangle\frac{\partial u_0}{\partial z}\right)+\eta_0\frac{\partial^2}{\partial z\partial x}\left(\langle\nu\rangle\frac{\partial v_0}{\partial z}\right)\right\rangle \tag{21}$$

$$\pi_2 = -g\frac{\partial}{\partial y}\left\langle\frac{1}{2}\xi_0\cdot\nabla\zeta_0\right\rangle + \frac{\partial}{\partial z}\left\langle\left(\frac{5}{2}\frac{\partial\eta_0}{\partial y}+\frac{\partial\xi_0}{\partial x}\right)\langle\nu\rangle\frac{\partial v_0}{\partial z}+\left(\frac{\partial\eta_0}{\partial x}+\frac{1}{2}\frac{\partial\xi_0}{\partial y}\right)\langle\nu\rangle\frac{\partial u_0}{\partial z}\right\rangle$$
$$-\left\langle\frac{\partial\xi_0}{\partial z}\cdot\nabla\left(\langle\nu\rangle\frac{\partial v_0}{\partial z}\right)\right\rangle - \frac{1}{2}\left\langle\frac{\partial^2\xi_0}{\partial z\partial y}\left(\langle\nu\rangle\frac{\partial u_0}{\partial z}\right)+\frac{\partial^2\eta_0}{\partial z\partial y}\left(\langle\nu\rangle\frac{\partial v_0}{\partial z}\right)\right.$$
$$\left.+\xi_0\frac{\partial^2}{\partial z\partial y}\left(\langle\nu\rangle\frac{\partial u_0}{\partial z}\right)+\eta_0\frac{\partial^2}{\partial z\partial y}\left(\langle\nu\rangle\frac{\partial v_0}{\partial z}\right)\right\rangle \tag{22}$$

$\boldsymbol{u}_\mathrm{L} = \boldsymbol{i}u_\mathrm{L} + \boldsymbol{j}v_\mathrm{L} + \boldsymbol{k}w_\mathrm{L}$ is the mass-transport velocity; $\langle\zeta_1\rangle$ is the residual elevation; Γ is the surface source; $\mathscr{K}$ is von Karman's constant; Δz is an arbitrary small distance from the bottom; h is the water depth; $\xi_0 = \boldsymbol{i}\xi_0 + \boldsymbol{j}\eta_0 + \boldsymbol{k}z_0$ is the zeroth-order displacement of water parcel; and the zeroth-order model describing astronomical tides is

$$\nabla\cdot\boldsymbol{u}_0 = 0 \tag{23}$$

$$\frac{\partial u_0}{\partial t} - fv_0 = -g\frac{\partial\zeta_0}{\partial x} + \frac{\partial}{\partial z}\left(\langle\nu\rangle\frac{\partial u_0}{\partial z}\right) \tag{24}$$

$$\frac{\partial v_0}{\partial t} + fu_0 = -g\frac{\partial\zeta_0}{\partial y} + \frac{\partial}{\partial z}\left(\langle\nu\rangle\frac{\partial v_0}{\partial z}\right) \tag{25}$$

$z = 0$:

$$w_0 = \frac{\partial\zeta_0}{\partial t} \tag{26}$$

$$\frac{\partial}{\partial z}(u_0, v_0) = 0 \tag{27}$$

$z = -h$:

$$\boldsymbol{u}_0 = 0 \tag{28}$$

A simple discussion about the generalized model of coastal circulation proposed above is of much interest. (i) The model proposed in Refs. [2–4] shows only a one-way effect of the nonlinearity of tides on the coastal circulation through the "tidal body force" (π_1,π_2), which should be due to the hypothesis on $\nu = \nu(x,y,z)$ as a known function, or as a result of the linearization of turbulent stress. The nonlinear coupling between tides and circulation shown in the present note is just a result of the nonlinear effect of turbulent stress. (ii) As the Prandtl's Mixing Length Hypothesis is applied to Eqs. (24) and (25), $\nu\approx\nu_0(x,y,z)\approx\langle\nu\rangle$ with an error of 20%, the problem of astronomical tides is not closed since the Prandtl's Mixing Length, l_m is unknown. Noting $l_\mathrm{m}\sim k^{3/2}/\varepsilon\sim\langle k\rangle^{3/2}/\langle\varepsilon\rangle$ and the Eqs. (9) and (10) satisfied by $\langle k\rangle$ and $\langle\varepsilon\rangle$, it seems that only the coupled problem of tidal and mean circulation in coastal seas is dynamically closed. (iii) $\langle\nu\rangle$ contained in the problem of astronomical tides (Eqs. (23)–(28)) might imply why generally exists a principal constituent, M_2, of tides in the coastal ocean although the turbulent stress terms are not weakly nonlinear. (iv) Eqs. (9)–(11) show that the astronomical tide affects not only directly the coastal circulation through the tidal body force but also indirectly the dynamics of circulation through the turbulent energy produced by the vertical shear of tidal current and the turbulent mixture. (v) The expression of the tidal body force derived in Refs. [2–4] exhibits a superfluous term, or $\left\langle \frac{\partial}{\partial z}\left[\nabla \nu \cdot \xi_0 \frac{\partial}{\partial z}(u_0, v_0) \right] \right\rangle$ (cf. expression (23) in Ref. [4]), when compared with (21) and (22). It is of interest that at least the superfluous term does not affect the vorticity dynamics of the depth-averaged coastal circulation[8].

References

[1] Csanady, G. T., Circulation in the Coastal Ocean, D. Reidel Pub., 1982.

[2] Feng, S. Proceedings of Qinghua University Conference on Engineering Mechanics and Thermophysics (in Chinese), Qinghua Univ. Press, Beijing, 1988, pp. 199–208.

[3] Feng, S. Acta Oceanologica Sinica, 1990, 9(3): 315.

[4] Feng, S., Lecture Notes on Coastal and Estuarine Studies, Springer-Verlag, New York, 1990, 38: 34.

[5] Rodi, W., Turbulence Models and Their Application in Hydraulics—A State of the Art Review, Book Pub. of IAHR, the Netherlands, 1984.

[6] Fang, G. & Ichiye, T., Geophys. J. R. Astr. Soc., 1983, 73(1): 62.

[7] Koutitas, C., Coastal and Estuarine Science, 4 (ed. Heaps, N.), AGU, Washington D. C., 1987, pp. 107–123.

[8] Feng, S., ChinesSe Science Bulletin, 1991, 36(24): 2043.

A Three-dimensional Numerical Calculation of the Wind-driven Thermohaline and Tide-induced Lagrangian Residual Current in the Bohai Sea *

Wang Hui, Su Zhiqing, Feng Shizuo and Sun Wenxin

Introduction

The development of the coastal industry and ocean exploitation brings considerably increasing quantity of polluting substance entering the coastal seas, and thus leads to serious pollution and damage to the environmental quality. This is harmful to the marine life, fishery resources and human health. So, ocean environment protection is widely concerned by the governments of the coastal countries. As those chemical, biological and physical processes occur continuously in the ocean, the polluting substance entering the ocean might be transferred and diffused, or, be oxidized reduced and concentrated biologically. Thus, to some extent, the sea water possesses purifications ability by itself. It should be pointed out that among the above-mentioned processes the physical aspect plays a very important role and circulation and long-term transport process is a basic research subject in this respect. However, former studies were mostly confined to the two-dimensional vertical integrated model of Eulerian residual current. In recent years, with the development of three-dimensional shallow sea hydrodynamical model and the Poetical requirement of attacking the environmental problems, studies on Lagrangian residual current and long-term transport process have been mostly concerned. In fact, the direction of the pollutant migration is closely related to the trajectory of water parcels and the Lagrangian residual current is of more direct importance in attacking the environmental problems (Feng, 1987).

Formulation

Derived from the non-dimensional form of the baroclinic shallow sea dynamic equations according to the weakly nonlinear approximation, a group of governing equations controlling the zeroth-order astronomical tide motion and controlling the wind-driven, thermohaline and tide-induced Lagrangian residual current are given separately as follows (Feng, 1987):

$$\nabla \cdot \boldsymbol{u}_0 = 0 \tag{1}$$

* Wang H, Su Z, Feng S, Sun W. 1993. A three-dimensional numerical calculation of the wind-driven thermohaline and tide-induced Lagrangian residual current in the Bohai Sea. Acta Oceanologica Sinica, 12(2): 169–182

$$\frac{\partial u_0}{\partial t} - fv_0 = -g\frac{\partial \zeta_0}{\partial x} + \frac{\partial}{\partial z}\left(\gamma \frac{\partial u_0}{\partial z}\right) \tag{2}$$

$$\frac{\partial u_0}{\partial t} + fu_0 = -g\frac{\partial \zeta_0}{\partial y} + \frac{\partial}{\partial z}\left(\gamma \frac{\partial v_0}{\partial z}\right) \tag{3}$$

$$z = 0, \quad w_0 = \frac{\partial \zeta_0}{\partial t}; \quad \frac{\partial u_0}{\partial z} = \frac{\partial v_0}{\partial z} = 0 \tag{4}$$

$$z = -h\,, \quad \boldsymbol{u}_0 = 0 \tag{5}$$

$$\nabla \cdot \boldsymbol{u}_{\mathrm{L}} = 0 \tag{6}$$

$$-fv_{\mathrm{L}} = -g\frac{\partial \langle \zeta_1 \rangle}{\partial x} + \frac{\partial}{\partial z}\left(\bar{\gamma}\frac{\partial u_{\mathrm{L}}}{\partial z}\right) + \pi_1 - g\frac{\partial}{\partial x}\int_z^0 \alpha \zeta_0 \mathrm{d}z' \tag{7}$$

$$fu_{\mathrm{L}} = -g\frac{\partial \langle \zeta_1 \rangle}{\partial y} + \frac{\partial}{\partial z}\left(\bar{\gamma}\frac{\partial v_{\mathrm{L}}}{\partial z}\right) + \pi_2 - g\frac{\partial}{\partial y}\int_z^0 \alpha \zeta_0 \mathrm{d}z' \tag{8}$$

$$z = 0, \; w_{\mathrm{L}} = 0, \; \bar{\gamma}\frac{\partial}{\partial z}(u_{\mathrm{L}}, v_{\mathrm{L}}) = (\bar{\tau}_{ax}, \bar{\tau}_{ay}) \tag{9}$$

$$z = -h, \quad \boldsymbol{u}_{\mathrm{L}} = 0 \tag{10}$$

where $\boldsymbol{u}_{\mathrm{L}}$ is the first-order Lagrangian residual current; $\langle \zeta_1 \rangle$ is the residual elevation; (π_1, π_2) indicates the nonlinear coupling between the astronomical tides and is named tide-induced body force; ζ_0 is the time-averaged apparent concentration; $(\bar{\tau}_{\mathrm{ax}}, \bar{\tau}_{\mathrm{ay}})$ are the mean wind stress components in (x,y) direction.

Equation of state shown below is a new one given by Millero *et al*. (1980).

$$\rho_{S,T,p} = \rho_{S,T,0} / [1 - p / \Gamma(S,T,p)] \tag{11}$$

$$\rho_{S,T,0} = \rho_{\mathrm{w}} + AS + BS^{3/2} + \mathrm{C}S^2 \tag{12}$$

where ρ_{w} is the density of fresh water. A and B are two coefficients related to temperature. C is a constant, $\Gamma(S,T,p)$ is the function of S, T and p. If only the salinity variation is considered, the approximate form of the equation of state is:

$$\rho - \rho_{\mathrm{w}} = \alpha S \tag{13}$$

with $\alpha = 8.2449 \times 10^{-4}$.

Richardson number is introduced into the eddy viscosity coefficient (Bowden, 1975). If $\xi = \frac{z}{h}$,

$$\gamma = \gamma_0 F_0(\xi) \tag{14}$$

$$\bar{\gamma} = \gamma_1 F_1(\xi)(1 + \beta Ri)^m \tag{15}$$

where

$$F_0(\xi)=\begin{cases} 0.1-2.7\xi & -\frac{1}{3}\leqslant\xi\leqslant 0 \\ 1 & -\frac{2}{3}\leqslant\xi\leqslant-\frac{1}{3} \\ 2.8+2.7\xi & -1\leqslant\xi\leqslant-\frac{2}{3} \end{cases} \tag{16}$$

$$F_1(\xi)=\left\{\begin{array}{ll} 0.1-5.4\xi & -\frac{1}{6}\leqslant\xi\leqslant 0 \\ 1 & -1\leqslant\xi\leqslant-\frac{1}{6} \end{array}\right\} \tag{17}$$

$Ri=g\alpha\frac{\Delta S_0}{\overline{u}^2}$ is the Richardson number of vertical integration. ΔS_0 is the difference of salinity between sea surface and bottom, $\overline{u}$ is the maximum value of vertical integrated tidal velocity. γ_0 and γ_1 are chosen to be 100 cm^2/s and 75 cm^2/s respectively. According to the observed data and numerical experiment, the two empirical constants are chosen to be:

$$\beta=8,\quad m=-\frac{1}{4} \tag{18}$$

The salinity data are given according to the General Chinese Oceanic Surrey data of 1958. As well known, the discharge of the Huanghe River has a considerably seasonal variation and its maximum value emerges in July or August. From the long-term statistical data of Lijin station, the runoff of the Huanghe River is chosen to be 2×10^9 cm^3/s, which is a typical value for the summer conditions (Pang and Si, 1980). The values of volumetric flux through the Bohai Sea Strait, namely a part of the Huanghai Warm Water coming from the Huanghai Sea is also derived from the GCOS data of 1958. The mean wind stresses are the same as those adopted by Zhang *et al.* (1984).

Numerical Method—the Velocity-splitting Model

The calculation of astronomical tide

Let $\xi=\frac{z}{h}$ and introduce the transformation of the constituent tidal wave method shown below

$$(u_0,v_0,w_0,\zeta_0)=\mathrm{Re}\left[(u,v,w,\zeta)\mathrm{e}^{-\mathrm{i}\sigma t}\right] \tag{19}$$

Equations corresponding to (1)–(5) are:

$$\frac{\partial u}{\partial x}+\frac{\partial v}{\partial y}-\frac{\xi}{h}\left(\frac{\partial u}{\partial \xi}\frac{\partial h}{\partial x}+\frac{\partial v}{\partial \xi}\frac{\partial h}{\partial y}\right)+\frac{1}{h}\frac{\partial w}{\partial \xi}=0 \tag{20}$$

$$-\mathrm{i}\sigma u-fv=-g\frac{\partial\zeta}{\partial x}+\frac{1}{h^2}\frac{\partial}{\partial\xi}\left(\gamma\frac{\partial u}{\partial\xi}\right)\tag{21}$$

$$-\mathrm{i}\sigma v+fu=-g\frac{\partial\zeta}{\partial y}+\frac{1}{h^2}\frac{\partial}{\partial\xi}\left(\gamma\frac{\partial v}{\partial\xi}\right)\tag{22}$$

$$\xi=0,\quad w=-\mathrm{i}\sigma\zeta;\quad \frac{\partial u}{\partial\xi}=\frac{\partial v}{\partial\xi}=0\tag{23}$$

$$\xi=-1,\quad \boldsymbol{u}=0\tag{24}$$

let

$$q=u+\mathrm{i}v=\frac{gh^2}{\gamma_0}\Pi\varphi\tag{25}$$

$$q=u-\mathrm{i}v=\frac{gh^2}{\gamma_0}\hat{\Pi}\hat{\varphi}\tag{26}$$

$$\Pi=\frac{\partial\zeta}{\partial x}+\mathrm{i}\frac{\partial\zeta}{\partial y}\tag{27}$$

$$\hat{\Pi}=\frac{\partial\zeta}{\partial x}-\mathrm{i}\frac{\partial\zeta}{\partial y}\tag{28}$$

the governing equation of profile functions φ, $\hat{\varphi}$ and water elevation ζ are derived as follows (Song, 1986):

$$\frac{\partial}{\partial\xi}\left(F_0(\xi)\frac{\partial\varphi}{\partial\xi}\right)+\mathrm{i}\mu_1\varphi=1\tag{29}$$

$$\xi=0,\quad \frac{\partial\varphi}{\partial\xi}=0\tag{30}$$

$$\xi=-1,\quad \varphi=0\tag{31}$$

$$\frac{\partial}{\partial\xi}\left(F_0(\xi)\frac{\partial\hat{\varphi}}{\partial\xi}\right)+\mathrm{i}\mu_2\hat{\varphi}=1\tag{32}$$

$$\xi=0,\quad \frac{\partial\hat{\varphi}}{\partial\xi}=0\tag{33}$$

$$\xi=-1,\quad \hat{\varphi}=0\tag{34}$$

$$m_{11}\left(\frac{\partial^2\zeta}{\partial x^2}+\frac{\partial^2\zeta}{\partial y^2}\right)+\left(\frac{\partial m_{11}}{\partial x}+\frac{\partial m_{12}}{\partial y}\right)\frac{\partial\zeta}{\partial x}+\left(\frac{\partial m_{11}}{\partial y}-\frac{\partial m_{12}}{\partial x}\right)\frac{\partial\zeta}{\partial y}-\mathrm{i}\sigma\zeta=0\tag{35}$$

where

$$\mu_1=\frac{(\sigma-f)h^2}{\gamma_0},\quad \mu_2=\frac{(\sigma+f)h^2}{\gamma_0},\quad m_{11}=\frac{gh^3}{2\gamma_0}(e+\hat{e})$$

$$m_{12} = \frac{gh^3}{\mathrm{i}z\gamma_0}(e - \hat{e}), \quad e = \int_{-1}^{0} \varphi \mathrm{d}\xi, \quad \hat{e} = \int_{-1}^{0} \hat{\varphi} \mathrm{d}\xi$$

First the equations of the profile function (29)–(31) and (32)–(34) are calculated, then the two dimensional elliptical governing equation of the residual elevation ξ. The tidal velocity is easily obtained from (25)–(26).

The calculation of Lagrangian residual current

Similar treatment as above is applied to the equation (6)–(10). The first-order Lagrangian residual current is:

$$q_{\mathrm{L}} = u_{\mathrm{L}} + \mathrm{i}v_{\mathrm{L}} = \frac{gh^2}{\gamma_1}(\Pi_1\varphi^{\Pi_1} + \varphi^{\theta} + \varphi^{\eta} + \varphi^{\tau}) + \frac{h\tau_a}{\gamma_1}\int_{-1}^{\xi} \frac{\mathrm{d}\xi'}{F_1(\xi')} \tag{36}$$

with the profile functions φ^{Π_1}, φ^{θ}, φ^{η} and φ^{τ} satisfying

$$\frac{\partial}{\partial \xi}\left(F_1(\xi)\frac{\partial \varphi^{\Pi_1}}{\partial \xi}\right) + \mathrm{i}\mu_3\varphi^{\Pi_1} = 1 \tag{37}$$

$$\xi = 0, \quad \frac{\partial \varphi^{\Pi_1}}{\partial \xi} = 0 \tag{38}$$

$$\xi = -1, \quad \varphi^{\Pi_1} = 0 \tag{39}$$

$$\frac{\partial}{\partial \xi}\left(F_1(\xi)\frac{\partial \varphi^{\theta}}{\partial \xi}\right) + \mathrm{i}\mu_3\varphi^{\theta} = \Theta \tag{40}$$

$$\xi = 0, \quad \frac{\partial \varphi^{\theta}}{\partial \xi} = 0 \tag{41}$$

$$\xi = -1, \quad \varphi^{\theta} = 0 \tag{42}$$

$$\frac{\partial}{\partial \xi}\left(F_1(\xi)\frac{\partial \varphi^{\eta}}{\partial \xi}\right) + \mathrm{i}\mu_3\varphi^{\eta} = \Phi \tag{43}$$

$$\xi = 0, \quad \frac{\partial \varphi^{\eta}}{\partial \xi} = 0 \tag{44}$$

$$\xi = -1, \quad \varphi^{\eta} = 0 \tag{45}$$

$$\frac{\partial}{\partial \xi}\left(F_1(\xi)\frac{\partial \varphi^{\tau}}{\partial \xi}\right) + \mathrm{i}\mu_3\varphi^{\tau} = T \tag{46}$$

$$\xi = 0, \quad \frac{\partial \varphi^{\tau}}{\partial \xi} = 0 \tag{47}$$

$$\xi = -1, \quad \varphi^{\tau} = 0 \tag{48}$$

where

$$\Pi_1 = \frac{\partial \langle \zeta_1 \rangle}{\partial x} + \mathrm{i}\frac{\partial \langle \zeta_1 \rangle}{\partial y}, \quad \Theta = \pi_1 + \mathrm{i}\pi_2, \quad \tau_a = \tau_{ax} + \mathrm{i}\tau_{ay}$$

$$T = \frac{\mathrm{i} f \tau_a}{\gamma_1} \int_{-1}^{\xi} \frac{\mathrm{d}\xi'}{F_1(\xi)}$$

$$\Phi = \left[g\frac{\partial}{\partial x}(h\int_{\xi}^{0} \alpha S_0 \mathrm{d}\xi') + g\alpha S_0 \xi \frac{\partial h}{\partial x} \right] + \mathrm{i}\left[g\frac{\partial}{\partial y}(h\int_{\xi}^{0} \alpha S_0 \mathrm{d}\xi') + g\alpha \xi \frac{\partial h}{\partial y} \right]$$

By integrating the continuous equation from sea surface to bottom and introducing stream function, ψ, its governing equation is derived as follows.

$$\eta_{11}\left(\frac{\partial^2 \psi}{\partial x^2} + \frac{\partial^2 \psi}{\partial y^2} \right) + \left(\frac{\partial n_{11}}{\partial x} - \frac{\partial n_{12}}{\partial y} \right)\frac{\partial \psi}{\partial y} + \left(\frac{\partial n_{11}}{\partial x} + \frac{\partial n_{12}}{\partial y} \right)\frac{\partial \psi}{\partial x} = \frac{\partial n_{23}}{\partial x} - \frac{\partial n_{13}}{\partial y} \tag{49}$$

where n_{11}, n_{12}, n_{13} and n_{23} are coefficients determined by the profile functions. Their formulas are not exhibited here, because of the complexity.

The residual elevation may be derived indirectly from Stream function ψ thus the first-order Lagrangian residual current can be obtained from (36).

In summary, two different types of equations must be resolved. One is ordinary differential equation of φ and $\hat{\varphi}$. The other is a two-dimensional elliptical equation of ψ. The β-spline method is adopted to solve the first type and the central differential scheme, the first kind of boundary condition and sequential over-relaxation iteration are adopted to solve the second one.

Numerical experiments

Besides the calculation of wind-induced, thermohaline and tide-induced Lagrangian residual current as a whole, the following numerical experiments have been conducted to determine separately the role that each factor plays individually in the general circulation of the Bohai Sea.

1. Tide-induced Lagrangian residual current;
2. thermohaline circulation;
3. wind-driven thermohaline and tide-induced Lagrangian residual current in constant depth;
4. wind-driven circulation;
5. wind-driven thermohaline and tide-induced Lagrangian residual current with or without discharge of the Huanghe River;
6. residual current caused only by the part of the Huanghai Warm Water coming from the Huanghai Sea through the Bohai Sea Strait is considered.

Analyses and Verification of Results

Basic feature of wind-driven, thermohaline and tide-induced, Lagrangian residual current

Figures 1(a) and (b) show respectively the surface and depth-averaged general circulation in summer induced by the wind-driven, thermohaline and tide-induced Lagrangian residual current with the runoff of the Huanghe River and a part of the Huanghai Warm Water coming from the Huanghai Sea through the Bohai Sea Strait.

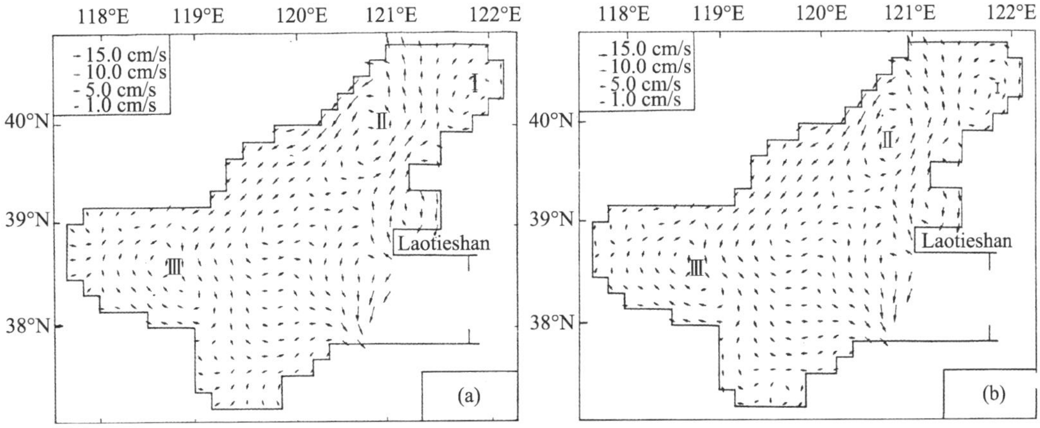

Fig. 1 Wind-driven, thermohaline and tide-induced Lagrangian residual current in the Bohai Sea
(a) Surface; (b) Depth-averaged

It can be seen from Fig. 1 that after a part of the Huanghai Warm Water coming from the Huanghai Sea enters the Bohai Sea through Laotieshan Waterway, it flows northward along the eastern coast of the Liaodong Gulf. Before arriving at the Changxing Island, a part of it turns westward, then also southward, then flows out of the basin through the southern part of the strait along the iso-depth contour of 30 m. The main branch of sea water flows northward along the east coast of Liaodong Gulf, then separated into two parts when it arrives at the head top of the Liaodong Gulf, one part turns eastward, forming one clockwise gyre, named Gyre I, the other turns westward and then southward along the western coast of the Liaodong Gulf, then flows out of the Bohai Sea through the southern part of the strait, forming one big counter-clockwise gyre in the central part of the Bohai Sea. On the way the second part flows southward, one branch of it turns northeastwards, forming one counter clockwise gyre in the central part of the Liaodong Gulf, named Gyre II. While inside the Bohai Gulf, there exists one distinct clockwise gyre, named Gyre III. The flow pattern in the central part of the Bohai Sea is comparatively simpler and no gyre emerges. The maximum value of the residual velocity is 25 cm/s, which is found at the deep, northern waterway of the Bohai Sea Strait.

Verification and analysis of results

Can the above numerical results give the basic features of general circulation in the Bohai Sea? Although no Lagrangian method has been used in the Bohai Sea for the field observation of currents, yet the results either obtained with Lagrangian method or Eulerian method, should be consistent with the basic features of the general circulation. So, two current charts① obtained in August and September respectively, are chosen to compare and verify the numerical results. The charts were drawn mainly according to observation data, and to the distribution of water masses in areas where no observation data were available. Thus, they should embody the main features of the general circulation in the Bohai Sea.

The consistency between calculated results and the theoretical extrapolation

There is one inference in the weakly nonlinear theory of Lagrangian residual current, i.e. the streamlines of depth-averaged mass-transport velocity coincide with the iso-concentration contours (Feng, 1988). A comparison between the depth-averaged circulation (Fig. 1 (b)) and the depth-averaged iso-salinity contours (Fig. 2) shows that the sea water flows approximately along the iso-salinity contours in the central part of the Bohai Sea, particularly in agreement near the mouth of the Bohai Sea Strait. The calculated results have a good coincidence with that inference.

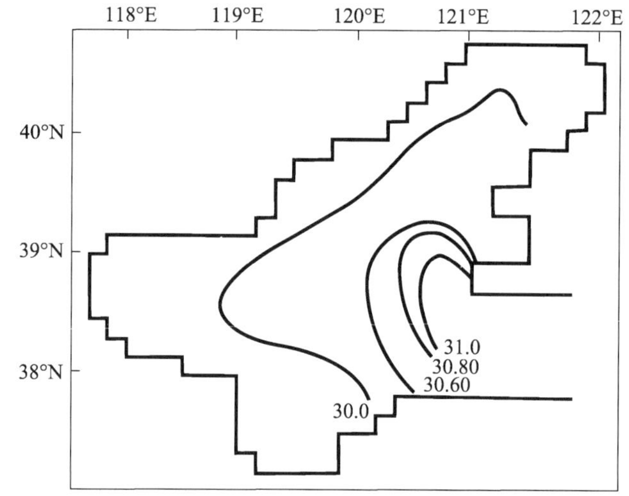

Fig. 2 Observed distribution of depth-averaged salinity (In summer)

Comparison between calculated results and observed surface current

A comparison between Fig. 1a and Fig. 3 shows that both exist a counterclockwise gyre in

① 原文注: Surface current charts of the Bohai Sea, Huanghai Sea and East China Sea. Institute of Oceanology, Academia Sinica. Qingdao (in Chinese)

the central-northern part of the Bohai Sea. This suggests that the calculated results are basically coincided with the observed data. The difference is that the calculated results show three distinct gyres while the observation dose not. The authors give a preliminary comment on these three gyres: (1) Owing to the lack of observation data at the head of the Liaodong Gulf. Gyre I cannot be verified directly yet. It may be formed perhaps by the effect of topography when the current arrives at the head of the Liaodong Gulf; (2) It is difficult to judge from Fig. 3 whether the Gyre II, III exit or not because the observed flow pattern is too coarse and the detail feature is not given. However, it is fairly confirmed to say that Gyre II is consistent with general tendency of the observed current of the Liaodong Gulf. From the fact that the gyres are not shown by the observed distribution of hydrological, chemical and geological factors, we also deduce that maybe the Bohai Sea is so shallow that the mixing induced by wind is extremely strong and may extend through out the whole depth, thus it would lead to the vanishment of the distribution features that might be induced by the gyres. So further studies must be performed to verify the existence of Gyres I, II and III. About the residual current direction in the Bohai Gulf, there exist two opposite opinions, one is that it flows clockwise, the other is counterclockwise. Our results fit the first one. Such a flow pattern is qualitatively consistent with the surface current field in the Bohai Gulf observed from ERTS imagery (Ping, 1983)(Fig. 4).

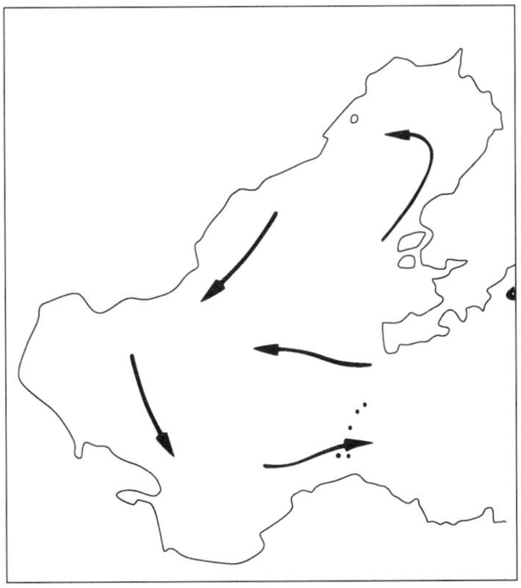

Fig. 3 Observed general circulation in the Bohai Sea (Surface, in summer)

In the present paper, except for the tide-induced Lagrangian residual current, the direction of the circulation in the Bohai Gulf computed with several schemes is counterclockwise. However, the general pattern of the summer circulation in the Bohai Gulf is clockwise. It shows: (1) the general pattern is mainly formed by the "tidal body force", (2) if wind-driven and the thermohaline circulation are emphasized in study the conclusion will be inversed.

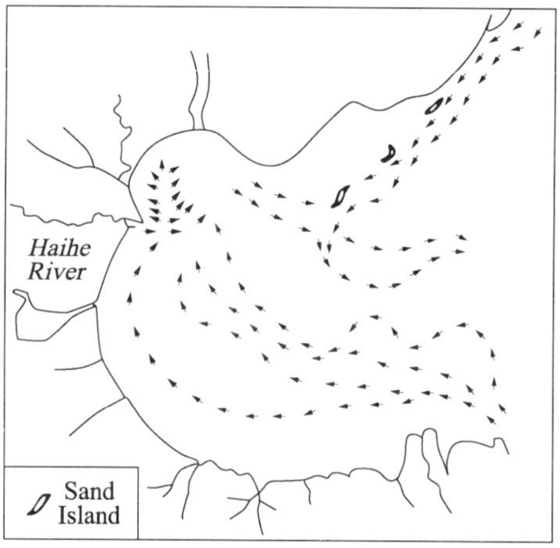

Fig. 4 The turbid water plume and interpretative pattern of flow at the surface of the Bohai Gulf from ERTS Photograph, on June 20, 1978

Effects of every factor in general summer circulation of the Bohai Sea

Tide-induced Lagrangian residual current. A comparison between the tide-induced Lagrangian residual current and the general pattern of the summer circulation shows that there is only small difference between them (Fig. 5). It suggests that the general pattern of the summer circulation in the Bohai Sea is mainly formed by the "tidal body force". The circulation at the water surface is hardly the same as depth-averaged circulation (not exhibited here).

Thermohaline circulation. Figure 6 shows the depth-averaged thermohaline circulation. It has three gyres. The counter clockwise gyre in the central part of the Liaodong Gulf (Fig. 6) has marked influence over gyre II. Thermohaline circulation is fairly small in the central part of the Bohai Sea. By the same reason mentioned above, the circulation at the water surface is not exhibited here.

Influence of the bottom topography. A flat bottom (depth for 25 m) circulation experiment carried out for the Bohai Sea in summer shows that the flow pattern is simple and the Gyres I, II and Ⅲ vanish (Fig. 7) Hence, the variable bottom topography is an important factor in forming the Gyres I, II and III.

Wind driven circulation. Figures 8(a) and (b) show the summer wind-driven circulation. The direction of the circulation at the water surface is hardly the same as the wind direction. The maximum value is 6.6 cm/s and emerges in the head of the Liaodong Gulf. The depth-averaged wind driven circulation has many small gyres. But, the velocity is so small that it has no marked influence over the general pattern of the circulation.

Influence of the runoff of the Huanghe Rirer. River runoff affects the current velocity only in the vicinity of the Huanghe Estuary. It does not change the summer circulation pattern throughout the Bohai Sea (Fig. 9).

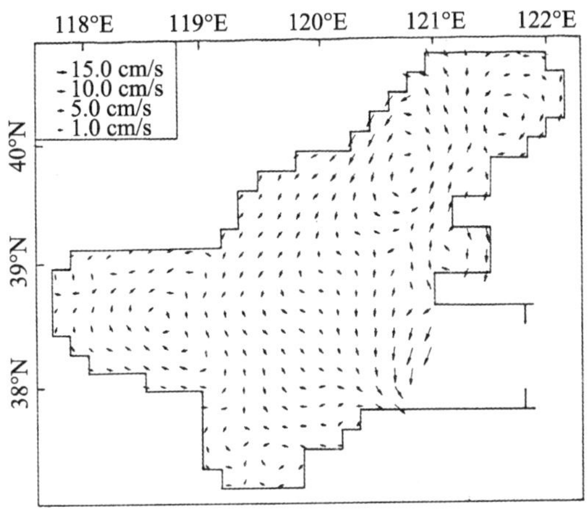

Fig. 5 Depth-averaged tide-induced Lagrangian residual current

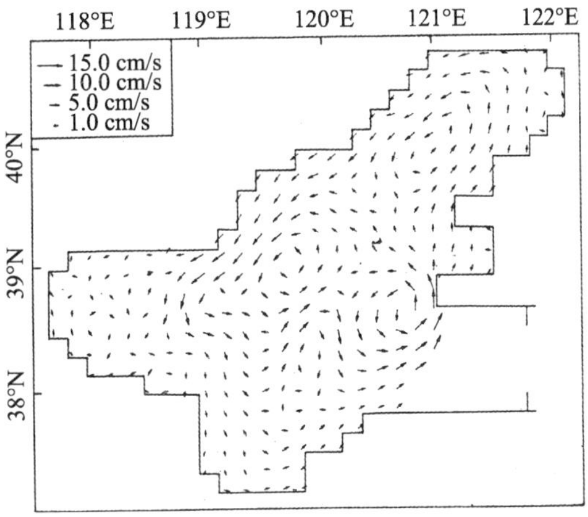

Fig. 6 Depth-averaged thermohaline circulation

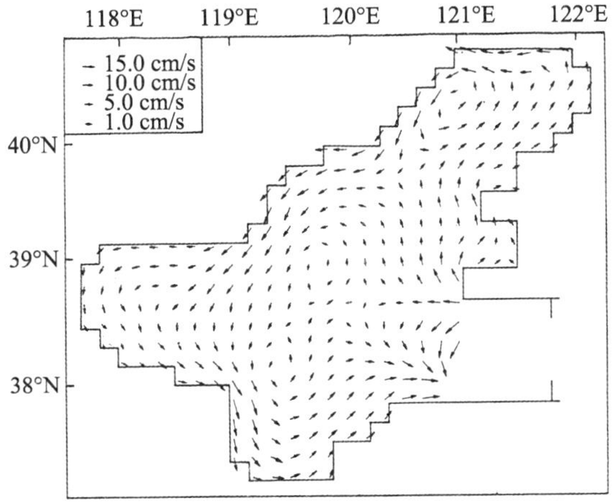

Fig.7 Depth-averaged wind-driven, thermohaline and tide-induced Lagrangian residual current (constant depth)

Influence of the Huanghai warm water. To study the influence of the Huanghai Warm Water coining from the Huanghai Sea through the Bohai Sea Strait over the general circulation, we have calculated the current velocity only induced by the Huanghai Warm Water (Fig. 10). It shows that the flow pattern is very simple and the current velocity is small in the central part of the Bohai Sea. However, the current velocity at the mouth of the Bohai Sea Strait is bigger. Its maximum value is 21.5 cm/s. Hence, the Huanghai Warm Water has marked influence at the mouth of the strait.

The distribution of vertical residual current velocity in Laotieshan cross-section

This cross-section is situated at latitude 38° N. The eastern apex is near the Laotieshan Waterway with 50 m in depth. The western apex is at the shallow water area and only 25 m in depth. Figure 11 shows that there is an upwelling climbing from deep water to shallow water. The maximum velocity of the upwelling current (28.1×10^{-4} cm/s) emerges in the east of this cross-section. We can see a downwelling current in the west of the cross-section. Its maximum value is only 4.1×10^{-4} cm/s.

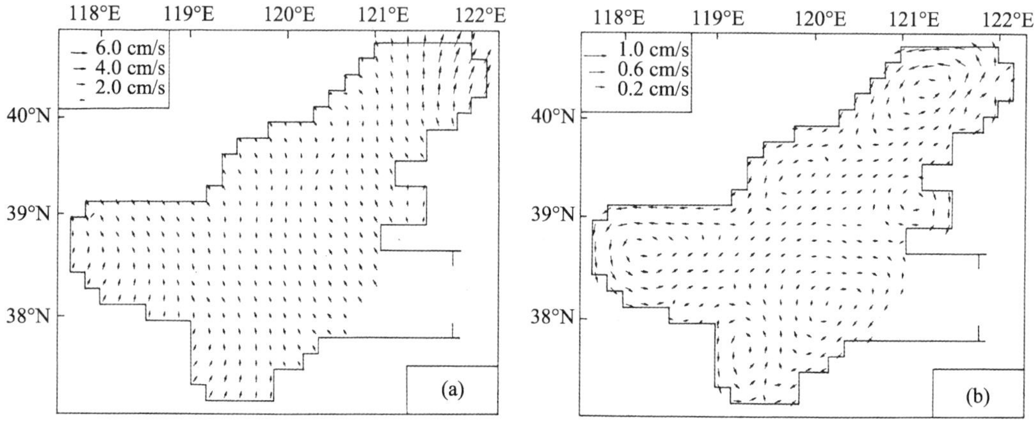

Fig. 8 Wind-driven circulation
(a) Surface; (b) Depth-averaged

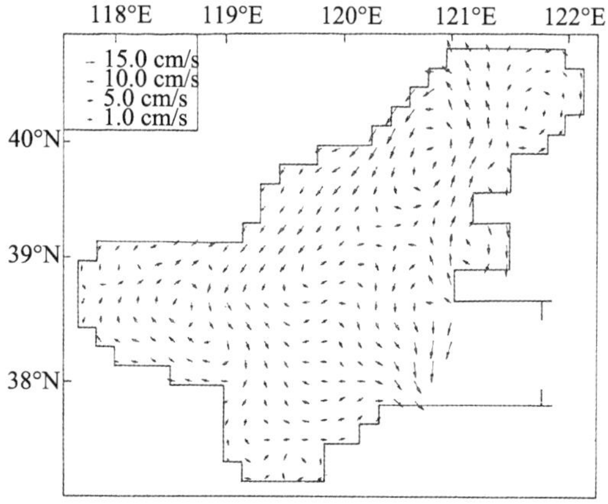

Fig. 9 Wind-driven, thermohaline and tide-induced Lagrangian residual current (no runoff of the Huanghe River)

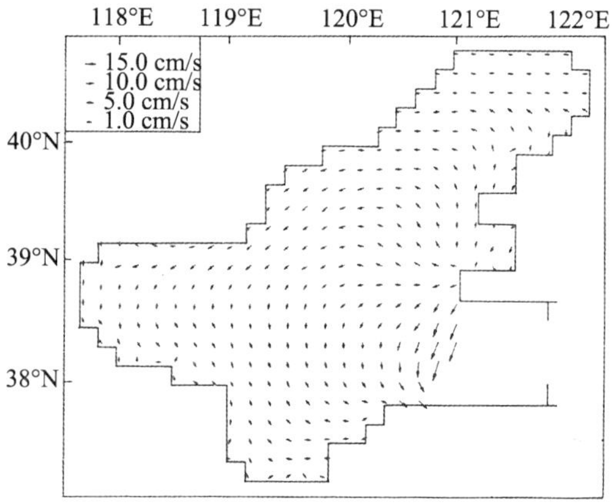

Fig. 10 Depth-averaged circulation, only associated with the part of the Huanghai Warm Water coming from the Huanghai Sea through the Bohai Sea Strait

Finally, the influence of the temperature has not been concluded in our calculation. Usually, the stratification is marked in summer of the Bohai Sea. We are going to study this question later.

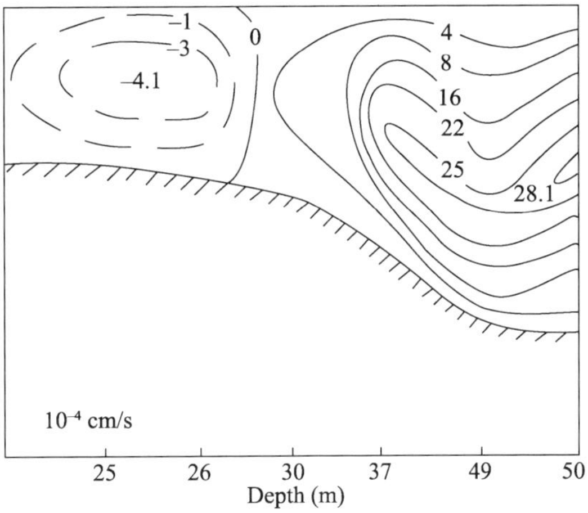

Fig. 11 Distribution of the residual current velocity at the cross-section of Laotieshan

Conclusion

This paper has provided summer wind-driven thermohaline and tide-induced Lagrangian residual current in the Bohai Sea. There exists a big counterclockwise gyre in the central of Bohai Sea and three small gyres situated at the Bohai Gulf and the Liaodong Gulf. Such flow patterns have been verified.

To sum up, the general pattern of the summer circulation in the Bohai Sea is mainly formed by the tide-induced Lagrangian residual current.

The bottom topography is very important for summer circulation in the Bohai Sea. The gyre I and gyre III are formed due to the nonlinear coupling of tidal effect and variable bottom topography. However, gyre II is formed due to the nonlinear common effects of the tide, wind, thermohaline and variable bottom topography.

Richardson number is introduced in eddy viscosity coefficient. On the basis of observed data and numerical experiments, the empirical constants have been obtained.

River runoff affects the current velocity only in the vicinity of the Huanghe Estuary.

The residual current caused by the part of the Huanghai Sea Warm Water from the Huanghai Sea through the Bohai Sea Strait is a single big counterclockwise gyre and its velocity is smaller.

References

Bowden K. F. and P. Hamilton (1975) Some experiment with a numerical model of circulation and mixing in a tide estuary. *Estuarine and Coastal Marine Science*, 3, 281–301.

Feng Shizuo (1987) A three-dimensional weakly nonlinear model of tide-induced Lagrangian residual current and mass-transport, with an application to Bohai Sea. *Three-dimensional Model of Marine and*

Estuarine Dynamics, J. C. J Nihoul & B. M. Jamart, Ed, Elsevier Oceanography series, 45, 99, 171–488.

Feng Shizuo (1988) Fundamentals of hydrodynamics of shallow water circulation and transport. In: *Proc of Conf. on Engineering Mechanics and Termophysics*,Tsinghua University Press (in Chinese), 99, 199–208.

Millero F. J., C. T. Chen, A. Bradshaw and K. Schlercher (1980) A new high pressure equation of state for sea water. *Deep-Sea Research*, 27A, 255–261.

Pang Jiazhen and Si Shuhen (1980) Fluvial process of the Huanghe River Estuary, II. Hydrographical character and region of sediment silting. *Oceanology & Limnology* (in Chinese), 11, 295–305.

Ping Zhongliang (1983) Surface current in Bohai Bay observed on ERTS imagery. *Oceanology & Limnology* (in Chinese), 14, 297–303.

Song Lina (1986) A hydrodynamic velocity-splitting model with depth-varying eddy viscosity in shallow seas, I, velocity-splitting model. *Journal of Shandong College of Oceanography* (in Chinese), 16, 15–24.

Zhang Shuzhen, Xi Pangen and Feng Shizuo (1984) Numerical modeling of steady circulation in the Bohai Sea. *Journal of Shandong College of Oceanography* (in Chinese), 14, 12–19.

An Inter-tidal Transport Equation Coupled with Turbulent K-ε Model in a Tidal and Quasi-steady Current System *

FENG Shizuo (冯士筰) and WU Dexing (吴德星)

This note attempts to derive an inter-tidal transport equation coupled with turbulent-closure model in a tidal and quasi-steady current system with convection dominant or in the order of turbulent diffusion.

In general, convection-diffusion equation describing the intra-tidal process of concentration c in shallow sea or estuary can be expressed as[1]

$$\frac{\partial c}{\partial t}+\boldsymbol{u}\cdot\nabla c=\frac{\partial}{\partial z}\left(\lambda\frac{\partial c}{\partial z}\right) \tag{1}$$

where t is time; ∇ is gradient operator; z is vertical coordinate; $\boldsymbol{u}$ denotes velocity vector; λ is coefficient of turbulent diffusivity. Nondimensional form of Eq. (1) is rendered as[2]

$$\frac{\partial c}{\partial \theta}+\kappa\boldsymbol{u}\cdot\nabla c=\mathscr{E}\frac{\partial}{\partial z}\left(\lambda\frac{\partial c}{\partial z}\right) \tag{2}$$

where both of nondimensional and dimensional variables have the same form, except $\theta=\omega t$ (ω is tidal frequency); κ is the ratio of current speed to long wave phase speed; $\sigma=\mathscr{E}/\kappa$ is the ratio of turbulent diffusion to convection transport.

Suppose that κ is a small parameter[2], that is,

$$\mathscr{O}(\kappa)<1, \tag{3}$$

and

$$\mathscr{O}(\sigma)\begin{cases}\kappa \ \text{(convection transport dominant)}^{[2]}\\ 1 \ \text{(convection transport having the order of turbulent diffusion)}\end{cases} \tag{4}$$

By means of a perturbation approach with the small parameter κ, the variables $(\boldsymbol{u},c,\lambda)$ and their time derivatives in the multiple time scale system[3–5] can be expanded in ascending series of κ. Substituting all the expanded variables and time derivatives into Eq. (2) as well as indicating the j-order solution by the subscript j, the zero-order equation becomes

$$\frac{\partial c_0}{\partial \theta_0}=0 \tag{5}$$

and the first-order equation is

---

* Feng S, Wu D. 1995. An inter-tidal transport equation coupled with turbulent K-ε model in a tidal and quasi-steady current system. Chinese Science Bulletin, 40(2): 136–139

$$\frac{\partial c_0}{\partial \theta_1}+\frac{\partial c_1}{\partial \theta_0}+\boldsymbol{u}_0\cdot\nabla c_0=\begin{cases}0 & (\mathscr{O}(\sigma)=\kappa)\\ \sigma\dfrac{\partial}{\partial z}\left(\lambda_0\dfrac{\partial c_0}{\partial z}\right) & (\mathscr{O}(\sigma)=1)\end{cases}\tag{6}$$

where

$$\boldsymbol{u}_0=\tilde{\boldsymbol{u}}_0+\bar{\boldsymbol{u}}_0\tag{7}$$

$\tilde{\boldsymbol{u}}_0$ is tidal current, the periodic function of θ_0; $\bar{\boldsymbol{u}}_0$ is quasi-steady current independent of θ_0. However, both of $\tilde{\boldsymbol{u}}_0$ and $\bar{\boldsymbol{u}}_0$ are the functions of $(\theta_1,\theta_2,\cdots)$ ($\theta_j=\omega_j t_j$, $\omega_j=\kappa^j\omega$, $j=0,1,2,\cdots$), representing the low frequency motion.

Further suppose that turbulence satisfies quasi-local-equilibrium state ($\mathscr{O}(\kappa^2)$), then $\dfrac{\partial\lambda_0}{\partial\theta_0}=0$ [6]. Integrating Eq. (6) with θ_0 and eliminating the secular term yields

$$\frac{\partial c_0}{\partial \theta_1}+\bar{\boldsymbol{u}}_0\cdot\nabla c_0=\begin{cases}0 & (\mathscr{O}(\sigma)=\kappa)\\ \sigma\dfrac{\partial}{\partial z}\left(\lambda_0\dfrac{\partial c_0}{\partial z}\right) & (\mathscr{O}(\sigma)=1)\end{cases}\tag{8}$$

and

$$c_1=\bar{c}_1+\tilde{c}_1\tag{9}$$

where $\tilde{c}_1=-\int\tilde{\boldsymbol{u}}_0\mathrm{d}\theta_0\cdot\nabla c_0$, $\bar{c}_1=\bar{c}_1(\theta_1,\theta_2,\cdots)$.

The second-order equation is

$$\frac{\partial c_0}{\partial \theta_2}+\frac{\partial c_1}{\partial \theta_1}+\frac{\partial c_2}{\partial \theta_0}+\boldsymbol{u}_1\cdot\nabla c_0+\boldsymbol{u}_0\cdot\nabla c_1=\begin{cases}\left(\dfrac{\sigma}{\kappa}\right)\dfrac{\partial}{\partial z}\left(\lambda_0\dfrac{\partial c_0}{\partial z}\right) & (\mathscr{O}(\sigma)=\kappa)\\ \sigma\dfrac{\partial}{\partial z}\left(\lambda_1\dfrac{\partial c_0}{\partial z}+\lambda_0\dfrac{\partial c_1}{\partial z}\right) & (\mathscr{O}(\sigma)=1)\end{cases}\tag{10}$$

where $\lambda_1-\bar{\lambda}_1\sim\tilde{\lambda}_1$ [6], besides $\boldsymbol{u}_1$ denoting the linear combination of tide-induced, wind-driven and thermohaline currents, it also includes an added θ_0-dependent periodic current induced by the nonlinear coupling between the quasi-steady current and tidal current as well as a θ_0-independent current induced by the auto-nonlinear coupling of quasi-steady current.

Integrating Eq. (10) with θ_0 and eliminating the secular term, we have

$$\frac{\partial \bar{c}_1}{\partial \theta_1}+\bar{\boldsymbol{u}}_0\cdot\nabla\bar{c}_1+\frac{\partial c_0}{\partial \theta_2}+\boldsymbol{u}_{\mathrm{L}}\cdot\nabla c_0=\begin{cases}\left(\dfrac{\sigma}{\kappa}\right)\dfrac{\partial}{\partial z}\left(\lambda_0\dfrac{\partial c_0}{\partial z}\right) & (\mathscr{O}(\sigma)=\kappa)\\ \sigma\dfrac{\partial}{\partial z}\left(\bar{\lambda}_1\dfrac{\partial c_0}{\partial z}+\bar{\lambda}_0\dfrac{\partial \bar{c}_1}{\partial z}\right) & (\mathscr{O}(\sigma)=1)\end{cases}\tag{11}$$

where $\boldsymbol{u}_{\mathrm{L}}=\boldsymbol{u}_{\mathrm{E}}+\boldsymbol{u}_{\mathrm{s}}$ is mass-transport velocity, $\boldsymbol{u}_{\mathrm{E}}$ denotes Eulerian time averaged velocity, $\boldsymbol{u}_{\mathrm{s}}$ is Stokes, drift velocity[3].

After multiplying Eq. (11) by κ and adding it to Eq. (8), a general equation for predicting the inter-tidal ($\omega_j^{-1}, j=1,2,\cdots$) evolution of concentration is obtained and its nondimensional form (correct to the order of $\mathcal{O}(\kappa)$) is

$$\frac{\partial \overline{c}}{\partial \Theta}+\overline{\boldsymbol{u}}\cdot\nabla\overline{c}=\sigma\frac{\partial}{\partial z}\left(\overline{\lambda}\frac{\partial \overline{c}}{\partial z}\right) \tag{12}$$

where $\overline{c}=c_0+\kappa\overline{c}_1$, $\frac{\partial}{\partial \Theta}=\frac{\partial}{\partial \theta_1}+\kappa\frac{\partial}{\partial \theta_2}$, $\overline{\boldsymbol{u}}=\overline{\boldsymbol{u}}_0+\kappa\boldsymbol{u}_{\mathrm{L}}=\boldsymbol{U}+\kappa\boldsymbol{u}_{\mathrm{s}}$, $\boldsymbol{U}=\overline{\boldsymbol{u}}_0+\kappa\overline{\boldsymbol{u}}_{\mathrm{E}}$, $\overline{\lambda}=\lambda_0+\kappa\overline{\lambda}_1$, $\mathcal{O}(\sigma)=\kappa$ or $\mathcal{O}(\sigma)=1$;

and its dimensional form is

$$\frac{\partial \overline{c}}{\partial \mathcal{T}}+\overline{\boldsymbol{u}}\cdot\nabla\overline{c}=\frac{\partial}{\partial z}\left(\overline{\lambda}\frac{\partial \overline{c}}{\partial z}\right) \tag{13}$$

where $\frac{\partial}{\partial \mathcal{T}}=\frac{\partial}{\partial t_1}+\frac{\partial}{\partial t_2}$. Note here that the inclusion of the terms with the order of $\mathcal{O}(\kappa^2)$ in Eq. (12) and Eq. (13) is only for variables' normalization and convenience in practical use.

To determine the turbulent diffusion coefficient λ, using the k-ε (k is the turbulent kinetic energy and ε is the rate of dissipation of k) turbulent closure model and the assumption of quasi-local equilibrium, we get (correct to the order of $\mathcal{O}(\kappa)$)[6]

$$\frac{\partial \overline{k}}{\partial \mathcal{T}}+\overline{\boldsymbol{u}}\cdot\nabla\overline{k}=\frac{\partial}{\partial z}\left(\overline{\lambda}_k\frac{\partial \overline{k}}{\partial z}\right)+\overline{P}+\overline{G}-\overline{\varepsilon} \tag{14}$$

and

$$\frac{\partial \overline{\varepsilon}}{\partial \mathcal{T}}+\overline{\boldsymbol{u}}\cdot\nabla\overline{\varepsilon}=\frac{\partial}{\partial z}\left(\overline{\lambda}_\varepsilon\frac{\partial \overline{\varepsilon}}{\partial z}\right)+c_{1\varepsilon}\frac{\overline{\varepsilon}}{\overline{k}}\left[(\overline{P}+\overline{G})(1-c_{3\varepsilon}Ri_{\mathrm{f}})-\frac{c_{2\varepsilon}}{c_{1\varepsilon}}\overline{\varepsilon}\right] \tag{15}$$

where

$$\overline{\lambda}=\frac{\overline{\gamma}}{Sc},\quad \overline{\lambda}_p=\frac{\overline{\gamma}}{Sc_p}(p=k,\varepsilon) \tag{16}$$

$$\overline{\gamma}=c_\mu\frac{\overline{k}^2}{\overline{\varepsilon}} \tag{17}$$

where $\overline{P}=\gamma\left[\left(\frac{\partial \tilde{u}_0}{\partial z}\right)^2+\left(\frac{\partial \tilde{V}_0}{\partial z}\right)^2+\left(\frac{\partial \overline{u}_0}{\partial z}\right)^2+\left(\frac{\partial \overline{V}_0}{\partial z}\right)^2\right]$ is the production of k, and $\overline{G}=g\beta\overline{\lambda}\frac{\partial \overline{c}}{\partial z}$ is the buoyancy production of k; g is the gravitational acceleration; β is the thermal expansion coefficient; (u_0, V_0) are the horizontal velocity components of $\boldsymbol{u}_0$; Sc is Schmidt number; Ri_{f} is the flux Richardson number; c_μ, $c_{1\varepsilon}$, $c_{2\varepsilon}$ and $c_{3\varepsilon}$ are nondimensional empirical coefficients[7].

If $\mathcal{O}(\sigma)=\kappa$, $\overline{\boldsymbol{u}}=0$ and $\boldsymbol{u}_{\mathrm{L}}$ is independent of θ_1 (for example, a pure M_2 nonlinear tidal system with convection transport dominant can meet these conditions), then from Eq. (8) it has $\frac{\partial c_0}{\partial \theta_1}=0$. Further noting that λ_0 is independent of θ_1 integrating Eq. (11) with θ_1 and eliminating the secular term yields

$$\frac{\partial c_0}{\partial \theta_2}+\boldsymbol{u}_{\mathrm{L}}\cdot\nabla c_0=\left(\frac{\sigma}{\kappa}\right)\frac{\partial}{\partial z}\left(\lambda_0\frac{\partial c_0}{\partial z}\right)\quad(\mathscr{O}(\sigma)=\kappa)\tag{18}$$

Eq. (18) is the degenerate form of Eq. (12) and represents an inter-tidal transport equation in a shallow sea system with only tidal convection transport dominant[3–6]. Eq. (18) also implies that inter-tidal transport of concentration evolves on θ_2 time scale[4–5].

The derived inter-tidal transport equation (12) shows that in a shallow sea or estuarine system in which tidal current and quasi-steady current (in the order of tidal current) are dominant, inter-tidal transport process of concentration evolves on θ_1 time scale. The convection-transport velocity is the sum of quasi-steady current $\bar{\boldsymbol{u}}_0$ and mass-transport velocity $\boldsymbol{u}_{\mathrm{L}}$, or is equivalent to the sum of quasi-steady current, Eulerian residual current ($\bar{\boldsymbol{U}}$) and tide-induced Stokes' drift velocity $\boldsymbol{u}_{\mathrm{s}}$. It is easy to verify that convection-transport velocity satisfies the continuity condition of flow field.

It is found by comparing Eq. (2) (or Eq. (1)) with Eq. (12) (or Eq. (13)) that if convection dominates the concentration transport during intra-tidal process on θ_0 time scale, convection also dominates the concentration transport during inter-tidal process on θ_1 time scale, and that if the contribution of convection to the concentration transport is in the order of that of turbulent diffusion during intra-tidal process on θ_0 time scale, it is also right during inter-tidal process on θ_1 time scale.

References

[1] Fischer, H. B., List, E. J., Koh, R. C. Y. *et al.*, *Mixing in Inland and Coastal Waters*, Orlando: Academic Press, 1979.

[2] Feng Shizuo, *Three-Dimensional Model of Marine and Estuarine Dynamics*, (eds. Nihoul, J. C. J., Jamart, B. M.), Amsterdam: Elsevier Publ, Co., 1987, 471–488.

[3] Feng Shizuo, *Physics of Shallow Estuaries and Bays*, (ed. Cheng, R. T.), New York: Springer-Verlag, 1990, 34–48.

[4] Hamrick, J. M., Time-averaged estuarine mass transport equations, in *Proc. of the 1987 National Conf. on Hydraulic Engineering*, New York: ASCE, 1987, 624–629.

[5] Lu Youyu, On the Lagrangian residual current and residual transport in a multiple time scale system of shallow seas, *Chin. J. Oceanol. Limnol*, 1991, 9(2): 184.

[6] Feng Shizuo, Lu Youyu, A turbulent closure model of coastal circulation, *Chinese Science Bulletin*, 1993, 38(20): 1737.

[7] Rodi, W., *Turbulence Models and Their Application in Hydraulics: A State of the Art Review*, Netherland: Book Pub. of IAHR, 1984.

On Circulation in Bohai Sea Yellow Sea and East China Sea *

Shizuo Feng

1. Introduction

As the development of economic hot spots in coastal region, oceanographers are paying more and more attention to the marine environmental problems. Estuarine and coastal management requires the evaluation of water quality over periods of several years. Studies of coastal oceanography in environmentally related problems have focused on transport processes which determine the long-term fate of dissolved and suspended matter, in particular of pollutants. In fact, an estuary or coastal sea, such as the Yellow Sea (the Huanghai Sea), can be considered as a marine ecosystem, in which the marine biological, chemical, physical, geochemical and geophysical processes occur and vary continuously, and they are interrelated and interacted on each other. There is hardly any need for further elaboration on the importance of shallow sea circulation as a carrier of matter in the processes.

The Yellow Sea is not an enclosed sea, but is a semi-enclosed shallow sea. The Yellow Sea is connected in the northwest corner with the Bohai Sea through the Bohai Strait and in the south with the East China Sea (the East Sea of China), an eqicontinental sea (Fig. 1). In fact, the water circulation in the Bohai Sea, the Yellow Sea and the East China Sea (BYES) is a continuous system. The general features of the circulation in BYES were sketched and exhibited by Guan (1994) (Fig. 2), where (a) for winter and (b) for summer; a, Kuroshio; b, Tsushima Warm Current (TSWC); c, Yellow Sea Warm Current (YSWC); d, Bohai Sea Circulation (BSC); e, Taiwan Warm Current (TWC); f, China Coastal Current (CCC); g, South China Sea Warm Current (SCSWC); h, West Korea Coastal Current (WKCC). The conceptual circulation features shown in Fig. 2 were inferred from the analyses of historical hydrographic data, and represent that the general circulation system of the Bohai Sea, the Yellow Sea and the East China Sea consists of two sub-systems: the warm and saline current system of oceanic origin (Kuroshio and its branches and extensions) and the less saline coastal current system. The former and the latter move northeastward and southward, respectively. With the exception of some local areas, the major circulation patterns of both summer and winter seasons in BYES are rather similar, roughly forming a respective cyclonic gyre (Guan, 1994). While the conceptual patterns of the general circulation of BYES shown in Fig. 2 are interesting and valuable, some of them are still controversial, such as the circulation pattern

---

* Feng S. 1998. On circulation in Bohai Sea Yellow Sea and East China Sea. In: G H Hong, J Zhang, B K Park (eds.) The Health of the Yellow Sea. Seoul: The Earth Love Publication Association. 43–77

in the Bohai Sea, YSWC and the summer circulation in the Yellow Sea, and the source of TWC, etc.

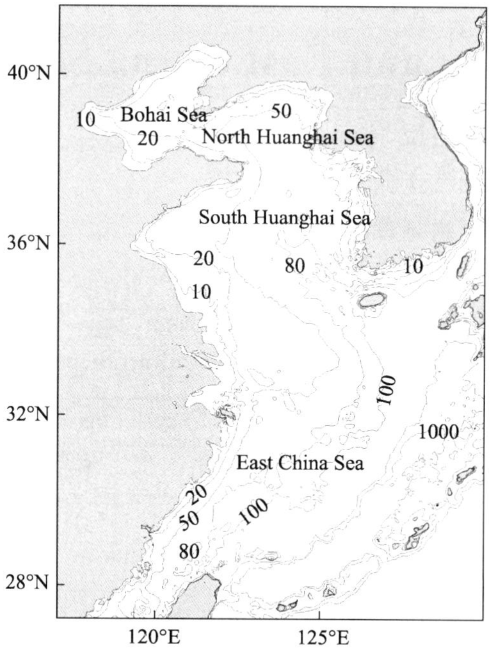

Fig. 1 Map of the Bohai Sea, the Yellow Sea and the East China Sea (BYES)

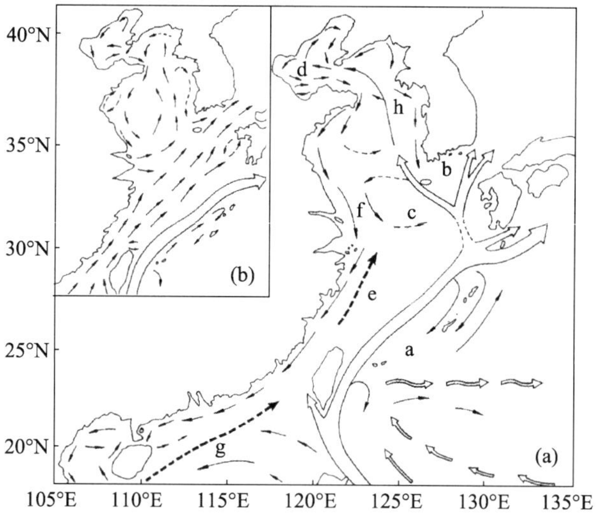

Fig. 2 Schematic representations of the major surface current systems in the Bohai, Yellow, East China seas, and adjacent areas for winter (a) and for summer (b)

a. Kuroshio; b. Tsushima Warm Current; c. YS Warm Current: d. BS Circulation; e. Taiwan Warm Current; f. China Coastal Current; g. SCS Warm Current; h. West Korea Coastal Current (Guan, 1994).

As an example, the controversies over differing opinions on YSWC and the related problems of circulation will be briefly introduced as follows. It has long been believed that YSWC was considered as a branch separated from TSWC and moved in general along the axis of the saline water tongue as a jet-like current (Guan and Chen, 1964). In fact, in summer, there exists a water mass in the central Yellow Sea deep layers, that is characterised by very low temperature, and the water mass is called the "Yellow Sea Cold Water Mass (YSCWM)" (Ho *et al.*, 1959). The YSCWM blocks the passage of YSWC flowing into the central part of the Yellow Sea, and thus the path of YSWC is deviated eastward and the water of oceanic origin moves northward chiefly around the boundary of the baroclinic circulation of the YSCWM (Guan and Chen, 1964). Another view point is that, in summer, cold waters occupy the middle region of the Yellow Sea, which means that there is no such a warm current mentioned above (Pang, 1996). The appearance of YSWC in winter was only considered as the compensative current of the wind-driven current by the monsoon (Hsueh and Romca, 1986). However, it is a warm and saline (salinity>32) water tongue, most remarkable in winter, and the northward flowing warm current with its extension appears also in the surface layer even in the period of weak winds in winter (Guan and Chen, 1964; Guan, 1994). Lee and Lee (1996) thought that since the detailed physical processes of the inflow (YSWC) has not been understood yet and no obvious observational evidence for the inflow has been obtained, one should raise a doubt about existence of the inflow.

Thus, it can be seen that there is a great need of doing the profound researches on the coastal ocean circulation, such as that in BYES, and collecting a vast amount of scientific data, in particular, field data for more thoroughly understanding the dynamics of the circulation and more exactly modeling and predicting the circulation field, although there are quite many contributions to the modeling and understanding of the circulation and its dynamics through the analyses of hydrographic data, numerical experiments and the studies of theoretical models (cf. Oceanology of China Seas, Vol. 1, edited by Zhou Di *et al.*, 1994; and the others).

To the author's knowledge, in fact, the physics of coastal water circulation could be called in an open question.

What is the definition of "circulation", or more generally, the problem of circulation in coastal seas, such as in the Bohai Sea, the Yellow Sea and the East China Sea, and how should it be described appropriately, particularly in dynamics?

Oceanic circulation is usually defined as time-averaged "climate" ocean water motions. However, in the tidal estuaries or coastal seas, such as the Bohai Sea, the Yellow Sea and the coastal shallow waters of the East China Sea, the dominant observable motions are periodical currents associated tides. The residual circulation is conventionally derived from current-meter records by using filter techniques or time averages of time series records and is defined as the Eulerian residual circulation. The Eulerian residual current is traditionally used to represent the circulation in tidal estuaries or coastal seas.

In the early stage of shallow sea circulation study, it was thought that the time-averaged circulation is driven only by the wind stress and the forces associated with the thermal stratification and salinity gradient. Until the middle of 1970s, Csanady (1975, 1976) and Nihoul and Ronday (1975) pointed out the incompleteness of this concept. In fact, in the tidal estuaries or coastal seas the dominant tidal motion is, to a certain extent, nonlinear and the nonlinear interactions among the dominant tidal variables, when time averaged over a tidal period or several tidal periods, will generate the "tidal stress". The "tidal stress", like as a wind stress, has also acted on the time-averaged circulation (Nihoul and Ronday, 1975). Furthermore, if one applied the above viewpoint to three-dimensional baroclinic model for the Eulerian residual circulation, it can be found that there exists not only a "tidal stress" but also a "tidal surface source" and "tide-induced body force" due to the nonlinear interactions between tides (Feng *et al.*, 1984). Since the tidal estuaries or the coastal seas are nonlinear systems, the time-averaged circulations in these systems are driven not only by the wind stress at the sea surface and the thermohaline effects but also by the "tide-induced forces" induced by the nonlinear couplings of astronomical tides. Unfortunately, the viewpoint of the Eulerian residual circulation representing the shallow sea circulation has been believed to be open to question (Feng, 1987, 1990, 1994). In fact, the distributions of important water properties, such as temperature and salinity, and the inter-tidal transport processes of pollutants or any tracer are closely related to the water circulation. It is becoming increasingly clear that the inter-tidal mass transport processes are not determined by the Eulerian residual current but determined by the net displacement of a marked water parcel divided by the lapsed time, that is the Lagrangian mean velocity (Zimmerman, 1979; Cheng and Casulli, 1982; Feng *et al.*, 1986a). The Lagrangian mean velocity can lead to the concept of the Lagrangian residual velocity or Lagrangian residual current (Feng, 1987).

The mass transport velocity (Longuet-Higgins, 1969) is derived as the first order approximation of Lagrangian residual velocity. It is worth notice that although the Eulerian residual velocity satisfies solenoidal field, it does not satisfy the conservation condition of material surface formed by water parcels in the flow field (Feng, 1987). For example, there exist "tidal surface source" at sea surface and "source" or "sink" in the solid lateral boundaries. On the other hand, the mass transport velocity satisfies not only solenoidal field, but also the conservation condition of a material surface in the water. From mentioned above, the mass transport velocity is more qualified for the description of the flow field than the Eulerian residual velocity is.

The favorable properties of the mass transport velocity for the description of the flow field of circulation are further confirmed in the inter-tidal transport equation. An inter-tidal transport equation, namely a convection-diffusion equation for the inter-tidal processes, describes the balance between convective and dispersive transports for the tidally averaged concentration of any conservative tracer. This equation is a key equation for studying the marine environmental problems. As it is well known in the conventional inter-tidal transport equation that the convection velocity is the Eulerian residual velocity, and an assumption of "tidal dispersion" has to be made (Fischer *et al.*, 1979). There are two weaknesses in the conventional inter-tidal transport equation. Firstly, it is questionable whether the Eulerian residual velocity describes the proper physics in an inter-tidal transport problem since for the inter-tidal transport processes the convective transport should be characterized by the Lagrangian residual velocity. Secondly, the physics and dynamics of the "tidal dispersion" are not well understood (Bowden, 1967; Fischer *et al.*, 1979). A new inter-tidal transport equation has already been proposed and extended (Feng *et al.*, 1986b; Cheng *et al.*, 1989; Feng, 1987, 1990; Hemrich, 1987; Lu, 1991). The characteristics of the new inter-tidal transport equation are that in the equation the convective velocity is the mass transport velocity and there is no need to introduce the "tidal dispersion". This new equation not only controls the time-averaged concentration of dissolved and suspended matter as well as any tracer but also controls the time- averaged field of salinity and temperature. In a baroclinic shallow sea or estuary, the velocity field of circulation with the time-averaged field of water properties, such as temperature and salinity, constructs a nonlinear thermo-hydrodynamic coupling system.

In view of the above-mentioned facts, the first-order Lagrangian residual velocity, or the mass-transport velocity, should be appropriate to the description of the lowest-order velocity field of (time-mean) circulation in coastal seas, such as in the Bohai Sea and the Yellow Sea. A set of field equations governing the mass-transport velocity, the residual elevation and the time-averaged apparent concentration has been proposed for a baroclinic shallow sea system (Feng, 1990). The set of field equations can be used to describe not only the wind-driven and thermohaline circulation but also the tide-induced (residual) circulation through a "Tidal body force" (Feng, 1987, 1990), and has, to a certain extent, been successfully applied to modeling the season circulation in the Bohai Sea and to studying the mechanisms of the formation of the same circulation (Wang *et al.*, 1993; Feng, 1994) and the related problems (cf. Wright *et al.*, 1990). Moreover, the new intertidal transport equation proposed by Feng (1987) and mentioned above has also been successfully applied to, for example, the simulations of the salinity transport processes for the entire year of 1985 and further, as a Lagrangian processor, to simulation of 22 (WQM) state variables transport processes for the years 1984–1986, and predicting and examining the long-term trends of eutrophication from 1959 to 1988 in Chesapeake Bay, U.S.A. (Dortch *et al.*, 1992; Cerco and Cole, 1993; Cerco, 1993, 1995) .

Based on the new theoretical frame of shallow sea circulation mentioned above, a generalized model, which can describe the circulation in a coastal sea system in which tidal currents and some quasi-steady flows, such as the Kuroshio in the East China Sea, are of the same order and dominant in the observable flow field, has been proposed and examined (Feng and Wu, 1995; Feng *et al.*, 1996). This model also can be used to describe the circulation in a tidal estuary with large runoff. In the present paper, the generalized model of coastal water circulation and its application to the problems of circulation in the BYES will be briefly introduced and discussed.

2. Model

A primary set of three-dimensional field equations for a thermo-hydrodynamic system in coastal seas is proposed as follows:
the continuity equation

$$\nabla \cdot \boldsymbol{u} + \frac{\partial w}{\partial z} = 0 \tag{1}$$

the momentum equation

$$\frac{\partial \boldsymbol{u}}{\partial t} + \kappa\left(\boldsymbol{u}\cdot\nabla\boldsymbol{u} + w\frac{\partial \boldsymbol{u}}{\partial z}\right) + f\boldsymbol{e}\times\boldsymbol{u} = -g\nabla\zeta + \frac{\partial}{\partial z}\left(\nu\frac{\partial \boldsymbol{u}}{\partial z}\right) - \kappa g\beta\nabla\int_z^{\kappa\zeta} c\mathrm{d}z' \tag{2}$$

the transport equation

$$TR(p) = \kappa^2 S_p \tag{3}$$

and the associated boundary conditions at $z = \kappa\zeta$

$$w = \frac{\partial \zeta}{\partial t} + \kappa\boldsymbol{u}\cdot\nabla\zeta \tag{4}$$

$$\nu\frac{\partial \boldsymbol{u}}{\partial z} = \kappa\boldsymbol{\tau}_{\mathrm{a}} \tag{5}$$

$$\frac{\partial c}{\partial z} = 0 \tag{6}$$

and at $z = -h$,

$$\boldsymbol{u} = w = 0 \tag{7}$$

$$\frac{\partial c}{\partial z} = 0 \tag{8}$$

in which t is time; ∇ is the horizontal gradient operator, z and $\boldsymbol{e}$ are the vertical coordinate and the corresponding unit vector respectively; $\boldsymbol{u}$ is the horizontal velocity vector, w is the vertical velocity; ζ and h are the water surface elevation and the water depth measured from mean sea level, respectively; f is the Coriolis parameter; g is the gravitational acceleration; β is the fractional increase in density per unit increase in salinity; $\boldsymbol{\tau}_{\mathrm{a}}$ is the wind stress vector at the water surface divide by the water density, ρ; ν is the eddy viscosity; and c is the "apparent concentration" (Feng, 1990) and is expressed by

$$c = (s - s_*) + \sum_{j=1}^{m}(\beta_j / \beta)(c_j - c_{j*}) - (\alpha / \beta)(\theta - \theta_*) \tag{9}$$

where c_j is the concentration of j-tracer, β_j is the corresponding constant, the notation "$*$" indicates the corresponding reference constant, and a generalized linearized equation of state containing salinity, s, temperature, θ and c_j, or $\rho - \rho_* = \beta c$ is used, and s, θ and c_j satisfy the same transport equation as Eq. (3).

At the left-hand side of Eq. (3), the transport operator, $TR(p)$, is defined as

$$TR(p) \equiv \left\{ \frac{\partial}{\partial t} + \kappa \left[\boldsymbol{u} \cdot \nabla + w \frac{\partial}{\partial z} - \frac{\partial}{\partial z} \left(\lambda_p \frac{\partial}{\partial z} \right) \right] \right\} (p) \tag{10}$$

where λ_p is the eddy diffusivity, and at the right-hand side of Eq. (3), S_p is the source. While $p = c$, $S_p = S_c$, and particularly the transport equation of a conservative tracer with the apparent concentration c is described by Eq. (3) for $S_p = S_c = 0$ (Feng, 1990).

Putting $p = k$ and $S_p = S_k = P + G - \varepsilon$, the turbulent kinetic energy, k, is described by Eq. (3), where the production of k, $P = \nu \frac{\partial \boldsymbol{u}}{\partial z} \cdot \frac{\partial \boldsymbol{u}}{\partial z}$ and the buoyancy production/destruction of k, $G = gbl_c \frac{\partial c}{\partial z}$. The rate of dissipation of k, ε, is also described by Eq. (3), if $p = \varepsilon$ and $S_p = S_\varepsilon = c_{1\varepsilon} \frac{\varepsilon}{k} \left[(P + G)(1 + c_{3\varepsilon} Ri_f) - \frac{c_{2\varepsilon}}{c_{1\varepsilon}} \varepsilon \right]$, where Ri_f is the flux Richardson number, $c_{j\varepsilon}$ $(j = 1,2,3)$ are nondimensional empirical coefficients (Rodi, 1984).

According to Kolmogorov-Prandtl, the eddy viscosity, ν, is expressed as

$$\nu = c_\mu \frac{k^2}{\varepsilon} \tag{11}$$

where c_μ is a nondimensional empirical coefficient.

Introducing the expression of the eddy diffusivity λ_p

$$\lambda_p = \frac{\nu}{\sigma_p}, \tag{12}$$

where σ_p is the turbulent Prandtl/Schmidt number. The boundary conditions (at $z = \kappa\zeta$ and $z = -h$) of k and ε are not exhibited here (Rodi, 1984).

κ, placed in front of some of the terms of Eqs. (1)–(12), is only a symbol used to point out that the order of magnitude of those terms marked by κ is equal to κ.

In coastal seas, such as the North Sea and the East China Sea, the oscillating currents associated with tides are dominant over the observable flow field, at least one of the dominant currents, which implies that κ should be, at least might be selected as

$$\kappa = \frac{u_c}{\sqrt{gh_c}} \tag{13}$$

where the quantity with the subscript "c" indicates the characteristic value of the corresponding quantity. Thus, the nondimensional parameter, κ, is essentially a nonlinear parameter characterizing and measuring the convective nonlinearity of the shallow water system. The turbulent stress and diffusion are the other origins of nonlinearity.

A scale analysis for the coastal ocean, such as the East China Sea, shows, according to the expression of κ in (13), $\kappa = \frac{1}{10} \sim \frac{1}{10\sqrt{10}}$, if taking the typical scales: u_c=1 m/s, g = 10 m/s^2 and h_c = (10~100) m. and $O\left(\left|g\beta\nabla\int_z^{\kappa\zeta} c\mathrm{d}z'\right|\right) = O(|\tau_a|) = \kappa$ (Feng *et al.*, 1984; Feng, 1990). Thus, let us assume

$$O(\kappa) < 1 \tag{14}$$

which shows that κ is a small parameter, thus convective nonlinearity measured by κ is small in the order of magnitude.

As is well known, in the coastal zone by the western boundary of the ocean. such as the East China Sea, there is a strong flow, such as the Kuroshio. In the Kuroshio's region, the flow is essentially a quasi-geostrophic motion, and the convective nonlinearity is characterized and measured by the Rossby number.

$$Ro = \frac{u_c}{f_c L}, \tag{15}$$

where L is the horizontal scale of the flow field. If taking the typical scales: u_c=1 m/s, f_c = 10^{-4} s^{-1} and L= (10^5~10^6) m, $Ro = \frac{1}{10} \sim \frac{1}{100}$.

Hence in the coastal ocean, such as the East China Sea, in which the tidal currents and the quasi-steady flow, such as the Kuroshio, are in the same order of magnitude and dominant over the flow field, it is relevant, or conservative at most, that only κ is used for measuring the convective nonlinearity.

Here a quasi-steady flow is defined as the flow that can be considered to be steady in the intra-tidal process at least, such as the Kuroshio varying with a period about 10 days (e.g. Sun and Su, 1994).

The expression for the transport operator, (10), has implied another assumption that both the convective and the diffusive transport are in the same order of magnitude. It means that the Schmidt number, σ_p, is the order of κ^{-1}. For example, $O(\sigma_p) = 10$ for the East China Sea dynamics.

In Eq. (3), we suppose formally

$$O(S_p) = \kappa^2 \tag{16}$$

or, more exactly speaking, $O(S_p) \leqslant \kappa^2$.

The assumption (16) is trivial for $p = c$, where c is the apparent concentration of a conservation tracer, because of $S_c \equiv 0$.

As for $p=k$ and $p=\varepsilon$, the assumption (16) implies a "quasi-local-budget" turbulent model in the intra-tidal process (Rodi, 1984; Feng and Lu, 1993). It seems more proper that the intra-tidal transport processes of turbulence can be approximately governed by a quasi-homogeneous transport equation for k or ε, or, correct to $O(\kappa)$, by a homogeneous transport equation for k or ε (Feng and Lu, 1993). The applicability of the proposed turbulent model to shallow sea problems has been discussed through the numerical experiments (Wei and Feng, 1997).

Eqs. (1)–(12) constitute the k-ε turbulent closure set of three-dimensional field equations for the thermo-hydrodynamic system in coastal seas, or the k-ε turbulent closure problem of a three-dimensional baroclinic coastal sea.

By means of a perturbation approach with the small parameter, κ, any variable A and its temporal derivative in a multiple time scale system (Hamrick, 1987; Lu, 1991; Feng and Wu, 1995) can be expanded in ascending series of κ, or

$$\left(A,\frac{\partial A}{\partial t}\right)=\sum_{j=0}^{2}\kappa^{j}\left(A_j,\sum_{m=0}^{j}\frac{\partial A_{j-m}}{\partial t_m}\right)+O(\kappa^3) \tag{17}$$

Substituting all the expanded variables and their temporal derivatives into Eqs. (1)–(12) and indicating the jth-order solution by the subscript j, the jth-order equations can be derived.

A substitution of p into Eq. (3) yields the zeroth-order equation, or

$$\frac{\partial p_0}{\partial t_0}=0,\ (p_0=k_0,\varepsilon_0,c_0) \tag{18}$$

which points out that the scalar p_0 is independent of t_0. (k_0,ε_0) is substituted into (11) and then ν_0 into (12), it is found that (ν_0,λ_{p_0}) is independent of t_0, or

$$\frac{\partial(\nu_0,\lambda_{p_0})}{\partial t_0}=0 \tag{19}$$

The first-order perturbation equation for p is derived as

$$\frac{\partial p_1}{\partial t_0}+\frac{\partial p_0}{\partial t_1}+\boldsymbol{u}\cdot\nabla p_0+w_0\frac{\partial p_0}{\partial z}=\frac{\partial}{\partial z}\left(\lambda_{p_0}\frac{\partial p_0}{\partial z}\right) \tag{20}$$

where

$$\begin{pmatrix}\boldsymbol{u}_0\\ w_0\end{pmatrix}=\begin{pmatrix}\boldsymbol{U}\\ W\end{pmatrix}+\begin{pmatrix}\boldsymbol{v}\\ \omega\end{pmatrix} \tag{21}$$

$(\boldsymbol{v},\omega)$ is the tidal current and is the tidally periodic function of t_0; $(\boldsymbol{U},W)$ is the quasi-steady current independent of t_0. However, both of the $(\boldsymbol{v},\omega)$ and $(\boldsymbol{U},W)$ are the function of $(t_1,t_2,\cdots)$, representing the low frequency motion.

Integrating Eq. (20) with t_0 and eliminating the secular term yields

$$p_1 = \overline{p}_1 + \tilde{p}_1 \tag{22}$$

where $\tilde{p}_1 = -\int \boldsymbol{v}\mathrm{d}t_0 \cdot \nabla p_0 - \int \omega \mathrm{d}t_0 \dfrac{\partial p_0}{\partial z}$; $\overline{p}_1 = \overline{p}_1(t_1, t_2, \cdots)$, and

$$\frac{\partial p_0}{\partial t_1} + \boldsymbol{U} \cdot \nabla p_0 + W \frac{\partial p_0}{\partial z} = \frac{\partial}{\partial z}\left(\lambda_{p_0} \frac{\partial p_0}{\partial z}\right) \tag{23}$$

Using the similar treatment to that of Eq. (20), a second-order perturbation equation for p is further derived, as

$$\frac{\partial \overline{p}_1}{\partial t_1} + \boldsymbol{U} \cdot \nabla \overline{p}_1 + W \frac{\partial \overline{p}_1}{\partial z} + \boldsymbol{u}_{\mathrm{L}} \cdot \nabla p_0 + w_{\mathrm{L}} \frac{\partial p_0}{\partial z} = \frac{\partial}{\partial z}\left(\overline{\lambda}_{p_1} \frac{\partial p_0}{\partial z} + \lambda_{p_0} \frac{\partial \overline{p}_1}{\partial z}\right) + \overline{S}_{p_0} \tag{24}$$

where the mass-transport velocity, $(\boldsymbol{u}_{\mathrm{L}}, w_{\mathrm{L}})$, as the sum of the Eulerian time-mean velocity, $(\boldsymbol{u}_{\mathrm{E}}, w_{\mathrm{E}})$ and the Stokes' drift velocity $(\boldsymbol{u}_{\mathrm{S}}, w_{\mathrm{S}})$ (Longuet-Higgins, 1969), or

$$\begin{pmatrix} \boldsymbol{u}_{\mathrm{L}} \\ w_{\mathrm{L}} \end{pmatrix} = \begin{pmatrix} \boldsymbol{u}_{\mathrm{E}} \\ w_{\mathrm{E}} \end{pmatrix} + \begin{pmatrix} \boldsymbol{u}_{\mathrm{S}} \\ w_{\mathrm{S}} \end{pmatrix} \tag{25}$$

$\overline{\lambda}_{p_1} = \dfrac{\nu_0}{\sigma_{\mathrm{p}}}\left(2\dfrac{\overline{\kappa}_1}{\kappa_0} - \dfrac{\overline{\varepsilon}_1}{\varepsilon_0}\right)$; $\overline{S}_{p_0} = 0$ ($p = c$), $\overline{S}_{p_0}$ $(p = k, \varepsilon)$ is expressed as $\overline{S}_{k_0} = \overline{P}_0 + \overline{G}_0 - \varepsilon_0$ and $\overline{S}_{\varepsilon_0} = c_{1\varepsilon}\dfrac{\varepsilon_0}{k_0}\left[(\overline{P}_0 + \overline{G}_0)(1 + c_{3\varepsilon} Ri_{\mathrm{f}}) - \dfrac{c_{2\varepsilon}}{c_{1\varepsilon}}\varepsilon_0\right]$, in which $\overline{P}_0 = \boldsymbol{v}_0\left[\dfrac{\partial \boldsymbol{U}}{\partial z} \cdot \dfrac{\partial \boldsymbol{U}}{\partial z} + \left\langle \dfrac{\partial \boldsymbol{v}}{\partial z} \cdot \dfrac{\partial \boldsymbol{v}}{\partial z} \right\rangle\right]$, and $\overline{G}_0 = g\beta\lambda_{\mathrm{c}_0}\dfrac{\partial c_0}{\partial z}$; <> denotes the Eulerian time-mean operator.

After multiplying Eq. (24) by κ and adding it to Eq. (23), an intertidal transport equation is, correct to the order of $O(\kappa)$ derived as

$$\frac{\partial \overline{p}}{\partial \tau} + \boldsymbol{v} \cdot \nabla \overline{p} + \overline{w}\frac{\partial \overline{p}}{\partial z} = \frac{\partial}{\partial z}\left(\lambda_p \frac{\partial \overline{p}}{\partial z}\right) + \kappa \overline{S}_p, \ (p = c, k, \varepsilon) \tag{26}$$

where

$$\overline{p} = p_0 + \kappa \overline{p}_1; \quad (\boldsymbol{v}, \overline{w}) = (\boldsymbol{U} + \kappa \boldsymbol{u}_{\mathrm{L}}, W + \kappa w_{\mathrm{L}})$$

$$\overline{\nu} = \nu_0 + \kappa \overline{\nu}_1; \quad \overline{\lambda}_p = \lambda_{p_0} + \kappa \overline{\lambda}_{p_1}$$

$$\frac{\partial}{\partial \tau} = \frac{\partial}{\partial t_1} + \kappa \frac{\partial}{\partial t_2}; \quad \overline{S}_{\mathrm{c}} = 0$$

$$\overline{S}_k = \overline{P} + \overline{G} - \overline{\varepsilon} \quad \text{and} \quad \overline{S}_\varepsilon = c_{1\varepsilon}\frac{\overline{\varepsilon}}{\overline{\kappa}}\left[(\overline{P} + \overline{G})(1 + c_{3\varepsilon} Ri_{\mathrm{f}}) - \frac{c_\varepsilon}{c_{1\varepsilon}}\overline{\varepsilon}\right]$$

$$\overline{P} = \nu\left[\frac{\partial \boldsymbol{U}}{\partial z} \cdot \frac{\partial \boldsymbol{U}}{\partial z} + \left\langle \frac{\partial \boldsymbol{v}}{\partial z} \cdot \frac{\partial \boldsymbol{v}}{\partial z} \right\rangle\right], \quad \overline{G} = g\beta\overline{\lambda}_{\mathrm{c}}\frac{\partial \overline{c}}{\partial z}$$

It should be pointed out that a substitution of (25) into the expression of $(\boldsymbol{v}, \overline{w})$ just defined yields

$$\begin{pmatrix} \boldsymbol{v} \\ \overline{w} \end{pmatrix} = \begin{pmatrix} \boldsymbol{U}_{\mathrm{E}} \\ W_{\mathrm{E}} \end{pmatrix} + \kappa \begin{pmatrix} \boldsymbol{u}_{\mathrm{S}} \\ w_{\mathrm{S}} \end{pmatrix} \tag{27}$$

where $(\boldsymbol{U}_{\mathrm{E}}, W_{\mathrm{E}}) = (\boldsymbol{U} + \kappa \boldsymbol{u}_{\mathrm{E}}, W + \kappa w_{\mathrm{E}})$ can be also considered as a Eulerian tidally time-mean velocity. Thus $(\boldsymbol{v}, \overline{w})$ can be also regarded as a mass transport velocity. Here we shall term $(\boldsymbol{v}, \overline{w})$ the generalized mass transport velocity in order to distinguish it from the traditional mass-transport velocity, $(\boldsymbol{u}_{\mathrm{L}}, w_{\mathrm{L}})$.

The inter-tidal transport equation (26) just derived reveals that in the coastal ocean, such as the East China Sea, in which both the tidal currents and the quasi-steady flow, such as the Kuroshio, are dominant and are supposed to have the same order of magnitude, the inter-tidal transport processes evolve on time scale and are convected by the generalized mass transport velocity, $(\boldsymbol{v}, \overline{w})$, expressed by (27) (Feng and Wu, 1995).

The generalized mass transport velocity, $(\boldsymbol{v}, \overline{w})$, can be easily verified to satisfy the continuity condition of flow field. In fact, in view of the linearity of the equation of continuity, Eq. (1), $\nabla \cdot \boldsymbol{U} + \partial W / \partial z = 0$ is tenable, and noting $\nabla \cdot \boldsymbol{u}_{\mathrm{L}} + \partial w_{\mathrm{L}} / \partial z = 0$ (Feng, 1987, 1990), we have $\nabla \cdot \boldsymbol{v} + \partial \overline{w} / \partial z = 0$.

Further, if a material surface in water is specified geometrically by the equation $F(\boldsymbol{x}, z, t) = \text{const.}$, where $\boldsymbol{x}$ is the horizontal coordinate vector, F is a quantity which is invariant for a water parcel on the material surface, so that at all points on the surface F satisfies the Eq. (3) if putting $p = F$, $S_p = S_F = 0$ and $\lambda_p = \lambda_F = 0$. Thus the equation governing $\overline{F} = F_0 + \kappa \overline{F}_1$ will be derived as

$$\frac{\partial \overline{F}}{\partial \tau} + \boldsymbol{v} \cdot \nabla \overline{F} + \overline{w} \frac{\partial \overline{F}}{\partial z} = 0 \tag{28}$$

This suggests that in the coastal water system, such as the East China Sea, the velocity of flow field of the inter-tidal circulation, or the tidally time mean circulation (called the "circulation" for short), should be embodied by the generalized mass transport velocity $(\boldsymbol{v}, \overline{w})$. It is evidently interesting to derive the continuity equation and the equation of motion governing the generalized mass transport velocity for the coastal circulation as well as the corresponding boundary conditions at the sea surface and at the sea bottom and join them with the inter-tidal transport equation, Eq. (26) for $p = c, k, \varepsilon$, for formulating a corresponding turbulent closure set of field equations for the coastal ocean circulation.

Omitting the tedious process of derivation, the governing equations, correct to the order of $O(\kappa)$, are the continuity equation

$$\nabla \cdot \boldsymbol{v} + \frac{\partial \overline{w}}{\partial z} = 0 \tag{29}$$

the momentum equation

$$\kappa\frac{\partial \boldsymbol{v}}{\partial \tau}+\kappa\left(\boldsymbol{v}\cdot\nabla\boldsymbol{v}+\overline{w}\frac{\partial \boldsymbol{v}}{\partial z}\right)+f\boldsymbol{e}\times\boldsymbol{v}=-g\nabla\overline{\zeta}+\frac{\partial}{\partial z}\left(\overline{\nu}\frac{\partial \boldsymbol{v}}{\partial z}\right)-\kappa g\beta\nabla\int_z^{\kappa\overline{\zeta}}\overline{c}\mathrm{d}z'+\kappa\boldsymbol{\pi} \tag{30}$$

$$z=\kappa\overline{\zeta}\;;\quad \overline{w}=\kappa\frac{\partial\overline{\zeta}}{\partial\tau}+\kappa\boldsymbol{v}\cdot\nabla\overline{\zeta} \tag{31}$$

$$\nu\frac{\partial \boldsymbol{v}}{\partial z}=\kappa\langle\boldsymbol{\tau}_a\rangle \tag{32}$$

$$\frac{\partial\overline{c}}{\partial z}=0 \tag{33}$$

$$z=-h,\quad \overline{\boldsymbol{v}}=\overline{w}=0 \tag{34}$$

$$\frac{\partial\overline{c}}{\partial z}=0 \tag{35}$$

where $\overline{\zeta}=\overline{\zeta}_0+\kappa\overline{\zeta}_1$, $\overline{\zeta}_0$ is the water surface elevation accompanying the dominant quasi-steady current, $(\boldsymbol{U},W)$; $\overline{\zeta}_1=\langle\zeta_1\rangle$; $\boldsymbol{\pi}$ is the tidal body force (Feng, 1987; Feng, 1990; Feng and Lu, 1993), which is induced by the tidally time averaged effects of the nonlinear couplings between tidal variables.

The set of equations governing astronomical tides are the continuity equation

$$\nabla\cdot\boldsymbol{v}+\frac{\partial\omega}{\partial z}=0 \tag{36}$$

the momentum equation

$$\frac{\partial \boldsymbol{v}}{\partial t_0}+f\boldsymbol{e}\times\boldsymbol{v}=-g\nabla\tilde{\zeta}_0+\frac{\partial}{\partial z}\left(\overline{\nu}\frac{\partial \boldsymbol{v}}{\partial z}\right) \tag{37}$$

the boundary condition at $z=0$

$$\omega=\frac{\partial\tilde{\zeta}_0}{\partial t_0} \tag{38}$$

$$\frac{\partial \boldsymbol{v}}{\partial z}=0 \tag{39}$$

$$z=-h\quad \boldsymbol{v}=\omega=0 \tag{40}$$

where ζ_0 is the tidal elevation.

Eqs. (29)–(35) and Eq. (26) constitute the set of field equations governing the generalized mass transport velocity $(\overline{\boldsymbol{v}},\overline{w})$ for the coastal ocean circulation.

3. Discussion

By using the generalized set of equations developed above (Feng and Wu, 1995; Feng *et al.*, 1996) with the assumption that the flow field is in a steady state and is compatible with the given eddy viscosity and density distribution, a diagnostic numerical experiment for the season circulation of the BYES, consisting of the Bohai Sea, the Yellow Sea and the East China Sea, has been made (Wang, 1996).

Shown in Figure 3 is the depth averaged circulation of the BYES in winter. As shown in Figure 3 that the circulation of the BYES is composed of two subsystems of flow, that is a warm and saline flow system of oceanic origin, or the Kuroshio and its branches and extension, and a less saline coastal flow system. The former flow northward and northeastward and the latter generally southward. The flow pattern of the circulation is roughly in the form of a cyclonic gyre, with the Kuroshio-TSWC-YSWC and its extension on the east side and the CCC, including the Southern Bohai Sea Coastal Current (SBSCC), the Yellow Sea Coastal Current (YSCC) and the East China Sea Coastal Current (ECSCC), on the west side. Along the western coast of Korea, the West Korea Coastal Current (WKCC) flows southward.

The extension of the YSWC enters the Bohai Sea through the northern part of the Bohai Strait. The less saline coastal water of the Bohai Sea, as the SBSCC, flows out of the Bohai Sea through the southern part of the Bohai Strait. The YSWC with its extension and the YSCC constitute grossly the Yellow Sea circulation. The Kuroshio enters the East China Sea east of Taiwan. It flows along the shelf edge until around north of 29°30'N where its main body leaves the shelf and re-enters the Pacific Ocean through the Tokara Strait. The mainstream of the Kuroshio has an anticyclonic meander and a swing south of 29°N. A small branch separated from the Kuroshio west of the Tokara Strait and the continental coastal current joint together to from the TSWC. After the separation of the YSWC from the TSWC west of Kyushu, the main body of the TSWC gradually turns northeastward, splits into two branches west of the Goto Islands, and eventually enters the Japan Sea. Part water separated from the YSWC west of Cheju Island combines with the WKCC and from Cheju Strait Current. The Changjiang Diluted Water (CDW) flows initially in the direction of its mouth. Main part of it flows southward along the coast. At a place not very far from the river mouth, the remainder of the CDW turns cyclonically to the northeast. Along the Fujian and Zhejiang Coast there is a weak southward ECSCC. There is the evidence of the intrusion of the Kuroshio water onto the shelf and as part of the TWC flows to the north round between the 50 m to 100 m isobath. The western part of the East China Sea is dominated by the influence of the TWC. It is obvious from Figure 3 that the East China Sea circulation is much less enclosed and essentially appears as a band with the dominant northeastward current. The general features of circulation in the BYES, mentioned above, agree well with those in Figure 2 inferred from the observed data analyses (Guan, 1994).

Shown in Figure 4 is the depth averaged circulation of the BYES in summer. Comparing Figure 3 with Figure 4, it can be found that the depth averaged circulation pattern in summer is similar to that of winter. However, still exist there some differences between them. The Kuroshio and the TWC are stronger in summer. It seems that the TWC can extend to the north of the Changjiang mouth and the ECSCC changes the direction to the north in summer. The CDW accedes to the TWC and flows northward and then turns to eastward.

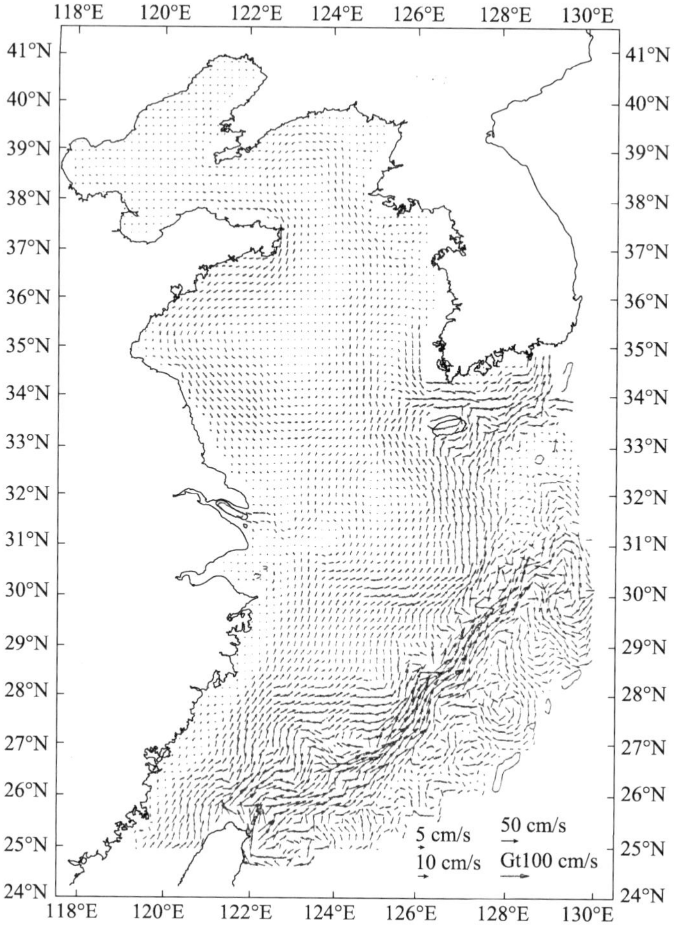

Fig. 3 Depth-averaged circulation of the BYES in winter

Shown in Figure 5 and Figure 6 are the flow patterns of circulation at 30 m layer of the BYES in winter and in summer, respectively. It is clear from comparing the two figures that circulation pattern at 30 m layer of the East China Sea in winter is almost the same as that in summer, excepting that the Kuroshio and the TWC are stronger in summer. However, the circulation pattern at 30 m layer of the Yellow Sea in winter is completely different to that in summer: In summer the flow pattern of circulation at 30 m layer of the Yellow Sea is roughly in the form of a cyclonic gyre. The core of the gyre coincides roughly with that of the YSCWM inferred by Ho *et al.* (1959) and discussed in detail by Su and Weng (1994) (Fig. 7). The YSWC, at least the most of it, turns to eastward into the Cheju Strait. In winter the flow pattern of circulation at 30 m layer of the Yellow Sea is much less enclosed and the northward current is dominant.

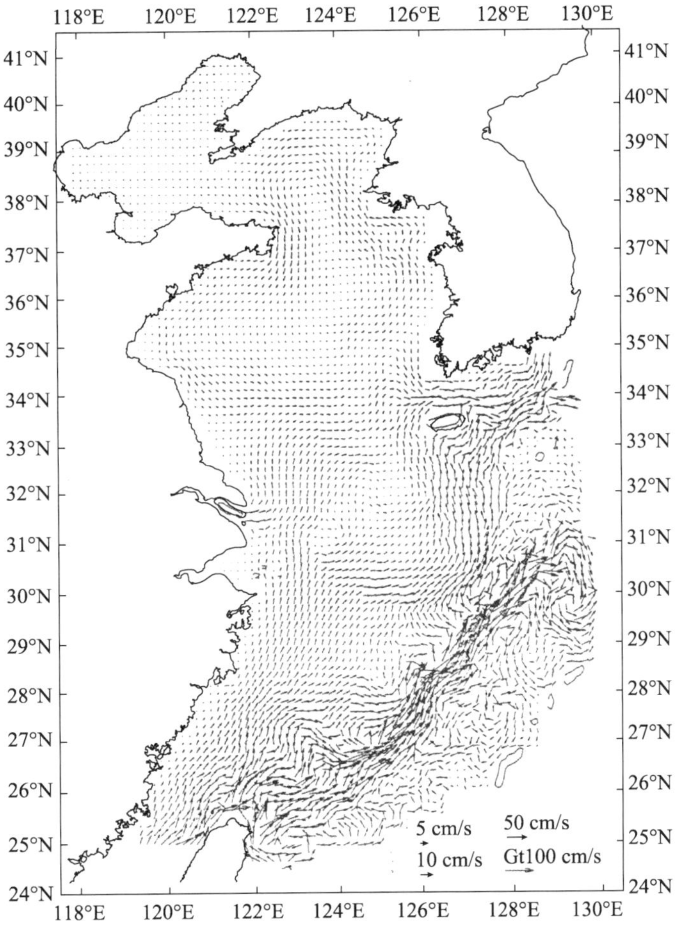

Fig. 4 Depth-averaged circulation of the BYES in summer

There seems no point in refusing the idea that the dominant northward current at 30 m layer of the central Yellow Sea in winter, just mentioned, should be considered as the compensative of the wind-driven current by the monsoon (Hsueh and Romea, 1986). In fact, firstly, a boundary forcing experiment is made (Fig. 8). For understanding the boundary forcing effects, there only are the water fluxes from the open boundaries considered as the driving forces of the BYES circulation. Under the given water fluxes in Table 1, the simulated flow system of oceanic origin, containing the Kuroshio, the TWC and the TSWC (Fig. 8), is similar to that in Fig. 5 and Fig. 6. This implies that while the boundary forces may be the dominant dynamic factor for the formation of flow system of oceanic origin, the generated currents are very weak to the north of 30°N and the west of 126°E. The latter suggests that the boundary forces hardly affect the circulation of the Yellow Sea and the Bohai Sea. Secondly, with the boundary forces, wind-driven circulation experiments are conducted and wind-driven circulation at 30 m layer is exhibited (Fig. 9). A comparison between Fig. 9a with Fig. 8a and Fig. 5 shows that the dominant northward current at 30 m layer (as a representative of deep layer circulations) of Yellow Sea in winter, mentioned above, is indeed of wind-driven but not from the oceanic origin.

The wind over the BYES is monsoonal, northeast in winter and southwest in summer. The main differences between the depth-averaged circulation in summer and that in winter are that there exists the relative stronger southward coastal current, along China coast in winter and the TWC in summer can reach to the north of the Changjiang River mouth (Figures cf. Wang, 1996). These differences may suggest that northerly wind of winter is a dominant factor for the formation of the relative stronger southward China Coastal Current and that the summer monsoon is a dominant factor for driving the TWC to reach farther north.

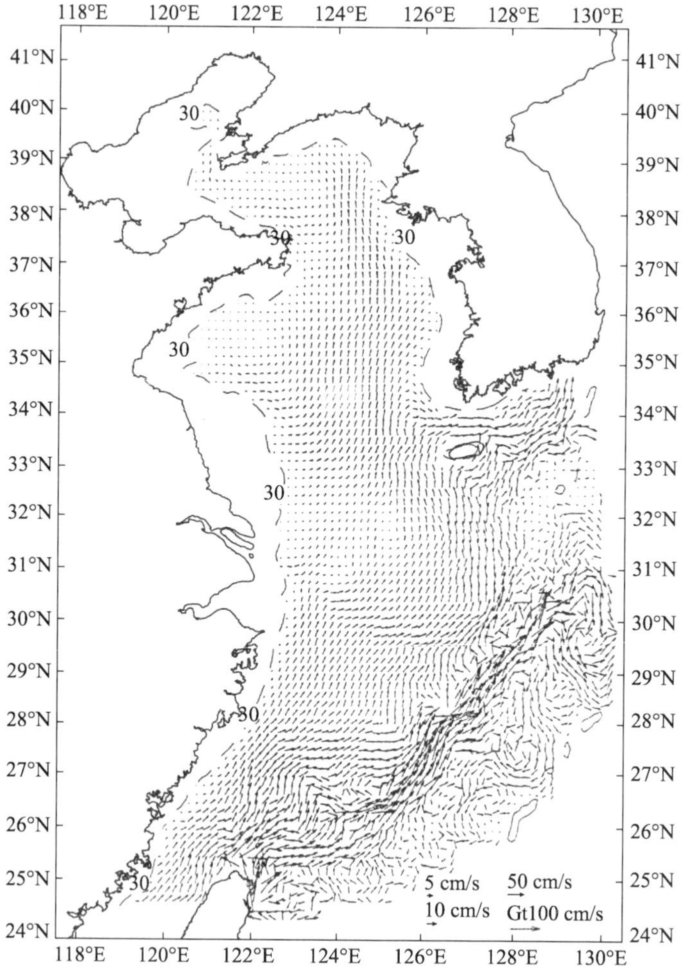

Fig. 5　Circulation at 30 m layer of the BYES in winter

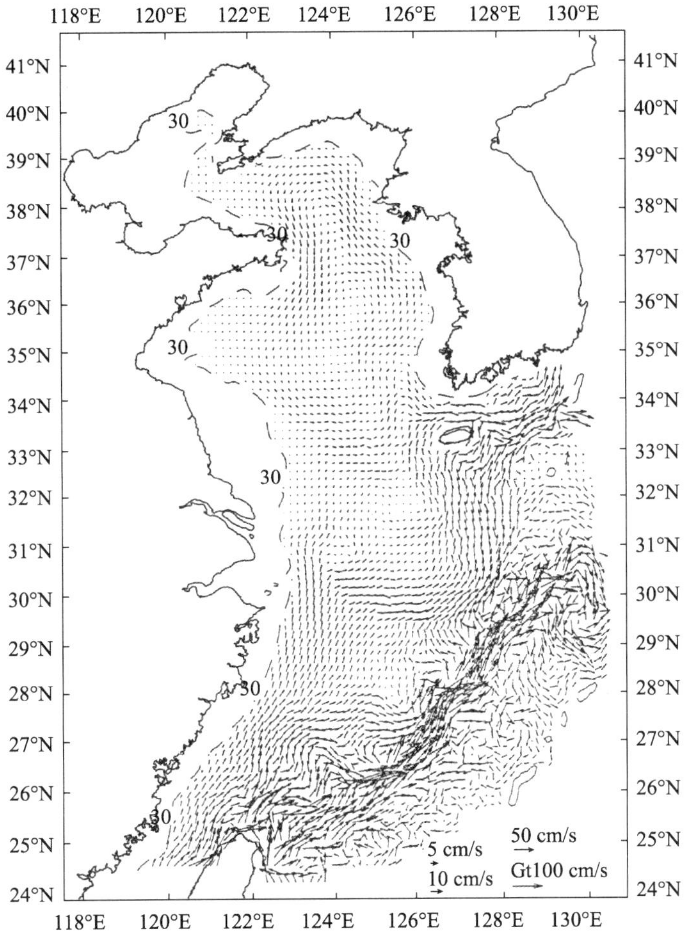

Fig. 6　Circulation at 30 m layer of the BYES in summer

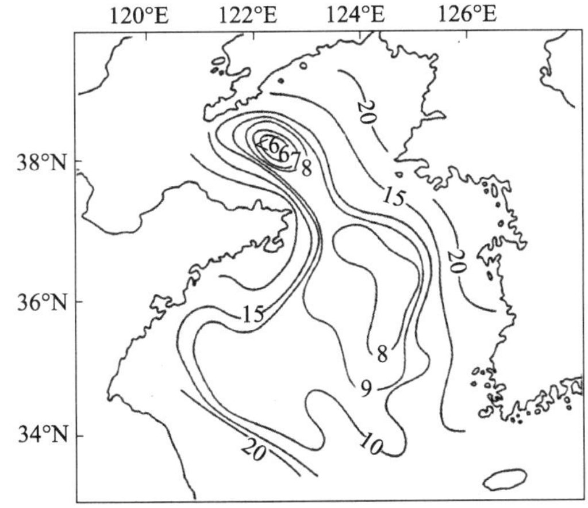

Fig. 7a　The distribution range and isothermal lines of the Yellow Sea Cold Water Mass (Su and Weng, 1994)

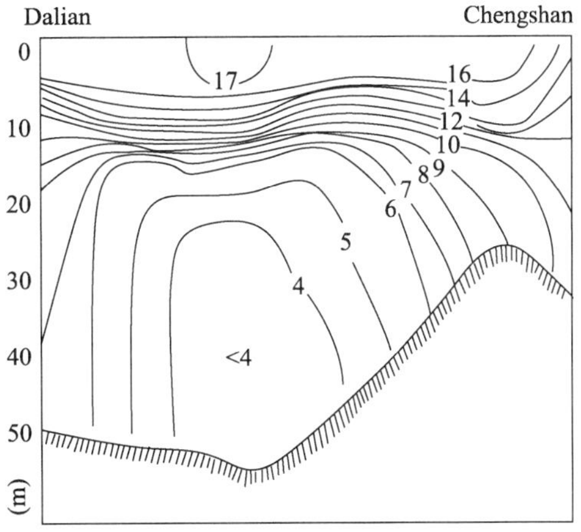

Fig. 7b　Temperature distribution along the Dalian-Chengshantou section in July 1977 (Hu, 1994)

Table 1 The water fluxes (Sv.) in different open boundaries (+ and – representing into and out, respectively)

Seasons	Taiwan Strait	Kuroshio region east of Taiwan	Tokara Strait	Tsushima Strait	Changjiang River	Huanghe River
Winter	+1.0	+22.5	–21.0	–2.53	+0.03	0.0
Summer	+3.0	+29.706	–29.2	–3.6	+0.09	0.004

With the boundary forces, thermohaline circulation experiments are carried out (Figures cf. Wang, 1996). In winter, the water in the Bohai Sea and the Yellow Sea and over the shelf in the East China Sea is basically vertically homogeneous. Corresponding to this kind of water condition, the thermohaline circulation is too weak to make contribution to the circulation of the BYES, except for the region of shelf break and east of it. In summer, the BYES

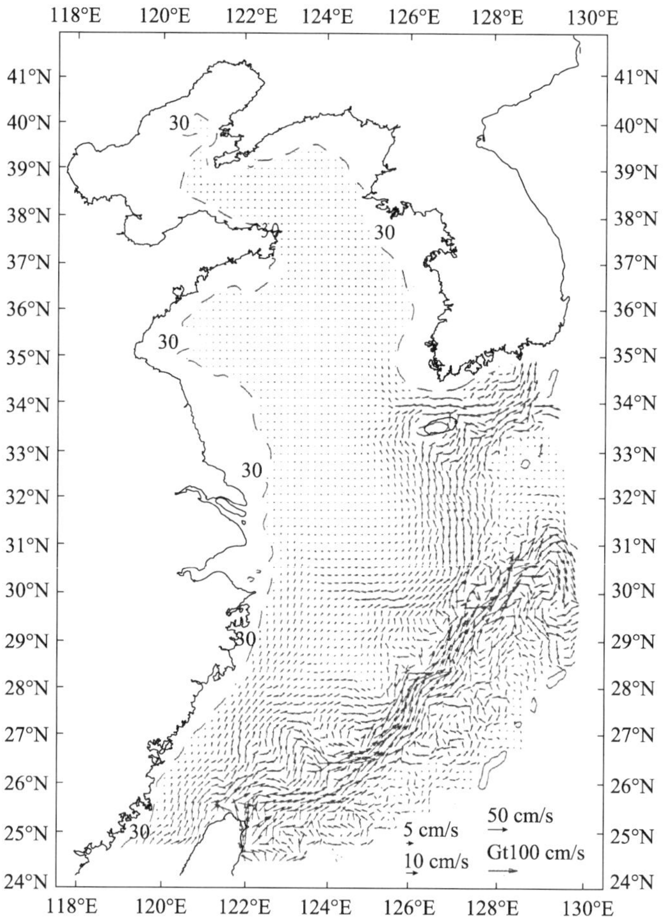

Fig. 8a Circulation at 30 m layer induced by the boundary force (such as the Kuroshio and the runoff of the Changjiang River, etc.) in the BYES: for winter

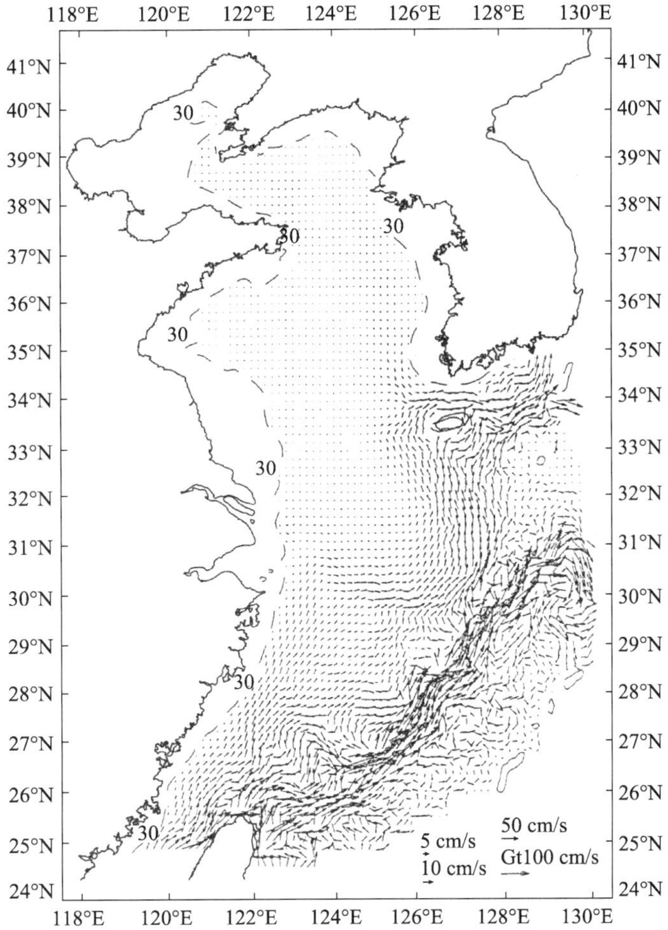

Fig. 8b Circulation at 30 m layer induced by the boundary force (such as the Kuroshio and the runoff of the Changjiang River, etc.) in the BYES: for summer

is well stratified. A basin-wide cyclonic gyre in the Yellow Sea is obvious at different layers. The different circulation structures between winter and summer reveal that thermohaline circulation may make an important contribution to the formation of the YSCWM. Fig. 9b implies that the summer wind-driven circulation makes little contribution to the cyclonic circulation pattern related to the YSCWM.

Shown in Figure 10 is the flow pattern of depth averaged M_2 tide-induced Lagrangian residual current of the BYES. The flow pattern reveals that M_2 tide-induced Lagrangian residual current forms a basin-wide cyclonic gyre in the Yellow Sea. The core of this gyre is almost consistent with that of the YSCWM. This suggests that tide-induced Lagrangian residual current may make an important contribution to the formation of the cyclonic circulation related to the YSCWM. Fig. 10 shows also that the M_2 tide-induced Lagrangian residual current enters the Bohai Sea through the northern part of the Bohai Strait and flows out of the

Bohai Sea through the southern part of the Bohai Strait. This flow pattern through the Bohai Strait is in keeping with that of the depth mean circulation in winter (Fig. 3). In addition, the east side current of the basin-wide cyclonic gyre stimulates the northward transport of the YSWC, and the west side current of the gyre tends to intensify the YSCC along the Shandong peninsula and off the northern Jiangsu coast. M_2 tide-induced Lagrangian residual current in the East China Sea forms a weak anticyclonic gyre in central part of the basin. The west part of this gyre tends to intensify the TWC and the east part of it tends to weaken the TSWC.

The origin of the TWC has long attracted much attention from both Chinese and Japanese oceanographers. Earlier hydrographic studies (Uda, 1950; Mao *et al.*, 1964) and bottom drifter studies (Inoue, 1975) argued that the TWC is a Kuroshio branch originated from the northeast of Taiwan, i.e., where shelf intrusion by Kuroshio takes place. However, Guan (1983) pointed out that summer GEK data obtained by Japanese scientists over 1972–1979 do not support the Kuroshio branch hypothesis, at least not in the upper and middle layers.

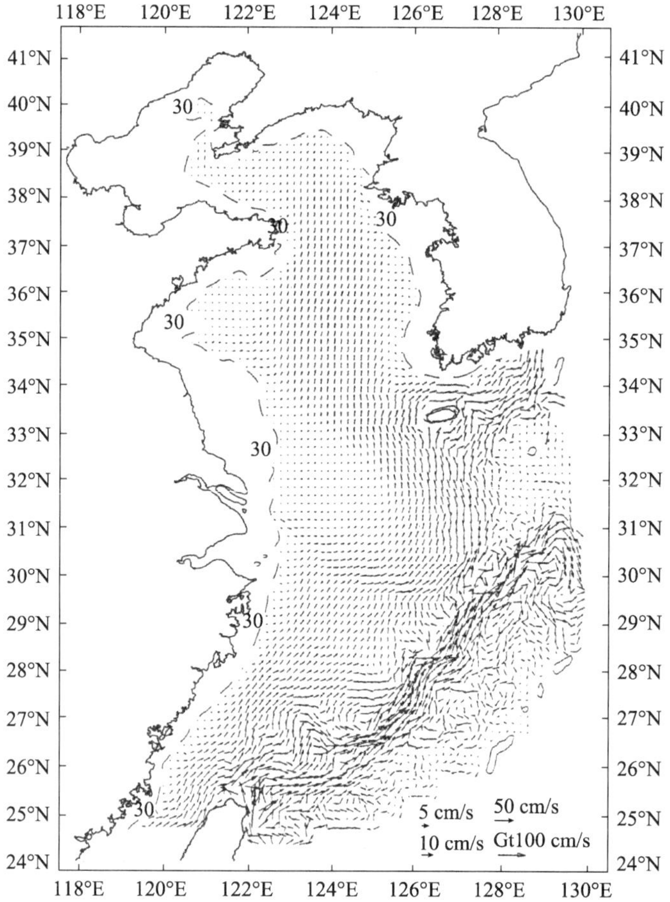

Fig. 9a Wind-driven circulation at 30 m layer in the BYES: for winter

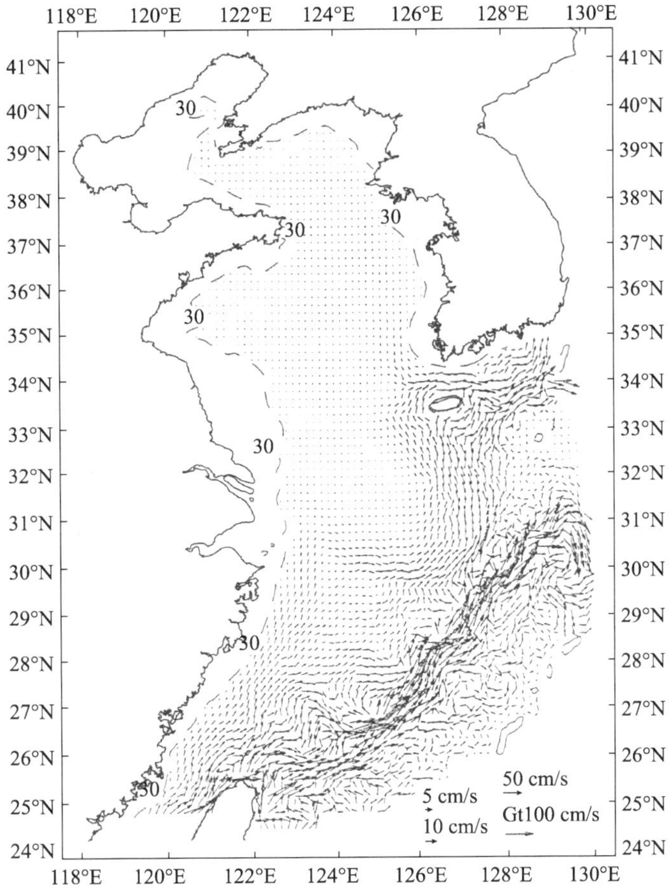

Fig. 9b Wind-driven circulation at 30 m layer in the BYES: for summer

For investigating the origin of the TWC, two comparable experiments of closing open boundaries are conducted (Figures cf. Wang, 1996). In the experiments of closing the Tokara Strait and no water entering the East China Sea from northeast of Taiwan, the flow patterns over the East China Sea shelf are very similar to those in Figure 3 and Figure 4, except for current being weak. In the experiment of only closing Taiwan Strait, still exists there the TWC, except that the TWC seems a little weaker than that shown in Figure 3 and Figure 4. From two experiments above, it can be concluded that the TWC water might originate partly from Kuroshio northeast of Taiwan and partly from the Taiwan Strait.

Finally, it should be pointed out that although the seasonal circulation in the BYES has been roughly simulated and some of striking features of circulation has been revealed, some of other important features related to the mechanism of the circulation variation are still unclear and, at least, controversial and thus further investigation on them should be made.

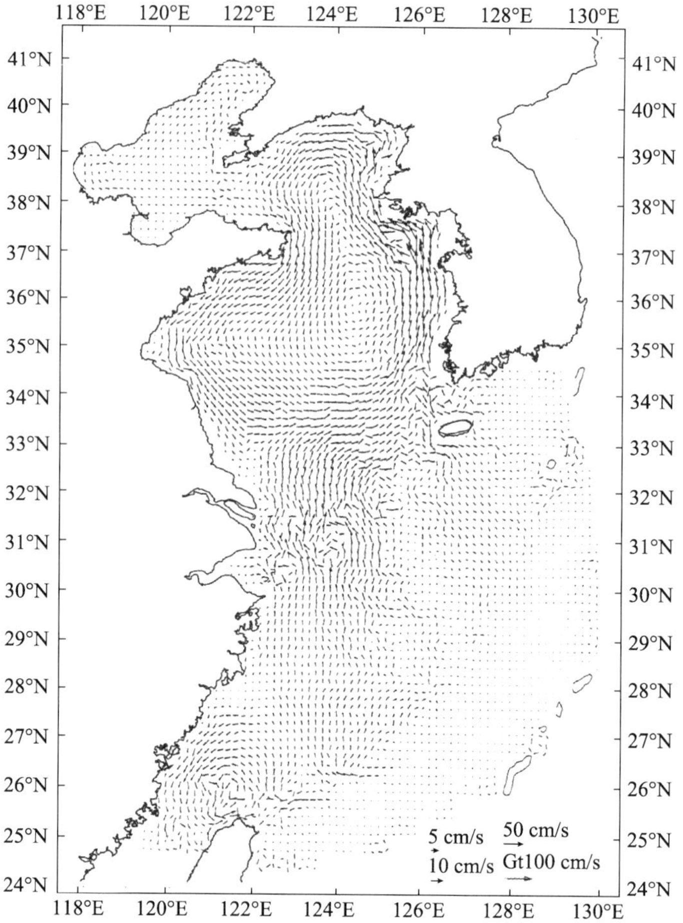

Fig. 10 Depth-averaged M_2 tide-induced Lagrangian residual current of the BYES

References

Bowden K. F. (1967) Circulation and diffusion. In: Estuaries. Publ., No. 83, G. H. Lauff (Editor), pp 15–36, AAAS, Washington, D. C.

Cerco C. (1995) Simulation of long-term trends in Chesapeake Bay eutrophication. J. Envir. Engrg., ASCE, 121(4), 398–410.

Cerco C. (1995) Response of Chesapeake Bay to nutrient load reductions. J. Envir. Engrg., ASCE, 121(8), 549–557.

Cerco C. and T. Cole (1993) Three-dimensional eutrophication model of Chesapeake Bay. J. Envir. Engrg., ASCE, 119(6), 1006–1025.

Cheng R. T. and V. Casulli (1982) On Lagrangian residual currents with applications in South San Francisco Bay, California. Water Resour. Res., 18, 1652–1662.

Cheng R. T., S. Feng and P. Xi (1989) On inter-tidal transport equation. In: Estuarine Circulation, by B. J. Neilson, J. Brubaker and A Kuo (Editors), pp 133–156,Humana Press, Inc.

Csanady G. T. (1975) Lateral momentum flux in boundary currents. J. Phys. Oceanogr., 5, 705–712.

Csanady G. T. (1976) Mean circulation in shallow seas. J. Geophys. Res., 81, 5389–5398.

Dortch M. S., R. S. Chapman and Steven R. Abt, Members, ASCE., (1992) Application of Three-Dimensional Lagrangian Residual Transport. J. Hydraulic Engineering, 118(6), 831–848.

Feng S. (1987) A three-dimensional weakly nonlinear model of tide-induced Lagrangian residual current and mass-transports, with an application to the Bohai Sea. In: Three-dimensional Models of Marine and Estuarine Dynamics, J. C. J. Nihoul and B. M. Jamart (Editors), Elsevier Oceanography Series, 45, pp 471–488, Elsevier.

Feng S. (1990) On the Lagrangian residual velocity and mass transport in a multi-frequency oscillatory system. In: Physics of Shallow Estuaries and Bays, R. T. Cheng (Editor), pp 18–34, Springer- Verlag, New York.

Feng S. and Y. Lu (1993) A turbulent closure model of coastal circulation. Chin. Sci. Bull., 38(20), 1737–1739.

Feng S. and D. Wu (1995) An inter-tidal transport equation coupled with turbulent k-ε model in a tidal and quasi-steady current system. Chin. Sci. Bull., 39(6), 136–139.

Feng S., R. T. Cheng and P. Xi (1986a) On tide-induced Lagrangian residual current and residual transport. Part I. Residual current. Water Resour. Res., 22(12), 1623–1634.

Feng S., R. T. Cheng and P. Xi (1986b) On tide-induced Lagrangian residual current and residual transport. Part II. Residual transport with application in South San Francisco Bay, California. Water Resour. Res., 22(12), 1635–1646.

Feng S., P. Xi and S. Zhang (1984) The baroclinic residual circulation in shallow seas. Chinese J. Oceanol. Limnol., 2, 49–60.

Feng S., S. Zhang and P. Xi (1994) A Lagrangian model of circulation in Bohai Sea. In: Oceanology of China Seas, Vol. 1, D. Zhou, Y B. Liang and C. K. Zeng (Editors), pp 83–90, Kluwer Academic Publishers.

Feng S., D. Wu, H. Wang and K. Wang (1996) A generalized set of equations for coastal ocean circulation. 8th Intern: Biennial Conf. on physics of Estuaries and coastal seas, the Hague, the Netherlands.

Fischer H. B., E. J. List, R. C. Y. Koh, J. Imberger and N. H. Brooks (1979) Mixing in Inland and Coastal Waters. Academic Press, New York, 483 pp.

Guan B. X. (1983) A sketch of the current structure and eddy characteristics in the East China Sea. In: Proceedings of International Symposium on Sedimentation on the Continental Shelf, pp 52–73, China Ocean Press.

Guan B. X. (1994) Patterns and structures of the currents in Bohai, Huanghai and East China Seas. In: Oceanology of China Seas, Vol. I, D. Zhou, Y. B. Liang and C. K. Zeng (Editors), pp 17–26, Kluwer Academic Publishers.

Guan. B. X. and S. J. Chen (1964) The Current Systems in Near-Sea Area of China Seas, pp 1–85 (in Chinese).

Hamrick J. M. (1987) Time averaged estuarine mass transport equations, Hydraulic Engineering. In: Proc. of the 1987 National Conf. on Hyaraulic Engineering, R. M. Ragan (Editor), pp 624–629, ASCE.

Ho C., Y. Wang, Z. Lei and S. Xu (1959) A preliminary study of the formation of the Yellow Sea mass and its properties, Oceanologia et. Limnoiogia Sinica, 2(1), 11–15 (in Chinese, with English abstract).

Hsueh Y. and R. D. Romca (1986) Winter wind and coastal sea-level fluctuations in the northeast China Sea, Part II. Numerical model. J. Phys. Oceanog., 16, 241–261.

Hu D. (1994) Some striking features of circulation in Huanghai Sea and East China Sea. In: Oceanology of China Seas. Vol. 1, D. Zhou, *et al.* (Editors), pp 27–38, Kluwer Academic Publishers.

Inoue N. (1975) Bottom current on the continental shelf of the East China Sea. Marine Science Monthly 2, 12–18.

Lee J. H. and S. Lee (1996) On the inflow of warm waters into the Yellow Sea in winter. A Symposium on the Yellow Sea Research (Abstracts), Qingdao, China, pp 3.

Longuet-Higgins M. S. (1969) On the transport of mass by time-varying ocean currents. Deep Sea Res., 16, 431–447.

Lu Y. (1991) On the Lagrangian residual current and residual transport in a multiple time scale system of shallow seas. Chin. J. Oceanol. Limnol, 9(2), 184–188.

Mao H. L., Y. W. Ren and K. M. Wan (1964) A preliminary investigation on the application of using T-S diagrams for a quantitative analysis of the water masses in the shallow water area. Oceanoiogica et Limnologica Sinica, 6, 1–22.

Nihoul J. C. J. and F. C. Ronday (1975) The influence of the "tidal stress" on the residual circulation. Tellus, 27, 484–489.

Pang I. C. (1996) Seasonal circulation around Cheju Island. A Symposium On The Yellow Sea Research (Abstracts), Qingdao, China, pp 13–14.

Rodi W. (1984) Turbulence models and their application in Hydraulics. A State of the Art Review. Netherlands, Book Pub. of IAHR.

Sun X. P. and Y. F. Su (1994) On the variation of Kuroshio in the East China Sea. In: Oceanology of China Seas, Vol. I, D. Zhou, Y. B. Liang and C. K. Zeng (Editors), pp 49–58, Kluwer Academic Publishers.

Su Y. S. and X. C. Weng (1994) Water masses in China seas. In: Oceanology of China Seas, Vol. I, D. Zhou, Y. B. Liang and C. K. Zeng (Editors), pp 3–16, Kluwer Academic Publishers.

Uda M. (1950) On the temperature variation in the East China Sea, I. Hydrography of the East China Sea and Yellow Sea, 2, 1–10.

Wang H., Z. Q. Su, S. Feng and W. X. Sun (1993) A three dimensional numerical calculation of the wind-driven thermohaline and tide-induced Lagrangian residual current in the Bohai Sea. Acta Oceanol. Sin., 12, 169–182.

Wang K. (1996) Study on the model and numerical method of coastal sea circulation and its application to the East China Sea. Ph. D. Dissertation, Ocean University of Qingdao (in Chinese).

Wei H and S. Feng (1997) Homogeneons κ-equation closure model and its application in the coastal sea. 29th Intern. Liege Colloquium on Ocean Hydrodynamics, Liege, Belgium.

Wright, L. D., W. J. Wisman Jr, Z. S. Yang, B. D. Bornhold, G. H. Keller, D. B. Prior and J. N. Suhayda (1990) Processes of marine dispersal and deposition of suspended silts off the modern mouth of the Huanghe (Yellow River), Cont. Shelf Res., 10(1), 1–40.

Zimmerman J. T. F. (1979) On the Euler-Lagrangian transformation and the Stokes' drift in the presence of oscillatory and residual currents. Deep Sea Res., 26A, 505–520.

Modelling Annual Cycles of Primary Production in Different Regions of the Bohai Sea*

Huiwang Gao, Shizuo Feng and Yuping Guan

Introduction

The Bohai Sea is a typical shallow sea in which the average depth is about 18 m, and which has a rich biodiversity and abundance of marine life. Studies have analysed its general circulation (Gu and Xiu, 1996), and summarized the general hydrographic features of the adjacent Yellow Sea (Guo, 1993). The nutrient status of the Bohai Sea has been examined (Cui and Song, 1996), as well as transparency and water colour (Fei, 1986) and chlorophyll a and primary productivity characteristics (Lu *et al.*, 1984; Fei *et al.*, 1988a, b). Relationships between the marine environment and the prawn fishery have been presented by Ge (1986).

The Bohai Sea is often divided into four regions (Fig. 1). These are Bohai Bay, Laizhou Bay, Liaodong Bay and the Central Bohai Sea. The water transparency, nutrient concentrations and primary production are quite different in the four regions (Lu *et al.*, 1984; Fei, 1986; Fei *et al.*, 1988b; Cui and Song, 1996). According to observations, the transparency is lowest in the Bohai Bay and highest in the Central Bohai Sea (Fei, 1986) (Fig. 2). Primary production in Laizhou Bay and the Central Bohai Sea is several times higher than in Bohai Bay (Fei *et al.*, 1988a; Fig. 3). Nutrient concentrations (total inorganic nitrogen) in Laizhou Bay are much higher than in the other regions, with the lowest being in Liaodong Bay (Cui and Song, 1996; Table 1). It is suggested that the spatial pattern of primary production is closely related to the distribution of nutrients and variations in water transparency (Fei *et al.*, 1988b).

Most of these studies have been based on observations. There have been few attempts to model the dynamics of the Bohai Sea in a way that includes the important physical, chemical and biological processes. Ecological modelling provides a useful method to relate the abundance and production of living organisms to variations in feeding conditions, predation, and to the abiotic environment. Many ecological models have been constructed (Fransz and Verhagen, 1985; Franks *et al.*, 1986; Frost, 1987; Fasham *et al.*, 1990; Taylor *et al.*, 1993; Sharples and Tett, 1994; Xu and Green, 1995; McCreary *et al.*, 1996; Chen *et al.*, 1997; Cui *et al.*, 1997) and used to simulate natural ecological systems.

A variety of models have been used to simulate the dynamics of ocean plankton ecosystems, ranging from the most basic nutrient-phytoplankton-zooplankton models to more

* Gao H, Feng S, Guan Y. 1998. Modelling annual cycles of primary production in different regions of the Bohai Sea. Fisheries Oceanography, 7(3–4): 258–264

complex formulations which include dissolved organic materials, detritus, bacteria and multiple size classes of organisms (Fasham *et al.*, 1990). The approach used in the present paper is to keep the model as simple as possible, yet to retain enough structure so that the basic functioning of the ecosystem is well represented. As a first step towards the development of a coupled physical-biological model, we present a biological model of the annual cycles of plankton dynamics and nutrient cycling in the Bohai Sea. In our initial development and exploration of the model, we have concentrated on modelling the annual cycle and comparing the values of primary production in the different regions of the Bohai Sea, using five components (nitrogen, phosphate, phytoplankton, zooplankton and detritus).

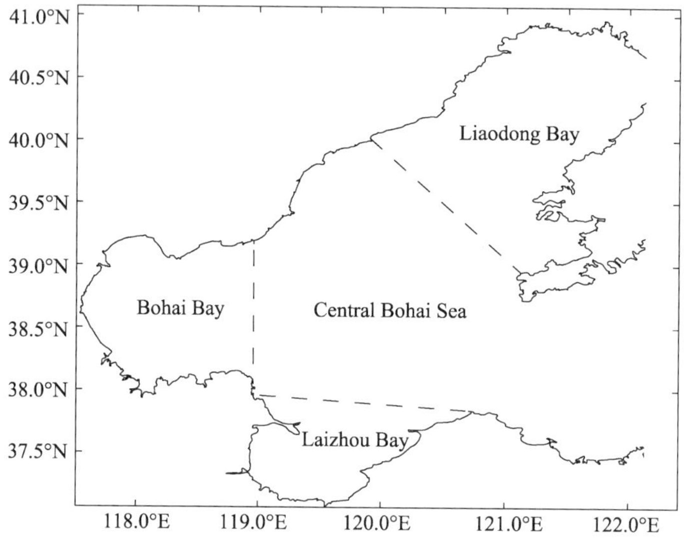

Figure 1 The four regions of the Bohai Sea, China

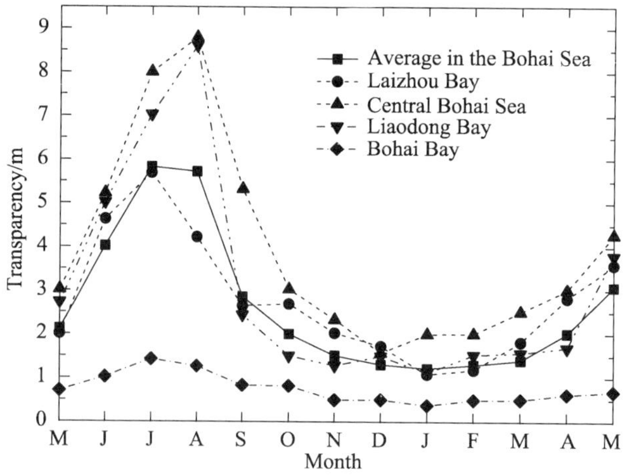

Figure 2 Observations by Secchi disk of water transparency in different regions of the Bohai Sea from 1982 to 1983

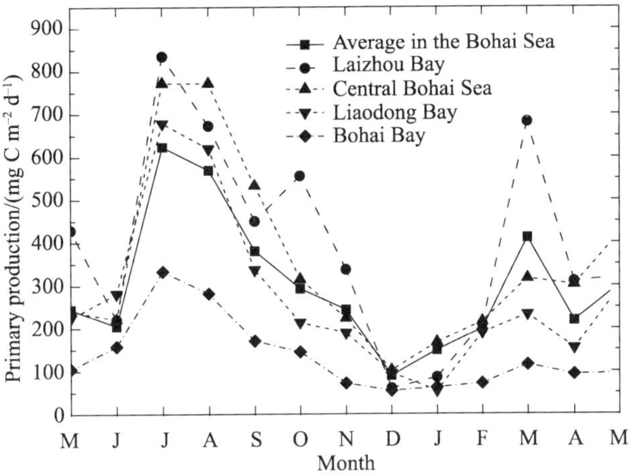

Figure 3 Observations of primary production in different regions of the Bohai Sea from 1982 to 1983

Model Equations and Parameters

Model equations

The model consists of two nutrient compartments (N_n, nitrogen and N_p, phosphate), with two reservoirs for living biomass (P, phytoplankton and Z, zooplankton) and one for detritus (D). In broad oceanic areas, nitrogen is generally the limiting nutrient, although in the coastal areas of the Bohai Sea, phosphate may be limiting (Chen *et al.*, 1991). Thus, it is necessary to consider both nitrogen and phosphate as the nutrient variables when modelling primary production of the Bohai Sea.

As with many other biological models, the model described in the present paper expresses the state variables in terms of nitrogen or phosphate. The model equations are as follows:

$$\frac{dP}{dt} = U_m \frac{N_i}{k_i + N_i} f(I)g(t)P - G_m h(t)Z(1 - e^{-\lambda P}) - d_P P, \quad i = n, p \tag{1}$$

$$\frac{dZ}{dt} = \gamma G_m h(t)Z(1 - e^{-\lambda P}) - d_Z Z \tag{2}$$

$$\frac{dN_n}{dt} = -U_m \frac{N_n}{k_n + N_n} f(I)g(t)P + \theta G_m h(t)Z(1 - e^{-\lambda P}) + eD \tag{3}$$

$$\frac{dN_p}{dt} = -U_m \frac{N_p}{k_p + N_p} f(I)g(t)P + \theta h(t)G_m Z(1 - e^{-\lambda P}) + eD \tag{4}$$

$$\frac{dD}{dt} = (1 - \gamma - \theta)G_m h(t)Z(1 - e^{-\lambda P}) + d_P P + d_Z Z - eD \tag{5}$$

Dissolved nutrients are taken up by Michaelis-Menten kinetics, while phytoplankton are grazed by zooplankton with an Ivlev functional response. The daily averaged growth rate of

Table 1 Observed nutrient concentrations in the Bohai Sea in May 1982 (μmol l^{-1}) (from Cui *et al.*, 1996)

Nutrient	Laizhou Bay	Bohai Bay	Liaodong Bay	Central area	Total Bohai Sea
NO_3-N	3.23	1.48	0.65	0.94	1.53
PO_4-P	0.68	1.17	0.97	1.0	0.94
IN[a]	6.36	2.91	1.28	1.85	3.01
N : P	9.35	1.97	1.97	1.97	3.2
$\frac{N_n}{k_n+N_n}$	0.87	0.74	0.56	0.65	0.75
$\frac{N_p}{k_p+N_p}$	0.91	0.95	0.94	0.94	0.94

[a]IN, Inorganic nitrogen.

phytoplankton is also controlled by the amount of photosynthetically available radiation (PAR) incident on a cell during a day, which in turn will be a function of time of the year, latitude, and the transparency of the water column. Mortality is assumed to be a proportion of the biomass of phytoplankton and zooplankton and is described by a specific natural mortality rate d_P and d_Z. The fraction of detritus that becomes dissolved nutrient is indicated by eD. In the model, it is assumed that phytoplankton growth is both light and nutrient limited. Following Steele (1962), the photosynthetic response to irradiance is:

$$f(I)=\frac{I}{I_o}e^{[1-(I/I_o)]} \tag{6}$$

Here, I is photosynthetically available radiation (PAR) with optimal light intensity I_o. I is considered to be a function of the water attenuation coefficient k_{ext} and the surface irradiance I_s :

$$I=I_s e^{-k_{ext}z} \tag{7}$$

where z is the depth below the sea surface. In the Bohai Sea, k_{ext} is a linear function of water transparency (S, in metres) as shown by Fei (1984):

$$k_{ext}=\frac{1.51}{S} \tag{8}$$

The temperature dependence of growth ($g(t)$, with t in °C) and grazing ($h(t)$, with t in °C) are parametrized according to the 'Q_{10}' rule (Fransz *et al.*, 1985). The same functional type is used for $g(t)$ and $h(t)$ but with a different Q_{10}:

$$g(t)=h(t)=Q_{10}^{(t-10.0)/t} \tag{9}$$

Primary production is defined as:

$$PP=\int_0^L U_m\frac{N_i}{k_i+N_i}f(I)g(t)P\mathrm{d}z, \tag{10}$$

where L is the depth of euphotic layer, U_m the maximum nutrient uptake rate (or maximum growth rate) with a half-saturation constant k_n (k_p), and i refers to the limiting nutrient, nitrogen or phosphate. For a specific region and time, only one type of nutrient is considered to be limiting. Before determining the limiting nutrient, the ratio of N : P was calculated. If the ratio was less than the Redfield ratio (16 : 1) and $N_n/(k_n+N_n)$ was less than $N_p/(k_p+N_p)$, then we assumed that nitrogen was the limiting nutrient, otherwise phosphate was limiting.

Zooplankton have a maximum grazing rate G_m with the grazing efficiency controlled by λ. Only a portion, γ, of the ingested phytoplankton is assimilated by the zooplankton, with part (θ) of the remainder being recycled into dissolved nutrients and the residue becoming detritus. The equations were solved using the software package Professional Dynamal Plus, with a time step of 0.05 day^{-1}.

Parameters

A successful simulation of the biological system requires a choice of parameters. Most of the parameters are poorly known from observations. In the present work, the temperature, surface incident irradiance and water attenuation coefficient were monthly averages from observations. Information in the literature provided the values for the other parameters. The set of parameters was then adjusted to reproduce the major observed features of the annual cycling of plankton and primary production through a series of trial simulations and studies. Table 2 lists the parameters and their values for the biological model in our main solution. In the final run, we used G_m = 1.5 day^{-1} and γ = 0.2, with the other parameter values as listed in Table 2. These parameters are all consistent with choices made in other modelling studies.

Table 2 Parameter values for the biological model

Parameter	Description	Value	Reference
U_m	Maximum nutrient uptake rate	2.0 day^{-1}	Franks *et al.* (1986)
k_n	Half-saturation constant for nitrogen uptake	1.0 μmol N l^{-1}	Franks *et al.* (1986)
k_p	Half-saturation constant for phosphate uptake	0.06 μmol P l^{-1}	Radach and Moll (1993)
G_m	Maximum grazing rate	1.0 day^{-1}	Fasham *et al.* (1990)
λ	Ivlev constant for grazing	0.2 (μmol N l^{-1})$^{-1}$	Franks *et al.* (1986)
γ	Zooplankton growth coefficient	0.1	McCreary *et al.* (1996)
θ	Zooplankton excretion coefficient	0.4	McCreary *et al.* (1996)
d_P	Phytoplankton death rate	0.1 day^{-1}	Franks *et al.* (1986)
d_Z	Zooplankton death rate	0.2 day^{-1}	Franks *et al.* (1986)
e	Detrital remineralization rate	0.05 day^{-1}	McCreary *et al.* (1996)
k_{ext}	Water attenuation coefficient	Variable	
I_o	Optimum light intensity	70 W m^{-2}	Raillard (1991)
Q_{10}	Temperature dependence coefficient for growth	2.08	Fransz and Verhagen (1985)
Q_{10}	Temperature dependence coefficient for grazing	3.1	Fransz and Verhagen (1985)

Annual Cycles of Nutrients and Plankton

We ran the model for 13 months from May 1982 to May 1983. To simulate the average nutrient and plankton cycles, the model was initialized with the average nutrient observations. The observed data used in running and testing of the model in this and the following sections are taken from the references of Fei (1984), Fei *et al.* (1988a, b) and Cui and Song (1996). In May 1982, the concentrations of PO_4-P, NO_3-N and total IN (inorganic nitrogen) were 0.94, 1.53 and 3.01 μmol L^{-1}, respectively (Cui and Song, 1996). The N : P ratio was about 3.2, which was much less than the Redfield ratio (16 : 1), and the value of nitrogen was less than the optimum concentration of 5.71 μmol L^{-1} for phytoplankton growth in the Bohai Sea (Cui *et al.*, 1996). The N : P ratios in the different regions of the Bohai Sea are shown in Table 1, and all were smaller than the Redfield ratio. In addition, $N_n / (k_n + N_n)$ was less than $N_p / (k_p + N_p)$, therefore nitrogen was selected as the limiting nutrient. The monthly averaged surface temperature, incident irradiance, and water transparency in the Bohai Sea are listed in Table 3.

Table 3 Observations of irradiance, surface temperature, transparency, zooplankton and simulated zooplankton in the Bohai Sea

Month	Irradiance/ (W m^{-2})	Temperature/℃	Transparency/m	Zooplankton/ (mg C m^{-3})	Simulated Zooplankton/(mg C m^{-3})
May 1982	225.7	12	2.1	60	16
June 1982	231.5	18	4.0	100	27.2
July 1982	208.3	24	5.8	50	36.8
Aug. 1982	196.8	28	5.7	70	50.4
Sept. 1982	187.5	22	2.8	96	50.8
Oct. 1982	138.9	20	2.0	36	44.8
Nov. 1982	104.2	14	1.5	38	29.6
Dec. 1982	92.6	9.0	1.3	30	18.4
Jan. 1983	98.4	5.0	1.2	38	12
Feb. 1983	144.7	2.5	1.3	24	13.6
Mar. 1983	185.2	3.0	1.4	28	40
Apr. 1983	202.5	6.0	2.0	88	52
May 1983	225.7	13	3.1		51.5

Simulated annual cycles of nitrogen, phytoplankton, zooplankton and detritus are presented in Fig. 4. Nitrogen has a peak value of 3.2 μmol N L^{-1}in December and January, and a minimum value of 0.5 μmol N L^{-1} in August. The annual cycles of phytoplankton, zooplankton and detritus are similar. There is a spring bloom in February and March with 1.2 μmol N L^{-1} of phytoplankton, corresponding to a chlorophyll *a* concentration of 1.9–4.8 mg C m^{-3}, depending on the choice of C : chlorophyll *a* ratio. From July to September, a second peak of plankton biomass occurs. Comparing the observed chlorophyll *a* concentrations with simulated values calculated using C : N of 106 : 16 (Redfield ratio) and C : chlo--rophyll *a* of 50 : 1 (Parsons *et al.*, 1984; Radach and Moll, 1993; Fig. 5), the model shows the main features of the annual phytoplankton cycle. However, the simulated minimum chlorophyll *a* of 0.09 mg m^{-3} is much lower than the observations. Figure 6 shows observed (Fei *et al.*, 1988a) and simulated values of zooplankton biomass. Generally, the simulation shows a pattern similar to the observations. There is a peak of simulated zooplankton biomass in September corresponding to an observed large increase, whereas the lowest simulated and observed zooplankton biomass was in February. Using zooplankton (mg m^{-3}, dry weight): C (mg C m^{-3}) of 1 : 0.4 (Wang *et al.*, 1988) and C : N of 106 : 16, the

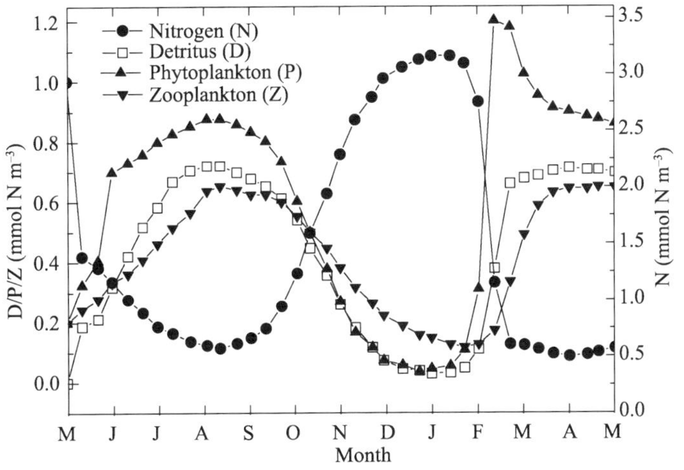

Figure 4 Simulated annual cycles of nitrogen, phytoplankton, zooplankton and detritus

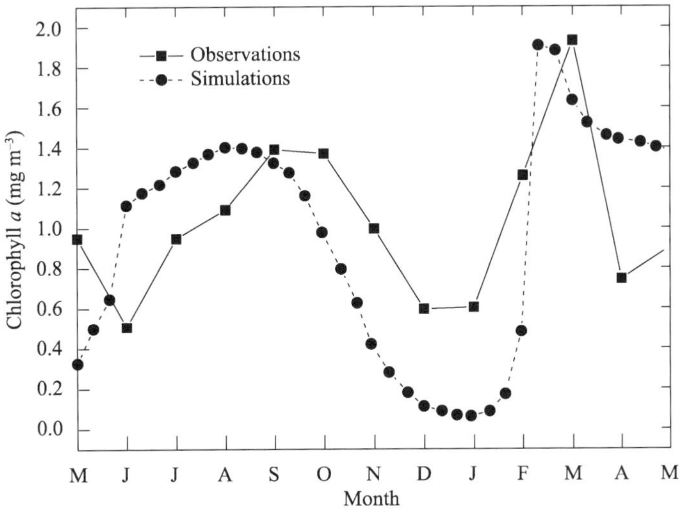

Figure 5 Comparison of simulated chlorophyll *a* and observations

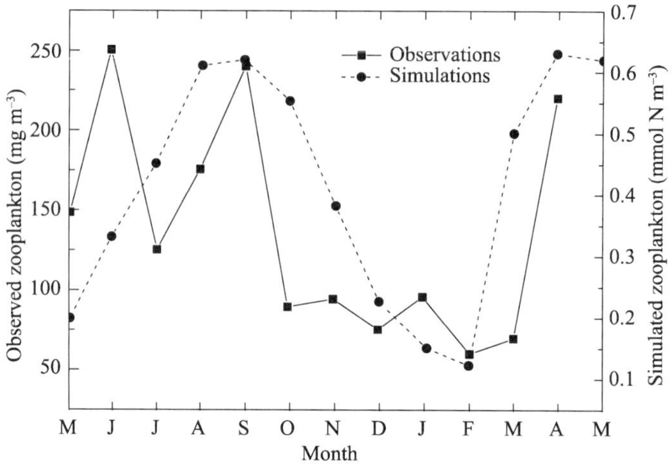

Figure 6 Comparison of simulated zooplankton and observations

observed and simulated zooplankton biomass (mg C m^{-3}) are listed in Table 3. This indicates that the simulated zooplankton biomass is smaller than the observations. Phytoplankton biomass is influenced by temperature and incident irradiance, so that in December and January the biomass of phytoplankton is significantly reduced due to lower temperatures and reduced irradiance.

Primary Production

In this section, simulated primary production values in different regions of the Bohai Sea are compared with observations. Because of limited data, it is assumed that there is no difference in incident irradiance and temperature in each part of the Bohai Sea, and that water transparency governs the available radiation for photosynthesis. The irradiance, temperature, and water transparency and nutrient concentrations used to drive the biological model are given in Table 3 and Fig. 2. The initial values of IN are listed in Table 1.

The patterns of variation are similar in each region of the Bohai Sea, with the highest levels found from July to September and the lowest from December to February (Fig. 7). Primary production is highest in Laizhou Bay and lowest in Bohai Bay. Variation of modelled primary production and water transparency is similar in Laizhou Bay, Bohai Bay, Liaodong Bay and the Central Bohai Sea. Transparency is lowest in winter because of strong wind mixing, whereas in summer the pycnocline prevents the mixing and resuspension of particles from the bottom, resulting in the highest transparency. Higher primary production corresponds to higher transparency in July to September, and lower primary production corresponds with lower transparency and lower incident radiation from December to February. The transparency in Bohai Bay is the lowest due to the particulate materials carried by the Yellow River and Haihe River. As a result, the primary production in this region is also the lowest. The initial concentration of nitrogen is another factor limiting primary production in this region (Table 1). Although the water transparency is much higher in Liaodong Bay and the central Bohai Sea than in Laizhou Bay, primary production is much higher in Laizhou Bay, because of much higher initial concentrations of nitrogen in the area (Table 1).

We correlated the observed and simulated monthly primary production in each part of the Bohai Sea. Figure 8 shows the regression with a linear correlation coefficient of more than 0.7. It indicates that the simulation is in reasonable agreement with the observations. However, the simulated values are lower than the observations, although the simulations are

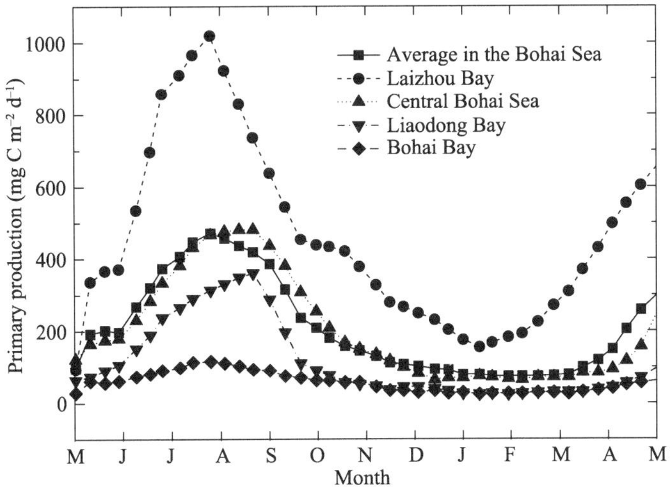

Figure 7 Simulated primary production in different regions of the Bohai Sea

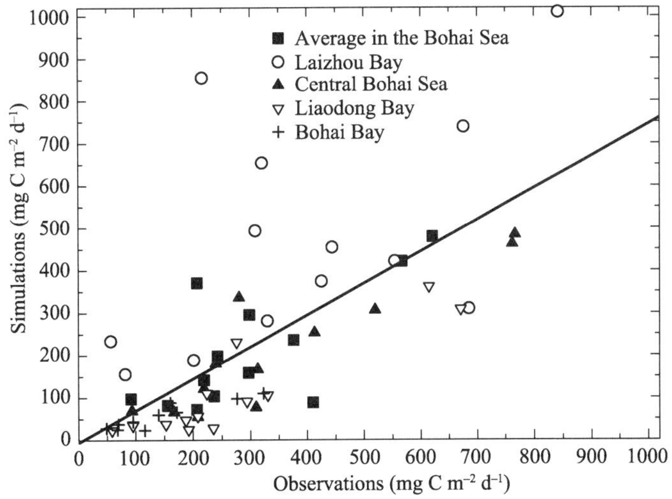

Figure 8 Comparison of simulated primary production and observations in different regions of the Bohai Sea; regression equation is $y = -7.2847 + 0.7512x$ ($r = 0.7037$)

somewhat larger than the observations in Laizhou Bay. This is probably because the model is a simple one that does not include processes such as the source of nutrients and predation by carnivores on zooplankton. Although the annual cycles of primary production are influenced by many factors, the differences of primary production in the different regions of the Bohai Sea indicate that it is mainly controlled by the water transparency and initial nutrient concentrations, confirming the hypotheses of Fei *et al.* (1988b).

In summary, a dynamically simple biological model is able to simulate the general features of the annual cycles of the plankton and primary production in the Bohai Sea forced by monthly observational data. Thus, our model provides a foundation for the hierarchy of coupled systems that will be required to understand better the complex biological dynamics of the Bohai Sea. Sensitivity analysis of the model and its coupling with a three-dimensional physical model are underway.

Acknowledgements

This research was supported by NSFC and the National Postdoctoral Science Foundation. We thank Professor Wang Jingyong and Sun Wenxin for their thoughtful comments and suggestions.

References

Chen, C., Wiesenburg, D. A. and Xie, L. (1997) Influences of river discharge on biological production in the inner shelf: a coupled biological and physical model of the Louisiana-Texas Shelf. J. Mar. Res. 55: 293–320.

Chen, S., Gu, Y., Liu, M. and Zhang, M. (1991) Nutrient distribution at the Huanghe River estuary. J. Ocean Univ. Qingdao 21: 32–37.

Cui, M., Wang, R. and Hu, D. (1997) Simple ecosystem model of the central part of the East China Sea in spring. Chin. J. Oceanol. Liminol. 15: 80–87.

Cui, Y. and Song, Y. L. (1996) Study on evaluation of nutrient status in the Bohai Sea. Mar. Fish. Res. 17: 57–62.

Fasham, M. J. R., Ducklow, H. W. and McKelvie, S. M. (1990) A nitrogen-based model of plankton dynamics in the oceanic mixed layer. J. Mar. Res. 48: 591–639.

Fei, Z. L. (1984) An estimation of the diffuse attenuation coefficient in offshore waters. J. Oceanogr. Huang Hai Bohai Sea 2: 26–29.

Fei, Z. L. (1986) Study on the water color and transparency in the Bohai Sea. J. Oceanogr. Huang Hai Bohai Sea 4: 33–40.

Fei, Z. L., Mao, X. H. and Zhu, M. (1988a) Studies on the production in Bohai Sea I: The distribution features of Chl. *a* and seasonal variation. Acta Oceanol. Sin. 10: 99–106.

Fei, Z. L., Mao, X. H. and Zhu, M. (1988b) Studies on the production in Bohai Sea II: evaluations of primary production and fish. Acta Oceanol. Sin. 10: 48–89.

Franks, P. J. S., Wroblewski, J. S. and Flierl, G. R. (1986) Behavior of a simple plankton model with food-level acclimation by herbivores. Mar. Biol. 91: 121–129.

Fransz, H. G. and Verhagen, J. H. G. (1985) Modelling research on the production cycle of phytoplankton in the Southern Bight of the North Sea in relation to river-borne nutrient loads. Neth. J. Sea Res. 19:

241–250.

Frost, B. W. (1987) Grazing control of phytoplankton stock in the open subarctic Pacific Ocean: a model assessing the role of mesozooplankton, particularly the large calanoid copepods, Neocalanus spp. Mar. Ecol. Prog. Ser. 39: 49–68.

Ge, C. (1986) On the relationship between marine environment and prawn fishery in Bohai Sea. J. Oceanogr. Huang Hai Bohai Sea 4: 77–83.

Gu, Y. and Xiu, R. (1996) On the current and storm flow in the Bohai Sea and their role in transporting deposited silt of the Yellow River. J. Oceanogr. Huang Hai Bohai Sea 14: 1–6.

Guo, B. H. (1993) Major features of the physical oceanography in the Yellow Sea. J. Oceanogr. Huang Hai Bohai Sea 11: 7–18.

Lu, P. D., Fei, Z. L. and Mao, X. (1984) Estimation of primary production and the distributions of Chl. a in Bohai Sea. Acta Oceanol. Sin. 6: 90–99.

McCreary, J. P., Kohler, K. E., Hood, R. R. and Olson, D. B. (1996) A four-component ecosystem model of biological activity in the Arabian Sea. Prog. Oceanogr. 37: 193–240.

Parsons, T. R., Takahashi, M. and Hargrave, B. C. (1984) Biological Oceanographic Processes. Oxford: Pergamon Press, 330 pp.

Radach, J. and Moll, A. (1993) Estimation of the variability of production by simulating annual cycles of phytoplankton in the central North Sea. Prog. Oceanogr. 31: 339–419.

Raillard, O. (1991) Etude des interaction entre les processus physiques et biologiques intervenant dans la production de l'huitre Crassostrea gigas du bassin de Marennes-Oleron. PhD thesis, University of Paris VI, 203 pp.

Sharples, J. and Tett, P. (1994) Modelling the effect of physical variability on the midwater chlorophyll maximum. J. Mar. Res. 52: 219–238.

Steele, J. H. (1962) Environmental control of photosynthesis in the sea. Limnol. Oceanogr. 7: 137– 150.

Taylor, A. H., Harbour, D. S. and Harris, R. P. (1993) Seasonal succession in the pelagic ecosystem of the North Atlantic and the utilization of nitrogen. J. Plank. Res. 15: 875–891.

Wang, X., Kang, J. and Li, S. (1988) Seasonal variation in body length, and weight and carbon, nitrogen and hydrogen content of Calanus sinicus in the Xiamen Harbour. J. Oceanog. Taiwan Strait 7: 173–179.

Xu, Y. and Green, J. S. A. (1995) Modeling of the seasonal variation of plankton. Acta Ecolog. Sin. 15: 246–250.

湍流局地平衡假设的新推论——齐次湍流动能输运方程封闭模型与应用(Ⅱ)*

魏　皓　冯士筰　武建平　张　平

在 Boussinesq 似黏性假设下，雷诺方程可由零方程模型（Prandtl 混合长理论）、一方程模型（k–方程封闭）、二方程模型（$k-\varepsilon$ 或 k–kl 模型）确定其湍流粘性来封闭。自 70 年代以来，这些模型在海洋中得以广泛应用①。普林斯顿海洋模式采用了 k–kl 封闭模型[1]，Leendertse 模式采用 k–方程来描述斜压过程的次网格湍流效应[2]，Johns 以 k–方程封闭计算了北海的风暴潮增水[3]，汉堡大学陆架海模式也增加了混合长封闭[4]研究北海环流。两方程模型在原有的海洋动力学模式中增加了两个预报变量，计算工作量大大增加，Prandtl 混合长模型在流速剪切为零处存在物质扩散系数为零的欠缺。我们考察了各种模型的性能与特点，在湍的局地平衡假设下建立了一个适用于浅海动力学的新的封闭模型——齐次湍能输运方程模型，它可以避免零方程模型的某些弱点，又可以比两方程模型减少计算工作量，性能较佳。

1　湍流局地平衡假设

1925 年 Prandtl 提出了半经验半理论的混合长假设 (Mixing Length Hypotheses 简称 MLH)[5]，直接比拟于气体分子运动论，类比于分子自由程定义一个涡不与其他涡碰撞的长度为混合长（1），湍粘性正比于混合长与速度剪切，这在当时是一个大胆推断，然而我们可以证明 MLH 是精确成立的 k–方程在局地平衡下的一个特例。

k–方程封闭模型为：

$$\frac{\partial k}{\partial t}+\boldsymbol{u}\cdot\nabla k=D+P+G-\varepsilon \tag{1}$$

其中 k 为湍流动能，D 为湍动能的扩散输运，P 为湍动能的剪切生成，G 为湍动能的浮力生成，ε 为湍动能的耗散。忽略湍的局地变化及对流扩散输运，则有：

$$P+G-\varepsilon=0 \tag{2}$$

其物理意义是：湍的生成等于其耗散，当地生成的湍能被就地消耗，这就是湍的局地

---

* 魏皓，冯士筰，武建平，张平．2000．湍流局地平衡假设的新推论——齐次湍流动能输运方程封闭模型与应用(Ⅱ)．青岛海洋大学学报，30(4): 557–562

① 编者注：魏皓，武建平，张平．2001．海洋湍流模式应用研究．青岛海洋大学学报，31(1)：7–13

平衡假设。

尽管没有实验直接证明此假设在机翼边界层中成立，几十年来空气动力学中 Prandtl 理论的成功应用，却说明至少在边界层中这种假设是可以接受的，因此它在海洋这一行星边界层运动中也应可以接受。

2 齐次湍能输运方程封闭模型

2.1 模型的建立

对（2）式的一种物理解释是认为固壁附近的湍，湍能的输运效应小到可以忽略，湍仅是一种局地效应，生成即被耗散。然而在远离壁界的湍流核心区，湍流一般要受到其他时空点上运动的影响，即湍动能具有输运的特性，此时仍保持湍的局地平衡假设，则（1）变为

$$\frac{\partial k}{\partial t}+\boldsymbol{u}\cdot\nabla k-\frac{\partial}{\partial z}\left(\frac{\nu}{\sigma_{\mathrm{k}}}\frac{\partial k}{\partial z}\right)=0 \tag{3}$$

其物理意义为：湍动能的局地变化率，由于湍动能的产生和耗散率相抵，只取决于对流和扩散输运过程。此时湍能输运方程是无源汇的，因此（3）称为齐次湍动能输运方程（Homogenous *k*–equation，以后简称 HKE），以 HKE 封闭的模型称为齐次湍动能输运方程封闭模型。

2.2 模型的检验

2.2.1 平板拖曳流

考虑相距为 h 的两块无限大平板之间，充满了不可压均质流体，上平板 $z=0$，上板在自身平面内向正 x 方向匀速移动，导致作用于流体上的切应力 $\boldsymbol{\tau}(0)=\tau_{\mathrm{a}}\boldsymbol{i}=\mathrm{const}>0$，这类似于平均风应力作用于海面，这种定常流动的 Reynolds 方程和边界条件为：$\frac{\partial}{\partial z}\left(\nu\frac{\partial u}{\partial z}\right)=0$; $z=0$, $\nu\frac{\partial u}{\partial z}=\tau_{\mathrm{a}}$; $z=-h$, $u=0$。此时 HKE 退化为：$\frac{\partial}{\partial z}\left(\frac{\nu}{\sigma_{\mathrm{k}}}\frac{\partial k}{\partial z}\right)=0$; $z=0$, $k=\lambda_{\mathrm{a}}\tau_{\mathrm{a}}$; $z=-h$, $k=\lambda_{\mathrm{b}}\tau_{\mathrm{b}}$, $\tau_{\mathrm{b}}=\nu\frac{\partial u}{\partial z}\Big|_{z=-h}$，其中 $\lambda_{\mathrm{a}}=C_{\mathrm{D}}^{-\frac{1}{2}}\big|_{z=0}$，$\lambda_{\mathrm{b}}=C_{\mathrm{D}}^{-\frac{1}{2}}\big|_{z=-h}$，$C_{\mathrm{D}}\big|_{z=0,-h}$ 分别为海底和海面的拖曳系数，此边界条件表明湍流动能在海底和海面是由风的混合及海底拖曳产生。再由 Kolmogorov-Prandtl 关系式，可以解得上述问题中运动及湍变量的解析解为

$$u=\sqrt{\frac{\tau_{\mathrm{a}}}{\lambda_{\mathrm{a}}}}\int_{-1}^{\sigma}\frac{\varPhi^{-\frac{1}{3}}}{L'(\sigma')}\mathrm{d}\sigma',\ k=\lambda_{\mathrm{a}}\tau_{\mathrm{a}}\varPhi^{\frac{2}{3}},\ \nu=h\sqrt{\lambda_{\mathrm{a}}\tau_{\mathrm{a}}}\,L'(\sigma)\varPhi^{\frac{1}{3}} \tag{4}$$

其中

$$\varPhi=\left[1-\left(\frac{\lambda_{\rm b}}{\lambda_{\rm a}}\cdot\frac{\tau_{\rm b}}{\tau_{\rm a}}\right)^{\frac{3}{2}}\right]\frac{\int_{-1}^{\sigma}\frac{1}{L'(\sigma')}{\rm d}\sigma'}{\int_{-1}^{0}\frac{1}{L'(\sigma')}d\sigma'}+\left(\frac{\lambda_{\rm b}}{\lambda_{\rm a}}\cdot\frac{\tau_{\rm b}}{\tau_{\rm a}}\right)^{\frac{3}{2}} \tag{5}$$

$L'=\dfrac{l}{h}$，$\sigma=\dfrac{z}{h}$，l 为湍流混合长度的经验剖面，$\varPhi$ 亦为一剖面函数。可见流速随拖曳力的增加而增加，湍流动能亦由拖曳做功产生，湍流粘性则随两板距离增加而增加，但它与流速的剖面却有很大不同。（5）式表明拖曳力不仅影响 u，k，ν 的量值，还会影响它们的剖面——空间分布状态。若利用 Prandtl 的混合长理论$\left(\nu=l_{\rm m}^2\dfrac{\partial u}{\partial z}\right)$求解，则有：

$$u=\sqrt{\tau_{\rm a}}\int_{-1}^{\sigma}\frac{1}{l'_{\rm m}(\sigma')}{\rm d}\sigma',\ \ \nu=h\sqrt{\tau_{\rm a}}\,l'_{\rm m}(\sigma)$$

比较两种方法得到的流速和湍流粘性系数剖面，发现只要令 $l_{\rm m}=\sqrt{\lambda_{\rm a}}\varPhi^{\frac{1}{3}}l$，则 HKE 模型与 MLH 理论给出了同样的结果。从而表明了：HKE 替代 MLH 是可行的；或者说它从逻辑推理出发，解释了 MLH 何以在远离固壁的湍流核心区（一般说存在湍动能输运过程的区域）有时也可使用的原因。

2.2.2　平板间梯度流

考虑两无限大平板间（相距 $2h$）关于 $z=0$ 对称、定常水平压力梯度（G）作用下的均质流体流动。其定解问题的控制方程和边界条件为：$\dfrac{\partial}{\partial z}\left(\nu\dfrac{\partial u}{\partial z}\right)=-G$，$z=0$，$\dfrac{\partial u}{\partial z}=0$；$z=-h, u=0$（运动关于 x 轴对称）。HKE 封闭的边界条件变为：$z=0$，$\dfrac{\partial k}{\partial z}=0$，加上 Kolmogorov-Prandtl 关系式可得其解为：$u=-\sqrt{\dfrac{Gh}{\lambda_{\rm b}}}\displaystyle\int_{-1}^{\sigma}\frac{\sigma'}{L'(\sigma')}{\rm d}\sigma'$，$k=\lambda_{\rm b}hG$，$\nu=h\sqrt{\lambda_{\rm b}hGL'}(\sigma)$，因为运动为定常水平压力梯度驱动，因而湍动能 k 为常数，全场均匀；但湍流粘性的垂直分布取决于混合长的剖面。

同样应用 MLH 假设，则其解为

$$u=\sqrt{Gh}\int_{-1}^{\sigma}\frac{\sqrt{-\sigma'}}{l'_{\rm m}(\sigma')}{\rm d}\sigma',\ \ \nu=h\sqrt{Ghl'_{\rm m}}(\sigma)\sqrt{-\sigma} \tag{6}$$

只要令

$$l_{\rm m}=\frac{\sqrt{\lambda_{\rm b}}}{-\sigma}l \tag{7}$$

两种流速和湍流粘性系数获得形式上相同的解。然而（7）式却表明，在 $\sigma=0$ 处对于一个有限的非零 $l(0)$ 混合长，将有一个无限的 $l_{\mathrm{m}}(0)$，即涡永不会与其他涡碰撞混合，显然这从物理上是不合理的。其实（6）式已表明，在 $\sigma=0$ 对于一有限的 $l_{\mathrm{m}}\neq 0$，必有 $\nu(0)=0$，此时物质扩散系数 $\lambda_{\mathrm{p}}(0)=\dfrac{\nu(0)}{\sigma_{\mathrm{p}}}=0$，即在流速剪切为零的地方湍粘性为零，且无物质扩散，这在物理上更加不合理，这一 MLH 的典型缺陷早已为人知。由 KHE 封闭所得的解可以看出，MLH 的这一缺陷被避免了。

Smith, T.[6]经过 k–方程封闭与 k - ε 模型的比较，找到一个较合理的混合长的经验公式：$l=C_{\mathrm{L}}\cdot h\cdot \operatorname{erf}\left[\alpha(l+\sigma)\right]$，$C_{\mathrm{L}}$ 和 α 为经验常数，erf 为误差函数，当 HKE 模型均选用此式时，得到的无因次速度剖面与 Laufer(1951)[7]对称平板湍流实验结果吻合良好（图 1），由此可见，HKE 比 MLH 具有更丰富而合理的内涵。

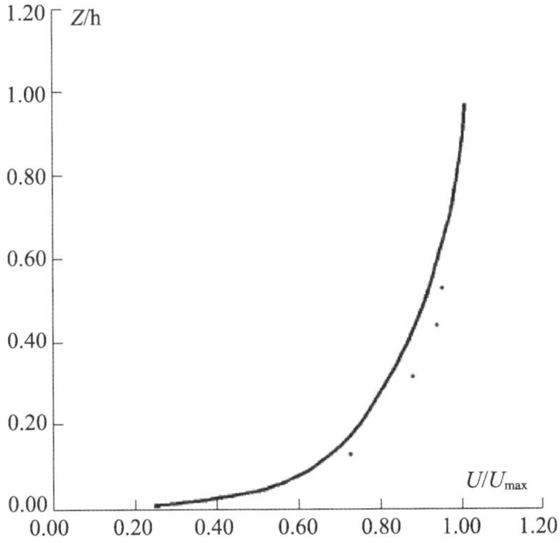

图 1　HKE 封闭模型计算的速度剖面与 Laufer 试验的比较

实线为 HKE 模型结果，黑点为 Laufer 实验结果

3　HKE 封闭的浅海动力学模型与运动分析

HKE 在经典流动中的成功应用，表明其在边界层运动中是可用的，我们进一步将其应用于浅海运动方程组的封闭。

3.1　HKE 封闭的浅海动力学方程组

一般浅海斜压系统模型的无因次化，是在湍的非线性参数化后，浅海运动中基本受力平衡为局地加速度与水位梯度力这一假设下获得的[8]。而湍流封闭问题中的非线性不是弱非线性，在浅海中湍流粘性效应充满整个研究海域的流场，水位梯度力正是近似地被这一效应所平衡，即二者构成了浅海流场中基本的力的平衡：

$$O(|g\nabla\zeta| / |\partial\tau/\partial z) = 0 \tag{8}$$

由此可推得 HKE 封闭的浅海流体动力学方程组：

$$\nabla\cdot\boldsymbol{u}=0,\ \varepsilon\left[\frac{\partial u}{\partial t}+k'\boldsymbol{u}\cdot\nabla u\right]-\phi fu=-g\frac{\partial\zeta}{\partial x}+\frac{\partial}{\partial z}\left(\nu\frac{\partial u}{\partial z}\right) \quad (9)(10)$$

$$\varepsilon\left[\frac{\partial v}{\partial t}+k'\boldsymbol{u}\cdot\nabla v\right]+\phi fu=-g\frac{\partial\zeta}{\partial y}+\frac{\partial}{\partial z}\left(\nu\frac{\partial v}{\partial z}\right) \quad (11)$$

$$\varepsilon\left[\frac{\partial c}{\partial t}+k'\boldsymbol{u}\cdot\nabla c\right]=\frac{\partial}{\partial z}\left(\frac{\nu}{\sigma_{\rm c}}\frac{\partial c}{\partial z}\right),\quad \varepsilon\left[\frac{\partial k}{\partial t}+k'\boldsymbol{u}\cdot\nabla k\right]=\frac{\partial}{\partial z}\left(\frac{\nu}{\sigma_k}\frac{\partial k}{\partial z}\right) \quad (12)(13)$$

$z=\kappa\zeta$

$$w=\left(\frac{\kappa}{\kappa'}\right)\left[\frac{\partial\zeta}{\partial t}+\kappa'\left(u\frac{\partial\zeta}{\partial x}+v\frac{\partial\zeta}{\partial y}\right)\right],\quad \nu\frac{\partial(u,v)}{\partial z}=F_{\rm S}(\tau_{\rm ax},\tau_{\rm ay}), \quad (14)(15)$$

$$k=\lambda_{\rm a}|\tau_{\rm a}|,\ \frac{\partial c}{\partial z}=k_{\varGamma}\varGamma \quad (16)(17)$$

$z=-h$

$$w=-u\frac{\partial h}{\partial x}-v\frac{\partial h}{\partial y},\quad \nu\frac{\partial(u,v)}{\partial z}=F_{\rm S}C_{\rm D}|\boldsymbol{u}|(u,v)=F_{\rm S}(\tau_{\rm bx},\tau_{\rm by}) \quad (18)(19)$$

$$k=\lambda_{\rm b}|\boldsymbol{\tau}_{\rm b}|,\ \frac{\partial c}{\partial z}=0 \quad (20)(21)$$

k 与 ν 满足 Kolmogorov-Prandtl 关系式，混合长剖面如 Smith[6]。式中 $\nabla\zeta_+$ 为广义压强梯度力，$\zeta_+=\zeta+F_{\rm e}\int_z^{\zeta}\beta c{\rm d}z'$。

3.2 运动分析

3.2.1 无量纲参数间的关系

（9）～（21）是一组有因次方程，只不过为了指出某项量阶而在该项前面标注了一个相关的无因次参数，无该参数者表明其量阶为 1，方程组中各量含义同文献①，无因次参数意义如下：$\varepsilon=\omega/\omega_\nu$，$k'=u_{\rm c}\omega^{-1}/L_{\rm c}$，$\kappa=\zeta_{\rm c}/h_{\rm c}$，$\phi=f_{\rm c}/\omega_\nu$，$F_{\rm s}=\tau_{\rm ac}/\tau_{\rm c}$，$F_{\rm c}=\delta_{\rm c}/\kappa$。式中 $\omega_\nu=v_{\rm c}/h_{\rm c}^2$ 为湍的阻尼频率。下标 c 表示该量的特征量，如 ω^{-1} 和 $L_{\rm c}$ 分

① 编者注：魏皓，武建平，张平. 2001. 海洋湍流模式应用研究. 青岛海洋大学学报，31(1)：7–13

别表示时间和水平空间尺度的特征量，τ_c 和 δ_c 分别为湍粘性力和密度相对增量的特征量。由（8）式有 $g\dfrac{\zeta_c}{L_c}\sim\dfrac{\tau_c}{h_c}$，可推得 $\zeta_c\sim\dfrac{\tau_c L_c}{g h_c}$。

3.2.2 运动方程中惯性项与湍应力项之比的估计

运动方程中惯性项与湍应力项之比为：$\varepsilon=\dfrac{\omega}{\nu_c/h_c^2}=\omega/\omega_v$。另外由 $\tau_c\sim\nu_c\dfrac{u_c}{h_c}$，有 $\dfrac{V_c}{h_c^2}\sim\dfrac{\tau_c}{h_c U_c}$。而在海底 $\tau=C_D|\boldsymbol{u}|\boldsymbol{u}$，应有 $\tau_a\sim C_D/u_c^2$，所以 $\dfrac{\nu_c}{h_c^2}\sim\dfrac{C_D u_c}{h_c}$。于是又可得到 ε 的另一表达 $\varepsilon=\omega/\omega_v=(k_c\cdot C_D)^{-1}$，$k_c=\dfrac{u_c}{\omega h_c}$ 称作 Keulegan-Carpenter 数。

当运动频率与阻尼频率接近时，$\varepsilon=O(1)$，此时运动的惯性效应不可忽略，惯性力亦是方程中一个基本平衡项。由 Johns 与 Odd[9]指出，$C_D=1.3\times10^{-3}$，汪景庸[10]亦指出底拖拽系数为$(2.5\sim4.3)\times10^{-3}$量级，因而当 $k_c\sim O(10^2)$ 时 $O(\varepsilon)=1$，对于 $O(h_c)=10^2$ m 的浅海，$\omega^{-1}\sim10^4$ s，这恰是全日潮或半日潮的周期的量级，因此潮波运动中惯性效应是一个基本特征。显然高频重力波需要考虑其惯性效应，而“气候式平均”的环流 $O(\varepsilon)<<1$，可以略去惯性项而作为定常态处理。

3.2.3 对流非线性项与惯性项之比的估计

对流非线性项与惯性项之比为：$\kappa'=\dfrac{u_c}{\omega L_c}=\dfrac{u_c}{\omega h_c}\cdot\dfrac{h_c}{L_c}=k_c\cdot\dfrac{h_c}{L_c}$。由于海洋为一薄层流体 $h_c<<L_c$，只有 k_c 数大到一定量级才能使 $O(\kappa')=1$。对于全日或半日潮波系统有：$O(\varepsilon)=1$, $u\sim1\ \text{ms}^{-1}$, $\omega\sim10^{-4}\ \text{s}^{-1}$, $L_c\sim100$ km, $\kappa'\sim10^{-1}$,因而在潮振荡占优的浅海，对流非线性项相对于惯性效应为小量。

3.2.4 科氏力与湍粘性力之比的估计

科氏力与湍粘性力之比为：$\phi\sim f_c u/\dfrac{\nu_c u_c}{h_c^2}=\dfrac{f_c}{\omega_v}$，此式表明当阻尼频率足够大时，科氏力亦为方程中的小项而可以忽略（如河口）。

3.2.5 运动学边界条件的因次估计

由连续方程（21），$w\sim u_c\dfrac{h_c}{L_c}$，海面运动学边界条件中水位的变化率与垂直流速之比为 $\dfrac{\partial\zeta}{\partial t}/w=\dfrac{\zeta_c\omega L_c}{h_c u_c}=\dfrac{\zeta_c}{h_c}/\dfrac{u_c}{\omega L_c}=\dfrac{\kappa}{\kappa'}$。由前分析知运动的对流项与惯性项之比为 κ'，所

以在海面用无因次参数标注的运动学边界条件为 $z=\kappa\zeta$，$w=\left(\frac{\kappa}{\kappa'}\right)\left[\frac{\partial\zeta}{\partial t}+\kappa'\left(u\frac{\partial\zeta}{\partial x}+v\frac{\partial\zeta}{\partial x}\right)\right]$。

当 $O(\kappa/\kappa')\ll 1$ 时，海面边界条件变为 $w=0$，即刚盖假定。对于浅海平均环流，其频率特征 $\omega\ll 10^{-5}\ \mathrm{s}^{-1}$，满足 $O(\kappa/\kappa')\ll 1$，显然刚盖假定是合理的；而对于潮波运动 $O(\kappa/\kappa')=1$，则不能用刚盖假定，而其对流非线性效应为小量。

以上运动分析是合乎实际的，因而以 HKE 封闭的浅海动力学模型及其运动基本平衡是合理的，我们将在下一部分介绍此模型在几种浅海运动模拟中的应用。

参考文献

[1] Blumerg A, Mellor G. A description of a three-dimensional coastal ocean circulation model. [J] In: Heaps N ed, Three-Dimensional Coastal Models. [M]Washington D C: American Geophysical Union, 1987. 208

[2] Leedertse J J, Liu S K. A Three-dimensional Model for Estuaries and Coastal Seas: Turbulent Energy Computation. [M] Santa Monica: Rand Corp, 1979. 1–58.

[3] Johns B. The modeling of tidal flow in a channel using a turbulence energy closure scheme. [J] J Phys Oceanogr, 1978, 8: 1042.

[4] Backhaus J O. A three-dimensional model for the simulation for shelf sea dynamics. [J] Dt Hydrogr Z, 1985, 38: 165.

[5] Rodi W. Turbulence models and their application in hydraulics —A state of the art review. [M] Netherlands: Book Pub of IAHR, 1980. 104.

[6] Smith Y J, Takhar H S. The calculations of oscillatory flow in open channels using mean turbulence energy models. [M] Rept Simon Engi Labs. England: University of Manchester, 1977. 70.

[7] Laufer J. NCCA Report 1053. [R] USA, 1951. 51.

[8] 冯士筰. 浅海环流物理及数值模拟. 见: 冯士筰, 孙文心主编. 物理海洋数值计算. [M] 郑州: 河南科技出版社. 1992. 610.

[9] Johns B, Odd N. On the vertical structure of tidal flow in river estuaries. [J] Geophys J R Astr Soc, 1966, 12: 103.

[10] 汪景庸. 风暴潮和天文潮条件下的底摩擦研究. [J] 青岛海洋大学学报, 1991, 21(4): 1–10.

Variability of the Bohai Sea Circulation Based on Model Calculations*

Dagmar Hainbucher, Hao Wei, Thomas Pohlmann,
Jürgen Sündermann, Shizuo Feng

1. Introduction

In the frame of the joint German/Chinese research programs Analysis and Modelling of Mass Fluxes in the Bohai Sea (AMBOS) and Analysis and Modelling Research of the Ecosystem in the Bohai Sea (AMREB) mass and energy fluxes in the Bohai Sea should be quantified and simulated. For this task, a baroclinic three-dimensional model (HAMSOM) was used, simulating a period of 14 years in order to get information about the long-term variability of the physical parameters. Not only that the data of these simulations can build the forcing for further models of the project like the SPM model (Jiang *et al.*, 2004) and the ecosystem model (Wei *et al.*, 2004a), the presented results also help to put the ship experiments and the further simulations of the cooperation in a general long-term context.

The Bohai Sea (Fig. 1) is a shallow, semi-enclosed marginal sea of the western Pacific Ocean located at the northern end of the Yellow Sea which is confronted with strong environmental pollution problems due to the high density of population at its coastlines.

The Bohai Sea is still under the control of the eastern Asia monsoon; mean wind directions from north to northwest prevail in winter, whereas in summer favoured wind directions are from south.

Since 1960 the hydrology of the Bohai Sea has been monitored at six stations around the Bohai Sea (State Ocean Information Center of China) and from this data it can be concluded that the hydrology is also influenced by climate variations. There is an increasing trend for both sea surface temperature (SST) and sea surface salinity (SSS). The SST increase seems to be correlated with the air temperature variations, even if the SST changes are somewhat delayed to the air temperature changes (Zou, 1998). This behaviour is the same as for the climate of Northern China (Chen *et al.*, 1998).

The increase of the sea surface salinity is rather strong, nowadays it is about 2.5 psu higher than 40 years ago. The salinity evolution is determined by evaporation, precipitation and river runoff in the coastal area. But fresh water discharge into the Bohai Sea decreases

---

* Hainbucher D, Wei H, Pohlmann T, Sündermann J, Feng S. 2004. Variability of the Bohai Sea circulation based on model calculations. Journal of Marine Systems, 44(3–4): 153–174

more and more due to the dam construction of the Huang He River, the main source for discharge in this waters. Furthermore, the annual precipitation of Northern China in the 1980s and 1990s is less than that of the 1950s (Chen *et al.*, 1998) while the sunshine duration in the area increased at the rate of 3 h/year.

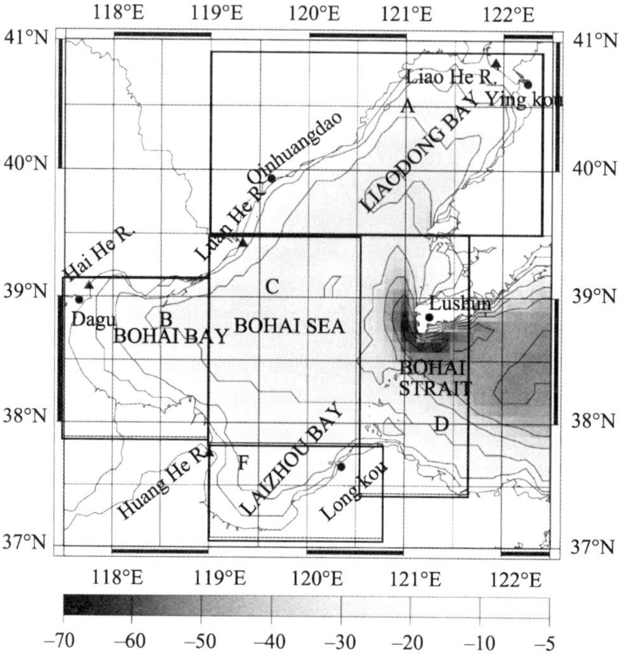

Fig. 1 Model area and bathymetry (m)

Triangles indicate the location of the three main river sources (Huang He, Hai He and Liao He) for the model. Flushing times and turnover times are calculated for the boxes A, B, C, D and F

Since about 1980 several model investigations of the Bohai Sea hydrodynamics had been carried out, most of them dealing with tides and tidal induced currents rather than with the thermohaline and atmospheric induced circulation. Tides and tidal currents have been reproduced quite well for the Bohai Sea using different theoretical and/or numerical approaches, like for example Leendertze's two-dimensional nonlinear long wave model (Dou *et al.*, 1981), Leendertze's three-dimensional model (Yu and Zhang, 1987), a two-dimensional ADI model (Huang and Wang, 1988), a finite element scheme (Wang, 1988), a three-dimensional current speed split method (Sun *et al.*, 1989) or a two-dimensional model using spherical coordinates (Zhao *et al.*, 1994). Additionally, numerical experiments had been carried out to investigate the influence of constant wind fields for typical winter or summer situations (Zhao and Shi, 1993; Miao and Liu, 1989). But all these model studies mentioned are just taking the barotropic signal into account. The first model studies also dealing with baroclinic processes have been carried out by Sun *et al.* (1990) and Huang *et al.* (1996).

The present investigation discusses results of a hydrodynamic model simulation which is not only fully baroclinic but which is also using time-dependent atmospheric forcing as input. Unfortunately, the time interval of the input data from 1980 to 1993 is not long enough to analyze really climatic variations. But the annual cycle is well resolved and high frequent fluctuations above the tidal cycle can be discussed.

2. Methods

2.1 The HAMSOM hydrodynamic model

The hydrodynamic Hamburg Shelf Ocean Model (HAMSOM) is a three-dimensional, baroclinic, leveltype model which solves the governing primitive equations of motion with finite differences on an Arakawa C-grid. The numerical scheme of HAMSOM was developed by Backhaus (1985). It is semi-implicit and therefore not dependent on the stability criteria required by explicit schemes. The implicit algorithms are applied for external gravity waves, vertical shear stresses and vertical diffusion of temperature and salinity. Furthermore, a stable second order approximation in the time domain is introduced for the Coriolis and the baroclinic pressure gradient terms in the equation of motion. Incompressibility and hydrostatic equilibrium are assumed for the pressure field, incorporating the Boussinesq approximation.

Vertical sub-grid scale turbulence is parameterized by means of a turbulent closure approach, proposed by Kochergin (1987) and later modified by Pohlmann (1996). The scheme is closely related to a Mellor and Yamada (1974) level-2-model where vertical eddy viscosity coefficients depend on stratification and vertical current shear. Convective overturning is parameterised by vertical mixing: an unstable stratification is turned into a neutral state through artificial enlargement of the vertical eddy viscosity coefficient. The horizontal diffusion of momentum is calculated using a constant, isotropic eddy viscosity coefficient.

The circulation model includes an Eulerian transport algorithm for temperature and salinity, based on the advection-diffusion equation within an upstream scheme. However, for the advection of momentum the Arakawa-J7 is used (Arakawa and Lamb, 1977). Additionally, a Lagrange approach (excluding diffusion) is used for the calculation of trajectories. Vertical eddy diffusivity coefficients (temperature and salinity) are calculated in the same way as vertical eddy viscosity coefficients, depending on stratification and vertical current shear. Horizontal eddy diffusion is neglected because of numerical diffusion stemming from the advection scheme.

Sea level and water mass properties are prescribed at the open boundaries of the model. Therewith, the inverse barometric effect is taken into account. At the sea surface and at the seabed, respectively, kinematic boundary conditions and quadratic stress laws apply.

Air temperature, relative humidity, cloud cover and wind speed determine the heat flux between the open ocean and the atmosphere. These values enter bulk formulae describing long-wave and global radiation and sensible and latent heat fluxes. The bulk formulae used are discussed in detail in Moll and Radach (1998).

Detailed descriptions of the model and its applications are given by Backhaus (1985), Backhaus and Hainbucher (1987), Pohlmann (1991), Stronach *et al.* (1993) and Schrum (1997).

The HAMSOM model was already applied to and validated for different shelf seas and also deep ocean areas with complex topographies all over the world. The results are documented in several publications like Hainbucher and Backhaus (1999), Harms *et al.* (1999), Schrum (1997), Alvarez Fanjul *et al.* (1997), Huang (1995), Carbajal (1993), Pohlmann (1991), and Hainbucher *et al.* (1987).

For the Bohai Sea (Fig. 1) a horizontal resolution of 5 min in latitude and longitude was chosen. In the vertical the model has 10 layers, the lower boundaries laying at 5, 10, 15, 20, 25, 30, 35, 40, 45 and 65 m. The model region covers the area from 37°N to 41°N and 117.5°E to 122.5°E. This model spacing was chosen because of the following reasons: a) the model results are comparable to those of Huang (1995) who made intensive tidal investigations of the Bohai Sea and used the same topography and spacing. This rendered possible a validation of our model for the tides. b) Under extreme wind conditions the sea surface elevation in the Bohai Sea can exceed heights more than 3 m. As a consequence the minimum vertical resolution of the model is about 5 m to avoid drying of grid cells but however, it still enables a representation of the vertical development of temperature and salinity. Furthermore, the horizontal spacing of less than 10 km is sufficient for resolving the baroclinic Rossby radius of deformation which ranges from about 25 to 30 km in this region (Chelton *et al.*, 1998).

2.2 Initial and forcing data

The topography of the model was obtained from the ETOPO5 database (Hirtzler, 1985). Amplitudes and phases of the five main tidal constituents (M_2, S_2, N_2, K_1, O_1) were made available by Huang (1995). He used the data of two opposite coastal tidal gauges at the entrance of the Bohai Strait (Dachangshandao, 39°16'N, 122°35'E and Jimingdao, 37°27'N, 122°29'E) to determine the tidal information. Additionally to the tidal information, the inverse barometric effect (calculated from the air pressure) and the dynamic heights (calculated from temperature and salinity) are prescribed at the open boundary.

The initial temperature and salinity distributions were extracted from the climatological Levitus data set (Levitus, 1982, seasonal values). This data is defined on a horizontal grid with a resolution of 1×1°. In the vertical plane, the data was available at hydrographic standard depths. The data was interpolated to the horizontal and vertical grid of the hydrodynamic model. Additionally, for the open boundary the data was interpolated in time. Daily values were used for the calculation of the dynamic heights.

Air temperature, humidity, cloud cover, precipitation, evaporation and wind speed data were used to calculate the sea surface heat fluxes; these originated from the European Centre for Medium Range Weather Forecast database (ECMWF, 1996). From the same source air pressure and wind stress data were used to force the model. The ECMWF reanalysed data covers a time period from 1980 to 1993 with a time step of 6 h and a horizontal resolution of approximately 100 km. Fig. 2 shows exemplary the evolution of air pressure and air temperature of this data set for special locations in the Bohai Sea. Both parameters show a distinguished seasonal signal. The time series show very clearly that spatial differences in the data are small especially for the air pressure evolution. Land/sea differences are obvious for the air temperature with more pronounced minima and maxima of the land nodes. However, differences between land and sea seemed not so distinguished that a special treatment of the data got necessary. A qualitative check of the data made obvious that the small area of the Bohai Sea is mainly land influenced which is not a consequence of the coarse resolution of the ECMWF grid but a meteorological fact due to the small size of the Bohai Sea compared to its surrounding land masses. Of course, a finer grid for the atmospheric data set would be desirable, which also resolves locally caused, spatial differences in the parameters but up to now it must be stated that the ECMWF database is the finest grid available for longer time scales.

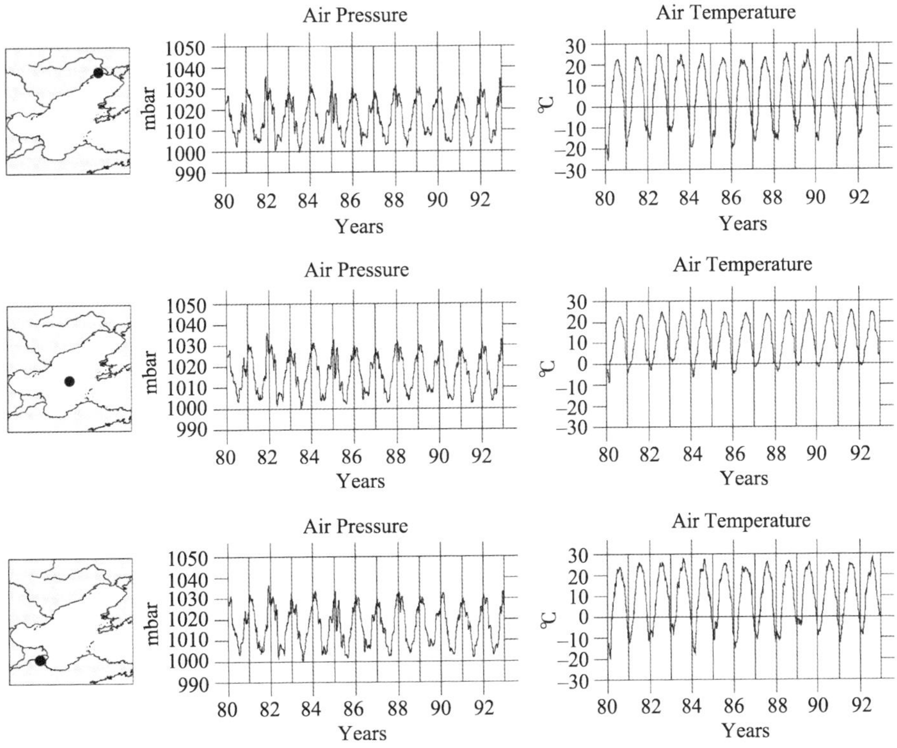

Fig. 2 Time series from 1980 to 1993 for three special locations in the Bohai Sea of air pressure and air temperature. The data source is the ECMWF (ECMWF, 1996) database. The data are filtered with a cut-off period of 30 days

Climatological monthly means of river run-off for three sources (Huang He River, Hai He River and Liao He River, Fig. 1) were taken from the Marine Atlas of Bohai Sea, Yellow Sea, East China Sea (Chen *et al.*, 1992).

The model was run with a time step of 10 min for a time period of 14 years (1980–1993) but the model output consists of daily values with the data averaged over two tidal cycles (M_2-tide). The daily means thus include the information of the tidal residual currents. In the calculations of the mean, the Stokes drift is taken into account (Wei *et al.*, 2004). For the barotropic tidal run, the data was stored hourly.

3. Results

The following description of our results concentrates on the analysis of fluctuations which are significantly longer than the tidal signal. The tides in the Bohai Sea have already been studied intensively. Furthermore, with the HAMSOM model, tidal simulations were already carried out and published (Huang, 1995). However, an additional barotropic tidal run had been performed for validation purposes.

3.1 Barotropic tides

Although the tides are only briefly discussed in this publication, they have been incorporated in all the simulations carried out. The influence of the tidal residual on the density or wind-induced flow is by all means not negligible for the Bohai Sea (Fang, 1994).

Fig. 3 shows the results for the additional barotropic tidal run. The co-tidal and co-range lines for the M_2-tide fit well with observations (Chen *et al.*, 1992) and the results of Huang

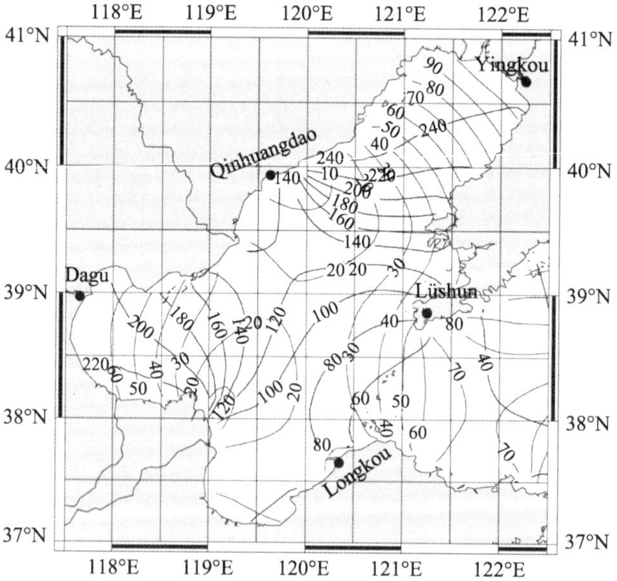

Fig. 3 Co-tidal lines (degrees) and co-range lines (cm) for the M_2-tide resulting from the barotropic tidal run

(1995). There are two amphidromic points in the Bohai Sea: one at the offshore area of Qinhuangdao and one near the mouth of the Huang He River. This agreement holds true also for the other four tidal constituents not shown here. The types of tides are mixed while the semi-diurnal signal predominates in most parts of the Bohai Sea. Also, this result agrees well with other publications (Fang, 1994).

3.2 Seasonal means

Monthly means of the sea surface elevation, of the flow field and of temperature and salinity have been calculated by averaging the data appropriately over the 14-year period.

Fig. 4a, b show the long-term means of the circulation for December and July represent ing the winter and summer season. The data is vertical integrated over the upper 15 m for reproducing the surface flow and integrated from 15 m to the bottom subsuming the bottom flow. This graduation fits with the mean location of the thermocline quite well (Fig. 5).

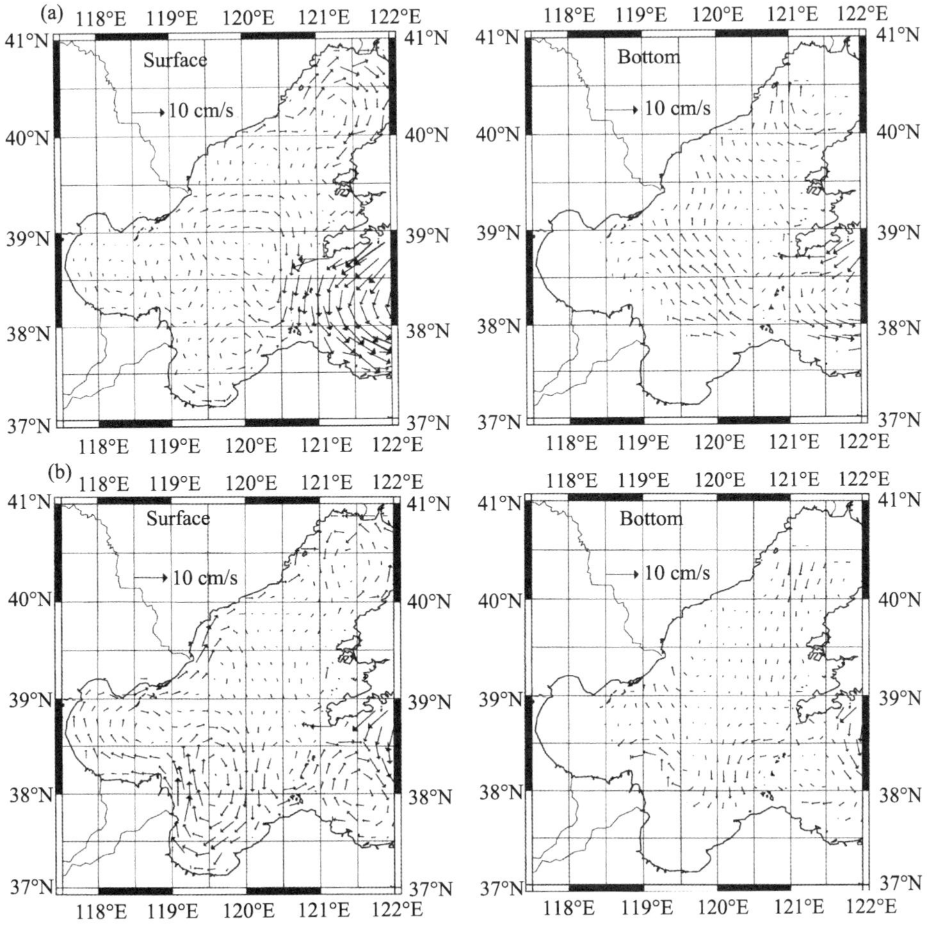

Fig. 4 Monthly mean flow fields

(a) December, (b) July. Left panel: surface flow (0–15 m). Right panel: bottom flow (15 m–bottom)

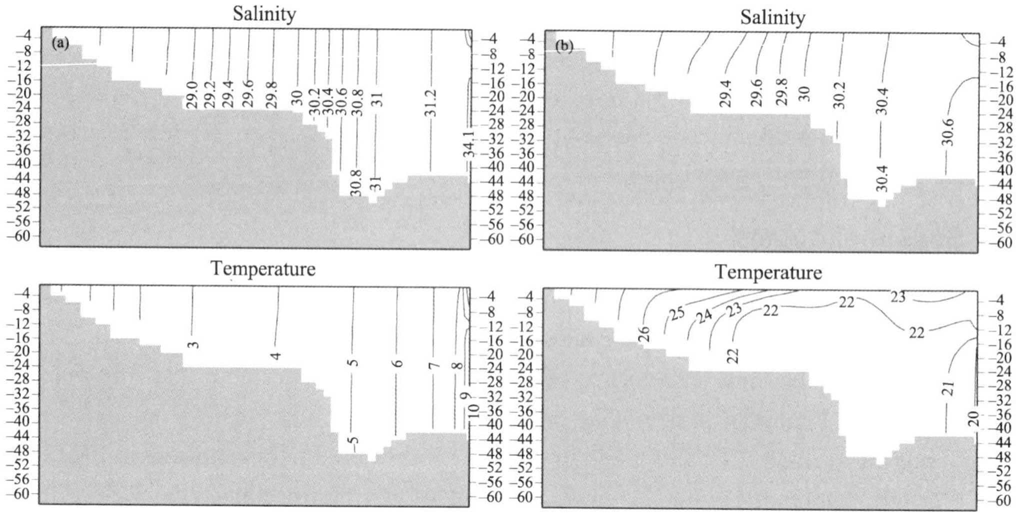

Fig. 5 Monthly mean distribution of temperature and salinity of a west-east section through the central Bohai Sea

Left side is located in the Bohai Bay; right side is located in the Bohai Strait. Upper panel: Salinity (psu), lower panel: temperature (°C). (a) December mean; (b) July mean

The summer and winter circulation distinguish obviously from each other. In winter, the circulation is strongest in the Bohai Strait region and east of it, presenting a close cyclonic gyre reaching from the surface to the bottom with few connections to the rest of the Bohai Sea except of the northern end of the Bohai Strait where some inflow to the Bohai Sea can be found. The surface flow during winter in the Bohai Sea itself is dominated by several small cyclonic and anticyclonic eddies, the main of these located in the three bays and the central Bohai Sea. The eddies of the bottom layer are modified by a northwestward flow. In summer, the cyclonic gyre east of the Bohai Strait has changed to a less pronounced anticyclonic gyre except of the northeast edge of the model area where a direct inflow to the Bohai Sea is still obtained. The eddies in the Bohai Sea itself are now stronger than in winter but changed position and partly direction. The central eddy is missing, whereas the eddy in the Laizhou Bay is now as pronounced as the eddy east of the Bohai Strait. There has also established a coastal current along the southern and western coastlines of the Bohai Sea. The eddies of the bottom regime in summer are modified by a mainly southward directed flow.

The development of the thermohaline distribution is exemplary presented by a section through the central Bohai Sea showing the vertical temperature and salinity profiles. The section starts in the west (Bohai Bay) and ends in the east (Bohai Strait). Again, the winter season (Fig. 5a) is represented by the mean distribution of December and the summer season (Fig. 5b) by the mean of July. In winter, the whole water column through all parts of the Bohai Sea is well mixed with warmest and most saline values to be found in the oceanic influenced region of the Bohai Strait and coldest and less saline values in the bay areas. An exception is the Bohai Strait area. Here weak vertical stratification exists. The situation changes in summer. The thermocline now exceeds from the Bohai Strait to the central Bohai Sea with a maximum vertical expansion of about 15 m. In contrast, the bay areas maintain the well mixed vertical structure. Altogether, the temperatures are higher now in the coastal areas and decline towards the oceanic Bohai Strait. For salinity only weak stratification developed.

3.3 Long-term fluctuations

The data of the 14 years run from 1980 to 1993 were analysed by means of time series and EOFs. The EOF analyses have been carried out for each year separately as well as over the whole simulation period. However, the timeframe of 14 years is too short to account for climatic signals and this is revealed in the 14 year EOF analysis which shows no significant variances. For this reason, a presentation of this EOF analysis has been skipped and only the year-by-year analyses have been interpreted in the following.

Due to the atmospheric forcing the model results reveal a pronounced annual cycle (Fig. 6). The time series of the sea surface elevation (SSE), the sea surface temperature (SST) and the sea surface salinity (SSS) for three selected locations in the Bohai Sea show only weak year-to-year differences and also the spatial differences between these locations are small. The SSE ranges between +40 cm and –40 cm with highest values in the middle of the year due to the evolution of the air pressure (Fig. 2). The SST reaches extreme minima of values lower 0 °C for the shallow bights where also the air temperature is lowest compared to the centre of the Bohai Sea. The evolution of salinity shows a decrease of values in the first years

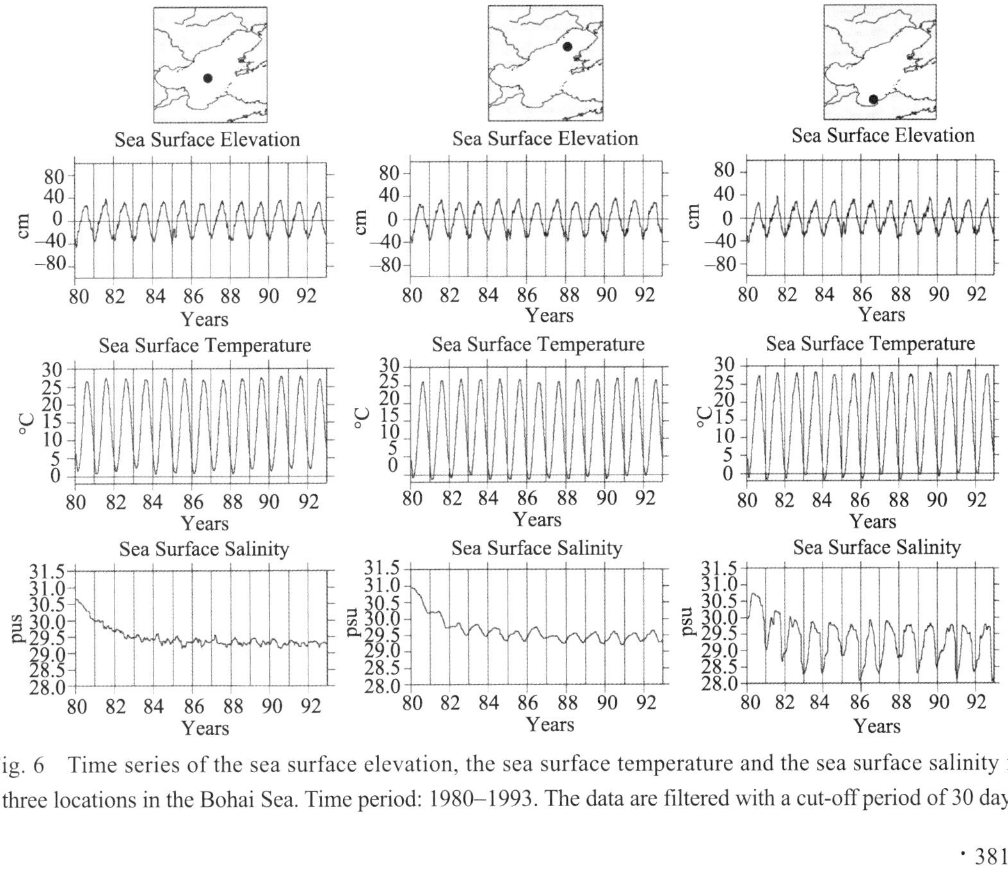

Fig. 6 Time series of the sea surface elevation, the sea surface temperature and the sea surface salinity for three locations in the Bohai Sea. Time period: 1980–1993. The data are filtered with a cut-off period of 30 days

of the simulation period all over the model area. This has an artificial reason. The Levitus data used for initialising the model are too coarse with a spatial resolution of 1×1° and due to the averaging procedure the deeper ocean parts are inappropriately influencing the shallow, coastal areas. The influence of the river runoff during the simulation leads then to the decrease of salinity until the values fluctuate around a mean value. The difference between minima and maxima are most pronounced for the third, most southerly location in the Laizhou Bay. This location is as well close to the mouth of the Huang He River and close to the oceanic boundary. So the gradients are bigger here.

Fig. 7 shows the first EOF and its associated principal component and explained variance of the SST for the years 1980, 1987 and 1990. These years are presented just as examples andare representative for the whole simulation period. It is striking that for all years the average explained variance of the first EOF is approximately 98%. Lower values only exist at the model boundary in the Bohai Strait. The associated principal component can be interpreted as the annual cycle with a maximum value during summer and a minimum value during winter. The pattern of the EOF anomaly reveals highest values in the three bights and lowest values in the Bohai Strait which can be explained by the fact that in the shallow areas the annual temperature differences are highest, whereas the Bohai Strait is the area which is compared to the rest of the Bohai Sea under strongest oceanic influence. This result agrees with observations (Chen *et al.*, 1992). Altogether, it can be concluded that the evolution of the SST in the Bohai Sea can be described by the annual cycle almost completely. This holds true also for the temperature distribution in deeper layers of the Bohai Sea, as most parts are vertically well mixed throughout the whole year, except of the central Bohai Sea area and the Bohai Strait where thermal stratification occurs in summer.

The evolution of the SSS is not thus clear (Fig. 8). However, more than 90% of the original signal is reproduced by the first and second EOF, and this holds also true for the first simulation years which are under control of unrealistic high oceanic influence (initial Levitus data). The associated principal components also reveal an annual cycle. For the first EOF a minimum can be found at the end of the summertime and the anomaly pattern shows the strongest gradients in the area of the mouth of the Huang He River. This may be related to the annual cycle of the river run-off where the maximum values occur in August (for all three run-off sources). For the second EOF the salinity reaches its minimum approximately in July and may be caused by the overall annual distribution of salinity as observed (Chen *et al.*, 1992) and prescribed by the eastern open boundary of the model. As in the case of the SST the SSS distribution is representative for the whole water column. Haline stratification is only low throughout the year.

The original SSE signal is also reproduced by the first and second EOF with an average of the explained variance of more than 95% for both EOFs together (Fig. 9). The principal component of the first EOF reveals the typical annual cycle with a maximum value during summer caused mainly by the air pressure forcing. Although the average of the explained variance for the first EOF lays between approximately 55% and 70% (for all simulated years), a value of more than 90% is always present for the central Bohai Sea. The explained variance decreases in direction to the coastlines. Also, the gradients of the anomaly pattern are small. The second EOF with an explained variance between 20% and 40% for all simulated years does not show any annual cyclic behaviour. The values of the principal component fluctuate around a mean with periods of the order of days or weeks, respectively. In contrast to the first EOF, the explained variance increases to the coastlines where also the gradients of the anomaly pattern are highest. This variability may be related to stochastic weather events. The closer to the coastlines and the shallower the region, the more dominant are the stochastic weather fluctuations.

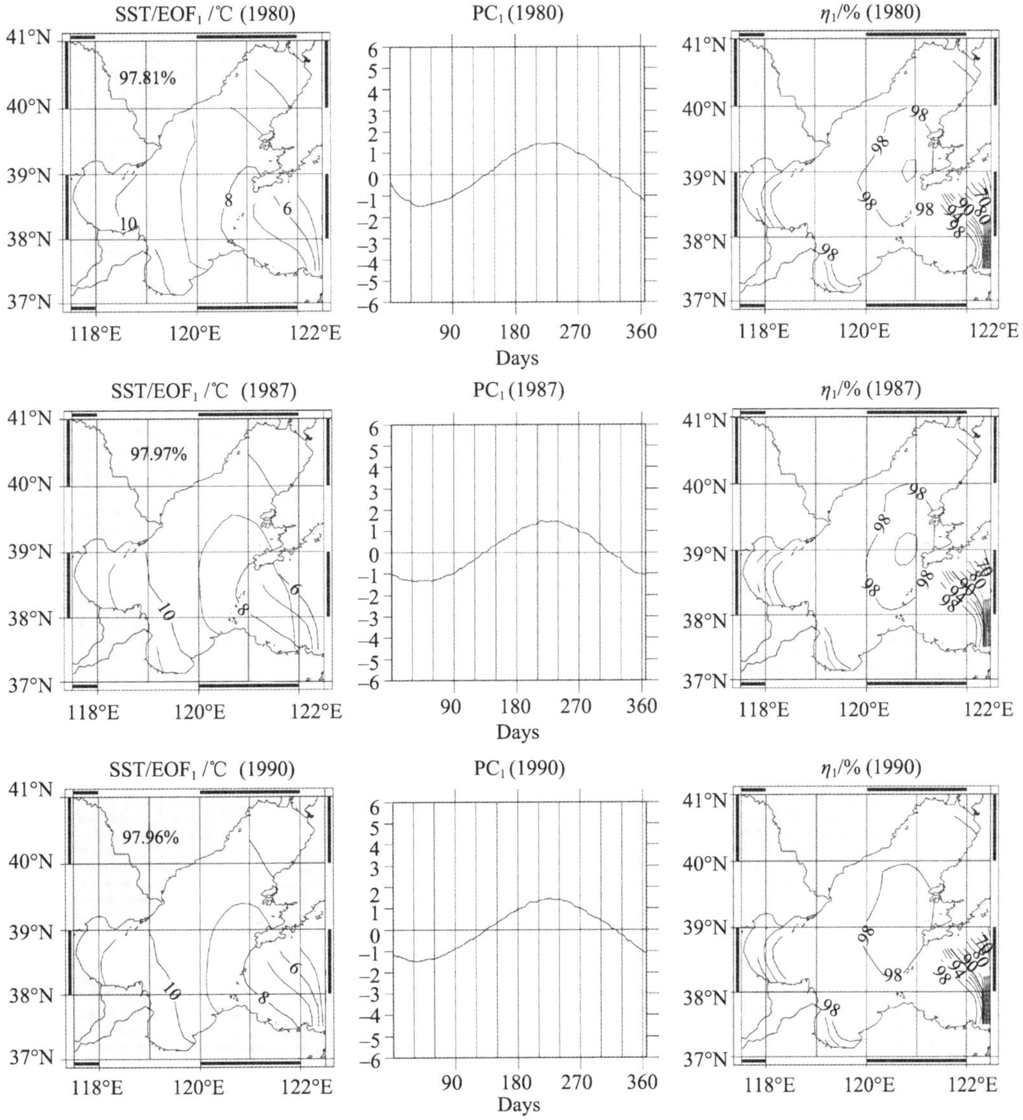

Fig. 7　EOF_1 of the sea surface temperature

From left to right: Pattern of the SST anomaly, principal component, explained variance. Number in the top left of the SST anomaly distribution: average of the explained variance. From top to bottom: EOF_1 for 1980, EOF_1 for 1987 and EOF_1 for 1990

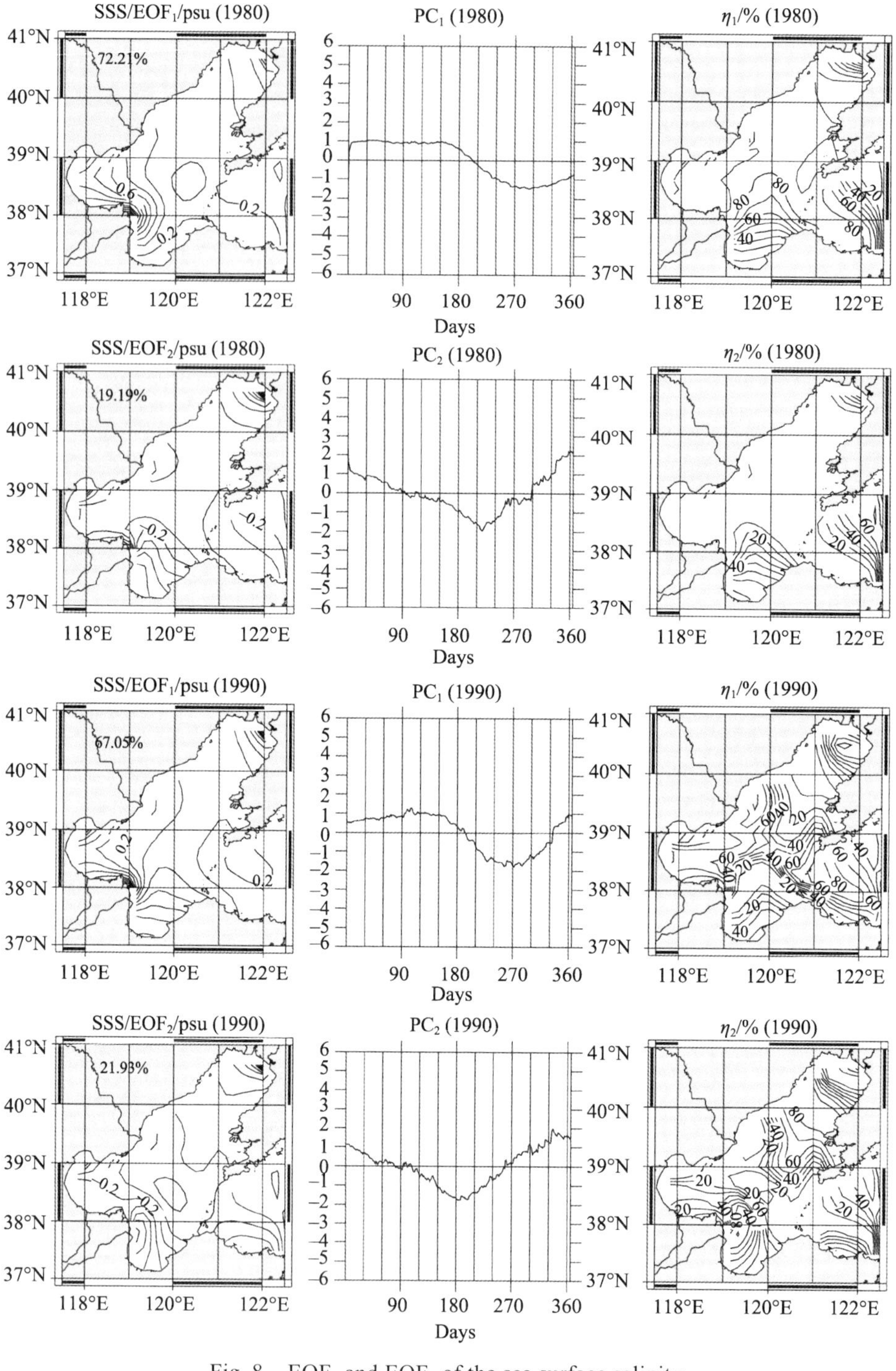

Fig. 8 EOF_1 and EOF_2 of the sea surface salinity

From left to right: Pattern of the SSS anomaly, principal component, explained variance. Number in the top left of the SSS anomaly distribution: average of the explained variance. From top to bottom: EOF_1 for 1980, EOF_2 for 1980 and EOF_1 for 1990, EOF_2 for 1990

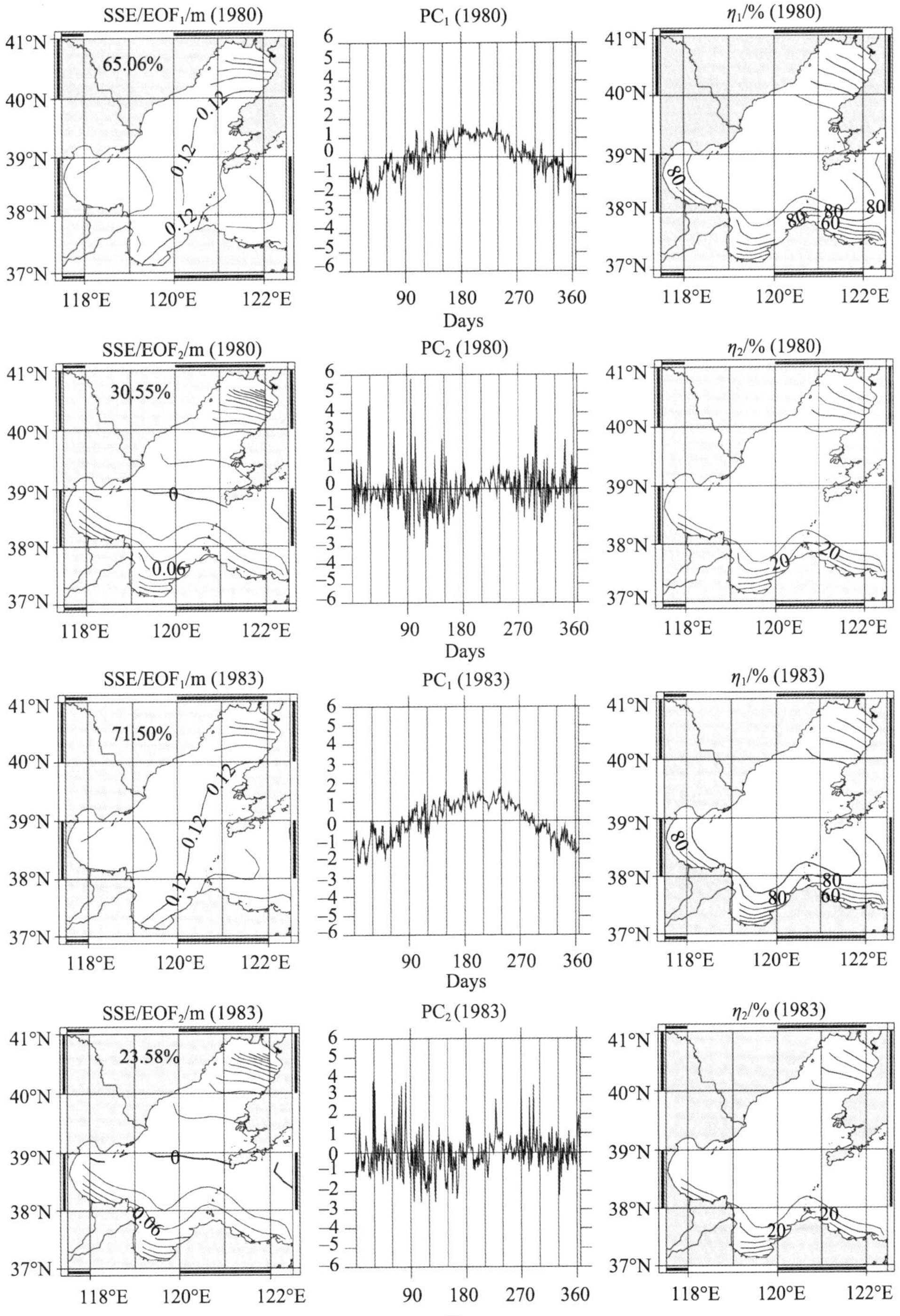

Fig. 9 EOF_1 and EOF_2 of the sea surface elevation

From left to right: Pattern of the SSE anomaly, principal component, explained variance. Number in the top left of the SSE anomaly distribution: average of the explained variance. From top to bottom: EOF_1 for 1980, EOF_2 for 1980 and EOF_1 for 1983, EOF_2 for 1983

3.4 Water exchange properties

In the frame of the project AMBOS/AMREB it is of special interest for the chemical and biological investigations to determine time scales for the renewal of water masses and to find out pathways which water masses may take to leave the Bohai Sea. For this investigation, three different methods had been utilised.

First, turnover times for different boxes in the Bohai Sea (Fig. 1) have been calculated. The turnover times have been defined by determining the time interval which the flow across the boundaries of a box needs to equal the volume of the box (Lenhart and Pohlmann, 1997).

Second, flushing times for the same boxes have been determined using the following procedure:

- The boxes were initialised with a 100% concentration of a fictive, passive tracer.
- The same Eulerian advection algorithm as for the calculation of temperature and salinity was applied using as input data the results of the 14 years simulation.
- The flushing time of a box was then defined as the time which is needed to reduce the concentration of the box to a value of 50% (Luff and Pohlmann, 1996).

Third, trajectories were calculated back in time in order to get some information from where water masses at special locations in the Bohai Sea may origin. The special locations were chosen at positions where measurements had been taken during the two cruises of AMBOS/AMREB.

Fig. 10 shows the turnover times for six boxes of the Bohai Sea for the year 1990 which has been chosen as an example but is representative also for the other years of the simulation period. The most remarkable outcome here is that the turnover times in late spring, early summer, respectively, are shorter than the turnover times of the other seasons. The volumes of the boxes are not equal compared to each other. But by dividing the turnover time by the corresponding volume it gets obvious that the renewal of water in the central Bohai Sea (box C) and the Bohai Strait (box D) are the fastest. The turnover times of the three bights (box A, box B, box F) are more or less of the same order. The turnover times for the special area boxes range between about half a month and 4 months. The whole Bohai Sea needs almost half a year for a renewal of water masses.

The calculation of flushing times was done for each year of the simulation period and for each box shown in Fig. 1. For each year, two cases have been defined, a winter case in which the concentration was initialised in January and a summer case with May as starting month. In case of box A (Figs. 11–13), shown here as an example, a relatively long time (about 240 days) is needed to reduce the total concentration in this box to 50% during the years 1983 (Fig. 11) and 1990 (Fig. 13). The month to month patterns of the dispersion (starting in January as month 1) look also similar for these two years. In the year 1987 (Fig. 12), on the contrary, the flushing is much faster, about 120 days and the exchange is stronger focused on the eastern side of the bight. It seems reasonable to conclude that changes in the

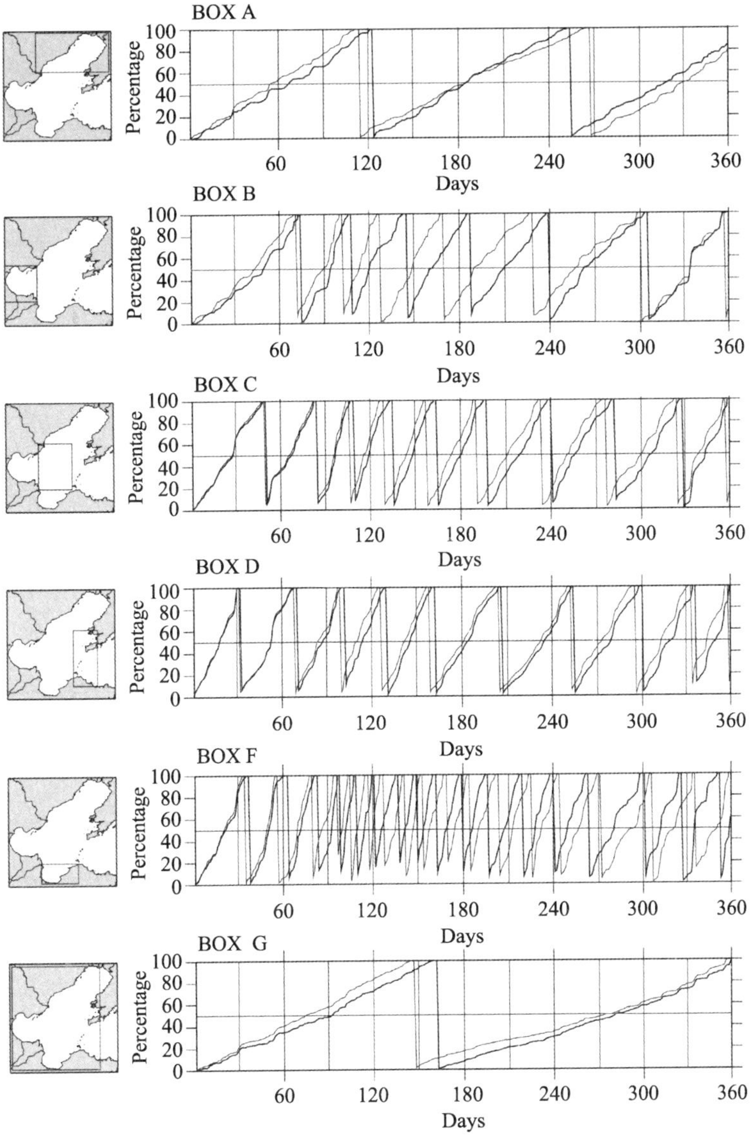

Fig. 10 Times which are needed to fill up the volume of boxes A, B, C, D, F and of the whole Bohai Sea in the year 1990 (turnover times)

Dark line: Turnover times calculated with the inflow through the box boundaries. Grey line: Turnover times calculated with the outflow through the box boundaries

flow field due to the stochastic weather fluctuations with periods of days to months cause these differences. Consequently, different patterns and flushing times are achieved for the same box when the dispersion is started on other months or years. The dispersion calculations for the other boxes reveal a similar variability due to the dependency of the flow field on the stochastic weather events.

In 1998 and 1999, two cruises in the Bohai Sea took place within the frame of the two

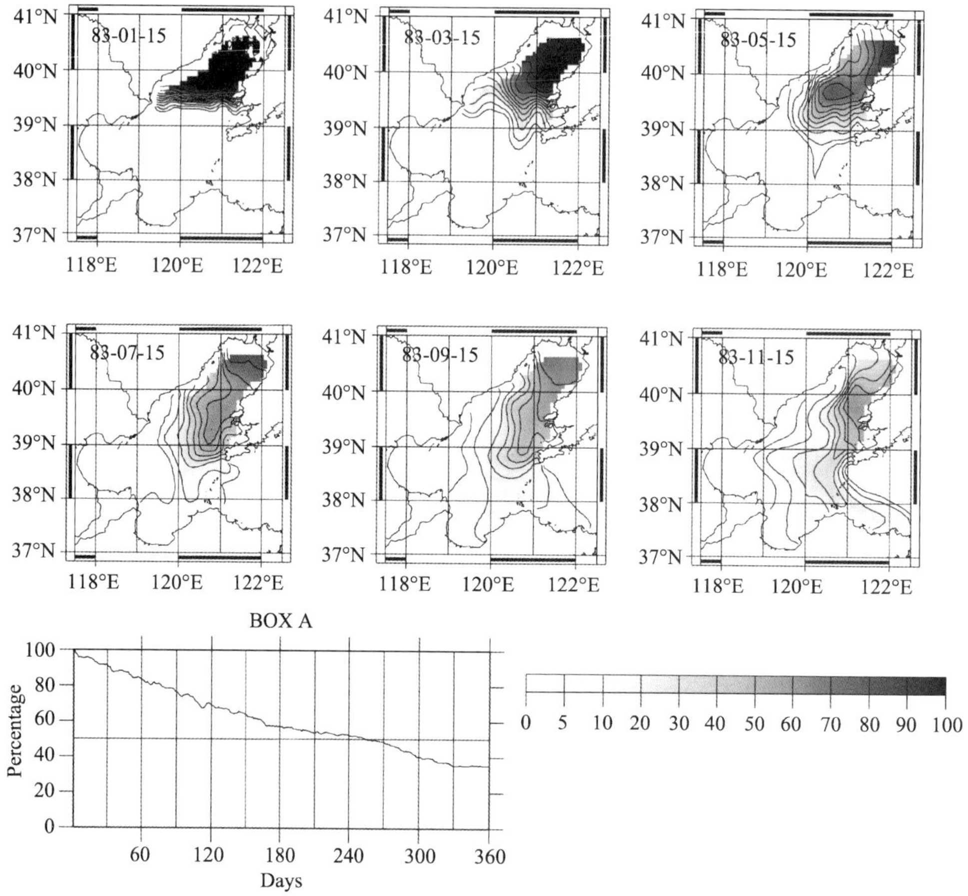

Fig. 11 Flushing time of box A for the year 1983

Above: Evolution of the concentration distribution, starting with January (month 1, day 15) and ending with November (month 11, day 15). Shown are the vertical integrated concentration patterns. Values given in %. Below: Evolution of the concentration integrated over the whole box volume for the year 1983 (day 1 to day 360)

projects AMBOS/AMREB. Unfortunately, both years are still not included in the ECMWF data set and therefore no simulations of these years have been carried out. However, at some locations of the station grid trajectories were calculated back in time for a time interval of 5 month for all years of the simulation period in order to conclude on the possible origin of the water masses at these locations. The trajectories were started during the month (May or September) in which the cruises took place. A selection of these calculations is shown in Fig. 14. It is obvious that pathways are not really predictable. For locations 1 and 3, the trajectories at least seem to follow up one direction even if the total length differs from year to year enormously. But for location 2, even the direction is strongly variable. This confirms the conclusion that for the evolution of the flow field in this shallow area and hence, for the water exchange high frequent weather fluctuations are dominant, whereas the annual cycle is of minor importance.

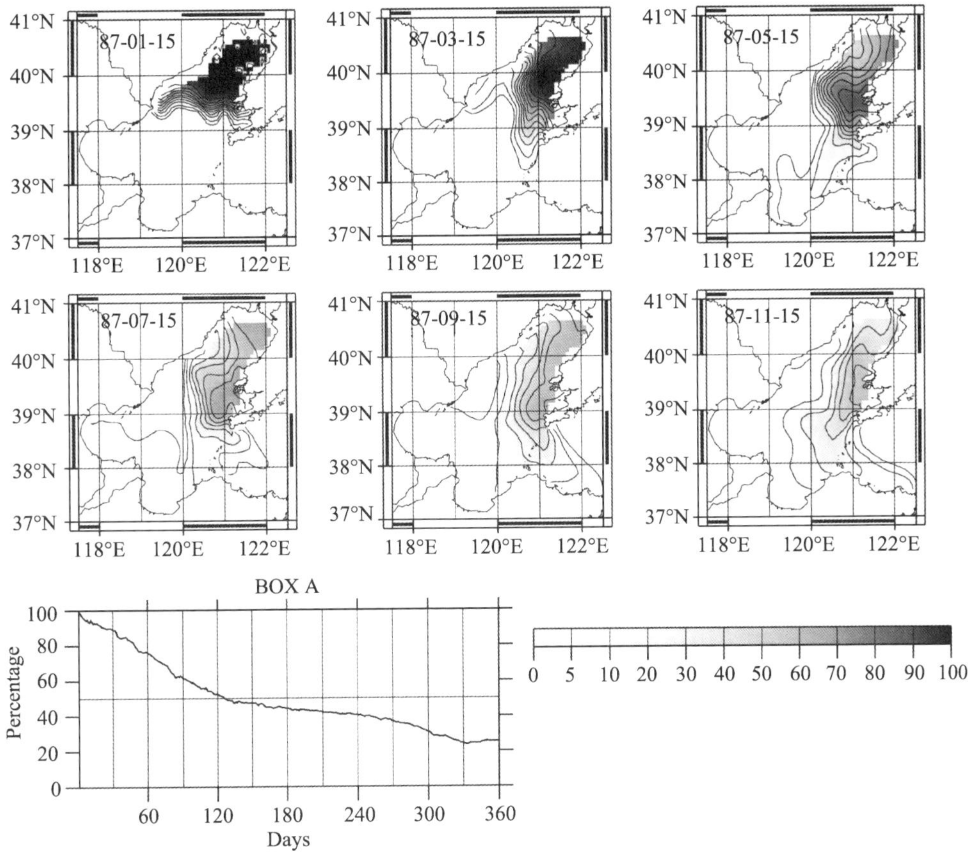

Fig. 12 Flushing time of box A for the year 1987

Above: Evolution of the concentration distribution, starting with January (month 1, day 15) and ending with November (month 11, day 15). Shown are the vertical integrated concentration patterns. Values given in %. Below: Evolution of the concentration integrated over the whole box volume for the year 1987 (day 1 to day 360)

4. Discussion and Conclusion

The Bohai Sea is a typical shallow shelf sea where the tidal, density and wind induced flows interact strongly with each other. The tides are mixed, mainly semi-diurnal and the residual tidal currents are generally rather weak (below 5 cm/s). Nevertheless, tidal turbulent mixing is influencing thermal and haline stratification as well residual currents are not negligible even if small. In winter, the whole Bohai Sea is vertically well mixed and in summer thermal stratification can be found, especially in the central Bohai Sea and the Bohai Strait. The temperature signal is strongly dominated by an annual cycle with sea surface temperatures higher than air temperatures in winter and vice versa in summer. The atmospheric forcing, still under Monsoon influence in this region, shows also a pronounced annual signal. The winds in spring and summer are prevailing north or northwest, while in summer and autumn the dominant direction is south.

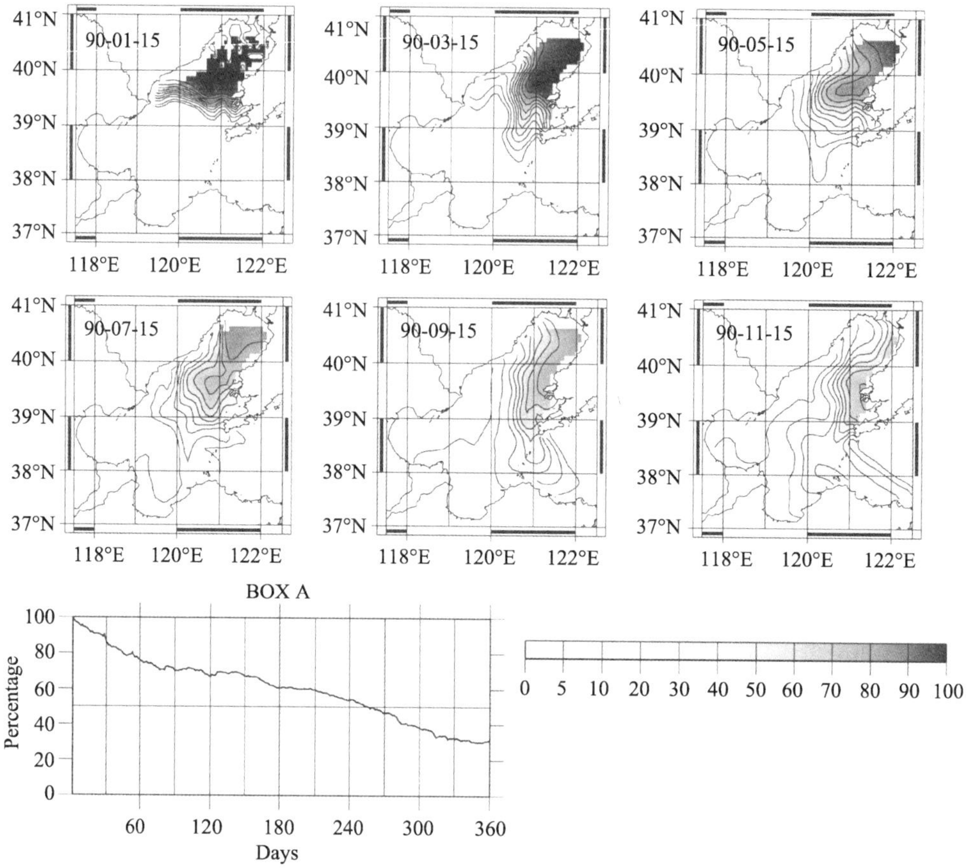

Fig. 13 Flushing time of box A for the year 1990

Above: Evolution of the concentration distribution, starting with January (month 1, day 15) and ending with November (month 11, day 15). Shown are the vertical integrated concentration patterns. Values given in %. Below: Evolution of the concentration integrated over the whole box volume for the year 1990 (day 1 to day 360)

A validation of the model was carried out for the tides by comparing the model results with the findings of Huang (1995), who carefully checked his results against observational data. Unfortunately, no comparison with other data was possible because of the failure to obtain observational data which would be suitable for this task. But in a qualitative manner, the results of the 14 years HAMSOM run reproduce the hydrography and the flow field of the Bohai Sea reasonably well.

The evolution of the temperature, of the salinity and of the sea surface elevation investigated by means of time series and EOFs reveal the typical pronounced annual cycle. Especially, the temperature is strongly dominated by it. Spatial horizontal fluctuations of these parameters are small but this might be caused by the coarse resolution of the atmospheric forcing and the initial temperature and salinity distribution. The vertical distribution of the hydrographic parameters is also reproduced reasonably by the model. In winter, the water column is well mixed, whereas in summer some thermal stratification exists in the central

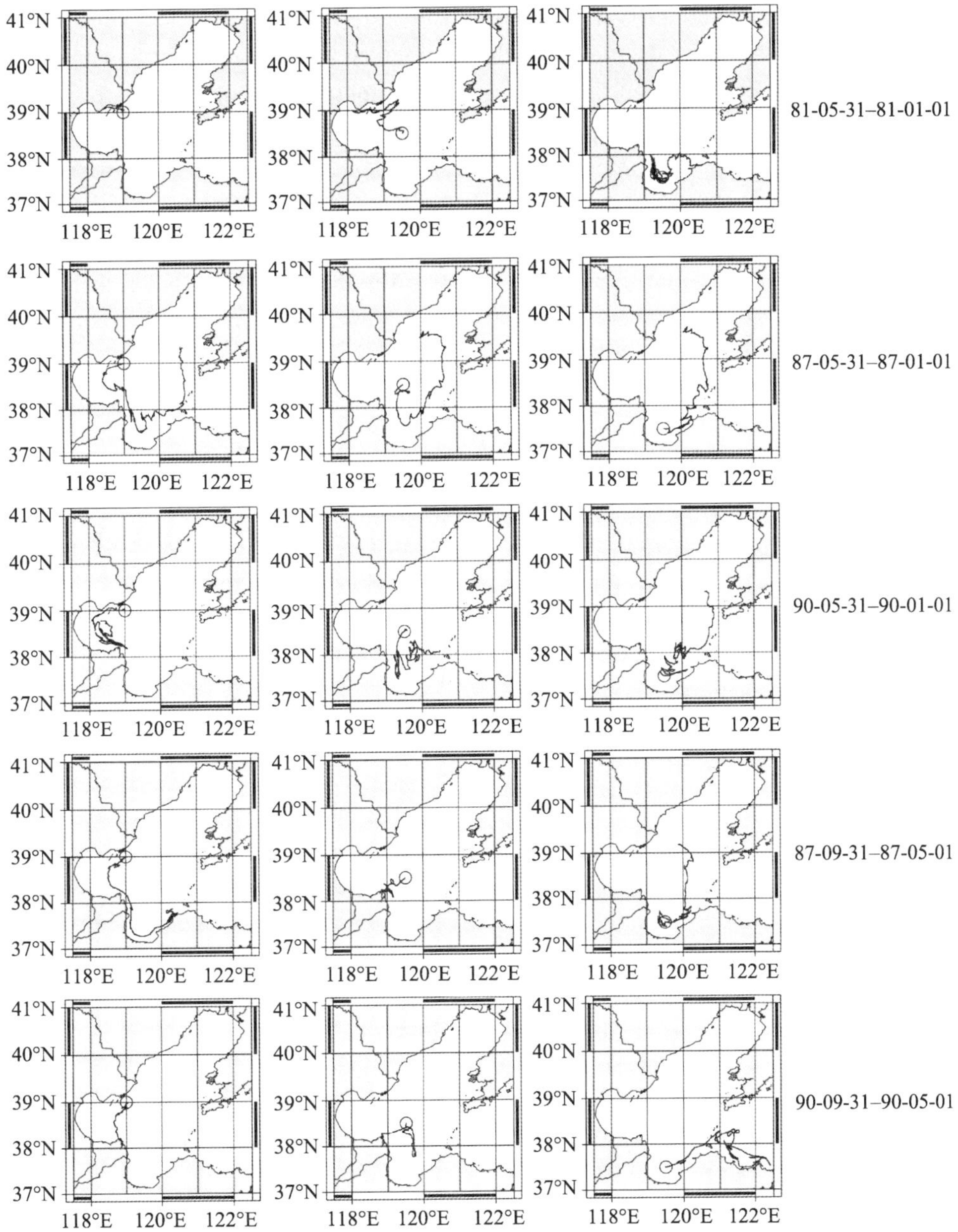

Fig. 14 Trajectories calculated back in time for three special locations in the Bohai Sea
The circle indicates the start position. From top to bottom: five different time intervals

Bohai Sea and the Bohai Strait. Although, the distribution of salinity and its seasonal evolution seem quantitatively to be correct, the results are not totally satisfying for two reasons. First, the boundary and initial values are taken from the Levitus database, which are overestimating the oceanic influence due to its coarse resolution; and in the case of salinity this causes unnatural high values especially at the beginning of the simulation. Second, the river runoff data used for the time-dependent simulation are climatological monthly mean values. The conclusion "the better the input data the better the model results" is not new but also valid for this investigation.

Water exchange properties, represented here by parameters like turnover or flushing times, are dominated by the flow field characteristics. In a shallow sea like the Bohai Sea, high frequent weather fluctuations in time scales of days to month prevail the circulation and consequently the water exchange. The annual cycle here is of less importance. The results show that representative turnover or flushing times for special areas of the Bohai Sea are hard to predict, the difference can be of the order of 50% for the same area but a different time window. Furthermore, pathways are also varying. Altogether, the turnover time for the whole Bohai Sea is in the range between half and 1 year. A statistical analysis of how different wind scenarios are influencing the flow field may give hints on favoured flushing times, turnover times and pathways, respectively, for specific areas in the Bohai Sea. But due to the high fluctuations, the most reliable way to determine water exchange properties in the Bohai Sea still remains to carry out simulations for the specific time range under consideration.

From the 40-years investigation of the hydrology of the Bohai Sea, it can be assumed that the Bohai Sea is also controlled by climatic weather fluctuations and climatic fluctuations resulting from changes of the hydrology in the Yellow Sea. These effects were not considered in the simulations. Except of the climatological Levitus temperature and salinity data, there were no further data sets available which could serve as boundary values at the transition zone between the Bohai Sea and Yellow Sea. This means that in this study fluctuations in the flow field and hydrography of the Bohai Sea simulations are controlled mainly by the atmosphere. The time series of the atmospheric forcing data are on the other hand too short for climatological investigations.

But supposed that these data sets will be available in future these simulations may be a beginning for extended climatic investigations. Fishing is an important economic aspect for the inhabitants around the Bohai Sea. Climatic changes may effect its ecosystem and consequently the availability of fish stocks in these waters. Hydrodynamic model simulations like those carried out here are a basis for additional SPM transport (Jiang *et al.*, 2004) and ecosystem model investigations (Wei *et al.*, 2004b), particle tracking models for biological purposes (Hainbucher and Backhaus, 1999; Bryant *et al.*, 1998) and statistical investigations of climatic changes (Janssen *et al.*, 2001).

Acknowledgements

We like to thank our colleague Frank Janssen for supporting us with the appropriate MATLAB routines for calculating EOFs.

We are indebted to the Federal Ministry of Education and Science (BMBF), Germany, which supported the present work under contract No. 03F0189A, to the National Natural Science Foundation of China (NSFC), which supported the work under contract No. 49576298, and to the State Oceanic Administration of China (SOA).

References

Alvarez Fanjul, E., Pérez Gómez, B., Rodríguez Sánchez-Arévalo, I., 1997. A description of the tides in the Eastern North Atlantic. Prog. Oceanogr. 40, 217–244.

Arakawa, A., Lamb, V. R., 1977. Computational design of the basic dynamical processes of the UCLA general circulation model. Methods Comput. Phys. 17, 174–267.

Backhaus, J. O., 1985. A three-dimensional model for the simulation of shelf sea dynamics. Dtsch. Hydrogr. Z. 38, 165–187.

Backhaus, J. O., Hainbucher, D., 1987. A finite-difference general circulation model for shelf seas and its application to low frequency variability on the North European shelf. In: Nihoul, J. C. J., Jamart, B. M. (Eds.), Three-Dimensional Models of Marine and Estuarine Dynamics. Elsevier Oceanogr. Ser., vol. 45, pp. 221–244.

Bryant, A. D., Hainbucher, D., Heath, M., 1998. Basin-scale advection and population persistence of Calanus finmarchicus. Fish. Oceanogr. 7 (3/4), 235–244.

Carbajal, N., 1993. Modelling of the circulation in the gulf of California. Rep. Centre Mar. Clim. Res. 3, 1–186.

Chelton, D. B., de Szoehe, R. A., Schlax, M. G., El Naggar, K., Siwertz, N., 1998. Geographical variability of the first-baroclinic Rossby radius of deformation. J. Phys. Oceanogr. 28, 433–460.

Chen, G., Niu, Y., Wen, S., Bao, C., Guan, D., Wu, B., Zhang, R., Gu, H. (Eds.), 1992. Marine Atlas of Bohai Sea, Yellow Sea, East China Sea (Hydrology). China Ocean Press, Beijing. 523 pp.

Chen, L., Zhu, W., Wang, W., Zhou, X., Li, W., 1998. Studies on climate change in China in recent 45 years. Acta. Meteorol. Sin. 56 (3), 257–271.

Dou, Z., Luo, Y., Huang, K., 1981. Numerical computation of tidal current and tide-induced residual circulation of the Bohai Sea. Acta Oceanol. Sin. 1 (3), 355–369 (in Chinese).

ECMWF, 1996. The ECMWF re-analysis (ERA) project. ECMWF Newsletter 73, 7–17.

Fang, G., 1994. Tides and tidal currents in East China Sea, Huanghai Sea and Bohai Sea. In: Zhou, D., Liang, Y., Tseng, C. (Eds.), Oceanology of China Sea. Kluwer Academic Publishing, Netherland, pp. 101–112.

Hainbucher, D., Backhaus, J. O., 1999. Circulation of the eastern North Atlantic and north-west European continental shelf—a hydrodynamic modelling study. Fish. Oceanogr. Trasnl. 8 (Suppl. 1), 1–12.

Hainbucher, D., Pohlmann, T., Backhaus, J. O., 1987. Transport of conservative passive tracers in the North Sea: first results of a circulation and transport model. Cont. Shelf Res. 7, 1161–1179.

Harms, I. H., Backhaus, J. O., Hainbucher, D., 1999. Modelling the seasonal variability of circulation and hydrography in the Iceland-Faeroe-Shetland overflow area. ICES CM 1999/L: 10, Annual Science Conference, 29 September to 2 October 1999, Stockholm, Sweden.

Hirtzler, J. D. (Ed.), 1985. Relief of the Surface of the Earth. MGG, 2. National Geographic Data Center, Boulder, CO, USA.

Huang, D., 1995. Modelling studies of barotropic and baroclinic dynamics in the Bohai Sea. Rep. Centre Mar. Clim. Res. 17, 1–126.

Huang, Z., Wang, X., 1988. ADI scheme: results and analysis. J. Shandong Coll. Oceanol. 18 (2(II)), 48–49 (in Chinese).

Huang, D., Su, J., Chen, Z., 1996. The application of three-dimensional shelf sea model in the Bohai Sea: II. Seasonal variation of temperature. Acta Oceanol. Sin. 18 (6), 8–17 (in Chinese).

Janssen, F., Schrum, C., Hübner, U., Backhaus, J. O., 2001. Uncertainty Analysis of a Decadal Simulation with a Regional Ocean Model for North Sea and Baltic Sea. Climate Research. 18(1/2), 55–62.

Jiang, W., Sun, J., Starke, A., 2004. SPM transport in the Bohai Sea: field experiments and numerical modelling. Journal of Marine Systems 44(3–4), 175–188.

Kochergin, V. P., 1987. Three-dimensional prognostic models. In: Heaps, N. (Ed.), Three-Dimensional Coastal Ocean Models. Coast. Estuar. Sci., vol. 4. AGU, Washington, D. C., pp. 201–208.

Lenhart, H. J., Pohlmann, T., 1997. The ICES-boxes approach in relation to results of a North Sea circulation model. Tellus 49A, 139–160.

Levitus, S., 1982. Climatological atlas of the world ocean. NOAA Prof. Pap. 13, 922–935.

Luff, R., Pohlmann, T., 1996. Calculation of water exchange times in the ICES-boxes with an Eulerian dispersion model using a half-life time approach. Dtsch. Hydrogr. Z. 47 (4), 287–299.

Mellor, G. L., Yamada, T., 1974. A hierarchy of turbulence closure models for planetary boundary layers. J. Atmos. Sci. 31, 1791–1806.

Miao, J., Liu, X., 1989. Numerical experiment on winter circulation of the North Huanghai Sea and the Bohai Sea. Acta Oceanol. Sin. 11 (1), 11–22 (in Chinese).

Moll, A., Radach, G., 1998. Advective contributions to the heat balance of the German bight (LV Elbe1) and the Central North Sea (OWS Famita). Dtsch. Hydrogr. Z. 50, 9–31.

Pohlmann, T., 1991. Evaluations of hydro- and thermodynamic processes in the North Sea with a 3-dimensional numerical model (in German). Rep. Centre Mar. Clim. Res. 23, 1–116.

Pohlmann, T., 1996. Calculating the annual cycle of the vertical eddy viscosity in the North Sea with a three-dimensional baroclinic shelf sea circulation model. Cont. Shelf Res. 16 (2), 147–161.

Schrum, C., 1997. Thermohaline stratification and instabilities at tidal mixing fronts: results of an eddy resolving model for the German bight. Cont. Shelf Res. 17, 689–716.

Stronach, J. A., Backhaus, J. O., Murty, T. S., 1993. An update on the numerical simulation of oceanographic processes in the waters between Vancouver Island and the Mainland: the GF8 model. Oceanogr. Mar. Biol., Annu. Rev. 31, 1–87.

Sun, W., Xi, P., Song, L., 1989. Numerical calculation of the three-dimensional tide-induced Lagrangian residual circulation in the Bohai Sea. J. Oceanol. Univ. Qingdao 19 (2(I)), 1–6.

Sun, Y., Chen, S., Zhao, K., 1990. A three-dimensional baroclinic model of the coastal water. J. Oceanol. Univ. Qingdao 18 (2(II)), 1–14 (in Chinese).

Wang, H., 1988. A finite element scheme: results and analysis. J. Shadong Colle. Oceanol. 18 (2(II)), 49–50 (in Chinese).

Wei, H., Hainbucher, D., Pohlmann, T., Feng, S., Sündermann, J., 2004. Tidal-Induced Lagrangian and Eulerian Mean Circulation in the Bohai Sea. Journal of Marine Systems 44(3–4), 141–151.

Wei, H., Sun, J., Moll, A., Zhao, L., 2004. Phytoplankton dynamics in the Bohai Sea—observations and modelling. Journal of Marine Systems 44(3–4), 233–251.

Yu, K., Zhang, F., 1987. A three-dimensional numerical model of the tidal motions in the Bohai Sea. Oceanol. Limnol. Sin. 8 (3), 227–236 (in Chinese).

Zhao, J., Shi, M., 1993. Numerical modelling of three-dimension characteristics of wind-driven current in the Bohai Sea. Chin. J. Oceanol. Limnol. 11 (1), 70–79.

Zhao, B., Fang, G., Cao, M., 1994. Simulation on tides and tidal currents of the Bohai Sea, Huanghai Sea and East China Sea. Acta Oceanol. Sin. 16 (5), 1–10 (in Chinese).

Zou, H., 1998. A preliminary exploration of climatic change in the Bohai region in recent hundred years. Mar. Forecasts 5 (4), 12–17.

Analysis and Modelling of the Bohai Sea Ecosystem—a Joint German-Chinese Study*

Jürgen Sündermann, Shizuo Feng

1. Introduction

The Bohai Sea located in Northeastern China is traditionally known as "fish storehouse", "salt storehouse" and "oil storehouse". Economists call the zone around the Bohai Sea "the golden necklace" of Northern China because its economical contribution amounts to 1/10 of the gross national product of China.

At the same time, there is a serious threat due to environmental pollution. Based on data of 1998, the Bohai Sea received 0.8 billion tons of industrial sewage water, which is 20.2% of total industrial sewage water discharging into all the seas of China that year (State Ocean Administration, 2000). Furthermore, as the second largest petroliferous base, the content of oil exceeds the water quality limit in about 30% of the Bohai Sea. In 1999, 2.044 million tons of oil-contaminated water were discharged into the Bohai Sea from eight oil fields and 42.3 tons of crude oil leaked into the sea (State Ocean Administration, 2000).

Aside from oil, the major pollution elements are inorganic nitrogen, organic phosphate, and lead. The inorganic nitrogen exceeded the water quality standard (0.4 mg/l) in some parts of the Liaodong Bay (State Environment Protection Agency of China, 1999). The eutrophication is closely related to red tides which occur more frequently in recent years and spread ubiquitously. The most serious red tide recorded up to 1998 in the Bohai Sea spread more than 5000 km^2 and the economic loss ran into 0.12 billion yuan (State Environment Protection Agency of China, 1999). The ecosystem of the Bohai Sea has degraded, the biological diversity is reduced. For the conception of sustainable management strategies, it is necessary to understand the capacity and reaction potential of the marine ecosystem. The joint German-Chinese research project is a step in this direction.

Several joint studies have been carried out in the Bohai Sea after the 1950s. The first basic study of China's Seas gave only a glimpse of the Bohai Sea ecosystem (Anonymous, 1977). A comprehensive environmental study in the Bohai Sea from 1985 to 1986 was named "Research on the Prediction of Water Quality and the Physical Self-Purification Capability of the Bohai Sea and the Ten Bays in China". In order to provide the hydrographic background of ecosystem studies as a key project of the National Science Foundation of China

---

* Sündermann J, Feng S. 2004. Analysis and modelling of the Bohai Sea ecosystem—a Joint German-Chinese study. Journal of Marine Systems, 44(3–4): 127–140

(NSFC), the "Study of China Shelf Sea Circulation and Its Dynamics" was carried out from 1992 to 1995. After that, there was no comprehensive research on the dynamics of the Bohai Sea ecosystem until the beginning of the present project in 1996.

At the same time, the Centre of Marine and Climate Research, University of Hamburg, had carried out three major field studies into the North Sea ecosystem and its variability: "Circulation and Contaminant Fluxes" —ZISCH (Sündermann, 1994), "Processes in the Contaminant Cycle Sea-Atmosphere" —PRISMA (Südermann and Radach, 1997), and "Matter Fluxes in Coastal Waters" —KUSTOS (Sündermann *et al.*, 1999). In all these projects, 3D models of circulation, transport of dissolved and particulate matter and primary production have been used to analyse, interpret and extrapolate field data. Since the Bohai Sea and the North Sea are hydrographically comparable and since both are facing similar environmental threats, e.g. eutrophication and pollution, it was natural to apply the same research methodology to the Bohai Sea. Correspondingly, a joint German-Chinese project "Analysis and Modelling of the Bohai Sea Ecosystem" (AMBOS/AMREB) was established.

The main objective of the project was the qualitative and quantitative description of the mass fluxes in the Bohai Sea. In addition to annual averages, seasonal variability as a consequence of meteorological forcing, riverine inputs and light availability has also been investigated. Particular emphasis has been placed on certain parameters which are characteristic for the local and global fluxes, such as salinity (freshwater), suspended particulate matter, nutrients (N,P) and carbon. As in the abovementioned North Sea projects, the fluxes are finally deduced from numerical model results. Two ship cruises of R/V 'Dong Fang Hong 2' in autumn 1998 and in spring 1999 delivered the database for process studies and model validation. The following series of articles present first the results of model investigations on circulation, transport and phytoplankton dynamics and the comparison with data (papers by Hainbucher *et al.*, 2004, Wei *et al.*, 2004 a, b, Sun *et al.*, 2004 and Jiang *et al.*, 2004). Next, the transfer of nutrient elements is discussed based on measurements (papers by Zhang *et al.*, 2004 and Raabe *et al.*, 2004).

2. The Bohai Sea: Hydrography and State Of Knowledge

The Bohai Sea is a shallow marginal sea of the western Pacific Ocean, extending about 550 km from north to south and about 350 km from east to west (Fig. 1); its mean depth is 18 m. The sea is structured into the inner region and the Liaodong, Bohai and Laizhou bays. The coastal line of the Bohai Sea has a length of nearly 3800 km. Its area is about 77000 km^2. The Bohai Strait in the east is the only passage connecting the Bohai Sea to the outer Yellow Sea.

The Bohai Sea is subject to strong tidal currents and to atmospheric forcing by wind stress and surface fluxes of heat and salt. A salient feature is the high concentration of suspended particulate matter (SPM) from rivers, particularly Huanghe River (the Yellow River). In its hydrography, the Bohai Sea and the North Sea exhibit remarkable similarities. This is also the case with regard to the environmental problems, although these are dramatically more severe in the Bohai Sea due to the shallow depth, the narrow connection with the ocean and the high population density in the region.

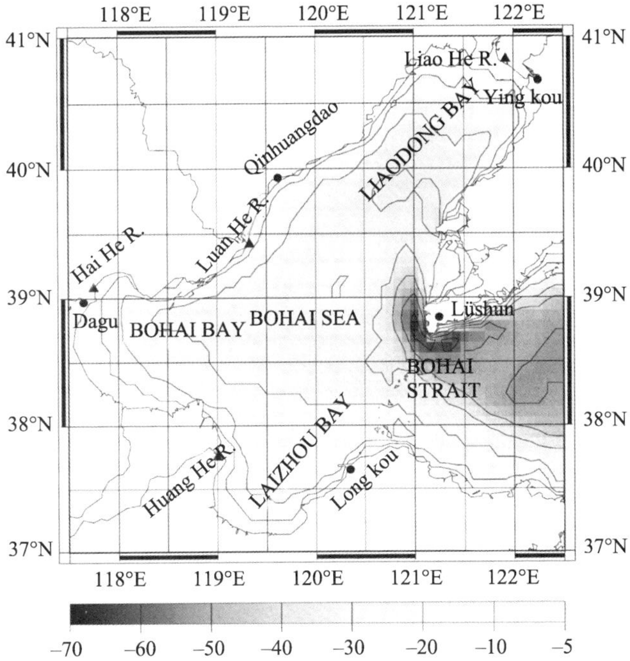

Fig. 1 Bathymetry of the Bohai Sea. The depths are in meters

The Bohai Sea is a typical region of freshwater influence (ROFI) and as such comparable with some coastal regions of the North Sea as the Rhine outflow or the German Bight. In a ROFI, the lateral input of freshwater buoyancy is equally important as the vertical input of heat buoyancy which occurs all over the shelf. Both processes lead to stratification within the water column. There is, however, the competing impact of stirring by tides, waves and wind which may destabilize the water mass and lead to strong horizontal fronts. All these processes are well known from the North Sea (Linden and Simpson, 1988; Simpson *et al.*, 1990). An appropriate indicator for frontal genesis is the Simpson-Hunter parameter (Simp son and Hunter, 1974)

$$\mathrm{SH} = \log_{10} h/u^3$$

which characterizes the relation of buoyancy and turbulent mixing. h is the water depth in m, u the mean tidal velocity in m/s. High values of SH support stable stratification. Fig. 2 shows the calculated horizontal distribution of SH in the Bohai Sea (Huang, 1995) with strong gradients near the river plumes. A comparable picture for the North Sea was calculated by Schrum (1994).

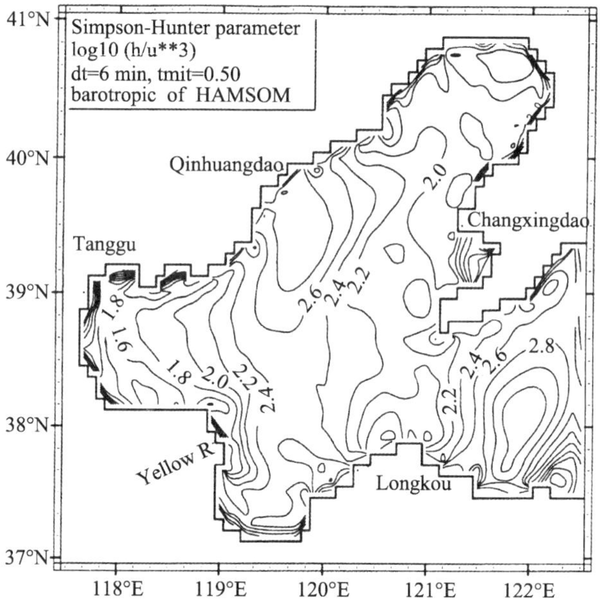

Fig. 2 Simpson-Hunter parameter in the Bohai Sea (Huang, 1995). *h* is in m, *u* in m/s

2.1 Meteorology

The East Asia Monsoon dominates the meteorology of this region. In winter, under the influence of Asian High Pressure and Aleutian Low Pressure, strong winds with a mean wind speed of 6–7 m/s frequently blow over the Bohai Sea from the north. In summer, southerly winds with a mean speed smaller than 4–6 m/s blow. The air temperature over the Bohai Sea reaches its lowest value of –4℃ to~0℃ in January and highest of 25℃ in July. Water temperature of the Bohai Sea is vertically homogeneous from November to March. A thermocline is gradually formed from April to May in the deep area of the Bohai Sea. The thermo-stratification intensifies and becomes greater than 1℃/m from June to August. Then it decreases rapidly in September. The averaged annual precipitation in the Bohai Sea is about 500–600 mm (Feng *et al.*, 1999).

2.2 Runoff

There are several rivers, such as the Yellow River (Huanghe), the Haihe River, the Luanhe River and the Liaohe River, discharging into the Bohai Sea. The total runoff is about 890×10^8 m^3 every year, among which the discharge of the Yellow River is 420×10^8 m^3.

The rivers discharge a large amount of solid matter into the sea, and sedimentation affects the estuaries, the coastal zone and the sea enormously. This effect is specifically great at the Yellow River's delta. Here each year, the coastline extends towards the sea 150–420 m and, in average, 23 km^2 land is created. It should be pointed out that in recent years, owing to the dry weather at the middle and low reaches of the Yellow River and the retaining and drawing of water from the Yellow River, the runoff of the river decreased dramatically, for example, in 1997 the lower river was dry for as many as 226 days (Feng *et al.*, 1999).

2.3 Tides and storm surges

Tides and tidal currents have been summarised by Fang (1994). The M_2 tide is the principal tidal constituent in the Bohai Sea. There are two amphidromic points, one is close to the coast near Qinhuangdao, the other off the Yellow River delta. The K_1 tide has an amphidromic point at the southern part of the Bohai. The maximum amplitudes of the $(M_2 + K_1)$ tide are about 2 m. The maximum velocity of $(M_2 + K_1)$ tidal current is about 1 m/s.

Fig. 3 shows the M_2 tidal ellipses in the surface layer as calculated by Huang (1995). There is a (left) rotating tidal flow entering the Bohai Sea from the Yellow Sea and propagating into the three inner bays almost as an alternating current.

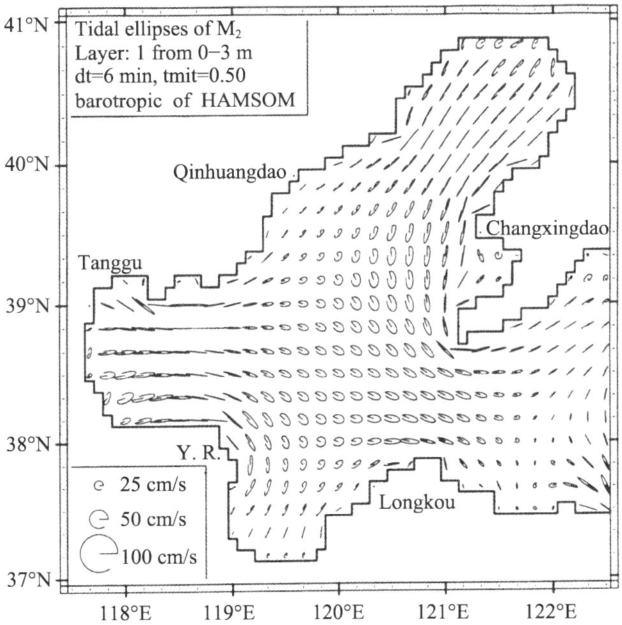

Fig. 3 Tidal ellipses of the M_2 in the surface layer (Huang, 1995)

Typhoons and cold atmospheric waves often strike the Chinese coastal zone. Then the water level rises, seawater intrudes. In summer and autumn, typhoon surges usually occur at the southeast and south Chinese coastal zones, but there are some typhoon surges in the Bohai Sea. In winter, there are strong wind surges in the Bohai Bay and the Laizhou Bay (Feng *et al.*, 1999).

2.4 Water masses, circulation and transport

In the Bohai Sea, five water masses are recognised. They are the Bohai Sea and the Yellow

Sea Mixed Water, the North Yellow Sea Bottom Cold Water, the Bohai Sea Central Water, the Bohai Sea Coastal Water and the Continental Diluted Water. Annual variations of temperature and salinity of three of these water masses are shown in Fig. 4 (Su and Weng, 1994). There is a strong seasonal signal in temperature with a smaller amplitude in the bottom cold water. Salinity exhibits a clear seasonality only for the Bohai Sea/Yellow Sea mixed water. In winter, the distribution of water masses at the sea surface and the bottom layer are basically identical (Fig. 5) (Su and Weng, 1994). Gradual modification occurs from November to April at the sea surface and from December to March at the bottom layer. Distributions of the water masses at the sea surface and the bottom layers in spring and autumn are transitional and patchy.

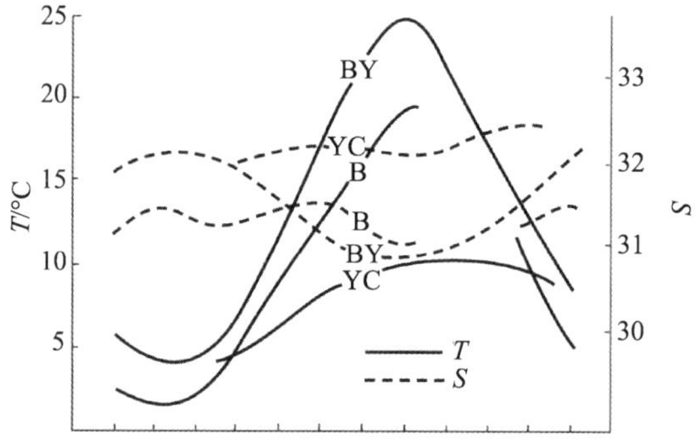

Fig. 4 Annual variations of temperature and salinity in the Bohai Sea and its subregions (Su and Weng, 1994)
B: Bohai Sea central water; BY: Bohai Sea and Yellow Sea mixed water; YC: North Yellow Sea bottom cold water

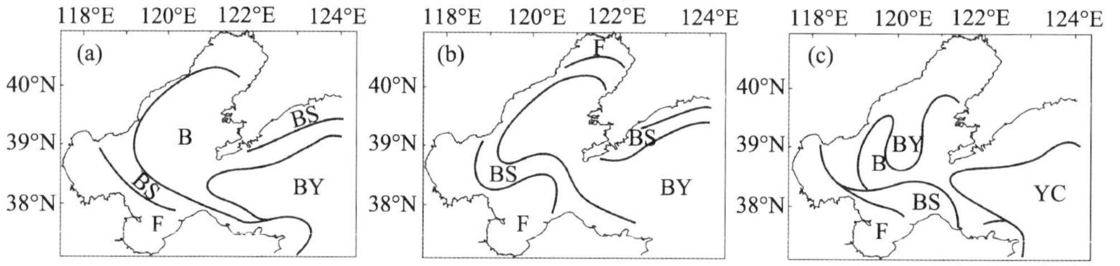

Fig. 5 Bohai Sea water masses at the sea surface and at the bottom (Su and Weng, 1994)
(a) winter, surface; (b) winter, bottom; (c) summer, bottom; B: Bohai Sea central water; BS: Bohai Sea coastal water; BY: Bohai Sea and Yellow Sea mixed water; F: Continental diluted water; YC: North Yellow Sea bottom cold water

Based upon the data from "Chinese National Comprehensive Oceanographic Survey (1959–1960)", Guan (1994) analysed the currents and the distributions of temperature and salinity and obtained the circulation pattern in the Bohai Sea (Fig. 6). The "Yellow Sea Warm Current Extension", like a jet, enters the Bohai Sea through the Bohai Strait, and moves westward along the central part until it meets the coast and there splits into two branches. One branch moves towards the Liaodong Bay, forming a clockwise gyre, and the other moves towards the Bohai Bay, forming a counterclockwise gyre.

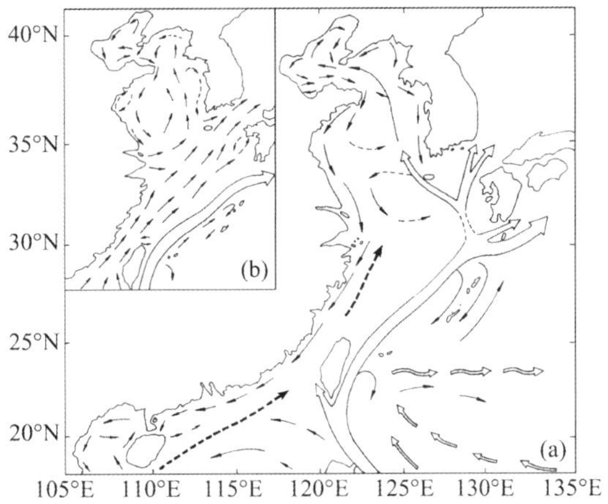

Fig. 6 Schematic diagram of the circulation in the Bohai Sea, Yellow Sea and East China Sea (Guan, 1994)
(a) winter; (b) summer

There are many descriptive, analytical and dynamic-numerical studies on the circulation in the Bohai Sea (for example, Dou *et al.*, 1981; Huang *et al.*, 1996; Jiang *et al.*, 1997; Shi and Du, 1994; Sun *et al.*, 1989; Wang *et al.*, 1993). The tide-induced residual circulation is an important component of the circulation in the Bohai Sea. Huang (1995) calculated the tidal residual currents using the five major tidal constituents. Generally, this current field is weaker than the wind-driven circulation. There are some regions in the Bohai Sea, however, where the tidal residuals are significant for the mean flow. Examples are the western boundaries, the whole Liaodong Bay and the northern Bohai Strait where a strong anti-cyclonic eddy occurs (see Fig. 7).

In a weakly nonlinear shallow water dynamical system, such as the Bohai Sea, the first-order Lagrangian mean velocity or the mass-transport velocity describes the lowest-order circulation. A set of field equations governing the mass transport by wind, tide and density action has been elaborated for a shallow water system (Feng, 1987, 1990, 1997; Feng and Lu, 1993). Then, the Lagrangian time mean circulation in the Bohai Sea is simulated and its dynamics is revealed (Sun *et al.*, 1989; Wang *et al.*, 1993). A comparison between the Lagrangian tide-induced residual circulation in the Bohai Sea and Eulerian one has been made through numerical simulations (Wei *et al.*, 2004a).

2.5 Suspended particulate matter

In the Bohai Bay, the sediment is very fine and it mainly consists of soft clay mud (sediment with sizes smaller than 0.01 mm amounts to a portion of more than 70%) and fine silt mud (sediment with sizes smaller than 0.01 mm amounts to a portion of 50%–70%). In the Liaodong Bay, the coarse silt (0.1–0.05 mm) and fine sand (0.25–0.1 mm) dominate in the sediment. In the Laizhou Bay the sediment consists of silt deposits, whereas in the central basin fine sand widely spreads. At the Laotieshan waterway (entrance to the Bohai Sea), the tidal flow is very strong and the area is often eroded; therefore, the sediment particles are coarse there.

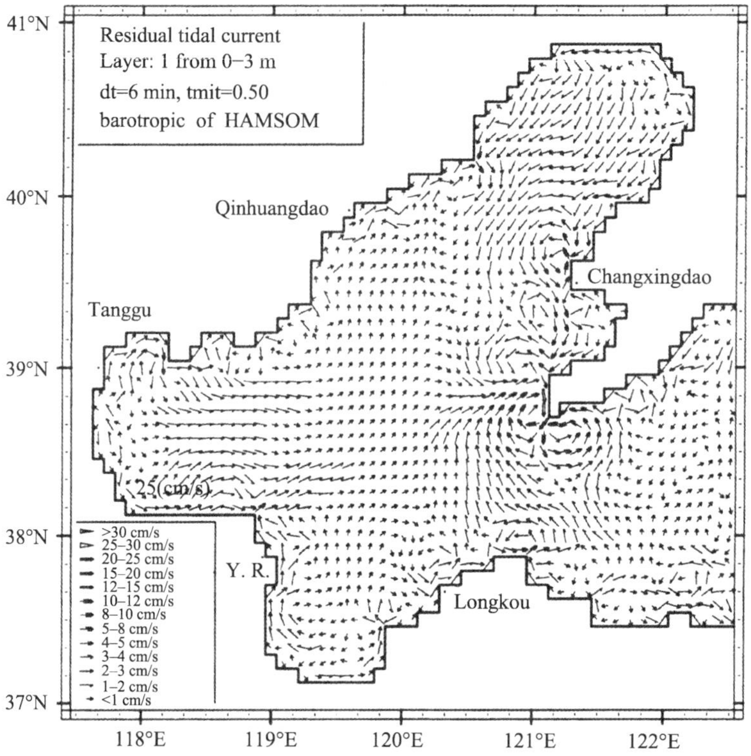

Fig. 7 Monthly mean tidal residuals in the surface layer (Huang, 1995)

SPM in the Bohai Sea mainly consists of nonbiotic particles from land. Most of the particle sizes are smaller than 0.01 mm (The Bohai Sea Geology, 1985). Sources of SPM are rivers and resuspension at the bottom. Atmospheric dust transported by the wind is a further source of SPM. The observed seasonal and regional variation of SPM concentration is closely related to the flow and the sediment concentration of the rivers.

Early descriptions of sediment types in the Bohai Sea are given by Shepard (1932), Niino and Emery (1961), and Qin and Liao (1962). Since then, investigations have continued so that now a comprehensive image of the sediment-type distribution has been achieved (Marine Atlas of Bohai Sea, Yellow Sea, East China Sea, 1992). Qin and Li (1982) analysed the distribution and source of the SPM in the Bohai Sea. This is the most detailed study concerning the SPM in the Bohai Sea up to now. They showed that the highest SPM concentrations are near the Yellow River and the Liaohe River mouths with sharp fronts towards the central Bohai Sea. There is a clear seasonal signal in SPM concentrations with maximum values in April and largest river plume extensions in July. A satellite picture is given by Jiang *et al.* (2004).

Ren and Shi (1986) focused on the influence of the sediment discharge from the Yellow River to the Bohai Sea and the Yellow Sea. According to their analysis, it is unlikely that the Bohai Sea will be filled up in a few thousand years, although there is a high sedimentation rate of 0.6 m ka^{-1}. They also analysed the transport process of fine SPM from the Bohai Sea to the Yellow Sea. Jiang (1987) studied the influence of the sand from the Yellow River on the Xiaoqing River mouth by analyzing satellite images. Martin *et al.* (1993) have studied the transport of SPM from the Bohai Sea to the western Pacific. Around the Yellow River mouth area, numerous studies, including international cooperation, were carried out. Wright *et al.* (1988) and Wiseman *et al.* (1986) studied the distribution and transport of SPM near the Yellow River delta.

Numerical modelling was started by Dong *et al.* (1989) when they simulated the erosion and deposition pattern in the Bohai Sea and the Yellow Sea by using a 2D numerical model of tidal currents. The first 3D SPM numerical model study in the Bohai Sea was done by Jiang and Mayer (1997). They simulated the transport of SPM from the Yellow River for 6 months.

2.6 Nutrients

The most striking feature of nutrient distribution in the Bohai Sea are the gradients of concentration from coastline towards the deep-water region. High concentration of nitrate, phosphate and dissolved silica can be found off the river mouths in this region. For example, freshwater input via rivers may have nitrate levels of up to 50–100 μM, 1.0–2.0 μM for phosphate and 100–200 μM for silica, respectively (Zhang, 1996a, b). Concentrations of nutrients fall rapidly with higher salinity in the estuary, and nutrient level approaches the marine end-members of Bohai and Yellow Sea when salinity reaches 25–30 psu. Among the various nutrient species, silica and nitrate have the tendency to be conservative along salinity profiles, whereas phosphate and ammonia can be regenerated in the estuary, which may significantly modify the nutrient fluxes from land to the Bohai Sea (Zhang *et al.*, 1997). Moreover, riverine concentrations for nutrients show high variability depending on time and space. Concentrations in the central Bohai Sea remain relatively low, and comparable to the Yellow Sea.

Only a few comprehensive investigations had been implemented before the present China-Germany joint project, e.g. the "Investigations of Bohai Sea environmental ecosystem and biological resources", carried out monthly from May 1982 to May 1983, in February, May, August and October 1992 and the Sino-France joint study in May and August 1985 (Diao *et al.*, 1985; Lin *et al.*, 1991; Chen *et al.*, 1991; Tang and Meng, 1997). These investigations showed that nutrients of the Bohai Sea have regular variations with seasons. Owing to the input from the Yellow River and the Haihe River, the levels are higher in the Bohai Bay and the Laizhou Bay located in the southern Bohai Sea. The increase of nitrogen and decrease of phosphate and silicate led to a dramatical increase of N/P ratio and a decrease of Si/N ratio (Cui and Song, 1996; Yu *et al.*, 2000). The situation of nitrogen being a limiting factor in the central Bohai Sea is gradually changing in the recent years towards phosphate and silicate. The decrease of the runoff of the Yellow River to the Bohai Sea may be responsible for this change. This, in turn, may limit the growth of diatoms and thus promote the development of dinoflagellates if other conditions are suitable (Tang and Meng, 1997; Yu *et al.*, 1999).

A model of biogeochemical interaction processes in the Bohai Sea was developed by Zhang and his group (Zhang *et al.*, 1990, 1997; Zhang, 1995; Dong *et al.*, 1998). It turned out that these processes can cause nutrient transport to the continental shelf more than 5–10 times higher than runoff does.

2.7 Phytoplankton

The observation of phytoplankton species composition has been carried out for more than four decades (Jin *et al.*, 1982a, b; Jin, 1988; Anonymous, 1977, 1992; Wang, 1936; Kang, 1986; Cheng and Cao, 1991). The research on primary production (Fei *et al.*, 1991, 1995; Li and Zhu, 1992) and some measurements (Diao, 1986; Shen *et al.*, 1987; Cui and Song, 1992) in the Bohai Sea have also been done for quite a long time. It has now become more and more clear that phytoplankton is the most important contributor to primary production in the Bohai Sea, and diatoms are the major phytoplankton component. The main feature of the annual variation of the phytoplankton growth are two peaks appearing in spring and autumn separately (Kang, 1991; Fei *et al.*, 1991; Yu and Li, 1993). Human-made changes are investigated by several authors (Zou and Dong, 1988; Xu *et al.*, 1993; Zhu and Xu, 1993).

3. Synthesis of Results

The results of the joint project are presented in detail in the following six articles. In the following some major findings are compiled. Although the venture was not exactly realised as intended in the beginning due to administrative constraints, to logistic difficulties, to weather conditions the general research strategy could be kept and turned out to work successfully. Two ship cruises with RV 'Dong Fang Hong 2' have been carried out during 23 September–8 October 1998 and 27 April–12 May 1999. Thus, for autumn and for spring conditions, comprehensive data sets of physical, chemical and biological parameters as current velocity, temperature, salinity, N, P, Si, pH, Chl–*a* on a 3D grid have been achieved. This data has been used differently: for quantification of processes, validation of model simulations, assessment of the actual situation against mean climatological conditions, indication of the quality status of the Bohai Sea. Two cruise reports are available (ZMK, 1998, 1999).

Simultaneously, a hierarchy of numerical models has been developed and applied to get a representation of the flow field, the transport of dissolved and particulate matter, the primary production depending on space and time. As the basic baroclinic circulation model, the Hamburg Shelf-Ocean Model (HAMSOM) was taken. It was complemented by the barotropic Lagrangian tidal residual flow model of Qingdao. The SPM transport calculations were done in Qingdao with the Lagrangian active tracer model of Hamburg using HAMSOM currents for forcing. Nutrient and phytoplankton dynamics have been simulated by the Ecosystem Model of Hamburg (ECOHAM), again using physical input from HAMSOM. Modelling suffered from the fact that the actual atmospheric forcing was not available on a basin-wide scale. So the circulation modelling was done for the period 1980–1993 and the SPM and phytoplankton simulations for 1982.

3.1 Circulation and transport

The circulation model has been used for a 14-year run in order to get the seasonal and inter-annual variability of physical parameters in the Bohai Sea. The period of 1980–1993 has been chosen due to the availability of atmospheric forcing data (ECMWF).

As a result, the flow field and the water mass distribution show the high seasonal variability as observed, especially the strong thermal stratification in summer and the perfect vertical mixing in winter. From the current velocities for different regions of the Bohai Sea, turnover times are deduced controlling the nutrient fluxes. They range from half to one year (Hainbucher *et al.*, 2004).

The tide-induced time mean circulation in the Bohai Sea is an important component of the general circulation in the Bohai Sea. A comparison between the patterns of the tide-induced Lagrangian and Eulerian mean circulation in the Bohai Sea through the numerical simulations reveals that they are quite different, at least in some areas of the Bohai Sea. While the Eulerian mean velocity has been traditionally used as the velocity field of circulation in the Bohai Sea, the mass-transport velocity, a first-order Lagrangian mean velocity, is appropriate for embodying the velocity field of circulation in the Bohai Sea. For example, in the Bohai Bay the tide-induced Lagrangian mean circulation has a clockwise gyre. Thus, it is able to bring sand from the Huanghe River into the Bohai Bay along the southern side of the bay. This explains why the Bohai Bay is a sink for sand coming from the Huanghe River. In addition, the Lagrangian residual transport concept has been successfully used in the modelling of the annual cycle of the primary production in the Bohai Sea in 1982 (Wei *et al.*, 2004b).

Simultaneously measurements of nutrients (nitrogen, phosphorus) and salinity with high correlations on timescales from hours to days indicate the high importance of advective transport, especially by the astronomical tides (Starke *et al.*, 2004).

3.2 Suspended particulate matter

Studies of SPM dynamics of the Bohai Sea are presented by Jiang *et al.* (2004). Generally, the SPM concentration at the surface is lower than that of the bottom layer. Four zones can be distinguished according to the horizontal SPM distributions in the survey areas, which are the Yellow River mouth area, the northeastern part of the survey area, the central Bohai Sea and its northwestern extension to the coast and the Bohai Strait area. The SPM concentration decreases in the order of the above four areas in both cruises. This shows the impact of the Yellow River, the water exchange through the Bohai Strait and the contribution of the Liaodong Bay.

Comparing the data for summer and spring the authors found that the seasonal variation is clearly evident. The SPM concentration changes from 10 mg/l in September and October to around 280 mg/l in spring near the Yellow River mouth area. For the other areas the SPM concentrations show small differences between different seasons. This reflects the fact that in spring the wind is stronger than that in summer and autumn, therefore, the resuspension is also stronger in spring. The measurements from the anchored station show that the SPM concentration is sensitive to wind, and this supports the above explanation of the seasonal variation. Similar findings are achieved by means of a 3D numerical model. In this model three major mechanisms are included: SPM transport by the water, sinking of SPM and its deposition and resuspension. More processes, which are related to these three mechanisms such as currents, waves, interaction between the SPM and water, etc., are involved. This model is also applied to the measuring campaigns and compared with the observed data. NCEP daily data is used as the meteorological forcing. The comparison between the simulation results and the observations gave similar patterns, and this suggests that the model performs well.

A SeaWifs satellite image during the 1999 cruise shows the same features as those obtained from the model results and the observations. Especially the high SPM concentration in the Liaodong Bay, shown in the satellite image, reveals the source of high SPM concentration in the northeastern part of the survey area.

3.3 Primary production

The observations in AMBOS/AMREB show, that the ecotype of phytoplankton in the Bohai Sea are temperate species and neritic species. The Bohai Sea flora is primarily made up of autochthonous meroplanktonic species; diatoms and dinoflagellates are the major components of phytoplankton in the Bohai Sea. The horizontal distribution of phytoplankton has a complex and close relation to the in situ physical-chemical parameters. The annual variation of phytoplankton shows two typical seasonal peaks with a main signal in spring and a weaker one in autumn. Compared with historical data, the phytoplankton and primary production obtained in this study were higher than that of the past. There is evidence for increasing frequency of red tide species such as *Ceratium* and *Protogonyaulax*.

The seasonal cycle of primary production was simulated with the HAMSOM/ECOHAM code for 1982 (Wei *et al.*, 2004b). The empirical input parameters are taken from the North Sea model, but they have been adapted to the Bohai Sea conditions by means of the AMBOS 1998/1999 data set. It turned out that the observed seasonal cycles of primary production could be only reproduced if two nutrient components (nitrogen and phosphorus) are considered. First model approaches with only one limiting nutrient (phosphorus) gained unsatisfactory results. Now, the observed seasonal cycle of biomass with two maxima in spring (April) and autumn (October) could be realistically simulated.

During the field campaigns inorganic and organic nutrient concentrations and fluxes have been determined and analysed (Zhang *et al.*, 2004; Raabe *et al.*, 2004). By in situ experiments the phytoplankton uptake and the benthic exchange fluxes have been quantified. Preliminary nutrient budgets have been established. It turned out that the input from the atmosphere gains increasing importance against riverine outflow and transport through the Bohai Strait, especially for nitrogen. Strong interactions between the bottom sediment and the water column enhance the biogeochemical turnover processes.

3.4 A composite view of the joint study

The present project AMBOS forms, in spite of all remaining deficits, a new step towards a holistic analysis of the Bohai Sea ecosystem, and that with respect to disciplines and methods. For the first time, physical, chemical and biological investigations have been carried out simultaneously. And for the first time, model simulations of the circulation, the SPM transport and the primary production have been done parallel to the observations. The currents in the Bohai Sea and correspondingly the transports of dissolved and suspended matter are mainly wind-driven, less by tides and again less by density gradients due to heat and salt fluxes. The wind field is dominated by the seasonal change of monsoons, but there are strong interannual and short-period changes on the El-Niño and the weather scale, respectively. Therefore, the field experiments cannot be considered as representative. They are realizations of a stochastic process, and their functions are mainly rough qualitative information on the global fields, quantification of local processes and getting data for model validation. The long-term simulations of Hainbucher *et al.* show that the chosen observation sites exhibit water masses of very different origin.

The physical model experiments, nevertheless, support some general statements. So the exchange with the ocean, especially the export of freshwater, dissolved substances and sediment, is relatively small compared with other marginal seas as the North Sea. Admittedly, there are strong tidal currents in the Bohai Strait, but the tidal residuals are (with the exception of a cyclonic eddy in the northern part) mainly directed into the Bohai Sea. The same is true for the wind and the thermohaline-driven circulation. The net outflow from the Bohai Sea due to river discharges becomes weaker due to water regulation works and climate change. That means a potential endangering of the marine ecosystem, because eutrophicated and contaminated waters and sediments remain within the system. The substances released from land and transported by river and atmospheric flow are spread close to the coast by tidal mixing and sea waves within the whole water column. Only in the central Bohai Sea, there appears a stratification from April to September.

The nutrient concentrations in the Bohai Sea, as AMBOS has shown, are presently still relatively small compared with European marginal seas as the North Sea or the Baltic. The contents of phosphate and silicate are even decreasing, but the nitrogen concentration is steadily increasing. This means eutrophication and at the same time a shift in the N : P : Si relations causing a relative decrease of diatom species. As a byproduct, the project has proven that in different regions of the Bohai Sea partly nitrogen, partly phosphorus is the limiting nutrient. Any realistic model of primary production in the Bohai Sea must therefore contain both nutrients.

4. Outlook

The following articles show that the joint German-Chinese project advanced the understanding and modelling of the Bohai Sea system. We know that the astronomical tides enhance all mixing processes and propagate dissolved and suspended material through the residual currents. Beyond the tides, the main external forcing is by the atmospheric fluxes of momentum and heat resulting in a strong seasonal signal. The hitherto dominating SPM and nutrient input by the Yellow River is weakening due to the decreasing freshwater discharge. This means that the atmospheric input of nutrients becomes more important. The models of the 3D hydrodynamics, of SPM transport, nutrient dynamics and primary production reproduce the natural situation as far as observational data are available. This is especially true for the simulation of the flow and mass fields for scales from hours to years.

Nevertheless, the existing database is too small for a rigorous validation of the models. Much more synoptic information is needed on SPM, nutrients and chlorophyll concentration for comprehensive model tests. Also, extended time series of ecosystem parameters at selected points are missing.

Summarizing, the circulation models will allow reliable scenario calculations in the decadal range provided the atmospheric forcing and the river discharges are available. For the biogeochemical fluxes and primary production modelling, more field data is necessary. Aside from long-term monitoring, specific multiship experiments are needed for identification and quantification of the major chemical and biological processes. These campaigns should be accompanied by high resolving remote sensing and modelling of ocean and atmosphere dynamics.

Acknowledgements

We are indebted to the Federal Ministry of Education and Science (BMBF), Germany, which supported the present work under contract no. O3F0189A, to the National Natural Science Foundation of China (NSFC) which supported the work under contract no. 49576298 and to the State Ocean Administration of China (SOA).

References

Anonymous, 1977. The coastal sea creatures of China Sea. The Report of Joint Investigation in China Seas, vol. 8. China Science Press, Beijing. Section 10, 159 pp. (In Chinese).

Anonymous, 1992. The marine biology report. Joint Investigation of Coastal Waters and Beach Waters in China Sea. China Ocean Press, Beijing (In Chinese).

Chen, Z., Gu, Y., Liu, M., Zhang, M., Yang, S., Li, J., 1991. Nutrient distribution of the Huanghe River estuary. Journal of Ocean University of Qingdao 21, 35–42.

Cheng, G., Cao, Y., 1991. The quantitative study of the diatoms in surface sediments of central southern part of Bohai Sea. Journal of Ocean University of Qingdao 21, 56–74.

Cui, Y., Song, Y., 1992. A preliminary study on the relationship between phytoplankton and physicochemical factors in Bohai Sea. Marine Environmental Science 11, 56–59.

Cui, Y., Song, Y., 1996. Study on evaluation of nutrient status in the Bohai Sea. Marine Fisheries Research 17(1), 57–62 (In Chinese).

Diao, H., 1986. The environmental assessment of temperature, phosphate and silicate on phytoplankton in Bohai Sea water. Transactions of Oceanology and Limnology, 35–40.

Diao, H., Shen, Z., Liu, X., Lu, J., 1985. The distribution of inorganic nitrogen in the southwest Bohai Sea. Studia Marina Sinica 25, 53–63 (In Chinese).

Dong, L., Su, J., Wang, K., 1989. The tidal current fields in the Bohai Sea and the Yellow Sea and their relation with the transport of the sediment. Acta Oceanologica Sinica 3, 355–369 (In Chinese).

Dong, L., Zhang, J., Xu, W., 1998. Modeling of nutrient dynamics in a macro-tidal bay from South China. In: Zhang, J. (Ed.), Land-Sea Interaction in Chinese Coastal Zones. China Ocean Press, Beijing, 91–131.

Dou, Z., Lou, Y., Huang, K., Zhang, C., Li, L., Cai, G., Ning, H., 1981. Numerical calculation of the three-dimensional tide-induced Lagrangian residual circulation in the Bohai Sea. Acta Oceanologica Sinica 3(3), 355–369 (In Chinese).

Fang, G., 1994. Tides and Tidal currents in East China Sea, Huanghai Sea and Bohai Sea. In: Zhou, D., Liang, Y., Tseng, C. (Eds.), Oceanography of China Sea, vol. 1. Kluwer Academic Publishing, Dordrecht, pp. 101–112.

Fei, Z., Mao, X., Zhu, M., Li, B., Li, B. H., Guan, Y., Zhang, X., Lu, R., 1991. The study of primary productivity in the Bohai Sea: chlorophyll a, primary productivity and potential fisheries resources. Marine Fisheries Research, 12, 55–69.

Fei, Z., Trees, C. C., Li, W., 1995. Analysis of CZCS data in coastal waters. Journal of Oceanography 13, 61–72.

Feng, S., 1987. A three-dimensional weakly nonlinear model of tide-induced Lagrangian residual current and mass-transport, with an application to the Bohai Sea. In: Nihoul, J. C. J., Jamart, B. M. (Eds.), Three-Dimensional Models of Marine and Estuarine Dynamics. Elsevier Oceanogranphy Series, vol. 45, 471–488.

Feng, S., 1990. On the Lagrangian residual velocity and mass-transport in a multi-frequency oscillatory system. In: Cheng, R. (Ed.), Physics of Shallow Estuaries and Bays. Springer, Berlin, pp. 34–48.

Feng, S., 1997. On circulation in Bohai Sea, Yellow Sea and East China Sea. In: Hong, G., Zhang, J., Park, B. (Eds.), The Health of the Yellow Sea. The Earth Love Publication Association, Seoul, Korea.

Feng, S., Lu, Y., 1993. A turbulent closure model of coastal circulation. Chinese Science Bulletin 38 (20), 1737–1741.

Feng, S., Li, F., Li, S., 1999. An Introduction to Marine Science, Higher Education Press, Beijing, 503 pp. (In Chinese).

Guan, B., 1994. Patterns and structures of the currents in Bohai, Huanghai and East China Seas. In: Zhou, D., Liang, Y., Tseng, C. (Eds.), Oceanography of China Sea, vol. 1. Kluwer Academic Publishing, Dordrecht, pp. 17–26.

Hainbucher, D., Wei, H., Pohlmann, T., Suendermann, J., Feng, S., 2004. Variability of the Bohai Sea circulation based on model calculations. Journal of Marine Systems 44(3–4), 153–174.

Huang, D., 1995. Modelling studies of barotropic and baroclinic dynamics in the Bohai Sea. Berichte aus dem Zentrumfür Meeres-und Klimaforschung der Universität Hamburg Reihe B: Ozeanographie 17, 126 pp.

Huang, D., Chen, Z., Su, J., 1996. The application of a 3D shelf sea model in the Bohai Sea. Acta Oceanologica Sinica 18(5), 1–13.

Jiang, W., Mayer, B., 1997. A study on the transportation of suspended particulate matter from Yellow River by using a 3D particle model. Journal of Ocean University of Qingdao 27(4), 439–445.

Jiang, W., Sun, J., Starke, A., 2004. SPM transport in the Bohai Sea: field experiments and numerical modelling. Journal of Marine Systems 44(3–4), 175–188.

Jiang, W., Wang, J., Zhao, J., Wang, H., 1997. The observation and analysis of circulation in the Bohai Gulf. Journal of Ocean University of Qingdao 27(1), 23–32.

Jin, D., 1988. The geographical distribution of diatom in China sea. Collection of Jin Dexiang. China Ocean Press, Beijing, 225–262.

Jin, D., Chen, Z., Lin, J., Liu, S., 1982a. The Benthonic Diatom of China Sea, vol. I. China Ocean Press, 233 pp.

Jin, D., Chen, Z., Liu, S., Ma, J., 1982b. The Benthonic Diatom of China Sea, vol. II. China Ocean Press, 437 pp.

Kang, Y., 1986. The ecological characteristics of phytoplankton and the relationship between phytoplankton and fisheries in the Yellow Sea. Marine Fisheries Research, 103–107.

Kang, Y., 1991. Distribution and seasonal variation of phytoplankton in the Bohai Sea. Marine Fisheries Research, 31–54.

Li, B., Zhu, M., 1992. The pigment species and distribution characteristics of phytoplankton in autumn in Bohai Sea. Journal of Oceanography 10, 32–37.

Lin, Q., Song, Y., Yang, Q., 1991. The hydrochemistry environment of marine enhancement in the Bohai Sea. Marine Fisheries Research 11(12), 11–30 (In Chinese).

Linden, P. F., Simpson, J. H., 1988. Modulated mixing and frontogenesis in shallow seas and estuaries. Continental Shelf Research 8, 1107–1127.

Marine Atlas of Bohai Sea, Yellow Sea, East China Sea, 1992. China Ocean Press, Beijing.

Martin, J.M., Zhang, J., Shi, M., Zhou, Q., 1993. Actual flux of Huanghe (Yellow River) sediment to the western Pacific Ocean. Netherlands Journal of Sea Research 31(3), 243–254.

Niino, H., Emery, K. O., 1961. Sediments of shallow portion of East China Sea and South China Sea. Bulletin of the Geological Society of America 72(5), 731–762.

Qin, Y., Li, F., 1982. Study of SPM in the Bohai Sea. Acta Oceanologica Sinica 4(2), 191–200 (In Chinese).

Qin, Y., Liao, X., 1962. The preliminary discussion of deposition processes in the Bohai Sea. Oceanologica et Limnologia Sinica 4(3–4), 15–20 (In Chinese).

Raabe, T., Yu, Z., Zhang, J., Sun, J., Starke, A., Brockmann, U., Hainbucher, D., 2004. Phase-transfer of nitrogen species within the water column of the Bohai Sea. Journal of Marine Systems, 44(3–4), 213–232.

Ren, M., Shi, Y., 1986. Sediment discharge of Yellow River (China) and its effect on the sedimentation of the Bohai and Yellow Sea. Continental Shelf Research 6, 785–810.

Shepard, F. P., 1932. Sediments of continental shelves. Bulletin of the Geological Society of America 43, 1017–1040.

Schrum, C., 1994. Numerische Simulation thermodynamischer Prozesse in der Deutschen Bucht. Berichte aus dem Zentrum für Meeres-und Klimaforschung der Universität Hamburg, Reihe B: Ozeanographie 15, 175 pp.

Shen, L., Huang, W., Zhu, L., 1987. Enclosure experiment for Bohai crude oil effects on the structure of phytoplankton community. Acta Oceanologica Sinica 6, 613–621.

Shi, Y., Du, B., 1994. A four layer diagnostic model of winter circulation in the East China Sea. Acta Oceanologica Sinica 16(6), 27–36 (In Chinese).

Simpson, J. H., Hunter, J. R., 1974. Fronts in the Irish Sea. Nature 250, 404–406.

Simpson, J. H., Brown, J., Matthews, J., Allen, G., 1990. Tidal straining, density currents and stirring control of estuarine stratification. Estuaries 13, 125–132.

State Environment Protection Agency of China, 1999. China Environmental State Bulletin: 1998. Environmental Protection 6, 3–9 (In Chinese).

State Ocean Administration, 2000. Bulletin of China Ocean Environmental State at the end of the 20th Century, Beijing (In Chinese).

Su, Y., Weng, X., 1994. Water Masses in China Seas. In: Zhou, D., *et al.* (Eds.), Oceanography of China Sea, vol. 1. Kluwer Academic Publishing, Dordrecht, pp. 3–16.

Sun, W., Xi, P., Song, L., 1989. Numerical modelling of 3D tidal residual current in the Bohai Sea. Journal of Ocean University of Qingdao 19(2), 27–33.

Sündermann, J. (Ed.), 1994. Circulation and Contaminant Fluxes in the North Sea. Springer, Berlin-Heidelberg-New York, 654 pp.

Sündermann, J., Radach, G., 1997. Fluxes and budgets of contaminants in the German Bight. Marine Pollution Bulletin 34, 395–397.

Sündermann, J., Hesse, K.-J., Beddig, S., 1999. Coastal mass and energy fluxes in the southern eastern North Sea. Deutsche Hydrographische Zeitschrift 51, 113–132.

Tang, Q., Meng, T., 1997. Atlas of the Ecological Environment and Living Resources in the Bohai Sea. Qingdao Press, Qingdao, China (In Chinese).

The Bohai Sea Geology, 1985. Marine Geology Division, Institute of Oceanography, Chinese Academic of Sciences. Science Press, Beijing. 260 pp.

Wang, C., 1936. Dinoflagellata of the Gula Bohai. Sinensia 7(2), 128–171.

Wang, H., Su, Z., Feng, S., Sun, W., 1993. Numerical calculation of the wind-driven, thermohaline and tide-induced Lagrangian residual current in the Bohai Sea. Acta Oceanologica Sinica 12(2), 169–182.

Wei, H., Hainbucher, D., Pohlmann, T., Feng, S., Sündermann, J., 2004a. Tidal-Induced Lagrangian and Eulerian Mean Circulation in the Bohai Sea. Journal of Marine Systems, 44(3–4), 141–151.

Wei, H., Sun, J., Moll, A., Zhao, L., 2004b. Phytoplankton dynamics in the Bohai Sea—observations and modelling. Journal of Marine Systems, 44(3–4), 233–251.

Wiseman Jr., W. J., Fan, Y., Bornhold, B. D., Keller, G. H., Su, Z., Prior, D. B., Yu, Z., Wright, L. D., Wang, F., Qian, Q., 1986. Suspended sediments advection by tidal currents off the Huanghe (Yellow River) delta. Geo-Marine Letters 6, 97–105.

Wright, L. D., Wiseman Jr., W. J., Bornhold, B. D., Prior, D. B., Suhayda, J. N., Keller, G. H., Yang, Z., Fan, Y. 1988. Marine dispersal and deposition of Yellow river silts by gravity-driven underflows. Nature 332, 629–632.

Xu, J., Zhu, M., Lu, R., 1993. The formation and environmental characteristics of the largest red tide in North China. 5 Int. Conf. on Toxic Marine Phytoplankton, Newport, RI, USA.

Yu, J., Li, R., 1993. Study on the phytoplankton ecology in the Bohai and Yellow Seas. Journal of Oceanography 11, 52–59.

Yu, Z., Zhang, J., Yao, Q., Mi, T., 1999. Nutrients in the Bohai Sea. In: Hong, G., Zhang, J., Chung, C. (Eds.), Biogeochemical Processes in the Bohai and Yellow Sea. The Dongjin Publication Association, Seoul.

Yu, Z., Mi, T., Xie, B., 2000. Changes of the environmental parameters and their relationship in the recent twenty years in the central Bohai Sea. Marine Environmental Science 19 (1), 15–19 (In Chinese).

Zhang, J., 1995. Geochemistry of trace metals in large Chinese estuaries—an overview. Estuarine, Coastal and Shelf Science 41, 631–658.

Zhang, J., 1996a. Nutrient elements in large Chinese estuaries. Continental Shelf Research 16, 1023–1045.

Zhang, J., 1996b. Nutrient elements from some selected North China Estuaries—Huanghe, Luanhe, Daliaohe and Yalujiang. In: Zhang, J. (Ed.), The Biogeochemical Processes Study for Major Estuaries in China. China Ocean Press, Beijing, 205–217 (In Chinese).

Zhang, J., Letolle, R., Martin, J. M., Jusser, C., Mouchel, J., 1990. Stable oxygen isotope distribution in the Huanghe (Yellow River) and Changjiang (Yangtze River) estuarine systems. Continental Shelf Research 10, 369–384.

Zhang, J., Yu, Z., Liu, S., 1997. Nutrient element dynamics of some North China estuaries (Luanhe, Shuangtaizihe and Yalujiang). Estuaries 20, 110–123.

Zhang, J., Yu, Z., Raabe, T., Liu, S., Starke, A., Zou, L., Gao, H., Brockmann, U., 2004. Dynamics of inorganic nutrient species in the Bohai seawaters. Journal of Marine Systems, 44(3–4), 189–212.

Zhu, M., Xu, J., 1993. Red tide in shrimp ponds along the Bohai Sea. 5. Int Conf. on Toxic Marine Phytoplankton, Newport, RI, USA.

ZMK, 1998. Cruise report 'Dong Fang Hong II' September/October 1998. Univ. of Hamburg, Hamburg.

ZMK, 1999. Cruise report 'Dong Fang Hong II' April 1999. Univ. of Hamburg, Hamburg.

Zou, J., Dong, L., 1988. Phytoplankton kinetics of the Haihe estuarine area and its interrelation with organic pollution. Studia Marina Sinica 29, 131–145.

Tidal-induced Lagrangian and Eulerian Mean Circulation in the Bohai Sea*

Hao Wei, Dagmar Hainbucher, Thomas Pohlmann,
Shizuo Feng, Jürgen Sündermann

1. Introduction

In a coastal sea such as the Bohai Sea (Bohai), the dominant observable motions are tidal oscillations. The M_2 tide is the principal tidal constituent. There are two amphidromic points off Qinhuangdao and the Huanghe River (Yellow River) estuary. The maximum tidal range is larger than 4 m and the maximum current is 2.0 m/s. The K_1 tide has an amphidromic point at the southern part of the Bohai Strait (Editorial Board Marine Atlas, 1994). For long-term transport processes, with time scales of weeks to seasons, the tidal residual currents, i.e. time mean circulation, are important (Delhez, 1996). The intrinsic time scales of the ecosystem dynamics are in the same range as the residual circulation. In winter, the surface current in the Bohai Sea is controlled by strong northern winds, but the tidal residual still has an important effect due to the nonlinear interaction of wind and tide. In summer, some parts of the Bohai Sea are stratified, but the horizontal gradients are small and the density circulation is weak. So the tide-induced mean circulation is even more important for the system.

The time mean circulation in a coastal sea can be obtained by averaging the current-meter data or averaging the simulation results of the instantaneous flow at a fixed point for several tidal periods, thus the Eulerian residual ($\boldsymbol{u}_E$) is obtained. Another way is dividing the net displacement of a water parcel in an averaged time period by the time it travels, resulting in the Lagrangian mean velocity. Longuet-Higgins (1969) has pointed out that it is more reasonable to use the Lagrangian mean velocity than the Eulerian mean velocity to determine the origin of a water mass in a time-varying flow. In the case of the Bohai Sea, the Lagrangian mean velocity can be taken as the Lagrangian residual ($\boldsymbol{u}_L$). The lowest order approximation of the Lagrangian residual is mass transport velocity (Feng and Cheng, 1987) which is the sum of the Eulerian residual and the Stokes drift ($\boldsymbol{u}_S$) that is induced by the nonlinear interactions (Longuet-Higgins, 1969; Zimmerman, 1979; Cheng and Xi, 1986; Feng *et al.*, 1986). $\boldsymbol{u}_L$ has been proven to be a solenoid vector satisfying the conservative condition at a material surface (Feng and Cheng, 1987; Feng, 1990). It has been shown that the Eulerian residual is not conserved while the Lagrangian residual is conserved at a section of the estu-

* Wei H, Hainbucher D, Pohlmann T, Feng S, Sündermann J. 2004. Tidal-induced Lagrangian and Eulerian mean circulation in the Bohai Sea. Journal of Marine Systems, 44(3–4): 141–151

ary (Ianniello, 1977). So the mass transport velocity, the lowest order approximation of the Lagrangian residual, should be appropriate for describing the lowest order mean circulation in coastal seas or tidal estuaries.

From the viewpoint of physics, it is easy to understand that the Lagrangian residual rather than the Eulerian residual can embody the coastal mean circulation. But in practice, this theory is not used widely now. It is taken for granted that the difference between the two kinds of residual is very small and can be neglected. This is not true in some real coastal sea situations. Delhez (1996) has calculated the Stokes drift on the Northwestern European Continental Shelf and compared the Lagrangian and Eulerian transports of passive tracers. He found significant differences.

In this paper, a numerical method for calculating the two kinds of tidal residuals with an existing hydrodynamic model (e.g. POM, HAMSOM) is introduced and the two residuals in the Bohai Sea are compared.

2. Circulation Studies in the Bohai Sea

Basing on "Chinese National Comprehensive Oceanographic Survey (1959–1960)", Guan (1994) analyzed the data of current, sea temperature and salinity, and then suggested the following scheme of the mean circulation in the Bohai Sea. The Yellow Sea Warm Current Extension enters the Bohai Sea through the deep trench of the Bohai Strait and moves westward into the central part of the Bohai Sea until it meets the coast where it splits into two branches. One branch moves toward the Liaodong Bay to form a clockwise gyre, and the other moves toward the Bohai Bay to form a counterclockwise gyre.

There were several numerical studies on the tidal-induced Eulerian mean and Lagrangian mean circulation in the Bohai Sea with different models. The circulation patterns derived were quite different. With regard to the mean Eulerian circulation, several authors concluded by using Leendertse's model (2- or 3-dimensional) that there are many small-scale local gyres and an obvious clockwise gyre in the Liaodong Bay and Laizhou Bay, while a counterclockwise gyre is located in the Bohai Bay (Dou *et al.*, 1981; Yu and Zhang, 1987; Sun *et al.*, 1990). Huang *et al.* (1999) using HAMSOM got a pair of significant eddies near the headland of Liaodong Peninsula. As for the mean Lagrangian circulation, Huang and Wang (1988) derived a counterclockwise gyre in the Bohai Bay and Liaodong Bay with a particle tracing scheme. Sun *et al.* (1989) calculated it by using the Stokes formula and got almost the same result as Huang and Wang (1988) except that the gyre in the western part of the Laizhou Bay points in the opposite direction. Feng *et al.* studied the circulation in the Bohai Sea using the weakly nonlinear theory in coastal seas. They pointed out that there exist a counterclockwise circulation in the central part of the Bohai Sea and two clockwise gyres, one in the Bohai Bay and one in the northeast corner of the Liaodong Bay (Feng, 1990; Zheng, 1992; Wang *et al.*, 1993). In most of the above-mentioned papers, the obtained circulation near the Bohai Strait is "inflow in the north and outflow at the southern part of the Bohai Strait." The magnitude of the tide-induced mean circulation in the Bohai Sea is less than 5 cm/s except in the area of the Laotieshan Channel where it exhibits 15–25 cm/s.

3. Schemes for Eulerian and Lagrangian Mean Circulation

3.1 Schemes based on HAMSOM

Hydrodynamic equations, computing domain, numerical scheme, spatial and temporal discretisation are the same as in Hainbucher *et al.* (2004). The open boundary is chosen on the 122.5°E line from Changshan Island to Jiming Island (Fig. 1). The harmonic constants of the M_2, S_2, N_2, K_1 and O_1 tides obtained from the ocean stations on these two islands are interpolated at the open boundary. The instantaneous water level and current speed $(\zeta,\boldsymbol{u})$ can be calculated. Based on this model, a scheme was developed to calculate the Eulerian and Lagrangian tidal-induced mean circulation at the same time.

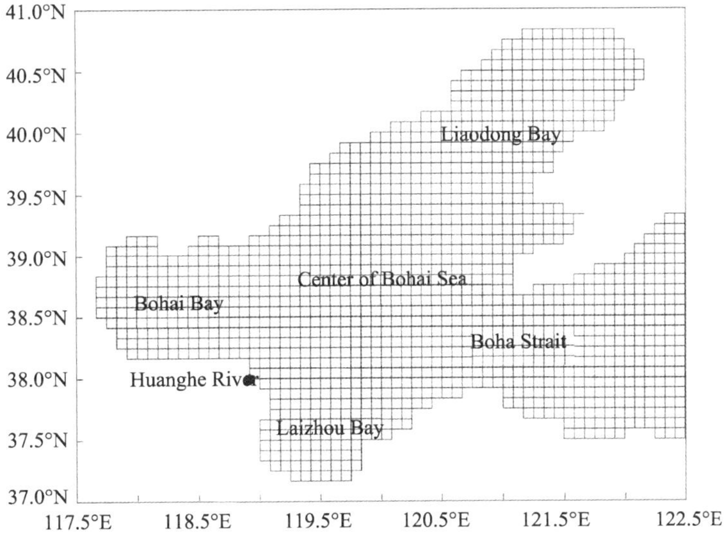

Fig. 1 Schematic grid for the computational domain

The definition of the time-mean operator is as follows:

$$\langle\rangle = \frac{1}{nT}\int_{t_0}^{t_0+nT} \mathrm{d}t \tag{1}$$

where T is the tidal period and n is the number of tidal cycles.

The Eulerian time-mean circulation can be calculated from

$$\boldsymbol{u}_{\mathrm{E}} = \langle \boldsymbol{u} \rangle \tag{2}$$

Stokes drift can also be derived from the instantaneous current by using the following formula:

$$\boldsymbol{u}_{\mathrm{S}} = \left\langle \int \boldsymbol{u} \mathrm{d}t \cdot \nabla \boldsymbol{u} \right\rangle \tag{3}$$

Then the Lagrangian residual is

$$\boldsymbol{u}_{\mathrm{L}} = \boldsymbol{u}_{\mathrm{E}} + \boldsymbol{u}_{\mathrm{S}} \tag{4}$$

Eq. (3) is Longuet-Higgins' original formula for the Stokes drift that is applicable to all kinds of oscillations, whereas Eq. (4) is the Stokes formula.

3.2 Theoretical analysis to the mechanism of Stokes drift

Supposing that there is a harmonic wave propagating only in the x-direction with phase speed c, the northern and eastern speed components for Stokes drift are derived from formula (3) as

$$u_{\mathrm{S}} = \frac{\left\langle u^2 + v^2 \right\rangle}{c} - \left\langle \int v \mathrm{d}\omega \right\rangle \tag{5}$$

$$v_{\mathrm{S}} = \left\langle \int v \mathrm{d}t \delta \right\rangle \tag{6}$$

In formulae (5) and (6), ω is the vorticity and δ is the divergence of the oscillating current,

$$\omega = \frac{\partial v}{\partial x} - \frac{\partial u}{\partial y} \tag{7}$$

$$\delta = \frac{\partial u}{\partial x} + \frac{\partial v}{\partial y} \tag{8}$$

Formula (5) and (6) mean that the Stokes drift is generated by the mean kinetic energy term $\left\langle u^2 + v^2 \right\rangle$, the mean nonlinear interaction of the net displacement $\int v \mathrm{d}t$ with vorticity ω and divergence δ. For the Bohai Sea, the average depth is 18 m so the phase speed ($c = \sqrt{gh}$) is less than 14 m/s. Tidal current in this shallow sea is quite strong (1 m/s in magnitude). So the Stokes drift could be as large as 7 cm/s if there is no rotation.

The transport associated with the Stokes drift in a water column can be written as

$$Hu_{\mathrm{s}} = \int_b^s \left\langle \int u \mathrm{d}t \frac{\partial u}{\partial x} + \int v \mathrm{d}t \frac{\partial u}{\partial y} \right\rangle \mathrm{d}z \tag{9}$$

For periodical functions A and B exists the relationship

$$\left\langle A \int B \mathrm{d}t \right\rangle = -\left\langle B \int A \mathrm{d}t \right\rangle \tag{10}$$

Applying Eq. (10) to Eq. (9) and using the continuity equation, it can be concluded that

$$Hu_{\mathrm{s}} = \left\langle u\zeta \right\rangle + \frac{\partial}{\partial y} \left\langle Hu \int v \mathrm{d}t \right\rangle \tag{11}$$

Vertically integrated mass flux in y-direction is

$$Hv_s = \langle v\zeta \rangle - \frac{\partial}{\partial x}\left\langle Hu \int v\mathrm{d}t \right\rangle \tag{12}$$

This transport is determined by the nonlinear interaction between the current and water elevation (term $\langle u\varsigma \rangle$), between topography, current and net displacement (second term of rhs).

So $\boldsymbol{u}_{\mathrm{L}}$ is different from $\boldsymbol{u}_{\mathrm{E}}$ as theoretical analysis shows. For the sake of mass conservation, Stokes drift transport should be concluded.

4. Results and Analysis

4.1 Tides and tidal currents simulation

Results of tidal waves simulated with this model are in good agreement with the observations. Two amphidromic points for M_2 tide and one for K_1 are simulated by using this model. The average error of the amplitude at 19 stations is 5 cm and that of the phase is about 10°. The simulated tidal level (solid line) of Tanggu, Qinhuangdao and Longkou which are representative for three bays are consistent with the monitoring data (dashed line) (Fig. 2).

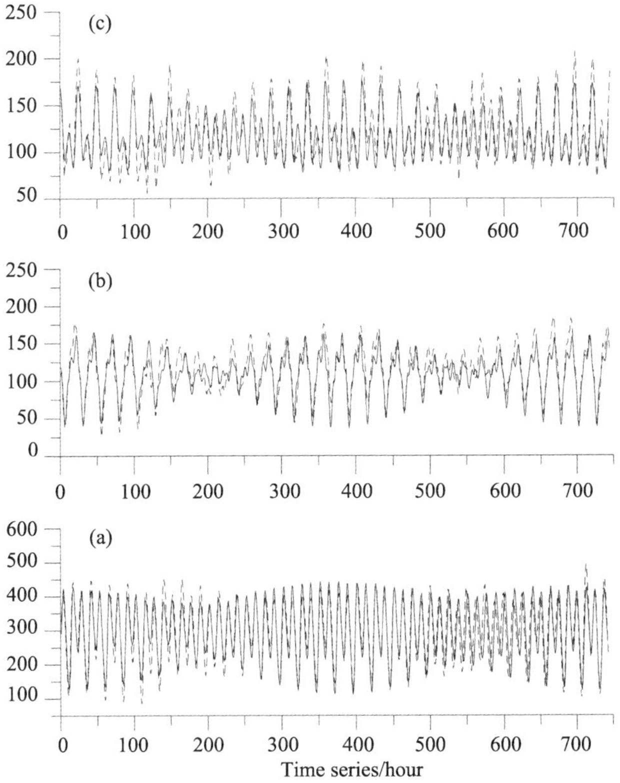

Fig. 2 Tidal level (cm) of (a) Tanggu, (b) Qinhuangdao, (c) Longkou, 1–31, July 1996

Monitored data are drawn in dashed line and simulated in solid line

Simulated tidal currents also agree with our observation during the R/V DongFangHong2 Spring Cruise in 1999 (Fig. 3). We have chosen stations where no stratification existed. E1 (118.5°E, 38.5°N, 22-m depth) is a station in the Bohai Bay, where the strongest current is about 80 cm/s and the tidal current is alternating in east-west direction. At station E3 (119.5°E, 38.5°N, 26-m depth), a station in the central basin, the current is quite weak (< 46 cm/s) and rotates clockwise. B1 (119.5°E, 37.74°N, 16-m depth), the shallowest station located in the Laizhou Bay, shows a little counterclockwise rotating current with a maximum value of 66 cm/s in north-south direction. The northeastern wind of 7 m/s blowing continuously at B1 is mainly responsible for the difference between simulated and observed data.

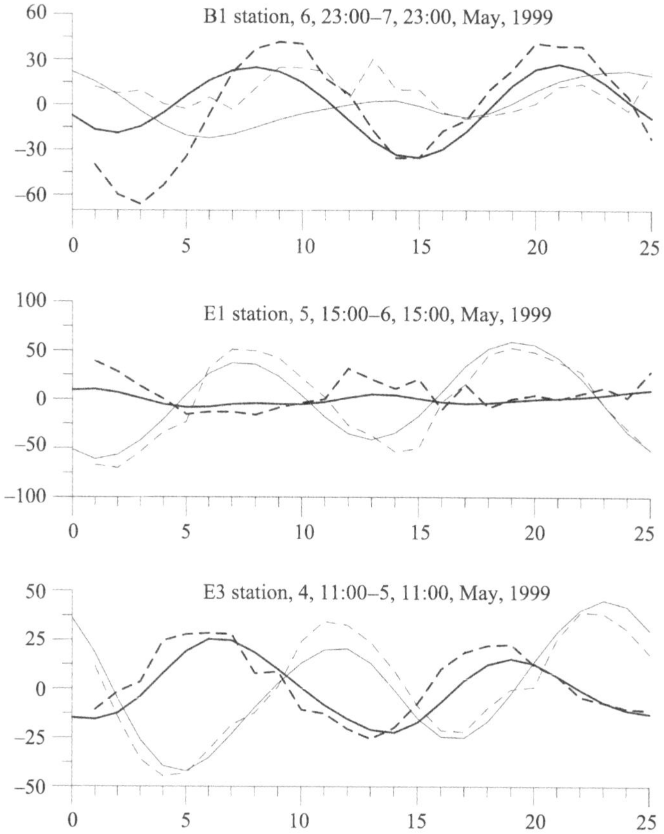

Fig. 3　Comparison between simulated tidal current and observations at three anchor stations in the Bohai Sea, May 1999. Thick solid line—v component simulated, thin solid line—u component simulated. Thick dashed line—v component observed, thin dashed line—u component observed

4.2 Comparison of Eulerian and Lagrangian time-averaged circulation in the Bohai Sea

The differences between Eulerian and Lagrangian time-averaged circulations are caused by the Stokes drift; therefore, the distribution of $\boldsymbol{u}_S$ are discussed first (Fig. 4). At the surface

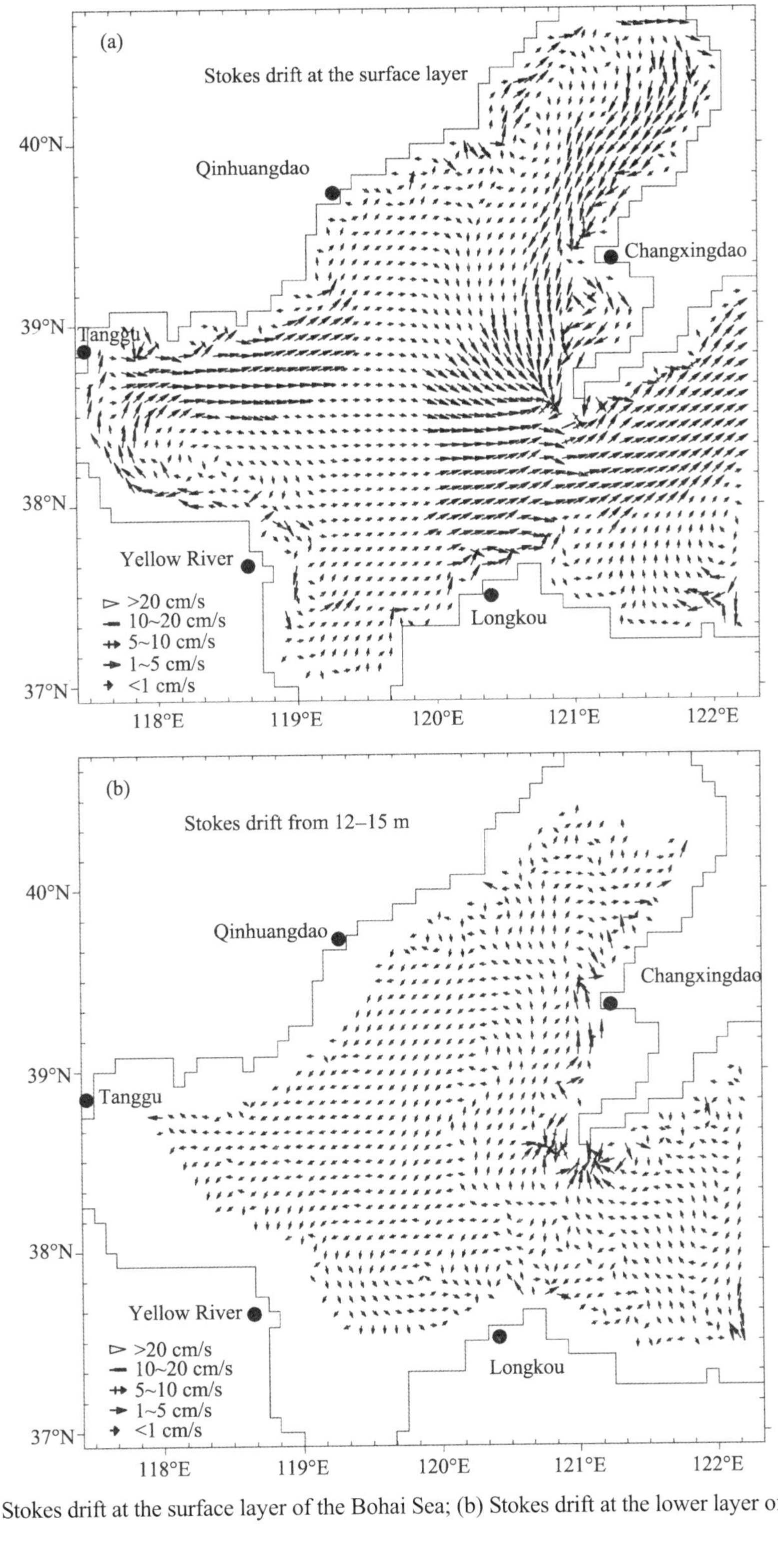

Fig. 4 (a) Stokes drift at the surface layer of the Bohai Sea; (b) Stokes drift at the lower layer of the Bohai Sea layer, $\boldsymbol{u}_S$ produces a clockwise flow in the Liaodong Bay and the Bohai Bay and a counter-clockwise flow in the Laizhou Bay. It flows out of the Bohai Strait. The areas of $|\boldsymbol{u}_S| >$ 1 cm/s are located near the eastern bank of the Liaodong Bay, the northern bank of the Bohai Bay and the Bohai Strait. Topography gradients in these areas are largest. Vorticity and divergence are also large due to the shape of the coastline (headland etc.). According to formulae (11) and (12), $|\boldsymbol{u}_S|$ in these areas is larger due to the strong nonlinear interaction of the net displacement with the vorticity and divergence. $|\boldsymbol{u}_S|$ is less than 1 cm/s in the lowest layer.

$\boldsymbol{u}_E$ at the surface layer is the same as that of the results of Huang *et al.* (1999). There are many pairs of eddies near the headland and $|\boldsymbol{u}_E|<1$ cm/s in the central basin and the Laizhou Bay (Fig. 5). The patterns of the Eulerian residual and the Stokes drift are obviously differrent. They even run in the opposite directions at the eastern bank of the Laizhou Bay, southern bank of the Bohai Bay and the northwest part of the Liaodong Bay. They are of the same order and make the Lagrangian residual pattern significantly different from the Eulerian one.

In the surface layer (Fig. 6), $\boldsymbol{u}_L$ does not exhibit as many eddies in the Bohai Bay and the Liaodong Bay as $\boldsymbol{u}_E$. The maximum Eulerian residual is about 8 cm/s, while the maximum Lagrangian residual is about 30 cm/s. In contrast to $|\boldsymbol{u}_E|$, $|\boldsymbol{u}_L|$ is larger than 1 cm/s in the central basin and the Laizhou Bay; where, the Stokes drift and the Eulerian residual are in the same direction. The two kinds of residuals are in opposite directions near the southwest coast of the Bohai Bay. $\boldsymbol{u}_L$ forms a clockwise flow in the Bohai Bay and the Liaodong Bay. The clockwise gyre in the Bohai Bay is also given in the results of Feng (1990) and Wang *et al.* (1993). The artificial jellyfish experiment supported this result (Jiang *et al.*, 1997). These currents can transport sand discharged from the Huanghe River into the Bohai Bay making the Bohai Bay a sand sink. This was proven by observations in the Sino-American joint investigations on the sediments of the Huanghe River estuary in 1985–1986 (Wright *et al.*, 1990; Shi, 1994) and its distributions of salinity and nutrients (Lue, 1985). In the satellite color images (Jiang *et al.*, 2004), the high turbidity area can be observed near the southern bank of the Bohai Bay.

$\boldsymbol{u}_L$ enters the Bohai Bay and the Liaodong Bay in their shallow parts and leaves these bights along the deep trench. This is consistent with observations of the surface current as given in the Marine Atlas of the Bohai Sea (Editorial Board, 1992). $\boldsymbol{u}_L$ is weak in the corner of the three bays. This means that there is an unfavorable condition for water exchange in these areas.

From the flow structure across sections of the Bohai Strait, it can be concluded that the transport of $\boldsymbol{u}_E$ and $\boldsymbol{u}_L$ are different. Transports by $\boldsymbol{u}_E$ enter the Bohai Sea by the eddy through the narrow northern part and central part of this strait. The net flux of $\boldsymbol{u}_E$ is 1.62×10^{-2} Sv. Transports by $\boldsymbol{u}_L$ run into the Bohai Sea through the northern deep channel with a flux of 0.16 Sv.

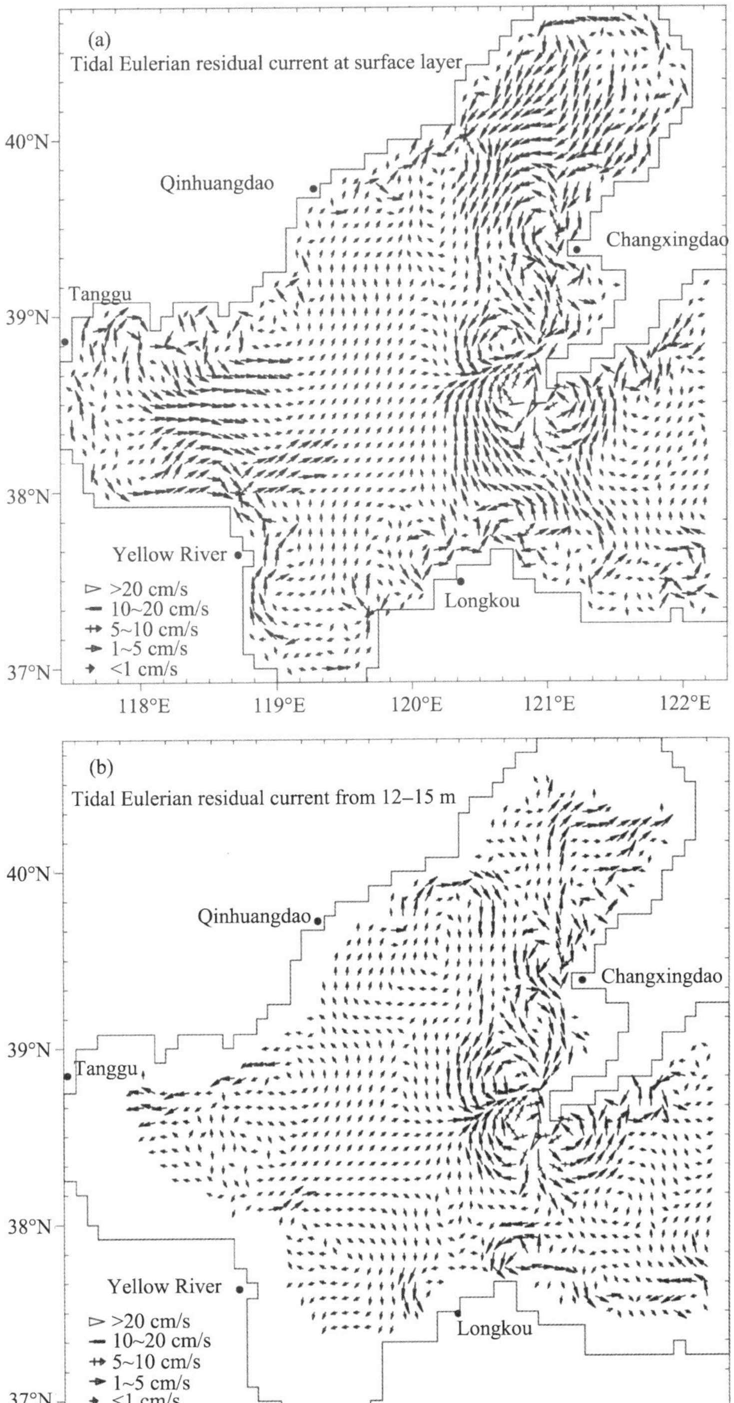

Fig. 5 (a) Eulerian tidal residual current at the surface layer of the Bohai Sea; (b) Eulerian tidal residual current at the lower layer of the Bohai Sea

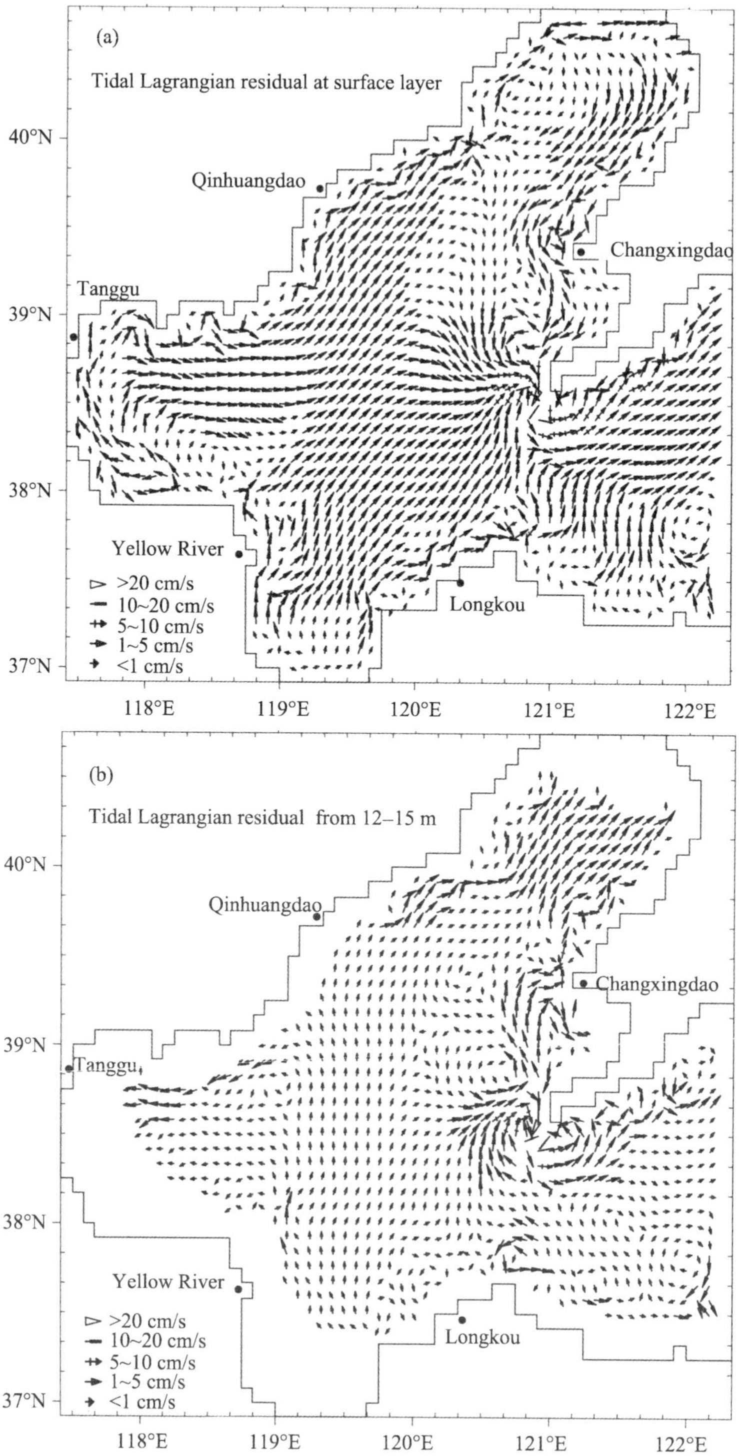

Fig.6 (a) Lagrangian tidal residual at the surface layer of the Bohai Sea; (b) Lagrangian tidal residual at the lower layer of the Bohai Sea

5. Conclusions

(1) The Eulerian and Lagrangian tidal-induced mean circulations are significantly different in coastal seas. This difference cannot be considered as negligible. The latter rather than the former can embody the total time-averaged circulation that a water mass experience.

(2) The Lagrangian tidal residual is stronger than the Eulerian one in most parts of the Bohai Sea and they can point in opposite directions in some areas. This will result in different transports, for example, of the sand discharged from the Yellow River into the Bohai Bay.

(3) The Lagrangian residual circulation should be considered as the base of the ecosystem dynamics in a coastal sea. We can deduct that it is easy to calculate the Lagrangian residual including tide-wind-thermohaline effects from an existing baroclinic model by using Stokes formula.

Since the tidal excursion is about 10 km, a grid of 1/12° in latitude and longitude may not resolve the net displacement in one tidal period reasonably. A finer grid is required for a better resolution of the tidal cycle.

Acknowledgements

We are indebted to the Federal Ministry of Education and Science (BMBF), Germany, which supported the present work under contract No. O3-F0189A, and to National Natural Science Foundation of China (NSFC) under contract No. 49576298 and State Oceanic Administration of China (SOA).

References

Cheng, R. T., Feng, S., Xi, P., 1986. On Lagrangian residual ellipse. In: Van de Kreede, J. (Ed.), Physics of Shallow Estuaries and Bays, Lecture Notes on Coastal and Estuarine Studies, vol. 16. Springer-Verlag, Berlin, pp. 102–113.

Delhez, E., 1996. On the residual advection of the passive constituents. J. Mar. Syst. 8, 147–169.

Dou, Z., Lou, Y., Huang, K., 1981. Numerical study of the tidal current and tidal residual current in the Bohai Sea. Acta Oceanol. Sin. 3(3), 355–369 (in Chinese).

Editorial Board Marine Atlas, K., 1994. Marine Atlas of Bohai Sea, Yellow Sea and East China Sea. China Ocean Press, Beijing (354 pp.).

Feng, S., 1990. On the Lagrangian residual velocity and the mass- transport in a multi-frequency oscillatory system. In: Cheng, R. T. (Ed.), Physics of Shallow Estuarine and Bays, Lecture Notes on Coastal and Estuarine Studies, vol. 20. Springer-Verlag, Berlin, pp. 18–34.

Feng, S., Cheng, R. T., 1987. A three-dimensional weakly nonlinear model of tide-induced Lagrangian residual current and mass-transport, with an application to the Bohai Sea. In: Nihoul, J., Jamart, B. M.

(Eds.), Three-Dimensional Models of Marine and Estuarine Dynamics, Elsevier Oceanography Series 45. Elsevier, The Netherlands, pp. 471–488.

Feng, S., Cheng, R. T., Xi, P. G., 1986. On tide-induced Lagrangian residual current and residual transport, 1. Lagrangian residual current. Water Resour. Res. 22(12), 1623–1634.

Guan, B., 1994. Patterns and structures of the currents in Bohai, Huanghai and East China Sea. In: Zhou, D., Liang, Y., Tseng, C. (Eds.), Oceanology of China Sea. Kluwer Academic Publishing, The Netherlands, pp. 17–26.

Hainbucher, D., Wei, H., Pohlmann, T., Suendermann, J., Feng, S., 2004. Variability of the Bohai Sea circulation based on model calculations. J. Mar. Syst. 44(3–4): 153–174.

Huang, Z., Wang, X., 1988. ADI scheme: results and analysis. J. Shandong Coll. Oceanol. 18 (2-II), 48–49 (in Chinese with English abstract).

Huang, D. J., Su, J. L., Backhaus, J. O., 1999. Modelling of the seasonal thermal stratification and baroclinic circulation in the Bohai Sea. Cont. Shelf Res. 19(11), 1485–1505.

Ianniello, J. P., 1977. Tidally-induced residual currents in estuaries of constant breadth and depth. J. Mar. Res. 35(4), 755–780.

Jiang, W. S., Wang, J. Y., Zhao, J. Z., Wang, H., 1997. An observation on the circulation in the Bohai Bay and its analysis. J. Ocean Univ. Qingdao 27(1), 13–21 (in Chinese with English abstract).

Jiang, W. S., Sun, J., Starke, A., 2004. SPM transport in the Bohai Sea: field experiments and numerical modelling. J. Mar. Syst. 44(3–4): 175–188.

Longuet-Higgins, M. S., 1969. On the transport of mass by time-varying currents. Deep-Sea Res. 16, 431–447.

Lue, X. Q., 1985. Nutrient distribution of the western Bohai Sea and near the Huanghe Estuary in summer. J. Shandong Coll. Oceanol. 15 (1), 21–26 (in Chinese with English abstract).

Shi, M., 1994. Analysis of the characteristic of the residual current field near the Huanghe River mouth. In: Wen, S. C. (Ed.), Collection of Bohai Ecosystem Dynamic Foundation. Ocean University of Qingdao, Qingdao, PR, China, pp. 70–93.

Sun, W., Xi, P., Song, L., 1989. Numerical calculation of the three-dimensional tide-induced Lagrangian residual circulation in the Bohai Sea. J. Ocean Univ. Qingdao 19(2-I), 1–6.

Sun, Y., Chen, S., Zhao, K., 1990. A three-dimensional baroclinic model of the coastal water. J. Ocean Univ. Qingdao 18 (2-II), 1–14 (in Chinese with English abstract).

Wang, H., Su, Z., Feng, S., Sun, W., 1993. Calculation of three-dimensional wind-driven, thermohaline and tide-induced Lagrangian residual in the Bohai Sea. Acta Oceanol. Sin. 15(1), 23–34.

Wright, L. D., Wisman, W. J., Yang, Z. S., 1990. Processes of marine dispersal and deposition of suspended silts off the modern mouth of the Huanghe (Yellow River). Cont. Shelf Res. 10(1), 1–40.

Yu, K., Zhang, F., 1987. A three-dimensional numerical model of the tidal motions in the Bohai Sea. Oceanol. Limnol. Sin. 18(3), 227–236 (in Chinese with English abstract).

Zheng, L. Y., 1992. Calculation of three-dimensional tidal Lagrangian residual and its application in the Bohai Sea. J. Ocean Univ. Qingdao 22(1), 1–14 (in Chinese with English abstract).

Zimmerman, J., 1979. On the Euler-Lagrangian transformation and the Stokes drift in the presence of oscillatory and residual currents. Deep-Sea Res. 26A, 505–520.

A Lagrangian Mean Theory on Coastal Sea Circulation with Inter-tidal Transports I. Fundamentals*

Feng Shizuo, Ju Lian, Jiang Wensheng

1. Introduction

The study on coastal water circulation with inter-tidal transports has a centrically important significance not only for underlying shallow sea dynamics but also for developing environmental oceanography, particularly including marine ecodynamics.

The time mean ocean circulation, or so called "ocean circulation" for short, and the time-mean fields of temperature and salinity associated with circulation, can be derived in practice by averaging current, temperature and salinity data sampled at a fixed point for periods of the order of a month or longer and in theory by solving a (quasi-) steady-state set of thermo-hydrodynamic equations governing the wind-driven and thermohaline ocean circulation. However, in coastal shallow seas such as the Bohai Sea in China, or in estuaries such as the San Francisco Bay and the Chesapeake Bay in U.S.A., or in the tidal areas of shelf seas, such as the Southern Bight of the North Sea and coastal shallow regions of the Huanghai Sea and the East China Sea, where dominant observable motions are, in general, tidal currents, there exist two open questions about the description of such mean circulation and the understanding of its dynamics. One of them is that, since the thermo-hydrodynamic system of shallow sea is essentially nonlinear, is there no influence of the dominant oscillating currents associated with tides on coastal tidally time-mean circulation and on the inter-tidal transports associated with circulation? The other is that, since any variable in flow field essentially has not only the local variation but also the convective variation, should the flow field variables of estuarine or coastal circulation be embodied by the Eulerian time-mean field used traditionally, as mentioned above?

The answer to the first question is that the (quasi-) steady-state set of equations adopted traditionally which only governs the wind-driven and thermohaline ocean circulation is not sufficient for describing the time-mean circulation in coastal seas. In fact, a set of depth-integrated two-dimensional field equations governing the Eulerian mean velocity of coastal sea circulation has been proposed by Nihoul and Ronday (1975), showing that the introduc-

* Feng S, Ju L, Jiang W. 2008. A Lagrangian mean theory on coastal sea circulation with inter-tidal transports I. Fundamentals. Acta Oceanologica Sinica, 27(6): 1–16

tion of the effects of nonlinear coupling between tides into the set of field equations leads to the generation of a so called "tidal stress" which combines with the wind stress at the sea surface to drive the water circulation. When the tidal stress model is applied to simulating the circulation in the Southern Bight of the North Sea, a comparison with the circulation obtained when neglecting the tidal stress implies the existence of determinant differences which enforces the tidal stress model (Nihoul and Ronday, 1975). Heaps (1978) has systematically derived the depth-averaged two-dimensional equations of motion for the circulation in coastal seas further showing the formal existence of the tidal stress, which seems to be in theory an important support to the tidal stress model proposed by Nihoul and Ronday (1975).

From the view point of physics, the answer to the second question is more important and essential for understanding the dynamics of the coastal water circulation and inter-tidal transport processes. In fact, it has been noticed that, when the methodology treating the Nihoul and Ronday's model mentioned above is extended to treat the three-dimensional problem for the circulation in coastal seas, in the set of equations governing the Eulerian mean velocity and corresponding boundary conditions there exists not only a "tidal stress" but also a so-called "tidal surface source" at the sea surface (Feng *et al.*, 1984). It should be valuable to notice that the "tidal surface source" shows a fictitious water mass source and/or sink at the sea surface, which implies that the conservation of water mass across the sea surface is broken! Furthermore, an inter-tidal transport equation describes essentially the balance between convective and diffusive/dispersive transports for the inter-tidal salinity, temperature and/or concentration of solutes or suspended matters, sometimes including sources and/or sinks. In the traditional inter-tidal transport equation, the convective transport velocity is the Eulerian tidal-mean velocity, and an assumption of "tidal dispersion" has to be made (Fischer *et al.*, 1979), of which the physics and dynamics are not well understood (Pritchard, 1954; Bowden, 1967; Fischer *et al.*, 1979). Longuet-Higgins (1969) has pointed out that it is more relevant to use a Lagrangian mean velocity rather than a Eulerian mean velocity to determine the origin of water mass. Csanady (1982) further considers that "monthly mean velocities determined at fixed locations in a shallow sea may bear no relationship with the Lagrangian velocity, and represent no more than a few statistics among many, characterizing local current climatology". It is becoming increasingly clear that the distribution of important water properties, such as temperature, salinity, the concentration of nutrients and fish eggs, and the long-term transport processes of these properties should depend on the Lagrangian rather than the Eulerian mean velocity, and on the turbulent diffusion and corresponding dispersion effects. Therefore, it is natural that the description on the problem of coastal water circulation with inter-tidal transport processes and the understanding of its dynamics require the development of appropriate theoretical models based on the concept of Lagrangian mean velocity and the related concepts.

The Lagrangian mean velocity can be equivalently defined as the net displacement of a labeled water parcel over an averaging period (one or a few of tidal cycles) divided by the averaging period (Zimmerman, 1979). As has been pointed out by Feng (1987), the Lagrangian mean velocity could lead to a concept of Lagrangian residual velocity, but the latter could be used as a Eulerian field variable and the aggregate of such local velocities may be specified as a Eulerian field of flow for the time-mean circulation in coastal seas or in tidal estuaries. In fact, the Lagrangian mean velocity cannot be expected unconditionally to embody the (Eulerian field) velocity of circulation unless the following three constrained conditions are satisfied. The first constrained condition is that the net displacement of the labeled water parcel over one or a few tidal cycles is negligibly small when compared in order of magnitude with the characteristic length of the flow field of circulation, which means that the averaging period mentioned above cannot be chosen to be too long. Another condition states that the Lagrangian mean velocity should be a solenoidal vector field of flow, since the shallow water circulation could be considered as an incompressible flow. The other is that the Lagrangian mean velocity must be proven to satisfy the conservation equation of a material surface in the flow field of circulation and in particular along the related boundaries, such as the free surface, based on the general principle of continuity of flow field. It is proposed that the Lagrangian mean velocity satisfying the three constrained conditions just stated can be termed as the "Lagrangian residual velocity", an instantaneous velocity in the flow field of circulation. Thus, a clear distinction between the Lagrangian mean velocity and (instantaneous) Lagrangian residual velocity is drawn. It should be, with emphasis, pointed out that the Eulerian mean velocity cannot and does not satisfy the conservation equation for a material surface, thus leading to a break in the continuity of flow field and boundary conditions. In fact, the "tidal surface source' at the sea surface, a fictitious water mass source/sink mentioned above, is a case in point. Also a related example is, in essence, the appearance of "tidal dispersion" in the traditional inter-tidal transport equation. Therefore, it can be seen that the velocity field for the time-mean circulation and particularly for the inter-tidal convective transport in coastal seas and tidal estuaries, where the tidal oscillating currents dominate in the observable intra-tidal flow field, should and could be rationally embodied by the Lagrangian residual velocity, or more exactly, by the aggregate of such local velocities, instead of the Eulerian mean velocity used conventionally.

As a matter of fact, among the several researches on tide-induced Lagrangian residual currents and long-term transports (Tee, 1976, 1987; Huang and Wang, 1986; Prandle, 1986; Cheng, 1983; Yu and Chen, 1983; Awaji, 1982; Cheng and Casulli, 1982; Awaji *et al.*, 1980; Dyke, 1980; Ianniello, 1977, 1979; Zimmerman, 1979; Longuet-Higgins, 1969), a Lagrangian time- mean theory on coastal/estuarine circulation and inter-tidal transports in a convectively weakly nonlinear tidal system has been systematically developed, of which a brief introduction is as follows.

The mass transport velocity, which is the sum of the Eulerian mean velocity and the Stokes' drift velocity (Longuet-Higgins, 1969), is a first order approximation of the Lagrangian mean velocity (Feng *et al.*, 1986a; Zimmerman, 1979), and has been proven to be a solenoidal vector and to satisfy the conservation equation of a material surface (Feng, 1987). It has been suggested that in coastal seas/tidal estuaries the mass transport velocity, as a first order Lagrangian residual velocity, should be appropriate to the description of tidally averaged circulation, or more exactly, may be used to embody the lowest order velocity field of circulation. Further, the field equations governing the mass transport velocity have been respectively proposed in a single frequency tidal system (Feng *et al.*, 1990; Feng, 1987) and in a multi-frequency oscillatory system (Feng, 1990), revealing clearly that the coastal water circulation is driven not only by the wind stress at free surface and the horizontal gradient of water density but also by the "tidal body force" resulting from the nonlinear coupling of tides. It should be noticed that the inter-tidal transport equations governing the Eulerian tidally-averaged temperature, salinity and/or concentration of any tracer, in which the convective transport velocity is the mass transport velocity and in particular no "tidal dispersion" appears (Cheng *et al.*, 1989; Feng, 1987; Feng *et al.*, 1986b), have been included in the complete set of field equations for circulation (Feng *et al.*, 1990; Feng, 1987, 1990).

Cheng and Cassulli (1982) pointed out that the Lagrangian residual velocity is a function of tidal phase when the water parcel is labeled and released, or of the initial tidal phase for short, but the mass transport velocity, the first order Lagrangian residual velocity, is independent of the initial tidal phase. Nevertheless, the tidal phase dependency of the Lagrangian residual velocity appears in the Lagrangian drift velocity, the second-order perturbation solution of the Lagrangian residual velocity, and in a weakly nonlinear single-frequency tidal system the Lagrangian drift velocity inscribes an ellipse on the hodograph plane over one tidal cycle for a depth-averaged two-dimensional tidal model (Cheng *et al.*, 1986; Feng *et al.*, 1986a), or so does the horizontal projection of a three-dimensional Lagrangian drift velocity (Feng, 1987). The Lagrangian residual velocity in a three-dimensional weakly nonlinear multi-frequency tidal system has been also analyzed (Feng, 1990). While the first-order Lagrangian residual velocity, or the mass-transport velocity, has been shown to be the sum of the mass-transport velocities derived from the respective tidal constituents, the second-order perturbation Lagrangian residual velocity, i.e., the Lagrangian drift velocity, has been shown to involve a series of nonlinear interactions between the products of the respective tidal constituents, and it reflects the periodicities of all the tidal constituents contained in the multi-frequency tidal system through the initial phases.

Further, it should be pointed out that in the field equations for the Lagrangian mean circulation, mentioned above, the eddy viscosity/diffusivity was supposed to be a known empirical function. This hypothesis on the eddy viscosity/diffusivity means that the nonlinear effects of turbulent stress and diffusion have been cancelled. In order to understand more thoroughly the dynamics of circulation and inter-tidal transports, a turbulent closure model of coastal/estuarine circulation has been developed (Feng and Lu, 1993). It is of much interest to notice that the nonlinear coupling between tide and circulation revealed in the model is just a result born of the nonlinear effect of turbulent stress/diffusion, but the corresponding model proposed by Feng (1987) showed only a one-way effect of the nonlinearity of tide on the tidally averaged circulation through the "tidal body force". Furthermore, the turbulent closure model also reveals that the astronomical tide affects the dynamics of circulation not only directly through the tidal body force but also indirectly through the turbulent energy produced by the vertical shear of tidal current and turbulent mixing.

Based on the above stated Lagrangian mean theoretical frame of coastal water circulation, a generalized model, which can describe the circulation in a coastal water-shelf sea system in which tidal currents and some quasi-steady flows, such as the Kuroshio in the East China Sea, are of the same order and dominant over the observable flow field, has been proposed and examined (Feng, 1998; Feng and Wu, 1995). This generalized model can also be used to describe the circulation and associated inter-tidal transports in a tidal estuary with large run-off. In the generalized model a generalized mass-transport velocity, which is the sum of quasi-steady flow velocity and mass transport velocity, has been verified to satisfy the continuity conditions of flow field and thus used to embody the velocity field of circulation.

It is interesting to mention that the analytical solutions of tidally induced residual currents in estuaries of constant/variable breadth and depth showed that the mass transport velocity does satisfy the conditions of continuity but the Eulerian tidally-averaged velocity does not (Ianniello, 1977, 1979).

These theoretical models have been applied to the simulations/predictions of circulation and inter-tidal transports in some realistic shallow seas/tidal estuaries and to the understanding of their dynamics (e.g., Zhou and Sun, 2006; Hainbucher *et al.*, 2004; Lei *et al.*, 2004; Wei *et al.*, 2004a, b; Jiang and Sun, 2002; Liu *et al.*, 2002; Zhao *et al.*, 2002; Zhou *et al.*, 2002; Sun *et al.* 2001; Feng, 1998; Wang, 1996; Wang *et al.*, 1993; Dortch *et al.*, 1992; Zheng, 1992; Sun *et al.*, 1989). And there are several related studies (e.g., Jiang *et al.*, 2002, 1997; Delhez, 1996; Ridderinkhof and Loder, 1993; Salomon and Breton, 1993; Foreman *et al.*, 1992; Feng, 1991; Jay, 1991; Lu, 1991; Prandle, 1991; Visser and Bowman, 1991; West *et al.*, 1990; Wright *et al.*, 1990; Orbi and Salomon, 1988; Tang and Tee, 1987).

Nevertheless, in realistic tidal estuaries, coastal waters, shallow bays and particularly embayments or firths, the basin geometry is usually quite complex. Not only the speed but also the direction of tidal currents is rectified by the basin geometry and bathymetry, which renders the structure of tidal flow field complicated. In the complex tidal flow field the velocity shear can become quite large and highly variable resulting in stronger nonlinearities of the flow field than that in regions of relatively smooth or weak velocity shears. Thus, the weakly nonlinear models mentioned above may not be valid for part or all of the estuary/embayment. In fact, a governing equation has been derived for the inter-tidal transport processes without invoking the weakly nonlinear assumption (Cheng and Casulli, 1991). In their inter-tidal transport equation governing the Eulerian tidal average of concentration, the convective transport velocity is a mean Lagrangian residual velocity, i.e., a Eulerian tidal average of Lagrangian residual velocity, and an "inter-tidal dispersion" term or a "tidally-averaged dispersion" term, appears. Unfortunately, the appearance of the inter-tidal dispersion term in the governing equation implies that the mean Lagrangian residual velocity cannot satisfy the conservation equation of material surface, and thus, according to our view, it cannot be used to embody the velocity field of circulation. A similar governing equation for the inter-tidal transports has been also proposed by Lu (1994) (personal communication). Consequently, a reformulation of the Lagrangian time-mean theory on circulation and inter-tidal transports in a general nonlinear tidally dominant shallow sea and estuarine system is highly desirable. An attempt is made to extend and reformulate this theory in the present paper.

2. Formulation

A coastal shallow sea, or a tidal estuary, in which tidal periodic currents generally dominate over the observable flow field, is a multiple time scale system. In this paper, a double time scale system is simply supposed. The independent time variables t and τ are respectively used to describe the time variations of dependent variables in the intra-tidal processes associated with the tidally dominated original motion and in the inter-tidal processes associated with the time-mean water circulation. The scale of intra-tidal time variable t, T, is supposed to be in the order of magnitude smaller than the scale of inter-tidal time variable τ, T_*, or

$$O(\delta) < 1 \tag{1}$$

where $\delta = T / T_* = O(t / \tau)$, and the period of principal tide, such as that of M_2-tide, can be selected as T.

Introducing a time mean operator, or a tidal cycle mean operator, $<\ \ >$

$$<\ \ > = \frac{1}{nT}\int_{t_0}^{t_0+nT} \mathrm{d}t \tag{2}$$

where n=1,2,⋯

Thus a Lagrangian mean velocity $\boldsymbol{u}_{\mathrm{L}}$ can be expressed as follows:

$$\boldsymbol{u}_{\mathrm{L}} = \left\langle \boldsymbol{u}\left[\ \boldsymbol{x}_0 + \xi\left(\ t;\tau\right), t;\tau\right]\right\rangle = \frac{\xi_{n\tau}}{nT} \tag{3}$$

where a water parcel, which is arbitrarily selected and thus can be viewed as the representative of all water parcels, occupies the spatial point $\boldsymbol{x} = \boldsymbol{x}_0$ at time $t = t_0$, and the water parcel begins its spatial traveling from $\boldsymbol{x}_0$ at time t_0 and then arrives at its terminal, $\boldsymbol{x}=\boldsymbol{x}_\mathrm{T}$, after n tidal cycles. $\boldsymbol{u}\left[\boldsymbol{x}_0+\xi(t;\tau),t;\tau\right]$ and $\xi(t;\tau)$ are the velocity and displacement of the water parcel, respectively, where $\xi=\boldsymbol{x}-\boldsymbol{x}_0$ and $\boldsymbol{u}=\dfrac{\partial\xi}{\partial t}$. $\xi_{nr}=\boldsymbol{x}_\mathrm{T}-\boldsymbol{x}_0$ is the net displacement of the water parcel after n tidal cycles.

First, the Lagrangian mean velocity, $\boldsymbol{u}_\mathrm{L}$, defined and expressed by Eq. (3), cannot be unconditionally postulated to be a continuous, smooth and differentiable function of $(\boldsymbol{x}_0,\tau;t_0)$, unless a necessary constrained condition (4), as a basic assumption, has been proposed and satisfied, or

$$O(\varepsilon)<1 \tag{4}$$

where $\varepsilon=\xi_{nrc}/L$, ξ_{nrc} is the scale of ξ_{nr} and L is the scale of the horizontal length of the flow field. Noting that the scale of $\boldsymbol{u}_\mathrm{L}$ should be selected to be L/T_*, i.e., $u_\mathrm{Lc}=L/T_*$, where u_Lc is the scale of $\boldsymbol{u}_\mathrm{L}$, and substituting the corresponding scales into Eq. (3) for forming the nondimensional form of Eq. (3), the necessary constrained condition, (4), can be written as

$$O(n\delta)<1 \tag{5}$$

The basic assumption (4), or (5), shows that the net displacement of the water parcel which moves through n tidal cycles, ξ_{nr}, should be in the order of magnitude smaller than the characteristic length of the flow field, L, i.e., $O(\varepsilon)=O(n\delta)<1$, where $\delta=T/T_*$ and $O(\delta)<1$ assumed above. Thus, it should be pointed out that the time, nT, which is used for averaging the velocity of a moving water parcel, cannot be chosen to be too long, or the number of tidal cycles, n, which is used for that same averaging, can not be too large.

The Lagrangian mean velocity can be further expressed by means of the differentiation of the water parcel displacement ξ with respect to τ. In fact, $\xi_{nr}=\dfrac{\partial\xi}{\partial\tau}nT$, being substituted into Eq. (3), we have

$$\boldsymbol{u}_\mathrm{L}=\frac{\partial\xi}{\partial\tau} \tag{6}$$

where $\xi=\xi(\tau;t_0)$. It is noticed that using $\boldsymbol{u}=\dfrac{\partial\xi}{\partial t}$, where $\xi=\xi(t;\tau)$, and $\boldsymbol{u}_\mathrm{L}=\left\langle\dfrac{\partial\xi}{\partial t}\right\rangle$, the same expression as Eq. (6) can be also obtained.

Next, it has to be proved that under the assumption (4) or (5), the Lagrangian mean velocity, $\boldsymbol{u}_\mathrm{L}$, satisfies both the conservation equation of material surface and the continuity equation for an incompressible fluid.

In fact, if a material surface in the water is specified geometrically by the equation $F\equiv F(\boldsymbol{x},t;\tau)=\text{const.}$, F is a quantity which is invariant for a water parcel on the material surface, then

$$\frac{\mathrm{D}F}{\mathrm{D}t}=0 \tag{7}$$

at all points on the surface, where

$$\frac{\mathrm{D}}{\mathrm{D}t}=\frac{\partial}{\partial t}+\boldsymbol{u}\cdot\nabla \tag{8}$$

and $\boldsymbol{u}=\boldsymbol{u}(\boldsymbol{x},t;\tau)$.

After Eq. (7) being substituted into the time mean operator < >, Eq. (2), its left hand term $\left\langle\frac{\mathrm{D}F}{\mathrm{D}t}\right\rangle=\frac{1}{nT}[F_{\mathrm{T}}-F_0]$, where $F_{\mathrm{T}}=F(\boldsymbol{x}_{\mathrm{T}},t_0+nT;\tau)=\frac{\partial F}{\partial\tau}nT+\xi_{n\mathrm{r}}\cdot\nabla F+F_0$ and $F_0=F(\boldsymbol{x}_0,t_0;\tau)$, and when Eq. (3) is introduced into it, then we get

$$\frac{\mathrm{D}F_{\mathrm{L}}}{\mathrm{D}\tau}=0 \tag{9}$$

where

$$\frac{\mathrm{D}}{\mathrm{D}\tau}=\frac{\partial}{\partial\tau}+\boldsymbol{u}_{\mathrm{L}}\cdot\nabla \tag{10}$$

and $F_{\mathrm{L}}=F_{\mathrm{L}}(\boldsymbol{x}_0,\tau;t_0)$ here.

The continuity equation of an incompressible fluid is

$$\nabla\cdot\boldsymbol{u}=0 \tag{11}$$

where $\boldsymbol{u}=\boldsymbol{u}(\boldsymbol{x},t;\tau)$, or equivalently $\frac{\mathrm{D}Q}{\mathrm{D}t}=0$ (since $\frac{1}{Q}\frac{\mathrm{D}Q}{\mathrm{D}t}=\nabla\cdot\boldsymbol{u}$), where Q is a volume of the corresponding moving water element and DQ/Dt=0 implies the water volume conservation. Following the operation process from Eq. (7) to Eq. (10), we have

$$\left\langle\frac{\mathrm{D}Q}{\mathrm{D}t}\right\rangle=\frac{\mathrm{D}Q_{\mathrm{L}}}{\mathrm{D}\tau}=0$$

or equivalently

$$\nabla\cdot\boldsymbol{u}_{\mathrm{L}}=0 \tag{12}$$

since $\frac{1}{Q_{\mathrm{L}}}\frac{\mathrm{D}Q_{\mathrm{L}}}{\mathrm{D}\tau}=\nabla\cdot\boldsymbol{u}_{\mathrm{L}}$, where $\boldsymbol{u}_{\mathrm{L}}=\boldsymbol{u}_{\mathrm{L}}(\boldsymbol{x}_0,\tau;t_0)$.

It can be found that the Lagrangian mean velocity $\boldsymbol{u}_{\mathrm{L}}$, under the assumption (4), $O(\varepsilon)<1$, does satisfy the conservation equation of material surface, Eq. (9), and the continuity equation for an incompressible fluid, Eq. (12). Since $(\boldsymbol{x}_0,t_0)$ is to be arbitrarily selected in the intra-tidal field of flow, $\boldsymbol{x}$ can be used instead of $\boldsymbol{x}_0$ in the inter-tidal field of flow, i.e., in the field of circulation, the Lagrangian mean velocity $\boldsymbol{u}_{\mathrm{L}}(\boldsymbol{x},\tau;t_0)$ can be viewed as a continuous, smooth and differentiable Eulerian field variable and the aggregate of such local velocities can be specified as a Eulerian field of flow, where t_0, as a parameter, varies in a tidal cycle, or the corresponding tidal phase, $\frac{2\pi}{T}t_0$, varies from 0 to 2π. Thus, such a Lagrangian mean velocity, $\boldsymbol{u}_{\mathrm{L}}(\boldsymbol{x},\tau;t_0)$, will be termed a "Lagrangian residual velocity", and the latter should and can be used to embody the velocity field of circulation in coastal seas or tidal estuaries.

It is, with emphasis, pointed out that the coastal water circulation described above, generally contains a set of infinite temporal-spatial fields of velocity, each of which is corresponding to a specific value of $\left(\frac{2\pi}{T}t_0\right)$ which varies continuously from 0 to 2π. It can be found that the concept on coastal water circulation formed and described here and the circulation pattern associated with it, as just mentioned, are very different from those defined and described traditionally.

The velocity field of circulation, i.e., $\boldsymbol{u}_{\mathrm{L}}(\boldsymbol{x},\tau;t_0)$ may be obtained through the following approach. The velocity field, $\boldsymbol{u}(\boldsymbol{x},t;\tau)$, associated with the intra-tidal processes, is previously derived from solving the original set of field equations. Then the net displacements of water parcels which are continuously released from spatial points $\boldsymbol{x}$ at instant $t=t_0$, $\xi_{nr}(\boldsymbol{x},\tau;t_0)$, can be derived from integrating the trajectory equation with respect to t from t_0 to t_0+nT, and thus $\boldsymbol{u}_{\mathrm{L}}(\boldsymbol{x},\tau;t_0)$ can be obtained by means of $\boldsymbol{u}_{\mathrm{L}}=\xi_{nr}/nT$ [Eq. (3)]. Finally, the infinite temporal-spatial fields of circulation velocity can be derived through varying the instant t_0, at which the water parcels are released, from 0 to T.

An inter-tidal transport process is closely related to, or even nonlinearly coupled with the circulation. Therefore, it should be noticed that the study on the problem of circulation must be, in general, combined with the study on the inter-tidal transport process.

A governing transport equation for intra-tidal processes is

$$\frac{\mathrm{D}c}{\mathrm{D}t}=S_{\mathrm{T}}+S \tag{13}$$

where $c=c(\boldsymbol{x},t;\tau)$ denotes the concentration of any tracer, the water temperature or salinity etc.; S_{T} is referred to the turbulent diffusion; S denotes the sources/sinks.

Just as the flow field the intra-tidal concentration c determined by Eq. (13) contains different temporal-spatial scales. The intra-tidal concentration is consistent with the intra-tidal flow field. In a tide dominant system the oscillatory tidal variation controls the main part of c. But because of the inter-tidal process there is also slowly varying part of concentration which can be called inter-tidal concentration accompanying the inter-tidal flow field-Lagrangian residual velocity.

Applying the time mean operator < >, Eq. (2), to Eq. (13) with $(S_{\mathrm{T}},S)=(S_{\mathrm{T}},S)$ $[\boldsymbol{x}_0+\xi(t;\tau),t;\tau]$ and going through the similar procedure from Eq. (7) to Eq. (10) yields an inter-tidal transport equation, i.e.,

$$\frac{\mathrm{D}c_{\mathrm{L}}}{\mathrm{D}\tau} = S_{\mathrm{TL}} + S_{\mathrm{L}} \tag{14}$$

with

$$\frac{\mathrm{D}}{\mathrm{D}\tau} = \frac{\partial}{\partial \tau} + \boldsymbol{u}_{\mathrm{L}} \cdot \nabla \tag{15}$$

Where $\boldsymbol{u}_{\mathrm{L}}$ is the Lagrangian residual velocity, defined above, and has been used to embody the velocity of circulation; S_{TL} and S_{L} are the Lagrangian tidally time averages of S_{T} and S, respectively.

Thus the inter-tidal concentration $c_{\mathrm{L}} = c_{\mathrm{L}}(\boldsymbol{x},\tau;t_0)$ is defined by Eq. (14) which describes the inter-tidal system combined with the Lagrangian residual velocity, $\boldsymbol{u}_{\mathrm{L}}$. $c_{\mathrm{L}}(\boldsymbol{x},\tau;t_0)$ is termed a "Lagrangian inter-tidal concentration".

The definition of $c_{\mathrm{L}} = c_{\mathrm{L}}(\boldsymbol{x},;t_0)$ is not straight forward and it cannot be understood as a result by filtering of the intra-tidal concentration c at one fixed point. While the idea is conventionally taken for granted that a Eulerian tidal average of intra-tidal concentration is naturally used to describe the inter-tidal transport process, it should be with emphasis pointed out that the Lagrangian inter-tidal concentration, $c_{\mathrm{L}}(\boldsymbol{x},\tau;t_0)$, obtained through solving

Eq. (14), is really appropriate to the description of the inter-tidal transport process, but not the Eulerian tidal average of intra-tidal concentration, at least not in a general nonlinear coastal/estuarine system.

The inter-tidal transport equation, Eq. (14), with the expression of a material derivative operator (15), states that the material derivative of $c_{\mathrm{L}}(\boldsymbol{x},\tau;t_0)$, or the material derivative of the Lagrangian inter-tidal concentration for the inter-tidal transport processes, is balanced by $(S_{\mathrm{TL}} + S_{\mathrm{L}})$ which is the sum of the Lagrangian time averages of S_{T} (referred to turbulent diffusion) and S (sources/sinks) for the intra-tidal process; the convective transport velocity is the Lagrangian residual velocity, $\boldsymbol{u}_{\mathrm{L}}$.

Further, it is revealed that the inter-tidal transport process described by Eq. (14), like the field of circulation revealed above, contains in general a set of infinite temporal-spatial fields of the Lagrangian inter-tidal concentration $c_{\mathrm{L}}(\boldsymbol{x},\tau;t_0)$, each of which is corresponding to a specific value of the tidal phase $\left(\frac{2\pi}{T}t_0\right)$, varying continuously from 0 to 2π.

The Lagrangian inter-tidal concentration, $c_{\mathrm{L}}(\boldsymbol{x},\tau;t_0)$, can be derived through the following approach: firstly solving the original set of field equations, particularly including the transport equation (13), we obtain the intra-tidal concentration, $c(\boldsymbol{x},t;\tau)$, and the intra-tidal velocity field, $\boldsymbol{u}(\boldsymbol{x},t;\tau)$. And then an arbitrary water parcel location at $\boldsymbol{x}$ when $t=t_0$, is tracked in the flow field until it reaches $\boldsymbol{x}_{\mathrm{T}}$ after one or a few of tidal periods. According to the definition the Lagrangian inter-tidal concentration $c_{\mathrm{L}}(\boldsymbol{x},\tau,t_0)$ equals to the instantaneous intra-tidal concentration c at $\boldsymbol{x}_{\mathrm{T}}$ when $t=t_0+nT$. Thus the inter-tidal concentration $c_{\mathrm{L}}(\boldsymbol{x},\tau,t_0)$ is got. By changing the starting point of t_0 from 0 to T all the infinite sets of inter-tidal concentration $c_{\mathrm{L}}(\boldsymbol{x},\tau;t_0)$ can be derived.

Let the equation of the intra-tidal free surface be denoted by

$$z-\zeta(x,y,t;\tau)=0 \tag{16}$$

where z is the vertical coordinate with its axis upwards; ζ is the displacement of the free surface from the undisturbed sea surface ($z=0$); $\boldsymbol{x}=(x,y,z)$.

Introducing $F(\boldsymbol{x},t;\tau)\equiv z-\zeta(x,y,t;\tau)$ into Eq. (7) and noticing Eq. (8), the kinematic boundary condition at the free surface is obtained, i.e., at $z=\zeta$:

$$w=\frac{\partial \zeta}{\partial t}+\boldsymbol{u}\cdot\nabla\zeta \tag{17}$$

where w is the vertical component of velocity $\boldsymbol{u}$.

It is obviously convenient that using Eq. (9) with Eq. (10) and $F_{\mathrm{L}}\equiv z-\zeta_{\mathrm{L}}$, where $\zeta_{\mathrm{L}}=\zeta_{\mathrm{L}}(x, y, \tau; t_0)$, termed a "Lagrangian inter-tidal free surface", and at

$$z=\zeta_{\mathrm{L}} \tag{18}$$

the kinematic boundary condition accompanying the flow field of circulation is derived

$$w_{\mathrm{L}}=\frac{\partial \zeta_{\mathrm{L}}}{\partial \tau}+\boldsymbol{u}_{\mathrm{L}}\cdot\nabla\zeta_{\mathrm{L}} \tag{19}$$

where w_{L} is the vertical component of the Lagrangian residual velocity $\boldsymbol{u}_{\mathrm{L}}$.

The kinematic boundary condition, Eq. (19) with Eq. (18), implies the existence of a set of infinite temporal-spatial fields of the Lagrangian inter-tidal free surface, ζ_{L}, each of which is corresponding to a specific value of the tidal phase $\left(\frac{2\pi}{T}t_0\right)$ varying continuously from 0 to 2π.

3. Discussion

A shallow water dynamic system, such as a shallow sea or a tidal estuary, is a nonlinear one. The birth of residual motion in the dynamic system is due to the wind stress at sea surface and the horizontal gradient of water density and, particularly due to the nonlinearity of the dynamic system. In view of the presence of tidal periodic currents generally dominated over the original observable flow field, there is a shallow water effect, i.e., a convectively nonlinear effect. The degree of the convective nonlinearity produced from the shallow water effect can be measured and characterized by a convectively nonlinear parameter κ. Turbulence may be considered as another nonlinear origin of the generation of residual motion.

3.1 The Nonlinearity due to convection

The convectively nonlinear parameter κ is a nondimentional one, and it may be expressed as (Feng *et al.*, 1986a)

$$\kappa = \frac{\zeta_c}{h_c} = \frac{\xi_c}{L} \tag{20}$$

where $\xi_c = u_c T$; h denotes the water depth measured from the undisturbed sea surface; the subscript “c” indicates the scales of corresponding variables.

If the net displacement of a water parcel through one tidal cycle is denoted by means of ξ_r, it is naturally supposed that $O(\xi_{nrc}) = n\xi_{rc}$. Noting $n\delta = O(\varepsilon) = O(n\xi_{rc}/L)$ and the equality for the order of magnitude, Eq. (20), $\kappa = \xi_c / L$, we have

$$\delta = O\left(\kappa\frac{\xi_{rc}}{\xi_c}\right) \tag{21}$$

and

$$\varepsilon = O\left(n\kappa\frac{\xi_{rc}}{\xi_c}\right) \tag{22}$$

In a convectively weakly nonlinear system, in which κ is a small parameter, i.e.,

$$O(\kappa) < 1 \tag{23}$$

the scale analysis shows $O(\xi_{rc}/\xi_c) = \kappa$. Substituting the latter into the scale relationships (21) and (22), there are

$$\delta = O(\kappa^2) \tag{24}$$

and

$$\varepsilon = O(n\kappa^2) \tag{25}$$

The solutions to this simplified model in the convectively weakly nonlinear system may be derived by using a perturbation method with the small parameter κ. All of the dependent variables, in particular including the velocity and concentration, are expanded in ascending series of κ, and taking the lowest-order terms of them, the Lagrangian residual velocity $\boldsymbol{u}_L$ expressed by Eq. (3) and the inter-tidal transport equation governing the Lagrangian inter-tidal concentration c_L, Eq. (14) with Eq. (15), will be respectively reduced to the expression of Eqs. (26) and (27) as follows:

$$\boldsymbol{u}_M = \boldsymbol{u}_E + \boldsymbol{u}_S \tag{26}$$

where $\boldsymbol{u}_M = \boldsymbol{u}_M(\boldsymbol{x}, \tau)$ is the mass-transport velocity, $\boldsymbol{u}_E = \boldsymbol{u}_E(\boldsymbol{x}, \tau)$ is the Eulerian mean velocity and $\boldsymbol{u}_S = \boldsymbol{u}_S(\boldsymbol{x}, \tau)$ is the Stokes’ drift velocity (Longuet-Higgins, 1969).

It should be pointed out that the mass transport velocity, $\boldsymbol{u}_M(\boldsymbol{x}, \tau)$, independent of t_0

seems convenient and proper to be used to embody the lowest-order velocity field of circulation in coastal seas, and the field equations governing the mass-transport velocity have been proposed (Feng, 1987; Feng and Lu, 1993). Therefore, in the convectively weakly nonlinear system the pattern of circulation, when compared with that in the general nonlinear system, has been reduced to a relatively simple one.

By the way, it is interesting to note that in the convectively weakly nonlinear system the second-order Lagrangian residual velocity, or the Lagrangian drift velocity, $\boldsymbol{u}_{\mathrm{LD}}(\boldsymbol{x},\tau;t_0)$ has revealed the dependence on t_0 by means of the analytical expressions (Cheng *et al.*, 1986; Feng *et al.*, 1986a; Feng, 1987, 1990), and thus predicted the relatively complex temporal-spatial field of circulations dependent on t_0 in the general nonlinear system, proposed in the present paper.

In the convectively weakly nonlinear system an inter-tidal transport equation is expressed as

$$\frac{\partial\langle c\rangle}{\partial\tau}+\boldsymbol{u}_{\mathrm{M}}\cdot\nabla\langle c\rangle=\langle S_{\mathrm{T}}\rangle+\langle S\rangle \tag{27}$$

where $\langle c\rangle$, $\langle S_{\mathrm{T}}\rangle$ and $\langle S\rangle$ are respectively the Eulerian tidal cycle averages of field variables $c(\boldsymbol{x},t;\tau)$, $S_{\mathrm{T}}(\boldsymbol{x},t;\tau)$ and $S(\boldsymbol{x},t;\tau)$.

The inter-tidal transport equation, Eq. (27), has been proposed based on an additional hypothesis: $O(S_{\mathrm{T}})\leqslant\kappa^2$ and $O(S)\leqslant\kappa^2$ (Lu, 1991; Feng, 1990; Cheng *et al.*, 1989; Feng 1987; Feng *et al.*, 1986b). The additional hypothesis may be in general valid at least for $O(S_{\mathrm{T}})\leqslant\kappa^2$ (Csanady, 1982).

Nevertheless, the following argument is of much interest.

Without invoking the additional hypothesis on $O(S_{\mathrm{T}},S)\leqslant\kappa^2$, a reduction of Eq. (14) to Eq. (27) can be realized only based on $O(S_{\mathrm{T}},S)\leqslant O(\kappa)<1$.

Analyzing $(c_{\mathrm{L}},\boldsymbol{u}_{\mathrm{L}})=(\bar{c}_{\mathrm{L}},\bar{\boldsymbol{u}}_{\mathrm{L}})+(\tilde{c}_{\mathrm{L}},\tilde{\boldsymbol{u}}_{\mathrm{L}})$, where $(\bar{c}_{\mathrm{L}},\bar{\boldsymbol{u}}_{\mathrm{L}})$ are independent of t_0 and $(\tilde{c}_{\mathrm{L}},\tilde{\boldsymbol{u}}_{\mathrm{L}})$ are the tidally periodic functions of t_0, and substituting them to Eq. (14) with Eq. (15), a Eulerian tidal average of Eq. (14) with respect to t_0 yields

$$\frac{\partial\bar{c}_{\mathrm{L}}}{\partial\tau}+\bar{\boldsymbol{u}}_{\mathrm{L}}\cdot\nabla\bar{c}_{\mathrm{L}}=\bar{S}_{\mathrm{TL}}+\bar{S}_{\mathrm{L}}+S_{\mathrm{D}} \tag{28}$$

where $\bar{A}=\frac{1}{T}\int_0^T A(\boldsymbol{x},\tau;t_0)\mathrm{d}t_0$, A is a Eulerian variable, such as c_{L} and $\boldsymbol{u}_{\mathrm{L}}$; $\left(\bar{S}_{\mathrm{TL}},\bar{S}_{\mathrm{L}}\right)=\frac{1}{T}\int_0^T(S_{\mathrm{TL}},S_{\mathrm{L}})\mathrm{d}t_0$; $S_{\mathrm{D}}=-\nabla\cdot\overline{(\tilde{c}_{\mathrm{L}}\tilde{\boldsymbol{u}}_{\mathrm{L}})}$.

If $O(\kappa)<1$, the Lagrangian residual velocity may be approximately expressed as

$$\boldsymbol{u}_{\mathrm{L}}(\boldsymbol{x},\tau;t_0)=\boldsymbol{u}_{\mathrm{M}}(\boldsymbol{x},\tau)+\kappa\boldsymbol{u}_{\mathrm{LD}}(\boldsymbol{x},\tau;t_0) \tag{29}$$

(Feng, 1987; Cheng *et al.*, 1986; Feng *et al.*, 1986a), thus $\bar{\boldsymbol{u}}_{\mathrm{L}}=\boldsymbol{u}_{\mathrm{M}}$, $\tilde{\boldsymbol{u}}_{\mathrm{L}}=\kappa\boldsymbol{u}_{\mathrm{LD}}$ and thus $S_{\mathrm{D}}=-\kappa\nabla\overline{(\tilde{c}_{\mathrm{L}}\tilde{\boldsymbol{u}}_{\mathrm{LD}})}=O(\kappa)$.

Noting $(S_{\mathrm{TL}}, S_{\mathrm{L}}) = \langle (S_{\mathrm{T}}, S)(\boldsymbol{x}_0 + \kappa \boldsymbol{\xi}(t;\tau), t;\tau) \rangle = \langle (S_{\mathrm{T}}, S)(\boldsymbol{x}_0, t;\tau) \rangle + O(\kappa)$ or $(\overline{S}_{\mathrm{TL}}, \overline{S}_{\mathrm{L}}) = \overline{\langle (S_{\mathrm{T}}, S)(\boldsymbol{x}_0, t;\tau) \rangle} + O(\kappa) = \langle (S_{\mathrm{T}}, S)(\boldsymbol{x}_0, t;\tau) \rangle + O(\kappa)$, and neglecting the terms of $O(\kappa)$, a substitution of Eq. (29) into Eq. (28) yields

$$\frac{\partial \overline{c}_{\mathrm{L}}}{\partial \tau} + \boldsymbol{u}_{\mathrm{M}} \cdot \nabla \overline{c}_{\mathrm{L}} = \langle S_{\mathrm{T}} \rangle + \langle S \rangle \tag{30}$$

In view of $O(S_{\mathrm{T}}, S) \leqslant O(\kappa) < 1$ and thus $c_{\mathrm{L}} = c_{\mathrm{L}}(\boldsymbol{x}, \tau; t_0) = c(\boldsymbol{x}, t_0; \tau)$, a comparison between $c_{\mathrm{L}} = c(\boldsymbol{x}, t_0; \tau)$ and $c = c(\boldsymbol{x}, t; \tau)$ implies $\overline{c}_{\mathrm{L}} = \langle c(\boldsymbol{x}, t; \tau) \rangle$. Thus, Eq. (30) becomes the form of Eq. (27).

From the view point of physics, the equality, $\overline{c}_{\mathrm{L}} = \langle c \rangle$, may be explained as follows: the tidally averaged concentration at a fixed point $\boldsymbol{x}$ in flow field, $\langle c(\boldsymbol{x}, t; \tau) \rangle$, is identical to an arithmetical average of values of the tracer concentration sampled from water parcels that traveled through the fixed point $\boldsymbol{x}$, in sequence, over the period of a tidal cycle, $\langle c(\boldsymbol{x}, t_0; \tau) \rangle$.

The inter-tidal transport equation, Eq. (27), when compared to the general inter-tidal transport equation Eq. (14), states that only in a convectively weakly nonlinear system with $O(S_{\mathrm{T}}, S) \leqslant O(\kappa)$ can the Eulerian tide-averaged concentration be used approximately to describe the inter-tidal concentration transport process.

In a convectively weakly nonlinear system, the kinematic boundary condition at the free surface of the flow field of circulation can be extremely simplified. In fact, in view of $\overline{u}_{\mathrm{L}} = \overline{u}_{\mathrm{M}} + O(\kappa)$ and $O(\zeta_{\mathrm{L}}) = O(\kappa)$, Eq. (19) with Eq. (18) becomes Eq. (32) with Eq. (31), i.e., at

$$z = \kappa \zeta_{\mathrm{L}} \tag{31}$$

$$w_{\mathrm{M}} = \kappa \left(\frac{\partial \zeta_{\mathrm{L}}}{\partial \tau} + \boldsymbol{u}_{\mathrm{M}} \cdot \nabla \zeta_{\mathrm{L}} \right) - \kappa w_{\mathrm{LD}} \tag{32}$$

where w_{M} is the vertical component of the mass transport velocity $\boldsymbol{u}_{\mathrm{M}}$; w_{LD} is the vertical component of the Lagrangian drift velocity $\boldsymbol{u}_{\mathrm{LD}}$; $\kappa^2 \boldsymbol{u}_{\mathrm{LD}}$ has been omitted as $O(\kappa^2)$.

Neglecting the terms of $O(\kappa)$, Eq. (32) with Eq. (31) is reduced to Eq. (34) with Eq. (33), i.e., at

$$z = 0 \tag{33}$$

$$w_{\mathrm{M}} = 0 \tag{34}$$

Eq. (34) with Eq. (33) shows that in the convectively weakly nonlinear system, the kinematic boundary condition at the free surface is simplified as $w_{\mathrm{M}} = 0$ at $z = 0$, thus it may be considered as a "rigid lid" hypothesis. This simplified condition has been derived and contained in the field equations governing the mass transport velocity (Feng, 1987, 1990; Feng and Lu, 1993).

A substitution of the ordering expression (25) into the necessary constrained condition (4) yields $n < O\left(\kappa^{-2}\right)$. If we assume $O(\kappa)=10^{-1}$, then $O(n)<10^{2}$, which implies that the theory given above cannot be applied to the cases of such long averaging period of time as a season or half a year in a convectively weakly nonlinear coastal sea or tidal estuary, as pointed out previously by Feng (1987). Moreover, in view of the scale equalities (1) and (24), we have $T / T_* = O(t / \tau) = O\left(\kappa^{2}\right)$, which is consistent with that obtained by Lu (1991) and in a previous paper (Hamrick, 1987).

In a convectively weakly nonlinear system with $O(\kappa)<1$, where tidal currents and some quasi-steady flows are of the same order and dominant over the observable flow field, the scale analysis shows $O(\xi_{\mathrm{rc}} / \xi_{\mathrm{c}})=1$. Further, substituting the latter to the scale Eqs. (22) and (21), and in view of Eqs. (4) and (1), we obtain respectively $n < O\left(\kappa^{-1}\right)$ and $\delta = T / T_* = O(t / \tau) = O(\kappa)$, which have been previously derived and deduced by Feng and Wu (1995) and by Feng (1998). Nevertheless, it should be pointed out that, in a general nonlinear system with tidal and some quasi-steady flows being of the same order and dominant over the observable flow field, the Lagrangian tidally averaged theory on shallow water circulation with inter-tidal transport processes, proposed in the present paper, is not valid, because the necessary constrained condition (4), a basic hypothesis, has been violated here.

3.2 The nonlinearity due to turbulence

The shallow coastal waters, such as shallow seas or tidal estuaries, containing the turbulence of several scales, are the dissipative systems. Turbulence is, in essence, an origin of nonlinearity. While the nonlinearity of turbulence is not a necessity for the generation of residual circulation, the turbulent effect has been believed to play a major role in the dynamics of shallow water circulation with inter-tidal transport processes (Feng, 1987, 1990). As we have noted in the previous paper (Feng and Lu, 1993), particularly, the introduction of a turbulent closure model reveals the nonlinear coupling between tides and circulation through the nonlinear effects of turbulent stress and diffusion, and it also reveals that the astronomical tide affects the dynamics of circulation not only directly through the tidal body force but also indirectly through the turbulent energy produced by the vertical shear of tidal current and turbulent mixing. The physics and dynamics on turbulence in the shallow coastal/estuarine system is undoubtedly the most complicated problem remaining to be solved. The further discussion about turbulence is beyond the scope of this paper. Nevertheless, it should be of great value to point out that, several turbulent closure models can be used for numerically solving the original fields of velocity/concentration present in the intra-tidal process, and then the fields of velocity/concentration in the circulation with the inter-tidal process, i.e., a set of fields of the Lagrangian residual velocity/the Lagrangian inter-tidal concentration depending on the tidal phase, will be obtained from the original fields, based on the Lagrangian tidally-averaged circulation theory for a general nonlinear tidally dominant shallow water system, presented in the present paper. It is quite evident that all of information on turbulence having been contained in the adopted turbulent closure model must be involved in the final solutions for circulation.

4. Conclusions

The fundamentals of a Lagrangian tidally-averaged dynamics on the circulation with inter-tidal transport processes has been proposed for a general nonlinear shallow coastal/estuarine system.

A Lagrangian residual velocity is strictly defined in terms of a Lagrangian mean velocity, based upon the following three constrained conditions having to be satisfied. One of them is a basic hypothesis that the net displacement of the labeled water parcel over one or a few tidal cycles is negligible small when compared with the characteristic length of the flow field of circulation, which constrains the number of tidal cycles used for averaging unable to be arbitrarily large. Another is that the Lagrangian mean velocity should be proven to be a solenoidal vector field of flow, for the shallow water circulation has been considered as an incompressible flow. The other is that the Lagrangian mean velocity must be proven to satisfy the conservation equation of a material surface in the flow field of circulation and in particular along the corresponding boundaries, based on the general principle of continuity of flow field. Thus, we define the Lagrangian residual velocity as the Lagrangian mean velocity satisfying the above three necessarily constrained conditions. Consequently, it is rationally proposed that the Lagrangian residual velocity should and can be used to embody the velocity of circulation.

A new concept of the concentration for inter-tidal transport processes associated with circulation is presented by means of a new inter-tidal transport equation derived through a Lagrangian tidal average of an intra-tidal transport equation. It thus seems proper to name this concentration describing the inter-tidal transport processes as a "Lagrangian inter-tidal concentration" according to its physical meaning. In the new inter-tidal transport equation governing the Lagrangian inter-tidal concentration derived, the convective transport velocity is the Lagrangian residual velocity, and here there is not any additional "dispersion" term.

It should be, with emphasis, pointed out that a unique property of the Lagrangian residual velocity/the Lagrangian inter-tidal concentration is its dependence on the tidal phase when the water parcels are labeled and released. The unique property reveals that the circulation described above contains a set of infinite temporal-spatial fields of velocity/concentration, each of which is corresponding to a specific value of tidal phases varying continuously over one tidal cycle.

As might be expected, when the convectively weakly nonlinear condition is approximately satisfied, a set of infinite temporal-spatial fields of velocity/concentration of circulation can be reduced to a single one: the Lagrangian residual velocity and the Lagrangian inter-tidal concentration will be approximately reduced to the mass-transport velocity and the Eulerian tidally-averaged concentration (with $O(S_T, S) \leqslant O(\kappa)$) exhibited traditionally, respectively.

In the general nonlinear system, analytical solutions for the problem of circulation do not exist, but the numerical researchers on it are effective. In the next paper, as the first example, the numerical researches in three types of rectangular model sea will illustrate the dependence of circulation on tidal phase due to the convectively nonlinear effect; and then, the present theory will be tested in the Bohai Sea.

References

Awaji T. 1982. Water mixing in a tidal current and the effect of turbulence on tidal exchange through a strait. Journal of Physical Oceanography, 12(6): 501–514

Awaji T, Imasato N, Kunishi H. 1980. Tidal exchange through a strait: a numerical experiment using a simple model basin. Journal of Physical Oceanography, 10(10): 1499–1508

Bowden K F. 1967. Circulation and diffusion. In: G H, Lauff, ed. Estuaries. Washington, D.C.: Amer Assoc Advance Sci Publ, 15–36

Cheng R T. 1983. Euler-Lagrangian computations in estuarine hydrodynamics. In: C Taylor, ed. Proc of the Third Intern Conf on Numerical Method in Laminar and Turbulent Flows. Pineridge Press, 341–352

Cheng R T, Casulli V. 1982. On Lagrangian residual currents with applications in South San Francisco Bay, California. Water Resources Research, 18(6): 1652–1662

Cheng R T, Casulli V. 1991. Dispersion in tidally-averaged transport equation. In: D Prandle, ed. Dynamics and Exchanges in Estuaries and the Coastal Zone. Coastal and Estuarine Studies, 40. American Geophysical Union, 409–428

Cheng R T, Feng Shizuo, Xi Pangen. 1986. On Lagrangian residual ellipse. In: J van de Kreeke ed. Physics of Shallow Estuaries and Bays. Lecture Notes on Coastal and Estuarine Studies, 16. Springer-Verlag, 102–113

Cheng R T, Feng Shizuo, Xi Pangen. 1989. On inter-tidal transport equation. In: B J Neilson, J Brubaker and A Kuo eds. Estuarine circulation. The Humana Press Inc, 133–156

Csanady G T. 1982. Circulation in the Coastal Ocean. Dordrecht-Boston-London: D Reidel Pub Com, 1–279

Delhez E J M. 1996. On the residual advection of passive constituents. Journal of Marine Systems, 8(3–4): 147–169

Dortch M S, Chapman R S, Abt S R. 1992. Application of three-dimensional Lagrangian residual transport. Journal of Hydraulic Engineering, 118(6): 831–848

Dyke P P G. 1980. On the Stokes' drift induced by tidal motions in a wide estuary. Estuarine and Coastal and Marine Science, 11(1): 17–25

Feng Shizuo. 1987. A three-dimensional weakly nonlinear model of tide-induced Lagrangian residual current and mass-transport, with an application to the Bohai Sea. In: J C J Nihoul, B M Jamart, eds. Three-Dimensional models of Marine and Estuarine Dynamics, Elsevier Oceanography Series, 45, Elsevier, 471–488

Feng Shizuo. 1990. On the Lagrangian residual velocity and the mass-transport in a multi-frequency oscillatory system. In: R T Cheng, ed. Residual Currents and Long-term Transport, Coastal and Estuarine Studies 38. Springer-Verlag, 34–48

Feng Shizuo. 1991. The dynamics on tidal generation of residual vorticity. Chinese Science Bulletin, 36(24): 2043–2046

Feng Shizuo. 1998. On circulation in Bohai Sea Yellow Sea and East China Sea. In: Hong G H, Zhang J, B K Park, eds. The Health of the Yellow Sea. Seoul, Korea: The Earth Love Publication Association, 43–77

Feng Shizuo, Cheng R T, Sun Wenxin, *et al*. 1990. Lagrangian residual current and long-term transport processes in a weakly nonlinear baroclinic system. In: Wang H, Wang J, Dai H, eds. Physics of Shallow Seas. Beijing: China Ocean Press, 1–20

Feng Shizuo, Cheng R T, Xi Pangen. 1986a. On tide induced Lagrangian residual current and residual transport, 1. Lagrangian residual current. Water Resources Research, 22(12): 1623–1634

Feng Shizuo, Cheng R T, Xi Pangen. 1986b. On tide induced Lagrangian residual current and residual transport, 2. Residual transport with application in South San Francisco Bay, California. Water Resources Research, 22(12): 1635–1646

Feng Shizuo, Lu Youyu. 1993. A turbulent closure model of coastal circulation. Chinese Science Bulletin, 38(20): 1737–1741

Feng Shizuo, Wu Dexing. 1995. An inter-tidal transport equation coupled with turbulent K-ε model in a tidal and quasi-steady current system. Chinese Science Bulletin, 40(2): 136–139

Feng Shizuo, Xi Pangen, Zhang Shuzhen. 1984. The baroclinic residual circulation in shallow seas. Chinese Journal of Oceanology and Limnology, 2(1): 49–60

Fischer H B, List E J, Koh R C Y, *et al*. 1979. Mixing in Inland and Coastal Waters. New York: Academic Press, 1–483

Foreman M G G, Baptista A M, Walters R A. 1992. Tidal model studies of particle trajectories around a shallow coastal bank. Atmosphere-Ocean, 30(1): 43–69

Hainbucher D, Wei Hao, Pohlmann T, *et al*. 2004. Variability of the Bohai Sea circulation based on model calculations. Journal of Marine System, 44 (3–4): 153–174

Hamrick J. 1987. Time-averaged estuarine mass transport equations. In: R M Ragen, ed. Proc. of the 1987 National Conf on Hydraulic Engineering. ASCE, 1–8

Heaps N S. 1978. Linearized vertical-integrated equations for residual circulation in coastal seas. Deut Hydrog Z, 31 (5): 147–169

Huang P S, Wang D P. 1986. Analysis of residual currents using a two-dimensional model. In: J van de Kreeke, ed. Physics of shallow Estuaries and Bays, Lecture Notes on Coastal and Estuarine Studies 16. Springer-Verlag, 71–80

Ianniello J P. 1977. Tidally-induced residual currents in estuaries of constant breadth and depth. Journal of

Marine Research, 35(4): 755–786

Ianniello J P. 1979. Tidally-induced residual currents in estuaries of variable breadth and depth. Journal of Physical Oceanography, 9(5): 962–974

Jay D A. 1991. Estuarine salt conservation—a Lagrangian approach. Estuarine, Coastal and Shelf Science, 32(6): 547–565

Jiang Wensheng, Sun Wenxin. 2002. Three dimensional tide-induced circulation model on a triangular mesh. International Journal of Numerical Methods in Fluids, 38(6): 555–566

Jiang Wensheng, Wang Jingyong, Zhao Jianzhong, *et al.* 1997. An observation of current in Bohai Gulf and its analysis. Journal of Ocean University of Qingdao (in Chinese), 27(1): 23–32

Jiang Wensheng, Wu Dexing, Gao Huiwang. 2002. The observation and simulation of bottom circulation in the Bohai Sea in summer. Journal of Ocean University of Qingdao (in Chinese), 32(4): 511–518

Lei Kun, Sun Wenxin, Liu Guimei. 2004. Numerical study of the circulation in the Yellow Sea and East China Sea: IV. Diagnostic calculation of the baroclinic circulation. Journal of Ocean University of Qingdao (in Chinese), 34(6): 937–941

Liu Guimei, Wang Hui, Sun Song, *et al.* 2002. Numerical study on density residual currents of the Bohai Sea in summer. Chinese Journal of Oceanology and Limnology, 21(2): 106–113

Longuet-Higgins M S. 1969. On the transport of mass by time-varying ocean currents. Deep-Sea Research, 16(5): 431–447

Lu Youyu. 1991. On the Lagrangian residual current and residual transport in a multiple time scale system of shallow seas. Chinese Journal of Oceanology Limnology, 9(2): 184–192

Nihoul J C J, Ronday F C. 1975. The influence of the "tidal stress" on the residual circulation. Tellus, 27(5): 484–490

Orbi A, Salomon J C. 1988. Dynamique de mare dans le golfe normndbreton. Oceanologica Acta, 11(1): 55–64

Prandle D. 1986. Generalized theory of estuarine dynamics. In: J van de Kreeke, ed. Physics of Shallow Estuaries and Bays. Lecture Notes on Coastal and Estuarine Studies, 16: 42–57

Prandle D. 1991. A new view of near-shore dynamics based on observations from HF radar. Progress in Oceanography, 27(3–4): 403–438

Pritchard D W. 1954. A study on the salt balance in a coastal plain estuary. Journal of Marine Research, 13(1): 133–144

Ridderinkhof H, Loder J W. 1993. Lagrangian characterization of circulation over submarine banks with application to the outer Gulf of Maine. Journal of Physical Oceanography, 24: 1184–1200

Salomon J C, Breton M. 1993. An atlas of long-term currents in the Channel. Oceanologicca Acta, 16(5–6): 439–448

Sun Wenxin, Liu Guimei, Lei Kun, *et al.* 2001. The numerical study of circulation in the Yellow and East China Sea, II. Numerical simulation of tide and tide-induced circulation. Journal of Ocean University of Qingdao (in Chinese), 31(3): 297–304.

Sun Wenxin, Xi Pangen, Song Lina. 1989. Numerical calculation of the three-dimensional tide-induced Lagrangian residual circulation in the Bohai Sea. Journal of Ocean University of Qingdao (in Chinese), 19(2): 27–36.

Tang Yuxiang, Tee K T. 1987. Effects of mean and tidal current interaction on the tidally induced residual current. Journal of Physical Oceanography, 17(2): 215–230

Tee K T. 1976. Tide-induced residual current: A 2-D nonlinear numerical model. Journal of Marine Research, 34(4): 603–628

Tee K T. 1987. Simple models to simulate three-dimensional tidal and residual currents. In: N S Heaps, ed. Three-dimensional Coastal Ocean Models. Washington D.C.: American Geophysical Union, 125–147

Visser A W, Bowman M J. 1991. Lagrangian tidal stress and basin-wind residual eddy dynamics in wide coastal sea straits. Geophys, Astrophys, Fluid Dynamics, 59: 113–145

Wang Kai. 1996. Study on model and numerical method of coastal sea circulation and its application to the East China Sea (in Chinese)[dissertation]. Qingdao, China: Ocean University of Qingdao

Wang Hui, Su Zhiqing, Feng Shizuo, *et al.* 1993. A three-dimensional numerical calculation of the wind driven thermohaline and tide-induced Lagrangian residual current in the Bohai Sea. Acta Oceanologica Sinica, 12: 169–182

Wei Hao, Hainbucher D, Pohlmann T, *et al.* 2004a. Tidal-induced Lagrangian and Eulerian mean circulation in the Bohai Sea. Journal of Marine System, 44(3–4): 141–151

Wei Hao, Sun Jun, Moll A, *et al.* 2004b. Phytoplankton dynamics in the Bohai Sea-observations and modeling. Journal of Marine Systems, 44(3–4): 233–251

West J R, Uncles R J, Stephens J A, *et al.* 1990. Longitudinal dispersion processes in the upper Tamar Estuary. Estuaries, 13(2): 118–124

Wright L D, Wiseman W J, Yang Zuosheng, *et al.* 1990. Processes of marine dispersal and deposition of suspended silts off the modern mouth of the Huanghe (Yellow River). Continental Shelf Research, 10(1): 1–40

Yu Guangyao, Chen Shijun. 1983. Numerical modeling of the circulation and the pollutant dispersion in Jiaozhou Bay: III. The Lagrangian residual current and the pollutant dispersion. Journal of Shandong College of Oceanology (in Chinese), 13(1): 1–13

Zhao Liang, Wei Hao, Feng Shizuo. 2002. Annual cycle and budgets of nutrients in the Bohai Sea. Journal of Ocean University of China, 1(1): 29–37

Zheng Lianyuan. 1992. Numerical simulation of three-dimensional tide-induced Lagrangian residual velocity and its application to Bohai Sea. Journal of Ocean University of Qingdao (in Chinese), 22(1): 39–49

Zhou Xubo, Sun Wenxin. 2006. Numerical simulation of Lagrangian circulations in the East China Sea: II. Simulation of the circulations. Journal of Ocean University of China (in Chinese), 36(1): 7–12

Zhou Xubo, Sun Wenxin, Su Jian. 2002. Self-adaptive numerical simulation of tide-induced residual current in the East China Sea. Journal of Ocean University of Qingdao (in Chinese), 32(2): 173–178

Zimmerman J T F. 1979. On the Euler-Lagrangian transformation and the Stokes' drift in the presence of oscillatory and residual currents. Deep-Sea Research, 26A (5): 505–520

A Lagrangian Mean Theory on Coastal Sea Circulation with Inter-tidal Transports II. Numerical Experiments*

Ju Lian, Jiang Wensheng, Feng Shizuo

1. Introduction

Oean circulation is a very important term in physical oceanography, which means the water flows in a closed circular pattern within the ocean. When people think of this term an image often appears in their minds that the water flows in a specific direction with a rather steady speed along a certain path, which is just like a river in the ocean. The circulation in the ocean is generally believed to be driven by both the wind forcing at the sea surface and horizontal pressure gradients resulting from the density stratification. Nevertheless, it is really another story in the coastal seas.

In the coastal seas normally the tidal currents contain most of the energy compared with movements of other types. So unlike the circulation in the open ocean the currents that can be seen are not the steady state uni-directional flow but are periodically oscillating currents. From the raw current speed data one can hardly tell any trends at a glance. It seems that there does not exist a long time scale circulation. Yet the water still has a long term moving tendency in the sea which can be sensed from the distribution of tracers no matter whether they are active tracers such as salinity or passive tracers such as contaminants. So the question is how to define, on the basis of the periodically tide dominated current data, a vector variable with the dimension of speed to describe the long term water moving tendency which can be regarded as the coastal sea circulation pattern. A straightforward way to do this is to eliminate the periodicity of tide by using a variety of time series analysis methods from the raw current data sequences sampled at a specific point. This is the core of the so-called Eulerian mean circulation theory which was proved improper by many researchers (Longuet-Higgins, 1969; Zimmerman, 1979; Cheng and Casulli, 1982; Csanady, 1982; Feng *et al.*, 1986a, b) and was followed by the proposition of the Lagrangian mean theory under weakly nonlinear conditions (e.g., Cheng *et al.*, 1986; Feng *et al.*, 1986a, b, 1990; Feng, 1987, 1990, 1998; Feng and Lu, 1993; Feng and Wu, 1995), on which a brief review has been made in the previous paper (Feng *et al.*, 2008).

* Ju L, Jiang W, Feng S. 2009. A Lagrangian mean theory on coastal sea circulation with inter-tidal transports II. Numerical experiments. Acta Oceanologica Sinica, 28(1): 1–14

In the previous paper (Feng *et al.*, 2008) a generalized Lagrangian residual circulation theory is proposed to describe the coastal circulation in a tide dominant system without restriction to weakly nonlinear system. Generally speaking, three requirements must be met to make a vector field to become a velocity field. The first requirement is that the vector field should be continuous and differentiable. Secondly, the material surface must be conservative and the third requirement is that it must satisfy the mass conservation equation of the incompressible fluid. Within the new concept framework the net transport velocity of the water parcel averaging over several tide cycles is proved to satisfy all the above three requirements if the net displacement of the water parcel is at least one order smaller compared with the scale of circulation. The so-constructed Lagrangian mean velocity, or a Lagrangian residual velocity, can represent the coastal sea circulation which is obviously not only the function of position and time but also the function of the initial tidal phase when the water parcel is released. So the circulation field is not a single field traditionally expected but a circulation set of infinite velocity fields corresponding to different initial phases. The dispersion of different velocity fields is influenced by the nonlinearity of the coastal sea hydrodynamic systems as discussed in Feng *et al.* (2008).

In addition to the wind forced part and the thermohaline part, the coastal sea circulation set also contains the residual current induced by the nonlinearity of oscillating tidal current. Of course there are also interactions between all these parts.

In order to demonstrate the characteristics of the newly constructed circulation field the numerical model is used with the single frequency tidal current considered only. Three idealized model areas are studied as examples to show the behaviors of the circulation. They are a flat bottom rectangular basin and a rectangular basin with staircase-like topography. The third basin, in addition to the characteristics of the second one, has an intrusion of the land into the sea area. Then the method is applied to the Bohai Sea in China.

2. Numerical Experiments in Model Sea Areas

2.1 Methods

As described in the previous paper (Feng *et al.*, 2008) the circulation field can be obtained from the net displacement of the water parcels after several tidal periods. This can be regarded as a data analysis method which is used to retrieve the circulation from the primitive current fields. So the tidal current should be simulated first.

The intra-tidal process is simulated by HAMSOM model developed by Professor Backhaus (1985) at Hamburg University and further developed by other researchers (Pohlmann, 1996, 2006). It is a 3D baroclinic prognostic model, of which only the barotropic part is used in the present study. In every example M_2 tide is imposed at the open boundary with amplitude of 1 m. The simulation results are of nonlinear single frequency tidal system. Although atmospheric forcing at the sea surface and the density stratification are not considered in the present study, turbulence cannot be omitted. For simplicity the eddy viscosity coefficient is taken as a constant value of 0.0075 m^2/s, though this is not a requirement for the new coastal sea circulation theory proposed here. The bottom friction is represented by a quadratic law.

After the tidal current is simulated a Lagrangian tracing program is applied to get the net displacement of the water parcel in the sea. As discussed in the previous paper the average time cannot be long, and only 1 tidal cycle is used in the study. Thus the Lagrangian residual velocity can be obtained for a specific initial tidal phase. In the study the initial tidal phase is also changed to see the dispersion of the whole circulation set.

Three kinds of model seas different in topography and shape are studied. The first one is the basic one and is a rectangular sea area with a flat bottom and an open boundary in the east (Case I). The topography changes to a staircase type in the second model (Case II). As for the third model a rectangular stretch of land extends into the model sea area (Case III).

In all model seas, the horizontal resolution is 1′ forming a grid of 180×100. The tidal waves propagate from the east boundary with the amplitude of 1 m. The time step is 360 s and the average time for the Lagrangian residual velocity is one tidal period.

2.2 Case I

The simulation results of the Lagrangian residual velocity in a rectangular basin with 10 m in depth (in view of Fig. 1) form an anti-clockwise gyre in the model area. The same results and analysis can be seen in Shi (1995), while the Eulerian mean velocity shows that the net water flux is out of the area which violates the mass conservation law.

As emphasized before, when the water parcels are released in the nonlinear coastal system, the Lagrangian residual velocity is not only the function of the position of the water parcel but also the function of the initial tidal phase. In Fig. 1 four circulation fields are displayed

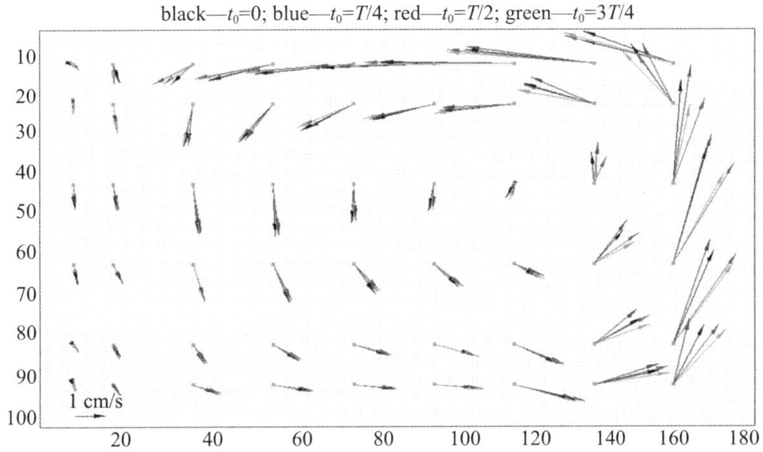

Fig. 1 Surface Lagrangian residual velocity at different initial tidal phase of constant depth model sea of 10 m

corresponding to four different initial tidal phases, which are t_0=0, $T/4$, $T/2$, $3T/4$. However there is nearly no obvious difference among the Lagrangian residual velocity fields at the four initial tidal phases.

It is pointed out in Feng *et al.* (1986a), Feng (1987) and also in the previous paper (Feng *et al.*, 2008) that in the coastal seas the ratio between the tidal range and the local water depth κ is an important parameter to evaluate the nonlinearity. The distribution of κ in this model sea area is shown in Fig. 2. It can be seen that the scale of the parameter κ is about $O\left(10^{-1}\right)$ in most part of the sea area, that is to say, the dynamic system here is a convectively weakly nonlinear one. The dispersion of the Lagrangian residual velocity fields here coincides with this result.

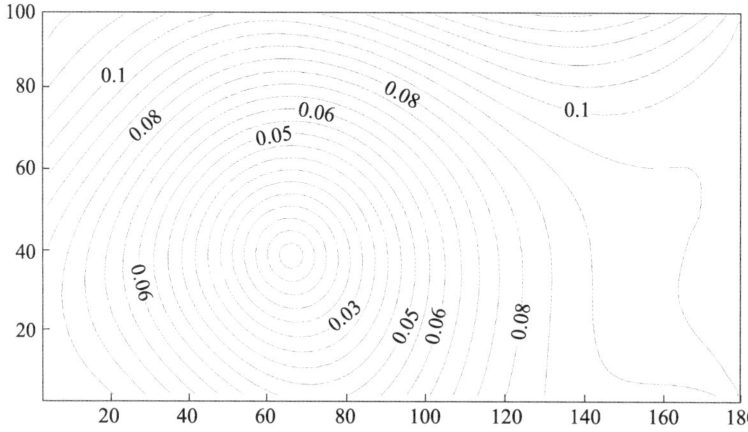

Fig. 2 The distribution of κ in the constant model sea of 10 m

The comparison between the distribution of the parameter κ and the Lagrangian residual velocity in this model sea shows an apparent conflict at the west end of the model sea. The parameter κ is comparatively big at that area, while the Lagrangian residual velocity and the dispersion among different velocity fields are small (Fig. 1 and Fig. 2). When the water depth of the model area is reduced from 10 m to 5 m (the figure is omitted in the paper) and the same tidal forcing is imposed at the open boundary, the Lagrangian residual velocity does not increase as expected. This can be partially due to the turbulent friction effect. It should be emphasized that friction plays an important role in the coastal sea system, and further attention should be paid on friction in the application (Feng, 1987; Feng and Lu, 1993).

2.3 Case II

In order to increase the nonlinearity of the system, the model sea with a staircase-like topography is constructed, with the depths ranging from 2 m at the end of the sea to 10 m at the open boundary. The Lagrangian residual velocity is shown in Fig. 3, and has the same anti-clockwise gyre as the model sea with the constant depth in Fig. 1. But in Fig. 3 the Lagrangian residual velocity is bigger than that in Fig. 1.

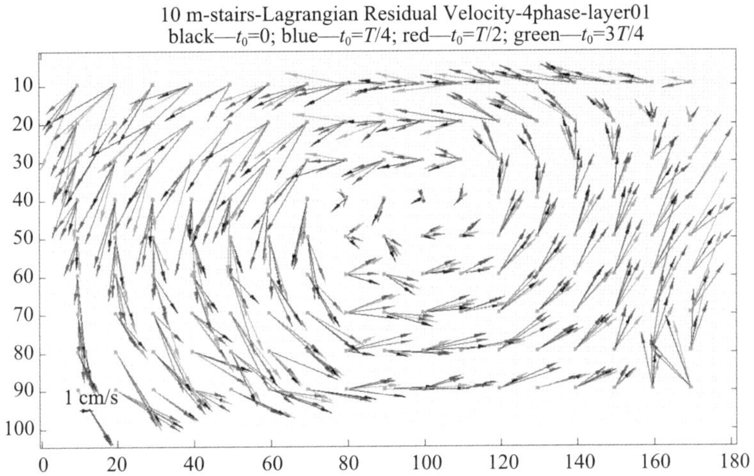

Fig. 3 Surface Lagrangian residual velocity at different initial tidal phase of stairs shape model sea

The Lagrangian residual velocity fields are also simulated at four different initial times, that is, t_0=0, T/4, T/2, 3T/4, corresponding to the initial tidal phases, $\theta_0 = 0, \pi/2, \pi, 3\pi/2$. The differences of the Lagrangian residual velocities are more prominent than those in the first model. This can be explained by comparing the distribution of κ in Fig. 2 and Fig. 4. It can be seen from both figures that the distribution of parameter κ in Fig. 4 is bigger than that in Fig. 2. According to the theory in the previous paper that the increase of parameter κ indicates that the nonlinear effect of the system is enhanced, the variation of the Lagrangian residual velocity with the initial tidal phase also increases.

The distribution of κ also shows different patterns in the two cases and in Fig. 4, the distribution of κ well corresponds to the staircase-like topography, while the corresponding pattern in the dispersion of different velocity fields is hardly seen. In order to quantitatively analyze the dispersion of the Lagrangian residual velocity with the initial tidal phase, a new index D, the drift dispersion index of the Lagrangian residual velocity, is introduced.

$$D = \frac{S(\xi_{\mathrm{r}})}{T}$$

where $\xi_{\mathrm{r}}(\boldsymbol{x}_0, \tau; t_0) = \int_{t_0}^{t_0+T} \boldsymbol{u}(\boldsymbol{x}_0 + \boldsymbol{\xi}(t,\tau), t; \tau) \mathrm{d}t$, which is the net displacement of the water parcel starting from $\boldsymbol{x}_0$ at time t_0 after one tidal cycle. With the change of t_0 from 0 to T, ξ_{r} can formulate an enclosed area which is exactly an eclipse when it is under convectively weakly nonlinear conditions (Feng, 1987). S is a functional of the enclosed curve ξ_{r}. The result of S is based on three steps, the first one is to get the projection of ξ_{r} to the horizontal plane, the second one is to have the projected ξ_{r} fitted to an eclipse and finally the area of the eclipse is obtained. Then the drift dispersion index D is obtained which has the same dimension of the turbulent eddy viscosity coefficient, m^2/s.

The distribution of the parameter D looks like several tongues in Fig. 5. Each local peak value appears at the stairs break where topography changes abruptly. A comparison between the distribution of D and κ shows the following similarity: where there is a big nonlinear parameter κ there is a large value of D which means the large dispersion of the different velocity fields.

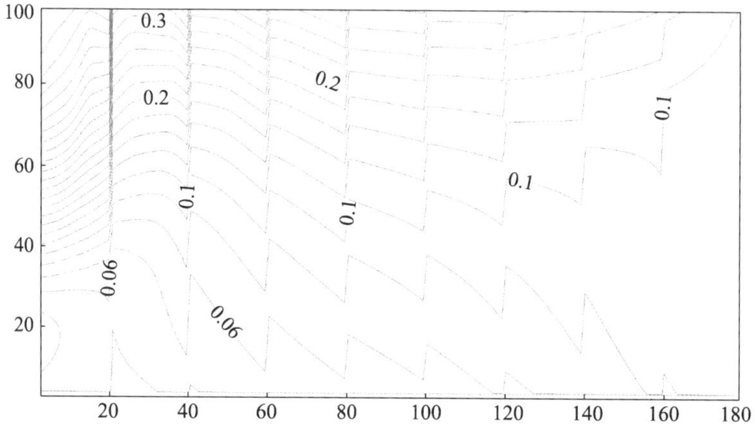

Fig. 4 The distribution of κ in the stairs shape model sea

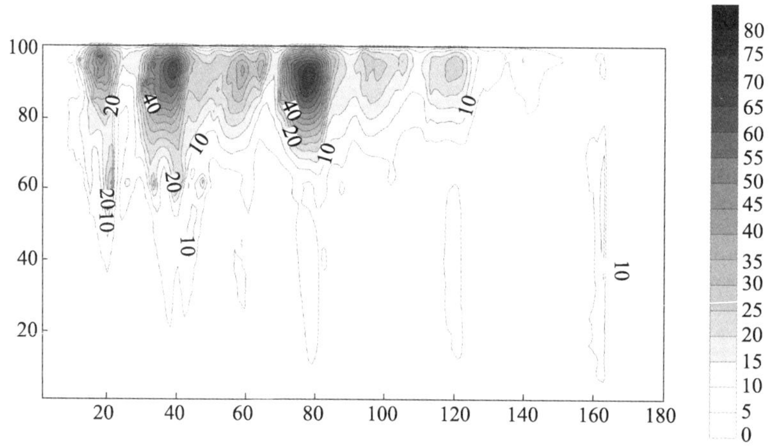

Fig. 5 The distribution of D at the surface of stair shape model sea

Through the comparison between the results of the model sea with the constant depth and those of the model sea with the varying topography, an important conclusion can be drawn, that is, with the increase of κ, the dependency of the Lagrangian residual velocity on the initial tidal phase becomes stronger and stronger. This supports the theoretical conclusion that the Lagrangian residual velocity, which can embody the coastal circulation, is the function of the time, space and the initial tidal phase and varies with the initial tidal phase on the circulation time scale.

In the weakly nonlinear system, the Lagrangian drift can be neglected. However, in the above two examples, although the scale of the parameter κ does not arrive at order 1, the variation of the Lagrangian residual velocity with the initial tidal phase cannot be neglected in some parts of the sea. Sun *et al.* (1989) has already noticed that the Lagrangian drift may not be neglected at some areas even when κ is only of $O(0.1)$ while applying the weakly nonlinear theory to the Bohai Sea. In fact, just as discussed in the previous paper (Feng *et al.*, 2008), κ is only an indicator parameter to define the nonlinearity. Some new parameters, such as D, which can capture more details of the local sea, may be introduced to define the effect of the nonlinearity.

2.4 Case III

The influences of the irregular coast line on the circulation pattern are studied in this part. A stretch of rectangular land extends into the model area described in the previous part. Thus the model sea is divided into three parts by the coast line, two bays with the constant depth of 5 m at the end of the model area and one outer bay with the maximum depth of 10 m in the shape of a staircase (Fig. 6).

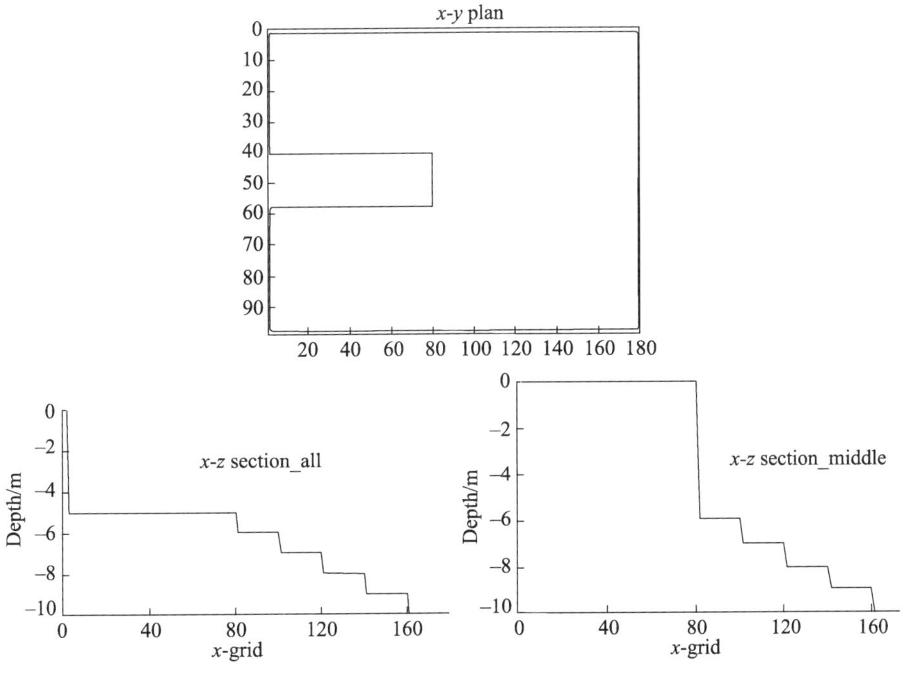

Fig. 6 The section of the model sea with a cape

The Lagrangian residual velocity fields are calculated as above and are displayed in Fig. 7. The results are similar to the above two examples. There is an anti-clockwise gyre in each of the three bays. The general Lagrangian residual velocity in the whole area is bigger than those in the above two cases which hints at the strong nonlinearity caused both by the topography and the coast line. The maximum Lagrangian residual velocity, which appears at the head of the cape, is nearly 20 cm/s. The pattern of the Lagrangian residual velocity is relatively smooth in the whole area in which no eddies can be found near the cape, just unlike the Eulerian residual circulation which always forms clear eddy pair. In view of the fact that the Eulerian residual circulation does not satisfy the conservation law of material surface that kind of eddy should be false one (Delhez, 1996; Wei *et al.*, 2004a; Foreman *et al.*, 1992).

The dispersion of the Lagrangian residual velocity fields with the initial tidal phase is more obvious in this model sea than that in the above two cases (Fig. 7). In the southern small bay with the constant depth, the nonlinear parameter κ is smaller than that in the rest of the sea area (Fig. 8), and the Lagrangian residual velocity fields for different initial tidal phases do not change much (Fig. 7). However, near the cape, the Lagrangian residual velocities for different initial tidal phases even run in the opposite direction because of the bigger κ there.

The quantitative estimation of the dispersion of Lagrangian residual velocity fields is shown in Fig. 9 through parameter D. It can be seen that larger dispersions occur both in the outer bay and in the northern small bay, which coincides with the distribution of κ very well. But a closer examination of Fig. 9 shows that the appearance of the large D near the cape cannot be explained by the distribution of κ. So κ is not fully appropriate to describe the nonlinearity at some local points and the nonlinear advection terms are examined below to give the answer.

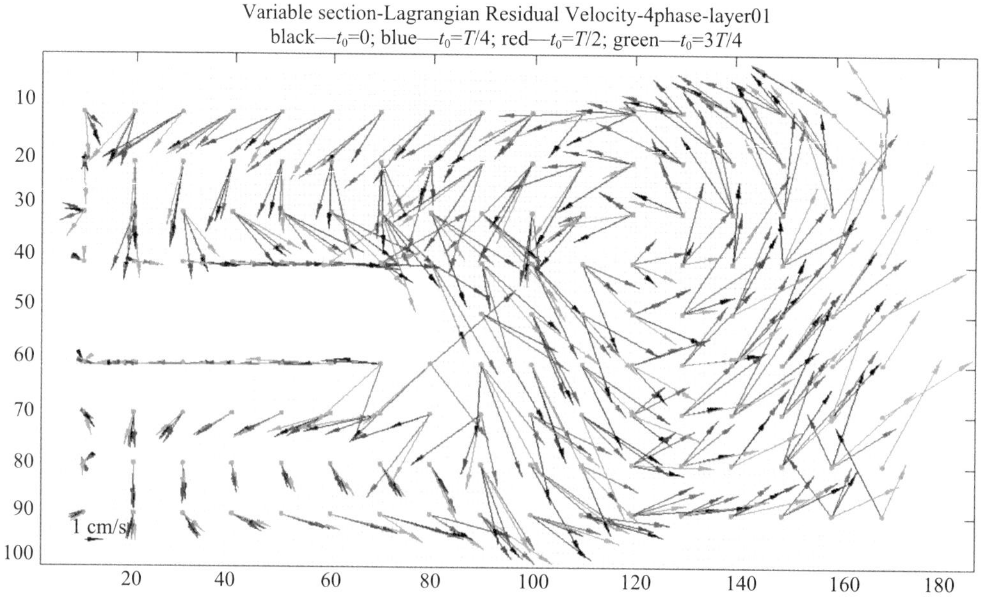

Fig. 7 Surface Lagrangian residual velocity at different initial tidal phase of model sea with a cape

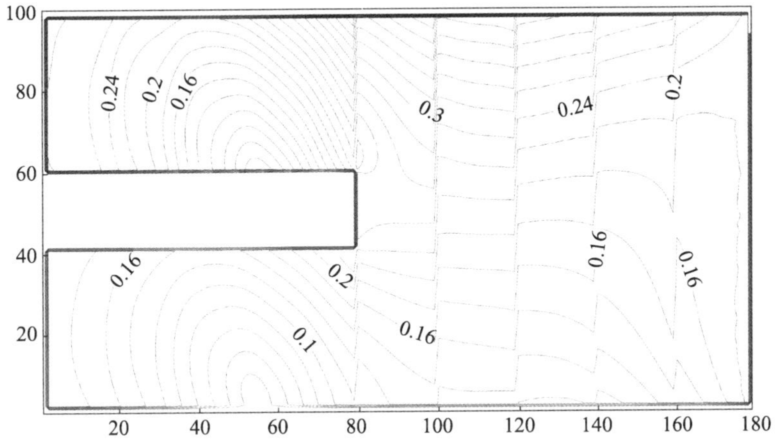

Fig. 8 The distribution of κ in the model sea with a cape

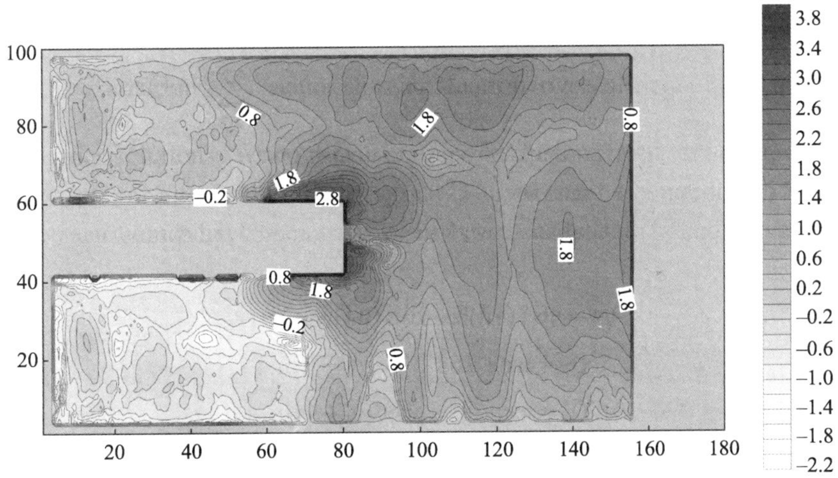

Fig. 9 The distribution of D at the surface of model sea with a cape (in the logarithm of 10)

According to the momentum equation in x-direction, the non-dimensional parameter ε to indicate the estimation of the ratio between nonlinear advection term over the local term is:

$$O(\varepsilon)=O\left(\frac{u\dfrac{\partial u}{\partial x}}{\dfrac{\partial u}{\partial t}}\right)=\frac{u_{\mathrm{c}}T}{L}$$

where u_{c} is the scale of the current speed, T is the tidal period or the characteristic eigen period of the system and L is the scale of the current field.

Assuming λ is the scale of wave length of the long gravity wave then the relation between ε and κ is as follows:

$$O(\varepsilon)=\frac{u_{\mathrm{c}}T}{L}=\frac{u_{\mathrm{c}}T}{\lambda}\left(\frac{\lambda}{L}\right)=\kappa\left(\frac{\lambda}{L}\right)$$

Supposing the horizontal scale of the current field, L, and the wave length of the gravity wave, λ, are in the same scale, then

$$O(\varepsilon)=\kappa \qquad (L\sim\lambda)$$

The parameter κ can be used to indicate the nonlinearity of the system. In certain cases L is smaller than λ in the scale. So the parameter κ itself cannot fully describe the convective nonlinearity. For example, when the study area is near a cape or with complicated geometry the horizontal scale of the current field becomes the horizontal scale of the local area which is much smaller than the wave length λ, and ε can approach unity while κ may be small. If κ is used to estimate the nonlinearity it may be underestimated.

3. Application to the Bohai Sea

The Bohai Sea is a semi-enclosed sea, which connects with the Yellow Sea only through the Bohai Strait. It is divided into four parts: the Liaodong Gulf, the Bohai Gulf, the Laizhou Bay and the Central Basin. It covers an area of 8×10^4 km^2 with a width of 300 km in east-west and a length of 500 km in northeast-southwest divection. The water depth increases gradually from the coastal area to the Central Basin with a mean depth of 18 m except the shallow Liaodong Bank and a canyon in the Liaodong Gulf. The deepest depth is 86 m in the northern part of the Bohai Strait.

The Bohai Sea is a typical shallow sea dynamic system in which the astronomical tidal movements are predominant. The circulation here is always a problem to the researchers since it is difficult to retrieve the weak circulation by eliminating the strong periodically tidal effect. Many researchers studied the circulation in the Bohai Sea and the results do not agree with each other very well. Some researchers also studied the ecosystem based on the circulation in the Bohai Sea (Wei *et al.*, 2003; Wei, 2004b; Liu and Yin, 2006; Wang *et al.*, 2007). The practical difficulty of acquiring the Bohai Sea circulation is that there is no strong inter-tidal current in the area and the fundamental problem is how to establish the conceptual framework of circulation in a tide dominant system. So here it is very interesting that the Lagrangian mean theory on coastal sea circulation is applied to the Bohai Sea after its application to the model seas in the previous section to see its applicability and feasibility.

3.1 Model scheme

M_2 tide is the major constituent in the tidal system in the Bohai Sea. So here the M_2-induced Lagrangian mean circulation is simulated to represent the tide-induced Lagrangian mean circulation of the Bohai Sea. The topography data of 2′ resolution is provided by Choi (personal communication) and is then interpolated to 1′. Because great changes in topography have taken place near the Yellow River estuary, which expands to the north, the topography is modified according to field data measured in 1998 and coastal lines based on satellite images (Fig. 10).

A grid with the resolution of $1' \times 1'$ both in longitude and latitude is taken when simulating the M_2 tidal current in the Bohai Sea. Ten layers are chosen vertically with the layer thickness of 3 m, 3 m, 3 m, 3 m, 3 m, 5 m, 5 m, 5 m, 10 m, 27 m respectively from the surface to the bottom. The time step is 360 s. The open boundary is placed away from the Bohai Strait at the 122.5° E in the north Yellow Sea to avoid unwanted open boundary effects.

The eddy viscosity $\nu = C_1 + C_2\sqrt{u^2 + v^2}$ is used according to the HAMSOM model when applying to the Bohai Sea, where $C_1 = 75.0 \times 10^{-4}\ \mathrm{m^2/s}$, $C_2 = 25.0 \times 10^{-5} \min\left(H, 30\ \mathrm{m}\right)$.

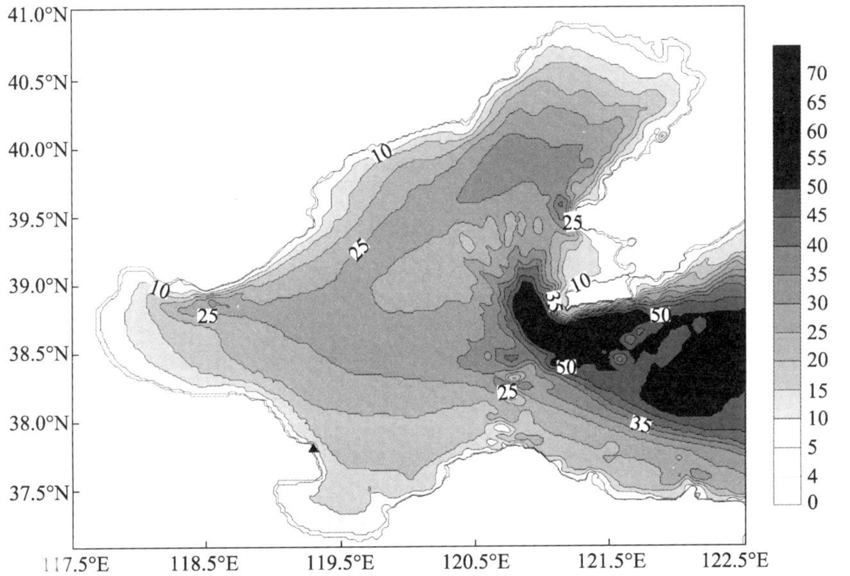

Fig. 10 The topography of the Bohai Sea (the black triangle represents the Yellow River)

3.2 Simulation results

Based on the one-month simulation, the co-tidal chart of M_2 is shown in Fig. 11. There are two amphidromic points for M_2 in the Bohai Sea. One is located at the offshore area of Qinhuangdao and the other is near the Yellow River mouth, which resembles the earlier studies.

The distribution of parameter κ is shown in Fig. 12 and is quite small in the Bohai Sea, most of which is of the scale of $O(10^{-2}–10^{-1})$. It is bigger only near the coast, especially at the head of the Liaodong Gulf and the Bohai Gulf, where the local maximum values rise to 0.5.

Because the Lagrangian residual velocity is directly calculated by its definition, that is to say, without using the mass transport velocity as in weakly nonlinear case, the nonlinearity of the system can be described appropriately in the general nonlinear area. Compared with what has been done by other researchers, more detailed circulation pattern in the Bohai Sea

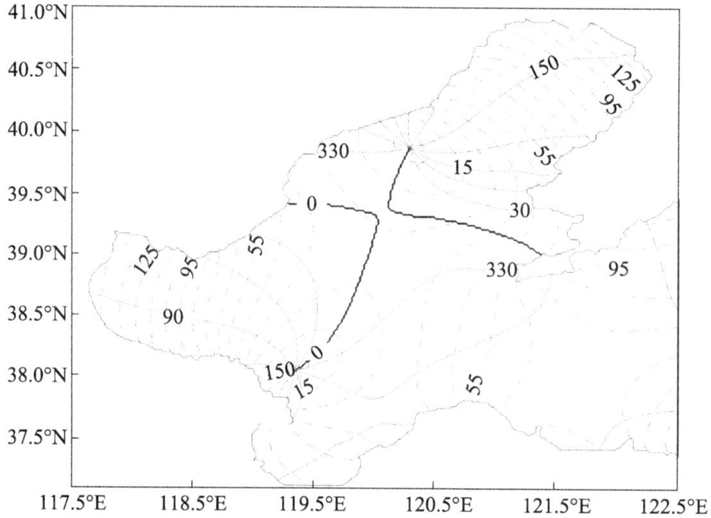

Fig. 11 The M_2 co-tidal chart of the Bohai Sea

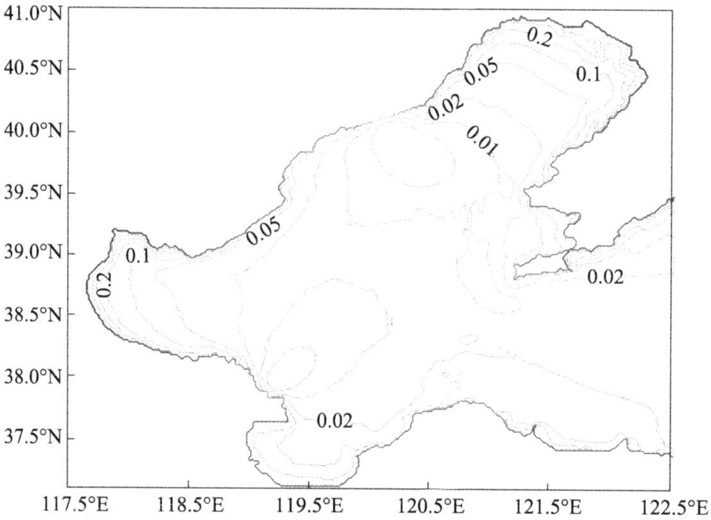

Fig. 12 The distribution of κ in the Bohai Sea

can be obtained because of the fine grid size (1′×1′) used in this study. Some new characteristics of the Lagrangian mean circulation in the Bohai Sea have been found (Fig. 13).

From the simulation results, it can be found that the Lagrangian residual velocity is 1–2 cm/s in most of the Bohai Sea and even smaller than 1 cm/s outside Qinhuangdao. There are three areas of relatively strong Lagrangian mean currents: one is in the Laotieshan waterway and its northern coastal area, another one is in the Liaodong Gulf and the third one is in the north part of the Bohai Gulf, where the maximum currents are over 10 cm/s.

As found in the model sea examples, the M_2 tide induced residual currents form an anti-clockwise gyre in the center of the semi-enclosed Bohai Sea and it can be found in every

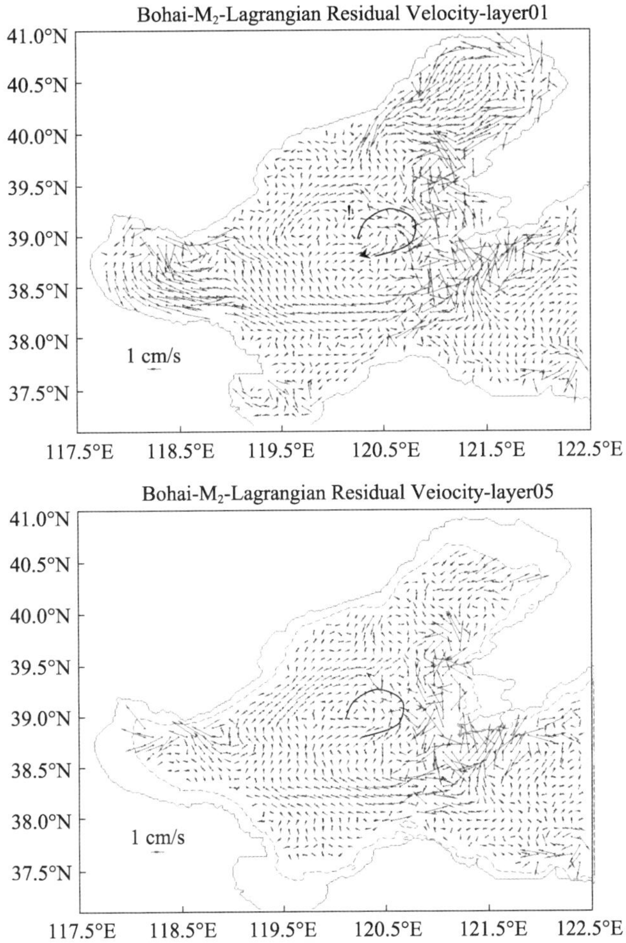

Fig. 13 The Lagrangian mean circulation in the Bohai Sea

vertical layer. The Lagrangian mean currents come in through the north of the Bohai Strait, then point to west at the mouth of the Liaodong Gulf and finally run to south along the coast from the west coast of the Liaodong Gulf. After passing the mouth of Bohai Gulf, the Lagrangian mean currents turn to east at the northeast of the Yellow River Delta and flow eastwards into the Bohai Strait.

This conclusion coincides with the early research work based on the weakly nonlinear theory (Sun *et al.*, 1989; Zheng, 1992; Wang and Zheng, 1992; Zhou *et al.*, 2002; Wang, 1996). Because of the weak nonlinearity in the center of the Bohai Sea, it is natural to get the same results generated by using either the new Lagrangian residual velocity concept (Feng *et al.*, 2007) or the early simulation model (Feng, 1987).

But a new interesting feature is found that a small clockwise ring with the diameter of 40 km exists inside the big anti-clockwise gyre mentioned above. Comparing the position of the ring with the topography of the Bohai Sea, one can find an underwater bank, the Bozhong

Bank, is located at the corresponding place. This ring is the around-bank flow induced by the nonlinear interaction between the bank and the tidal currents (Lorder, 1980). Similar research can be found in Huthnance (1973) and Zimmerman (1978a, b).

The Lagrangian mean currents of the Liaodong Gulf show an obvious three-dimensional structure. The surface Lagrangian mean currents flow outside the Liaodong Gulf, with the maximum value near the coast except for a small anti-clockwise ring in the northeast corner. Then the surface currents go into the flow which comes in from the Laotieshan waterway and form the north part of the anti-clockwise gyre in the central Bohai Sea. In the lower layer of the water, the Lagrangian residual velocity becomes smaller and its direction changes from the outwards to inwards of the Liaodong Gulf and forms a clockwise ring near the coast.

The 3D structure can also be found in the Bohai Gulf. At the surface of the Bohai Gulf the water flows out in most part of the area. There is only a small anti-clockwise gyre near the northeast part of the Bohai Gulf. But when the current field of the deeper layer is examined the current becomes smaller and smaller and there is also a direction change. The anticlockwise gyre becomes a clockwise one and the center of the gyre moves further to the northeast and finally moves out of the Bohai Gulf. This 3D structure is different from the results of previous studies that a clockwise gyre exists in the whole water volume of the Bohai Gulf when it is considered as a convectively weakly nonlinear case (Zheng, 1992; Wang *et al.*, 1992; Wei *et al.*, 2004a).

In the Laizhou Bay, the structure of the Lagrangian mean circulation is simpler than those in the other two bays. The Laizhou Bay can be divided into two parts. The flow in its west part forms a clockwise circulation, while in its deeper northeast part the flow is part of the anticlockwise gyre in the central basin of the Bohai Sea.

The circulation structure derived by the numerical model in the new conceptual framework explains some phenomena observed before. It can be seen from the salinity distribution map in Ocean Atlas (1992) that the fresh water coming from the Yellow River will go out of the Bohai Sea along its southern coast. This can be explained by the big anticlockwise gyre in the central Bohai Sea. The fresh water does not go into the Bohai Gulf and this means that there is no net water flux going into the Bohai Gulf. In consideration of the direction of the surface flow in the Bohai Gulf, the stronger current speed and the low salinity in the surface layer, this result is obvious. According to Wright *et al.* (1985), there are modern sediments in the Bohai Gulf based on the sediment composition analysis which means there should be mass transport going into the Bohai Gulf from the Yellow River mouth area. Indeed Jiang and Mayer (1997) and Jiang *et al.* (2002) found the same bottom circulation pattern based on the artificial jellyfish. Jiang and Mayer (1997) and Jiang *et al.* (2002) got the same results by numerical models when studying the transport of suspended particulate maters. All these facts may suggest the bottom flow direction is towards west in the Bohai Gulf. Some early results (Zheng, 1992; Wang *et al.*, 1992; Wei *et al.*, 2004a) mentioned above also got the similar flow trend in this part of the sea by using the weakly nonlinear theory but they cannot explain some phenomena at the surface.

The Lagrangian mean currents in the Bohai Strait are relatively complicated due to the complicated topography and coastal lines. In general the Lagrangian mean currents are larger in the north part than in the south part. The water flows in the Bohai Sea through the north part and out through the south part in the whole depth. The inflow becomes concentrated in the deeper part of the water.

The above discussed Lagrangian mean circulation is derived at one initial tidal phase. The surface Lagrangian mean circulation fields from 4 different initial tidal phases (0, $T/4$, $T/2$, $3T/4$) are displayed in Fig. 14. In general the overall pattern of the circulation fields do not change. But the Lagrangian residual velocity changes dramatically in the area where the coastline or the topography is complicated. It is even in opposite directions at some points near the Yellow River mouth which is also observed by artificial jellyfish for bottom currents in Jiang *et al.* (2002). This shows the ability of the present theory to describe the nonlinearity. In Fig. 14 another feature can be noticed that although the circulation pattern is complicated near the complicate coastline there is no eddies near the headland which always occurs in Eulerian mean velocity distribution map.

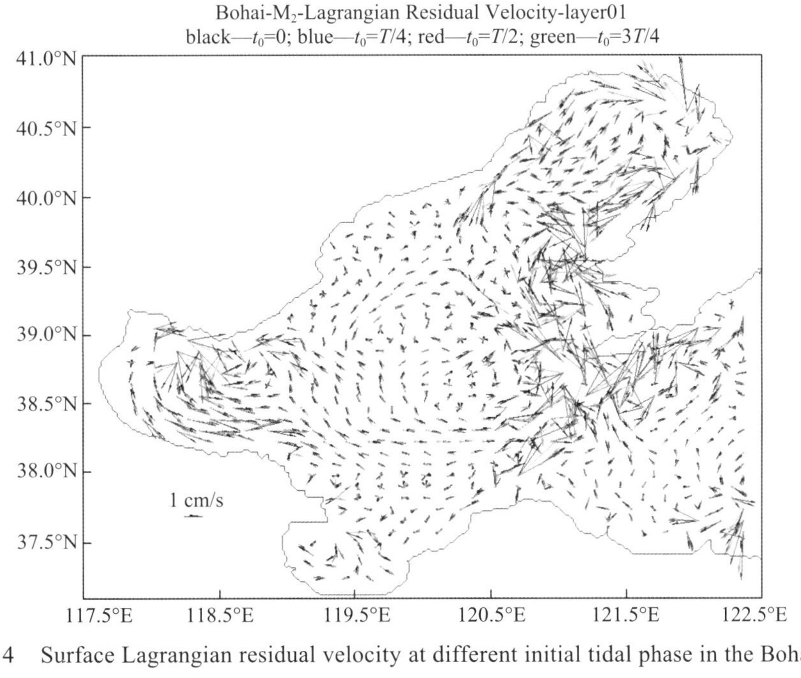

Fig. 14 Surface Lagrangian residual velocity at different initial tidal phase in the Bohai Sea

4. Conclusions

The new concept of the Lagrangian mean theory on coastal circulation is applied in a single frequency tide induced case for the first time in the present paper. The circulation of three model areas and the Bohai Sea are simulated based on the new theory proposed in the previous paper (Feng *et al.*, 2008). The results show that the application of the theory does not add too much work to the simulation of calculation but can really give insights into the structure of the tide induced coastal circulation. In the three model areas the circulation pattern is an anti-clockwise gyre in all rectangular basins which shows the conservation of the material surface.

In the generally nonlinear case the coastal circulation can not be described by a single field but a set of infinite fields. In the present study the infinite circulation fields are derived by changing the initial tidal phase from 0 to 2π. It is found that the complicated coastline and the topography are the two causes of nonlinearity in the coastal area. But the dispersion of the fields which can be accounted for quantitatively by a drift dispersion index introduced in this study concentrates only at some parts where the geometry changes abruptly. That is to say, the convectively weakly nonlinear theory can be used in most cases defined by the parameter κ but the effects of the local geometry should be sufficiently taken into account.

The application to the Bohai Sea also shows similar results. The overall circulation speed is small with only several cm/s. There is also an anticlockwise gyre in the central Bohai Sea which resembles the model sea cases. A new interesting feature is that a clockwise gyre exists in the center of the above mentioned anticlockwise gyre which can be explained by the shallowness of the Bozhong Bank. In the Bohai Gulf the tide induced residual circulation shows a 3D structure with outflow at the surface and inflow at the bottom. This circulation pattern partly explains the spread of the Yellow River fresh water out of the Bohai Gulf and in the meanwhile the transport of the sediment from the same river into the Bohai Gulf.

The dispersion of the circulation fields is large at some parts of the Bohai Sea, for example, the Bohai Strait and the Yellow River mouth area, where the current direction even reverses. The strong dependence of the initial tidal phases of the circulation shows that the theory can cope with the nonlinearity well. And the false eddies associated with capes do not occur as in the Eulerian mean velocity.

Of course the application is confined to the single frequency tidal system only, without including many complicated processes such as the interaction with the baroclinic effect. All of these works need further development in the future.

References

Backhaus J O. 1985. A three-dimensional model for the simulation of shelf sea dynamics. Dt. Hydrogr. Z., 38(4): 165–187

Cheng R T, Casulli V. 1982. On Lagrangian residual currents with applications in South San Francisco Bay, California. Water Resources Research, 18(6): 1652–1662

Cheng R T, Feng Shizuo, Xi Pangen. 1986. On Lagrangian residual ellipse. In: J van de Kreeke ed., Physics of Shallow Estuaries and Bays, Lecture Notes on Coastal and Estuarine Studies, 16, Springer-Verlag, 102–113

Csanady G T. 1982. Circulation in the Coastal Ocean. Dordrecht-Boston-London: D. Reidel Pub. Com. 1–279

Delhez E J M. 1996. On the residual advection of passive constituents. Journal of Marine Systems, 8 (3–4): 147–169

Feng Shizuo. 1987. A three-dimensional weakly nonlinear model of tide-induced Lagrangian residual current and mass-transport, with an application to the Bohai Sea. In: J C J Nihoul and B M Jamart eds., Three-Dimensional models of Marine and Estuarine Dynamics. Elsevier Oceanography Series, 45, Elsevier, 471–488

Feng Shizuo. 1990. On the Lagrangian residual velocity and the mass-transport in a multi-frequency oscillatory system. In: R T Cheng ed., Physics of shallow Estuaries and Bays. Lecture Notes on Coastal and Estuarine Studies, Springer-Verlag, 18–34

Feng Shizuo. 1998. On circulation in Bohai Sea Yellow Sea and East China Sea. In: G H Hong, J Zhang and B K Park eds., The Health of the Yellow Sea. Seoul The Earth Love Publication Association, 43–77

Feng Shizuo, Cheng R T, Xi Pangen. 1986a. On tide induced Lagrangian residual current and residual transport, 1. Lagrangian residual current. Water Resources Research, 22(12): 1623–1634

Feng Shizuo, Cheng R T, Xi Pangen. 1986b. On tide induced Lagrangian residual current and residual transport, 2. residual transport with application in South San Francisco Bay, California. Water Resources Research, 22(12): 1635–1646

Feng Shizuo, Cheng R T, Sun Wenxin, *et al.* 1990. Lagrangian residual current and longterm transport processes in a weakly nonlinear baroclinic system. In: H Wang, J Wang, and H Dai eds., Physics of Shallow Seas. Beijing: China Ocean Press, 1–20

Feng Shizuo, Ju Lian, Jiang Wensheng. 2008. A Lagrangian mean theory on coastal sea circulation with inter-tidal transports, I. fundamentals. Acta Oceanologica Sinica, 27(6): 1–16

Feng Shizuo, Lu Youyu. 1993. A turbulent closure model of coastal circulation. Chinese Science Bulletin, 38(20): 1737–1741

Feng Shizuo, Wu Dexing. 1995. An inter-tidal transport equation coupled with turbulent K-ε model in a tidal and quasi-steady current system. Chinese Science Bulletin (in Chinese), 40(2): 136–139

Foreman M G G, Baptista A M, Walters R A. 1992. Tidal model studies of particle trajectories around a shallow coastal bank. Atmosphere-Ocean, 30(1): 43–69

Huthnance J M. 1973. Tidal current asymmetries over the Norfolk Sandbanks. Estuarine Coastal Marine Science, 1: 89–99

Jiang Wensheng, Pohlmann T, Suendermann J, *et al.* 2000. A modeling study of SPM transport in the Bohai Sea. Journal of Marine System, 24(3–4): 175–200

Jiang Wensheng, Mayer B. 1997. A study on the transportation of suspended particulate matter from Yellow River by using a 3D particle model. Journal of Ocean University of China, 27(4): 439–445

Jiang Wensheng, Wang Jingyong, Zhao Jianzhong, *et al.* 1997. An observation of current in Bohai gulf and its analysis. Journal of Ocean University of Qingdao (in Chinese), 27(1): 23–32

Jiang Wensheng, Wu Dexing, Gao Huiwang. 2002. The observation and simulation of bottom circulation in the Bohai Sea in summer. Journal of Ocean University of Qingdao (in Chinese), 32(4): 511–518

Liu Hao, Yin Baoshu. 2006. Model study on Bohai ecosystem: 1. Model description and primary productivity. Acta Oceanologica Sinica, 25(4): 77–90

Loder J W. 1980. Topographic rectification of tidal currents on the sides of Georges Bank. Journal of

Physical Oceanography, 10: 1399–1416
Longuet-Higgins M S. 1969. On the transport of mass by time-varying ocean currents. Deep-Sea Research, 16(5): 431–447
Pohlmann T. 1996. Predicting the thermocline in a circulation model of the North Sea: Part I: model description, calibration and verification. Continental Shelf Research, 16(2): 131–146
Pohlmann T. 2006. A meso-scale model of the central and southern North Sea: Consequences of an improved resolution. Continental Shelf Research, 26(19): 2367–2385
Shi Fengyan. 1995. On Moving Boundary Numerical Models of Coastal Sea Dynamics: [dissertation] (in Chinese). Qingdao: Ocean University of Qingdao
Sun Wenxin, Xi Pangen, Song Lina. 1989. Numerical calculation of the three-dimensional tide-induced Lagrangian residual circulation in the Bohai Sea. Journal of Ocean University of Qingdao (in Chinese), 19(2): 27–36
Wang Kai. 1996. Study on model and numerical method of coastal sea circulation and its application to the East China Sea: [dissertation] (in Chinese). Qingdao: Ocean University of Qingdao
Wang Hui, Liu Guimei, Sun Song, *et al.* 2007. A three-dimensional coupled physical-biological model study in the spring of 1993 in the Bohai Sea of China. Acta Oceanologica Sinica, 26(6): 1–12
Wang Hui, Zheng Lianyuan. 1992. The numerical calculation of the three-dimensional trajectory of the pollutants in the Bohai Sea. Journal of Ocean University of Qingdao (in Chinese), 22(3): 1–10
Wei Hao, Hainbucher D, Pohlmann T, *et al.* 2004a. Tidal-induced Lagrangian and Eulerian mean circulation in the Bohai Sea. Journal of Marine Systems, 44(3–4): 141–151
Wei Hao, Sun Jun, Andreas Moll, *et al.* 2004b. Phytoplankton dynamics in the Bohai Sea-observations and modelling. Journal of Marine Systems, 44: 233–251
Wei Hao, Zhao Liang, Feng Shizuo. 2003. A model study on carbon cycle and phytoplankton dynamical processes in the Bohai Sea. Acta Oceanologica Sinica, 22(1): 47–55
Wright L D, Wiseman W J, Yang Zuosheng, *et al.* 1990. Processes of marine dispersal and deposition of suspended silts off the modern mouth of the Huanghe (Yellow River). Continental Shelf Research, 10(1): 1–40
Zheng Lianyuan. 1992. Numerical simulation of three-dimensional tide-induced Lagrangian residual velocity and its application to Bohai Sea. Journal of Ocean University of Qingdao (in Chinese), 22(1): 39–49
Zhou Xubo, Sun Wenxin, Su Jian. 2002. Self-adaptive numerical simulation of tide-induced residual current in the East China Sea. Journal of Ocean University of Qingdao (in Chinese), 32(2): 173–178
Zimmerman J T F. 1978a. Dispersion by tide-induced residual current vortices. Hydrodynamics of Estuaries and Fjords. Amsterdam: Elsevier, 207–216
Zimmerman J T F. 1978b. Topographic generation of residual circulation by oscillatory (tidal) currents. Geophysical and Astrophysical Fluid Dynamics, 11: 35–47
Zimmerman J T F. 1979. On the Euler-Lagrangian transformation and the Stokes' drift in the presence of oscillatory and residual currents. Deep-Sea Research, 26A(5): 505–520
Marine Atalas of the Bohai Sea, Yellow Sea and East China Sea. 1992. Beijing: Ocean Press

Analytical Solution for the Tidally Induced Lagrangian Residual Current in a Narrow Bay*

Wensheng Jiang, Shizuo Feng

1. Introduction

In shallow seas, the tidal current is normally the dominant component of the total movement of the sea water. Because of nonlinearity, the tidal flow induces an aperiodic current that is called the residual current. It has been long recognized that the residual current plays an important role in the inter-tidal mass transport in the shallow seas (e.g., Nihoul and Ronday, 1975).

The meaning of "residual current" is rather vague, which is related to the various ways to derive it from the directly measured current fields. The residual current and the accompanying temperature and salinity fields should constitute a slowly varying, aperiodic dynamic system in shallow seas. The prerequisite for the retrieved aperiodic movement to become the residual current is that it can describe the net displacement of the water parcel after one or a few tidal periods. The residual current should also be in accordance with the basic law of fluid motion.

A straightforward definition of the residual current is the average of the measured velocity at a fixed location over multiple tidal periods (e.g., Abbott, 1960). This is called the Eulerian residual velocity, $\boldsymbol{u}_{\mathrm{E}}$. However, if $\boldsymbol{u}_{\mathrm{E}}$ is used to study the inter-tidal movement in a 3D case, there is fictitious source (or sink) of water mass at the sea surface (Feng *et al.*, 1984). There is also an extra-dispersion term in the inter-tidal transport equation (Fischer *et al.*, 1979). These are questionable aspects related to describing the subtidal current with the Eulerian residual velocity.

An alternative approach to define the residual velocity was inspired by the idea of relating the mass-transport velocity to the sediment transport in tidal estuaries by Hunt (1961). Longuet-Higgins (1969) first related the mass-transport velocity to the large-scale ocean circulation. He pointed out clearly that the mass-transport velocity at a fixed location is not solely controlled by the mean velocity at that point alone, and the mass-transport velocity, $\boldsymbol{u}_{\mathrm{L}}$, equals the sum of the Eulerian residual velocity ($\boldsymbol{u}_{\mathrm{E}}$) and the Stokes' drift velocity ($\boldsymbol{u}_{\mathrm{S}}$). For the case of the tidal flow, Zimmerman (1979) obtained the analytic expression of $\boldsymbol{u}_{\mathrm{S}}$ and found that $\boldsymbol{u}_{\mathrm{L}}$ and $\boldsymbol{u}_{\mathrm{E}}$ are rather different; Loder (1980) and Cheng and Casulli (1982)

* Jiang W, Feng S. 2011. Analytical solution for the tidally induced Lagrangian residual current in a narrow bay. Ocean Dynamics, 61(4): 543–558

demonstrated similar results based on numerical modeling results.

In a weakly nonlinear tidal system, Feng *et al.* (1986a) systematically applied the perturbation method to the depth-averaged tidal dynamic equations and obtained the Lagrangian residual velocity (LRV, here-after), $\boldsymbol{u}_{\mathrm{LR}}$, to the second-order expressed by the zeroth-order tidal solutions. They showed that the first-order $\boldsymbol{u}_{\mathrm{LR}}$ is the mass-transport velocity, $\boldsymbol{u}_{\mathrm{L}}$; and defined the second-order $\boldsymbol{u}_{\mathrm{LR}}$ as the Lagrangian drift velocity, $\boldsymbol{u}_{\mathrm{ld}}$. They analytically revealed the dependence of $\boldsymbol{u}_{\mathrm{ld}}$ on the tidal phase, which is related to the initial timing when the water parcels are released. This demonstrates that $\boldsymbol{u}_{\mathrm{LR}}$ has a specific Lagrangian nature (Cheng, 1983). Feng *et al.* (1986b) proposed the use of $\boldsymbol{u}_{\mathrm{L}}$ to replace $\boldsymbol{u}_{\mathrm{E}}$ to act as the advective transport velocity in the inter-tidal transport equation, in which the tidal dispersion term disappears.

Feng (1987) extended the analysis to the 3D case and established a set of dynamic equations governing $\boldsymbol{u}_{\mathrm{L}}$. A tidal body force appears in the momentum equations, which represents the nonlinear coupling of the zeroth-order tide. The water mass is continuous at the sea surface, i.e., the fictitious "tidal surface source" in Feng *et al.* (1984) vanishes, which keeps the conservation of the mass. The inter-tidal mass-transport equation using $\boldsymbol{u}_{\mathrm{L}}$ as the advective velocity was also derived, with the tidal dispersion term in Fischer *et al.* (1979) disappearing. The analysis clearly demonstrated that $\boldsymbol{u}_{\mathrm{L}}$ is the more appropriate residual circulation in the tide-dominant shallow seas.

The distinction between the Lagrangian and the Eulerian residual velocity has been theoretically recognized for more than 30 years. The Lagrangian residual current framework has been successfully applied to many cases (e.g., Jay, 1991; Foreman *et al.*, 1992; Dortch *et al.*, 1992; Cerco and Cole, 1993; Wang *et al.*, 1993; Ridderinkhof and Loder, 1994; Delhez, 1996; Wei *et al.*, 2004; Hainbucher *et al.*, 2004; Muller *et al.*, 2009). However, the Lagrangian framework is still not widely accepted; and it is not uncommon in literature that the Eulerian residual velocity ($\boldsymbol{u}_{\mathrm{E}}$) is used to explain the mass transport in shallow seas. This is partly because the residual current is at least one order smaller than the tidal current, causing it to be often contaminated by the errors in observations or model simulations. This leads to difficulty in demonstrating the differences among various definitions of the residual current.

In some special cases, $\boldsymbol{u}_{\mathrm{E}}$ was questioned explicitly based on the physics. For example, when studying the depth-averaged 2D case, Robinson (1983) replaced $\boldsymbol{u}_{\mathrm{E}}$ with the Eulerian residual transport velocity, u_{T}, as the tidal mean flux at one specific location divided by the averaged water depth. He stated that u_{T} should be used to quantify the mean mass transport at a specific location. This statement was proven wrong by Feng *et al.* (1986b) who showed it is $\boldsymbol{u}_{\mathrm{L}}$ rather than u_{T} that is related to the inter-tidal transport. The product of the depth-averaged LRV and the mean water depth, the so-called Lagrangian residual transport, is the net mass transport in a depth-averaged 2D case. Since $\boldsymbol{u}_{\mathrm{T}}$ is of the same order as $\boldsymbol{u}_{\mathrm{E}}$ and u_{L}, it is also difficult to distinguish them from observations or modeling results.

Under such a circumstance, the analytical solutions for idealized cases are the most ideal for assessing the differences among different residual velocities. For example, Dyke (1980) gave an analytical solution of the Stokes' drift in a two-layer rectangular basin with a flat bottom, based on a specific solution of the linear tide. Ianniello (1977) analytically calculated $\boldsymbol{u}_{\mathrm{E}}$ and $\boldsymbol{u}_{\mathrm{L}}$ in a length-depth section of a narrow bay, and found that $\boldsymbol{u}_{\mathrm{E}}$ is always seaward while $\boldsymbol{u}_{\mathrm{L}}$ is of a two-layer structure. This clearly demonstrated that the Eulerian residual velocity violates mass conservation while the LRV does not. It should be pointed out that the expression of $\boldsymbol{u}_{\mathrm{L}}$ in Ianniello (1977) is indeed $\boldsymbol{u}_{\mathrm{T}}$, later defined by Robinson (1983). It happens that in the case of Ianniello (1977) $\boldsymbol{u}_{\mathrm{T}}$ is a valid simplification of $\boldsymbol{u}_{\mathrm{L}}$.

For general cases $\boldsymbol{u}_{\mathrm{T}}$ and $\boldsymbol{u}_{\mathrm{L}}$ are different, but in recent years $\boldsymbol{u}_{\mathrm{T}}$ has still been used to represent the inter-tidal mass transport even in some theoretical studies. Li and O'Donnell (2005) gave an analytic solution of $\boldsymbol{u}_{\mathrm{T}}$ in a narrow bay and used it to represent the mass-transport velocity. They studied the dependence of the Eulerian residual transport velocity on the length (Li and O'Donnell, 2005) and bending of the bay (Li *et al.*, 2008). Winant (2007) found a solution for a 3D case in a bay and then studied the first-order LRV, $\boldsymbol{u}_{\mathrm{L}}$ (Winant, 2008). However, he used the Eulerian residual transport to represent the depth integrated residual transport even though he noticed the discrepancy between $\boldsymbol{u}_{\mathrm{T}}$ and $\boldsymbol{u}_{\mathrm{L}}$ in his results.

The purpose of the present study is to clarify the different definitions of the residual current, through quantitative comparison of $\boldsymbol{u}_{\mathrm{E}}$, $\boldsymbol{u}_{\mathrm{T}}$ and $\boldsymbol{u}_{\mathrm{L}}$ calculated for a specific case of a narrow bay. The governing equations for the Lagrangian residual flow in a depth-averaged 2D case will be derived for the first time, based on the concept developed by Feng *et al.* (1986a, b). By confining to a narrow bay, the equations will be solved analytically to obtain the first- and second-order LRV. The behavior of the LRV will be analyzed and some factors influencing it will be discussed. The analysis results will help to draw the attention of researchers when studying the circulation dynamics in shallow coastal waters.

2. Definitions and Governing Equations

2.1 Definitions of residual velocities

For clarity, Table 1 provides a list of symb ols related to the major definitions of the residual currents. The definitions of these residual currents are provided below, and more details can be found in Feng *et al.* (1986a, 2008) and Delhez (1996).

The Lagrangian residual velocity $\boldsymbol{u}_{\mathrm{LR}}$ is commonly defined as the net displac ement of the water parcel divided by the elapsed time of n tidal periods (e.g., Zimmerman, 1979):

$$\boldsymbol{u}_{\mathrm{LR}}=\frac{\xi(t_0+nT;\boldsymbol{x}_0,t_0)}{nT}$$

where $\xi(t;\boldsymbol{x}_0,t_0)=\int_{t_0}^{t}\boldsymbol{u}(\boldsymbol{x}_0+\xi,t')\mathrm{d}t'$ defines the displacement of a water parcel with the initial position at $\boldsymbol{x}_0$ when $t=t_0$.

Table 1 Symbols for major definitions of the residual current

Symbol	Meaning
$\boldsymbol{u}_{\mathrm{LR}}$	Lagrangian residual velocity (LRV)
$\boldsymbol{u}_{\mathrm{L}}$	First-order LRV, mass-transport velocity
$\boldsymbol{u}_{\mathrm{ld}}$	Second-order LRV, Lagrangian drift velocity
$\boldsymbol{u}_{\mathrm{E}}$	Eulerian residual velocity
$\boldsymbol{u}_{\mathrm{S}}$	Stokes' drift velocity
$\boldsymbol{u}_{\mathrm{T}}$	Eulerian residual transport velocity

The tidal oscillation is removed from the net displacement of the water parcel after n tidal periods, but a slowly varying part with an inter-tidal time scale τ may still exist. Feng *et al.* (2008) proved that if n is not too large to ensure that the water parcel's net displacement after n tidal periods is at least one order of magnitude smaller than the tidal excursion, $\xi(t_0+nT;\boldsymbol{x}_0,t_0)$ can be a smooth function of $\boldsymbol{x}_0$ and τ. Thus, assuming $\xi_{n\mathrm{r}}(\boldsymbol{x}_0;\tau,t_0)=\xi(t_0+nT;\boldsymbol{x}_0,t_0)$, the LRV is defined as

$$\boldsymbol{u}_{\mathrm{LR}}(\boldsymbol{x}_0;\tau,t_0)=\frac{\partial\xi_{n\mathrm{r}}(\boldsymbol{x}_0,\tau;t_0)}{\partial\tau}$$

The nonlinearity of the shallow sea tidal system can be quantified by a nondimensional number κ, the ratio between the tidal amplitude and the water depth. In the weakly nonlinear case, i.e., $O(\kappa)<1$, all the state variables such as u, v and ζ can be expanded into the power series accord ing to κ,

$$u=u_0+\kappa u_1+\kappa^2u_2+\cdots$$

and similar expansion can be applied to v and ζ.

Feng *et al.* (1986a) proved that the mass-transport velocity $\boldsymbol{u}_{\mathrm{L}}$ defined by Longuet-Higgins (1969) is the first-order approximation to $\boldsymbol{u}_{\mathrm{LR}}$ and is independent of the initial phase t_0, i.e.,

$$\boldsymbol{u}_{\mathrm{LR}}(\boldsymbol{x}_0,\tau;t_0)=\boldsymbol{u}_{\mathrm{L}}(\boldsymbol{x}_0,\tau)+O(\kappa) \tag{1}$$

where the mass-transport velocity is defined as

$$\boldsymbol{u}_{\mathrm{L}}(\boldsymbol{x}_0,\tau)=\boldsymbol{u}_{\mathrm{E}}+\boldsymbol{u}_{\mathrm{S}} \tag{2}$$

The Eulerian residual velocity is defined as

$$\boldsymbol{u}_{\mathrm{E}}=(u_{\mathrm{E}},v_{\mathrm{E}})=(\langle u_1\rangle,\langle v_1\rangle) \tag{3}$$

and the Stokes' drift velocity is defined as

$$\boldsymbol{u}_{\mathrm{S}}=(u_{\mathrm{S}},v_{\mathrm{S}})=\langle\xi_0\cdot\nabla\boldsymbol{u}_0\rangle \tag{4}$$

where the tidal average operator is defined as follows:

$$\langle f \rangle = \frac{1}{nT}\int_{t_0}^{t_0+nT} f(\boldsymbol{x},t)\mathrm{d}t \tag{5}$$

and

$$\boldsymbol{\xi}_0 = (\xi_0,\eta_0) = \left(\int_{t_0}^{t} u_0(\boldsymbol{x}_0,t')\mathrm{d}t', \int_{t_0}^{t} v_0(\boldsymbol{x}_0,t')\mathrm{d}t'\right)$$

thus

$$\boldsymbol{u}_0 = (u_0, v_0) = \frac{\partial \boldsymbol{\xi}_0}{\partial t} \tag{6}$$

The definition of the Eulerian residual transport velocity $\boldsymbol{u}_\mathrm{T}$ is based on the idea that the calculation of the residual transport should take the water surface oscillation into consideration, thus,

$$\boldsymbol{u}_\mathrm{T} = (u_\mathrm{T}, v_\mathrm{T}) = \boldsymbol{u}_\mathrm{E} + \frac{\langle \boldsymbol{u}_0 \zeta_0 \rangle}{h} \tag{7}$$

It is proved in Eqs. 53 and 54 in the Appendix (or Feng *et al.*, 1986b) that

$$\boldsymbol{u}_\mathrm{S} = \boldsymbol{u}_\mathrm{D} + \frac{\langle \boldsymbol{u}_0 \zeta_0 \rangle}{h}$$

where

$$\boldsymbol{u}_\mathrm{D} = (u_\mathrm{D}, v_\mathrm{D}) = \frac{1}{h}\left(\frac{\partial \langle h u_0 \eta_0 \rangle}{\partial y}, \frac{\partial \langle h v_0 \xi_0 \rangle}{\partial x}\right) \tag{8}$$

We thus get $\boldsymbol{u}_\mathrm{L} = \boldsymbol{u}_\mathrm{T} + \boldsymbol{u}_\mathrm{D}$, which means that $\boldsymbol{u}_\mathrm{T}$ is only a part of $\boldsymbol{u}_\mathrm{L}$. In general, even to the first-order approximation, $\boldsymbol{u}_\mathrm{T}$ cannot describe the water parcel movement as the LRV does, except in cases when $\boldsymbol{u}_\mathrm{D} = 0$ such as the section-averaged one-dimensional or breadth-averaged two-dimensional cases of a bay.

The second-order $\boldsymbol{u}_\mathrm{LR}$ is related to the Lagrangian drift velocity $\boldsymbol{u}_\mathrm{ld} = (u_\mathrm{ld}, v_\mathrm{ld})$,

$$\boldsymbol{u}_\mathrm{LR} = \boldsymbol{u}_\mathrm{L} + \boldsymbol{u}_\mathrm{ld} + O(\kappa^2) \tag{9}$$

$\boldsymbol{u}_\mathrm{ld}$ is dependent on the initial phase of a tidal period and its expression can be found in Feng *et al.* (1986a). The Lagrangian drift velocity represents the interaction between the tidal current and $\boldsymbol{u}_\mathrm{L}$ and is one order smaller than $\boldsymbol{u}_\mathrm{L}$ in the weakly nonlinear case.

2.2 The tidal equations

For a narrow bay, the Cartesian coordinates are used with the *x* and *y*-axes oriented along and across the axis of the bay respectively. The bay has a length of *L* and width of *B*. A single-frequency tide is imposed at the open boundary at *x*=0. The nondimensional, depth-averaged governing equations describing this system of long gravity waves can be found in literature, e.g., in Feng *et al.* (1986a). Here we use the linearized quadric bottom friction based on Proudman (1953). The influence of the nonlinear bottom friction on the tide-induced residual current will be discussed in Section 4.4. The governing equations are

$$\frac{\partial \zeta}{\partial t}+\frac{\partial(h+\kappa\zeta)u}{\partial x}+\frac{\partial(h+\kappa\zeta)v}{\partial y}=0 \tag{10}$$

$$\frac{\partial u}{\partial t}+\kappa\left(u\frac{\partial u}{\partial x}+v\frac{\partial u}{\partial y}\right)-\delta Fv=-\frac{\partial \zeta}{\partial x}-\beta_{\mathrm{T}}\frac{u}{h+\kappa\zeta} \tag{11}$$

$$\delta^2\left[\frac{\partial v}{\partial t}+\kappa\left(u\frac{\partial v}{\partial x}+v\frac{\partial v}{\partial y}\right)\right]+\delta Fu=-\frac{\partial \zeta}{\partial y}-\delta^2\beta_{\mathrm{T}}\frac{v}{h+\kappa\zeta} \tag{12}$$

where t, $\boldsymbol{u}=(u,v)$, ζ and h are the nondimensional time, velocity and its components in x and y directions, sea surface elevation and water depth, respectively. For brevity, they take the same form as the dimensional variables.

The characteristic scales for the various variables are $t_{\mathrm{c}}=T$, $x_{\mathrm{c}}=\lambda$, $y_{\mathrm{c}}=B$, $u_{\mathrm{c}}=\sqrt{g/h_{\mathrm{c}}}\zeta_{\mathrm{c}}$ and $v_{\mathrm{c}}=B\zeta_{\mathrm{c}}/(h_{\mathrm{c}}T)$, where h_{c} is the averaged water depth, ζ_{c} is the typical tidal amplitude at the open boundary, g is the acceleration due to gravity, and $\lambda=T\sqrt{gh_{\mathrm{c}}}$ is the wave length.

Equations 10–12 contain four nondimensional parameters, i.e., $\kappa=\zeta_{\mathrm{c}}/h_{\mathrm{c}}$, $\delta=B/\lambda$, $F=fT$ and $\beta_{\mathrm{T}}=\beta T/h_{\mathrm{c}}$, where β is the bottom friction coefficient and f is the Coriolis parameter. The present study considers the semi-diurnal or diurnal tides and the bay is assumed to be located at the mid-latitude, so $O(F)=1$. For tides in the shallow seas, the bottom friction is significant and should be included in the lowest-order tide equations, hence $O(\beta_{\mathrm{T}})=1$. The third parameter κ, the ratio of the tidal amplitude over the water depth, characterizes the nonlinear effect in the system. The fourth parameter δ is the aspect ratio of the cross and along-channel length scales.

In this study, we consider the case of $O(\kappa)<1$ and assume δ to be of the same order as κ, i.e., $O(\delta)=O(\kappa)$. This describes a system of weakly nonlinear tide in a narrow bay. The perturbation method can be applied to Eqs. 10–12 based on the small parameter κ.

Substituting the expansion of the state variables according to κ into Eqs. 10–12 yields the zeroth-order equations

$$\frac{\partial \zeta_0}{\partial t}+\frac{\partial hu_0}{\partial x}+\frac{\partial hv_0}{\partial y}=0 \tag{13}$$

$$\frac{\partial u_0}{\partial t}=-\frac{\partial \zeta_0}{\partial x}-\beta_{\mathrm{T}}\frac{u_0}{h} \tag{14}$$

$$0=\frac{\partial \zeta_0}{\partial y} \tag{15}$$

$$(u_0,v_0)\cdot\boldsymbol{n}=0 \tag{16}$$

at the fixed boundary,

$$\zeta_0=\zeta_{\mathrm{open}},\ x=0 \tag{17}$$

In the above equations, $\boldsymbol{n}$ is the unit vector normal to the land boundary and ζ_{open} is the nondimensional water elevation at the open boundary.

According to Eq. 15, ζ_0 is a function of x and t but independent of y. This is an obvious inference caused by the narrow nature of the bay, which does not need to be an assumption as proposed in Li and O'Donnell (2005). According to Eqs. 14 and 15, the Coriolis force does not influence the zeroth-order tide. If the mean water depth is 10 m, $f = 10^{-4}\,\text{s}^{-1}$ and a semi-diurnal tide has a characteristic amplitude of 1 m, then $O(\delta) = O(B/FR) = O(B/100\,\text{km})$, where $R = \sqrt{gh_c}/f$ is the barotropic Rossby deformation radius. This means that B should be less than 10 km for the bay to be considered as a narrow one.

The equations for the first-order motions are

$$\frac{\partial \zeta_1}{\partial t} + \frac{\partial h u_1}{\partial x} + \frac{\partial h v_1}{\partial y} = -\left(\frac{\partial u_0 \zeta_0}{\partial x} + \frac{\partial v_0 \zeta_0}{\partial y}\right) \tag{18}$$

$$\frac{\partial u_1}{\partial t} - F v_0 = -\frac{\partial \zeta_1}{\partial x} - \beta_{\text{T}} \frac{u_1}{h} - \left(u_0 \frac{\partial u_0}{\partial x} + v_0 \frac{\partial u_0}{\partial y} - \beta_{\text{T1}} \frac{u_0 \zeta_0}{h^2}\right) \tag{19}$$

$$F u_0 = \frac{\partial \zeta_1}{\partial y} \tag{20}$$

$$(u_1, v_1) \cdot \boldsymbol{n} = 0 \tag{21}$$

at the fixed boundary

$$\zeta_1 = 0,\ x = 0 \tag{22}$$

In the above equations, $\beta_{\text{T1}} = \beta_1 T / h_c$ with β_1 being equal to β. If β_1 is assumed to be 0, then the bottom friction term in Eqs. 11 and 12 retreats to the linear form case (see Feng *et al.*, 1986a, Eqs. 3 and 4). This case is referred to as the "linear form of the bottom friction term" for short hereafter. Otherwise, it is called the "nonlinear form of the bottom friction term". The effect of these two forms on the residual current will be discussed in Section 4.4.

2.3 Equations for u_{L}

In the present study, only a single frequency tide is imposed at the open boundary, i.e.,

$$\zeta_{\text{open}} = \zeta_{\text{openR}} \cos 2\pi t + \zeta_{\text{openI}} \sin 2\pi t = \text{Re}\left[\zeta_{\text{open}}^{\text{c}} \text{e}^{-\text{i}2\pi t}\right]$$

where $\zeta_{\text{open}}^{\text{c}} = \zeta_{\text{openR}} + i\zeta_{\text{openI}}$.

The solution of the zeroth-order tide takes the form of

$$\zeta_0 = \zeta_{0\text{R}} \cos 2\pi t + \zeta_{0\text{I}} \sin 2\pi t = \text{Re}\left[\zeta_0^{\text{c}} \text{e}^{-\text{i}2\pi t}\right] \tag{23}$$

$$u_0 = u_{0\text{R}} \cos 2\pi t + u_{0\text{I}} \sin 2\pi t = \text{Re}\left[u_0^{\text{c}} \text{e}^{-\text{i}2\pi t}\right] \tag{24}$$

$$v_0 = v_{0R}\cos 2\pi t + v_{0I}\sin 2\pi t = \mathrm{Re}\left[v_0^c e^{-i2\pi t} \right] \tag{25}$$

where $\zeta_0^c = \zeta_{0R} + i\zeta_{0I}$, $u_0^c = u_{0R} + iu_{0I}$ and $v_0^c = v_{0R} + iv_{0I}$.

The nondimensional equivalence of Eq. 5 is

$$\langle f \rangle = \frac{1}{n}\int_{t_0}^{t_0+n} f(\boldsymbol{x},t)\mathrm{d}t \tag{26}$$

By applying the operator (Eq. 26) to Eqs. 18–20 and taking into consideration the forms of the zeroth- and first-order solutions in Feng *et al.* (1986a), the equations governing the first-order LRV for the weakly non-linear tide can be obtained according to the methodology proposed by Feng (1987). The detailed procedure to obtain them is presented in the Appendix.

The equations governing the first-order LRV equations in a narrow bay are

$$\frac{\partial hu_L}{\partial x} + \frac{\partial hv_L}{\partial y} = 0 \tag{27}$$

$$u_L = \frac{h}{\beta_T}\left(-\frac{\partial \zeta_E}{\partial x} + \pi_x \right) \tag{28}$$

$$\frac{\partial \zeta_E}{\partial y} = 0 \tag{29}$$

where

$$\pi_x = \pi_{xN} + \pi_{x\beta} \tag{30}$$

$$\pi_{xN} = -\frac{1}{2}\frac{\partial}{\partial x}\left\langle \xi_0 \frac{\partial \zeta_0}{\partial x} \right\rangle \tag{31}$$

$$\pi_{x\beta} = \beta_{T1}\frac{\langle u_0\zeta_0 \rangle}{h^2} + \frac{\beta_T}{h}\left(\frac{1}{h}\frac{\partial \langle hu_0\eta_0 \rangle}{\partial y} + \langle \xi_0 \nabla \cdot \boldsymbol{u}_0 \rangle + \left\langle -\eta_0 \frac{\partial u_0}{\partial y} \right\rangle \right) \tag{32}$$

$$\boldsymbol{u}_L \cdot \boldsymbol{n} = 0 \tag{33}$$

at the fixed boundary,

$$\zeta_E = 0, x = 0 \tag{34}$$

Equation 28 tells that the longitudinal LRV can be obtained directly. The terms contributing to $\boldsymbol{u}_L$ include the gradient of the residual water elevation and a depth-averaged version of the tidal body force π_x, which was first defined by Feng (1987). The irrotational term π_{xN} represents the horizontal gradient of the tidally averaged coupling of the parcel displacement and the gradient of the water elevation. The rotational term $\pi_{x\beta}$ is more complex and is controlled by the bottom friction. It includes four terms corresponding to four mechanisms to drive the LRV. The first represents the tidally averaged interaction between the velocity and the water elevation. It diminishes when the tidal wave is a standing wave or when the tidal bottom friction term is in the linear form. The second is u_D defined in Eq. 8 and will be discussed in detail in Section 4.1. It is the gradient of the coupling of the water depth, the tidal current velocity in one direction and the water parcel displacement in the other direction. The third is the tidally averaged product of the along-channel water parcel displacement and the divergence of the velocity. The fourth is the tidally averaged product of the lateral water parcel displacement and the vorticity of the tidal current.

After the solution of the first-order LRV $(\boldsymbol{u}_{\mathrm{L}})$ is obtained, the second-order LRV $\boldsymbol{u}_{\mathrm{ld}} = (u_{\mathrm{ld}}, v_{\mathrm{ld}})$ can be formulated as

$$u_{\mathrm{ld}} = u'_{\mathrm{ld}} \cos 2\pi t_0 + u''_{\mathrm{ld}} \sin 2\pi t_0 \tag{35}$$

$$v_{\mathrm{ld}} = v'_{\mathrm{ld}} \cos 2\pi t_0 + v''_{\mathrm{ld}} \sin 2\pi t_0 \tag{36}$$

$$u'_{\mathrm{ld}} = u_{0\mathrm{I}} \frac{\partial u_{\mathrm{L}}}{\partial x} + v_{0\mathrm{I}} \frac{\partial u_{\mathrm{L}}}{\partial y} - u_{\mathrm{L}} \frac{\partial u_{0\mathrm{I}}}{\partial x} - v_{\mathrm{L}} \frac{\partial u_{0\mathrm{I}}}{\partial y} \tag{37}$$

$$u''_{\mathrm{ld}} = -u_{0\mathrm{R}} \frac{\partial u_{\mathrm{L}}}{\partial x} - v_{0\mathrm{R}} \frac{\partial u_{\mathrm{L}}}{\partial y} + u_{\mathrm{L}} \frac{\partial u_{0\mathrm{R}}}{\partial x} + v_{\mathrm{L}} \frac{\partial u_{0\mathrm{R}}}{\partial y} \tag{38}$$

$$v'_{\mathrm{ld}} = u_{0\mathrm{I}} \frac{\partial v_{\mathrm{L}}}{\partial x} + v_{0\mathrm{I}} \frac{\partial v_{\mathrm{L}}}{\partial y} - u_{\mathrm{L}} \frac{\partial v_{0\mathrm{I}}}{\partial x} - v_{\mathrm{L}} \frac{\partial v_{0\mathrm{I}}}{\partial y} \tag{39}$$

$$v''_{\mathrm{ld}} = -u_{0\mathrm{R}} \frac{\partial v_{\mathrm{L}}}{\partial x} + v_{0\mathrm{R}} \frac{\partial v_{\mathrm{L}}}{\partial y} + u_{\mathrm{L}} \frac{\partial v_{0\mathrm{R}}}{\partial x} + v_{\mathrm{L}} \frac{\partial v_{0\mathrm{R}}}{\partial y} \tag{40}$$

3. Solutions in a Narrow Bay

For a rectangular bay, its constant width B can be taken as the length scale in y direction. Thus, $x = L_{\mathrm{N}}$, $y = 0$ and $y = 1$ constitute the fixed boundary of the bay, where $L_{\mathrm{N}} = L / \lambda$ is the dimensionless length of the bay which influences the solution together with β_{T}. The solution of Eqs. 13–17 for the zeroth-order tide in the above narrow bay is given in Li and Valle-Levinson (1999), but the assumption that the transverse gradient of the tidal elevation is zero in Li and Valle-Levinson (1999) is not needed here. According to Eq. 15, ζ_0 is independent of y, i.e., $\zeta_0 = \zeta_0(x,t)$, which is a natural deduction for a narrow bay system.

3.1 General solutions

If the water depth varies only along the bay, i.e., $h = h(x)$, then v is always 0 according to Eqs. 13–17. For nonzero v the water depth h should at least depend on y, the across-bay direction. Here, the case of $h = h(y)$ is considered.

The solution procedure of the linear tide follows Li and Valle-Levinson (1999) and the time invariant parts of the solution defined in Eqs. 23–25 are

$$\zeta_0^{\mathrm{c}}(x) = \zeta_{\mathrm{open}}^{\mathrm{c}} \frac{\cos\left[\Omega(x - L_{\mathrm{N}})\right]}{\cos(\Omega L_{\mathrm{N}})} \tag{41}$$

$$u_0^c(x,y) = -\zeta_{\text{open}}^c \frac{\Omega}{\mathrm{i}2\pi - \beta_T / h} \frac{\sin[\Omega(x - L_N)]}{\cos(\Omega L_N)} \tag{42}$$

$$v_0^c(x,y) = \frac{\zeta_{\text{open}}^c}{h} \frac{\cos[\Omega(x - L_N)]}{\cos(\Omega L_N)} \cdot \left[\mathrm{i}2\pi y + \Omega^2 \int_0^y \frac{h}{\mathrm{i}2\pi - \beta_T / h} \mathrm{d}y' \right] \tag{43}$$

where

$$\Omega^2 = -\frac{\mathrm{i}2\pi}{\mathcal{H}}$$

$$\mathcal{H} = \int_0^1 \left(\frac{h}{\mathrm{i}2\pi - \beta_T / h} \right) \mathrm{d}y$$

The first-order LRV is obtained in the following procedure. By applying the operator $\int_0^1 \mathrm{d}y$ to Eq. 27 with the boundary conditions taken into consideration, the following equation can be obtained,

$$\int_0^1 h u_L \mathrm{d}y = 0$$

Then the substitution of Eq. 28 into the above relation followed by a rearrangement yields

$$\frac{\partial \zeta_E}{\partial x} = \frac{\int_0^1 h^2 \pi_x \mathrm{d}y}{\int_0^1 h^2 \mathrm{d}y} \tag{44}$$

Thus, ζ_E can be obtained by integrating the above equation from 0 to x with $\zeta_E = 0$ at $x = 0$. By substituting Eq. 44 into Eq. 28, one gets

$$u_L = \frac{h}{\beta_T \int_0^1 h^2 \mathrm{d}y} \left(\pi_x \int_0^1 h^2 \mathrm{d}y - \int_0^1 h^2 \pi_x \mathrm{d}y \right) \tag{45}$$

Then v_L can be calculated by integrating Eq. 27 from 0 to y. Thus, the first-order LRV, $\boldsymbol{u}_L$, is expressed explicitly with the solution of the zeroth-order tide.

The Lagrangian drift velocity $\boldsymbol{u}_{ld}$ can be obtained with Eqs. 35–40 after the zeroth-order tidal current and the first-order LRV are obtained.

The explicit expression of each term in the solution is obtained by using Maxima, a system for the manipulation of symbolic and numerical expressions. It evolves from Macsyma, a computer algebra system developed in the late 1960s at Massachusetts Institute of Technology (Maxima.sourceforge.net 2009). The expression is exported as a segment of a Fortran code and inserted in a Fortran program directly. In the solution the integration concerning the water depth cannot always be solved analytically, so a numerical integration software package Quadpack is used (Piessens *et al.*, 1983). To ensure accuracy, the double precision real number is used.

3.2 Solution for a specific bottom profile

The water depth across the bay is assumed to taken the dimensional form of

$$\tilde{h}(y) = \tilde{h}_1 + \tilde{h}_2 e^{-[(y/B-1/2)/\alpha]^2} \tag{46}$$

Where α is an adjusting parameter.

In the nondimensional form the bottom topography is

$$h(y) = h_1 + h_2 e^{-[(y-1/2)/\alpha]^2} \tag{47}$$

where $h_1 = \tilde{h}_1 / h_c$ and $h_2 = \tilde{h}_2 / h_c$, The nondimensional form y is written the same as its dimensional counterpart for simplicity, as stated in Section 2.

Because the width of the bay, B, is neither in the depth profile nor in the residual circulation equations explicitly, the nondimensional results are independent of B if the narrow bay condition $O(\delta) \leqslant O(\kappa) < 1$ is satisfied.

The dimensional parameters in Eq. 46 are set the same as those in Li and O'Donnell (2005), with $\alpha = 7/40$ and the width B = 2000 m. The water depth in the cross section of the bay (in meters) becomes

$$\tilde{h}(y) = \tilde{h}_1 + \tilde{h}_2 e^{-[(y-2000/2)/350]^2}$$

If $\tilde{h}_1$ = 5 m and $\tilde{h}_2$ = 10 m, the nondimensional depth profile across the bay is displayed in Fig. 1.

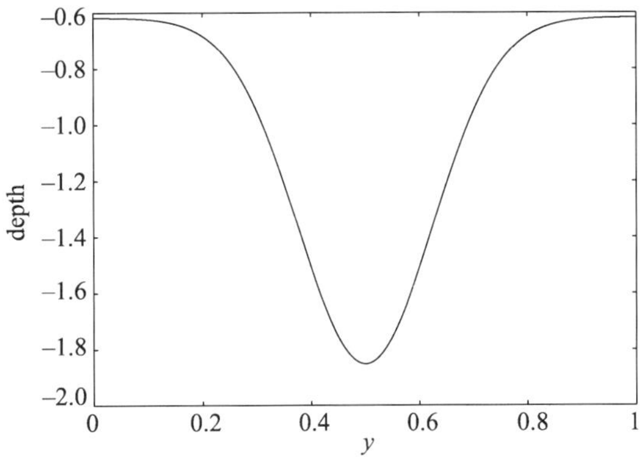

Fig. 1 The nondimensional depth profile across the bay

A single-frequency tide with the period T =12 h is imposed with the amplitude of 1 m at the open boundary. The bottom friction coefficient is taken to be $\beta = \beta_1$ = 0.00176 m s^{-1} as in Li and O'Donnell (2005). All the above parameters determine the nondimensional parameters in the narrow bay system (Eqs. 13–17 and 27–34), except the length of the bay. This configuration is referred to as the standard case in the remaining parts of the paper.

Based on the above configuration, the bay with a length L = 105 km is first considered. The Eulerian residual transport velocity $\boldsymbol{u}_T$ is displayed in Fig. 2 with the variables being in the nondimensional form. It can be seen that the flow pattern is symmetric with respect to the central line of the bay. Two mirror-symmetric gyres are formed at the head of the bay with the inward flow in the deep water area. At the outer part of the bay, two semi-gyres are formed with openings at the open boundary. The Eulerian residual transport velocity is outward at the deep water area and inward at the shoal area in the outer half of the bay. This result is consistent with that of Li and O'Donnell (2005).

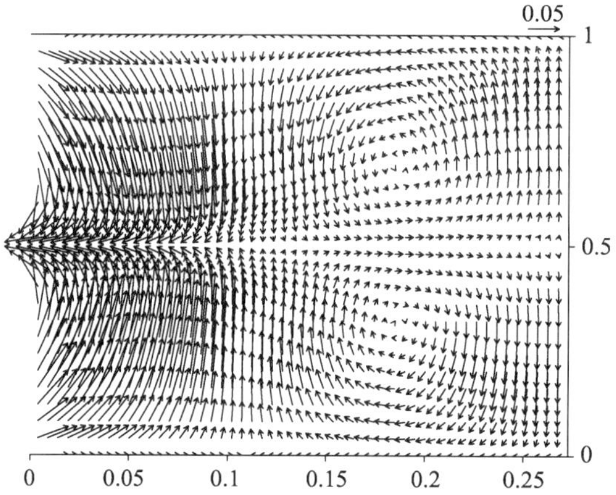

Fig. 2 The vectors of $\boldsymbol{u}_T$ in a bay of 105 km long and 2 km wide
All the labels and vectors are in the nondimensional form

4. Analysis of Solutions in a Narrow Bay

4.1 u_L and its comparison with u_E and u_T

The solutions of $\boldsymbol{u}_L$, $\boldsymbol{u}_E$, $\boldsymbol{u}_S$, $\boldsymbol{u}_T$ and $\boldsymbol{u}_D$ are displayed in Fig. 3 for the standard case with the length of the bay being the same as the tidal wave length.

The pattern of $\boldsymbol{u}_L$ shown in Fig. 3a is rather complex. A pair of mirror-symmetric gyres exist near the head of the bay, with the cyclonic one located in the upper half of the bay. Adjacent to these two gyres, there exists another pair of gyres with opposite rotation directions and they are compressed by four semi-gyres all extending to the open boundary. The four semi-gyres can be divided into two symmetric groups. In each group, the flow is inward along the lateral boundary and along the central axis of the bay; the two semi-gyres join together and flow out at the slope of the bay.

Distinct differences between $\boldsymbol{u}_L$ and $\boldsymbol{u}_E$ are found in the outer part of the bay. Fig. 3b shows that the Eulerian residual velocity $\boldsymbol{u}_E$ flows outward along the whole open boundary, violating mass conservation law (Feng *et al.*, 2008). However, Fig. 3c shows that $\boldsymbol{u}_S$ is in the opposite direction of $\boldsymbol{u}_E$ along the open boundary. As a result, the magnitude of $\boldsymbol{u}_L$, shown in Fig. 3a, has a smaller magnitude than that of $\boldsymbol{u}_E$ for the present case. This difference has been noticed in previous studies such as Loder (1980), but a more general theoretical explanation is given in this study. Fig. 3b shows that there are two mirror-symmetric gyres near the head of the bay as in Fig. 3a, but the sizes of the gyres are different. The gyres of $\boldsymbol{u}_{\mathrm{L}}$ occupy larger areas than those of $\boldsymbol{u}_{\mathrm{E}}$.

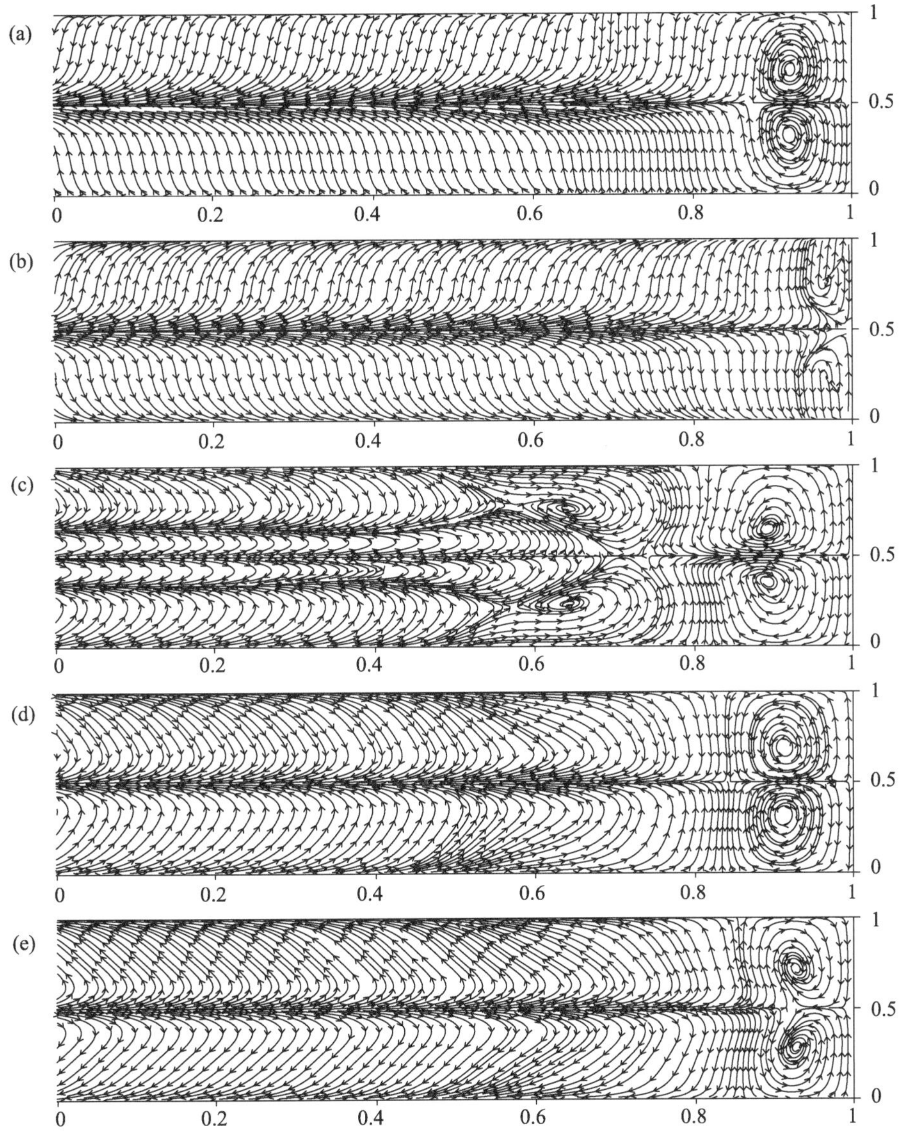

Fig. 3 Streamlines of nondimensional (a)$\boldsymbol{u}_{\mathrm{L}}$, (b)$\boldsymbol{u}_{\mathrm{E}}$, (c)$\boldsymbol{u}_{\mathrm{S}}$, (d)$\boldsymbol{u}_{\mathrm{T}}$, and (e)$\boldsymbol{u}_{\mathrm{D}}$ in a bay with length equal to one tidal wavelength

$\boldsymbol{u}_{\mathrm{T}}$ shown in Fig. 3d is different from $\boldsymbol{u}_{\mathrm{L}}$ shown in Fig. 3a. Two semi-gyres, which are adjacent to the two gyres near the head of the bay, are located at the outer part. The two semi-gyres combine to form an outflow in the deep part of the bay and an inflow along the sides of the bay. At the open boundary $\boldsymbol{u}_{\mathrm{T}}$ flows outward near the central axis of the bay while $\boldsymbol{u}_L$ flows inward; near the sides of the bay both $\boldsymbol{u}_T$ and $\boldsymbol{u}_L$ flow inward. For $\boldsymbol{u}_L$, we have noted that there exist two branches of outflows at the slope of the bay to ensure the mass conservation. Clearly, the LRV differs from the Eulerian residual transport velocity not only in magnitudes but also in patterns.

$\boldsymbol{u}_D$, the missing part in $\boldsymbol{u}_T$ for $\boldsymbol{u}_L$, is shown in Fig. 3e. Clearly, $\boldsymbol{u}_D$ countervails with $\boldsymbol{u}_T$ nearly everywhere. This countervailing feature makes the magnitude of $\boldsymbol{u}_L$ smaller than $\boldsymbol{u}_T$; and even makes the directions of $\boldsymbol{u}_L$ and $\boldsymbol{u}_T$ opp osite. This difference may suggest, for instance, that the "residence time of water inside the bay" estimated from $\boldsymbol{u}_L$ is longer than that estimated from $\boldsymbol{u}_T$.

$\boldsymbol{u}_D$ results from the interaction between the along-bay tidal velocity and the across-bay displacement, but the bottom topography still plays a role in determining it according to

$$u_D = \frac{\partial \ln h}{\partial y}\langle u_0\eta_0\rangle + \frac{\partial \langle u_0\eta_0\rangle}{\partial y} \tag{48}$$

The tidally averaged product of the along-bay velocity and the across-bay displacement can be expressed as the product of u_0^c and the conjugate of v_0^c, v_0^{c*}, or the product of v_0^c and the conjugate of u_0^c, u_0^{c*}, i.e.,

$$\langle u_0\eta_0\rangle = -\langle v_0\xi_0\rangle = \frac{1}{4\pi}\mathrm{Im}(u_0^c v_0^{c*}) = -\frac{1}{4\pi}\mathrm{Im}(v_0^c u_0^{c*}) \tag{49}$$

Equation 49 is rather complex and Fig. 4 shows its distribution for a bay with length equal to 0.3 tidal wavel ength since the major charac teristics are confined to the head of the bay. The bay can be approximately split into four areas, with two bigger areas occupying the outer part of the bay and two smaller areas located near the head of the bay. $\langle u_0\eta_0\rangle$ is the biggest at the mouth of the bay and decreases towards the head of the bay. The variation of $\langle u_0\eta_0\rangle$ is also larger at the mouth than near the head of the bay. Across the outer part of the bay from $y = 0$ to $y = 1$, $\langle u_0\eta_0\rangle$ is negative for $y < 1/2$ and positive for $y > 1/2$. Meanwhile, $\partial \ln h / \partial y$ is positive in the lower half of the bay and negative in the upper half of the bay. Thus, the first term in Eq. 48 is negative across the bay. The second term in Eq. 48 is negative at the lateral boundary of the bay where the water is shallow, thus u_D is negative. u_D

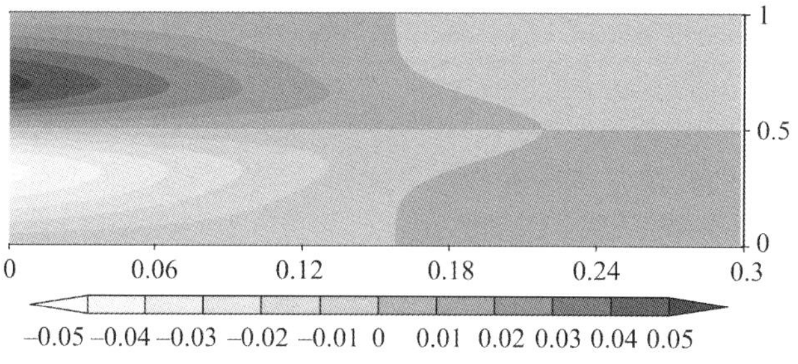

Fig. 4 The nondimensional form of $\langle u_0\eta_0\rangle$ in a narrow bay

should be positive at the deep part of the bay because it obeys mass conservation. The condition is opposite near the head of the bay where two symmetric eddies exist. As a result, the pattern of $\boldsymbol{u}_{\mathrm{D}}$ contrasts with that of $\boldsymbol{u}_{\mathrm{T}}$ shown in Fig. 3d, e.

4.2 Dependence on L_{N}

The nondimensional length of the bay, L_{N}, is a free parameter in the present case of a narrow bay. Here we examine the patterns of the first-order LRV by varying L_{N} from 0.12 to 0.5 tidal wavelength. Figure 5a shows that for the length of the bay being 0.5 wavelength, there are two mirror-symmetric gyres near the head of the bay as mentioned in Section 4.1 (Fig. 3a). The diameter of the gyres is about 0.18 wavelength. Again, in the outer part of the bay there are two gyres with different rotation directions, which are separated by four semi-gyres at the open boundary.

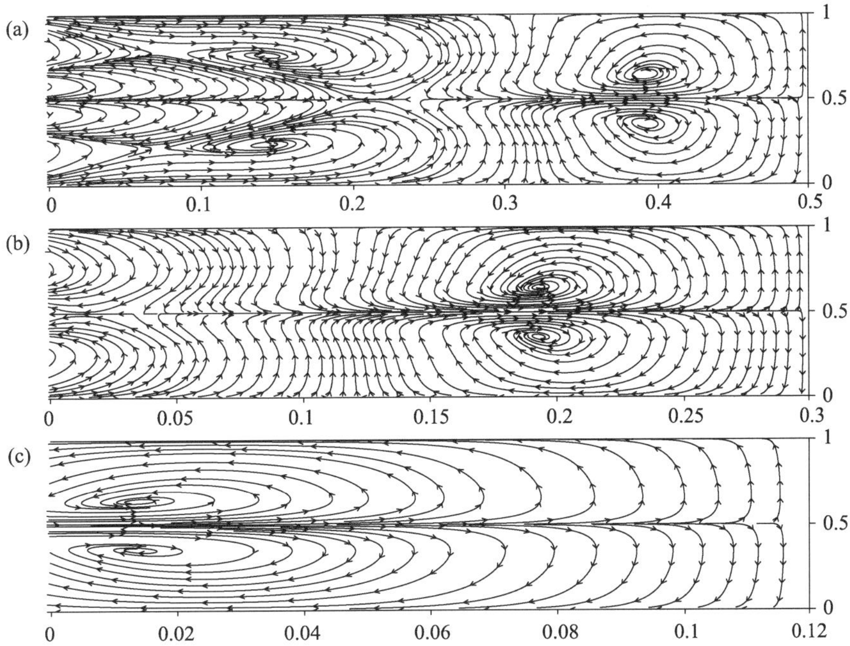

Fig. 5 The streamlines of nondimensional $\boldsymbol{u}_{\mathrm{L}}$ in a bay with lengths of (a) 0.5, (b) 0.3, and (c) 0.12 tidal wavelength

Figure 5b, c show the first-order LRV for the length of the bay being 0.3 and 0.12 wavelength respectively. The combination of Fig. 5 and Fig. 3a tells that the first-order LRV streamline of a shorter bay is a part of that of the longer bay. Therefore, the pattern of $\boldsymbol{u}_{\mathrm{L}}$ is determined by the dimensionless distance away from the head of the bay, according to which the bay can be divided into three zones. In the inner zone, which is within 0.18 wave length away from the head of the bay, the water flows inward in the deep part and outward in the shallow part. In the outer zone, with distance larger than 0.5 wavelength, the water flows inward along the sides and in the deep part, and flows outward in the slope area. The transition zone is located with distances between 0.18 and 0.5 wavelength away from the head of the bay. Here, the first-order LRV has a more complex pattern than the other two zones. The water flows inward along the sides and bounces back when encountering the two inner gyres; then the two branches will converge and flow outward for a short distance in the deep central part of the bay before detaching because of the inflow from the open boundary, and finally join together and flow outward in the slope area of the bay.

4.3 Dependence on β_T

According to Eq. 28, the nondimensional parameter $\beta_T = \beta T / h_c$ can influence the solution of LRV. Figures 6 and 7 show the streamlines of $\boldsymbol{u}_L$ and $\boldsymbol{u}_T$ with β_T being half or twice

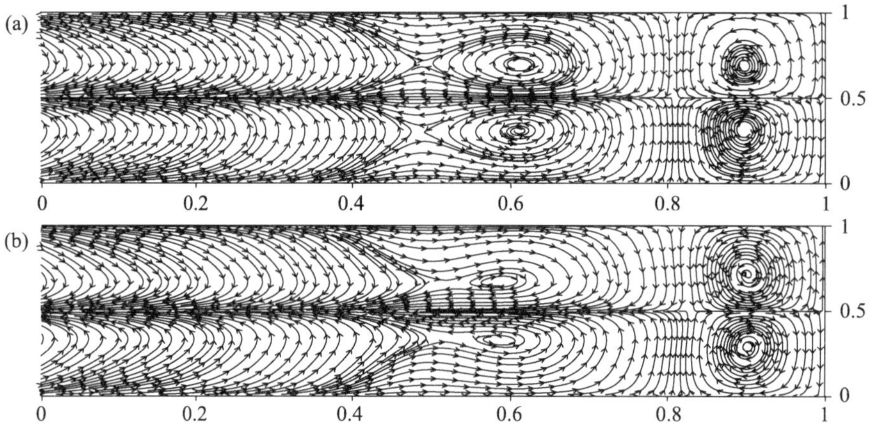

Fig. 6 The streamlines of nondimensional (a) $\boldsymbol{u}_L$ and (b) $\boldsymbol{u}_T$ with β_T being half of that used in creating Fig. 3

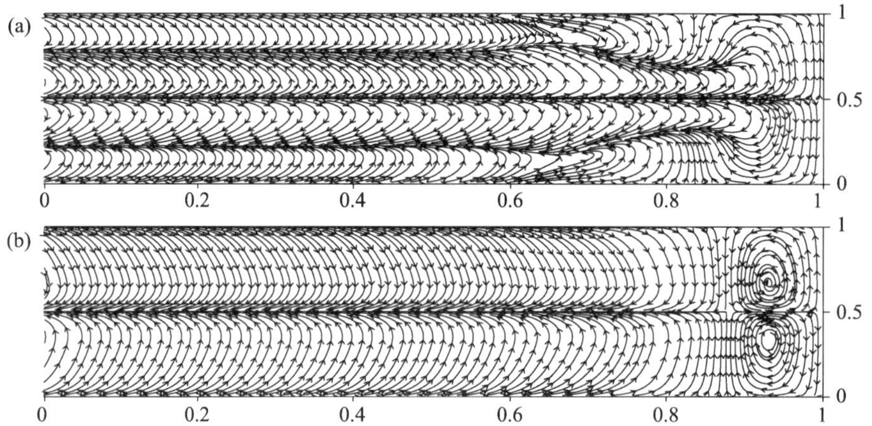

Fig. 7 The streamlines of nondimensional (a) $\boldsymbol{u}_L$ and (b) $\boldsymbol{u}_T$ with β_T being twice of that used in creating Fig. 3

of that used for creating Fig. 3, respectively. With a reduced β_T, $\boldsymbol{u}_L$ is closer to $\boldsymbol{u}_T$ and the two semi-gyres located at the deep central part disappear. This corresponds to a reduced contribution of the rotational part of the tidal body force, $\pi_{x\beta}$. With an increased β_T, the gyres of $\boldsymbol{u}_L$ at the head of the bay do not appear but join the two semi-gyres in the deep central part of the bay. The structure of $\boldsymbol{u}_T$ does not change much with a changing β_T.

In the real world, the values of β_T differ for different bays with different water depths. For a single bay, the values of β_T differ for different tidal constitutes, for example, β_T for a semi-diurnal tide is only half of that for a diurnal tide. For a specific bay if the semi-diurnal and diurnal tides are both important, the residual current is induced by the combination of these tidal constituents. Under the weakly nonlinear condition, Feng (1990) proved that the Lagrangian residual current induced by multi-frequency tides is the linear summation of $\boldsymbol{u}_L$ generated by each tidal component. Therefore, the difference of the LRV caused by different values of β_T can add more complexity to the gross result than the single tide case. This suggests that the feature of the LRV may not be fully described by the bay length only.

4.4 Dependence on the form of the bottom friction

As stated in Section 2.2, the linearized bottom friction based on Proudman (1953) is used in this study. However, the bottom friction term in Eqs. 11 and 12 takes the nonlinear form, whereas in other studies the linear form, i.e., $\beta_{T1}=0$, was used (Feng *et al.*, 1986a).

It can be seen from Eqs. 13 to 15 that the zeroth-order tide is not influenced by the different forms of the bottom friction term. However, the nonlinear bottom friction term contributes to the first-order LRV through $\pi_{x\beta}$ in Eq. 28. The difference in $\boldsymbol{u}_L$ between the linear and nonlinear forms of the bottom friction is denoted as $\boldsymbol{u}_{L\beta}=\langle u_{L\beta}, v_{L\beta}\rangle$ and can be expressed as

$$u_{L\beta}=-\frac{h\int_0^1\langle u_0\zeta_0\rangle \mathrm{d}y}{\beta_T\int_0^1 h^2\mathrm{d}y}+\frac{\langle u_0\zeta_0\rangle}{\beta_T h} \tag{50}$$

$$v_{L\beta}=\frac{\int_0^y h^2\mathrm{d}y'}{\beta_T h\int_0^1 h^2\mathrm{d}y}\int_0^1\frac{\partial\langle u_0\zeta_0\rangle}{\partial x}\mathrm{d}y-\frac{1}{\beta_T h}\int_0^y\frac{\partial\langle u_0\zeta_0\rangle}{\partial x}\mathrm{d}y' \tag{51}$$

The streamline of $\boldsymbol{u}_L$ for the linear form of the bottom friction term is shown in Fig. 8a. Compared with Fig. 3a, it can be seen that the structure of $\boldsymbol{u}_L$ becomes much simpler. The two gyres near the head of the bay now disappear, and the four semi-gyres in the outer part of the bay are reduced to two semi-gyres with one occupying the upper and the other occupying the lower half of the bay.

The part of the LRV associated with the nonlinear part of the bottom friction term, $\boldsymbol{u}_{L\beta}$, is shown in Fig. 8b. Interestingly, the two gyres near the head of the bay in Fig. 8b are very similar to those in Fig. 3a. This means that the two gyres near the head of the bay in Fig. 3a are mainly induced by the nonlinear part of the bottom friction term. This explains why the

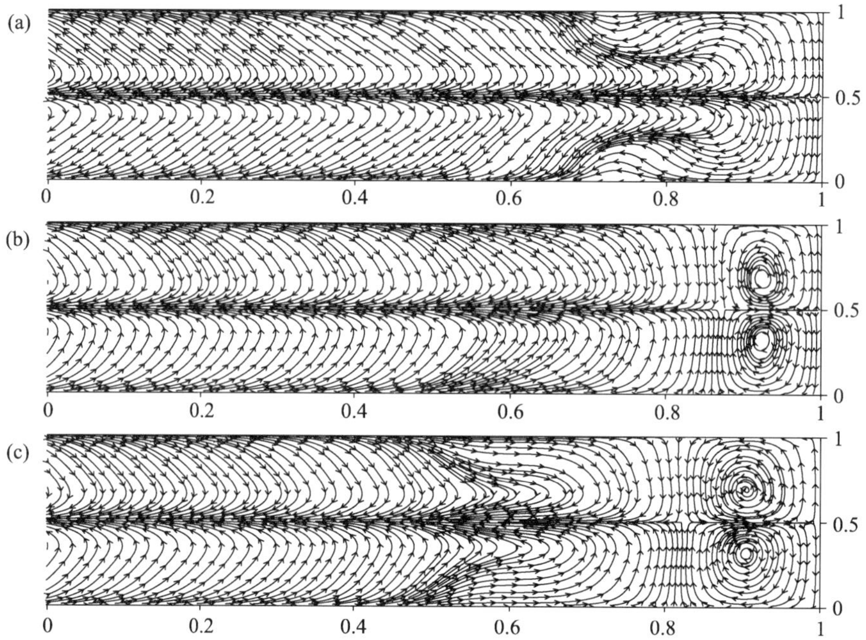

Fig. 8 (a) The streamline of $\boldsymbol{u}_{\mathrm{L}}$ for linear bottom friction. (b) The streamline of $\boldsymbol{u}_{\mathrm{L}\beta}$. (c) The streamline of $\boldsymbol{u}_{\mathrm{T}}$ for linear bottom friction

pattern of $\boldsymbol{u}_{\mathrm{L}}$ becomes two simple semi-gyres when the linear form of the bottom friction term is used. For the outer part, $\boldsymbol{u}_{\mathrm{L}}$ in Fig. 8a is similar to $\boldsymbol{u}_{\mathrm{D}}$ in Fig. 3e, both showing inward flow in the deep part and outward flow in the shallow shoal. If the nonlinear form of the bottom friction term is used, $\boldsymbol{u}_{\mathrm{L}}$ beco mes much more complex as displayed in Fig. 3a.

The pattern of $\boldsymbol{u}_{\mathrm{L}\beta}$ shown in Fig. 8b is similar to that of $\boldsymbol{u}_{\mathrm{T}}$ (Fig. 3d), but differs from that of $\boldsymbol{u}_{\mathrm{D}}$ (Fig. 3e). Beca use $\boldsymbol{u}_{\mathrm{L}\beta}$ is the difference in $\boldsymbol{u}_{\mathrm{T}}$ between the cases using the linear and the nonlinear forms of the bottom friction, the structure of $\boldsymbol{u}_{\mathrm{T}}$ for the linear form (Fig. 8c) does not show much difference with that of $\boldsymbol{u}_{\mathrm{T}}$ for the nonlinear form of the bottom friction term (Fig. 3d). For the case of the linear form of the bottom friction term, the pattern of $\boldsymbol{u}_{\mathrm{T}}$ differs from that of $\boldsymbol{u}_{\mathrm{L}}$ in the outer part of the bay. $\boldsymbol{u}_{\mathrm{L}}$ flows inward at the deep central part of the bay and flows outward in the shallow part and along the sides. $\boldsymbol{u}_{\mathrm{T}}$ is in the opposite direction of $\boldsymbol{u}_{\mathrm{L}}$. Despite of the above differences, $\boldsymbol{u}_{\mathrm{T}}$ still shows two gyres near the head of the bay. This means that for the case of the linear form of the bottom friction term, the structure of the first-order LRV is smoother than that of the Eulerian residual transport velocity.

The bottom friction has two different effects on the residual current. In the case where the nonlinear form of the bottom friction term is used, only the nonlinear part associated with β_1 contributes to the generation of the residual velocity. The linear part associated with β

only dissipates the residual current. In numerical modeling studies the bottom friction itself can be nonlinear, thus generating additional complexity.

4.5 Residual current and water exchange

One of the important applications of the residual transport is to estimate the water exchange. Many methods have been used to define the water exchange across a section. If the steady-state residual transport is obtained, the water volume bounded by a specific cross section divided by the water exchange flux at the same section is used to describe the residence time, denoted as "the residence time at that cross section" hereafter.

Since $h\boldsymbol{u}_{\mathrm{L}}$ satisfies the mass conservation equation, the integration of $h\boldsymbol{u}_{\mathrm{L}}$ over a section across the bay should be 0. The integration of hu_{L} over half of the section across the bay is the water exchange flux at that section. The water exchange flux is calculated at the consecutive cross sections in a bay with a length of one tidal wavelength. The same procedure is also applied to $h|u_{\mathrm{T}}|$. The resulting estimates are displayed in Fig. 9a.

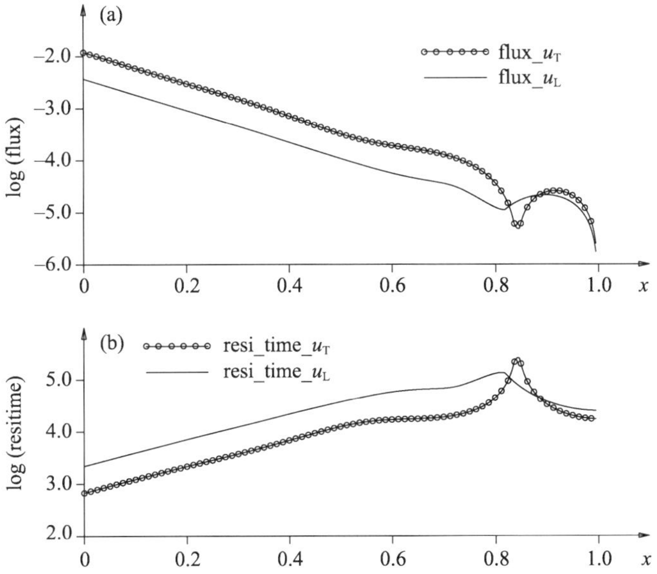

Fig. 9 (a) The water exchange flux at every cross section in a bay with length of one tidal wavelength based on hu_{L} and hu_{T}. (b) The residence time of water inside every cross section. The vertical axes are in logarithmic scales

It can be seen that the water exchange flux generally decreases from the open boundary towards the head of the bay; but at 0.18 wavelength away from the head of the bay, the water exchange flux reaches a minimum value. Accor ding to Fig. 3a, this is where the boundary of the two inner gyres is located. This location corres ponds to a "barrier" for the water exch ange in the bay. Similarly, Fig. 9b shows that the residence time incre ases sharply from the head of the bay and reaches the maximum at 0.18 wavelength away from the head. It then gradually decreases until toward the open boundary. Clearly, the variation of residence time along the bay corresponds to the two rather isolated circulation systems separated at 0.18 wavelength away from the head.

The Eulerian residual transport, hu_T, is also used to estimate the water exchange and the resulting estimates are shown in Fig. 9a, b along with the estimates with the first-order Lagrangian residual transport hu_L. The characteristics of the estimates based on hu_T and hu_L are rather similar, but differ quantitatively. Firstly, the minimum water exchange flux based on hu_T occurs at around 0.16 wavelength away from the head, further inside the bay than that based on hu_L. This corresponds to the slight difference in the positions of the two gyres between the hu_L and hu_T cases. Secondly, the residence time based on hu_T is generally shorter than that based on hu_L, except near where the residence time reaches the maximum. At the open boundary, the residence time based on hu_L is three times longer than that based on hu_T. Therefore, if hu_T is used to examine the water exchange, the capacity of the physical self-purification of the bay will be significantly exaggerated.

The residence time for the bays with lengths ranging from 0.05 to 1 tidal wavelength is shown in Fig. 10. Clearly, the residence time increases with the length of the bay. The difference in the residence time between estimates based on hu_L and hu_T is nearly constant for the bays with lengths longer than 0.2 tidal wavelength.

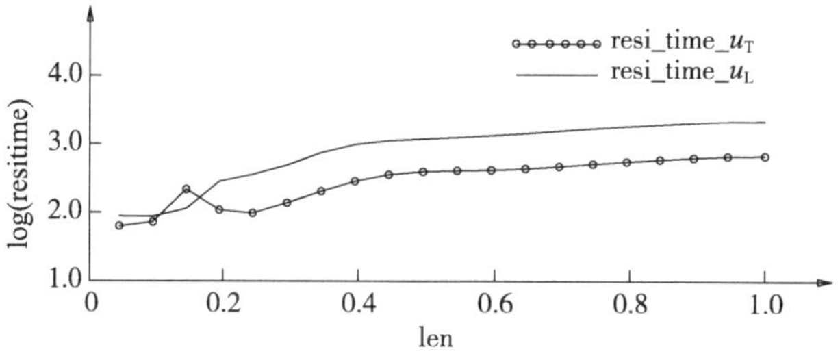

Fig. 10 The residence time based on hu_L and hu_T for the bays with different lengths
The vertical axis is in logarithmic scale

The water exchange fluxes at each cross section for the bays with lengths ranging from 0.3 to 1 tidal wavelength are plotted in Fig. 11. By aligning the head of the bay at $x = 1$, the end of each curve line corresponds to the position of the open boundary. It can be seen that the water exchange flux increases as the length of the bay decreases. For the same tidal energy input, the energy is spread further in a longer bay than in a shorter bay. At the open boundaries, the water exchange fluxes are similar; while the rates of variation along the axes of the bays are similar for the bays with various lengths. Interestingly, for bays with different lengths the minimum water exchange fluxes occur at the same location, i.e., at around 0.18 tidal wavelength away from the head of the bay, corresponding to the outer boundary of the two inner gyres.

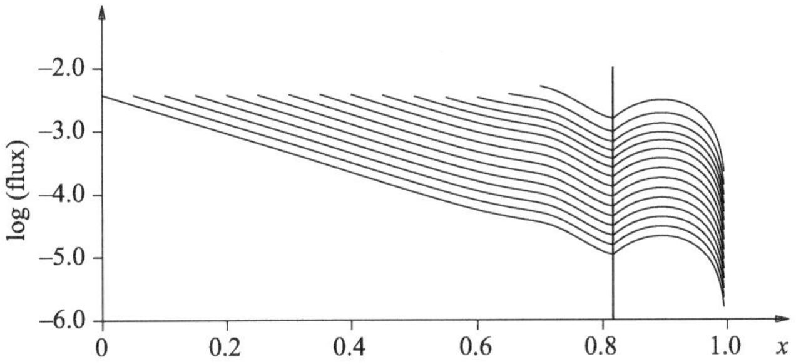

Fig. 11 The water exchange flux based on hu_L at each cross section of the bay with lengths ranging from 0.3 to 1 wave length, corresponding to curves displaced from top to bottom. The vertical line denotes the position where the minimum water exchange flux occurs. The vertical axis is in logarithmic scale

The water exchange flux for a bay with different values of β_T is displ yed in Fig. 12. The water exchange flux increases at most cross sections of the bay as β_T decreases. The minimum water exchange flux occurs where the inner and outer gyres are separated. That point is located at around 0.18 wavel ngth away from the head of the bay and shifts towards it as β_T increases. Moreover, the separation becomes less obvious as β_T gets bigger and finally the separation nearly disappears, corre pond-ing to the merging of the inner and outer gyres as displayed in Fig. 7a.

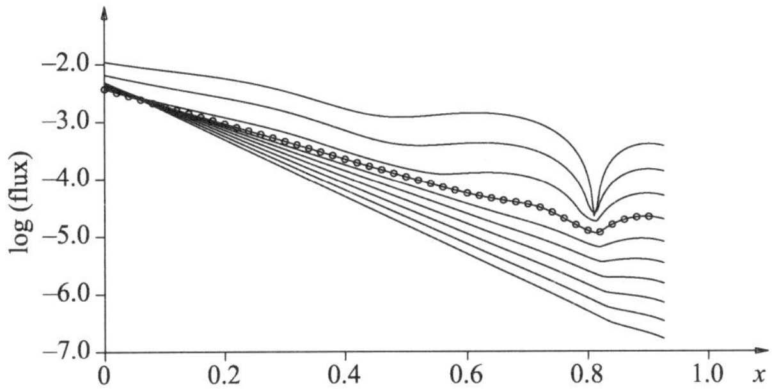

Fig. 12 The water exchange flux based on $h\boldsymbol{u}_L$ with different values of β_T, at each cross section of a bay with a length of one wave length. The curves from top to bottom correspond to the values of β_T being half and twice of that in the standard case. The curve with circles shows the standard case. The vertical axis is in logarithmic scale

The water exchange flux in bays with different depth profiles is exa ined by varying α from 2×7/40 to 0.5×7/40 in Eq. 47 while keeping h_1 and h_2 unchanged. A constant is added to the depth profile to adjust the averaged water depth, so that β_T is kept the same as in the standard case. For $\alpha = 2\times7/40$, $1.4\times7/40$, and $1.0\times7/40$, the water exchange fluxes are displayed in Fig. 13. The results of $\alpha < 1\times7/40$ are not shown because they are very similar to that of $\alpha = 1\times7/40$. Fig. 13 shows that the water exchange flux becomes bigger if α gets smaller, i.e., the bottom is flatter. The location of the minimum flux shifts away from the head of the bay when α gets bigger.

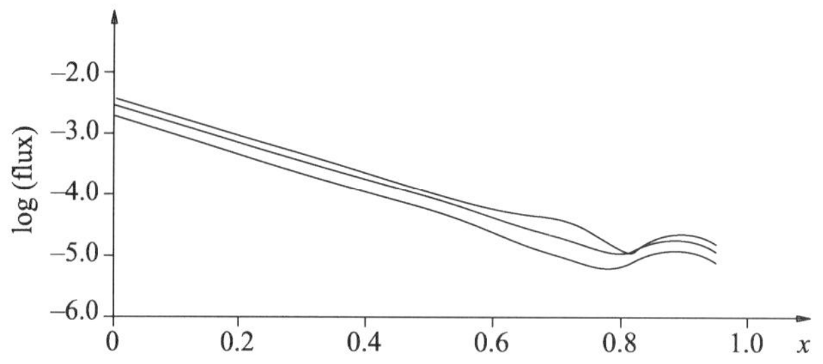

Fig. 13 The water exchange flux based on hu_L at each cross section of the bay with length of one wave length but different values of α. The curves from bottom to top correspond to $\alpha = 2\times7/40$, $1.4\times7/40$, and $1.0\times7/40$, respectively. The vertical axis is in the logarithmic scale

4.6 The second-order LRV

Equations 35 and 36 tell that the Lagrangian drift velocity depends on the initial tidal phase. At each location, the vectors of the Lagrangian drift velocity constitute an ellipse, as shown in Fig. 14. The ellipses are bigger at the open boundary and in the slope area of the bay. The ellipses are one order of magnitude smaller compared with the mass-transport velocity. Therefore, in this weakly nonlinear case the mass-transport velocity is a good approximation of the LRV. On the other hand, the dependence of the LRV on the initial tidal phase is a Lagrangian feature, which is important in many applications such as choosing the optimal time for sewage discharge.

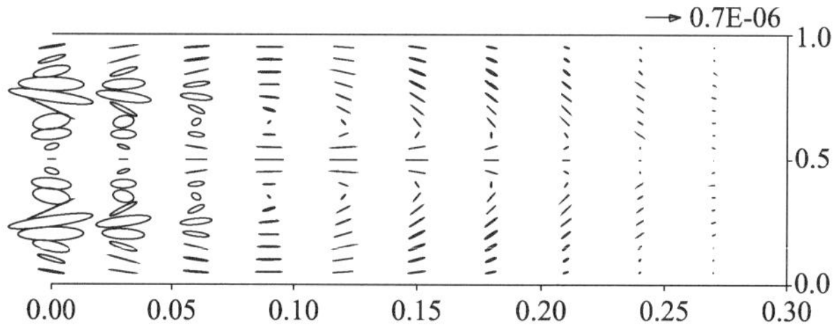

Fig. 14 Ellipses of the nondimensional $\boldsymbol{u}_{\mathrm{ld}}$ in a narrow bay

5. Conclusions

In the present study, the depth-averaged equations for the single frequency tidal currents and the first-order LRV, and the expressions for the second-order LRV, are derived systematically based on the weakly nonlinear assumption. The equations are simplified for the case of a narrow bay and are solved analytically. The results for a specific axisymmetric bottom topography are discussed in detail.

Generally, according to the streamlines of the first-order LRV, the bay can be divided into an inner part near the head of the bay, the transitional zone, and the outer part near the open boundary. Two mirror-symmetric gyres exist in the inner part with the inflow in the central deep region. Adjacent to the two gyres in the inner part, two gyres with opposite rotation directions exist in the transitional zone. These two gyres are compressed by four semi-gyres in the outer part, where the water flows inward along the sides and the deepest part of the bay and flows outward along the slope.

The dependence of the first-order LRV on the length of the bay length, L_N, is examined. It is proved that the streamline of the nondimensional current velocity in a shorter bay is only part of that in a longer bay. The nondimensional parameter $\beta_T = \beta T / h_c$ includes the combined influences of the bottom friction parameter, the tidal period and the averaged water depth on the first-order LRV. For smaller values of β_T, the first- order LRV is closer to the Eulerian residual transport velocity. For larger values of β_T, the two gyres in the inner part of the bay tend to merge with two of the four semi-gyres in the outer part. A practical significance in that the semi-diurnal or diurnal tides can change β_T by a factor of two. The form of the bottom friction also plays an important role in the pattern of the firstorder LRV. This suggests that the parameterization of bottom friction in numerical models should be treated carefully.

The water exchange at each cross section of the bay is controlled by the LRV. The water exchange flux is higher at the open boundary and decreases towards the head of the bay, but the minimum flux occurs at around 0.18 wave length away from the head of the bay. This location can be regarded as where the inner and the outer parts of the bay separate. The position of this separation location is not fixed; it changes with both β_T and the bottom topography.

The second-order LRV is shown to be dependent on the initial tidal phase to take the Lagrangian averaging. This shows the Lagrangian feature of the LRV.

Different definitions of the residual current are different, not only in magnitudes but also in patterns and directions. The Eulerian residual transport velocity, that can be regarded as the partially modified Eulerian residual velocity, differs from the LRV. For the specific case considered here, the pattern of the Eulerian residual transport velocity is simpler than that of the first-order LRV; and in the deep part of the bay, the two velocities are in opposite directions. This leads to significant difference in the water residence time calculated according to different residual currents.

It is noteworthy that the symmetry of the depth profile about $y = 0.5$ is not a prerequisite to get the solution. Indeed there is a streamline along $y = 0.5$ so the narrow bay can be split into two separate bays without affecting the solutions. Thus, the depth profile can be of any form if it only varies in the across-bay direction.

In general the LRV represents the water mass inter-tidal transport in the tide dominant area. The residual current derived by other methods may cause the misunderstanding of the mass-transport mechanism in the study area. Yet, this belief is not easy to be proved in the field because in most tide dominant areas the residual current, which is of the order of cm s^{-1}, is a small part of the current. This value is even beyond the precision of the current meter. The traditional Eulerian observation cannot obtain the LRV, which makes the proof more difficult.

In this paper the depth-averaged LRV obtained analytically in the narrow bay gives a good example. Although it is confined to a single-frequency tidal case, the analysis can be extended to cases with wind driven, baroclinic, and multi-frequency tides, because the residual currents generated by the above factors can be linearly combined following Feng (1987, 1990). River runoff can also be included at the head of the bay based on the analyses of Feng and Wu (1995) and Feng (1998). Further steps to verify these theoretical analyses include laboratory experiments and the development of innovative methods of Lagrangian observations.

Acknowledgements

This study was supported by project 40976003 from National Science Foundation of China and National Basic Research Program of China (2010CB428904). We thank the two anonymous reviewers for constructive comments on the original manuscript.

Appendix: The Derivation of Equations for $\boldsymbol{u}_{\mathrm{L}}$

By applying the operator (Eq. 26) to Eq. 18 and making use of the definition of $\boldsymbol{u}_{\mathrm{E}}$ in Eq. 3, one gets

$$\frac{\partial hu_{\mathrm{E}}}{\partial x}+\frac{\partial hv_{\mathrm{E}}}{\partial y}=-\left(\frac{\partial\langle u_0\zeta_0\rangle}{\partial x}+\frac{\partial\langle v_0\zeta_0\rangle}{\partial y}\right) \tag{52}$$

The substitution of Eqs. 23–25 into Eq. 4 gives

$$hu_{\mathrm{S}}=\langle u_0\zeta_0\rangle+\frac{\partial}{\partial y}\langle hu_0\eta_0\rangle \tag{53}$$

$$hv_{\mathrm{S}}=\langle v_0\zeta_0\rangle+\frac{\partial}{\partial x}\langle hv_0\xi_0\rangle \tag{54}$$

By performing $\partial/\partial x$ (Eq. 53)+ $\partial/\partial y$ (Eq. 54) and taking into consideration of Eq. 52, the continuity equation of $\boldsymbol{u}_{\mathrm{L}}$, Eq. 27, can be obtained.

The momentum equations of the LRV are derived as follows. By applying Eq. 26 to Eq. 19 and making use of Eq. 2, one gets

$$0=-\frac{\partial\zeta_{\mathrm{E}}}{\partial x}-\beta_{\mathrm{T}}\frac{u_{\mathrm{L}}}{h}+\beta_{\mathrm{T1}}\frac{\langle u_0\zeta_0\rangle}{h^2}-\left\langle u_0\frac{\partial u_0}{\partial x}+v_0\frac{\partial u_0}{\partial y}\right\rangle+\beta_{\mathrm{T}}\frac{1}{h}\left\langle\xi_0\frac{\partial u_0}{\partial x}+\eta_0\frac{\partial u_0}{\partial y}\right\rangle \tag{55}$$

By applying the operator $(\xi_0\partial/\partial x+\eta_0\partial/\partial y)$ to Eq. 14 and making use of Eq. 15, followed by applying Eq. 26, one yields

$$\left\langle\xi_0\frac{\partial^2 u_0}{\partial x\partial t}+\eta_0\frac{\partial^2 u_0}{\partial y\partial t}\right\rangle=-\left\langle\xi_0\frac{\partial^2\zeta_0}{\partial x^2}\right\rangle-\beta_{\mathrm{T}}\left\langle\xi_0\frac{\partial}{\partial x}\left(\frac{u_0}{h}\right)+\eta_0\frac{\partial}{\partial y}\left(\frac{u_0}{h}\right)\right\rangle \tag{56}$$

The left side of Eq. 56 is rearranged with integration by parts. Furthermore, by making use of Eq. 6 and the periodicity of the variables, one gets

$$\begin{aligned}\left\langle\xi_0\frac{\partial^2 u_0}{\partial x\partial t}+\eta_0\frac{\partial^2 u_0}{\partial y\partial t}\right\rangle&=\left\langle\frac{\partial}{\partial t}(\xi_0\frac{\partial u_0}{\partial x}+\eta_0\frac{\partial u_0}{\partial y})\right\rangle-\left\langle\frac{\partial\xi_0}{\partial t}\frac{\partial u_0}{\partial x}+\frac{\partial\eta_0}{\partial t}\frac{\partial u_0}{\partial y}\right\rangle\\&=-\left\langle u_0\frac{\partial u_0}{\partial x}+v_0\frac{\partial u_0}{\partial y}\right\rangle\end{aligned} \tag{57}$$

Then by changing Eq. 56 to the following form

$$\begin{aligned}-\left\langle u_0\frac{\partial u_0}{\partial x}+v_0\frac{\partial u_0}{\partial y}\right\rangle=&-\left\langle\xi_0\frac{\partial^2\zeta_0}{\partial x^2}\right\rangle-\beta_{\mathrm{T}}\frac{1}{h}\left\langle\xi_0\frac{\partial u_0}{\partial x}+\eta_0\frac{\partial u_0}{\partial y}\right\rangle\\&-\beta_{\mathrm{T}}\left\langle\xi_0 u_0\frac{\partial}{\partial x}\left(\frac{1}{h}\right)+\eta_0 u_0\frac{\partial}{\partial y}\left(\frac{1}{h}\right)\right\rangle\end{aligned}$$

and by adding Eq. 55 to the above equation, one gets

$$0=-\frac{\partial\zeta_{\mathrm{E}}}{\partial x}-\beta_{\mathrm{T}}\frac{u_{\mathrm{L}}}{h}+\beta_{\mathrm{T1}}\frac{\langle u_0\zeta_0\rangle}{h^2}-\left\langle\xi_0\frac{\partial^2\zeta_0}{\partial x^2}\right\rangle-\beta_{\mathrm{T}}\left\langle\xi_0 u_0\frac{\partial}{\partial x}\left(\frac{1}{h}\right)+\eta_0 u_0\frac{\partial}{\partial y}\left(\frac{1}{h}\right)\right\rangle \tag{58}$$

By applying $\xi_0\partial/\partial x$ to Eq. 14, one obtains

$$\xi_0\frac{\partial^2\zeta_0}{\partial x^2}=-\xi_0\left(\frac{\partial^2 u_0}{\partial x\partial t}+\beta_{\mathrm{T}}\frac{\partial}{\partial x}\left(\frac{u_0}{h}\right)\right) \tag{59}$$

and then $\partial/\partial x[\xi_0\cdot(\text{Eq.}14)]$ leads to

$$\frac{\partial}{\partial x}\left(\xi_0\frac{\partial\zeta_0}{\partial x}\right)=-\frac{\partial}{\partial x}\left(\xi_0\frac{\partial u_0}{\partial t}+\beta_{\mathrm{T}}\xi_0\frac{u_0}{h}\right) \tag{60}$$

Then, Eq. 59 is multiplied with 2 and subtracted with Eq. 60; and the resulting equation is averaged over one tidal period. By making use of Eq. 6 and the periodicity of the zeroth-order solution, one obtains

$$\left\langle\xi_0\frac{\partial^2\zeta_0}{\partial x^2}\right\rangle=\left\langle\frac{1}{2}\frac{\partial}{\partial x}\left(\xi_0\frac{\partial\zeta_0}{\partial x}\right)\right\rangle+\beta_{\mathrm{T}}\frac{1}{h}\left\langle u_0\frac{\partial\xi_0}{\partial x}\right\rangle \tag{61}$$

Substituting Eq. 61 into Eq. 58 leads to

$$0=-\frac{\partial \zeta_{\mathrm{E}}}{\partial x}-\beta_{\mathrm{T}} \frac{u_{\mathrm{L}}}{h}+\beta_{\mathrm{T1}} \frac{\left\langle u_0 \zeta_0\right\rangle}{h^2}-\left\langle\frac{1}{2} \frac{\partial}{\partial x}\left(\xi_0 \frac{\partial \zeta_0}{\partial x}\right)\right\rangle \\ -\beta_{\mathrm{T}}\left\langle\xi_0 u_0 \frac{\partial}{\partial x}\left(\frac{1}{h}\right)+\eta_0 u_0 \frac{\partial}{\partial y}\left(\frac{1}{h}\right)\right\rangle-\beta_{\mathrm{T}} \frac{1}{h}\left\langle u_0 \frac{\partial \xi_0}{\partial x}\right\rangle \tag{62}$$

For the case of single tidal frequency $\langle \xi_0 u_0\rangle=\langle\eta_0 v_0\rangle=0$, thus

$$-\beta_{\mathrm{T}}\left\langle\xi_0 u_0 \frac{\partial}{\partial x}\left(\frac{1}{h}\right)+\eta_0 u_0 \frac{\partial}{\partial y}\left(\frac{1}{h}\right)\right\rangle=\beta_{\mathrm{T}} \frac{1}{h^2} \frac{\partial}{\partial y}\left\langle h u_0 \eta_0\right\rangle+\beta_{\mathrm{T}} \frac{1}{h}\left\langle\xi_0 \frac{\partial v_0}{\partial y}-\eta_0 \frac{\partial u_0}{\partial y}\right\rangle$$

$$-\beta_{\mathrm{T}} \frac{1}{h}\left\langle u_0 \frac{\partial \xi_0}{\partial x}\right\rangle=\beta_{\mathrm{T}} \frac{1}{h}\left\langle\xi_0 \frac{\partial u_0}{\partial x}\right\rangle$$

By applying the above results to Eq. 62, Eq. 28 can be obtained. It is trivial to obtain the momentum equation in the *y* direction.

References

Abbott M R (1960) Boundary layer effects in estuaries. J Mar Res 18: 83–100

Cerco C F, Cole T (1993) Three-dimensional eutrophication model of Chesapeake Bay. J Environ Eng 119: 1106–1125

Cheng R T (1983) Euler-Lagrangian computations in estuarine hydrodynamics. In: Taylor J, Johnson A, Smith R (eds) Proceedings of Third International Conference on Numerical Methods in Laminar and Turbulent Flow. Pineridge, Swansea. pp 341–352

Cheng R T, Casulli V (1982) On Lagrangian residual currents with applications in South San Francisco Bay, California. Water Resour Res 18: 1652–1662

Delhez E J M (1996) On the residual advection of passive constituents. J Mar Syst 8: 147–169

Dortch M S, Chapman R S, Abt S R (1992) Application of three-dimensional Lagrangian residual transport. J Hydraul Eng 118: 831–848

Dyke P P G (1980) On the Stokes' drift induced by tidal motions in a wide estuary. Estuarine Coastal Mar Sci 11: 17–25

Feng S (1987) A three-dimensional weakly nonlinear model of tide-induced Lagrangian residual current and mass-transport, with an application to the Bohai Sea. In: Nihoul J C J, Jamart B M (eds) Three-dimensional Models of Marine and Estuarine Dynamics, Elsevier Oceanography Series, 45. Elsevier, Amsterdam. pp 471–488

Feng S (1990) On the Lagrangian residual velocity and the mass-transport in a multi-frequency oscillatory system. In: Cheng R T (ed) Residual Currents and Long-term Transport, Coastal and Estuarine Studies 38. Springer, Berlin, pp 34–48

Feng S (1998) On circulation in Bohai Sea Yellow Sea and East China Sea. In: Hong G H, Zhang J, Park B K (eds) Health of the Yellow Sea. The Earth Love Publication Association, Seoul, pp 43–77

Feng S, Wu D (1995) An inter-tidal transport equation coupled with turbulent K-ϵ model in a tidal and quasi-steady current system. Chin Sci Bull 40: 136–139

Feng S, Xi P, Zhang S (1984) The baroclinic residual circulation in shallow seas. Chin J Oceanol Limnol 2: 49–60

Feng S, Cheng R T, Xi P (1986a) On tide-induced Lagrangian residual current and residual transport, 1. Lagrangian residual current. Water Resour Res 22: 1623–1634

Feng S, Cheng R T, Xi P (1986b) On tide-induced Lagrangian residual current and residual transport, 2. residual transport with application in South San Francisco Bay. Water Resour Res 22: 1635–1646

Feng S Z, Ju L, Jiang W S (2008) A Lagrangian mean theory on coastal sea circulation with inter-tidal transports, I. fundamentals. Acta Oceanol Sin 27: 1–16

Fischer H B, List E J, Koh R, Imberger J, Brooks N H (1979) Mixing in Inland and Coastal Waters. Academic, New York

Foreman M G G, Baptista A M, Walters R A (1992) Tidal model studies of particle trajectories around a shallow coastal bank. Atmos Ocean 30: 43–69

Hainbucher D, Wei H, Pohlmann T, Sündermann J, Feng S (2004) Variability of the Bohai Sea circulation based on model calculations. J Mar Syst 44: 153–174

Hunt J N (1961) Oscillations in a viscous liquid with an application to tidal motion. Tellus 13: 79–84

Ianniello J P (1977) Tidally induced residual currents in estuaries of constant breadth and depth. J Mar Res 35: 755–786

Jay D A (1991) Estuarine salt conservation: a Lagrangian approach. Estuarine, Coastal Shelf Sci 32: 547–565

Li C, O'Donnell J (2005) The effect of channel length on the residual circulation in tidally dominated channels. J Phys Oceanogr 35: 1826–1840

Li C, Valle-Levinson A (1999) A two-dimensional analytic tidal model for a narrow estuary of arbitrary lateral depth variation: the intratidal motion. J Geophys Res 104: 23525–23543

Li C, Chen C, Guadagnoli D, Georgiou I Y (2008) Geometry induced residual eddies in estuaries with curved channels: observations and modeling studies. J Geophys Res 113: C01, 005. doi: 10.1029/2006JC004031

Loder J W (1980) Topographic rectification of tidal currents on the sides of Georges Bank. J Phys Oceanogr 10: 1399–1416

Longuet-Higgins M S (1969) On the transport of mass by time-varying ocean currents. Deep-Sea Res 16: 431–447

Maxima.sourceforge.net (2009) Maxima, a computer algebra system version 5.18.1. Available at: http://maxima.sourceforge. net. Accessed 14 July 2010

Muller H, Blanke B, Dumas F, Lekien F, Mariette V (2009) Estimating the Lagrangian residual circulation in the Iroise sea. J Mar Syst 78: S17–S36. doi: 10.1016/j.jmarsys.2009.01.008

Nihoul I C J, Ronday F C (1975) The influence of the tidal stress on the residual circulation. Tellus A 27: 484–489

Piessens R, de Doncker-Kapenga E, Überhuber C, Kahaner D (1983) Quadpack: A Subroutine Package for Automatic Integration, Springer Series in Computational Mathematics, vol 1. Springer, Berlin

Proudman J (1953) Dynamical Oceanography. Methuen, London

Ridderinkhof H, Loder J W (1994) Lagrangian characterization of circulation over submarine banks with application to the outer Gulf of Maine. J Phys Oceanogr 24: 1184–1200

Robinson I S (1983) Tidally induced residual flows. In: Johns B (ed) Physical Oceanography of Coastal and Shelf Seas. Elsevier, Amsterdam, pp 321–356

Wang H, Su Z, Feng S, Sun W (1993) A three dimensional numerical calculation of the wind driven thermohaline and tide-induced Lagrangian residual current in the Bohai Sea. Acta Oceanol Sin 12: 169–182

Wei H, Hainbucher D, Pohlmann T, Feng S, Sündermann J (2004) Tidal-induced Lagrangian and Eulerian mean circulation in the Bohai Sea. J Mar Syst 44: 141–151

Winant C D (2007) Three-dimensional tidal flow in an elongated, rotating basin. J Phys Oceanogr 37: 2345–2362

Winant C D (2008) Three-dimensional residual tidal circulation in an elongated, rotating basin. J Phys Oceanogr 38: 1278–1295

Zimmerman J T F (1979) On the Euler-Lagrange transformation and the Stokes' drift in the presence of oscillatory and residual currents. Deep-Sea Res 26A: 505–520

Simulation of the Lagrangian Tide-induced Residual Velocity in a Tide-dominated Coastal System: A Case Study of Jiaozhou Bay, China *

Guangliang Liu, Zhe Liu, Huiwang Gao, Zengxiang Gao and Shizuo Feng

1. Introduction

The inter-tidal transport of dissolved and suspended constituents, particularly nutrients, sediments and pollutants, is of great importance to the physical, chemical and biological states of marine ecosystems. It has been well documented that the inter-tidal mass transport is closely related to the residual velocities at tidal and subtidal frequencies, i.e. the tidally time-averaged velocities (Tee, 1976; Zimmerman, 1979; Cheng and Casulli, 1982; Feng *et al.*, 1986a, b, 2008; Feng, 1987; Dortch *et al.*, 1992; Delhez, 1996; Wei *et al.*, 2004; Ju *et al.*, 2009; Muller *et al.*, 2009; Jiang and Feng, 2011). However, as the usual dominant seawater movement in coastal seas, tidal currents vary much faster than the characteristic time scale of the residual velocity. It is necessary to extract the residual velocity from the oscillating ocean currents to study the inter-tidal mass transport.

The Eulerian mean velocity is a convenient method of rectifying transiently oscillating ocean velocities at a fixed position using time-averaging or other filtering techniques (e.g. Abbott, 1960; Nihoul and Ronday, 1975). However, it has been theoretically recognised for a long time that the Eulerian mean velocity lacks a well-defined physical meaning. Feng *et al.* (1984) developed the equations for the three-dimensional baroclinic Eulerian mean velocity and found artificial mass exchange across the sea surface (named the Tidal Surface Source). If the Eulerian mean velocity is used as the advection velocity in the inter-tidal mass transport equation, an assumptive dispersion (named the Tidal Dispersion) must be made, but its physical meaning is ambiguous (Fischer *et al.*, 1979). In addition, for the intermittent presence of water between the highest and lowest water levels, the Eulerian mean velocity cannot even be calculated (Delhez, 1996) because the Eulerian mean velocity should be calculated at a fixed point in the Eulerian coordinate system (in x, y and z spaces).

To avoid these problems, the Eulerian residual transport velocity was defined based on the residual transport (Feng *et al.*, 1986b; Delhez, 1996; Li and O'Donnell, 2005; Jiang and Feng, 2011). The Eulerian residual transport velocity can be calculated by averaging the transient

---

* Liu G, Liu Z, Gao H, Gao Z, Feng S. 2012. Simulation of the Lagrangian tide-induced residual velocity in a tide-dominated coastal system: A case study of Jiaozhou Bay, China. Ocean Dynamics, 62(10–12): 1443–1456

velocity weighted by the transient depth (see Eq. 1 in Section 3.2). Because the Eulerian residual transport velocity conserves mass and material surface, it has been widely used. For instance, the dependence of the Eulerian residual transport velocity on the length (Li and O'Donnell, 2005) and bending of the bay (Li *et al.*, 2008) was studied using an analytical solution of the Eulerian residual transport velocity in a narrow bay. In an elongated rotating bay, Winant (2007, 2008) used the Eulerian residual transport as the depth-integrated residual transport.

An alternative method is to define residual velocities in the Lagrangian mean framework. The Lagrangian mean velocity can be defined as the net displacement of the labelled water parcel over one or a few tidal cycles divided by the corresponding time interval (Zimmerman, 1979; Cheng and Casulli, 1982). Recently, the Lagrangian residual velocity was strictly defined by Feng *et al.* (2008). They proposed a theoretical basis for the Lagrangian residual velocity that applies to inter-tidal transport processes (Feng *et al.*, 2008). If the net displacement of a labelled water parcel over one or a few tidal cycles is negligibly smaller than the characteristic length of the flow field of circulation, the Lagrangian mean velocity field is continuous, smooth, differentiable and subject to the conservation of mass and material surface. Under such circumstances, the Lagrangian mean velocity can be called the Lagrangian residual velocity (Feng, 1987; Feng *et al.*, 2008). In general nonlinear systems, the Lagrangian residual velocity depends on the initial tidal phase. Previous theoretical and numerical studies indicated that it is more appropriate to use the Lagrangian residual velocity to determine the origin of a water mass (e.g. Longuet-Higgins, 1969; Zimmerman, 1979; Cheng and Casulli, 1982; Feng *et al.*, 1986a, b; Feng, 1987; Foreman *et al.*, 1992; Delhez, 1996; Ju *et al.*, 2009; Muller *et al.*, 2009).

As an approximation of the Lagrangian residual velocity in weakly nonlinear systems, the mass transport velocity, which is the sum of the Eulerian mean velocity and the Stokes drift velocity (Longuet-Higgins, 1969), is the first-order Lagrangian residual velocity (Zimmerman, 1979; Feng *et al.*, 1986a). According to Feng (1987), under weakly nonlinear assumptions, the first-order Lagrangian residual velocity is proven to satisfy the conservation of mass and material surface. Many studies have revealed that the first-order Lagrangian residual velocity is useful for studying the inter-tidal mass transport dynamics of weakly nonlinear systems (e.g. Dortch *et al.*, 1992; Cerco and Cole, 1993; Wang *et al.*, 1993; Cerco, 1995a, b; Delhez, 1996; Sun *et al.*, 2001; Hainbucher *et al.*, 2004; Wei *et al.*, 2004; Zou *et al.*, 2009).

In theory, the distinction between the Eulerian and Lagrangian residual velocities has been recognised for a long time, but the application of various residual velocities is confusing in the literature. This confusion may lead to the misunderstanding and misinterpretation of inter-tidal mass transport. For a specific, narrow, weakly nonlinear tidal bay, Jiang and Feng (2011) quantitatively compared the analytical solutions of the Eulerian mean velocity, the Eulerian residual transport velocity and the first-order Lagrangian residual velocity to clarify these residual velocities. They demonstrated that the use of the Eulerian mean velocity and the Eulerian residual transport velocity may lead to the misunderstanding of inter-tidal mass transport in shallow seas.

The purpose of the present study is to simulate the Lagrangian residual velocity of a real nonlinear tidal system (i.e. Jiaozhou Bay (JZB)) and compare it with the Eulerian residual transport velocity and the first-order Lagrangian residual velocity. Section 2 describes the state of the residual velocity study on JZB. The model configuration and the method used to calculate the residual velocities are described in Section 3. In Section 4, the residual velocities are compared with the observed net surface sediment transport pattern. The flushing time and its relationship to the Lagrangian residual velocity are also analyzed. Section 5 presents the comparison of the residual velocities. The hydrodynamic mechanisms generating the nonlinearity are discussed. Finally, in Section 6, we summarise our findings and offer some conclusions.

2. Study Area and Unresolved Problems

JZB is a semi-enclosed shallow water body off the southern coast of the Shandong Peninsula in East China (Fig. 1). The average water depth of JZB is approximately 7 m, with a maximum depth of 60~70 m near the narrow inlet (~3 km) connecting it to the Yellow Sea. Tidal flats cover approximately one third of the total area. JZB possesses some important indications of nonlinearity because of its coastline irregularities, topographical complexity and strong tidal currents. The currents can reach 2 m/s during the spring tide (Lü *et al.*, 2010) and be dominated by the semidiurnal (M_2) tide, which provides over 80% of the kinetic and potential energy of the seawater (Ding, 1992). The wind over JZB is relatively weak, with an annual average wind speed of 5.4 m/s (Editorial Board of Annals in China, 1993). Several small seasonal rivers discharge into JZB, with the Dagu River contributing over 90% of the freshwater discharge. However, the total annual river discharge is rather limited ($<1.0\times10^9$ m^3 $year^{-1}$). Even in summer, the stratification is not strong.

Like many other coastal water bodies, the environmental problems in JZB call for a better understanding of inter-tidal mass transport. The Eulerian mean velocity has been extensively used to study the inter-tidal mass transport in JZB. The Eulerian mean velocity pattern in JZB is characterized by several eddies (Chen *et al.*, 1999). A large anticlockwise eddy appears between Tuandao and Xuejiadao, and a clockwise eddy dominates the region between Tuandao and Huangdao. In the east side of the Inner Bay, the Eulerian mean velocity is characterized by a weak clockwise gyre. However, a significant discrepancy exists between the Eulerian mean velocity and the inter-tidal mass transport indicated by the net surface sediment transport pattern in JZB. The net surface sediment transport pattern is opposite to the clockwise flow indicated by the Eulerian mean velocity. In the west side of the Inner Bay, there are also apparent discrepancies. The applicability of the first-order Lagrangian residual velocity in JZB is also questionable. Due to the strong tidal motion, complex topography, irregular coastline and other factors, JZB is more likely to be a general nonlinear system than a weakly nonlinear system. The transport in JZB is further complicated by the fact that the flushing time for water parcels is closely related to the initial tidal phase when the water parcels are released (Wang *et al.*, 2009). Because the Eulerian residual transport velocity and the first-order Lagrangian residual velocity are both independent of the initial tidal phase, their validity in JZB is questionable.

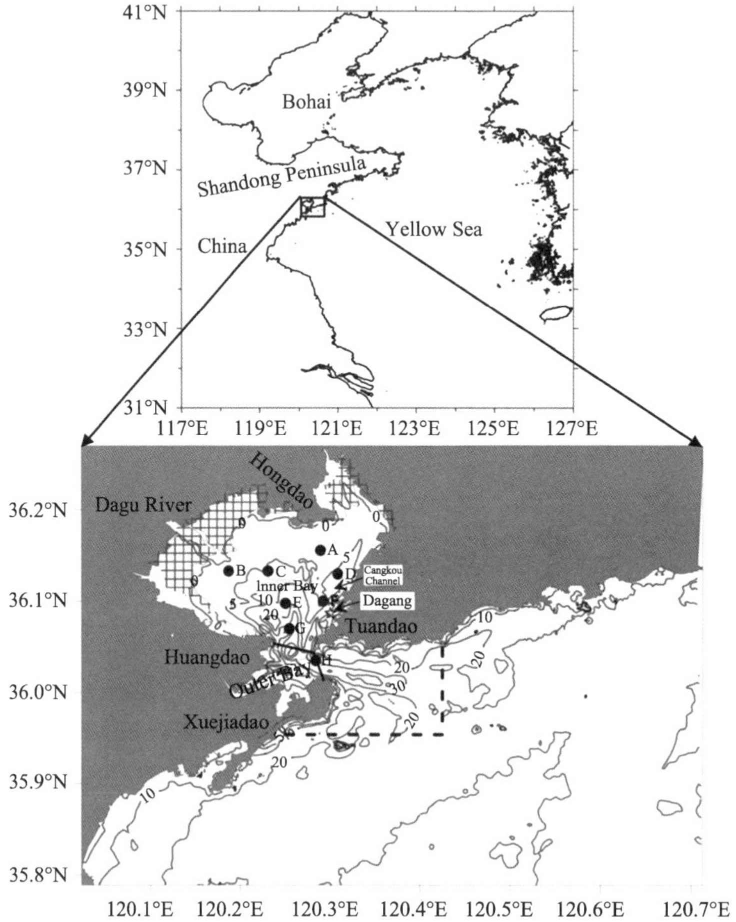

Fig. 1 Topography of JZB (in metres). JZB can be divided into the Inner and Outer Bays by the line between Tuandao and Huangdao. The bay channel connecting JZB to the Yellow Sea is located between Tuandao and Xuejiadao. The gridded region is the tidal flat. The water parcels are released initially at each grid point within the bold dashed black lines. The solid circles indicate the current measurement stations

Thus, it is important to understand the inter-tidal mass transport pattern and its intensity in JZB from the general nonlinear Lagrangian perspective.

3. Numerical Model and Methods

3.1 Numerical model description

The hydrodynamic model used in this study was based on Princeton Ocean Model with Wetting and Drying Capability (POM-WAD) (Blumberg and Mellor, 1987; Oey, 2005, 2006). The vertical mixing coefficients were provided by the Mellor-Yamada level 2.5 turbulence closure model (Mellor and Yamada, 1982), while the horizontal mixing coefficients were calculated using a Smagorinsky-type formulation (Smagorinsky *et al.*, 1965). The calculation domain covering the entire JZB and its adjacent area was divided into a 312×213 grid with horizontal grid size of approximately 200×250 m (Fig. 1). The computational domain included the tidal flat using the WAD technique. A total of 17 vertical sigma levels provided an average vertical resolution of approximately 0.5 m in the bay area. Due to the weak wind stress and stratification, the wind effect and baroclinic processes could be neglected in most cases (Weng *et al.*, 1992). Because of the dominant role of the M_2 tidal component, the barotropic current was derived from the POM-WAD driven only by the M_2 tidal component at the open boundary (Ding, 1992). The amplitudes and phases used in the simulation at the coastal sea open boundaries were obtained from *Marine Atlas of Bohai Sea, Yellow Sea, East China Sea* (Chen, 1992). At the open boundary, the tidal elevation was specified; the radiation condition was used for the normal component of the velocity. The initial values of the currents and elevation were set to zero. The time steps for the external and internal modes were 2.5 s and 40 s, respectively.

To validate the model in this study, in-situ observational data on the tidal currents were obtained during August 2009. There were a total of eight anchor stations (Fig. 1). The observation duration was over 25 h, with an average sampling interval of 10 min. Through harmonic analysis, the M_2 tidal currents were extracted, and the tidal ellipse parameters were calculated (Table 1).

Table 1 Observed and modelled tidal ellipse parameters for the surface M_2 currents at the stations in Fig. 1

Stations	Observed				Modelled				Difference			
	Semi-major axis/ (m s^{-1})	Semi-minor axis/ (m s^{-1})	Inclination/(°)	Phase/ (°)	Semi-major axis/ (m s^{-1})	Semi-minor axis/ (m s^{-1})	Inclination/(°)	Phase/ (°)	Semi-major axis/ (m s^{-1})	Semi-minor axis/ (m s^{-1})	Inclination/(°)	Phase/ (°)
A	0.36	0.08	43.93	59.35	0.27	0.04	61.00	40.18	0.09	0.04	−17.07	19.17
B	0.30	0.00	110.51	45.13	0.31	0.01	133.21	41.52	−0.01	0.00	−22.70	3.61
C	0.31	0.04	97.06	65.20	0.30	0.04	103.77	43.60	0.02	0.00	−6.71	21.60
D	0.30	0.02	54.14	30.07	0.31	0.03	64.93	38.11	0.00	−0.01	−10.79	−8.04
E	0.24	0.01	128.91	15.49	0.33	0.03	102.13	48.20	−0.09	−0.02	26.77	−32.72
F	0.29	0.05	59.74	24.50	0.32	0.08	70.94	21.65	−0.03	−0.03	−11.20	2.85
G	1.09	0.01	97.24	65.39	0.74	0.05	99.44	57.44	0.35	−0.04	−2.20	7.95
H	0.80	0.03	154.58	54.90	0.81	0.01	157.46	47.49	0.00	0.02	−2.88	7.41
Mean	0.46	0.03	93.26	45.00	0.42	0.04	99.11	42.27	0.04	−0.01	−5.85	2.73
r. m. s.									0.13	0.03	15.10	16.23

3.2 Calculation method for residual velocities

The water parcel tracking module developed according to Zhang (1995) was embedded into the hydrodynamic model. The Lagrangian residual velocity can be calculated as the net displacement over one or more tidal circles divided by the corresponding time interval (Zimmerman, 1979; Cheng and Casulli, 1982; Feng *et al.*, 2008). Initially, the water parcels were deployed at each wet cell inside the bold dashed black lines shown in Fig. 1. In this study, the labelled water parcels were tracked over one tidal period. Then, the Lagrangian residual velocities were calculated and deployed at the initial positions of the labelled water parcels.

Let t and τ represent the intra-tidal process-independent time variable and the inter-tidal process-independent time variable, respectively. T is the tidal period, n is the number of tidal cycles and $\langle\ \rangle = \frac{1}{nT}\int_{t_0}^{t_0+nT} \mathrm{d}t$ is defined as the tidal cycle mean operator.

Then, the Eulerian residual transport velocity (Feng *et al.*, 1986b; Delhez, 1996) can be defined as:

$$\bar{\boldsymbol{u}}_{\mathrm{T}} = \frac{\langle (h+\zeta)\,\bar{\boldsymbol{u}}(\bar{\boldsymbol{x}},t)\rangle}{\langle h+\zeta\rangle} \tag{1}$$

As a comparison with the Eulerian residual transport velocity, the depth-averaged Stokes drift velocity (Longuet-Higgins, 1969) can be calculated from

$$\bar{\boldsymbol{u}}_{\mathrm{S}} = \left\langle \int_{t_0}^{t} \bar{\boldsymbol{u}}(\bar{\boldsymbol{x}},t')\mathrm{d}t' \cdot \nabla \bar{\boldsymbol{u}}(\bar{\boldsymbol{x}},t') \right\rangle \tag{2}$$

The first-order Lagrangian residual velocity (Longuet-Higgins 1969; Feng *et al.*, 1986a) is

$$\bar{\boldsymbol{u}}_{\mathrm{M}} = \bar{\boldsymbol{u}}_{\mathrm{E}} + \bar{\boldsymbol{u}}_{\mathrm{S}} \tag{3}$$

$\bar{\boldsymbol{u}}_{\mathrm{E}}$ is the depth-averaged Eulerian mean velocity.

$\bar{\boldsymbol{u}}_{\mathrm{L}}$, the Lagrangian residual velocity (Zimmerman, 1979; Feng *et al.*, 2008), can be expressed as follows:

$$\boldsymbol{u}_{\mathrm{L}}(\boldsymbol{x},\tau;t_0) = \langle \boldsymbol{u}(\boldsymbol{x}_0 + \boldsymbol{\xi}(t;\tau),t;\tau)\rangle = \frac{\boldsymbol{\xi}_{n\mathrm{r}}}{nT} \tag{4}$$

where $\boldsymbol{x}(x, y, z)$ is a three-dimensional position vector in the Cartesian coordinate system, $\bar{\boldsymbol{x}}(x,y)$ is the horizontal two-dimensional position vector, t_0 is the initial time for an arbitrary water parcel to be tracked, and $\boldsymbol{x}_0$ is the initial position vector. Note that $\boldsymbol{x}$ is used for the independent space variable of $\boldsymbol{u}_{\mathrm{L}}$ instead of $\boldsymbol{x}_0$ since $\boldsymbol{x}_0$ is any given initial position. $\boldsymbol{u}(\boldsymbol{x}_0 + \boldsymbol{\xi}(t;\tau),t;\tau)$ and $\boldsymbol{\xi}(t;\tau)$ are the tidal velocity and the displacement of the water parcel, respectively, where $\boldsymbol{\xi} = \boldsymbol{x} - \boldsymbol{x}_0$ and $\boldsymbol{u} = \frac{\partial \boldsymbol{\xi}}{\partial t} \cdot \bar{\boldsymbol{u}}(\bar{\boldsymbol{x}},t)$ is the depth-averaged velocity. $\zeta(\bar{\boldsymbol{x}},t)$ is the fluctuating position of the water surface, and $h(\bar{\boldsymbol{x}})$ is the depth of the undisturbed water column. $\boldsymbol{\xi}_{n\mathrm{r}}$ is the net displacement of the water parcel over n tidal cycles (in this study, n=1).

To investigate the dependence of the Lagrangian residual velocity on the initial tidal phase, 12 different initial times were selected as the initial release time.

$$t_i = \{t_0 + (i-1) \times T_{M_2} / N\} \qquad (i = 1, 2, \cdots, N; N = 12) \tag{5}$$

Referring to the Eulerian residual transport velocity, to infer the net transport, the Lagrangian residual transport velocity can be calculated as follows:

$$\bar{\boldsymbol{u}}_{LT} = \frac{\langle (h + \zeta_L) \bar{\boldsymbol{u}}_L \rangle}{\langle h + \zeta_L \rangle} \tag{6}$$

where $\bar{\boldsymbol{u}}_L$ is the depth-averaged Lagrangian residual velocity and ζ_L is termed the Lagrangian inter-tidal free surface (Feng *et al.*, 2008).

4. Results

4.1 Tides and tidal currents

The simulated M_2 tidal coamplitudes, cophases (Fig. 2a) and current ellipses (Fig. 2b) are in good agreement with previous studies (e.g. Chen *et al.*, 2011).

During flood tide, seawater comes from the northeast. Part of the flood tide water enters JZB, and the rest flows southwest along the coastline. During ebb tide, seawater from the southwest mixes with the outflow water of JZB and flows northeast. When the tidal wave propagates into JZB, the amplitude gradually increases with decreasing water depth.

Inside JZB, the tidal current is characterized by a reversing current (Fig. 2b). At high and low tides, the currents reverse direction, and the velocity magnitudes are at a minimum. The currents reach their maximum when the tidal elevations approach zero. To calibrate the simulated tidal currents, field observations were collected from a ship survey performed in August 2009. The comparison results shown in Table 1 indicate that the modelled M_2 tidal ellipses are in reasonable agreement with the observed data. The mean difference in the semimajor axis is 0.04 m^{-1}. The mean differences in ellipse inclination and phase are both less than 10°. All of the rotational directions of the simulated tidal currents are consistent with the observations.

4.2 Eulerian residual transport velocity, the first-order Lagrangian residual velocity and Lagrangian residual velocity

Several eddies in the Eulerian mean velocity (Chen *et al.*, 1999) also existed in the Eulerian residual transport velocity (Fig. 3a). The location and rotation of the simulated eddies are similar to those found by Chen *et al.* (1999). The first-order Lagrangian residual velocity, defined as the sum of the Eulerian mean velocity and the Stokes drift velocity, maintains a multi-vortex structure. However, except for the multi-vortex region, the first-order Lagrangian residual velocity (Fig. 3b) differs considerably from the Eulerian residual transport velocity (Fig. 3a).

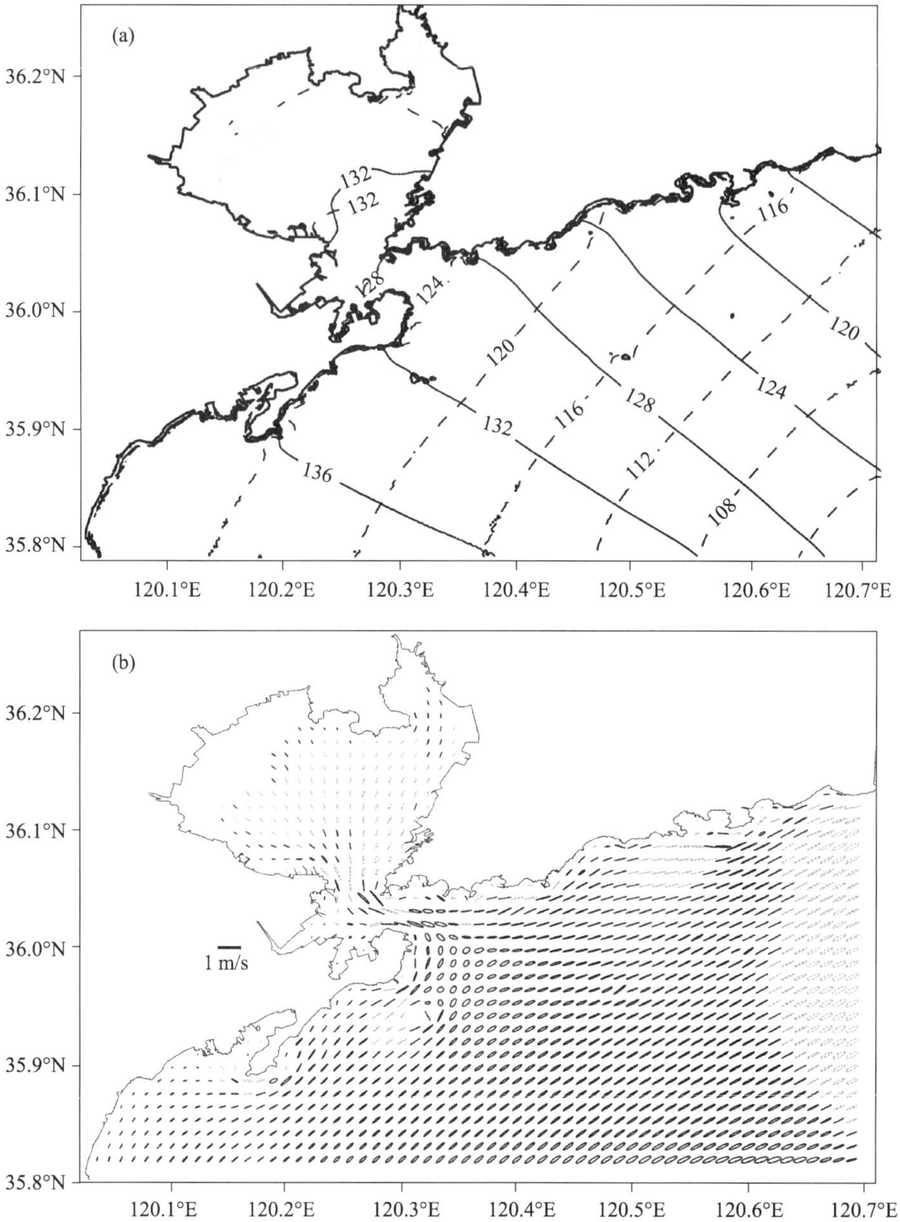

Fig. 2 Computed M_2 tidal coamplitudes (dashed lines) and cophases (solid lines) (a) and the tidal current ellipse (b) (grey, clockwise; black, anticlockwise)

The Lagrangian residual transport velocity shows some remarkable differences compared with the Eulerian residual transport velocity and the first-order Lagrangian residual velocity: (I) a large anticlockwise flow originates north of the dividing line between the Outer and Inner Bays and then flows toward the northern boundary of JZB along the Cangkou Channel. The flow turns anticlockwise and flows straight along the bank. To the south of western Hongdao, the flow splits into two branches: one flows along the central axis of JZB, and the other forms a small anticlockwise eddy covering the northwestern part of JZB; (II) in the northeastern part of JZB, one small anticlockwise eddy dominates; and (III) the Lagrangian residual velocity flows into the bay in the south and away from the bay outside the bay mouth (Fig. 3c).

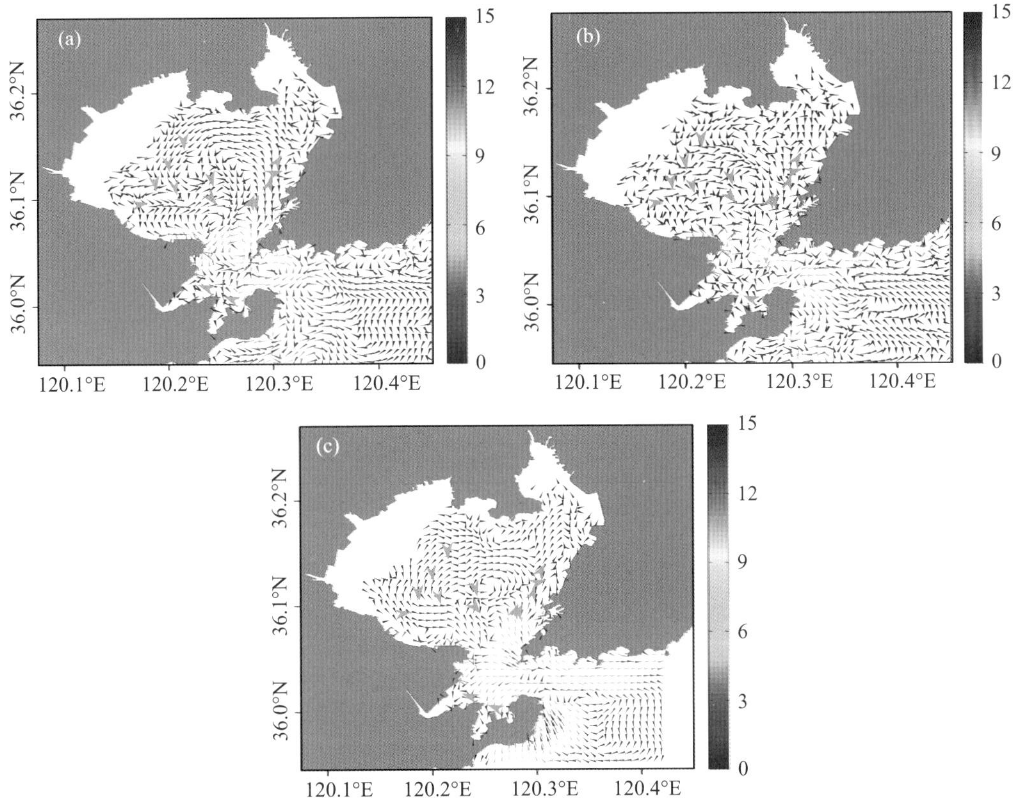

Fig. 3 Distributions of the Eulerian residual transport velocity (a), the first-order Lagrangian residual velocity (b) and the Lagrangian residual transport velocity (c) (in centimetres per second) induced by the M_2 tidal constituent.

The Lagrangian residual velocities were deployed at the initial positions of the labelled water parcels. The colours indicate the magnitude, and the arrows indicate the direction. The orange arrows represent the surface sediment transport pattern described by Liu *et al.* (2008)

It should be emphasised that even in a purely periodic, single tidal constituent-driven barotropic system, the Lagrangian residual velocity is three-dimensional in nature and depends on the initial time at which the water parcels are released. The patterns of the tidal composition of the Lagrangian residual velocity in the surface (Fig. 4a) and middle layers (Fig. 4b) both maintain the anticlockwise eddy in the central bay and do not show much vertical difference. However, in the bottom layer (Fig. 4c), the Lagrangian residual velocity shows an opposite flow direction in the north bay compared with those in the upper layers due to bottom friction. Both the magnitude and direction of the Lagrangian residual velocity vary significantly with the initial tidal phase in JZB (Fig. 5). For example, the depth-averaged Lagrangian residual velocity near the bay mouth exhibits two small anticlockwise eddies in the central bay during high tide (Fig. 5a). However, during low tide, the two eddies move southward and one of them nearly disappears (Fig. 5c).

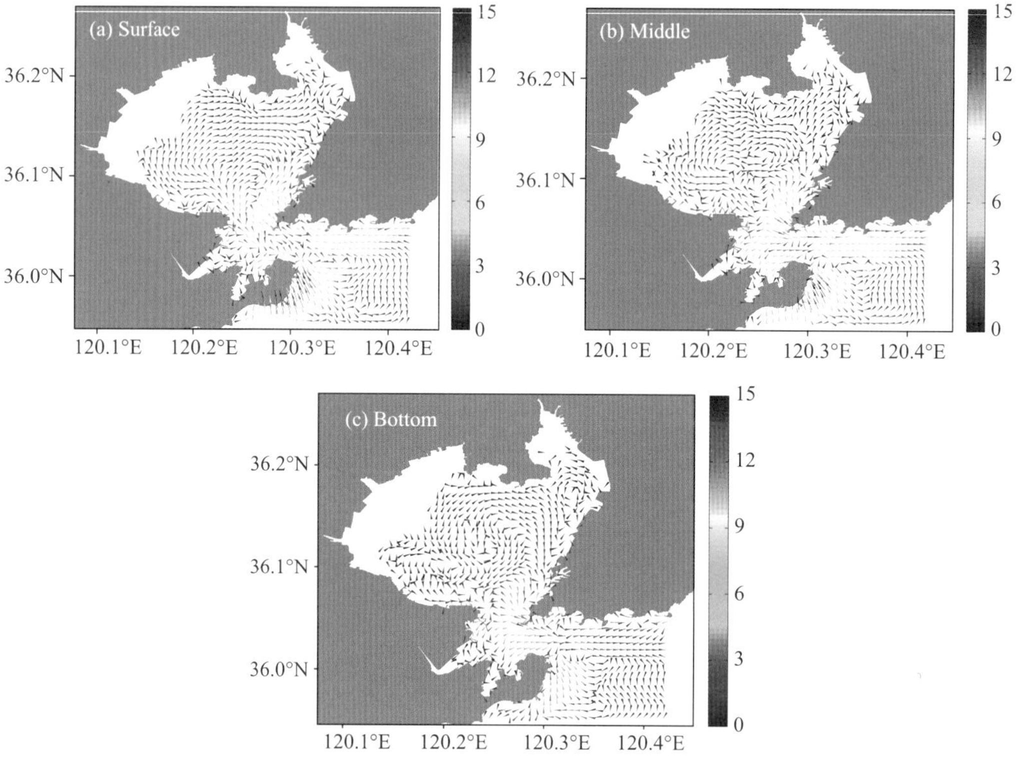

Fig. 4 The tidal composition of the Lagrangian residual velocity (in centimetres per second) in different layers

The Lagrangian residual velocities were deployed at the initial positions of the labelled water parcels. The colours indicate the magnitude, and the arrows indicate the direction

4.3 Net surface sediment transport pattern and its comparison with residual velocities

The tidal residual velocity is an important determinant of the net sediment transport pattern in tide-dominated regions (e.g. Fiechter *et al.*, 2006; Sanay *et al.*, 2007; De Swart and Zimmerman, 2008; Wu *et al.*, 2011). The net surface sediment transport pattern may be closely related to the tideinduced residual velocity in JZB for the following reasons: (1) because of weak stratification, the density-induced residual velocity is weak in most regions of JZB (Ding, 1992); (2) the wind is relatively weak, and extreme events (such as typhoon-induced storm surges) are rare (Editorial Board of Annals in China 1993); and (3) the horizontal dimension of JZB is only approximately 20 km, which is not long enough for wind-induced waves to grow strong (Editorial Board of Annals in China, 1993; Liu *et al.*, 2008). Thus, the sediment transport is closely related to the tide-induced mass transport. The net surface sediment transport pattern was derived from grain size analysis of surface (0–4 cm) sediments using the Gao-Collin model (Liu *et al.*, 2008).

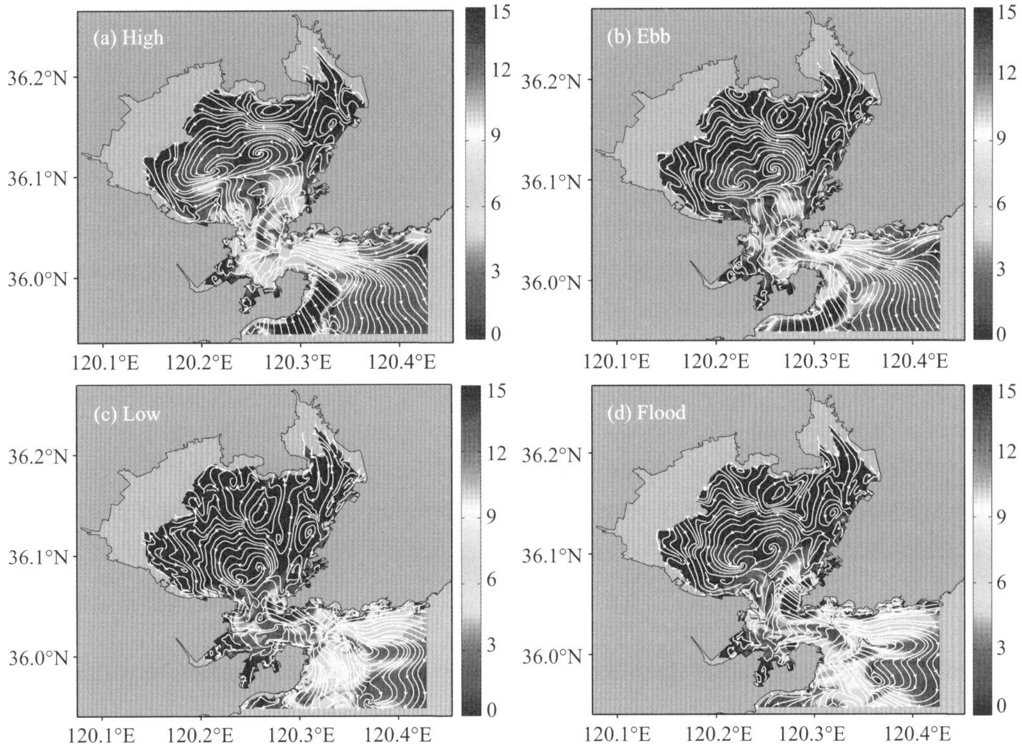

Fig. 5 The streamline of the depth-averaged Lagrangian residual velocity (in centimetres per second) corresponding to the initial tidal phases when the labelled water parcels were released: high tide (a), maximum ebb tide (b), low tide (c) and maximum flood tide (d). The colours indicate the magnitude

Overall, the spatial structure of the Lagrangian residual transport velocity is consistent with the net surface sediment transport pattern (Fig. 3c) while the Eulerian residual transport velocity (Fig. 3a) and the first-order Lagrangian residual velocity (Fig. 3b) are not. Although the pattern of the streamline varies significantly over one tidal cycle, an anticlockwise eddy can be observed in the central area of JZB (Fig. 5). In agreement with the net surface sediment transport pattern, the Lagrangian residual transport velocity indicates the existence of an anticlockwise circulation over the central area of JZB. The Lagrangian residual transport velocity is also consistent with the net southeastward surface sediment transport in the northwest bay and northeastward transport along the Cangkou Channel. In contrast, the Eulerian residual transport velocity flows opposite to the net surface sediment transport pattern on the east side of the Inner Bay. Obvious disagreements also exist in the west side of the Inner Bay. The use of the first-order Lagrangian residual velocity reduces the difference between the transport pattern and the sediment pattern in the northwest bay, but large differences still exist. This finding implies that it is the Lagrangian residual velocity that determines the inter-tidal mass transport in JZB.

4.4 Flushing time and Lagrangian residual velocity

The flushing time is a valid tool with which to quantify water exchange capability. It was noted by Wang *et al.*, (2009) that the flushing time of the water in JZB is closely related to the initial release time. For instance, the flushing time for the water parcels released at high tide is 28 days, which is significantly shorter than that at low tide (i.e., 43 days) (Wang *et al.*, 2009).

The Eulerian residual transport velocity and the first-order Lagrangian residual velocity cannot explain such a phenomenon because they remain invariant with different initial tidal phases. In contrast, the streamline of the Lagrangian residual velocity in JZB indicates that the transport of pollutants discharged at different tidal phases should follow different pathways. Some pathways, corresponding to certain tidal phases, are more effective in flushing pollutants out of the bay. The major anticlockwise eddy in the central bay may be responsible for the different flushing time scales. During high tide, the major anticlockwise eddy moves northward and is far from the bay mouth. The streamline near the bay mouth points toward the outside of the bay (Fig. 5a). As a result, it is easier for water parcels to be flushed out. At low tide, the major anticlockwise eddy approaches the bay mouth, which makes it unfavorable for the flushing out of water parcels (Fig. 5c).

Under the assumption that the internal bay water can be mixed immediately, the flushing time of JZB can be roughly calculated as the water volume of the entire bay divided by the flux obtained from the Lagrangian residual velocity at the bay mouth. The flushing time in JZB is apploxomately 5 days at high tide and 9 days at low tide, which is consistent with previous box model results (Wu *et al.*, 1992; State Oceanic Administration of China (SOA) and Qingdao Aquiculture Bureau (QAB), 1998). The flushing time in this study is shorter than that found by Wang *et al.*, (2009) because our model did not consider the spatial difference in mixing process and therefore overestimated the water exchange capability of JZB. However, our study is helpful for understanding why the water exchange time depends on the initial tidal phase in a pure tidal system.

5. Discussion

5.1 Comparison of the Lagrangian residual velocity with the Eulerian residual transport velocity and the first-order Lagrangian residual velocity

As shown in Fig. 3b, the first-order Lagrangian residual velocity preserves the multi-vortex structure near the capes and is similar to the Eulerian residual transport velocity outside JZB. However, the difference between the first-order Lagrangian residual velocity and the Eulerian residual transport velocity cannot be neglected inside JZB because, in theory, the Eulerian residual transport velocity is only part of the first-order Lagrangian residual velocity

(Feng *et al.*, 1986b; Jiang and Feng, 2011).

The obvious difference between the Eulerian residual transport velocity and Lagrangian residual velocity is that several eddies in the Eulerian residual transport velocity field disappear in the Lagrangian residual velocity field.

The Lagrangian residual velocity also shows considerable differences from the first-order Lagrangian residual velocity. The first-order Lagrangian residual velocity is derived based on the weakly nonlinear hypothesis of tidal systems, which means that κ, the ratio of the displacement of the free surface to the water depth, both measured from the undisturbed sea surface, is small (κ should usually be equal to or less than less than 0.1). The average water depth of JZB is approximately 7 m, and the typical elevation is greater than 1 m, so the value of κ is slightly above 0.1. Thus, it seems that JZB may not be too far from a weakly nonlinear system. However, the order of the local κ is much more than 1 in the tidal flats that occupy one third of the total area of JZB. The water depth in another one third of the total area is less than 5 m, while the mean tidal range is greater than 2.5 m, which also leads to κ values much larger than 0.1. The deepest water region of JZB has strong tidal currents and an irregular coastline. Ju (2007) noted that the nonlinearity may be underestimated if the κ principle is applied to regions such as JZB because the nonlinearity can be strong due to large depth gradients or complicated coastlines. The nonlinearity should not be weak over all of JZB. Thus, the variations in κ may be the reason why the Lagrangian residual velocity differs substantially from the first-order Lagrangian residual velocity in JZB.

5.2 Separation of advection and viscosity regimes

In a general nonlinear system, the Lagrangian residual velocity is characterized by a strong dependence on the tidal phase at the time of the initial release of the water parcels. In general, horizontal advection, vertical advection, horizontal viscosity and vertical viscosity are the major hydrodynamic processes directly contributing to the nonlinearity of the tidal system within the flow field. Bottom friction results in a vertical gradient of the current, affecting advection in the bottom layer and the vertical viscosity.

To identify the key process(es) among the four processes, that is (are) responsible for the nonlinearity of the tidal system in JZB, several statistical indexes are defined (see Appendix A). *LN* (Eq. 7) is used to quantify the dependence of the Lagrangian residual velocity on the initial tidal phase or the nonlinearity of the tidal system. The index *ha* (Eq. 10) measures the nonlinear effects of horizontal advection. By scale analysis, *ha* can be expressed as the inverse of the Strouhal number $\left(Sr=\frac{L_c}{u_c T}\right)$, which is the ratio of the inertial forces to the unsteadiness of the flow or to local acceleration. u_c, L_c and T are, respectively, the characteristic values of the instantaneous horizontal velocity, the spatial length and the tidal period. In a tidally dominated system, $\frac{u_c}{T} \sim \frac{g\zeta_c}{L_c}$ and $L_c \sim T\sqrt{gh_c}$. ζ_c and h_c are the characteristic values of water elevation and depth, respectively. Thus, the inverse of the Strouhal number is just κ ($\frac{\zeta_c}{h_c}$ or $\frac{u_c}{\sqrt{gh_c}}$), a widely used parameter used to measure the horizontal advection-induced nonlinearity (e.g. Heaps, 1978; Feng *et al.*, 1986a; Pedlosky, 1987). The index *va* (Eq. 11) is similar to *ha*, but it is used to measure the vertical advection-induced nonlinearity. The index *hv* (Eq. 12) measures the nonlinear effects of horizontal viscosity. By scale analysis, *hv* can be expressed as $\left(=\frac{K_c T}{L_c^2}\right)$, the ratio of the horizontal viscosity terms to the local acceleration. K_c is the characteristic kinematic viscosity. The index *hv* can also be expressed as $\frac{\kappa}{Re}$ or $(SrRe)^{-1}$. $Re\left(=\frac{u_c L_c}{K_c}\right)$ is the Reynolds number. The index *vv* (Eq. 13) is similar to *hv* but is used to measure the vertical viscosity-induced nonlinearity.

Figure 6 shows the distribution of *LN* the index of nonlinearity of the tidal system. The irregular coastline and complex topography lead to strong nonlinearity near the bay mouth. For instance, the maximum values of *LN* can be found in the vicinity of the two capes, i.e. Huangdao and Tuandao, in JZB. Overall, the nonlinearity of the tidal system in the surface layer is stronger than that in the bottom layer.

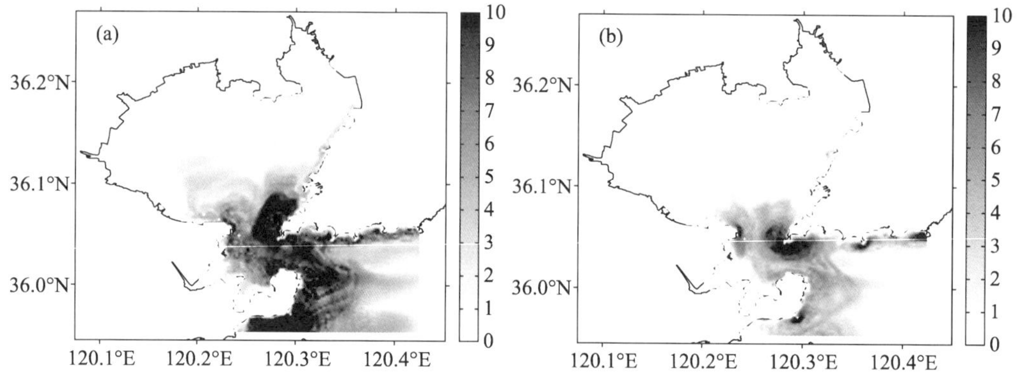

Fig. 6 The nonlinearity of the M_2 tidal system, *LN* (in centimetres per second), in the surface layer (a) and the bottom layer (b)

Next, the indexes (*ha*, *va*, *hv* and *vv*) were calculated. To separate the advection and the viscosity regimes in JZB, the indexes were normalised by their summation and multiplied by 100% to express them as percentages (Appendix A). These percentages were used to quantify their respective roles in the tidal system (Fig. 7). In the surface layer, horizontal advection dominates near the bay mouth where tidal currents are strong. Vertical viscosity plays a key role in the shallow waters of the northern area of JZB. The effects of vertical advection and horizontal

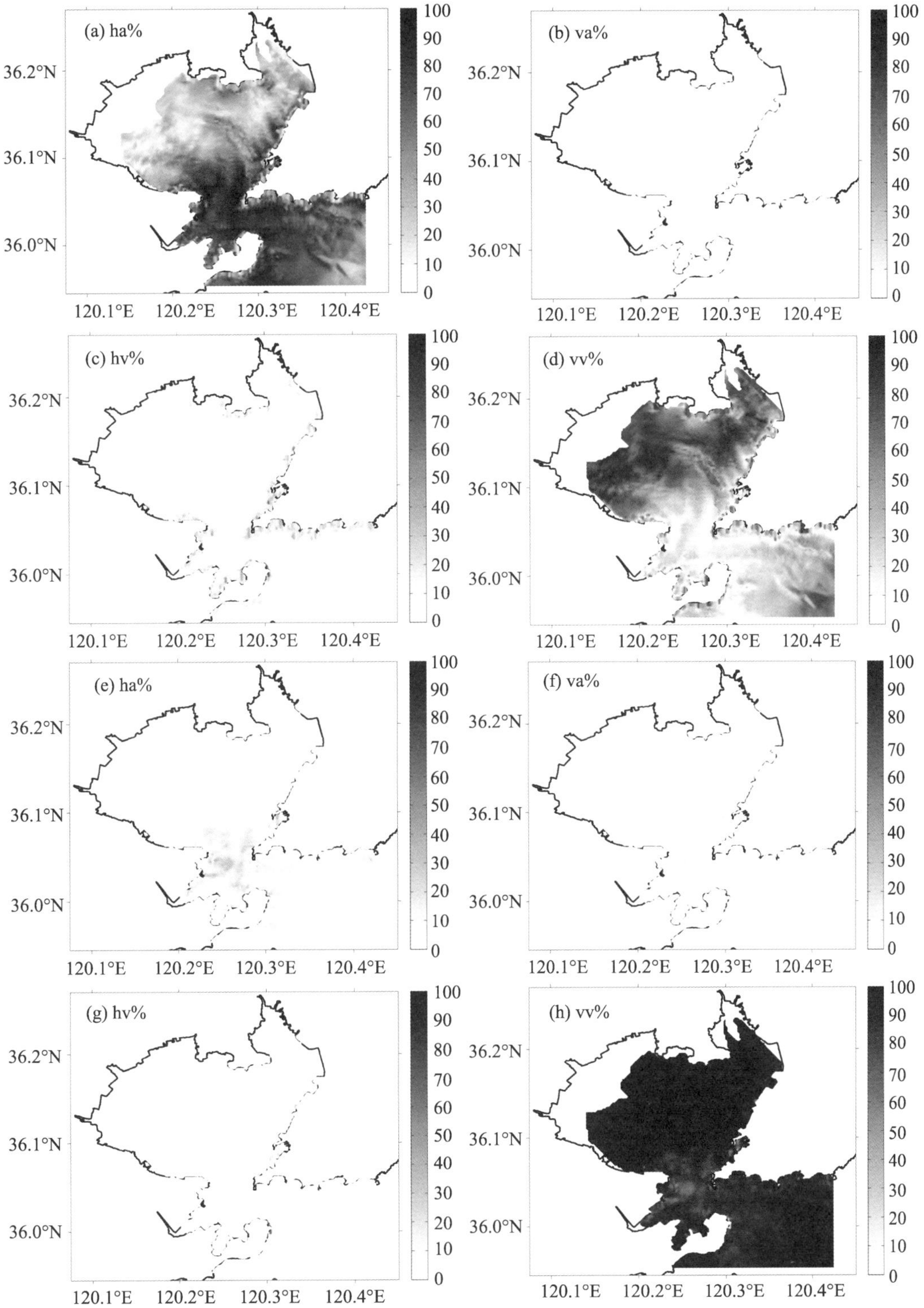

Fig. 7 The percentage contribution of the advection and the viscosity to the nonlinearity of the tidal system in the surface (a)–(d) and the bottom layers (e)–(h)

viscosity are relatively weak. The effect of horizontal advection decreases with reduced depth due to bottom friction, while vertical viscosity plays an increasing role. In the bottom layer, vertical viscosity dominates over almost the entire bay, except near the bay mouth. The effect of horizontal advection is weak due to bottom friction, but it still cannot be neglected near the bay mouth. Horizontal viscosity and vertical advection can be neglected in the bottom layer.

Comparing the nonlinearity metric of the tidal system *LN* (Fig. 6) with the advection and viscosity regimes (Fig. 7), it can be concluded that horizontal advection is the most important contributor to *LN* in the surface layer, while the large values of *LN* in the bottom layer may result primarily from vertical viscosity and horizontal advection. Except near the bay mouth in the bottom layer, vertical viscosity does not cause great tidal system nonlinearity. The reason for this result may be the double role of vertical viscosity: on one hand, vertical viscosity makes the system nonlinear, and, on the other hand, vertical viscosity also leads to strong dissipation. The dissipation by vertical viscosity may also explain why the nonlinearity in the bottom layer is weaker than that in the surface layer.

6. Conclusion

The significant intra-tidal variation of the Lagrangian residual velocity in JZB indicates that the M_2 tidal system in JZB is a general nonlinear, rather than a weakly nonlinear system. Unlike the Eulerian residual transport velocity and the first-order Lagrangian residual velocity, the Lagrangian residual velocity can be used to describe the net surface sediment transport pattern and the sensitivity of flushing time to the initial tidal phase. Thus, the Lagrangian residual velocity represents the inter-tidal mass transport process more accurately than the Eulerian residual transport velocity and the first-order Lagrangian residual velocity. In the surface layer, horizontal advection and vertical viscosity are both important hydrodynamic processes. Vertical viscosity plays a dominant role in the bottom layer, but horizontal advection still cannot be neglected near the bay mouth. Because of the complex hydrodynamics, the M_2 tidal system in JZB shows strong nonlinearity near the bay mouth. In the surface layer, horizontal advection is the main contributor to the large nonlinearity near the bay mouth, while the strong nonlinearity near the bay mouth in the bottom layer may result from vertical viscosity and horizontal advection.

Acknowledgements

This study was financially supported by the Special Fund for Public Welfare Industry (Oceanography; grant No. 200805011 and 201205018) and the Youth Foundation of the State Oceanic Administration of the People's Republic of China (grant No. 2012206). Discussion with Prof. Lian Xie of North Carolina State University and Dr. Xueqing Zhang of the Ocean University of China are highly appreciated. We appreciate the two anonymous reviewers for their critical comments, which were very helpful for improving the manuscript. We are also grateful to the High Performance Center of OUC for providing technical support to our programs running in clusters.

Appendix A. Definition of Indexes

LN in Eq. 7 measures the intra-tidal variation of the Lagrangian residual velocity. It can also be used to represent the degree of nonlinearity of the M_2 tidal system.

$$LN=\sqrt{\frac{\sum_{p=1}^{N}(\Delta u_{lp}^{2}+\Delta v_{lp}^{2})}{N-1}} \tag{7}$$

where p is the initial tidal phase, N (=12) is the total number of tidal phases, Δu_l and Δv_l represent the deviation between the Lagrangian residual velocity at different tidal phases and the time-averaged Lagrangian residual velocity, respectively.

$$\Delta u_l=u_l-\overline{u}_{la} \tag{8}$$

$$\Delta v_l=v_l-\overline{v}_{la} \tag{9}$$

where u_l and v_l are, respectively, the east and north components of the Lagrangian residual velocity (i.e., $\boldsymbol{u}_{\mathrm{L}}$) at a given initial tidal phase; $\boldsymbol{u}_{\mathrm{L}a}=(\sum_{p=1}^{N}\boldsymbol{u}_{\mathrm{L}p})/N$, p=1,2⋯N is the time average of $\boldsymbol{u}_{\mathrm{L}}$; and $\overline{u}_{la}$ and $\overline{v}_{la}$ are, respectively, the east and north components of $\boldsymbol{u}_{\mathrm{L}a}$.

The indexes, *ha*, *va*, *hv* and *vv* quantitatively measure the effect of horizontal advection, vertical advection, horizontal viscosity and vertical viscosity, respectively. In fact, in the indexes *va* and *vv*, the effect of bottom friction has been taken into account due to the vertical gradient of the current induced by bottom friction.

$$ha=\frac{1}{nT_{\mathrm{M}_2}}\int_{t_0}^{t_0+nT_{\mathrm{M}_2}}\frac{\sqrt{\left(u\frac{\partial u}{\partial x}+v\frac{\partial u}{\partial y}\right)^2+\left(u\frac{\partial v}{\partial x}+v\frac{\partial v}{\partial y}\right)^2}}{\sqrt{\left(\frac{\partial u}{\partial t}\right)^2+\left(\frac{\partial v}{\partial t}\right)^2}}\mathrm{d}t \tag{10}$$

$$va=\frac{1}{nT_{\mathrm{M}_2}}\int_{t_0}^{t_0+nT_{\mathrm{M}_2}}\frac{\sqrt{\left(\frac{\omega}{D}\frac{\partial u}{\partial \sigma}\right)^2+\left(\frac{\omega}{D}\frac{\partial v}{\partial \sigma}\right)^2}}{\sqrt{\left(\frac{\partial u}{\partial t}\right)^2+\left(\frac{\partial v}{\partial t}\right)^2}}\mathrm{d}t \tag{11}$$

$$hv=\frac{1}{nT_{\mathrm{M}_2}}\int_{t_0}^{t_0+nT_{\mathrm{M}_2}}\frac{\sqrt{\left(\frac{\partial}{\partial x}\left(2A_{\mathrm{M}}\frac{\partial u}{\partial x}\right)+\frac{\partial}{\partial y}\left(A_{\mathrm{M}}\left(\frac{\partial u}{\partial y}+\frac{\partial v}{\partial x}\right)\right)\right)^2+\left(\frac{\partial}{\partial y}\left(2A_{\mathrm{M}}\frac{\partial u}{\partial x}\right)+\frac{\partial}{\partial x}\left(A_{\mathrm{M}}\left(\frac{\partial u}{\partial y}+\frac{\partial v}{\partial x}\right)\right)\right)^2}}{\sqrt{\left(\frac{\partial u}{\partial t}\right)^2+\left(\frac{\partial v}{\partial t}\right)^2}}\mathrm{d}t \quad (12)$$

$$vv=\frac{1}{nT_{\mathrm{M}_2}}\int_{t_0}^{t_0+nT_{\mathrm{M}_2}}\frac{\sqrt{\left(\frac{1}{D}\frac{\partial}{\partial\sigma}\left(\frac{K_M}{D}\frac{\partial u}{\partial\sigma}\right)\right)^2+\left(\frac{1}{D}\frac{\partial}{\partial\sigma}\left(\frac{K_{\mathrm{M}}}{D}\frac{\partial v}{\partial\sigma}\right)\right)^2}}{\sqrt{\left(\frac{\partial u}{\partial t}\right)^2+\left(\frac{\partial v}{\partial t}\right)^2}}\mathrm{d}t \quad (13)$$

where T_{M_2} is the M_2 tidal period; n is the number of tidal periods (in this study, n=1); u and v are horizontal tidal current components of the velocities; ω is the vertical velocity in the sigma coordinate system; D is the transient total water depth; A_{M} is the horizontal diffusivity; K_{M} is the vertical diffusivity; and t_0 is the initial time at which the water parcels are released.

To separate the advection and the viscosity regimes, the above indexes are divided by their summation values and multiplied by 100% to obtain percentage values.

$$ha\%=\frac{ha}{ha+va+hv+vv}\times 100\% \quad (14)$$

$$va\%=\frac{va}{ha+va+hv+vv}\times 100\% \quad (15)$$

$$hv\%=\frac{hv}{ha+va+hv+vv}\times 100\% \quad (16)$$

$$vv\%=\frac{vv}{ha+va+hv+vv}\times 100\% \quad (17)$$

References

Abbott MR (1960) Boundary layer effects in estuaries. J Mar Res 18: 83–100

Blumberg AF, Mellor GL (1987) A description of a three-dimensional coastal ocean circulation model. In: NS H (ed) Three-dimensional coastal ocean models, vol. 4. AGU, Washington, DC, pp 1–17

Cerco CF (1995a) Response of Chesapeake Bay to nutrient load reductions. J Environ Eng 121: 549–557

Cerco CF (1995b) Simulation of long-term trends in Chesapeake Bay eutrophication. J Environ Eng 121: 298–310

Cerco CF, Cole T (1993) Three-dimensional eutrophication model of Chesapeake Bay. J Environ Eng 119: 1006–1025

Chen DX (1992) Marine Atlas of Bohai Sea, Yellow Sea, East China Sea: hydrology. China Ocean Press, Beijing (in Chinese)

Chen CS, Ji RB, Zheng LY, Zhu MY, Rawson M (1999) Influences of physical processes on the ecosystem in Jiaozhou Bay: a coupled physical and biological model experiment. J Geophys Res 104:

29925–29949. doi: 10.1029/1999JC900203

Chen JR, Chen XE, Yu HM, Yan YW, Shan SL, Zhao J (2011) Three dimensional high resolution numerical study of the tide and tidal current in the Jiaozhou Bay. Period Ocean Univ China 41: 29–35 (in Chinese, with English abstract)

Cheng RT, Casulli V (1982) On Lagrangian residual currents with applications in south San Francisco Bay California. Water Resour Res 18: 1652–1662

De Swart HE, Zimmerman JTF (2008) Morphodynamics of tidal inlet systems. Annu Rev Fluid Mech 41(1): 203–229

Delhez EJM (1996) On the residual advection of passive constituents. J Mar Syst 8: 147–169

Ding WL (1992) Tides and tidal currents. In: Liu RY (ed) Ecology and Living Resources of Jiaozhou Bay. Science Press, Beijing, pp 39–57 (in Chinese)

Dortch MS, Chapman RS, Abt SR, ASCE Members (1992) Application of three dimensional Lagrangian residual transport. J Hydraul Eng 118: 831–848

Editorial Board of Annals in China (1993) Jiaozhou Bay. In: Wang JL (ed) Annals of Bays in China. China Ocean Press, Beijing, pp 157–260 (in Chinese)

Feng SZ (1987) A three-dimensional weakly nonlinear model of tide-induced Lagrangian residual current and mass-transport with an application to the Bohai Sea. In: Nihoul JCJ, Jamart BM (eds) Three-dimensional Models of Marine and Estuarine Dynamics. Elsevier Oceanography Series 45. Elsevier, Amsterdam, pp 471–488

Feng SZ, Xi PG, Zhang SZ (1984) The baroclinic residual circulation in shallow seas. Chin J Oceanol Limnol 2: 49–60

Feng SZ, Cheng RT, Xi PG (1986a) On tide-induced Lagrangian residual current and residual transport 1. Lagrangian residual current. Water Resour Res 22: 1623–1634

Feng SZ, Cheng RT, Xi PG (1986b) On tide-induced Lagrangian residual current and residual transport 2. Residual transport with application in south San Francisco Bay California. Water Resour Res 22: 1635–1646

Feng SZ, Ju L, Jiang WS (2008) A Lagrangian mean theory on coastal sea circulation with inter-tidal transports I. Fundamentals. Acta Oceanol Sin 27: 1–16

Fiechter J, Steffen LK, Mooers NKC, Haus KB (2006) Hydrodynamics and sediment transport in a southeast Florida tidal inlet. Estuar Coast Shelf Sci 70: 297–306

Fischer HB, List EJ, Koh RCY, Imberger J, Brooks NH (1979) Mixing in Inland and Coastal Waters. Academic, New York

Foreman MGG, Baptista AM, Walters RA (1992) Tidal model studies of particle trajectories around a shallow coastal bank. Atmos-Ocean 30: 43–69

Hainbucher D, Wei H, Pohlmann T, Sündermann J, Feng SZ (2004) Variability of the Bohai Sea circulation based on model calculations. J Mar Syst 44: 153–174

Heaps NS (1978) Linearized vertically-integrated equations for residual circulation in coastal seas. Ocean Dyn 31: 147–169

Jiang WS, Feng SZ (2011) Analytical solution for the tidally induced Lagrangian residual current in a narrow bay. Ocean Dyn 61: 543–558

Ju L (2007) Lagrangian Residual Current and inter-tidal transport in the nonlinear estuary or coastal sea system with an application to the Bohai Sea. Dissertation, Ocean University of China (in Chinese, with English abstract)

Ju L, Jiang WS, Feng SZ (2009) A Lagrangian mean theory on coastal sea circulation with inter-tidal transports II. Numerical Experiments. Acta Oceanol Sin 28: 1–14

Li C, O'Donnell J (2005) The effect of channel length on the residual circulation in tidally dominated channels. J Phys Oceanogr 35 (10): 1826–1840

Li C, Chen C, Guadagnoli D, Georgiou IY (2008) Geometry-induced residual eddies in estuaries with curved channels: observations and modeling studies. J Geophys Res 113(C1): C1005. doi: 10.1029/2006JC004031

Liu YL, Wang YP, Gao JH, Xia XM, Jia JJ (2008) The sediment distribution pattern in spatio-temporal scales and recent sediment transport pathway in the Jiaozhou Bay. Mar Sci Bull 27: 57–66 (in Chinese, with English abstract)

Longuet-Higgins MS (1969) On the transport of mass by time-varying ocean currents. Deep-Sea Res 16: 431–447

Lü XG, Zhao C, Xia CS, Qiao FL (2010) Numerical study of water exchange in the Jiaoozhou Bay and the tidal residual currents near the bay mouth. Acta Oceanol Sin 32: 20–30 (in Chinese, with English abstract)

Mellor GL, Yamada T (1982) Development of a turbulence closure model for geophysical fluid problems. Rev Geophys 20(4): 851–875

Muller H, Blanke B, Dumas F, Lekien F, Mariette V (2009) Estimating the Lagrangian residual circulation in the Iroise Sea. J Mar Syst 78: 17–36

Nihoul JCJ, Ronday FC (1975) The influence of the "tidal stress" on the residual circulation. Application to the Southern Bight of the North Sea. Tellus 48: 484–490

Oey L (2005) A wetting and drying scheme for POM. Ocean Model 9: 133–150

Oey L (2006) An OGCM with movable land–sea boundaries. Ocean Model 13: 176–195

Pedlosky J (1987) Geophysical Fluid Dynamics. Springer, New York

Sanay R, Voulgaris G, Warner JC (2007) Tidal asymmetry and residual circulation over linear sandbanks and their implication on sediment transport: a process-oriented numerical study. J Geophys Res 112: C12015. doi: 10.1029/2007JC004101

Smagorinsky J, Manabe S, Holloway JL Jr (1965) Numerical results from a nine-level general circulation model of the atmosphere. Mon Weather Rev 93(12): 727–768

State Oceanic Administration of China (SOA), Qingdao Aquiculture Bureau (QAB) (1998) Report on total drainage amount control of contaminant from land in Jiaozhou Bay (in Chinese)

Sun WX, Liu GM, Lei K, Jiang WS, Zhang P (2001) A numerical study on circulation in the Yellow and East China Sea II: numerical simulation of tide and tide-induced circulation. J Ocean Univ Qingdao 31: 297–304 (in Chinese, with English abstract)

Tee K (1976) Tide-induced residual current, a 2-D nonlinear numerical tidal model. J Mar Res 34: 603–628

Wang H, Su ZQ, Feng SZ, Sun WX (1993) A three dimensional numerical calculation of the wind driven thermohaline and tide induced Lagrangian residual current in the Bohai Sea. Acta Oceanol Sin 12:

169–182

Wang C, Zhang XQ, Sun YL (2009) Numerical simulation of water exchange characteristics of the Jiaozhou Bay based on a three-dimensional Lagrangian model. China Ocean Eng 23: 277–290

Wei H, Hainbucher D, Pohlmann T, Feng SZ, Suendermann J (2004) Tidal-induced Lagrangian and Eulerian mean circulation in the Bohai Sea. J Mar Syst 44: 141–151

Weng XC, Zhu LB, Wang YF (1992) Physical oceanography. In: Liu RY (ed) Ecology and Living Resources of Jiaozhou Bay. Science Press, Beijing, pp 39–57 (in Chinese)

Winant CD (2007) Three-dimensional tidal flow in an elongated, rotating basin. J Phys Oceanogr 37: 2345–2362. doi: 10.1175/JPO3122.1

Winant CD (2008) Three-dimensional residual tidal circulation in an elongated, rotating basin. J Phys Oceanogr 38: 1278–1295. doi: 10.1175/2007JPO3819.1

Wu YC, Wang CM, Zhang YK, Zhang QL, Meng XC (1992) Water exchange and mixing diffusion. In: Liu RY (ed) Ecology and Living Resources of Jiaozhou Bay. Science Press, Beijing, pp 57–72 (in Chinese, with English abstract)

Wu YS, Chaffey J, Greenberg AD, Colbo K, Smith CP (2011) Tidally-induced sediment transport patterns in the upper Bay of Fundy: a numerical study. Contin Shelf Res 31: 2041–2053

Zhang XY (1995) Ocean outfall modeling-interfacing near and far field models with particle tracking method. Dissertation, Massachusetts Institute of Technology

Zimmerman JTF (1979) On the Euler-Lagrange transformation and the Stokes' drift in the presence of oscillatory and residual currents. Deep-Sea Res Part A Oceanogr Res Pap 26: 505–520

Zou T, Gao HW, Sun WX, Liu Z (2009) Numerical simulation of Lagrange Residual Current in the Changjiang Estuary Hangzhou Bay and their adjacent sea I: barotropic circulation. Period Ocean Univ China 39: 153–159 (in Chinese, with English abstract)

Acquisition of the Tide-induced Lagrangian Residual Current Field by the PIV Technique in the Laboratory*

Tao Wang, Wensheng Jiang, Xu Chen, and Shizuo Feng

1 Introduction

In bays and estuaries, the tide-induced residual current, which is generated by the nonlinearity of the tidal system, is an important component in coastal and estuarine circulations (Nihoul and Ronday, 1975; Feng, 1990). Although the discussion of the tide-induced residual current is inevitable when studying the dynamics of the tide dominant shallow sea systems, its definition is far from unified. As a result, different methods have been proposed to derive it from the tidal current fields. The most frequently used definition is the Eulerian mean velocity, the so-called Eulerian residual velocity, which is defined as the time average or the steady part in the harmonic analysis of the tidal current at one fixed location for one or several tidal periods. The Eulerian mean method was used in many analytical or numerical studies to get the inter-tidal motion (e.g., Lopes and Dias, 2007; Huijts *et al.*, 2009; Vaz *et al.*, 2009). However, it is not an appropriate representative of the inter-tidal current because of its violation of the mass conservation (Feng *et al.*, 1984) and the ambiguity of the fictitious dispersion term (Fischer *et al.*, 1979).

Another definition of the tidally averaged velocity is the quotient of the net displacement of a water parcel after several periods and the elapsed time, namely, the Lagrangian residual velocity. Longuet-Higgins (1969) was the first to put forward the concept of the Lagrangian mean velocity in large-scale circulations. He pointed out that the mass transport velocity equaled the sum of the Eulerian residual velocity and the Stokes' drift velocity. Zimmerman (1979) examined the approximation of the rigorous Euler-Lagrangian transformation and defined the Lagrangian residual velocity as the net displacement of a labeled water parcel over one or several tidal periods. Cheng and Casulli (1982) proposed that the Lagrangian residual velocity was a function of the tidal phase when the water parcel was labelled. Feng *et al.*, (1986a) systematically analyzed the depth-averaged dynamic system in the weakly nonlinear shallow sea. They divided the Lagrangian residual velocity into different orders under the assumption of the weakly nonlinear convective terms and found that the first-order Lagrangian residual velocity was the mass transport velocity and the second order of the

* Wang T, Jiang W, Chen X, Feng S. 2013. Acquisition of the tide-induced Lagrangian residual current field by the PIV technique in the laboratory. Ocean Dynamics, 63(11–12): 1181–1188

Lagrangian residual velocity was the Lagrangian residual drift velocity, which displayed the Lagrangian nature of the Lagrangian residual velocity. Furthermore, they established a new inter-tidal mass transport equation, in which the convective velocity was the mass transport velocity, making the "tidal dispersion" term in the traditional inter-tidal mass transport equation vanish (Feng *et al.*, 1986b; Cheng *et al.*, 1989). Feng (1987) set the field equations for the Lagrangian residual circulation induced by an M_2 tidal system and developed a three-dimensional Lagrangian residual current model.

Dortch *et al.* (1992) used the model developed by Feng (1987) to simulate the annual salinity distribution in 1985 in the Chesapeake Bay, whose results coincided with the field observation results. This model solved the problem that with the use of the Eulerian residual velocity, the simulated salinity was lower than the observational salinity. Also, based on this model, several studies (Cerco and Cole, 1993; Cerco, 1995a, b) simulated the multiyear nutrients distribution and the eutrophication distribution trend and found that the results were reasonable. Delhez (1996), after discussing the different approaches used to simulate the long-term advection of passive constituents on tidal shelves in the framework of large-scale hydrodynamic modeling, found that the Eulerian residual transport velocities failed to represent the long-term motions when the tidal nonlinearities were important and concluded that the first-order Lagrangian residual velocity was a very good solution. Wei *et al.* (2004) used the Hamburg Shelf Ocean Model (HAMSOM) to simulate the Eulerian and Lagrangian residual current in the Bohai Sea. Jiang and Feng (2011) used the analytical method to study the tide-induced Lagrangian residual current in a narrow bay and compared the difference among the Eulerian residual velocity, the Eulerian residual transport velocity and the Lagrangian residual velocity, demonstrating that the Lagrangian residual current was the appropriate representative of the water mass inter-tidal transport in the tide-dominant area. The above results clearly show that it is appropriate to use the first-order Lagrangian residual velocity to describe the circulation fields in the shallow sea.

All the above results are based on the weakly nonlinear assumptions. Feng *et al.*, (2008) broke this limit and studied the tide-induced circulation and mass transport in the general nonlinear system. They proposed a new concept of the concentration for inter-tidal transport processes, named as the "Lagrangian inter-tidal concentration".

The studies mentioned above prove that the Lagrangian residual current is the appropriate representative of the inter-tidal current, but it has not been widely accepted up to now. One of the possible reasons is that the mathematics of the Lagrangian residual current is more complex than that used in the Eulerian residual velocity and the accurate acquisition of the difference between them is difficult due to the small value of the residual currents compared with the tidal current.

Generally, there are two ways to get the Lagrangian residual current. One way is to get the sum of the Eulerian residual velocity and the Stokes' drift velocity, and the other is to analyze the trajectories of water parcels. However, it is not an easy task to obtain the Lagrangian residual current field by either method. As mentioned above, the tide-induced residual current is very small and is often of the same order with the instrument measurement error. The direct measurement error of the time series of tidal currents will contaminate the Lagrangian residual current. The trajectory method, which arises from the original definition of the Lagrangian residual current, is reliable, but it is not possible to get enough trajectories to make a satisfying spatial coverage.

Things may be easier in the laboratory because of long history of the experimental fluid dynamics studies. By now, there have been a number of laboratory experimental studies on the tide-induced residual current, but most of them are confined to the obtaining of the Eulerian residual velocities. Yanagi (1976, 1978) did a series of laboratory experiments on the Eulerian residual velocity in a square bay model. Yanagi and Yoshikawa (1983) investigated the effects of the Coriolis force and bottom slope on the tide-induced residual current, but they also only acquired the Eulerian residual velocity. Yasuda (1984) obtained the surface Eulerian residual velocity and Lagrangian residual current in a bay with a sloping bed. In all of these studies, the method used to acquire the tide-induced residual current is the surface-floats tracing method, i.e., some floats are spread on the surface of the water and the trajectories of the floats are acquired by taking a series of photos in several tidal periods. However, in order to distinguish the floats in the frames acquired by the camera, the distances among these floats must be large enough. Therefore, the velocity can only be obtained at sporadic points and the velocity field can hardly be obtained.

In the 1980s, a method to obtain the high-resolution velocity field, the particle image velocimetry (PIV) technique, was developed in experimental fluid dynamics (Adrian, 2005). This technique makes the acquisition of accurate and high-resolution instantaneous velocity fields possible. Recently, it has been used in many laboratory studies on ocean phenomena (e.g., Dossmann *et al*., 2011; Mercier *et al*., 2011; Wang *et al*., 2012). In the present study, based on the tide-induced Lagrangian residual current theory, a method that enables the acquisition of the high-resolution Lagrangian residual current field through the PIV technique is proposed. The paper is organized as follows: Section 2 describes the measurement principles, the experimental setup, and the similarity analysis. The results and evaluation are discussed in section 3. Section 4 analyzes the impact of measurement error on the measurement of the residual current. Conclusions are presented in section 5.

2 Measurement Principles, Experimental Setup, and Similarity Analysis

2.1 Measurement principles

With the classical PIV technique, the particles with a diameter of tens of microns are suspended in the fluid. Illuminating one slice of the fluid with a laser beam and taking photos of the slice with a high-speed charge-coupled device (CCD) camera, a sequence of images that reflects the distribution of the particles can be obtained. Then, every frame is split into an array of interrogation windows and a displacement vector for each window can be obtained by processing two consecutive frames with autocorrelation or cross-correlation techniques, which is called the pattern-matching algorithm. Every interrogation window can be regarded as a fluid parcel, and the size of the interrogation window should be chosen to have at least 6 pixels per window on average. As a result, the instantaneous velocity of each fluid parcel is obtained to formulate the high-resolution velocity field.

To accurately measure the fluid motion, three conditions should be satisfied. Firstly, the PIV particles should follow the flow of the fluid very well. Secondly, the distances that fluid parcels move during the time interval between two neighboring frames should not be too long to hinder identifying the movement of the fluid parcel defined by the interrogation window. Thirdly, the distribution of the particles in a fluid parcel, i.e., an interrogation window, should remain almost unchanged between the two frames used for calculating. The third problem is called the particle pattern decorrelation which is of major concern in acquiring the velocity field by applying the PIV technique. Two mechanisms contribute to this phenomenon: one is the effect of the diffusion, which can change the particle distribution pattern in one water parcel; the other is the velocity normal to the plane being studied.

According to the definition, the Lagrangian residual velocity can be obtained by processing the two images separated by one tidal period with pattern-matching algorithm if the above three conditions are satisfied. In the present-study, the PIV particles are chosen as those used in similar studies; then, the first condition is satisfied easily. According to the behavior of the tide-dominant movement, the fluid parcels will come back to the positions near their initial positions after one or several tidal periods. Therefore, the net displacements of water parcels in one or several tide periods are small and may be measured by the PIV technique, which will be discussed in the following part. As for the particle pattern decorrelation, velocity normal to the illuminated plane can be estimated to evaluate its influence, but the effect of the turbulent diffusion can only be discussed after the experiment is carried out.

2.2 Experimental setup

This experiment setup is based on a narrow bay model with an analytical solution. Winant (2008) first gave the analytical solution of the first-order Lagrangian residual velocity for a parabolic depth profile by calculating the Eulerian residual velocity and the Stokes' drift velocity separately. The present author established the governing equations for the first-order Lagrangian residual velocity and solved it analytically (Jiang and Feng, 2014). In the above models, the momentum equation of the vertical direction is reduced to hydrostatic equilibrium because of the small ratio of depth to horizontal spatial scale when studying the tidal currents in a bay. This ratio is of the order of 10^{-3} at most which is difficult to reproduce in the laboratory without changing its dynamics.

In the present study, the laboratory experiments were conducted in a rectangular tank of 5000 mm long (x-axis), 150 mm wide (y-axis) and 400 mm high (z-axis). At the left-hand side of the tank, a wave generator was used to generate gravity waves with a period of 7.6 s to simulate tides. At the right-hand side, a bottom slope was put in the tank which was 1000 mm in length and made the water depth decreasing linearly along the y-coordinate from 120 mm to 40 mm (Fig. 1), i.e., the mean depth of the water on the slope was 80 mm. The measurement area was chosen at the right-hand side of the tank which was 420 mm long. The laser beam illuminated a horizontal plane just below the water surface. The PIV particles with a diameter of about 30 μm were suspended in the water. At about 1m above the water surface, a camera with a frame rate of 30 fps was set to record the particle distributions of the illuminated plane with a resolution of 1392×1040 pixels.

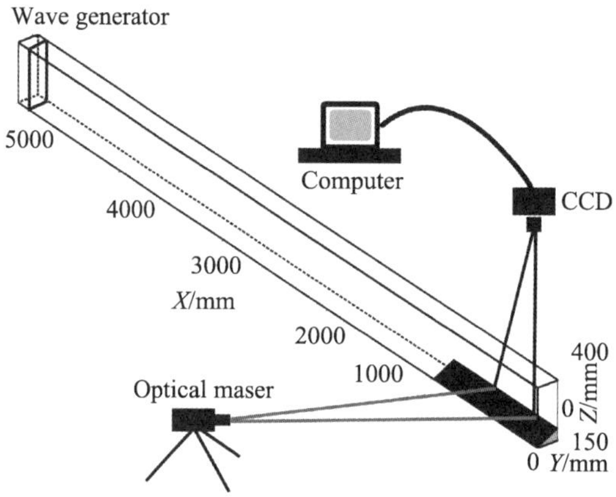

Fig. 1 The experimental setup

2.3 Similarity analysis

First, the properties of the wave in this experiment are diagnosed. In this experiment, the elevation of the water ζ is of the order of 1 mm, which is small compared with the water depth, so the dispersion relations can be written as $\omega^2 = gk\,\mathrm{th}(hk)$, where ω, k, and h represent the angular frequency, the wave number, and the mean water depth, respectively. According to the equation $k = \frac{2\pi}{\lambda}$, where λ denotes the wavelength, it can be deduced that the ratio of the water depth to the wavelength is of the order of 0.01. Therefore, the wave generated in this experiment is a long gravity wave, whose vertical velocity is much smaller than longitudinal velocity. This ensures that the vertical velocity will not affect the distribution of the PIV particles in any horizontal plane, which partly satisfies the third condition mentioned in Section 2.1.

Second, based on the experimental setup, the independent characteristic scales are presented, namely, the basin length scale L~10 m, the basin width scale B~0.1 m, the basin depth scale H~0.1 m, and the time scale T~10 s. According to the measurement results, the peak longitudinal current is about 10 mm/s, i.e., the longitudinal velocity scale U~10 mm/s. Because $B/L = 0.01$ and $H/L = 0.01$, the experimental area is geometrically similar to a narrow shallow bay.

According to the above analysis, the second condition mentioned in Section 2.1 is satisfied in this experiment: because $UT/L << 1$, the experiment setup defines a weakly nonlinear system, where UT means that the characteristic value of water parcel excursion is about 0.1 m. In addition, in the weakly nonlinear system, the ratio of the residual current to the tidal current is also equal to UT/L (Ianniello, 1977; Feng *et al.*, 1986a). Therefore, the net displacement of the water parcel over one period is of the order of 0.001 m. The fluid parcels will be back to the positions near their initial positions after one tidal period. According to the measurement area and the CCD resolution, the net displacement of the order of 0.001 m can be measured through the PIV technique.

As to the third condition, it is assumed that the horizontal convection dominates in the experiment, i.e., the PIV particles' distribution in a horizontal interrogation window holds well in one tidal period. If this assumption is right, the high-resolution horizontal Lagrangian residual current field on a certain layer can be obtained rigorously by applying the pattern-matching algorithm on the initial and final frames over one tidal period. Yet the direct estimation of the importance of the turbulence diffusion in an interrogation window is beyond the scope of the present study because it is of the subgrid scale here. So in the following section, this method will be evaluated by analyzing the experiment results.

3 Results and Evaluation

3.1 Results

In this experiment, 228 frames can be acquired in one tidal period. The phase when the ebb tidal velocity reaches its maximum is assumed to be $\varphi = 0$. In the experiment, the images of three periods are taken. Then, if the pattern-matching algorithm is applied to each pair of frames separated by one tidal period, the high-resolution Lagrangian residual current fields with different initial phases are obtained. Replicating experiments have also been done, and Lagrangian residual current fields are nearly the same. Averaging these results with the same initial phase, the Lagrangian residual current fields with the different phases are displayed in Fig. 2.

It can be seen in Fig. 2 that when the tidal current flows over the narrow bay with the bottom slope decreasing linearly along the width, the Lagrangian residual current flows in at the shallow side and flows out at the deep side. The residual current is around 0.1 mm/s, and the tidal velocity amplitude is about 10 mm/s. Therefore, the residual current is one to two orders in magnitude smaller than the tidal velocity. The pattern of the Lagrangian residual current field does not change with the initial phase. This is because the experiment setup defines a weakly nonlinear system; the second-order Lagrangian residual current which is dependent on the phase of tidal current is much smaller than the first-order Lagrangian residual current (Feng *et al.*, 1986a). The Lagrangian residual current field for a tidal period of 5 s is also obtained and is similar to that in Fig. 2 (figures are omitted for brevity).

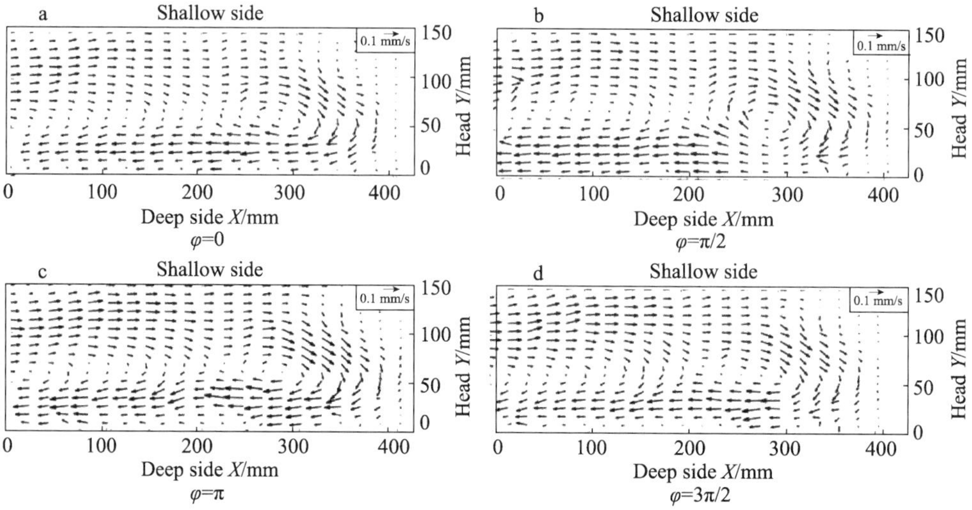

Fig. 2 The Lagrangian residual current fields obtained by the PIV technique

a. The initial phase is 0; b. The initial phase $\pi/2$; c. The initial phase is π; d. The initial phase is $3\pi/2$.

This is in accordance with the result of the analytical model when the viscosity coefficient is taken to be very small to make the turbulent viscosity force two orders smaller than the inertial force (see Fig. 3). It should be noted that the equations of the analytical model are derived based on the shallow-water approximation because the water depth is at least three orders smaller than the horizontal spatial scale in a real bay. But limited by the experimental technique, the water depth is of the same order as the bay width in the present experiment though it is one to two orders smaller than the bay length. Furthermore, the analytical result in Fig. 3 is only the first-order approximation to the Lagrangian residual current which is independent of the initial phase. Thus, the experimental results can only be compared with the analytical results qualitatively.

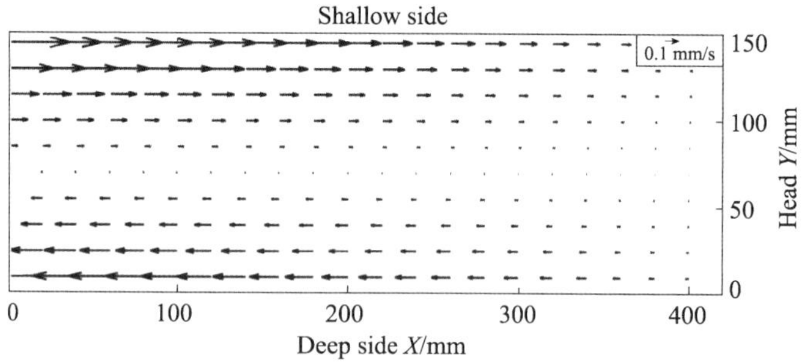

Fig. 3 The first-order Lagrangian residual current field obtained by an analytical method

The Lagrangian residual velocity can also be obtained by processing the two images separated by two or more tidal periods. The results show that the Lagrangian residual currents cannot be determined at some interrogation windows which means that the pat tern decorrelation becomes important if the time interval between the frame of images is too long.

3.2 Error analysis

Although the spatial resolution of the camera is 1392×1040, the spatial resolution of the measurement area in the tank is 1392×485 because the measurement area covers only part of the frame. Considering the geometry scale of the measurement area, it can be found that a pixel is equal to about 0.31 mm in the present experiment setup. The software for calculation in this paper is DigiFlow (Dalziel Research Partners, 2008). In this experiment, the accuracy is set to best when processing the data by using DigiFlow, in which the calculated accuracy of the net displacement through PIV is about 1/100 of a pixel which is 3.1×10^{-3} mm. Because the tidal period in the experiment is 7.6 s, and there are 228 frames in one period, the error is about 9.3×10^{-2} mm/s of the instantaneous velocities and is about 4.1×10^{-4} mm/s of the Lagrangian residual velocities. Therefore, the measurement error of the PIV method in the present experiment is two orders smaller than the measured variables.

3.3 Comparison with the surface-float tracing method

According to its definition, the Lagrangian residual current can describe the net displacement of the water parcel after one or several tidal periods. To evaluate the Lagrangian residual current field obtained by the PIV technique, the traditional surface-float tracing method was used as a reference. The surface-floats tracing method has been used by some researchers (Yanagi, 1976, 1978; Yanagi and Yoshikawa, 1983; Yasuda, 1984), and it was employed again in this study to obtain the Lagrangian residual current at some specific points with the same experimental conditions as in the PIV experiment. When the PIV technique was applied to get the current field, 30 surface-floats with a diameter of about 2–3 mm were spread on the water surface simultaneously. Just considering the initial and final positions of these surface floats over one period, the net-displacements of the floats over one period can be obtained. Taking $\varphi=0$ as the initial time and dividing these net-displacements by one tidal period, the Lagrangian residual current at these initial points is presented in Fig. 4.

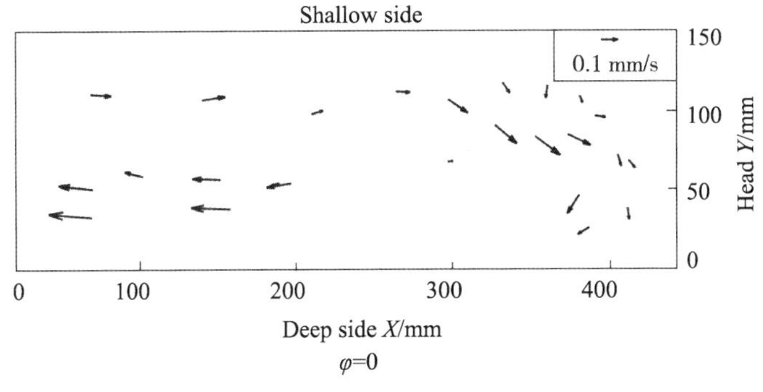

Fig. 4 The Lagrangian residual current obtained by the surface-floats tracing method. The initial phase is 0.

The Lagrangian residual current in Fig. 4 exhibits the same flow pattern as that in Fig. 2. This coincides with that of Yasuda (1984), who also used the surface-float tracing method to obtain the Lagrangian residual current in a narrow bay model with a sloping bed along the transverse direction.

However, the velocities in Fig. 4 are not exactly equal to those in Fig. 2. According to the surface-floats tracing method, the positions of these floats are determined by identifying the high-brightness area in the two frames as shown in Fig. 5a. The contiguous pixels with a brightness value greater than 0.5 are regarded as being occupied by the floats. The geometrical center of the float can be obtained by averaging the pixel positions occupied by each float. Assuming the float to be a sphere, as shown in Fig. 5a, the geometrical center of the float can be regarded as the center of a circle, and the largest distance between the geometrical center and the edge can be regarded as the radius. Actually, the floats are not exactly spherical and one float only occupies several pixels, which means that the shape of the same float may not be the same in different frames. So there exits positioning errors for center positions of these floats in different frames, which is of the same order as the difference between the radii of the floats in the initial frame and final frame.

In Fig. 5b, one float starting from the pixel (575, 114) is taken as an example to analyze the net displacement over one tidal period obtained by both the surface-float tracing method and the PIV method. Firstly, the angle difference between the directions of both displacements is approximately 1°, which is small enough to demonstrate that there exists little difference between the directions of the Lagrangian residual currents obtained by these two methods. Secondly, the difference in magnitude between the net displacements is nearly 0.2 pixels. Fig. 5a shows that the positioning error of the float is 0.3 pixels, which is of the same order as the difference between these two methods. Therefore, the difference in magnitude between the net displacements can be attributed to the positioning error of the surface float.

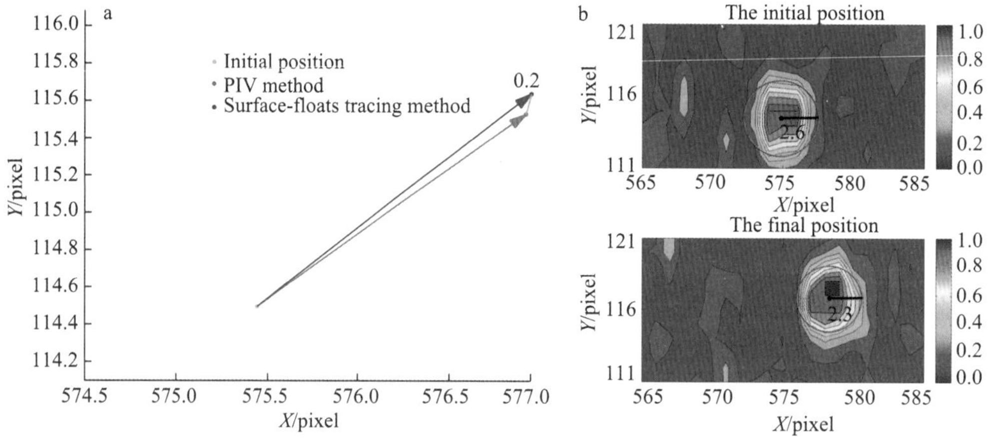

Fig. 5　a. Net-displacements over one tidal period; b. Positioning error for the center position of the float in the surface-floats tracing method

The angle and displacement differences between these two methods and the positioning errors in surface-float tracing method were calculated for all the 30 floats. The averaged angle and displacement difference between these two methods are 6° and 0.4 pixels respectively, and the averaged positioning error is 0.3 pixels. This shows that the angle is very small and the mean displacement difference is of the same order as the mean positioning error. Therefore, the difference between these two methods is mainly caused by the positioning errors of the center positions of the floats. In addition, the floats are too large to follow the flow well, which may also explain why the surface-float tracing method resulted in a larger measurement error than the PIV method.

Therefore, it is feasible to use the PIV method to acquire the high-resolution Lagrangian residual current field in the laboratory. Moreover, the PIV method ensures greater accuracy than the surface-floats tracing method. Thirdly, with the PIV method, the particles suspend in the fluid, which makes the acquisition of the three-dimensional structure of the Lagrangian residual current possible.

4 The Impact of Measurement Error on the Residual Current

It has been proved that under weakly nonlinear conditions, the first-order Lagrangian residual velocity equals the sum of the Eulerian residual velocity and the Stokes' drift velocity (Longuet-Higgins, 1969; Feng *et al.*, 1986a). This proposes another approach to obtain the first-order Lagrangian residual current, which is often used in the numerical studies (e.g., Dortch *et al.*, 1992; Delhez, 1996; Jiang and Sun, 2002; Hainbucher *et al.*, 2004; Wei *et al.*, 2004). The Eulerian residual velocity (Fig. 6a) in turn can be obtained by averaging the tidal velocities in one tidal period, and the Stokes' drifts velocity (Fig. 6b) can be obtained according to the formula $\boldsymbol{U}_{\mathrm{s}}=\overline{(\int_{t_0}^{t}\boldsymbol{U}\mathrm{d}t\cdot\nabla)\boldsymbol{U}}$, which depends on the time series of the tidal current at one fixed point and the time series of the gradient of the velocity at the same point. The over bar represents the tidal average operator defined as $\overline{\boldsymbol{f}}=\frac{1}{nT}\int_{t_0}^{t_0+nT}\boldsymbol{f}(x,y,t)\mathrm{d}t$. The present experiment, however, uses the PIV technique to get the instantaneous tidal velocity fields at a time interval of 1/30 s. The first-order Lagrangian residual current acquired by this approach is shown in Fig. 6c.

It can be seen that the pattern of the Lagrangian residual current in Fig. 6c is similar to that in Fig. 2, but it is discontinuous in the left part of the shallow side and the velocity field is more irregular than that in Fig. 2. This is because the measurement errors in the tidal velocities seriously contaminate the residual current in this approach.

In the PIV technique, the instantaneous tidal velocity is obtained by the quotient of the displacement of one interrogation window and the time interval. Obviously, there are measurement errors in determining the displacements. Because the instrument and the experimental setup are fixed, it is assumed that all the displacement vector measurement errors are

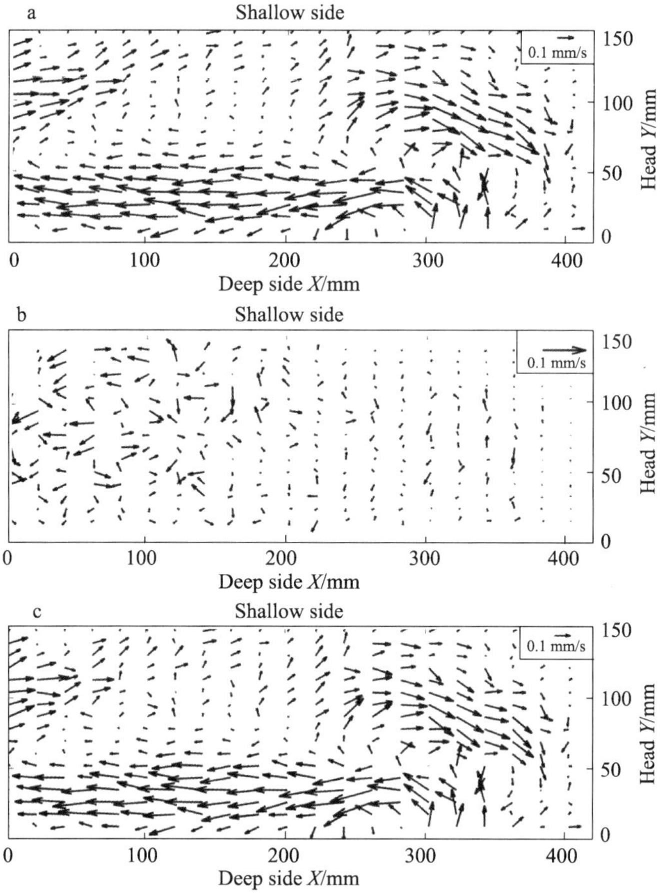

Fig. 6 a. The Eulerian residual velocity; b. The Stokes drift velocity field; c. The Lagrangian residual current filed obtained by calculating the sum of the Eulerian residual velocity and the Stokes' drift velocity

within the same range and can be represented by (σ_x, σ_y). It is assumed that the number of instantaneous tidal velocity fields in one tidal period is N and the displacement can be denoted as $(\Delta x_k, \Delta y_k)$ for the kth time step. Then, the Eulerian residual velocity can be obtained through the equation $(u_{\mathrm{E}}, v_{\mathrm{E}}) = \dfrac{1}{N}\sum_{k=1}^{N}\dfrac{(\Delta x_k, \Delta y_k) \pm (\sigma_x, \sigma_y)}{\Delta t}$ with the error being $(\sigma_x, \sigma_y)/\Delta t$. Therefore, the minimal error for the Lagrangian residual current obtained through the equation $(u_{\mathrm{L}}, v_{\mathrm{L}}) = (u_{\mathrm{E}}, v_{\mathrm{E}}) + (u_{\mathrm{S}}, v_{\mathrm{S}})$ is at least $(\sigma_x, \sigma_y)/\Delta t$.

However, if the Lagrangian residual velocity is calculated through dividing the net displacement in one tidal period by the tidal period and the net displacement of a fluid parcel is assumed as $(\Delta x_{\mathrm{L}}, \Delta y_{\mathrm{L}})$, the formula will be $(u_{\mathrm{L}}, v_{\mathrm{L}}) = \dfrac{(\Delta x_{\mathrm{L}}, \Delta y_{\mathrm{L}}) \pm (\sigma_x, \sigma_y)}{N\Delta t}$ with the error being $(\sigma_x, \sigma_y)/(N\Delta t)$, whose magnitude is $1/N$ of the error in the former approach. When N=228, which is the number of frames in the present experiment, the error arising from the former approach is two orders larger than the error resulting from the approach dividing the net displacement by the tidal period. Moreover, the influence of the measurement error caused by the former approach will be fatal, especially when the residual current is very small.

In in situ observation analysis, this problem may also arise. Generally, the traditional current measurement error is about 0.01 m/s, so if the Eulerian residual velocity and the Stokes' drift velocity are added together to get the Lagrangian residual current, the minimal error in the Lagrangian residual current will be about 0.01 m/s. However, in most cases, the order of the Lagrangian residual current is 0.01 m/s. Therefore, the measurement error will influence the results vitally. On the other hand, if the Lagrangian residual current is obtained by tracking drifters that fully behave like the surrounding water particles, the major error will come from the positioning process. It is quite common to get a positioning accuracy at the order of 10 m, and this will make the Lagrangian residual current error fall at the order of 10^{-4} m/s if M_2 tide is assumed, causing the error to be at least one order smaller than the Lagrangian residual current itself. Therefore, in the field observation, the drifter tracking method is a good practice to get the Lagrangian residual current.

5 Conclusions

In the present experiment, a long gravity wave is generated in a 5-m-long water tank to simulate the tide. According to the tidal oscillatory behavior that the fluid parcels will be back to the vicinity of their initial positions after one tidal period, a method to obtain the high-resolution tide-induced Lagrangian residual current by the PIV technique in the laboratory is proposed. A comparison with the traditional surface-float tracing method demonstrates that this method is feasible and better than the traditional method. Furthermore, it is found that in laboratory experiments and in situ observation analysis, the method of direct acquisition of the Lagrangian residual current may reduce the error by at least one order compared with those acquisition methods that require the detailed information of the tidal cycle. Although it seems difficult to extend this idea to the field research, the error analysis of the present method hints that the surface drifter tracing is better than adding up the Eulerian residual velocity and the Stokes' drift velocity to acquire the first-order Lagrangian residual velocity in the field.

The result in this paper proves that the analytical solution under similar condition as the laboratory experiment is reasonable. In future, the measurement method will be used in laboratory to study the Lagrangian residual current under more realistic conditions.

Acknowledgements

This work is supported by the National Natural Science Foundation of China (projects

40976003 and 40906001) and the Scholarship Award for Excellent Doctoral Student granted by Ministry of Education of China. Tao Wang acknowledges the financial support from Chinese Scholarship Council.

References

Adrian RJ (2005) Twenty years of particle image velocimetry. Exp Fluids 39(2): 159–169. doi: 10.1007/s00348-005-0991-7

Cerco CF (1995a) Simulation of long-term trends in Chesapeake Bay eutrophication. J Environl Eng 121: 298–310

Cerco CF (1995b) Response of Chesapeake Bay to nutrient load reductions. J Environ Eng 121: 549–557

Cerco CF, Cole T (1993) Three-dimensional eutrophication model of Chesapeake Bay. J Environ Eng 119: 1106–1125

Cheng RT, Casulli V (1982) On Lagrangian residual currents with applications in South San Francisco Bay, California. Water Resour Res 18(6): 1652–1662

Cheng RT, Feng S, Xi P (1989) On inter-tidal transport equation. In: Neilson BJ, Kuo A, Brubaker J (eds) Estuarine Circulation. Humana, Clifton, pp 133–156

Dalziel Research Partners (2008) DigiFlow user guide. Dalziel Research Partners, Cambridge.

Delhez EJM (1996) On the residual advection of passive constituents. J Mar Syst 8: 147–169

Dortch MS, Chapman RS, Abt SR (1992) Application of three-dimensional Lagrangian residual transport. J Hydraul Eng 118: 831–848

Dossmann Y, Paci A, Auclair F, Floor JW (2011) Simultaneous velocity and density measurements for an energy-based approach to internal waves generated over a ridge. Exp Fluids 51(4): 1013–1028. doi: 10.1007/s00348–011–1121–3

Feng S (1987) A three-dimensional weakly nonlinear model of tide-induced Lagrangian residual current and mass-transport, with an application to the Bohai sea. In: Nihoul JCJ, Jamart BM (eds) Three Dimensional Models of Marine and Estuarine Dynamics, vol 45, Elsevier Oceanography Series. Elsevier, Amsterdam, pp 471–488

Feng S (1990) On the Lagrangian residual velocity and the mass-transport in a multi-frequency oscillatory system. In: Cheng RT (ed) Coastal and Estuarine Studies, vol 38. Springer, Berlin, pp 34–48

Feng S, Xi P, Zhang S (1984) The baroclinic residual circulation in shallow seas. Chin J Oceanol Limnol 2: 49–60

Feng S, Cheng RT, Xi P (1986a) On tide-induced Lagrangian residual current and residual transport: 1. Lagrangian residual current. Water Resour Res 22: 1623–1634

Feng S, Cheng RT, Xi P (1986b) On tide-induced Lagrangian residual current and residual transport: 2. Residual transport with application in South San Francisco Bay. Water Resour Res 22: 1635–1646

Feng S, Ju L, Jiang W (2008) A Lagrangian mean theory on coastal sea circulation with Inter-tidal transports I. Fundamentals. Acta Ocean Sin 27(6): 1–16

Fischer HB, List EJ, Koh R, Imberger J, Brook NH (1979) Mixing in Inland and Coastal Waters. Academic, New York

Hainbucher D, Wei H, Pohlmann T, Sündermann J, Feng S (2004) Variability of the Bohai Sea circulation based on model calculations. J Mar Syst 44(3–4): 153–174. doi: 10.1016/j.jmarsys. 2003.09.008

Huijts KMH, Schuttelaars HM, de Swart HE, Friedrichs CT (2009) Analytical study of the transverse distribution of along-channel and transverse residual flows in tidal estuaries. Cont Shelf Res 29(1): 89–100. doi: 10.1016/j.csr.2007.09.007

Ianniello JP (1977) Tidally induced residual currents in estuaries of constant breadth and depth. J Mar Res 35(4): 755–786

Jiang W, Feng S (2011) Analytical solution for the tidally induced Lagrangian residual current in a narrow bay. Ocean Dyn 61(4): 543–558. doi: 10.1007/s10236–011–0381–z

Jiang W, Feng S (2014) 3D Analytical solution to the tidally induced Lagrangian residual current equations in a narrow bay. to Ocean Dyn 64(8): 1073–1091. doi: 10.10071s10236–014–0738–1

Jiang W, Sun WX (2002) Three dimensional tide-induced circulation model on a triangular mesh. Int J Numer Methods Fluids 38: 555–566. doi: 10.1002//d.231

Longuet-Higgins MS (1969) On the transport of mass by time-varying ocean currents. Deep Sea Res 16: 431–447

Lopes JF, Dias JM (2007) Residual circulation and sediment distribution in the Ria de Aveiro lagoon, Portugal. J Mar Syst 68(3–4): 507–528. doi: 10.1016/j.jmarsys.2007.02.005

Mercier M, Peacock T, Saidi S, Viboud S, Didelle H, Gostiaux L, Sommeria J, Dauxois T, Helfrich K (2011) The Luzon Strait experiment. Paper presented at the 7th International Symposuim on Strait, Rome

Nihoul I, Ronday F (1975) The influence of the tidal stress on the residual circulation. Tellus A 27: 484–489

Vaz N, Miguel Dias J, Chambel Leitão P (2009) Three-dimensional modelling of a tidal channel: The Espinheiro Channel (Portugal). Cont Shelf Res 29(1): 29–41. doi: 10.1016/j.csr.2007.12.005

Wang T, Chen X, Jiang W (2012) Laboratory experiments on the generation of internal waves on two kinds of continental margin. Geophys Res Lett 39(4): L04602. doi: 10.1029/2011gl049993

Wei H, Hainbucher D, Pohlmann T, Feng S, Suendermann J (2004) Tidal-induced Lagrangian and Eulerian mean circulation in the Bohai Sea. J Mar Syst 44(3–4): 141–151. doi:10.1016/j.jmarsys. 2003.09.007

Winant CD (2008) Three-dimensional residual tidal circulation in an elongated, rotating basin. J Phys Oceanogr 38(6): 1278–1295. doi: 10.1175/2007jpo3819.1

Yanagi T (1976) Fundamental study on the tidal residual circulation I. J Oceanogr Soc Jpn 32: 199–208

Yanagi T (1978) Fundamental study on the tidal residual circulation II. J Oceanogr Soc Jpn 34: 67–72

Yanagi T, Yoshikawa K (1983) Generation mechanisms of tidal residual circulation. J Oceanogr Soc Jpn 39: 156–166

Yasuda H (1984) Horizontal circulations caused by the bottom oscillatory boundary layer in a bay with a sloping bed. J Oceanogr Soc Jpn 40: 124–134

Zimmerman JTF (1979) On the Euler-Lagrange transformation and the stokes drift in the presence of oscillatory and residual currents. Deep-Sea Res 26A: 505–520

3D Analytical Solution to the Tidally Induced Lagrangian Residual Current Equations in a Narrow Bay *

Wensheng Jiang • Shizuo Feng

1 Introduction

The tidal current is the principal movement in most shallow seas. Yet, the time scale of the tidal period is often not of major concern to people dealing with environment-related problems. People tend to pay much more attention to the fate of the pollutants on a time scale of days or even months. As a result, the movement of the tidal frequency needs to be filtered out of the total movement, making the residual current represent the mass transport in an intertidal time scale. In order to eliminate the tidal signal, the early work such as Abbott (1960) used a very direct and simple way by taking just the average of the current velocity over one or several tidal periods at a specific location. The so processed velocity, known as the Eulerian residual velocity ($\boldsymbol{u}_{\mathrm{E}}$), is still widely used to study the intertidal circulation nowadays (The definitions related to the residual currents used in this paper are summarized in Section 2.1). For example, Huijts *et al.* (2009) decomposed $\boldsymbol{u}_{\mathrm{E}}$ into several terms based on different mechanisms, and Burchard and Schuttelaars (2012) specifically studied in detail the mechanism of the tidal straining effect on $\boldsymbol{u}_{\mathrm{E}}$ which was used to represent the estuarine circulation in their study.

In all of these studies, it was implied a priori that $\boldsymbol{u}_{\mathrm{E}}$ could represent the circulation on an intertidal time scale, but it turns out that this is not the case. Nihoul and Ronday (1975) deduced the depth-averaged governing equations of $\boldsymbol{u}_{\mathrm{E}}$ from a tide-resolving primitive equation model and found that the so-called 'tidal stress' term is responsible for generating the residual current because of the system's nonlinearity. Heaps (1978) reexplored this problem and systematically derived the depth-averaged coastal sea circulation equations, which supported the results in Nihoul and Ronday (1975). These studies showed the researchers' attempts to separate the intertidal residual circulation system from the tidal current system. Feng *et al.* (1984) expanded this methodology to a 3D baroclinic case and found that an additional term called 'tidal surface source' exists at the sea surface even with no water supply imposed, which violates the mass conservation law. Moreover, in Fischer *et al.* (1979), it was found that there must be a tidal dispersion term in the intertidal mass transport equations for the closure of the equations, in which the dispersion coefficient is often of several orders' variations in different cases.

* Jiang W, Feng S. 2014. 3D analytical solution to the tidally induced Lagrangian residual current equations in a narrow bay. Ocean Dynamics, 64(8): 1073–1091

Another approach to remove the tidal signal to get the coastal sea circulation is to use the definition of the Lagrangian residual current. As a matter of fact, this concept originated from the mass transport velocity ($\boldsymbol{u}_{\mathrm{L}}$), which was first made related to the large-scale ocean circulation by Longuet-Higgins (1969). He pointed out that the mass transport velocity ($\boldsymbol{u}_{\mathrm{L}}$) equals the sum of the Eulerian residual velocity ($\boldsymbol{u}_{\mathrm{E}}$) and the Stokes' drift velocity ($\boldsymbol{u}_{\mathrm{S}}$). Zimmerman (1979) loosely defined the Lagrangian residual velocity ($\boldsymbol{u}_{\mathrm{LR}}$) as the net displacement of a labeled water parcel over one or several tidal periods. Under the weakly nonlinear condition, Feng *et al.* (1986a) proved rigorously that $\boldsymbol{u}_{\mathrm{L}}$ is a first-order approximation to $\boldsymbol{u}_{\mathrm{LR}}$ and the second-order of $\boldsymbol{u}_{\mathrm{LR}}$ is the Lagrangian residual drift velocity ($\boldsymbol{u}_{\mathrm{ld}}$), which displayed the Lagrangian nature of $\boldsymbol{u}_{\mathrm{LR}}$. Then in Feng *et al.* (1986b), $\boldsymbol{u}_{\mathrm{L}}$ was used to replace $\boldsymbol{u}_{\mathrm{E}}$ to act as the advective velocity in the intertidal transport equation, in which the tidal dispersion term disappears. Feng (1987) deduced the dynamic equations governing $\boldsymbol{u}_{\mathrm{L}}$ in a 3D case with the tidal body force appearing in the inter-tidal momentum equations but with the fictitious 'tidal surface source' in Feng *et al.* (1984) vanishing to keep the conservation of the mass. $\boldsymbol{u}_{\mathrm{L}}$ is also applied to the intertidal mass-transport equation, and the tidal dispersion term in Fischer *et al.* (1979) disappears naturally. These studies clearly show that $\boldsymbol{u}_{\mathrm{L}}$ is the appropriate representative of the residual circulation in tide-dominated shallow seas. A series of studies extended the idea to the study of the wind-driven baroclinic multi-frequency tide-induced system (Feng, 1990), the system which takes into account the turbulence closure (Feng and Lu, 1993) and the system with a zeroth-order strong background current (Feng and Wu, 1995; Feng, 1998). Recently, the Lagrangian mean theory without the restriction of the weak nonlinearity was proposed in Feng *et al.* (2008).

Although the application of the Lagrangian residual velocity to coastal seas is not rare (e.g., Jay and Smith, 1990; Foreman *et al.*, 1992; Ridderinkhof and Loder, 1994; Delhez, 1996; Wei *et al.*, 2004; Hainbucher *et al.*, 2004; Muller *et al.*, 2009), and the dynamical equations proposed in Feng (1987) were also applied successfully under weakly nonlinear conditions (e.g., Dortch *et al.*, 1992; Wang *et al.*, 1993; Cerco and Cole, 1993; Cerco, 1995), this concept is not widely used in dealing with coastal circulation problems. This may be due to the fact that the residual current is often more than one-order smaller than the tidal current and is often of the order of centimeter per second. It is just of the same order with the measurement error and may be even smaller than the error in numerical modeling. Therefore, the difference between $\boldsymbol{u}_{\mathrm{L}}$ and $\boldsymbol{u}_{\mathrm{E}}$ cannot be clearly defined in all cases. As a result, the analytical solution becomes very important to the understanding of the mechanisms.

Ianniello (1977) gave the first analytical solution to the residual current in a breadth-averaged narrow tidal bay, with the solution confined to a vertical-longitudinal plane. In his solution, he clearly showed that $\boldsymbol{u}_{\mathrm{E}}$ flowed out of the bay at all depths, which indicates that the bay will be empty if $\boldsymbol{u}_{\mathrm{E}}$ is assumed to be the intertidal residual velocity, while $\boldsymbol{u}_{\mathrm{L}}$ has a two-layer structure with the inflow at the surface and the outflow near the bottom, which ensures the mass conservation. Li and O'Donnell (2005) solved the depth-averaged tidal system analytically by using the Eulerian residual transport velocity ($\boldsymbol{u}_{\mathrm{T}}$) proposed by Robinson (1983) to describe the intertidal movement. However, $\boldsymbol{u}_T$ is different from the mass transport velocity ($\boldsymbol{u}_L$) defined by Longuet-Higgins (1969). An interesting point is that $\boldsymbol{u}_L$ used in Ianniello (1977) is of the same form as $\boldsymbol{u}_T$ defined by Robinson (1983), but this is a mere coincidence because under general conditions an additional term is missing for $\boldsymbol{u}_T$, as pointed out in Feng *et al.* (1986b). Jiang and Feng (2011) proposed and analytically solved the depth-averaged Lagrangian residual current equations, which were the 2D counterpart of the 3D equations established by Feng (1987). This enhances the understanding of the differences between $\boldsymbol{u}_L$, $\boldsymbol{u}_T$, and $\boldsymbol{u}_E$ pointed out in Feng *et al.* (1986b).

It should be noted that $\boldsymbol{u}_L$ has a 3D nature (Feng 1987). Because the tidal current is generally of 3D, a water column at the initial time cannot be kept as the same water column during the whole tidal period, which means that the derivation of $\boldsymbol{u}_L$ should start from the 3D primitive equations. Winant (2008) first gave the 3D $\boldsymbol{u}_L$ in a narrow bay. He acquired $\boldsymbol{u}_E$ first and obtained $\boldsymbol{u}_L$ by adding up $\boldsymbol{u}_E$ and $\boldsymbol{u}_S$ following Longuet-Higgins (1969). The results showed that $\boldsymbol{u}_E$ violates the mass conservation but $\boldsymbol{u}_L$ does not, though Winant (2008) did not point this out explicitly. However, when the residual water transport was studied in Winant (2008), $\boldsymbol{u}_T$ was used again which cannot represent the real exchange flow situation.

In the present study, the 3D governing equations for $\boldsymbol{u}_L$ are deduced from the 3D primitive equations by applying the perturbation method and are then solved in a narrow bay directly. The feature and generating mechanisms of $\boldsymbol{u}_L$ are discussed, and the exchange flow in the bay is obtained from the 3D $\boldsymbol{u}_L$. The structure of the paper is as follows: in Section 2, the formulation of the model is given. The solution procedure of $\boldsymbol{u}_L$ is described in Section 3. And in Section 4, the discussion of the solution under several specific bottom profiles is made. Finally, the conclusions are drawn in Section 5.

2 Formulation

2.1 Definitions of residual velocities

As mentioned in the introduction, there are several ways to define the residual current. For clarity, the definitions used in this paper are summarized in Table 1 and more detailed discussion can be found in Jiang and Feng (2011).

The Lagrangian residual velocity $\boldsymbol{u}_{LR}$ is loosely defined by Zimmerman (1979) as

$$\boldsymbol{u}_{LR} = \frac{\boldsymbol{\xi}(t_0 + nT; \boldsymbol{x}_0, t_0)}{nT}$$

where $\boldsymbol{\xi}(t; \boldsymbol{x}_0, t_0) = \int_{t_0}^{t} \boldsymbol{u}(\boldsymbol{x}_0 + \boldsymbol{\xi}, t')\mathrm{d}t'$ defines the displacement of a water parcel with its initial position at $\boldsymbol{x}_0$ when $t = t_0$, $\boldsymbol{u}$ is the tidal current velocity. Because $\boldsymbol{x}_0$ and t_0 can be arbitrarily selected, $\boldsymbol{u}_{LR}$ is well defined in the whole domain for any time which constitutes a Eulerian field. Thus, $\boldsymbol{u}_{LR}$ is a Eulerian quantity.

In the weakly nonlinear case, the velocity can be expanded into the power series according to a small nondimensional number κ, i.e.,

$$\boldsymbol{u} = \boldsymbol{u}_0 + \kappa \boldsymbol{u}_1 + O(\kappa^2)$$

with κ being the ratio of the tidal amplitude to the water depth.

Feng *et al.* (1986a) proved that

$$\boldsymbol{u}_{\mathrm{LR}} = \boldsymbol{u}_{\mathrm{L}} + O(\kappa) \tag{1}$$

where the mass transport velocity $\boldsymbol{u}_{\mathrm{L}}$ defined by Longuet-Higgins (1969) is

$$\boldsymbol{u}_{\mathrm{L}} = (u_{\mathrm{L}}, v_{\mathrm{L}}, w_{\mathrm{L}}) = \boldsymbol{u}_{\mathrm{E}} + \boldsymbol{u}_{\mathrm{S}} \tag{2}$$

The Eulerian residual velocity $\boldsymbol{u}_{\mathrm{E}}$ is defined as

$$\boldsymbol{u}_{\mathrm{E}} = (u_{\mathrm{E}}, v_{\mathrm{E}}, w_{\mathrm{E}}) = \langle \boldsymbol{u}_1 \rangle \tag{3}$$

and the Stokes' drift velocity $\boldsymbol{u}_{\mathrm{S}}$ is defined as

$$\boldsymbol{u}_{\mathrm{S}} = (u_{\mathrm{S}}, v_{\mathrm{S}}, w_{\mathrm{S}}) = \langle \boldsymbol{\xi}_0 \cdot \nabla \boldsymbol{u}_0 \rangle \tag{4}$$

where the tidal-averaging operator is defined as follows with T being the tidal period:

$$\langle \cdot \rangle = \frac{1}{nT} \int_{t_0}^{t_0+nT} \cdot \mathrm{d}t \tag{5}$$

and

$$\boldsymbol{\xi}_0 = (\xi_0, \eta_0, t_0) = \int_{t_0}^{t} \boldsymbol{u}_0(\boldsymbol{x}, t')\mathrm{d}t' \tag{6}$$

Thus,

$$\boldsymbol{u}_0 = (u_0, v_0, w_0) = \frac{\partial \boldsymbol{\xi}_0}{\partial t} \tag{7}$$

The definition of the Eulerian residual transport velocity $\boldsymbol{u}_{\mathrm{T}}$ is based on the idea that the calculation of the residual transport should take the water surface oscillation into consideration, thus,

$$\boldsymbol{u}_{\mathrm{T}} = \frac{1}{h}\left\langle \int_{-h}^{0} \boldsymbol{u}_1 \mathrm{d}z \right\rangle + \frac{1}{h}\left\langle \boldsymbol{u}_0 \big|_{z=0} \zeta_0 \right\rangle \tag{8}$$

where ζ_0 is the zeroth-order water elevation and h is the undisturbed water depth.

Table 1 Symbols for major definitions of the residual current

Symbol	Meaning	Definition
$\boldsymbol{u}_{\mathrm{LR}}$	Lagrangian residual velocity (LRV)	$\frac{1}{nT}\xi(t_0 + nT; \boldsymbol{x}_0, t_0)$
$\boldsymbol{u}_{\mathrm{E}}$	Eulerian residual velocity	$\langle \boldsymbol{u} \rangle$
$\boldsymbol{u}_{\mathrm{S}}$	Stokes' drift velocity	$\langle \xi_0 \cdot \nabla \boldsymbol{u}_0 \rangle$
$\boldsymbol{u}_{\mathrm{L}}$	first-order LRV, mass transport velocity	$\boldsymbol{u}_{\mathrm{E}} + \boldsymbol{u}_{\mathrm{S}}$
$\boldsymbol{u}_{\mathrm{T}}$	Eulerian residual transport velocity	$\frac{1}{h}\left\langle \int_{-h}^{0} \boldsymbol{u}_1 \mathrm{d}z \right\rangle + \frac{1}{h}\left\langle \boldsymbol{u}_0 \big|_{z=0} \zeta_0 \right\rangle$

2.2 The non-dimensional tidal equations

The 3D single frequency (e.g., the M_2 tide) tidal current equations will be solved in a semi-enclosed rectangular bay. The x and y coordinates are along the two horizontal sides of the bay with $x = 0$ at the open boundary and $x = L$ at the end of the bay. $y = 0$ and $y = B$ are the two lateral boundaries. $z = 0$ is set at the surface of the still water, while $z = -h$ is set at the sea bottom. The free surface is at $z = \zeta(x, y, t)$ to represent the tidal elevation. The single frequency tidal signal with the period being T and the tidal amplitude being ζ_c is imposed at the open boundary and the Coriolis force is omitted in the present study.

In this system, there are five basic characteristic values which are the spatial scales $x_c=\lambda_c$, $y_c=B$, $z_c=h_c$, the temporal scale $t_c=T/2\pi$, and the scale for the tidal amplitude ξ_c, with h_c being the average depth of the sea area and $\lambda_c = \sqrt{gh_c}T$ denoting the typical tidal wavelength.

In this paper, the water density is assumed to be constant to study the barotropic case only. Because the barotropic tidal wave is a long gravity wave, it holds that $\lambda_c \gg h_c$. Then in the vertical direction, the hydrostatic condition is a natural inference. It is also clear that the major balance of the tidal wave system is the local acceleration term and the pressure gradient force term. The 3D non-dimensional governing equations for the tidal wave can be deduced from the normalization of the dimensional equations based on the notion above. The non-dimensional governing equations are

$$\nabla \cdot \boldsymbol{u} = 0 \tag{9}$$

$$\frac{\partial u}{\partial t} + \kappa \boldsymbol{u} \cdot \nabla u = -\frac{\partial \zeta}{\partial x} + \beta \frac{\partial}{\partial z}\left(\nu \frac{\partial u}{\partial z}\right) \tag{10}$$

$$\delta^2 \frac{\partial v}{\partial t} + \delta^2 \kappa \boldsymbol{u} \cdot \nabla v = -\frac{\partial \zeta}{\partial y} + \beta \delta^2 \frac{\partial}{\partial z}\left(\nu \frac{\partial v}{\partial z}\right) \tag{11}$$

At the sea surface, $z = \kappa \zeta$,

$$w = \frac{\partial \zeta}{\partial t} + \kappa \left(u \frac{\partial \zeta}{\partial x} + v \frac{\partial \zeta}{\partial y} \right) \tag{12}$$

$$\frac{\partial (u, v)}{\partial z} = 0 \tag{13}$$

At the sea bottom, $z = -h$,

$$\boldsymbol{u} = 0 \tag{14}$$

At the open boundary, $x = 0$,

$$\zeta = \zeta_{\text{open}} \tag{15}$$

where t, $\boldsymbol{u}=(u, v, w)$, ζ and h are the non-dimensional time, the velocity and its components in x, y, and z directions, the sea surface elevation, and the water depth, respectively. L refers to the non-dimensional form of the bay length which is normalized by the tidal wavelength λ_c. ζ_{open} denotes the non-dimensional tidal wave imposed at the open boundary. $\nu = \nu(x,y,z)$ is the non-dimensional eddy viscosity coefficient with ν_c being its characteristic value. For brevity, they take the forms identical to their dimensional counterparts. The velocity scales are, $u_c = \zeta_c\sqrt{g/h_c}$, $v_c = \zeta_c B/(h_c t_c)$, and $w_c = \zeta_c / t_c$, respectively.

Three non-dimensional numbers exist in the system, which are $\beta = \nu_c T_c / h_c^2$, $\kappa = \zeta_c / h_c$ and $\delta = B/\lambda_c$. β reflects the importance of the eddy viscosity force. Since in the shallow sea, the eddy viscosity can not be neglected, $O(\beta) = 1$ is assumed in general in the present study. The other two non-dimensional numbers are basic in the whole system. One is κ, the ratio of the tidal amplitude scale (ζ_c) to the depth scale (h_c), which reflects the advective nonlinearity in the system, and the other is δ, the aspect ratio, which reflects the asymmetric feature of the horizontal geometry of the sea area.

In the present study, the case for $O(\kappa) < 1$ is studied, which means that the tidal motion is of a weakly nonlinear case if the eddy viscosity coefficient is a known value. $O(\delta) < 1$ is also assumed so that the system describes a weakly nonlinear tide in a narrow bay.

2.3 The perturbation in a narrow bay

In reality, $O(\kappa) = 0.1$ is very common for a typical bay and $O(\delta) = 0.1$ is also not uncommon in bays around the world. Therefore, in the weakly nonlinear case in a narrow bay considered in the present study, κ and δ are two small independent parameters. The perturbation method with two parameters is applied to the system of Eqs. 9–15.

Since in the system of Eqs. 9–15, only κ and δ^2 are present, $\boldsymbol{u}$ and ζ can be expressed in terms of a power series in small parameters κ and δ^2 to the order of $O(\kappa\delta^2)$,

$$\boldsymbol{u} = \boldsymbol{u}_0 + \kappa\boldsymbol{u}_1 + \delta^2\boldsymbol{u}_0' + \kappa\delta^2\boldsymbol{u}_1' + \cdots \tag{16}$$

$$\zeta = \zeta_0 + \kappa\zeta_1 + \delta^2\zeta_0' + \kappa\delta^2\zeta_1' + \cdots \tag{17}$$

The system of Eqs. 9–15 can then be decomposed into different subsystems based on the different orders of κ and δ^2.

It should be noticed that at the sea surface $z=\kappa\zeta$, the velocity $\boldsymbol{u}$ can be expanded in a Taylor series about z=0. Then, the surface boundary condition Eqs. 12 and 13 can be changed to their equivalent forms at z=0 after neglecting the high-order terms from the Taylor expansion. If we substitute the perturbation series Eqs. 16 and 17 into them, the surface boundary conditions of different orders can be obtained at z=0. For brevity, only the equations used in the deduction procedure are listed as follows. All the details of the deduction procedure are supplied in the Supplementary material.

2.3.1 The O(1) order equations

$$\nabla \cdot \boldsymbol{u}_0 = 0 \tag{18}$$

$$\frac{\partial u_0}{\partial t} = -\frac{\partial \zeta_0}{\partial x} + \beta \frac{\partial}{\partial z}\left(\nu \frac{\partial u_0}{\partial z}\right) \tag{19}$$

$$0 = -\frac{\partial \zeta_0}{\partial y} \tag{20}$$

At the sea surface, $z = 0$,

$$w_0 = \frac{\partial \zeta_0}{\partial t} \tag{21}$$

$$\frac{\partial (u_0, v_0)}{\partial z} = 0 \tag{22}$$

At the sea bottom, $z = -h$,

$$\boldsymbol{u}_0 = 0 \tag{23}$$

At the open boundary, $x = 0$,

$$\zeta_0 = \zeta_{\text{open}} \tag{24}$$

2.3.2 The O(κ) order equations

$$\nabla \cdot \boldsymbol{u}_1 = 0 \tag{25}$$

$$\frac{\partial u_1}{\partial t} + \boldsymbol{u}_0 \cdot \nabla u_0 = -\frac{\partial \zeta_1}{\partial x} + \beta \frac{\partial}{\partial z}\left(\nu \frac{\partial u_1}{\partial z}\right) \tag{26}$$

$$0 = -\frac{\partial \zeta_1}{\partial y} \tag{27}$$

At the sea surface, $z = 0$,

$$w_1 = \frac{\partial \zeta_1}{\partial t} + \frac{\partial u_0 \zeta_0}{\partial x} + \frac{\partial v_0 \zeta_0}{\partial y} \tag{28}$$

$$\frac{\partial (u_1, v_1)}{\partial z} = -\zeta_0 \frac{\partial^2 (u_0, v_0)}{\partial z^2} \tag{29}$$

At the sea bottom, $z = -h$,

$$\boldsymbol{u}_1 = 0 \tag{30}$$

At the open boundary, $x = 0$,

$$\zeta_1 = 0 \tag{31}$$

2.3.3 The O(δ^2) order equations

$$\frac{\partial v_0}{\partial t} = -\frac{\partial \zeta_0'}{\partial y} + \beta \frac{\partial}{\partial z}\left(\nu \frac{\partial v_0}{\partial z}\right) \tag{32}$$

2.3.4 The $O(\kappa\delta^2)$ order equations

$$\frac{\partial v_1}{\partial t}+\boldsymbol{u}_0\cdot\nabla v_0=-\frac{\partial \zeta_1'}{\partial y}+\beta\frac{\partial}{\partial z}\left(\nu\frac{\partial v_1}{\partial z}\right) \tag{33}$$

2.4 The deduction of Lagrangian residual circulation equations

The mass transport velocity $\boldsymbol{u}_{\mathrm{L}}$ is a first-order accurate approximation to the Lagrangian residual velocity $\boldsymbol{u}_{\mathrm{LR}}$ which represents the water parcel's net transport in one or several tidal periods. The physics of $\boldsymbol{u}_{\mathrm{L}}$ is revealed by its governing equations derived by Feng (1987) in general. In this section, the governing equations of $\boldsymbol{u}_{\mathrm{L}}$ will be given in a narrow bay. The non-dimensional form of the operator defined in Eq. 5 is listed as follows. It is used in the following part of the paper.

$$\langle\cdot\rangle=\frac{1}{2n\pi}\int_{t_0}^{t_0+2n\pi}\cdot\mathrm{d}t \tag{34}$$

2.4.1 The continuity equation of $\boldsymbol{u}_{\mathrm{L}}$

If the divergence operator is applied to both sides of Eq. 2 with noticing Eqs. 18 and 25 to get

$$\nabla\cdot\boldsymbol{u}_{\mathrm{L}}=\left\langle\frac{\partial\xi_0}{\partial x}\cdot\nabla u_0\right\rangle+\left\langle\frac{\partial\xi_0}{\partial y}\cdot\nabla v_0\right\rangle+\left\langle\frac{\partial\xi_0}{\partial z}\cdot\nabla w_0\right\rangle \tag{35}$$

Because of Eq. 7, the method in Appendix A is applied to Eq. 35 and noticing the periodicity of the zeroth-order variables, Eq. 35 can be changed to

$$\nabla\cdot\boldsymbol{u}_{\mathrm{L}}=-\left\langle\frac{\partial\boldsymbol{u}_0}{\partial x}\cdot\nabla\xi_0\right\rangle-\left\langle\frac{\partial\boldsymbol{u}_0}{\partial y}\cdot\nabla\eta_0\right\rangle-\left\langle\frac{\partial\boldsymbol{u}_0}{\partial z}\cdot\nabla\iota_0\right\rangle \tag{36}$$

If we add Eqs. 35 and 36 and expand the right-hand side, it can be found that

$$\nabla\cdot\boldsymbol{u}_{\mathrm{L}}=0 \tag{37}$$

2.4.2 The x-direction momentum equation

Equation 34 is applied to Eq. 26, with the periodicity of u_1 taken into consideration and based on the assumption $\nu=\nu(x,y,z)$, the following equation can be obtained

$$\langle\boldsymbol{u}_0\cdot\nabla u_0\rangle=-\frac{\partial\langle\zeta_1\rangle}{\partial x}+\beta\frac{\partial}{\partial z}\left(\nu\frac{\partial\langle u_1\rangle}{\partial z}\right) \tag{38}$$

According to the definition of $\boldsymbol{u}_{\mathrm{L}}$ in Eq. 2, Eq. 38 is changed to

$$\langle\boldsymbol{u}_0\cdot\nabla u_0\rangle=-\frac{\partial\langle\zeta_1\rangle}{\partial x}+\beta\frac{\partial}{\partial z}\left(\nu\frac{\partial u_{\mathrm{L}}}{\partial z}\right)-\beta\frac{\partial}{\partial z}\left(\nu\frac{\partial\langle\xi_0\cdot\nabla u_0\rangle}{\partial z}\right) \tag{39}$$

Based on the method in Appendix A and Eq. 7, the following relation can be obtained,

$$\left\langle \boldsymbol{u}_0 \cdot \nabla u_0 \right\rangle = -\left\langle \xi_0 \cdot \nabla \frac{\partial u_0}{\partial t} \right\rangle \tag{40}$$

Then Eq. 39 can be written as

$$0 = -\frac{\partial \zeta_{\mathrm{E}}}{\partial x} + \beta \frac{\partial}{\partial z}\left(\nu \frac{\partial u_{\mathrm{L}}}{\partial z} \right) + \pi_1 \tag{41}$$

where $\zeta_{\mathrm{E}} = \left\langle \zeta_1 \right\rangle$, which is independent of y according to Eq. 27, and

$$\pi_1 = \left\langle \xi_0 \cdot \nabla \frac{\partial u_0}{\partial t} \right\rangle - \beta \frac{\partial}{\partial z}\left(\nu \frac{\partial \left\langle \xi_0 \cdot \nabla u_0 \right\rangle}{\partial z} \right) \tag{42}$$

π_1 is the x-direction tidal body force named in Feng (1987) which is composed of two terms. The first term means the transport of the local inertia, and the other term is the eddy viscosity term of the Stokes' drift velocity, u_{S}.

2.4.3 The y-direction momentum equation

Equation 34 is applied to Eq. 33 while noticing the periodicity of v_1 and the assumption $\nu = \nu(x, y, z)$ to get

$$\left\langle \boldsymbol{u}_0 \cdot \nabla v_0 \right\rangle = -\frac{\partial \left\langle \zeta_1' \right\rangle}{\partial y} + \beta \frac{\partial}{\partial z}\left(\nu \frac{\partial \left\langle v_1 \right\rangle}{\partial z} \right) \tag{43}$$

According to the definition of v_{L} in Eq. 2, Eq. 43 can be changed to

$$\left\langle \boldsymbol{u}_0 \cdot \nabla v_0 \right\rangle = -\frac{\partial \left\langle \zeta_1' \right\rangle}{\partial y} + \beta \frac{\partial}{\partial z}\left(\nu \frac{\partial v_{\mathrm{L}}}{\partial z} \right) - \beta \frac{\partial}{\partial z}\left(\nu \frac{\partial \left\langle \xi_0 \cdot \nabla v_0 \right\rangle}{\partial z} \right) \tag{44}$$

Based on the method in Appendix A and Eq. 7, the following relation can be obtained,

$$\left\langle \boldsymbol{u}_0 \cdot \nabla v_0 \right\rangle = -\left\langle \xi_0 \cdot \nabla \frac{\partial v_0}{\partial t} \right\rangle \tag{45}$$

Then Eq. 44 can be written as

$$0 = -\frac{\partial \zeta_{\mathrm{E}}'}{\partial y} + \beta \frac{\partial}{\partial z}\left(\nu \frac{\partial v_{\mathrm{L}}}{\partial z} \right) + \pi_2 \tag{46}$$

where $\zeta_{\mathrm{E}}' = \left\langle \zeta_{\mathrm{E}}' \right\rangle$ and

$$\pi_2 = \left\langle \xi_0 \cdot \nabla \frac{\partial v_0}{\partial t} \right\rangle - \beta \frac{\partial}{\partial z}\left(\nu \frac{\partial \left\langle \xi_0 \cdot \nabla v_0 \right\rangle}{\partial z} \right) \tag{47}$$

2.4.4 The kinematic boundary condition at the sea surface

When Eq. 34 is applied to Eq. 28 and based on the periodicity of ζ_1, Eq. 21, Eq. 7 and the method in Appendix A, the following equation can be obtained

$$\langle w_1 \rangle = -\frac{\partial \langle \xi_0 w_0 \rangle}{\partial x} - \frac{\partial \langle \eta_0 w_0 \rangle}{\partial y} \tag{48}$$

When Eq. 48 is inserted into Eq. 2, using the method in Appendix A repeatedly, the kinematic boundary condition at the sea surface can be obtained by noticing the zeroth-order continuity equation (Eq. 18).

$$\begin{aligned} w_L &= -\left\langle \frac{\partial \xi_0}{\partial x} w_0 \right\rangle - \left\langle \frac{\partial \eta_0}{\partial y} w_0 \right\rangle + \left\langle \frac{\partial w_0}{\partial z} \iota_0 \right\rangle \\ &= \left\langle \frac{\partial u_0}{\partial x} \iota_0 \right\rangle + \left\langle \frac{\partial v_0}{\partial y} \iota_0 \right\rangle + \left\langle \frac{\partial w_0}{\partial z} \iota_0 \right\rangle \\ &= 0 \end{aligned} \tag{49}$$

2.4.5 The dynamic boundary condition at the sea surface

Equation 34 is applied to Eq. 29 with its right-hand side further manipulated by applying the method in Appendix A repeatedly, and based on Eqs. 18, 22, 21, and 6, the following can be obtained

$$\begin{aligned} -\frac{\partial^2}{\partial z^2}\langle \zeta_0(u_0, v_0) \rangle &= \frac{\partial^2}{\partial z^2}\langle w_0(\xi_0, \eta_0) \rangle \\ &= -\frac{\partial}{\partial z}\left\langle \left(\frac{\partial u_0}{\partial x} + \frac{\partial v_0}{\partial y} \right)(\xi_0, \eta_0) \right\rangle \\ &= -\frac{\partial}{\partial z}\langle \xi_0 \cdot \nabla u_0, \xi_0 \cdot \nabla v_0 \rangle \end{aligned} \tag{50}$$

Then Eq. 29 is changed to

$$\frac{\partial \langle u_1, v_1 \rangle}{\partial z} = -\frac{\partial}{\partial z}\langle \xi_0 \cdot \nabla u_0, \xi_0 \cdot \nabla v_0 \rangle \tag{51}$$

Based on the definition in Eq. 2, Eq. 51 is changed to

$$\frac{\partial \langle u_L, v_L \rangle}{\partial z} = 0 \tag{52}$$

2.4.6 The boundary condition at the sea bottom

Because at the sea bottom, $z = -h$, $\boldsymbol{u}_0=0$ and $\boldsymbol{u}_1=0$, according to Eq. 6, $\boldsymbol{\xi}_0=0$. Therefore, according to the definition in Eq. 2, $\boldsymbol{u}_L = 0$.

2.5 The tide-induced Lagrangian residual current equations

In summary, the tide-induced Lagrangian residual current equations are listed as follows by rewriting Eqs. 35, 41, and 47 here,

$$\frac{\partial u_{\mathrm{L}}}{\partial x}+\frac{\partial v_{\mathrm{L}}}{\partial y}+\frac{\partial w_{\mathrm{L}}}{\partial z}=0 \tag{53}$$

$$0=-\frac{\partial \zeta_{\mathrm{E}}}{\partial x}+\beta\frac{\partial}{\partial z}\left(\nu\frac{\partial u_{\mathrm{L}}}{\partial z}\right)+\pi_1 \tag{54}$$

$$0=-\frac{\partial \zeta'_{\mathrm{E}}}{\partial y}+\beta\frac{\partial}{\partial z}\left(\nu\frac{\partial v_{\mathrm{L}}}{\partial z}\right)+\pi_2 \tag{55}$$

with the boundary conditions specified as follows:
At the surface, $z = 0$

$$w_{\mathrm{L}}=0 \tag{56}$$

$$\frac{\partial u_{\mathrm{L}}}{\partial z}=0 \tag{57}$$

$$\frac{\partial v_{\mathrm{L}}}{\partial z}=0 \tag{58}$$

At the bottom, $z = -h$,

$$u_{\mathrm{L}}=v_{\mathrm{L}}=w_{\mathrm{L}}=0 \tag{59}$$

At the open boundary, $x = 0$, the residual water elevation $\zeta_{\mathrm{E}}=0$ and $\zeta'_{\mathrm{E}}=0$.
At the fixed boundary, because of the no water inflow at the banks of the bay,

$$U_{\mathrm{L}}=\int_{-h}^{0}u_{\mathrm{L}}\mathrm{d}z=0,\ \ x=L \tag{60}$$

$$V_{\mathrm{L}}=\int_{-h}^{0}v_{\mathrm{L}}\mathrm{d}z=0,\ \ y=0,1 \tag{61}$$

It can be seen that the governing equations of the Lagrangian residual current are in linear form. The Lagrangian residual current is driven by the tidal body force $\pi=(\pi_1,\pi_2)$ which is defined by the zeroth-order linear tide.

This set of equations clearly shows the dynamics of the tide-induced Lagrangian residual current. Compared with the so-called Eulerian residual velocity, the Lagrangian residual current satisfies the conservation law of the material surface. For example, at the sea surface, there is no fake source/sink term in Eq. 56, while in the case of the Eulerian residual velocity, such a term is inevitable (Feng *et al*. 1984).

3 The Solution to the Tide-induced Residual Current Equations

In order to find an analytical solution to the Lagrangian residual current, the eddy viscosity coefficient is assumed to be a constant. Since ν_{c} is the characteristic value of the eddy viscosity coefficient, the non-dimensional ν should be 1. It is assumed that water depth h varies only with y.

Integrate Eq. 54 from z to 0 first and then from $-h$ to z with noticing the boundary conditions Eqs. 57 and 59, we get

$$u_{\mathrm{L}} = \frac{z^2 - h^2}{2\beta}\frac{\partial \zeta_{\mathrm{E}}}{\partial x} + \frac{\int_{-h}^{z}\int_{z_1}^{0}\pi_1(x, y, z_2)\mathrm{d}z_2\mathrm{d}z_1}{\beta} \tag{62}$$

Then, the longitudinal volumic transport U_{L} is

$$U_{\mathrm{L}} = \int_{-h}^{0} u_{\mathrm{L}}\mathrm{d}z = \frac{\Pi_1}{\beta} - \frac{h^3}{3\beta}\frac{\partial \zeta_{\mathrm{E}}}{\partial x} \tag{63}$$

where

$$\Pi_1 = \int_{-h}^{0}\int_{-h}^{z}\int_{z_1}^{0}\pi_1(x, y, z_2)\mathrm{d}z_2\mathrm{d}z_1\mathrm{d}z \tag{64}$$

The same procedure, when applied to Eq. 55, results in

$$v_{\mathrm{L}} = \frac{z^2 - h^2}{2\beta}\frac{\partial \zeta'_{\mathrm{E}}}{\partial y} + \frac{\int_{-h}^{z}\int_{z_1}^{0}\pi_2(x, y, z_2)\mathrm{d}z_2\mathrm{d}z_1}{\beta} \tag{65}$$

Then, the latitudinal volumic transport V_{L} is

$$V_{\mathrm{L}} = \int_{-h}^{0} v_{\mathrm{L}}\mathrm{d}z = \frac{\Pi_2}{\beta} - \frac{h^3}{3\beta}\frac{\partial \zeta'_{\mathrm{E}}}{\partial y} \tag{66}$$

where

$$\Pi_2 = \int_{-h}^{0}\int_{-h}^{z}\int_{z_1}^{0}\pi_2(x, y, z_2)\mathrm{d}z_2\mathrm{d}z_1\mathrm{d}z \tag{67}$$

When Eq. 53 is integrated from $-h$ to 0 in the vertical direction, taking into account the boundary condition Eqs. 56–59, it becomes

$$\frac{\partial U_{\mathrm{L}}}{\partial x} + \frac{\partial V_{\mathrm{L}}}{\partial y} = 0 \tag{68}$$

The substitution of Eqs. 63 and 66 into Eq. 68 makes,

$$\frac{\partial}{\partial y}\left(h^3\frac{\partial \zeta'_{\mathrm{E}}}{\partial y}\right) + h^3\frac{\partial^2 \zeta_{\mathrm{E}}}{\partial x^2} - 3\frac{\partial \Pi_1}{\partial x} - 3\frac{\partial \Pi_2}{\partial y} = 0 \tag{69}$$

If Eq. 69 is integrated across the width of the bay, in view of the lateral boundary condition Eq. 61 with the substitution of Eq. 66 into it gives

$$\frac{\partial^2 \zeta_{\mathrm{E}}}{\partial x^2} = \frac{3\frac{\partial}{\partial x}\int_0^1 \Pi_1\mathrm{d}y}{\int_0^1 h^3\mathrm{d}y} \tag{70}$$

The integration of Eq. 70 from L to x with the boundary condition Eq. 60 with the substitution of Eq. 63 into it gives

$$\frac{\partial \zeta_{\mathrm{E}}}{\partial x} = \frac{3\int_0^1 \Pi_1\mathrm{d}y}{\int_0^1 h^3\mathrm{d}y} \tag{71}$$

When Eq. 70 is substituted into Eq. 69 and integrated from 0 to y with the lateral boundary condition Eq. 61 taken into consideration,

$$\frac{\partial \zeta_E'}{\partial y} = \frac{3\left(\Pi_2 - \dfrac{\frac{\partial}{\partial x}\int_0^1 \Pi_1 \mathrm{d}y \int_0^y h^3 \mathrm{d}y}{\int_0^1 h^3 \mathrm{d}y} + \frac{\partial}{\partial x}\int_0^y \Pi_1 \mathrm{d}y \right)}{h^3} \tag{72}$$

Then, according to Eqs. 63 and 66, the volumic transport in the x and y directions are

$$U_L = \frac{\Pi_1}{\beta} - \frac{h^3 \int_0^1 \Pi_1 \mathrm{d}y}{\beta \int_0^1 h^3 \mathrm{d}y} \tag{73}$$

$$V_L = \frac{\frac{\partial}{\partial x}\int_0^1 \Pi_1 \mathrm{d}y \int_0^y h^3 \mathrm{d}y}{\beta \int_0^1 h^3 \mathrm{d}y} - \frac{\frac{\partial}{\partial x}\int_0^y \Pi_1 \mathrm{d}y}{\beta} \tag{74}$$

The substitution of Eqs. 71 and 72 into Eqs. 62 and 65 gives

$$u_L = \frac{3(z^2 - h^2)\int_0^1 \Pi_1 \mathrm{d}y}{2\beta \int_0^1 h^3 \mathrm{d}y} + \frac{\int_{-h}^{z}\int_{z_1}^{0} \pi_1(x,y,z_2)\mathrm{d}z_2 \mathrm{d}z_1}{\beta} \tag{75}$$

$$v_L = \frac{3(z^2 - h^2)\left(\Pi_2 - \frac{\partial}{\partial x}\int_0^1 \Pi_1 \mathrm{d}y \dfrac{\int_0^y h^3 \mathrm{d}y}{\int_0^1 h^3 \mathrm{d}y} + \frac{\partial}{\partial x}\int_0^y \Pi_1 \mathrm{d}y \right)}{2\beta h^3} + \frac{\int_{-h}^{z}\int_{z_1}^{0} \pi_2(x,y,z_2)\mathrm{d}z_2 \mathrm{d}z_1}{\beta} \tag{76}$$

Then the vertical velocity w_L can be deduced by inserting Eqs. 75 and 76 into the continuity equation Eq. 53, with the result integrated from $-h$ to z based on the boundary condition Eq. 59 to obtain

$$\begin{aligned} w_L = & -\frac{z(z^2 - 3h^2)}{2\beta h^3} \times \int_{-h}^{0}\int_{-h}^{z}\int_{z_1}^{0}\left[\frac{\partial \pi_1(x,y,z_2)}{\partial x} + \frac{\partial \pi_2(x,y,z_2)}{\partial y} \right] \mathrm{d}z_2 \mathrm{d}z_1 \mathrm{d}z \\ & + \frac{1}{\beta}\int_{z}^{0}\int_{-h}^{z}\int_{z_1}^{0}\left[\frac{\partial \pi_1(x,y,z_2)}{\partial x} + \frac{\partial \pi_2(x,y,z_2)}{\partial y} \right] \mathrm{d}z_2 \mathrm{d}z_1 \mathrm{d}z + \frac{3z(z^2 - h^2)}{2\beta h^4}\frac{\partial h}{\partial y} \\ & \left(\Pi_2 - \frac{\frac{\partial}{\partial x}\left(\int_0^1 \Pi_1 \mathrm{d}y \right)\int_0^y h^3 \mathrm{d}y}{\int_0^1 h^3 \mathrm{d}y} + \frac{\partial}{\partial x}\int_0^y \Pi_1 \mathrm{d}y \right) - \frac{z(z^2 - h^2)}{2\beta h^2}\frac{\partial h}{\partial y}\int_{-h}^{0} \pi_2(x,y,z)\,\mathrm{d}z \end{aligned} \tag{77}$$

Equations 73–77 constitute the 3D LRV solution in a narrow bay, and they are expressed by the zeroth-order solution of the linear tide which is listed in Appendix B.

4 Results and Discussions

In order to demonstrate the features of the residual current, two types of topography are used, the exponential type and the parabolic type. The non-dimensional depth profile is as follows exponential:

$$h = \left[5 + 10\mathrm{e}^{-(40y-20)^2/49}\right] / 8.1 \tag{78a}$$

parabolic:

$$h = \left[10 - 9(2y-1)^2\right] / 7 \tag{78b}$$

In the exponential profile, there is a deep channel in the middle of the bay and a shoal along either bank of the bay. The water depth in the channel decreases sharply before it reaches the shoals. This kind of depth profile can represent the typical topography in an estuary.

For the parabolic topography, no shoals exist along the banks of the bay and the water depth decreases smoothly from the center of the bay towards the banks. This is used as a comparison to the exponential topography.

$\beta=1$ is selected to reflect the fact that the vertical turbulence effect is always present in the zeroth-order basic balance. It is found that the bay length is an important factor in determining the pattern of the residual current (e.g., Li and O'Donnell, 2005; Winant, 2008; Jiang and Feng, 2011). In the present study, the same conclusion can also be drawn. So the non-dimensional bay length L of 0.3, 1.0, and 1.5 are chosen to reflect the effect of the bay length in the model.

4.1 The 3D structure of the results

4.1.1 The results of the cross section

In a bay with the exponential topography, the solutions of $\boldsymbol{u}_{\mathrm{L}}$ for different bay length L are displayed in Fig. 1, which exhibits the $\boldsymbol{u}_{\mathrm{L}}$ pattern at different cross sections. For all the cross sections, the magnitude of the first-order LRV is approximately the same, that is of $O(1)$. This reflects the correctness of the nondimensionalization procedure.

It can be seen in all cross sections that the longitudinal current flows towards the head of the bay in the upper part of the deep channel. The area with a positive longitudinal residual velocity at different sections becomes larger and larger from the open boundary of the bay to the head of the bay and in the mean time the speed of the current gets weaker and weaker.

In the shallow part of the bay, that is along both banks of the bay, the first-order LRV is negative when the cross section is less than one wavelength away from the head of the bay.

This indicates that the water flows out of the bay through these areas. However, if the cross section is more than 1.1 wavelengths away from the head of the bay, two areas with a positive longitudinal velocity will evolve from both banks of the bay.

The solutions of $\boldsymbol{u}_{\mathrm{L}}$ with parabolic cross sections defined in Eq. 78b are displayed in Fig. 2. This cross section is similar to that used in Winant (2008) defined in Eq. 79.

$$h = 0.1 + 0.9(1 - y^2) \tag{79}$$

Since the nondimensionalization procedure in the present study is different from that in Winant (2008), the pattern of the first-order LRV in Fig. 2 is only similar to Fig. 7 in Winant (2008). However, it can be easily proved that if $y \in [-1,1]$, $\beta = 0.5$, and $L = 1.5$, with the maximum water depth taken as the characteristic depth and the characteristic value of the longitudinal velocity taken as $L\zeta_{\mathrm{c}}\sqrt{g / h_{\mathrm{c}}}$, then the present solution is the same as that in Winant (2008) neglecting the rotation of the earth (figures are omitted). These two mutually

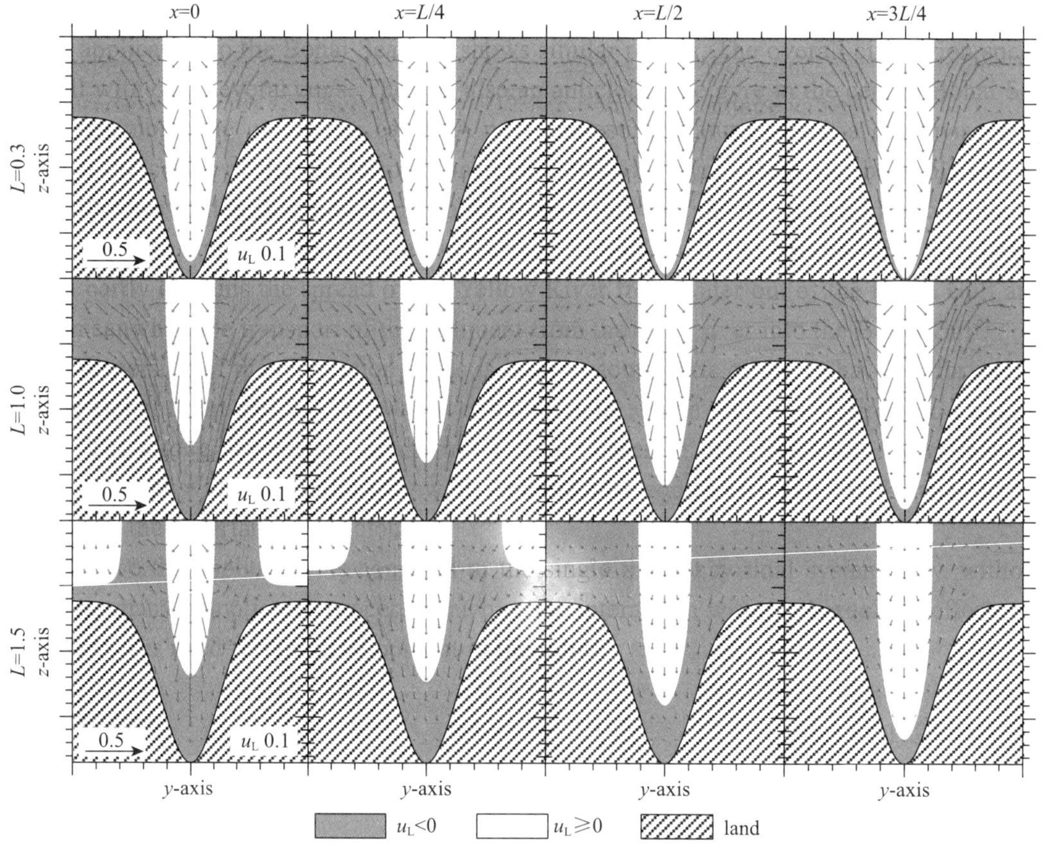

Fig. 1 The Lagrangian residual current at four cross sections for three different bay lengths with the depth profile being exponential. The contour lines denote the magnitude of the axial velocity $\boldsymbol{u}_{\mathrm{L}}$ and the contour interval is 0.1. The axial velocity is negative in the shaded area, which is towards the mouth of the bay. The arrows represents the velocity in y-z plane with the scale at the lower-left corner

demonstrate the validity of the results without considering the effects of the rotation.

The pattern of the first-order LRV in bays with the parabolic bottom topography displayed in Fig. 2 is generally similar to that in Fig. 1. The water flows in through the upper layer of the central deeper part and flows out at the areas either adjacent to the sea bottom or the bank of the bay. One striking difference occurs near the open boundary when the bay is 1.5 wavelengths long. Almost no inflow area can be found in Fig. 2 near the open boundary as in Fig. 1. This reflects the importance of the shoals along the banks defined in Eq. 78a.

4.1.2 The discussion of w_L

The acquisition of w_L is important in the study of the sediment transport and ecosystem modeling. Because normally w_L is too small to get both in the field research and in numerical modeling, the analytical solution becomes an important method in discussing the upwelling and the downwelling process especially for the residual current study.

It is noticed that the water flows in through the upper half of the deep channel in Fig. 1. In order to keep the mass conservation, the water goes out either through the horizontal circulation or through the vertical circulation. The horizontal circulation was discussed in the above subsection. The vertical circulation seems to have a rather stable feature. The water

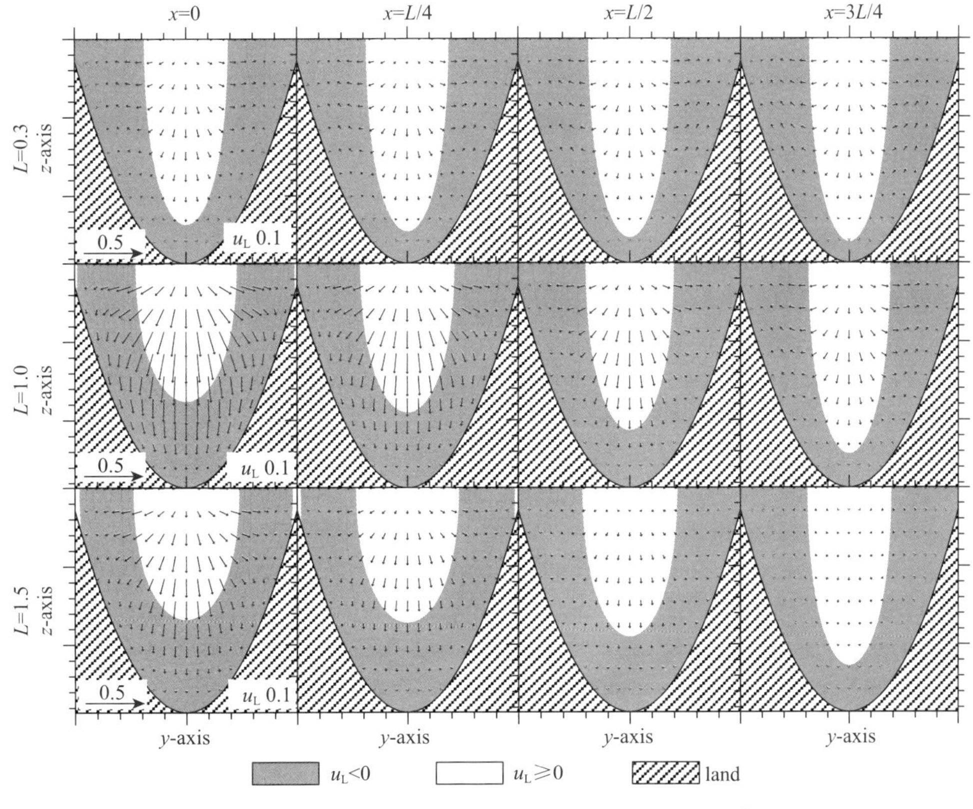

Fig. 2 As in Fig. 1, but with the parabolic depth profile

sinks at the center of the deep channel from the surface down to the bottom as displayed by the arrows in Fig. 1 for different bay lengths and at different depths. On either side of the downwelling area, there exists an upwelling area, which expands from the bottom to the surface and extends to a little more than 0.5 wavelengths from the head of the bay in the longitudinal direction in Fig. 1. In the latitudinal direction, the expansion stops at the distance of 0.1 to 0.4 bay widths away from each bank. Another two upwelling areas also present themselves at the upper half of the water column and are more than 1.2 wavelengths away from the head of the bay in Fig. 1.

If the results of the parabolic depth profile defined in Eq. 78b are compared with those of the exponential bottom profile, it can be found that the depth profile plays an important role in determining the upwelling and downwelling areas. For the parabolic one, the upwelling areas are expanded horizontally to the banks of the bay and vertically to the surface of the water, but are confined within 0.5 wavelengths away from the head of the bay. Another difference between the two depth profiles is that there is no upwelling area near the open boundary in the parabolic bottom case. This shows the importance of the bottom topography.

4.2 The results of the water transport

In a bay or an estuary, the horizontal water transport, i.e., the exchange flow, is important in understanding the environmental problems associated with the water exchange and the mass transport, such as salt, nutrients, or pollutants. Since the 3D Lagrangian residual current is obtained in Section 3, the residual horizontal water transport can be obtained by integrating it across the depth as defined in Eqs. 73 and 74.

4.2.1 The feature of the water transport

The water transport is displayed in Fig. 3 for the exponential cross section. In Fig. 3, it can be obviously seen that the water flows in through the deep channel of the bay until it strikes the head, where the water is split into two branches which flow out along the shoals. However, if the bay is longer than one wavelength, the outflowing water will detach from the bank of the bay and flows towards the deep channel, forming a front with the inward flow in the deep channel. At the outer part of the bay, the water flows in along the banks of the bay and turns towards the deep channel before joining the outflow from the inner bay at around one wavelength away from the head.

The pattern of the water transport in a bay whose length is shorter than one wavelength is similar to the result in Fig. 8a of Jiang and Feng (2011) when the 2D depth-averaged Lagrangian residual current equations were solved with the linear bottom friction term. However, when the bay is 1.5 wavelengths long, the water transport shows a more complex feature in the outer part of the bay than that in Fig. 8a of Jiang and Feng (2011) but is similar to that in Fig. 3a of Jiang and Feng (2011). If the parabolic depth profile is taken as in Eq. 78b, the transport pattern is simpler than that in Fig. 3 (figures are omitted). For all the bay lengths, the water goes in from the deep channel to the head of the bay and goes back along the banks towards the entrance of the bay. This again demonstrates the importance of the bottom topography.

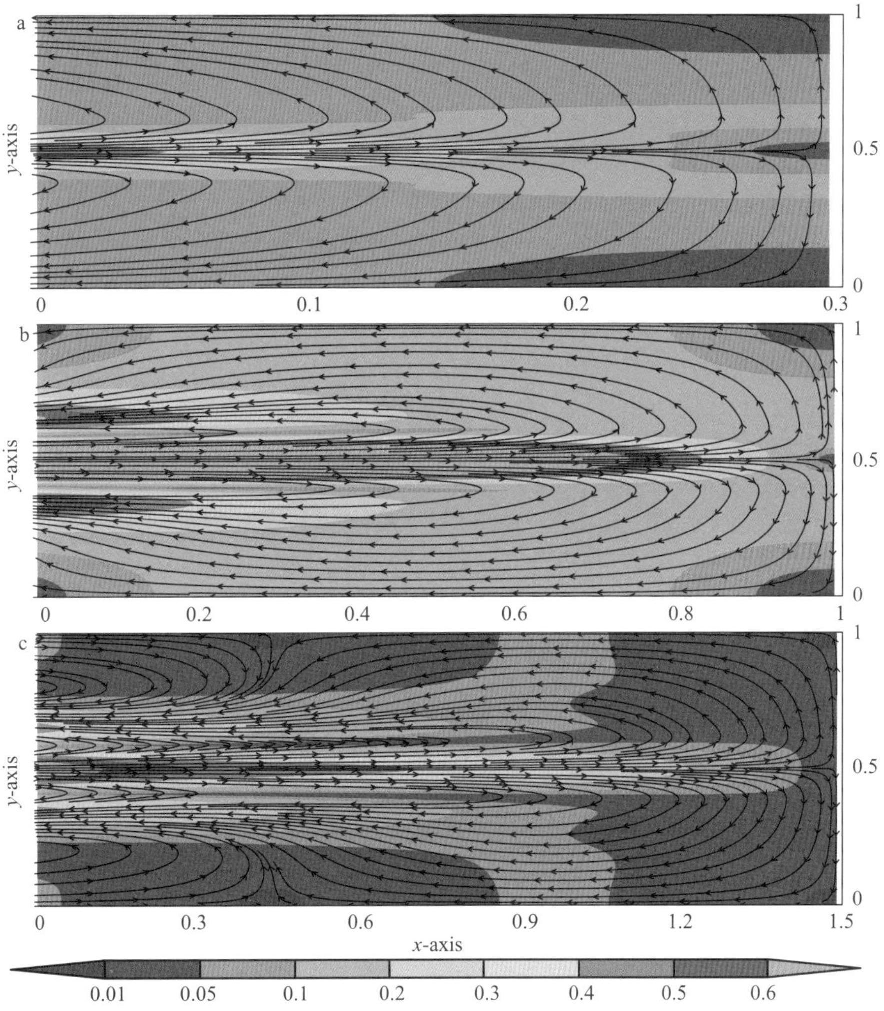

Fig. 3 The streamline of the depth-integrated Lagrangian residual current in bays with different lengths (a. $L = 0.3$; b. $L = 1$; c. $L = 1.5$). The depth profile is exponential and the colorbar represents the magnitude.

This transport pattern, however, is contrary to Fig. 4 in Winant (2008), in which the transport result was obtained when $f = 0$ is assumed, though the solution of the 3D LRV is the same as that described in the previous text in this section. Indeed, the transport obtained according to (35) and (36) in Winant (2008) is the Eulerian residual transport defined by Robinson (1983). It was pointed out in Feng *et al.* (1986b) theoretically that the Eulerian residual transport is different from the Lagrangian residual transport, the depth-integrated LRV. Therefore, it cannot represent the net transport behavior of the water and the corresponding material in it after removing the tidal effect. Li and O'Donnell (2005) calculated the Eulerian residual transport in a 2D narrow bay, which explains why the similar results are obtained as declared in Winant (2008). Jiang and Feng (2011) solved the depth-averaged LRV equations analytically and after comparing their results with those in Li and O'Donnell (2005), concluded that the Eulerian residual transport is different from the depth-integrated LRV. In this study, the same conclusion can also be reached from a 3D analytical solution. Indeed, this conclusion can also be made if Figs. 4 and 7 in Winant (2008) are examined against each other. According to the distribution of the contour lines in the lower-left figure in Fig. 7 of Winant (2008), the depth-integrated LRV, i.e., the Lagrangian residual transport, should be positive, which is contrary to that in Fig. 4 of Winant (2008). Therefore, the Eulerian residual transport is different from the Lagrangian residual transport and cannot represent the net transport effect after getting rid of the tide.

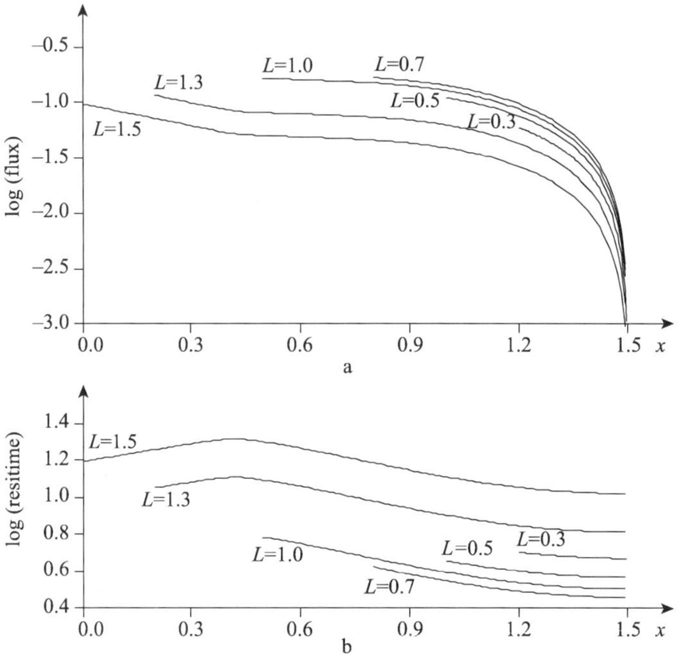

Fig. 4 The water flux (a) and residence time (b) along the section of the bay with the depth profile being exponential. All the lines are aligned with each other at $x = 1.5$ where the head of the bay is located. The tag denotes the length of the bay and is placed near the open boundary of the bay.

4.2.2 The effect of the bottom friction

If Eq. 54 is integrated from $-h$ to 0, and the boundary condition at sea surface Eq. 57 is taken into consideration, then

$$0 = -h\frac{\partial \zeta_E}{\partial x} - \beta\nu\frac{\partial u_L}{\partial z}\bigg|_{z=-h} + \int_{-h}^{0}\pi_1 dz \tag{80}$$

When Eq. 63 is inserted into Eq. 80,

$$\beta\nu\frac{\partial u_{\mathrm{L}}}{\partial z}\bigg|_{z=-h}=-\frac{3\beta\nu}{h^2}U_{\mathrm{L}}-\left(\frac{3\Pi_1}{h^2}+\int_{-h}^{0}\pi_1\mathrm{d}z\right) \tag{81}$$

This shows that the bottom friction for $\boldsymbol{u}_{\mathrm{L}}$ is in the linear form. Actually, if the zeroth-order tide is examined, the bottom friction is also in the linear form. From Eq. 86 in Appendix B, it can be seen that

$$\frac{\partial U_0}{\partial z}\bigg|_{z=-h}=\frac{\mathrm{i}+1}{\sqrt{2\beta\nu}}\tan\left(\frac{(\mathrm{i}+1)h}{\sqrt{2\beta\nu}}\right)\times\left[\frac{\sqrt{2\beta\nu}}{\mathrm{i}+1}\tan\left(\frac{(\mathrm{i}+1)h}{\sqrt{2\beta\nu}}\right)-h\right]^{-1}\int_{-h}^{0}U_0\mathrm{d}z \tag{82}$$

This indicates that the vertical shear of the horizontal velocity at the sea bottom is proportional to the depth-integrated velocity. This result may be due to the fact that the eddy coefficient ν is regarded as a constant and this simplification filters out some complexity of the real nature of the Lagrangian residual current.

4.2.3 The water exchange in the bay

Water exchange is one of the important processes governing the environmental problems in a coastal sea, and the tidal effect is always an important factor in determining the water exchange. Since $\boldsymbol{u}_{\mathrm{L}}$ is obtained in the present elongated narrow bay, the tide-induced water exchange can be obtained naturally. The water exchange flux can be calculated at each latitudinal cross section by integrating the residual current of either positive or negative values. The water flux calculated on the basis of the positive residual currents is equal in the magnitude but is opposite in signs to the water flux calculated on the basis of their negative counterparts, which again reflects the fact that the mass conservation of the Lagrangian residual current is kept. The residence time of the water is also calculated at each section by dividing the volume of the water between the section and the head of the bay by the water exchange flux.

The water exchange flux and the residence time for different bay lengths ranging from 0.3 to 1.5 wavelengths with 0.1 wavelengths apart are calculated and six of them are plotted in Fig. 4. The heads of the bays are aligned with each other for bays with different lengths. The horizontal coordinate of the leftmost point of each line in Fig. 4 can be used as the indicator of different bay lengths. It can be seen in the upper panel of Fig. 4 that the water exchange flux increases smoothly with the distance away from the head of the bay and the increasing trend becomes sharpen at around one wavelength away from the head of the bay. The water exchange flux is also related to the length of the bay. In the present case, the water exchange flux increases with the bay length if the latter is less than around 0.8 wavelengths. If the bay is longer than 0.9 wavelengths long, the water exchange flux will decrease with the bay length. This feature is different from that in a 2D model in Jiang and Feng (2011), which demonstrates the Lagrangian residual currents' 3D nature and its complex 3D feature.

The residence time shown in the lower panel of Fig. 4 exhibits a similar feature in accordance with the water exchange flux, but the change of the residence time along the bay displays a prominent point at around one wavelength away from the head of the bay. This point is determined by the corresponding point in the water exchange flux line. This means that two separate systems of $\boldsymbol{u}_L$ occur with the separating point at around one wavelength away from the head of the bay, which is also noted in the previous section of the present paper.

4.3 The results of the breadth-averaged currents

In a narrow bay or an estuary, the circulation in the longitudinal-vertical section is important to the transport of the materials. The breadth-averaged model is a classical simplification when studying the estuary circulation, e.g., the analytical study by Ianniello (1977). But this kind of simplification ignored the lateral variation of the depth profile. In the present study, 3D $\boldsymbol{u}_L$ is obtained directly with taking into consideration of the lateral variation of the topography. Then, the 3D $\boldsymbol{u}_L$ can be averaged across the bay to get the breadth-averaged residual current. The results are compared with those in Ianniello (1977), and the effects of the lateral variation of the sea bottom are discussed between flat and non-flat bottom.

4.3.1 The flat bottom

The breadth-averaged $\boldsymbol{u}_L$ in a flat bottom bay is displayed in Fig. 5 for different bay lengths. It can be found that the water flows in at the upper half of the bay and turns back at the lower half, which is just contrary to the gravitational flow in an estuary. The speed of the current is higher at the open mouth of the bay than inside the bay. The pattern remains the same with different bay lengths. The results are in accordance with those in Ianniello (1977).

4.3.2 The non-flat bottom

Normally, the bottom of a bay cannot be flat and the results show different features in such non-flat bottom conditions. The breadth-averaged $\boldsymbol{u}_L$ in a bay with the exponential bottom profile as that in Eq. 78a is shown in Fig. 6. The pattern shown in Fig. 6 is different from that in Fig. 5, the flat bottom case. It can be seen in Fig. 6b, c that an anti-clockwise gyre exists in the inner part of the bay. At the surface, the gyre extends to the point where it is 0.8 wavelengths away from the head of the bay. In the vertical direction, the gyre does not occupy the whole water column but stops at around 0.1 water depth above the bottom. In the outer part of the bay, there is a clockwise semi-gyre, which intrudes into the bay and squeezes the inner gyre before submerging beneath it. In Fig. 6a, because the bay length is less than 0.8 wave-lengths, only two semi-gyres exist. Therefore, in the area which is less than 0.8 wavelengths away from the head, the water flows out of the upper layer of the bay, which will enhance the gravity flow if fresh water is discharged from the head of the bay. Then, the water goes downward and flows inside at the near bottom layer. However, in the outer part of the bay, the water flows inside at the surface layer and flows outside at the bottom layer, similar to the pattern shown in Fig. 5. It should be noticed that there is always an outward flow at the bottom.

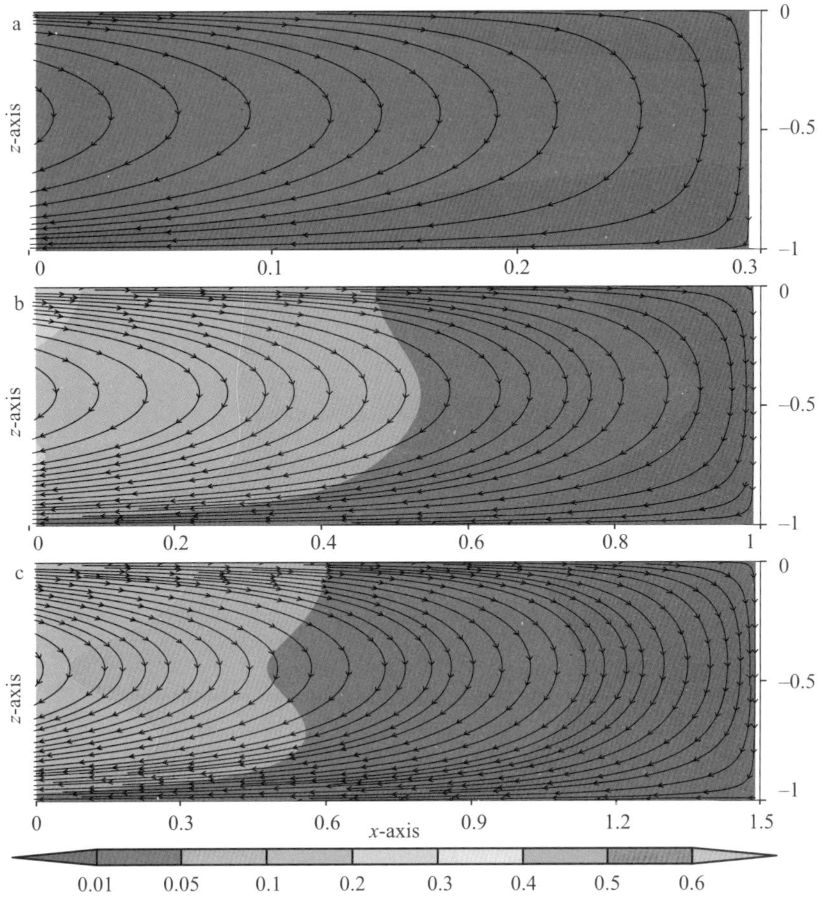

Fig. 5 The streamline of the breadth-averaged Lagrangian residual current in bays with different lengths (a. $L = 0.3$; b. $L = 1$; c. $L = 1.5$). The depth profile is flat and the colorbar represents the magnitude

The pattern of the breadth-averaged $\boldsymbol{u}_{\mathrm{L}}$ for the parabolic profile Eq. 78b is also studied (figures are omitted). Because the parabolic bottom is flatter than the exponential bottom, the result is more like that in Fig. 5. The clockwise semi-gyre occupies almost the whole section with a small anti-clockwise gyre huddled at the inner surface corner of the bay. Therefore, the results in Ianniello (1977) are valid only when the bottom profile is flat. When it is in a complex form, the breadth-averaged LRV becomes complex accordingly.

4.4 The contribution from each component

In Eqs. 54 and 55, $\boldsymbol{\pi} = (\pi_1, \pi_2)$ is the force to generate the residual current caused by the tidal movement, which was called the tidal body force in Feng (1987). Based on the different dynamic

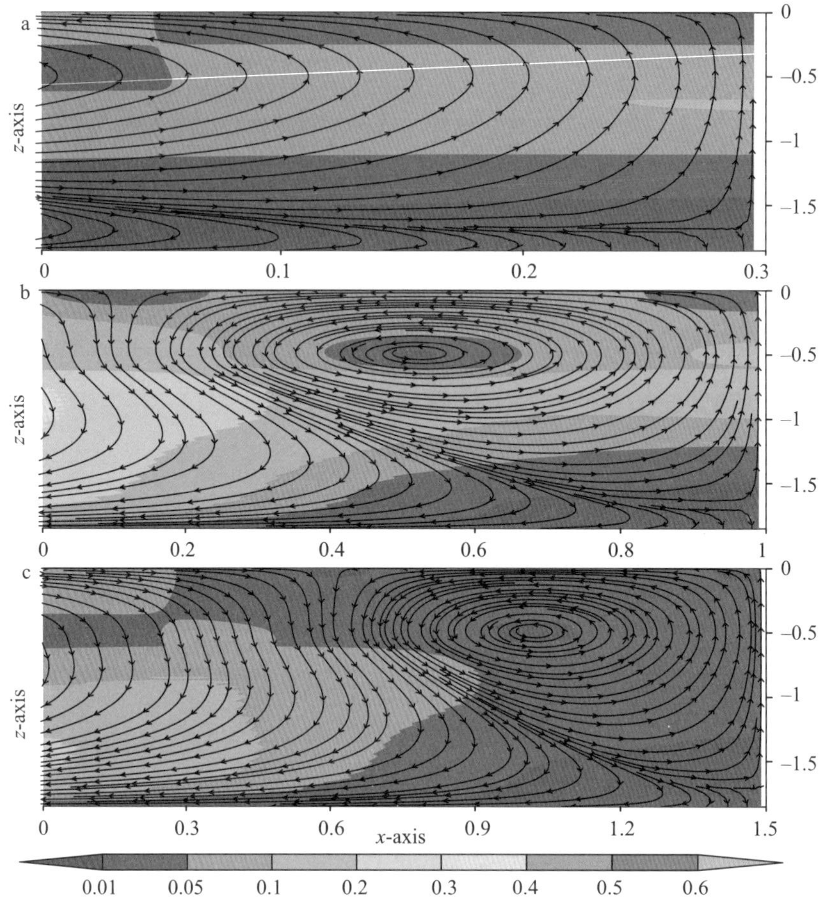

Fig. 6 As in Fig. 5, but with the exponential depth profile

features, the tidal body force can be decomposed into three components, which are $\boldsymbol{\pi}_{\mathrm{N}}=(\pi_{\mathrm{N}1},\pi_{\mathrm{N}2})$, the inviscid term, $\boldsymbol{\pi}_{\mathrm{b}}=(\pi_{\mathrm{b}1},\pi_{\mathrm{b}2})$, the bottom friction related term, and $\boldsymbol{\pi}_{v}=(\pi_{v1},\pi_{v2})$, the turbulent viscosity related term, respectively. The expression for them is in Appendix C and D, and the detailed decomposition procedures are described in the supplementary material. The set of the residual current equations (Eqs. 53–61) describes a linear system, so each component of $\boldsymbol{\pi}$ can act separately. The contribution of each component to the depth-integrated and breadth-averaged residual currents is discussed as follows.

4.4.1 The contribution to the depth-integrated currents

It can be seen that the depth-integrated $\boldsymbol{\pi}$ drives the depth-integrated residual currents directly if Eqs. 54 and 55 are integrated from the bottom to the surface. It can be seen in Fig. 7 that the depth-integrated residual currents driven by each of the three components have a rather stable pattern.

Among the three components, $\boldsymbol{\pi}_{\mathrm{b}}$ is the dominant one, and the pattern of the residual currents

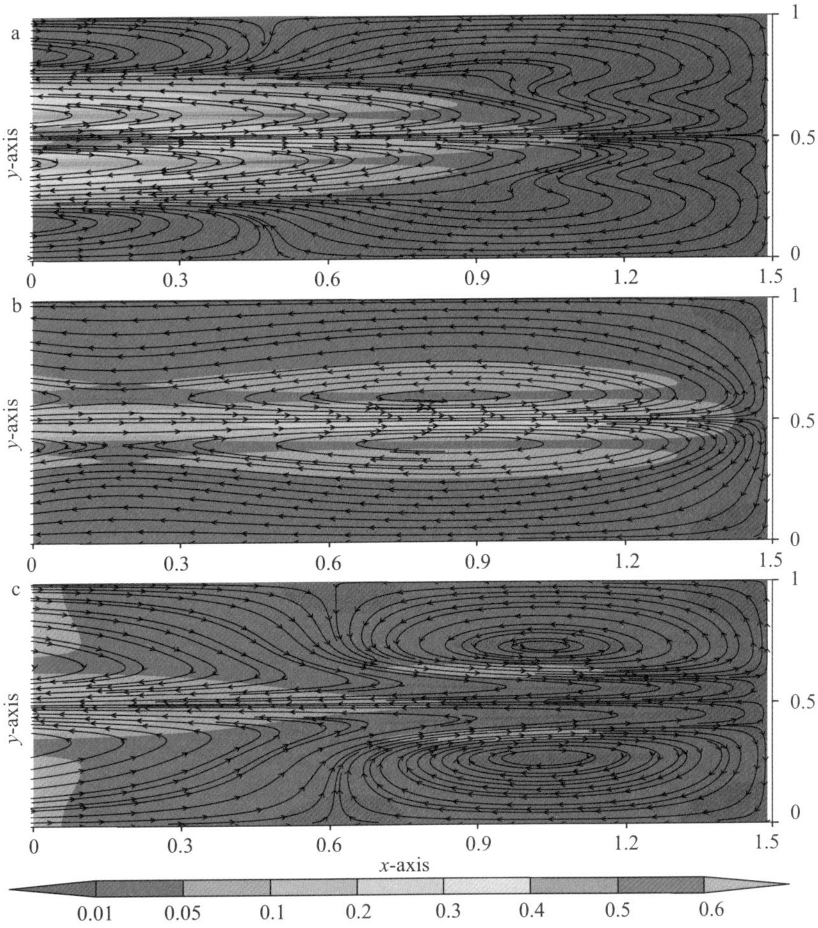

Fig. 7 The streamline of the depth-integrated Lagrangian residual current driven by each component of the tidal body force (a. π_b; b. π_N; c. π_v). The depth profile is exponential and the colorbar represents the magnitude

is mostly determined by it. According to the expression of π_b, its value at the bottom alone contributes to the depth-integrated residual currents. It is shown in Fig. 7 that the water goes in from the deep channel till the head of the bay. Then, the water is split into two branches before returning towards the head of the bay along the banks. After that, it encounters two reversing gyres, detaches from the banks, and flows out of the bay at the slope of the channel. In the case of the parabolic bottom profile, the reversing gyres stay at the corner of the outer part of the bay (the figure is omitted). This reflects its sensitivity to the bottom topography.

The general pattern of the residual currents driven by π_N is similar to that driven by π. The water goes in from the deep part and goes out from the shallow part. But the residual currents driven by π_N is smoother than those of the other two components.

The residual currents driven by π_v is generally reverse to the other two and the total one.

The water goes out from the deep channel till the head of the bay. Two gyres of 0.9 wavelengths long exist at the inner part of the bay. Two semi-gyres with opposite direction exist at the outer half of the bay. The water goes in along the banks of the bay and is compressed by the two inner gyres to formulate two intrusion gyres. The velocity is higher at the open boundary of the deep channel. But, in general, the value is small, and this may hint that the detailed structure of the eddy viscosity is not important to the residual transport circulation.

4.4.2 The contribution to the breadth-averaged currents

The three components of the tidal body force also play different roles in the breadth-averaged Lagrangian residual currents displayed in Fig. 8. The component driven by $\boldsymbol{\pi}_{\mathrm{b}}$ always goes in at the surface and goes out at the bottom, no matter what the bottom topography is like. The component driven by $\boldsymbol{\pi}_{\mathrm{v}}$ dominates the general pattern. The result in Fig. 8c looks similar to that in Fig. 6c, but there is a difference in the location of the contacting line between the two gyres, with the line in Fig. 6c stretching further into the bay.

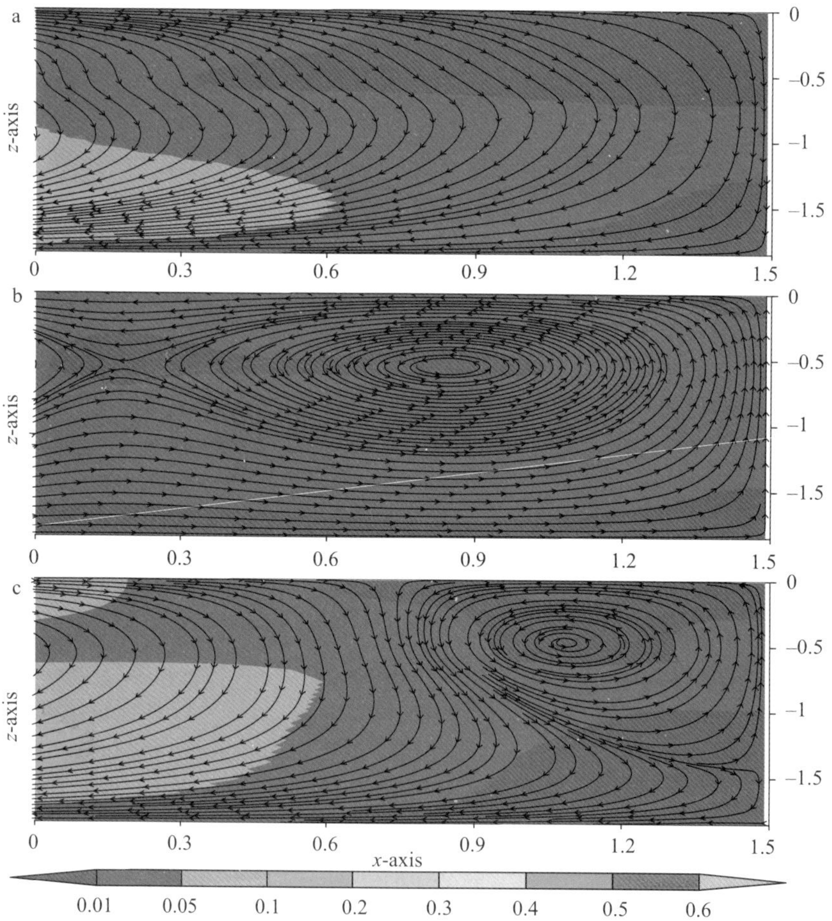

Fig. 8 As in Fig. 7, but for the breadth-averaged Lagrangian residual current

The component driven by $\boldsymbol{\pi}_{\mathrm{N}}$ is sensitive to the bottom topography, and the results are totally different between the flat and non-flat bottom profiles. When the bottom is not flat, the currents caused by $\boldsymbol{\pi}_{\mathrm{N}}$ will go out at the surface and go in at the bottom, with the direction of the residual current components opposite to the direction of those driven by $\boldsymbol{\pi}_{\mathrm{b}}$ and $\boldsymbol{\pi}_{\nu}$.

5 Conclusions

In this paper, the governing equations for the 3D mass transport velocity ($\boldsymbol{u}_{\mathrm{L}}$), the first-order approximation of the Lagrangian residual velocity, are deduced by the perturbation method. This set of equations defines an independent system on an intertidal time scale, with the tidal effect reflected in the tidal body force. The analytical solution to $\boldsymbol{u}_{\mathrm{L}}$ is given in a narrow bay in this paper, and the results for several bottom profiles are presented.

The results show clearly the 3D structure of $\boldsymbol{u}_{\mathrm{L}}$. The water flows in through the deep channel from the open boundary directly to the head of the bay. The inflow does not expand to the whole water column, but occupies the upper half of the water only, and the inflow layer becomes thicker when it approaches the head of the bay. When the inflow reaches the head of the bay, the water will be squeezed to the bottom and the two sides of the bay. This makes the water flow out of the bay through the layer just above the bottom. However, when the bay with shoals is longer than one wavelength, there are also inflows from the shoal area at the outer bay which is more than one wavelength away from the head.

If the results are integrated vertically, the intertidal water transport can be obtained. An obvious feature is that the water flows in at the deep bay until it strikes the head of the bay and flows out along the two banks within one wavelength away from the head of the bay. If the bay length is greater than one wavelength, the outward flow detaches from the banks from the point more than one wavelength away from the head and goes out at the slope with two branches of water flowing in along the banks. This pattern is generally similar to the results in Jiang and Feng (2011) in which the 2D depth-averaged Lagrangian residual current equations were solved. It can be found that in the present 3D case when the vertical eddy viscosity is assumed as a constant, the bottom friction is proportional to the depth-integrated velocity. This may explain the similarity. However, the disappearance of the two small gyres in Jiang and Feng (2011) may be due to the vertical movement in the 3D case.

The water exchange flux is then calculated in this paper. It can be found that the water exchange flux increases smoothly with the distance to the head of the bay increasing until about one wavelength away from the head, and the water exchange flux is also related to the bay length. In the present case, the water exchange flux increases with bay length until the bay length reaches around 0.8 wavelengths before the trend reverses. This feature is different from that in a 2D model in Jiang and Feng (2011), which further demonstrates the Lagrangian residual currents' 3D nature and its complex 3D feature. However, as far as the residence time is concerned, another feature can be detected. Generally, the residence time increases with the distance away from the head of the bay, but when the distance is more than one wavelength, the residence time decreases with the distance away from the head of the bay. The relationship between the residence time and the bay length is in accordance with the relationship between exchange flux and the bay length. When the bay length is around 0.8 wavelengths, the residence time is the smallest.

In the vertical direction, the downwelling area covers most of the bay. In the deep channel area, in particular, the water sinks down to the bottom. The upwelling area is centered around the deep channel with two sub-areas near the head of the bay and another two near the open boundary when the bay length is greater than one wavelength.

As to the breadth-averaged Lagrangian residual current, if a flat bottom profile is assumed, the feature is the same as that in Ianniello (1977), which is against the traditional density gradient flow. On the other hand, if the bottom profile is non-flat, one may find that the pattern is dependent on the bay length. In the inner bay which is within 0.8 wavelengths, the surface flow is outward as the density gradients flow, while in the outer bay it is again contrary to that of the density gradient flow.

The tidal body force can be divided into three terms, which are $\boldsymbol{\pi}_{\mathrm{N}}$, the inviscid term, $\boldsymbol{\pi}_{\mathrm{b}}$, the bottom friction related term, and $\boldsymbol{\pi}_{\nu}$, the turbulent viscosity-related term. It is found that the depth-integrated LRV is mainly determined by $\boldsymbol{\pi}_{\mathrm{b}}$, the bottom friction-related term and the breadth-averaged LRV are mainly determined by $\boldsymbol{\pi}_{\nu}$, the turbulent viscosity-related term.

Acknowledgements

This study was supported by project 40976003 from National Natural Science Foundation of China and National Basic Research Program of China (2010CB428904). Helpful comments and suggestions provided by two anonymous reviewers are greatly appreciated.

Appendix A: The Method Frequently Used in the Paper

If $f(x,y,z,t)$ and $g(x,y,z,t)$ are two periodic functions of t and satisfy the relations in Eq. 83,

$$f=\frac{\partial F}{\partial t},\ g=\frac{\partial G}{\partial t} \tag{83}$$

Then

$$\langle Fg\rangle=\left\langle\frac{\partial}{\partial t}(FG)-\frac{\partial F}{\partial t}G\right\rangle=-\langle fG\rangle \tag{84}$$

Appendix B: The Solution to the Zeroth-order Tidal Currents

If ν is a constant, the zeroth-order tidal current system can be solved analytically by using the method similar to that in Winant (2007). The solution can be assumed in the following forms.

$$u_0 = \mathrm{Re}\left[U_0 \mathrm{e}^{-\mathrm{i}t}\right],\ v_0 = \mathrm{Re}\left[V_0 \mathrm{e}^{-\mathrm{i}t}\right],\ w_0 = \mathrm{Re}\left[W_0 \mathrm{e}^{-\mathrm{i}t}\right],$$
$$\text{and } \zeta_0 = \mathrm{Re}\left[N_0 \mathrm{e}^{-\mathrm{i}t}\right],\ N_0 = \frac{\cos\left[\mu(L-x)\right]}{\cos(\mu L)} \tag{85}$$

where $\mu^{-2} = -\int_0^1 P_0 \mathrm{d}y$ is a constant, and $P_0 = \dfrac{\sqrt{2\beta\nu}\sin\dfrac{(\mathrm{i}+1)h}{\sqrt{2\beta\nu}}}{(\mathrm{i}+1)\cos\dfrac{(\mathrm{i}+1)h}{\sqrt{2\beta\nu}}} - h$.

$$U_0 = \frac{\mathrm{i}\mu\left[\cos\dfrac{(\mathrm{i}+1)z}{\sqrt{2\beta\nu}} - \cos\dfrac{(\mathrm{i}+1)h}{\sqrt{2\beta\nu}}\right]\sin\left[\mu(L-x)\right]}{\cos\left[\dfrac{(\mathrm{i}+1)h}{\sqrt{2\beta\nu}}\right]\cos(\mu L)} \tag{86}$$

$$V_0 = -\frac{\mathrm{i}\mu^2\left[\cos\dfrac{(\mathrm{i}+1)z}{\sqrt{2\beta\nu}} - \cos\dfrac{(\mathrm{i}+1)h}{\sqrt{2\beta\nu}}\right]G\cos\left[\mu(L-x)\right]}{\cos\left[\dfrac{(\mathrm{i}+1)h}{\sqrt{2\beta\nu}}\right]\cos(\mu L)P_0} \tag{87}$$

where $G = -\int_0^y P_0(y')\mathrm{d}y' - y\mu^{-2}$.

$$\begin{aligned} W_0 = \frac{\mathrm{i}\mu^2\cos\left[\mu(L-x)\right]}{\cos(\mu L)}&\left[\left(\frac{1}{\mu^2 P_0} + \frac{\partial h}{\partial y}\frac{G}{P_0^2}\tan^2\frac{(\mathrm{i}+1)h}{\sqrt{2\beta\nu}}\right)\right.\\ &\times\left(\frac{\sqrt{2\beta\nu}\left(\sin\dfrac{(\mathrm{i}+1)z}{\sqrt{2\beta\nu}} + \sin\dfrac{(\mathrm{i}+1)h}{\sqrt{2\beta\nu}}\right)}{-(1+\mathrm{i})\cos\dfrac{(\mathrm{i}+1)h}{\sqrt{2\beta\nu}}} + z + h\right)\\ &\left.+\frac{\partial h}{\partial y}\sin\frac{(\mathrm{i}+1)h}{\sqrt{2\beta\nu}}\frac{\left(\sin\dfrac{(\mathrm{i}+1)z}{\sqrt{2\beta\nu}} + \sin\dfrac{(\mathrm{i}+1)h}{\sqrt{2\beta\nu}}\right)G}{\cos^2\dfrac{(\mathrm{i}+1)h}{\sqrt{2\beta\nu}}P_0}\right] \end{aligned} \tag{88}$$

Appendix C: The Decomposition of the Tidal Body Force π_1

$$\pi_1 = \pi_{\mathrm{N}1} + \pi_{\mathrm{v}1} + \pi_{\mathrm{b}1} \tag{89}$$

where

$$\pi_{\mathrm{N1}} = -\frac{1}{2}\frac{\partial}{\partial x}\left\langle \xi_0 \frac{\partial \zeta_0}{\partial x}\right\rangle \tag{90}$$

$$\pi_{\mathrm{v1}} = -\frac{\beta}{2}\left\langle \xi_0 \frac{\partial^2}{\partial x \partial z}\left(\nu \frac{\partial u_0}{\partial z}\right)\right\rangle - \frac{\beta\nu}{2}\left\langle \frac{\partial u_0}{\partial z}\frac{\partial^2 \xi_0}{\partial x \partial z}\right\rangle - \beta\left\langle \frac{\partial \boldsymbol{\xi}_0}{\partial z}\cdot\nabla\left(\nu\frac{\partial u_0}{\partial z}\right)\right\rangle \tag{91}$$

$$\pi_{\mathrm{b1}} = \beta\frac{\partial}{\partial z}\left(\nu\left\langle \frac{\partial u_0}{\partial z}\left(\frac{5}{2}\frac{\partial \xi_0}{\partial x} + \frac{\partial \eta_0}{\partial y}\right) + \frac{\partial v_0}{\partial z}\frac{\partial \xi_0}{\partial y}\right\rangle\right) + \beta\frac{\partial}{\partial z}\left\langle \frac{\partial u_0}{\partial z}\boldsymbol{\xi}_0\cdot\nabla\nu\right\rangle \tag{92}$$

Appendix D: The Decomposition of the Tidal Body Force π_2

$$\pi_2 = \pi_{\mathrm{N2}} + \pi_{\nu 2} + \pi_{\mathrm{b2}} \tag{93}$$

where

$$\pi_{\mathrm{N2}} = -\frac{1}{2}\frac{\partial}{\partial y}\left\langle \eta_0 \frac{\partial \zeta_0'}{\partial y}\right\rangle - \left\langle \xi_0 \frac{\partial^2 \zeta_0'}{\partial x \partial y}\right\rangle \tag{94}$$

$$\pi_{\mathrm{v2}} = -\frac{\beta}{2}\left\langle \eta_0 \frac{\partial^2}{\partial y \partial z}\left(\nu \frac{\partial v_0}{\partial z}\right)\right\rangle - \frac{\beta\nu}{2}\left\langle \frac{\partial v_0}{\partial z}\frac{\partial \eta^2{}_0}{\partial y \partial z}\right\rangle - \beta\left\langle \frac{\partial \boldsymbol{\xi}_0}{\partial z}\cdot\nabla\left(\nu\frac{\partial v_0}{\partial z}\right)\right\rangle \tag{95}$$

$$\pi_{\mathrm{b2}} = \beta\frac{\partial}{\partial z}\left(\nu\left\langle \frac{\partial v_0}{\partial z}\left(\frac{\partial \xi_0}{\partial x} + \frac{5}{2}\frac{\partial \eta_0}{\partial y}\right) + \frac{\partial u_0}{\partial z}\frac{\partial \eta_0}{\partial x}\right\rangle\right) + \beta\frac{\partial}{\partial z}\left\langle \frac{\partial v_0}{\partial z}\boldsymbol{\xi}_0\cdot\nabla\nu\right\rangle \tag{96}$$

References

Abbott MR (1960) Boundary layer effects in estuaries. J Mar Res 18: 83–100

Burchard H, Schuttelaars HM (2012) Analysis of tidal straining as driver for estuarine circulation in well-mixed estuaries. J Phys Oceanogr 42: 261–271

Cerco CF (1995) Simulation of long-term trends in Chesapeake Bay eutrophication. J Environ Eng 121: 298–310

Cerco CF, Cole T (1993) Three-dimensional eutrophication model of Chesapeake Bay. J Environ Eng 119: 1106–1125

Delhez EJM (1996) On the residual advection of passive constituents. J Mar Syst 8: 147–169

Dortch MS, Chapman RS, Abt SR (1992) Application of three-dimensional Lagrangian residual transport. J Hydraul Eng 118: 831–848

Feng SZ (1987) A three-dimensional weakly nonlinear model of tide-induced Lagrangian residual current and mass-transport, with an application to the Bohai Sea. In: Nihoul JCJ, Jamart BM (eds) Three-dimensional Models of Marine and Estuarine Dynamics, Elsevier Oceanography Series, vol 45. Elsevier, pp 471–488

Feng SZ (1990) On the Lagrangian residual velocity and the mass transport in a multi-frequency oscillatory system. In: Cheng RT (ed) Residual Currents and Long-term Transport, Coastal and Estuarine Studies, vol 38. Springer, pp 34–48

Feng SZ (1998) On circulation in Bohai Sea, Yellow Sea and East China Sea. In: Hong GH, Zhang J, Park BK (eds) Health of the Yellow Sea. The Earth Love Publication Association, Seoul, pp 43–77

Feng SZ, Lu YY (1993) A turbulent closure model of coastal circulation. Chin Sci Bull 38: 1737–1741

Feng SZ, Wu DX (1995) An inter-tidal transport equation coupled with turbulent K-ε model in a tidal and quasi-steady current system. Chin Sci Bull 40: 136–139

Feng SZ, Xi PG, Zhang SZ (1984) The baroclinic residual circulation in shallow seas. Chin J Oceanol Limnol 2: 49–60

Feng SZ, Cheng RT, Xi PG (1986a) On tide-induced Lagrangian residual current and residual transport, 1. Lagrangian residual current. Water Resour Res 22: 1623–1634

Feng SZ, Cheng RT, Xi PG (1986b) On tide-induced Lagrangian residual current and residual transport, 2. residual transport with application in South San Francisco Bay. Water Resour Res 22: 1635–1646

Feng SZ, Ju L, Jiang WS (2008) A Lagrangian mean theory on coastal sea circulation with inter-tidal transports, I. fundamentals. Acta Oceanol Sin 27: 1–16

Fischer HB, List EJ, Koh R, Imberger J, Brooks NH (1979) Mixing in Inland and Coastal Waters. Academic Press, New York

Foreman MGG, Baptista AM, Walters RA (1992) Tidal model studies of particle trajectories around a shallow coastal bank. Atmos Ocean 30: 43–69

Hainbucher D, Wei H, Pohlmann T, Sundermann J, Feng SZ (2004) Variability of the Bohai Sea circulation based on model calculations. J Mar Syst 44: 153–174

Heaps N (1978) Linearized vertically-integrated equations for residual circulation in coastal seas. Dtsch Hydrogr Z 31: 147–169

Huijts KMH, Schuttelaars HM, de Swart HE, Friedrichs CT (2009) Analytical study of the transverse distribution of along-channel and transverse residual flows in tidal estuaries. Cont Shelf Res 29: 89–100

Ianniello JP (1977) Tidally induced residual currents in estuaries of constant breadth and depth. J Mar Res 35: 755–786

Jay DA, Smith JD (1990) Residual circulation in shallow estuaries 1. highly stratified, narrow estuaries. J Geophys Res 95(C1): 711–731

Jiang WS, Feng SZ (2011) Analytical solution for the tidally induced Lagrangian residual current in a narrow bay. Ocean Dyn 61: 543–558. doi: 10.1007/s10236-011-0381-z

Li CY, O'Donnell J (2005) The effect of channel length on the residual circulation in tidally dominated channels. J Phys Oceanogr 35: 1826–1840

Longuet-Higgins MS (1969) On the transport of mass by time-varying ocean currents. Deep-Sea Res 16: 431–447

Muller H, Blanke B, Dumas F, Lekien F, Mariette V (2009) Estimating the Lagrangian residual circulation in the Iroise sea. J Mar Syst 78: S17–S36. doi: 10.1016/j.jmarsys.2009.01.008

Nihoul ICJ, Ronday FC (1975) The influence of the tidal stress on the residual circulation. Tellus A 27: 484–489

Ridderinkhof H, Loder JW (1994) Lagrangian characterization of circulation over submarine banks with application to the outer Gulf of Maine. J Phys Oceanogr 24: 1184–1200

Robinson IS (1983) Tidally induced residual flows. In: Johns B (ed) Physical oceanography of coastal and shelf seas. Elsevier, Amsterdam, pp 321–356

Wang H, Su ZQ, Feng SZ, Sun WX (1993) A three dimensional numerical calculation of the wind driven thermohaline and tide-induced Lagrangian residual current in the Bohai Sea. Acta Oceanol Sin 12: 169–182

Wei H, Hainbucher D, Pohlmann T, Feng SZ, Sundermann J (2004) Tidal-induced Lagrangian and Eulerian mean circulation in the Bohai Sea. J Mar Syst 44: 141–151

Winant CD (2007) Three-dimensional tidal flow in an elongated, rotating basin. J Phys Oceanogr 37: 2345–2362

Winant CD (2008) Three-dimensional residual tidal circulation in an elongated, rotating basin. J Phys Oceanogr 38: 1278–1295

Zimmerman JTF (1979) On the Euler-Lagrange transformation and the Stokes' drift in the presence of oscillatory and residual currents. Deep-Sea Res 26A: 505–520

Numerical Study on Inter-tidal Transports in Coastal Seas*

Mao Xinyan, Jiang Wensheng, Zhang Ping, and Feng Shizuo

1 Introduction

In tide-dominant shallow seas, the water properties, such as the concentration of nutrients, pollutants and other tracers, have obvious oscillatory variation at semidiurnal or diurnal tidal frequencies. There is also a slowly varying part of the property, termed as inter-tidal (sub-tidal) concentration due to long term processes, such as residual motion, turbulent diffusion and relevant mass sources/sinks.

In conventional inter-tidal transport equations, inter-tidally averaged concentration is expressed as the Eulerian (tidally-) mean concentration (EMC), the employed convective transport velocity is the Eulerian mean velocity (EMV), and a term of "tidal dispersion" has to be included (Pritchard, 1954; Bowden, 1967; Officer, 1976). This "tidal dispersion" term arises from correlation between tidal fluctuations of current and concentration. However, the physics of "tidal dispersion" is not understood until now. Many studies hammered at parameterizing this term in the form of a diffusion coefficient ("tidal diffusivity") acting on the mean concentration gradient, and then quantified the "diffusivity" with field observations (e.g., Geyer *et al.*, 2000; Ralston *et al.*, 2008; Scully *et al.*, 2009; Scully and Geyer, 2012; Aristizabal and Chant, 2014).

It has become clear that the long term transport should depend on the Lagrangian mean rather than the Eulerian mean velocity since the 1960s. A Lagrangian tidally-mean theory under convectively weakly nonlinear assumptions has been established by means of perturbation expansion since the late 1980s (Feng *et al.*, 1986a, b; Feng, 1987, 1990; Feng and Lu, 1993; Feng and Wu, 1995; Feng, 1998). In this theory, EMC is considered proper to represent inter-tidal concentration as before, while the mass transport velocity (MTV, Longuet-Higgins, 1969) is deemed as the convective transport velocity, which is the first order approximation of the Lagrangian residual velocity (LRV). It is noticeable that the blurry "tidal dispersion" term no more appears in the transport equation. This weakly nonlinear Lagrangian tidally-mean theory has been successfully applied in a number of tide-dominant coastal and estuarine seas (e.g., Dortch *et al.*, 1992; Cerco, 1995; Wei *et al.*, 2004b; Jiang and Feng, 2011, 2014; Quan *et al.*, 2014), as well as laboratory investigations (Wang *et al.*, 2013).

* Mao X, Jiang W, Zhang P, Feng S. 2016. Numerical study on inter-tidal transports in coastal seas. Journal of Ocean University of China, 15(3): 379–388

Nevertheless, the weakly nonlinear assumption, i.e., the ratio of the tidal elevation range to the water depth being at least one order of magnitude smaller than $O(1)$, may not always be valid for the shallow seas, especially the regions with tortured geometry or complicated bathymetry. For that reason, Feng *et al.* (2008) reformulated the Lagrangian mean theory in a generally nonlinear tide-dominant system, where the tidal range could match the water depth. A new concept of inter-tidal concentration was proposed, associated with an updated inter-tidal transport equation.

Feng *et al.* (2008) and Ju *et al.* (2009) focused respectively on illustrating this improved theory and on the numerical application aiming at coastal circulation. Subsequent to their studies, the present study puts emphases on inter-tidal transport. The numerical experiments in this study are configured similar to that in Ju *et al.* (2009). For simplicity, only the M_2 tide-induced effect will be considered, with the diurnal tide, wind-induced and baroclinic effects being ignored for now and left for future work.

A short introduction of the general nonlinear theory is given in the next section. Section 3 provides a model description. Results for three idealized model seas are presented in Section 4, followed with the application in the Bohai Sea in Section 5. While we pay more attention to the inter-tidal concentration, the relevant residual current will also be presented. Section 6 presents a summary of conclusions.

2 Outline of the Theory

Here, we give a brief outline of the Lagrangian tidally-mean theory in a general nonlinear tide-dominated system. For details, the reader is referred to Feng *et al.* (2008).

2.1 Lagrangian residual velocity

Under the assumption that the net displacement of a water parcel which moves through '*n*' tidal cycles (nT) be at least one order of magnitude smaller than the characteristic length of the flow field, the Lagrangian mean velocity $\boldsymbol{u}_{\mathrm{L}}$ will have the following properties:

(1) $\boldsymbol{u}_{\mathrm{L}}(\boldsymbol{x}_0,\tau;t_0)$ is a continuous, smooth and differentiable function of spatial and temporal variables $(\boldsymbol{x}_0,\tau)$. $\boldsymbol{x}_0$ denotes the initial position of any water parcel; τ is the temporal variable of inter-tidal scale. Being a parameter rather than an independent variable, t_0 is the releasing moment of water parcel during one tidal cycle.

(2) The conservation law of the material surface is satisfied.

(3) The continuity equation for an incompressible fluid is satisfied.

Such a constricted vector, the Lagrangian mean velocity, is called the Lagrangian residual current (or velocity, LRV). Noticing that LRV varies with the tidal phase $(2\pi t_0/T)$ over one tidal period (T), and the coastal circulation is not a single one but a set of fields.

2.2 Lagrangian inter-tidal concentration

The inter-tidal transport processes of tracer concentration can be expressed by

$$\frac{\partial C_{\mathrm{L}}}{\partial \tau}+\boldsymbol{u}_{\mathrm{L}}\cdot\nabla C_{\mathrm{L}}=S_{\mathrm{TL}}+S_{\mathrm{L}} \tag{1}$$

where $C_{\mathrm{L}}(\boldsymbol{x}_0,\tau;t_0)$ is termed as "Lagrangian inter-tidal concentration" (LIC), and the transport velocity in the equation is LRV. S_{TL} and S_{L} are respectively the Lagrangian mean of turbulent diffusion and sources/sinks. For Lagrangian tracking calculation, LIC is the final concentration of water parcel over the tracking period, and the water parcel is denoted by its initial position.

The inter-tidal transport process also contains a set of temporal-spatial fields of C_{L}, each one of which corresponds to a specific value of tidal phase. When the convectively weakly nonlinear condition is satisfied, a set of infinite temporal-spatial fields of LRV/LIC can be approximately reduced to a single one (mass transport velocity, MTV/Eulerian mean concentration, EMC) as exhibited previously.

3 Model Description

A 3-D hydrodynamic model, HAMburg Shelf Ocean Model (HAMSOM), is employed in this study to simulate the intra-tidal processes, including variables of elevation, velocity, and passive tracer concentration. HAMSOM is a primitive equation, *z*-coordinate, ocean model developed at the University of Hamburg by Backhaus (1985). The vertical viscosity and diffusivity are parameterized by a modified model of Munk and Anderson (1948), with a quadratic law for bottom stress. HAMSOM has been widely applied and shown to be a robust model for studying continental shelf sea hydrodynamics (e.g., Pohlmann, 1996; Wei *et al.*, 2004a; Simionato *et al.*, 2006; Marinone *et al.*, 2009; O'Driscoll *et al.*, 2011, 2013; Santiago-Garcia *et al.*, 2014).

Trajectory tracking method in the SPM (Suspended Particulate Matter Model, Jiang *et al.*, 2004) is used here to track an arbitrary water parcel for complete tidal cycles. The net displacement is divided by the tracking time to obtain the LRV. At the same time, tracer concentration of the tracked parcel is redistributed from its ending position to its starting one, which accords with the definition of LIC. By changing the starting time t_0 from 0 to T, a set of inter-tidal concentrations can be derived. If one water parcel runs out of the open boundary during the cycles, its residual current will not be calculated, nor for the inter-tidal concentration. For the Eulerian mean results (EMV and EMC), we simply average the intra-tidal velocity and concentration time-series in each computing grid for one tidal cycle.

4 Results of Idealized Models

In Ju *et al.* (2009), the LRVs of three idealized models were studied according to the general nonlinear Lagrangian tidally-mean theory. Because inter-tidal transport process is closely related, even nonlinearly coupled, to the residual motion, here we study the inter-tidal passive tracer concentration in the same idealized settings.

4.1 Configuration of idealized models

Three idealized model seas are constructed with different topography or coastline, termed as flat-bottom, stairs and cape cases, respectively, similar to that given in Ju *et al.* (2009). They are all semi-enclosed bays, with 150×100 horizontal grids, 1' (approximately 1.8 km) resolution, and the east boundary being set as open. The maximum depth in each case is 10 m, and the water body is evenly divided into 5 layers. Fig. 1 shows the bathymetry of stairs and cape cases. In all the three cases, wind forcing and baroclinic effect are ignored; a single-frequency tidal wave M_2 is introduced at the open boundary, with amplitude of 1 m. The Orlanski radiation condition is used for velocity. Initial concentration of a passive tracer is prescribed to be 2.0 (unit) over the model domain. Along the open boundary, the concentration linearly decreases from 3.0 at north to 1.0 at south, and a restoring condition is adopted. At the lateral land boundary, no normal flux condition is applied to both velocity and concentration.

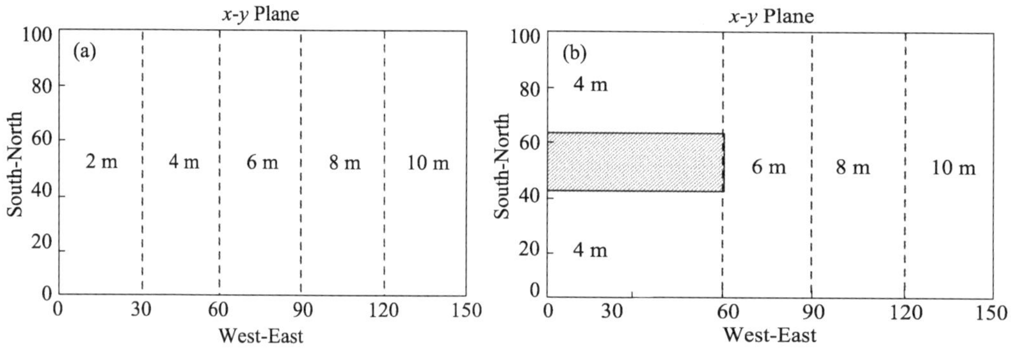

Fig. 1 Model sea design of stairs (a) and cape (b) cases: dashed lines denotes different water depth subarea

4.2 Discussion on LRV

The analyses on the LRV have been presented in Ju *et al.* (2009). Here we first discuss the residual current from some different points of views.

Firstly, considering that the trajectories of water parcels may tell the long-term trend of transport, the paths of 126 particles over 60 tidal cycles (almost one month) are shown in Fig. 2. The particles are initially distributed evenly in the surface layer of the flat-bottom case, and tracked from the same moment. Fig. 3 shows streamlines of surface LRV and EMV. It is clear that particles move closely along the streamline of LRV, suggesting that the LRV field accurately depicts the coastal circulation.

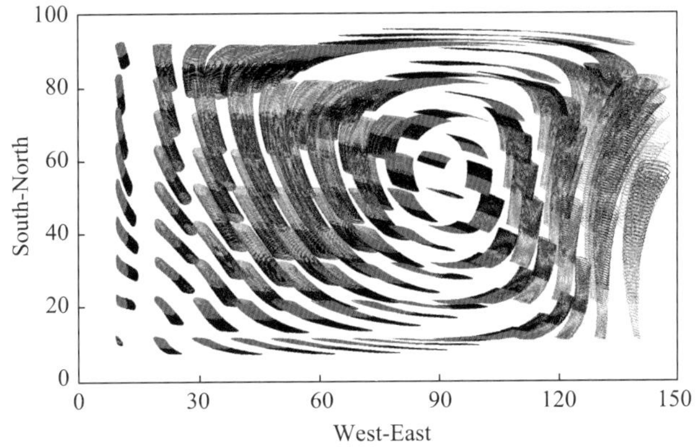

Fig. 2 Trajectories of 126 particles for 60 tidal cycles at surface in flat-bottom case (purple line is for first 30 cycles, and black one for later 30 cycles)

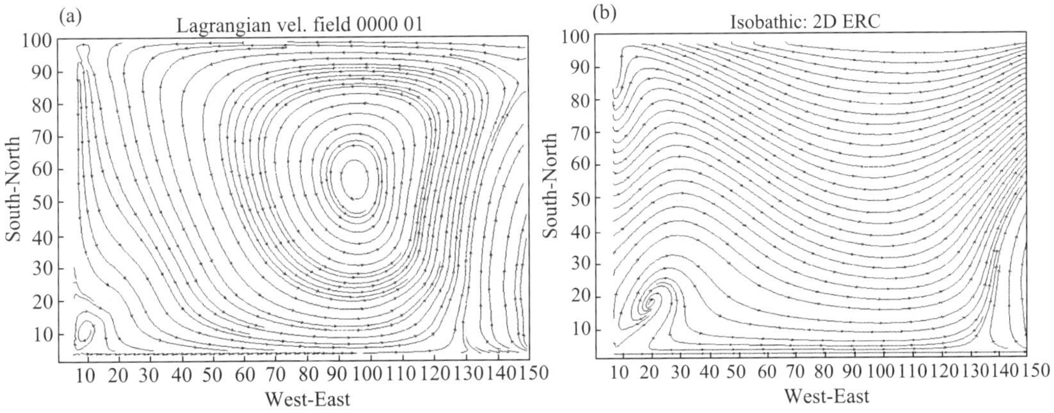

Fig. 3 Streamlines of vertically averaged LRV (a) and EMV (b) in flat-bottom case

Ju *et al.* (2009) proposed an index to estimate the dependence of LRV on initial tidal phase. Here we try to quantify the dependence in terms of both magnitude and direction. Fig. 4 shows the surface LRV vectors at four initial mments $(0,T/4,T/2,3,T/4)$ in the three idealized models. The four initial moments represent the times of maxi mum flood, slack water, maximum ebb and another slack water, respectively.

The variation of LRV magnitude with initial phase is obtained with the following expression:

$$U_{\text{coefficient}}=\max(|\boldsymbol{u}_\text{L}|)-\min(|\boldsymbol{u}_\text{L}|)/\text{mean}(|\boldsymbol{u}_\text{L}|) \tag{2}$$

where $\max(|\boldsymbol{u}_\text{L}|)$ denotes the maximum LRV among four initial tidal phases of a water parcel, $\min(|\boldsymbol{u}_\text{L}|)$is the minimum one, and $\text{mean}(|\boldsymbol{u}_\text{L}|)$ is the mean value of the four. It implies that the larger $U_{\text{coefficient}}$ is, the LRV deviates the more from the mean value, and the stronger is the convective nonlinearity. With regard to the variation of LRV direction, two vectors with the largest difference in directions are picked out to get α of each water parcel. Table 1 and Table 2 list the statistically piecewise results of $U_{\text{coefficient}}$ and α in the surface layer for flat-bottom, stairs, and cape cases.

In all three cases, over 85% of α is less than 45°. For α between 45° and 90° and between 90° and 180°, the percentages for the stairs and cape cases are about four times and twice larger than that for the flat-bottom case, respectively. It can be concluded from Fig. 4 and the two tables that the convective nonlinearity is much stronger in the stairs and cape cases than that in the flat-bottom case, mainly in regions with significant topographic variation.

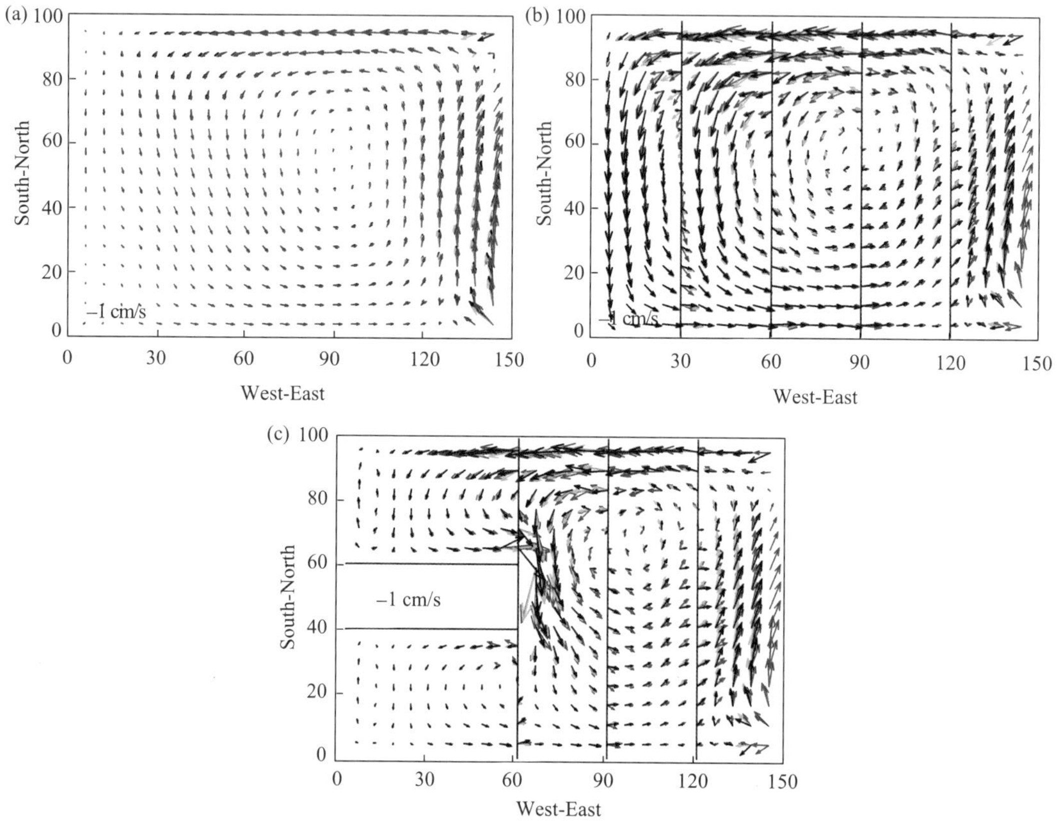

Fig. 4 Surface LRV of four initial tidal phases: (a) flat-bottom case; (b) stairs case; (c) cape case. Blue—$t_0 = 0$, red—$t_0 = T/4$, green—$t_0 = T/2$, black—$t_0 = 3T/4$, and lines in (b) and (c) indicate positions where bottom topography or coastline changes

Table 1 Piecewise statistics of $U_{\text{coefficient}}$ in three cases

Case	0.0–0.2	0.2–0.4	0.4–0.6	0.6–1.0	1.0–
Flat-bottom	59.2%	30.6%	5.7%	4.1%	0.4%
Stairs	39.8%	31.1%	12.9%	9.9%	6.3%
Cape	33.6%	35.4%	15.3%	10.2%	5.5%

Table 2 Piecewise statistics of α in three cases

Case	0°–45°	45°–90°	90°–35°	135°–180°
Flat-bottom	97.4%	1.9%	0.5%	0.2%
Stairs	87.8%	10.3%	1.2%	0.7%
Cape	86.3%	10.4%	1.9%	1.4%

4.3 Variation of Lagrangian inter-tidal concentration (LIC) with tidal phase

Similar to the analyses of LRV, 0, $T/4$, $T/2$ and $3T/4$ are still chosen as initial moments to study the dependence of the LIC on tidal phase. Fig. 5–Fig. 7 show the results of the flat-bottom, stairs and cape cases. Analyses mainly focus on the inner regions, far away from the influence of open boundary condition. This also avoids the situation that particles released close to the open boundary may leave the domain.

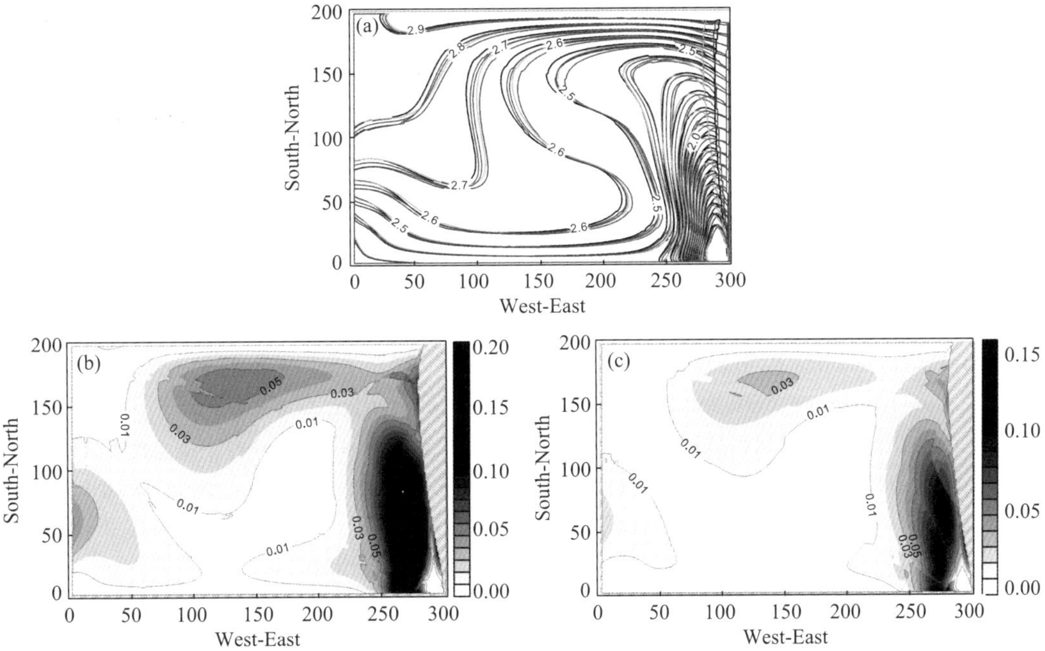

Fig. 5 Results of LIC and EMC in flat-bottom case: (a) Distribution of EMC (purple) and LIC with four tidal phases (blue—$t_0 = 0$, red—$t_0 = T/4$, green—$t_0 = T/2$, black—$t_0 = 3T/4$); (b) Distribution of maximum difference among four LICs; (c) Distribution of maximum difference between LIC and EMC. Hatched area indicates where particles left model domain during tracking periods.

In the flat-bottom case, the surface LIC varies from 2.3 to 2.9, and contours of the same concentration at different initial phases stay close in most areas, even overlap with each other, and with the Eulerian mean concentration (EMC, Fig. 5a). The LIC shows obvious dependence on tidal phase in the northern and southwestern parts of the model sea, where it varies from 2.5 to 2.9 and from 2.4 to 2.7, respectively.

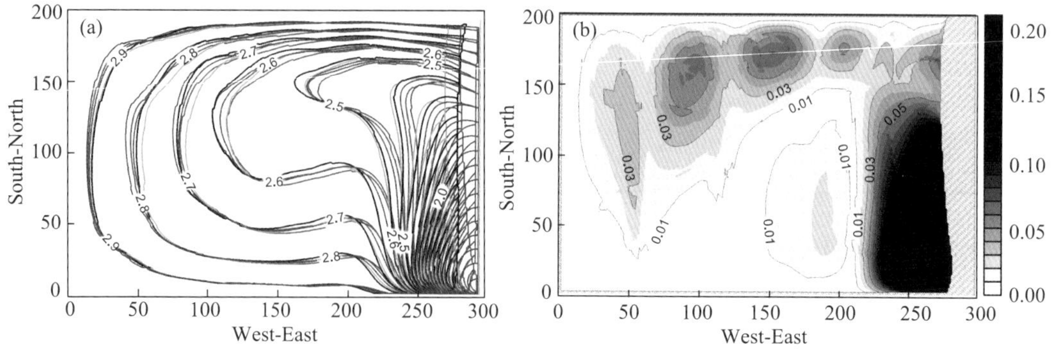

Fig. 6　Similar to Fig. 5(a, b) but for stairs case

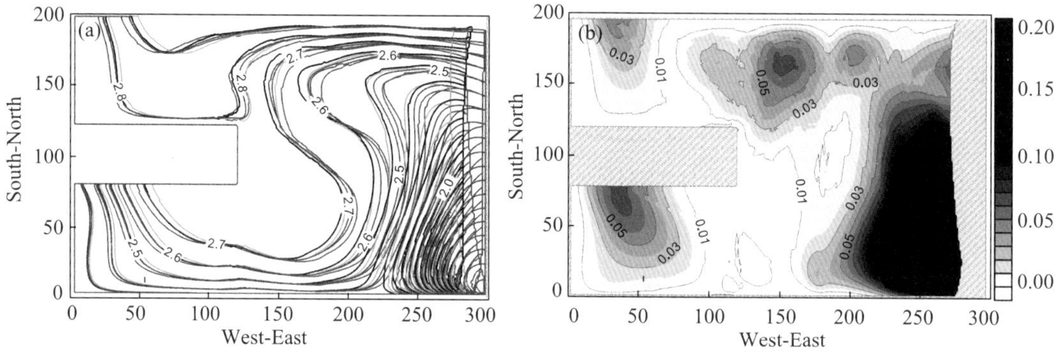

Fig. 7　Similar to Fig. 5(a, b) but for cape case

The difference of the LIC due to different tidal phases is quantified and shown in Fig. 5b. The difference reaches its maximum (~0.05) in the northern part of the model sea. By com parison, the difference between the EMC and LIC is smaller, about 0.03 (Fig. 5c). Both ofthese two differences are one order of magnitude smaller than either the whole-field or local variations of concentration. Therefore, the flat-bottom model sea can be regarded as a convectively weakly nonlinear system. The same conclusion is obtained from the above LRV analysis. In this case, the EMC can approximately substitute LIC.

In Fig. 6a and Fig. 7a, the water with high concentration from the northern part of the open boundary intrudes much more into the other water body than in the flat-bottom case. This is mainly due to the shallower mean depth in stairs and cape cases. In any rectangular bay of the three idealized models, either a large one like the whole domain of the flat-bottom case, or a smaller one like the northern (or southern) bay close to the headland in the cape case, the variation of LIC with tidal phase is much more significant in the northern part of the bay than in the other parts (Figs. 5b, 6b, and 7b).

Also in this northern part of the stairs and cape cases, the variation of LIC with tidal phase is no more an order of magnitude smaller than the local concentration variation (Fig. 6, Fig. 7). The same applies to the difference between the LIC and EMC in both cases (figures not shown here). This is attributed to the increasing convective nonlinearity, because the area with large LIC difference fits to that with distinct LRV dependence on the tidal phase (as in Fig. 4). Therefore, in the stairs and cape cases, the EMC cannot replace the LIC in describing the inter-tidal transport.

5 Application to the Bohai Sea

The Bohai Sea (BS) is a semi-enclosed shallow sea with an average depth of 18 m. It consists of five parts: Liaodong Bay, Bohai Bay, Laizhou Bay, Bohai Strait and the central area. The Bohai Strait is the only passage from the inner Bohai Sea to the Yellow Sea (Fig. 8).

Over the last several decades, studies of coastal circulation and transport in the BS have been carried out based on field investigation (e.g. Kuang *et al.*, 1991; Jiang *et al.*, 2002) and numerical simulation (e.g., Wang *et al.*, 1993; Hainbucher *et al.*, 2004). In Ju *et al.* (2009), the general nonlinear Lagrangian mean theory has been first applied to the BS for a numeri cal study of coastal circulation. The present work will take the Lagrangian inter-tidal salinity (LIS) in the BS as an example to study the inter-tidal transport. It should be pointed out that salinity is treated as a passive tracer by neglecting its contribution to the water motions, through changing the water density.

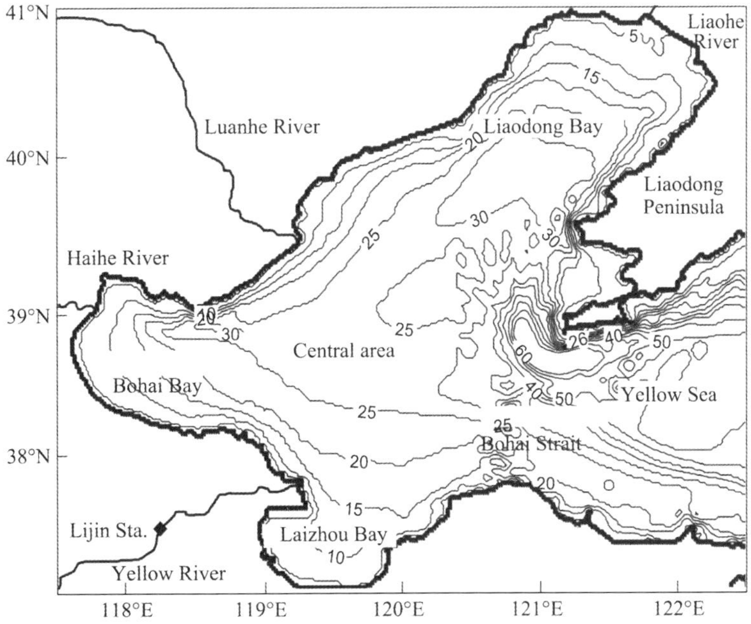

Fig. 8 Bathymetry of the Bohai Sea

5.1 Model setup

Same as the configuration in Mao *et al.* (2008), the BS model has a grid resolution of 1' in both zonal and meridional directions. There are 10 layers in the vertical, and the maximum depth is set to be 65 m. The harmonic constants of M_2 tide at the open boundary (122°30′E) are obtained by linearly interpolating the values from coastal gauges at both sides of the Bohai Strait. A restoring condition is adopted for salinity at the open boundary. Runoffs from four rivers, i.e., Yellow River, Liaohe River, Haihe River and Luanhe River, are taken into account. The freshwater flux from the ERA-40 dataset is applied at the sea surface (Uppala *et al*., 2005).

The model is run in a hydro-thermo decoupled mode, as salinity behaves just like a general dissolved passive tracer concentration that is transported and diffused with the flow and has a source/sink (freshwater flux at sea surface). The February of 2002 and August of 2004 are chosen as two typical months with low and high rates of runoff. February represents the dry season and it is before the Yellow River flow-sediment regulation. August represents the flood season and it is after the regulation (Mao *et al*., 2008). The runoff rates of the four rivers used in simulation are shown in Table 3.

Table 3 Runoff rates (m^3/s) of four rivers in February 2002 and August 2004

	Yellow River	Liaohe River	Haihe River	Luanhe River
2002.02	10.0	0.5	0.6	1.2
2004.08	1570.0	168.0	14.9	29.8

The simulated tidal elevation and current, obtained with harmonic analysis, show excellent agreement with the results of previous studies, in terms of the two M_2 amphidromic points in the BS, directing of the two 0-cophase lines, and the distribution of tidal current ellipses (Fang, 1986; Huang, 1991; Wei *et al*., 2004a) (figures not shown).

5.2 LRV in the Bohai Sea

By tracking water parcels in February 2002 and August 2004, the distributions of the LRV in the two months are obtained. They are almost the same, even near the river mouths in the BS, mainly because the simulation is set to be hydro-thermal decoupled. The consistency also indicates that estuarine freshwater input has little effect on the BS circulation. Therefore, only the LRV field of February 2002 is shown in Fig. 9 for the analysis of the circulation structure.

There are four regions with strong LRV in the BS: around the southern headland of the Liaodong Peninsula, at the top of the Liaodong Bay, in the northern part of the Bohai Bay, and in the vicinity of the Yellow River mouth. This result is consistent with that of Ju *et al*. (2009). This indicates the successful reproduction of the tide-induced residual current in our study.

In these four regions, where topography and coastline are complicated, local convective nonlinearity increases, resulting in strong residual current, as well as the remarkable dependence of the LRV on initial tidal phase. The maximum residual velocity reaches 10 cm/s,

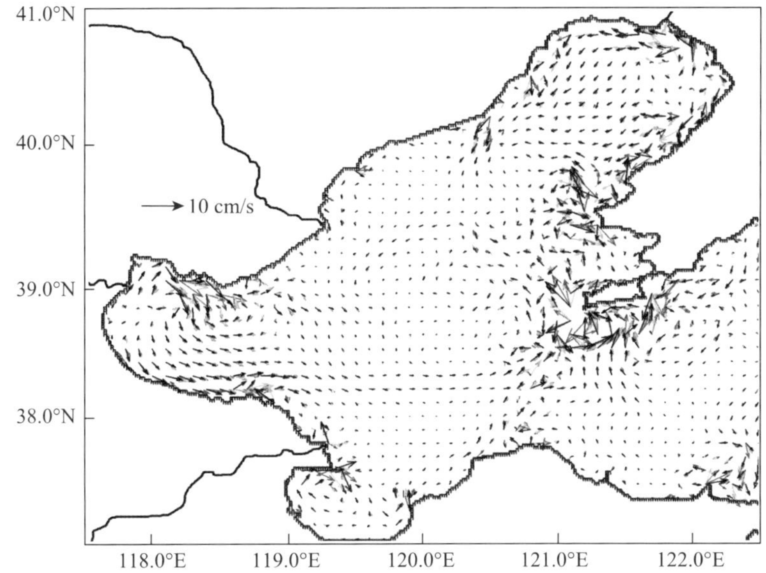

Fig. 9 Surface LRV at four initial moment in the BS (blue—$t_0 = 0$, red—$t_0 = T/4$, green—$t_0 = T/2$, black—$t_0 = 3T/4$)

and even 20 cm/s at the southern headland of the Liaodong Peninsula. In the central part of the BS, the residual velocity is much weaker (~1 cm/s), and hardly varies with the tidal phase. For the salinity distribution, it can be expected that the LIS in the four strong LRV regions should vary distinctly with the tidal phase, and the LIS in the rest parts of BS, particularly in the central area, can be approximated by the Eulerian mean salinity (EMS) due to the weak nonlinearity there.

5.3 Lagrangian inter-tidal salinity (LIS) in the Bohai Sea

The intra-tidal salinity fields in February 2002 and August 2004 are processed by the traditional time-average method to obtain the EMS. It is shown in Fig. 10 that the freshwater

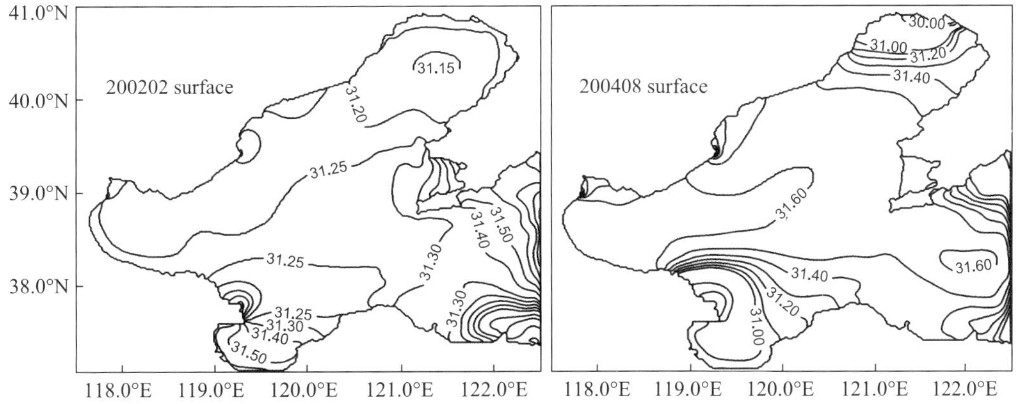

Fig. 10 Surface EMS in the BS in February 2002 (left) and August 2004 (right)

plume of each river in 2004 occupies larger area than that in 2002, due to the difference in runoff rates. Because the Yellow River and the Liaohe River discharge larger quantity of freshwater into the BS, salinity in these two estuaries shows a decreasing trend. However, salinity in the Bohai Bay and the central part increases by 0.3–0.4 due to the persistent net evaporation at the sea surface.

Fig. 11 shows the LIS at four tidal phases in February 2002. Isohaline groups can be distinguished in most areas of the BS. Especially in the central area, isohalines of the same salinity are close to, even overlap each other. Consequently, both the variation of the LIS with tidal phase and the difference between the LIS and EMS are very small (figures not shown). This is consistent with result of the LRV, and verifies the weak nonlinearity in this area. Therefore, the EMS in the central area of the BS can substitute the LIS in describing the long-term transport. It also suggests that the variation of monthly-mean salinity in this area in Mao *et al*. (2008) is reasonable and convincing.

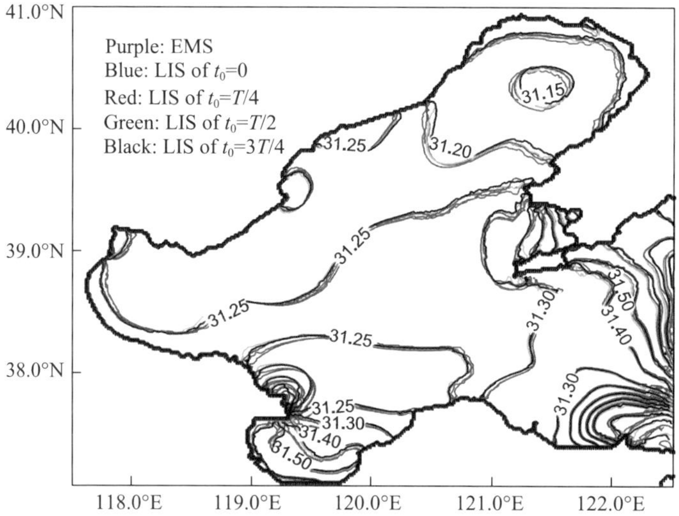

Fig. 11 Surface EMS (purple) and LIS of February 2002 in the BS: LIS of t_0 =0 (blue), t_0 = T/4 (red), t_0 =T/2(green) and t_0 =3T/4 (black)

Additionally, it can be seen from Fig. 11 that the isohaline groups near the southern headland of the Liaodong Peninsula and in the vicinity of the Yellow River mouth are intersected and chaotic. The LIS around the headland varies significantly (0.05–0.10) in a tidal cycle, while the local spatial variation of the salinity is only 0.20, i.e., in the range of 31.30–31.50 (Fig. 12). In this case, variation of the LIS with tidal phase and the difference between the LIS and EMS cannot be regarded as small quantities at all. This indicates the locally strong transport nonlinearity, consistent with the strong LRV in this area.

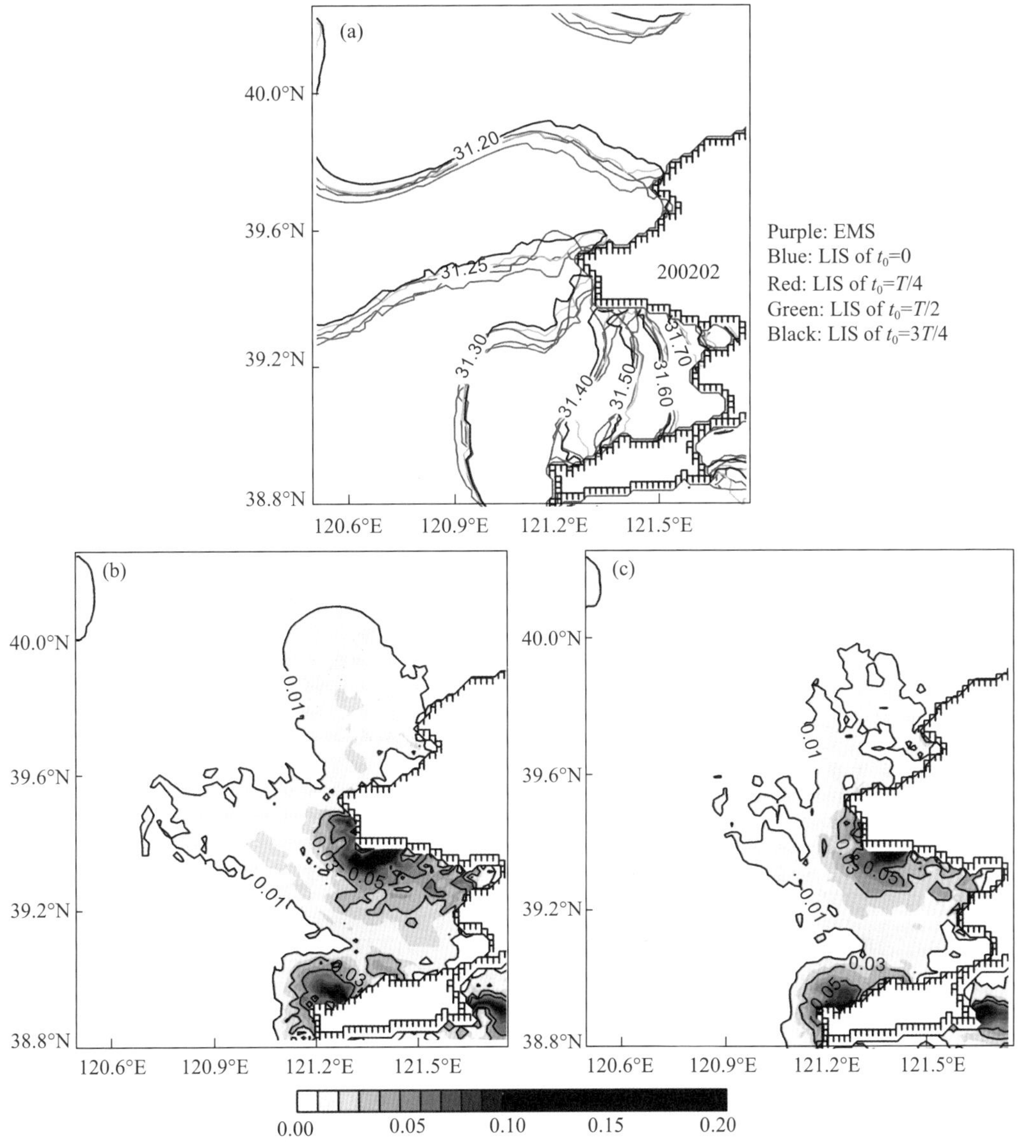

Fig. 12 Surface EMS and LIS at southern headland of Liaodong Peninsula in February 2002: (a) Distribution of EMS (purple) and LIS of four tidal phases (blue—$t_0 = 0$, red—$t_0 = T/4$, green—$t_0 = T/2$, black—$t_0 = 3T/4$); (b) Distribution of maximum difference in the four LIS with tidal phase; (c) Distribution of maximum difference between LIS and EMS

Besides the southern headland of the Liaodong Peninsula, the LIS in the Laizhou Bay also shows remarkable variation with the initial tidal phase, mainly caused by the Yellow River freshwater input and its influence on the salinity gradient (Fig. 13). In August of 2004, the flood season, largest variation of the LIS with tidal phase occurs near the river mouth and reaches almost 6.00 psu. Broadly, variation of the LIS with tidal phase is between 0.10 and

1.00 in the estuarine area, while the spatial distribution has a range of 29.00–31.00. The LIS depends evidently on the initial tidal phase, manifesting the tidal influence on the estuarine salinity front. Even in the dry season (February of 2002), the variation of the LIS with tidal phase (0.03–0.10) can still match the spatial variation (31.00–31.20). In another word, the EMS cannot substitute the LIS around the southern headland of the Liaodong Peninsula or around the Yellow River mouth. The long-term salinity variations in these areas, shown by the EMS in Mao *et al.* (2008), need be re-examined.

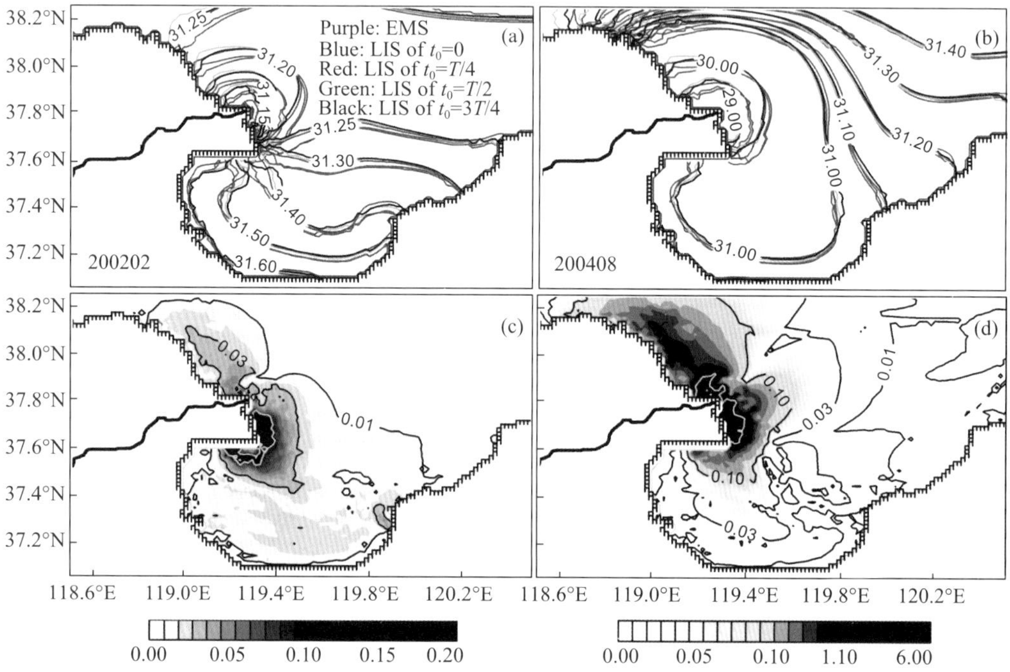

Fig. 13 Similar to Fig. 12a, b, but for the Laizhou Bay. February 2002 (left), August 2004 (right); light blue contour in lower panel indicates 0.10 psu (left) and 1.00 psu (right)

In the abovementioned areas, the dependence of the LIS on tidal phase is quite significant, and agrees well with the behavior of the LRV in these areas. However, it should be noted that not all regions with strong LRV show distinct LIS variations. For example, the LRV is strong in the northern part of the Bohai Bay (see Fig. 9), but variation of the LIS with tidal phase is very small, showing much weaker transport nonlinearity due to the uniform-distributed salinity in this area. This suggests that the transport nonlinearity is linked with both the residual velocity and salinity gradient.

6 Conclusions

Based on the Lagrangian tidally-mean theory in a general nonlinear coastal/estuarine system, this paper investigates the Lagrangian inter-tidal concentration (LIC) of a passive tracer, associated with the M_2 tide-induced Lagrangian residual velocity (LRV), first in three idealized model seas, and then in the realistic setting of the Bohai Sea.

The three idealized model seas are constructed with different topography or coastline. The results suggest that the nonlinearities in the stairs and cape cases are stronger than that in the flat-bottom case as shown by the significant dependence of the LIC and LRV on the initial tidal phase. It is the tortuous and complicated coastline or topography that increases the local nonlinearity of tidal system.

For the application of the Lagrangian mean theory to study the inter-tidal transport of salinity in the Bohai Sea, the model is run in a hydro-thermo decoupled mode. The results show that in most parts of the BS, especially in the central area, the Eulerian mean salinity can substitute the Lagrangian inter-tidal salinity in describing the long-term transport. The dependence of the LIS on the initial tidal phase is remarkable near the southern headland of the Liaodong Peninsula and in the vicinity of the Yellow River mouth. In the former case, this is mainly due to the complicated coastline that augments the local nonlinearity. In the latter case, this is resulting from the estuarine plume and salinity front. The long-term salinity variations in these areas in previous studies were generally shown by EMS, a point that warrants re-examination.

Acknowledgements

This study is supported by National Basic Research Program of China (No. 2010CB428904), National Science Foundation of China (Nos. 41106006, 40976003). We are grateful to the anonymous reviewers for comments on the original manuscript.

References

Aristizabal M and Chant R, 2014. Mechanisms driving stratification in Delaware Bay estuary. Ocean Dynamics, 64, 1615–1629.

Backhaus J, 1985. A three-dimension model for the simulation of shelf sea dynamics. Deutsche Hydrographische Zeitschrift, 38, 165–187.

Bowden K, 1967. Circulation and diffusion. In: Lauff G ed. Estuaries, American Association for the Advancement of Science, Washington D.C., 15–36.

Cerco C, 1995. Simulation of long-term trends in Chesapeake Bay eutrophication. Journal of Environmental Engineering, 121(4), 298–310.

Dortch M, Chapman R, and Abt S, 1992. Application of three-dimensional Lagrangian residual transport. Journal of Hydraulic Engineering, 118(6), 831–848.

Fang G, 1986. Tide and tidal current charts for the marginal seas adjacent to China. Chinese Journal of Oceanology and Limnology, 4(1), 1–16.

Feng S, 1987. A three-dimensional weakly nonlinear model of tide-induced Lagrangian residual current and mass-transport, with an application to the Bohai Sea. In: Nihoul J, Jamart B, eds. Three-dimensional

Models of Marine and Eestuarine Dynamics, Elsevier Oceanography Series, 45, Elsevier, 471–488.

Feng S, 1990. On the Lagrangian residual velocity and the mass-transport in a multi-frequency oscillatory system. In: Cheng R, ed. Residual Currents and Long-term Transport, Coastal and Estuarine Studies, Springer-Verlag, 18–34.

Feng S, 1998. On circulation in Bohai Sea, Yellow Sea and East China Sea. In: Hong G, Zhang J, Park B, eds. The Health of the Yellow Sea. The Earth Love Publication Association, Seoul, 43–77.

Feng S and Lu Y, 1993. A turbulent closure model of coastal circulation. Chinese Science Bulletin, 38(20), 1737–1741.

Feng S and Wu D, 1995. An inter-tidal transport equation coupled with turbulent k-ε model in a tidal and quasi-steady current system. Chinese Science Bulletin, 40(2), 136–139.

Feng S, Cheng R and Xi P, 1986a. On tide-induced Lagrangian residual current and residual transport, I Lagrangian residual current. Water Resources Research, 22(12), 1623–1634.

Feng S, Cheng R, and Xi P, 1986b. On tide-induced Lagrangian residual current and residual transport, II. Residual transport with application in South San Francisco Bay, California. Water Resources Research, 22(12), 1635–1646.

Feng S, Ju L and Jiang W, 2008. A Lagrangian mean theory on coastal sea circulation with inter-tidal transports, I. Fundamental. Acta Oceanologica Sinica, 27(6), 1–16.

Geyer W, Trowbridge J, and Bowen M, 2000. The dynamics of a partially mixed estuary. Journal of Physical Oceanography, 30, 2035–2048.

Hainbucher D, Wei H, Pohlmann T, *et al.*, 2004. Variability of the Bohai Sea circulation based on model calculations. Journal of Marine Systems, 44, 153–174.

Huang Z, 1991. Tidal waves in the Bohai Sea and their variations. Journal of Ocean University of Qingdao, 21(2), 1–12 (in Chinese with English abstract).

Jiang W and Feng S, 2011. Analytical solution for the tidally induced Lagrangian residual current in a narrow bay. Ocean Dynamics, 61, 543–558.

Jiang W and Feng S, 2014. 3D analytical solution to the tidally induced Lagrangian residual current equations in a narrow bay. Ocean Dynamics, 64, 1073–1091.

Jiang W, Pohlmann T, Sun J, *et al.*, 2004. SPM transport in the Bohai Sea: field experiments and numerical modeling. Journal of Marine Systems, 44, 175–188.

Jiang W, Wu D and Gao H, 2002. The observation and simulation of bottom circulation in the Bohai Sea in summer. Journal of Ocean University of Qingdao, 32(4), 511–518 (in Chinese with English abstract).

Ju L, Jiang W and Feng S, 2009. A Lagrangian mean theory on coastal sea circulation with inter-tidal transports, II. Numerical Experiments. Acta Oceanologica Sinica, 28(1), 1–14.

Kuang G, Zhang Q and Dai Y, 1991. Observation of long-term currents and analysis of residual currents in central Bohai Sea. Transactions of Oceanology and Limnology, 2, 1–11 (in Chinese with English abstract).

Longuet-Higgins M, 1969. On the transport of mass by time-varying ocean currents. Deep Sea Research, 16(5), 431–447.

Mao X, Jiang W, Zhao P, *et al.*, 2008. A 3-D numerical study of salinity variations in the Bohai Sea during the recent years. Continental Shelf Research, 28(19), 2689–2699.

Marinone S, Gonzalez J and Figueroa J, 2009. Prediction of currents and sea surface elevation in the Gulf of California from tidal to seasonal scales. Environmental Modelling & Software, 24, 140–143.

Munk W and Anderson E, 1948. Notes on a theory of the thermocline. Journal of Marine Research, 7, 276–295.

O'Driscoll K, Ilyina T, Pohlmann T, *et al.*, 2011. Modelling the fate of persistent organic pollutants (POPs) in the North Sea system. Procedia Environmental Sciences, 6, 169–179.

O'Driscoll K, Mayer B, Ilyina T, *et al.*, 2013. Modelling the cycling of persistent organic pollutants (POPs) in the North Sea system: Fluxes, loading, seasonality, trends. Journal of Marine Systems, 111–112, 69–82.

Officer C, 1976. Physical Oceanography of Estuaries. Wiley, New York, 465pp.

Pohlmann T, 1996. Calculating the development of the thermal vertical stratification in the North Sea with a three-dimensional baroclinic circulation model. Continental Shelf Research, 16(2), 163–194.

Pritchard D, 1954. A study of the salt balance in a coastal plain estuary. Journal of Marine Research, 13, 133–144.

Quan Q, Mao X and Jiang W, 2014. Numerical computation of the tidally induced Lagrangian residual current in a model bay. Ocean Dynamics, 64, 471–486.

Ralston D, Geyer W and Lerczak J, 2008. Subtidal salinity and velocity in the Hudson River estuary: Observations and modeling. Journal of Physical Oceanography, 38, 753–770.

Santiago-Garcia M, Marinone S and Velasco-Fuentes O, 2014. Three-dimensional connectivity in the Gulf of California based on a numerical model. Progress in Oceanography, 123, 64–73.

Scully M and Geyer W, 2012. The role of advection, straining, and mixing on the tidal variability of estuarine stratification. Journal of Physical Oceanography, 42, 855–868.

Scully M, Geyer W and Lerczak J, 2009. The influence of lateral advection on the residual estuarine circulation: a numerical modeling study of the Hudson River estuary. Journal of Physical Oceanography, 39, 107–124.

Simionato C, Meccia V, Dragani W, *et al.*, 2006. On the use of the NCEP/NCAR surface winds for modeling barotropic circulation in the Rio de la Plata Estuary. Estuarine, Coastal and Shelf Science, 70, 195–206.

Uppala S M, Kallberg P W, Simmons A J, *et al.*, 2005. The ERA-40 re-analysis. Quarterly Journal of the Royal Meteorological Society, 131, 2961–3012.

Wang H, Su Z, Feng S and Sun W, 1993. A three-dimensional numerical calculation of the wind driven thermohaline and tide-induced Lagrangian residual current in the Bohai Sea. Acta Oceanologica Sinica, 12(2), 169–182.

Wang T, Jiang W, Chen X, *et al.*, 2013. Acquisition of the tide-induced Lagrangian residual current field by the PIV technique in the laboratory. Ocean Dynamics, 63, 1181–1188.

Wei H, Hainbucher D, Pohlmann T, *et al.*, 2004a. Tidal-induced Lagrangian and Eulerian mean circulation in the Bohai Sea. Journal of Marine Systems, 44, 141–151.

Wei H, Sun J, Moll A, *et al.*, 2004b. Phytoplankton dynamics in the Bohai Sea—observations and modeling. Journal of Marine Systems, 44, 233–251.

附录：冯士筰院士学术著作列表

学术著作

1. 冯士筰. 1982. 风暴潮导论. 北京：科学出版社. 241pp
2. 冯士筰，孙文心 主编. 1992. 物理海洋数值计算. 郑州：河南科技出版社. 610pp
3. 冯士筰，石广玉，高会旺 主编. 2006. 上层海洋与低层大气研究的前沿科学问题. 北京：气象出版社. 303pp
4. 冯士筰，张经，魏皓，等著. 2007. 渤海环境动力学导论. 北京：科学出版社. 281pp

学术文章（其中中国科学等期刊中英文都有的只保留了英文的）

5. Chin T, Feng S. 1975. A preliminary study on the mechanism of shallow water storm surges. Science in China, XVII(2): 242–261
6. Feng S. 1977. A three-dimensional nonlinear model of tides. Science in China, XX(4): 436–446
7. 冯士筰. 1978. 风暴潮的理论模化. 山东海洋学院学报, 8(2): 72–79
8. 冯士筰. 1978. 湍流速度分布对数律的一点修正. 山东海洋学院学报, 8(1): 51–53
9. 冯士筰. 1979. 大洋风生-热盐环流模型. 山东海洋学院学报, 9(2): 1–14
10. 冯士筰. 1979. f-平面上的宽陆架诱导阻尼波. 海洋学报, 9(2): 177–192
11. 孙文心，冯士筰，秦曾灏. 1979. 超浅海风暴潮的数值模拟(一)——零阶模型对渤海风潮的初步应用. 海洋学报, 1(2): 193–211.
12. 冯士筰，施平. 1980. 含变涡动系数的超浅海风暴潮模型. 海洋学报, 2(3): 1–11
13. 孙文心，秦曾灏，冯士筰. 1980. 超浅海风暴潮的数值模拟(Ⅱ)——渤海风潮的一阶模型. 山东海洋学院学报, 10(2): 7–19
14. 奚盘根，张淑珍，冯士筰. 1980. 东中国海环流的一种模型Ⅰ、冬季环流的数值模拟. 山东海洋学院学报, 10(3): 13–25
15. 冯士筰，张淑珍，奚盘根. 1981. 东中国海环流的一种模型Ⅱ、夏季环流和相似准则. 山东海洋学院学报, 11(2): 8–26
16. 冯士筰. 1981. 常底坡有限宽陆架诱导阻尼波的一种模型. 海洋与湖沼, 12(1): 1–8
17. 孙文心，陈宗镛，冯士筰. 1981. 一种三维空间非线性潮波的数值控拟(Ⅰ)——渤海 M_4 和 MS_4 分潮波的试算. 山东海洋学院学报, 11(1): 23–31
18. 冯士筰. 1982. 论f-和β-坐标系. 山东海洋学院学报, 12(3): 1–10
19. Sun W, Feng S, Qin Z. 1982. Numerical study on the Bohai Sea wind surges—the zeroth-order dynamical model. Acta Oceanologica Sinica, 1(2): 175–188
20. Feng S, Sun W. 1983. A tidal three-dimensional nonlinear model with variable eddy viscosity(I)—A dynamic model. Chinese Journal of Oceanology and Limnology, 1(2): 166–170
21. 冯士筰. 1984. 论大洋环流的尺度分析及风旋度-热盐梯度方程式. 山东海洋学院学报, 14(1): 33–43

22. 张淑珍，奚盘根，冯士筰. 1984. 渤海环流数值模拟. 山东海洋学院学报, 14(2): 12–19
23. Feng S, Xi P, Zhang S. 1984. The baroclinic residual circulation in shallow seas I. The hydrodynamic models. Chinese Journal of Oceanology and Limnology, 2(1): 49–60
24. Feng S. 1984. A three-dimensional nonlinear hydrodynamic model with variable eddy viscosity in shallow seas. Chinese Journal of Oceanology and Limnology, 2(2): 177–187
25. Xi P, Zhang S, Feng S. 1985. An investigation on numerical modeling of circulation in the East China Sea. Acta Oceanologica Sinica, 4(4): 510–514
26. Cheng R T, Feng S, Xi P. 1986. On Lagrangian residual ellipse. In: J van de Kreeke (ed.) Physics of Shallow Estuaries and Bays. Berlin, New York: Springer-Verlag. 102–113
27. Zhang S, Wang H, Feng S, Xi P. 1986. A Simulation of the Residual Flow in the Bohai Sea. In: J van de Kreeke (ed.) Physics of Shallow Estuaries and Bays, Berlin, New York: Springer-Verlag. 114–119
28. Feng S, Cheng R T, Xi P. 1986. On tide-induced Lagrangian residual current and residual transport: 1. Lagrangian residual current. Water Resources Research, 22(12): 1623–1634
29. Feng S, Cheng R T, Xi P. 1986. On tide-induced Lagrangian residual current and residual transport: 2. Residual transport with application in south San Francisco Bay, California. Water Resources Research, 22(12): 1635–1646
30. Feng S. 1986. A three-dimensional weakly nonlinear dynamics on tide-induced Lagrangian residual current and mass-transport. Chinese Journal of Oceanology and Limnology, 4(2): 139–158
31. Feng S. 1987. A three-dimensional weakly nonlinear model of tide-induced Lagrangian residual current and mass-transport, with an application to the Bohai Sea. In: J C J Nihoul, B M Jamart (eds.) Three-Dimensional Models of Marine and Estuarine Dynamics. Amsterdam, Oxford, New York, Tokyo: Elsevier. 471–488
32. 冯士筰. 1988. 论浅海环流及其输运的流体动力学基础. 见：张维 编. 清华大学工程力学与工程热物理学术会议论文集. 北京：清华大学出版社. 199–208
33. 冯士筰. 1988. 第一章 海洋环境流体动力学研究的焦点. 山东海洋学院学报, 18(2(II)): 1–7
34. 孙文心，奚盘根，冯士筰. 1988. 第三节 三维非线性潮波边值问题模型计算结果及分析. 山东海洋学院学报, 18(2(II)): 51–52+81–100
35. 奚盘根，冯士筰，孙文心. 1988. 第五章 非线性三维潮波边值问题模型. 山东海洋学院学报, 18(2(II)): 37–47
36. Cheng R T, Feng S, Xi P. 1989. On inter-tidal transport equation. In: B J Neilson, A Kuo, J Brubaker (eds.) Estuarine Circulation. Clifton: Humana Press. 133–156
37. 唐永明，孙文心，冯士筰. 1990. 三维浅海流体动力学模型的流速分解法. 海洋学报, 12(2): 149–158
38. Feng S, Cheng R T, Sun W, Xi P, Song L. 1990. Lagrangian residual current and longterm transport processes in a weakly nonlinear baroclinic system. In: H Wang, J Wang, H Dai (eds.) Physics of Shallow Seas. Beijing: China Ocean Press. 1–20
39. Feng S. 1990. On the fundamental dynamics of barotropic circulation in shallow seas. Acta Oceanologica Sinica, 9(3): 315–329
40. Feng S. 1990. On the Lagrangian residual velocity and the mass-transport in a multi-frequency oscillatory system. In: R T Cheng (ed.) Residual Currents and Long-term Transport. New York: Springer-

Verlag. 34–48

41. Feng S. 1991. The dynamics on tidal generation of residual vorticity. Chinese Science Bulletin, 36(24): 2043–2046
42. 冯士筰. 1992. 第九章 浅海环流物理及数值模拟. 见: 冯士筰, 孙文心 编. 物理海洋数值计算. 郑州: 河南科学技术出版社. 543–610
43. 冯士筰, 鹿有余. 1993. 浅海 Lagrange 余流和长期输运过程的研究——一种三维空间弱非线性理论. 自然科学进展, 3(2): 126–132
44. 王辉, 苏志清, 冯士筰, 孙文心. 1993. 渤海三维风生–热盐–潮致 Lagrange 余流数值计算. 海洋学报, 15(1): 9–21
45. Feng S, Lu Y. 1993. A turbulent closure model of coastal circulation. Chinese Science Bulletin, 38(20): 1737–1741
46. Wang H, Su Z, Feng S, Sun W. 1993. A three-dimensional numerical calculation of the wind-driven thermohaline and tide-induced Lagrangian residual current in the Bohai Sea. Acta Oceanologica Sinica, 12(2): 169–182
47. 冯士筰. 1994.《海洋环境流体动力学在海洋环境保护中的应用》序. 青岛海洋大学学报, 24(S1): 11–12
48. Feng S, Zhang S, Xi P. 1994. A Langrangian model of circulation in Bohai Sea. In: D Zhou, Y Liang, C Zeng (eds.) Oceanology of China Seas. Dordrecht, Boston, London: Kluwer Academic Publishers. 83–90
49. 王辉, 冯士筰. 1995. 河口 Si 输运过程研究. 海洋与湖沼, 26(2): 161–168
50. Feng S, Wu D. 1995. An inter-tidal transport equation coupled with turbulent K-ε model in a tidal and quasi-steady current system. Chinese Science Bulletin, 40(2): 136–139
51. Wu D, Feng S. 1995. The dynamic effects of a finite depth ambient fluid on bottom eddy. Chinese Journal of Oceanology and Limnology, 13(2): 124–133
52. 管玉平, 冯士筰, 丑纪范, 孙德田. 1996. 定性数学的若干基本特征. 青岛海洋大学学报, 26(4): 67–74
53. 吴德星, 王彩霞, 冯士筰. 1996. 热带海区温跃层深度空间变化——对波致环流的影响. 青岛海洋大学学报, 26(4): 5–9
54. 管玉平, 高会旺, 冯士筰, 林一骅. 1997. 海洋生态系统动力学浅说. 地球科学进展, 12(5): 50–53
55. Feng S, Xi P. 1997. Physical transport of contaminants in bays, estuaries and coastal environment. In: J Zhou (ed.) Sources Transport and Environmental Impact of Contaminants in the Coastal and Estuarine Area of China. Beijing: China Ocean Press. 21–38
56. 冯士筰. 1998. 风暴潮的研究进展. 世界科技研究与发展. 20(4): 44–47
57. 管玉平, 高会旺, 冯士筰. 1998. 海洋生态信息系统 I. 海洋生态系统的数据处理问题. 青岛海洋大学学报, 28(1): 27–32
58. 王凡, 吴德星, 冯士筰, 侍茂崇. 1998. 赤道海洋波动弱非线性动力学系统浅析. 海洋科学, 22(3): 40–42
59. Feng S. 1998. On circulation in Bohai Sea Yellow Sea and East China Sea, the health of the Yellow Sea. In: G H Hong, J Zhang, B K Park (eds.) The Health of the Yellow Sea. Seoul: The Earth Love Publication Association

60. Gao H, Feng S, Guan Y. 1998. Modelling annual cycles of primary production in different regions of the Bohai Sea. Fisheries Oceanography, 7(3–4): 258–264
61. 冯士筰，高会旺．1999．发展海洋科学技术 促进资源的可持续利用．中国科学院院刊，14(1): 42–44
62. 管玉平，高会旺，冯士筰．1999．海洋生态调控智能决策技持系统的结构模型．海洋环境科学，18(1): 1–5
63. 管玉平，高会旺，冯士筰．1999．海洋生态模型库系统初探．海洋通报，18(3): 63–67
64. 王凡，吴德星，冯士筰，侍茂崇．1999．赤道海洋波致 Lagrange 余流的弱非线性动力学模型及其解．海洋与湖沼，30(6): 701–710
65. Wang K, Feng S. 1999. Numerical experiments on the wind-driven circulation in the Bohai, Huanghai and East China Seas in winter. Chinese Journal of Oceanology and Limnology, 17(1): 10–18
66. 高会旺，冯士筰，管玉平．2000．海洋浮游生态系统动力学模式的研究．海洋与湖沼，31(3): 341–348
67. 王凡，吴德星，冯士筰，侍茂崇．2000．混合 Rossby 惯性重力波致 Lagrange 余流．海洋与湖沼，31(2): 168–177
68. 魏皓，冯士筰，武建平，张平．2000．湍流局地平衡假设的新推论——齐次湍流动能输运方程封闭模型与应用(Ⅱ)．青岛海洋大学学报，30(4): 557–562
69. Jiang W, Pohlmann T, Sündermann J, Feng S. 2000. A modelling study of SPM transport in the Bohai Sea. Journal of Marine Systems, 24(3–4): 175–200
70. 冯士筰，李凤岐，顾育翘．2001．海洋科学发展对教育改革的要求．中国地质教育，10(2): 6–11
71. 王凯，冯士筰，施心慧．2001．渤、黄、东海夏季环流的三维斜压模型．海洋与湖沼，32(5): 551–560
72. 魏皓，冯士筰，武建平，刘桂梅．2001．齐次湍流动能输运方程封闭模型在浅海动力学中的应用．青岛海洋大学学报，31(6): 821–827
73. 赵领娣，冯士筰．2001．大力提升中国企业的绿色国际竞争力．经济界，(6): 48–52
74. Ren L, Zhang M, Lu X, Feng S. 2001. Simple simulation of the annual variation of the specific photosynthesis rate in Jiaozhou Bay. Chinese Journal of Oceanology and Limnology, 19(1): 63–72
75. Wei H, Zhao L, Feng S. 2001. Comparison of the Eulerian and Lagrangian tidal residuals in the Bohai Sea. Chinese Journal of Oceanology and Limnology, 19(2): 119–127
76. 冯士筰，王修林．2002．适应新形势 加快海洋科学教育的发展．中国大学教学，(Z1): 23–25
77. 赵亮，魏皓，冯士筰．2002．渤海氮磷营养盐的循环和收支．环境科学，23(1): 78–81
78. Zhao L, Wei H, Feng S. 2002. Annual cycle and budgets of nutrients in the Bohai Sea. Journal of Ocean University of Qingdao, 1(1): 29–37
79. 孙军，刘东艳，冯士筰．2003．近海生态系统动力学研究中浮游植物采样及分析策略．海洋与湖沼，34(2): 224–232
80. 魏皓，赵亮，冯士筰．2003．渤海浮游植物生物量与初级生产力变化的三维模拟．海洋学报，25(S2): 66–72
81. 魏皓，赵亮，冯士筰．2003．渤海碳循环与浮游植物动力学过程研究．海洋学报，25(S2): 151–156
82. Ren L, Zhang M, Brockmann U, Feng S. 2003. Pelagic nitrogen cycling in Jiaozhou Bay, a model study I: the conceptual model. Chinese Journal of Oceanology and Limnology, 21(4): 358–367
83. Wei H, Zhao L, Feng S. 2003. A model study on carbon cycle and phytoplankton dynamical processes

in the Bohai Sea. Acta Oceanologica Sinica, 22(1): 47–55
84. Hainbucher D, Wei H, Pohlmann T, Sündermann J, Feng S. 2004. Variability of the Bohai Sea circulation based on model calculations. Journal of Marine Systems, 44(3–4): 153–174
85. Sündermann J, Feng S. 2004. Analysis and modelling of the Bohai sea ecosystem—a joint German-Chinese study. Journal of Marine Systems, 44(3–4): 127–140
86. Wei H, Hainbucher D, Pohlmann T, Feng S, Sündermann J. 2004. Tidal-induced Lagrangian and Eulerian mean circulation in the Bohai Sea. Journal of Marine Systems, 44(3–4): 141–151
87. Zhao L, Wei H, Xu Y, Feng S. 2005. An adjoint data assimilation approach for estimating parameters in a three-dimensional ecosystem model. Ecological Modelling, 186(2): 235–250
88. Feng S, Ju L, Jiang W. 2008. A Lagrangian mean theory on coastal sea circulation with inter-tidal transports I. Fundamentals. Acta Oceanologica Sinica, 27(6): 1–16
89. 冯士筰. 2008. 海洋灾害学及劳动经济学交叉研究的新探索——评价《海洋灾害及海洋收入的经济学研究》. 现代渔业信息, 23(11): 9–10+15
90. 冯士筰. 2008. 交叉科学研究的新探索. 管理观察, (21): 36–37
91. 高会旺, 祁建华, 石金辉, 石广玉, 冯士筰. 2009. 亚洲沙尘的远距离输送及对海洋生态系统的影响. 地球科学进展, 24(1): 1–10
92. Ju L, Jiang W, Feng S. 2009. A Lagrangian mean theory on coastal sea circulation with inter-tidal transports II. Numerical Experiments. Acta Oceanologica Sinica, 28(1): 1–14
93. Jiang W, Feng S. 2011. Analytical solution for the tidally induced Lagrangian residual current in a narrow bay. Ocean Dynamics, 61(4): 543–558
94. Liu G, Liu Z, Gao H, Gao Z, Feng S. 2012. Simulation of the Lagrangian tide-induced residual velocity in a tide-dominated coastal system: a case study of Jiaozhou Bay, China. Ocean Dynamics, 62(10–12): 1443–1456
95. Wang T, Jiang W, Chen X, Feng S. 2013. Acquisition of the tide-induced Lagrangian residual current field by the PIV technique in the laboratory. Ocean Dynamics, 63(11–12): 1181–1188
96. Jiang W, Feng S. 2014. 3D analytical solution to the tidally induced Lagrangian residual current equations in a narrow bay. Ocean Dynamics, 64(8): 1073–1091
97. Wang T, Geyer W R, Engel P, Jiang W, Feng S. 2015. Mechanisms of tidal oscillatory salt transport in a partially stratified estuary. Journal of Physical Oceanography, 45(11): 2773–2789
98. Mao X, Jiang W, Zhang P, Feng S. 2016. Numerical study on inter-tidal transports in coastal seas. Journal of Ocean University of China, 15(3): 379–388
99. Deng F, Jiang W, Feng S. 2017. The nonlinear effects of the eddy viscosity and the bottom friction on the Lagrangian residual velocity in a narrow model bay. Ocean Dynamics, 67(9): 1105–1108